PRACTICAL APPLICATIONS OF MICRORESONATORS IN OPTICS AND PHOTONICS

OPTICAL SCIENCE AND ENGINEERING

Founding Editor
Brian J. Thompson
University of Rochester
Rochester, New York

PRACTICAL APPLICATIONS OF MICRORESONATORS IN OPTICS AND PHOTONICS

Edited by
ANDREY B. MATSKO

CRC Press
Taylor & Francis Group
Boca Raton London New York

CRC Press is an imprint of the
Taylor & Francis Group, an **informa** business

Cover images reprinted from:

Central figure: Savchenkov, A.A., Matsko, A. B., Grudinin, I., Savchenkova, E. A., Strekalov, D., and Maleki, L., Optical vortices with large orbital momentum: generation and interference. *Opt. Express,* 14, 2888-2897 (2006).

Top and bottom figures: Savchenkov, A.A., Matsko, A. B., Ilchenko, V. S., Strekalov, D., and Maleki, L. Direct observation of stopped light in a whispering-gallery-mode microresonator. *Phys. Rev.,* A 76, 023816 (2007).

CRC Press
Taylor & Francis Group
6000 Broken Sound Parkway NW, Suite 300
Boca Raton, FL 33487-2742

First issued in paperback 2019

© 2009 by Taylor & Francis Group, LLC
CRC Press is an imprint of Taylor & Francis Group, an Informa business

No claim to original U.S. Government works

ISBN-13: 978-1-4200-6578-7 (hbk)
ISBN-13: 978-1-138-38139-1 (pbk)

Visit the Taylor & Francis Web site at
http://www.taylorandfrancis.com

and the CRC Press Web site at
http://www.crcpress.com

Contents

Preface

The race for compactness and scalability of optical and photonic devices calls for the development of efficient micro- and nano-optical elements. The optical resonators are important here because they can be used in the optical signal processing systems as modulators, filters, delay lines, switches, sensors, and so on. The number of the different types of resonators increases every day, and the basic research of their properties is gradually and steadily substituted with applied research. Such practical issues as efficient packaging and robust coupling, as well as integration of the resonators into complex optical systems, become especially important when one tries to bridge the gap between the fundamental research and practical implementation.

There are many scientific books and reviews discussing the properties of optical microresonators, and I believe that at this stage it is important to have a collection reviewing the basic directions in the development of the practical applications of the microresonators, which is my goal with this book. Though it is practically impossible to cover the whole field with several contributions, I hope that this collection will provide readers with the flavor of the applied studies in the field and will convince them that systems containing microresonators will soon become as common and widespread as electronic devices containing quartz oscillators. I also hope that this book will attract the attention of a general audience dealing with R&D in broadly defined physics/electrical engineering areas to the fascinating world of the microresonators. The chapters are written by brilliant scientists and engineers working in the field and can be understood by any graduate student in the field.

Traditional mirrored optical resonators are utilized in all branches of optics where, for example, multiple recirculation of optical power is required to maintain laser oscillation, to increase the effective path length in spectroscopic or resolution in interferometric measurements, and to enhance wave mixing interactions. Crucial properties of the resonators, such as high quality (Q) factor and finesse, can be achieved with the highest reflectivity and low-loss mirrors. Despite their versatility, these resonators have remained fairly complex devices. They are prone to vibration instabilities because of relatively low-frequency mechanical resonances. Stability and small modal volume are of great importance for practical applications; however, miniaturization of conventional Fabry–Perot resonators is either complicated and expensive, or yields rather low Q-factors.

This book contains several reports on the progress in the rapidly growing field of monolithic micro- and nano-resonators. Such resonators do not have localized mirrors as such. The light is confined inside these resonators due to their morphology. The monolithic resonators are characterized by the unique combination of properties unreachable in other resonator structures. They have tiny volumes along with huge finesse and Q factors. The modal spectrum of the resonators can be efficiently engineered. These properties make the resonators extremely efficient in multiple applications.

The first chapter in this book, authored by Takasumi Tanabe et al. (NTT Corporation, Japan), is devoted to photonic crystal-based resonators (nanocavities). Among various microresonators, photonic crystal nanocavities have the smallest mode volume (V) and nearly the highest Q/V value. High Q/V devices are attracting considerable attention because they enable multiple quantum and nonlinear optics applications. Recent progress on these ultrahigh-Q photonic crystal nanocavities is also discussed in the first chapter.

Various designs of photonic crystal nanocavities, fabrication and the characterization technologies are reviewed. In addition, various applications like light buffering, slow light propagation, all-optical switching, and bistable memory operation are discussed by the authors.

The second chapter, authored by Charles Santori et al. (Hewlett-Packard Laboratories, USA), is devoted to the discussion of applications of a particular type of distributed feedback microresonators called "pillar microcavities". These microcavities are well suited for efficient coupling of dipole emitters to a single mode in free space and thus are suitable for generation of photons on demand. The design, fabrication and characterization of single-photon devices based on single InAs quantum dots coupled to pillar microcavities formed from AlAs/GaAs distributed-Bragg-reflector mirrors are described in this chapter. Several applications including quantum cryptography and entanglement formation through two-photon interference are presented. Future applications that could be developed as the devices improve are also discussed.

Chapters 3 through 11 deal with the resonators in which the closed trajectories of light are supported by a variety of total internal reflection in curved and polygonal transparent dielectric structures. The circular optical modes in such resonators, frequently called whispering gallery modes (WGMs), can be understood as closed circular beams supported by total internal reflections from the boundaries of the resonators. High values of Q-factor can be achieved in WGMs of very small volume, in certain cases as small as cubic wavelength, with appropriately designed dielectric interface and with use of transparent materials. Applications of the microresonators made of various materials, including silicon, fused silica, fluorite, lithium niobate, and polymers are discussed in these chapters.

These resonators have cylindrical, spherical, spheroidal, toroidal, ring, and other shapes and topologies. When the reflecting boundary has high index contrast, and radius of curvature exceeds several wavelengths, the radiative losses, similar to bending losses of a waveguide with high refractive index contrast, become very small, and the Q factor of the resonators becomes limited only by and material attenuation scattering caused by geometrical imperfections (e.g. surface roughness).

Fabrication of the open dielectric resonators can be simple and inexpensive, and they lend themselves to integration. The unique combination of very high Q (as high as 10^{11}) and very small volume has attracted interest in the applications of the resonators in fundamental science and engineering. Small size also results in excellent mechanical stability and easy control of the resonator parameters. The authors describe applications of the resonators for filtering and modulating light, for detecting chemical and biological substances. Various lasers and oscillators based on the resonators are also discussed.

Namely, Lute Maleki et al. (OEwaves Inc., USA) discuss application of crystalline WGM resonators in filtering and laser stabilization in Chapter 3. Applications of polygonal-shaped microdisk resonators are studied in Chapter 4, authored by Andrew W. Poon et al. (The Hong Kong University of Science and Technology, People's Republic of China). Applications of electro-optic polymer ring resonators for millimeter-wave modulation and optical signal processing are reviewed by William H. Steier et al. (University of Southern California, USA) in Chapter 5. Chapter 6, authored by Melanie Lebental et al. (Ecole Normale Superieure de Cachan, France), is devoted to the discussion of properties of organic micro-lasers. Practical applications of optical microfiber loop and coil resonators are described in Chapter 7 by Misha Sumetsky (OFS Laboratories, USA). Chapter 8, authored by Xudong Fan (University of Missouri, Columbia, USA), deals with optofluidic ring resonator biological and chemical sensors. An application of crystalline microresonators for fabrication of a non-electronic wireless receiver with immunity to damage by electromagnetic pulses is introduced in Chapter 9 by Bahram Jalali et al. (the University of California, Los Angeles, USA). Properties and applications of cavity enhanced opto-mechanics are studied in Chapter 10, authored by Tobias Kippenberg (Max Planck Institut für Quantenoptik, Garching, Germany) and Kerry Vahala (California Institute of Technology, USA). Generation of optical frequency combs in optical microresonators and applications of the combs are described in Chapter 11, authored by Oliver Arcizet et al. (Max Planck Institut für Quantenoptik, Garching, Germany).

The last two chapters are devoted to the theoretical discussion of the properties of long chains of coupled microresonators. Though the fabrication of such chains is still problematic because of technological immaturity, the theoretical studies shed light on the problems and phenomena one needs to expect once the chain fabrication becomes feasible. Chapter 12, written by Jacob Khurgin (Johns Hopkins University, USA), is focused on the two most important factors limiting the performance of linear and nonlinear optical devices based on coupled resonator structures. These factors are, respectively, dispersion of loss and dispersion of group velocity. Chapter 13, authored by Shayan Mookherjea (University of California, San Diego, USA), deals with linear and nonlinear localization of light in chains of nearly identical resonators.

I hope that this book will help accelerate the already rapid pace of the research and developments in the exciting field of the applications of optical microresonators. I would like to thank the authors for their contributions making this book a success. This book would not have been possible without the assistance of my colleagues Lute Maleki, Anatoliy Savchenkov, and Vladimir Ilchenko. I am also thankful to Allyson Beatrice for her assistance.

Andrey B. Matsko

Editor

Andrey B. Matsko (MS, 1994 and PhD, 1996, Moscow State University, Russia) has been a principal engineer with OEwaves Inc. since 2007. He joined the company after six year employment as a senior/principal member of technical staff at Jet Propulsion Laboratory (JPL) and four year post-doctoral training at the Department of Physics, Texas A&M University. He has numerous publications in the field and holds several patents. His current research interests include, but are not restricted to, applications of whispering gallery mode resonators in quantum and nonlinear optics, and photonics; coherence effects in resonant media; and quantum theory of measurements. He is a member of the *Optical Society of America* and a member of the *Program Committee of Photonics West: Laser Resonators and Beam Control Conference.* He received JPL's Lew Allen Award for excellence in 2005.

Contributors

Olivier Arcizet
Max Planck Institut für Quantenoptik
Garching, Germany

Ali Ayazi
Electrical Engineering
University of California
Los Angeles, California

Gary Betts
Photonic Systems Inc.
Burlington, Massachusetts

Eugene Bogomolny
Laboratoire de Physique Théorique et
    Modéles Statistiques
Université Paris-Sud
Orsay, France

Bart Bortnik
Department of Electrical Engineering
University of California Los Angeles
Los Angeles, California

Hui Chen
Photonic Device Laboratory
Department of Electronic and Computer
    Engineering
The Hong Kong University of Science and
    Technology
Clear Water Bay, Hong Kong, People's
    Republic of China

Pascal Del'Haye
Max Planck Institut für Quantenoptik
Garching, Germany

Xudong Fan
Biological Engineering Department
University of Missouri
Columbia, Missouri

David Fattal
Hewlett-Packard Laboratories
Palo Alto, California

Harold R. Fetterman
Department of Electrical Engineering
University of California Los Angeles
Los Angeles, California

Ronald Holzwarth
Max Planck Institut für Quantenoptik
Garching, Germany

Nick K. Hon
Photonic Device Laboratory
Department of Electronic and Computer
    Engineering
The Hong Kong University of Science and
    Technology
Clear Water Bay, Hong Kong, People's
    Republic of China

Rick C. J. Hsu
Broadcom Corporation
Irvine, California

Yu-Chueh Hung
Department of Electrical Engineering
University of California Los Angeles
Los Angeles, California

Vladimir S. Ilchenko
OEwaves Inc.
Pasadena, California

Bahram Jalali
Electrical Engineering
University of California
Los Angeles, California

Jacob Khurgin
Department of Electrical and Computer
Engineering
Johns Hopkins University
Baltimore, Maryland

Seongku Kim
Department of Electrical Engineering
University of California Los Angeles
Los Angeles, California

Tobias Jan Kippenberg
Max Planck Institut für Quantenoptik
Garching, Germany

Eiichi Kuramochi
Optical Science Laboratory
NTT Basic Research Laboratories
NTT Corporation
Atsugi, Japan

Melanie Lebental
Laboratoire de Photonique Quantique et
Moléculaire
Ecole Normale Supérieure of Cachan
Cachan, France
and
Laboratoire de Physique Théorique et
Modéles Statistiques
Université Paris-Sud
Orsay, France

Jonathan Y. Lee
Photonic Device Laboratory
Department of Electronic and Computer
Engineering
The Hong Kong University of Science and
Technology
Clear Water Bay, Hong Kong, People's
Republic of China

Chao Li
Photonic Device Laboratory
Department of Electronic and Computer
Engineering
The Hong Kong University of Science and
Technology
Clear Water Bay, Hong Kong, People's
Republic of China

Xianshu Luo
Photonic Device Laboratory
Department of Electronic and Computer
Engineering
The Hong Kong University of Science and
Technology
Clear Water Bay, Hong Kong, People's
Republic of China

Lute Maleki
OEwaves Inc.
Pasadena, California

Andrey B. Matsko
OEwaves Inc.
Pasadena, California

Shayan Mookherjea
Department of Electrical Engineering
University of California
San Diego, California

Masaya Notomi
Optical Science Laboratory
NTT Basic Research Laboratories
NTT Corporation
Atsugi, Japan

Matthew Pelton
Center for Nanoscale Materials
Argonne National Laboratory
Argonne, Illinois

Andrew W. Poon
Photonic Device Laboratory
Department of Electronic and Computer
Engineering
The Hong Kong University of Science and
Technology
Clear Water Bay, Hong Kong,
People's Republic of China

David Press
Edward L. Ginzton Laboratory
Stanford University
Stanford, California

Charles Santori
Hewlett-Packard Laboratories
Palo Alto, California

Anatoliy A. Savchenkov
OEwaves Inc.
Pasadena, California

Albert Schliesser
Max Planck Institut für Quantenoptik
Garching, Germany

Byoung-Joon Seo
Department of Electrical Engineering
University of California Los Angeles
Los Angeles, California

Akihiko Shinya
Optical Science Laboratory
NTT Basic Research Laboratories
NTT Corporation
Atsugi, Japan

Siyka I. Shopova
Biological Engineering Department
University of Missouri
Columbia, Missouri

Glenn S. Solomon
Joint Quantum Institute
National Institute of Standards and
    Technology and University of
    Maryland
Gaithersburg, Maryland

William H. Steier
Department of Electrical Engineering
University of Southern California
Los Angeles, California

Misha Sumetsky
OFS Laboratories
Somerset, New Jersey

Yuze Sun
Biological Engineering Department
University of Missouri
Columbia, Missouri

Jonathan D. Suter
Biological Engineering Department
University of Missouri
Columbia, Missouri

Takasumi Tanabe
Optical Science Laboratory
NTT Basic Research Laboratories
NTT Corporation
Atsugi, Japan

Hidehisa Tazawa
Department of Electrical Engineering
University of Southern California
Los Angeles, California

Kerry J. Vahala
California Institute of Technology
Pasadena, California

Jelena Vučković
Edward L. Ginzton Laboratory
Stanford University
Stanford, California

Edo Waks
Institute for Research in Electronics and
    Applied Physics
University of Maryland
College Park, Maryland

Ian M. White
Biological Engineering Department
University of Missouri
Columbia, Missouri

Fang Xu
Photonic Device Laboratory
Department of Electronic and Computer
    Engineering
The Hong Kong University of Science and
    Technology
Clear Water Bay, Hong Kong, People's
    Republic of China

Yoshihisa Yamamoto
Edward L. Ginzton Laboratory
Stanford University
Stanford, California

Gilmo Yang
Biological Engineering Department
University of Missouri
Columbia, Missouri

Andrew Yick
Department of Electrical
    Engineering
University of Southern
    California
Los Angeles, California

Linjie Zhou
Photonic Device Laboratory
Department of Electronic and Computer
    Engineering
The Hong Kong University of Science and
    Technology
Clear Water Bay, Hong Kong, People's
    Republic of China

Hongying Zhu
Biological Engineering Department
University of Missouri
Columbia, Missouri

Joseph Zyss
Laboratoire de Photonique Quantique et
    Moléculaire
Ecole Normale Supérieure of Cachan
Cachan, France

1

Ultrahigh-Q Photonic Crystal Nanocavities and Their Applications

Takasumi Tanabe, Eiichi Kuramochi, Akihiko Shinya, and Masaya Notomi

NTT Basic Research Laboratories

NTT Corporation

CONTENTS

1.1 Introduction

Light is fast and thus can carry large amounts of data in a very short time, which makes photonic technology a promising communication tool. In fact, photonic technologies, such as optical fiber, that support a high transmittance speed[1] are becoming more important in our lives. However, these technologies have had limited practical applications in data transmission, and signal processing has yet to find a commercial use. Although all-optical signal processing has been widely studied for several decades,[2] it has been difficult to employ in practical systems, often because the required operating energy was too large.[3,4] This is due to the fundamental nature of light. In other words, light is fast but difficult to store or confine in a small space. This makes photonic approaches difficult to handle. In contrast, there is a growing demand for a practical all-optical signal processor because today the network system bandwidth is often limited by the speed of the electronics used at network nodes.

Optical nonlinearities, which can change such material characteristics as refractive index or absorption, are key phenomena in terms of achieving optical signal processing.[5–7] But their coefficients are usually small. As a result, a high input power is required, which makes the device impractical. However, the input power can be significantly reduced if we can achieve strong light confinement.

Photonic crystals[8,9] are attracting considerable attention because of their strong light confinement and small structure. It has already been shown that photonic crystals can confine light in a very small space if we construct ultrahigh-Q (quality factor) photonic crystal nanocavities.[10,11] Since the photon density in these small cavities is extremely high, various optical nonlinearities occur at a small input energy, which enables the fabrication of practical all-optical switches[12] and logic gates.[13] Indeed all-optical switching at an input energy of less than 100 fJ has been demonstrated using a two-dimensional (2D) silicon photonic crystal nanocavity.[12]

In this chapter, we describe recent progress on 2D photonic crystal nanocavities, and introduce some applications. These applications include add-drop filters for wavelength division multiplexing, optical buffers, slow light, all-optical switching, bistable memory operation, and logic gates such as flip-flop and pulse retiming circuits.

1.2 Small Optical Cavities Fabricated on 2D Photonic Crystal Slabs

1.2.1 2D and 3D Photonic Crystals

Artificial dielectric structures with periodically modulated refractive indexes at a structural dimension close to an optical wavelength can alter the density of photonic states, thus allowing the creation of unique photonic band diagrams in terms of frequency and wavevector.[14] For those with 2D or 3D structures whose refractive index modulation contrast is relatively large (i.e. between air ($n = 1$) and semiconductors such as silicon ($n = 3.4$), GaAs, or InP) are called photonic crystals,[15] by analogy with the electrical property of natural crystals. Photonic crystals enable us to control the spontaneous emission or propagation of light.[8] For instance, light rapidly decays exponentially, as evanescent wave, when the photon energy (light frequency) is within the photonic forbidden band. In other words, light cannot penetrate a photonic crystal, which enables the fabrication of perfect mirrors. If we surround transparent material with photonic crystals, light cannot escape towards the outside, and this allows us to confine light securely in a tiny space. From this point of view, 3D photonic crystals are ideal because light propagation can be altered or prohibited in all dimensions. However, such crystals are difficult to fabricate, because complex woodpile structure[16] or highly sophisticated novel 3D material processing[17] is required. Fortunately, 2D semiconductor photonic crystals are relatively straightforward to fabricate, because we can utilize mature planar semiconductor processing technologies. Although 3D photonic crystals have been considered essential in terms of taking full advantage of the various unique properties of photonic crystals, recent studies have revealed that a 2D structure is sufficient for various applications such as optical switching[12] or even for spontaneous emission control[18] if they are designed carefully. Low-loss waveguides and ultrahigh-Q nanocavity fabrication play important roles with respect to these applications.

Figure 1.1 shows an example of 2D photonic crystals fabricated on a silicon slab, which was designed for telecom wavelengths. It should be noted that silicon is transparent at telecom wavelengths. The photonic crystals consist of hexagonal arrays with air rods. The calculated energy band diagram for a hexagonal photonic crystal is shown in Figure 1.2. The graph shows a clear forbidden band where no light propagation is allowed. Indeed, the experiment in Figure 1.3 shows low optical transmittance in the Γ–M and Γ–K directions for wavelengths of several hundred nanometers, which is clear evidence of the presence of a photonic band gap.

A low-loss waveguide is essential if we are to use photonic crystals as a platform for on-chip photonic integration. For photonic crystals with an air-rod array with a hexagonal lattice, one row with no holes can act as an optical waveguide, because light can propagate along this line defect. Figure 1.4a shows an example illustration of a photonic crystal line-defect waveguide fabricated on a silicon-on-insulator wafer, and Figure 1.4b shows an electron scanning microscope image of a fabricated air-bridge type photonic crystal waveguide. Figure 1.4c shows the band diagram for these structures. It should be noted that the vertical confinement of a 2D photonic crystal is achieved by total reflection at the slab surface. The critical angle of the total reflection is given by Snell's law. Starting from Snell's law, the condition for a 1D case is described as:

$$\omega = \frac{c}{n} k \tag{1.1}$$

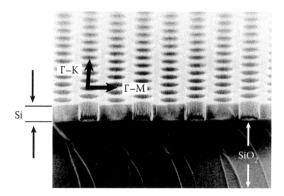

FIGURE 1.1
Scanning electron microscope image of a 2D photonic crystal fabricated on a silicon slab. Hexagonal arrays of air holes were fabricated by using electron beam lithography and dry etching. The diameter of the air hole is 200 nm and the lattice constant is 400 nm. The silicon slab is about 200 nm thick.

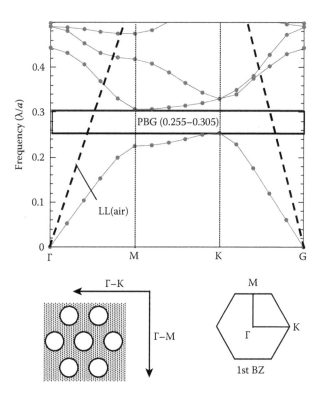

FIGURE 1.2
Energy band diagram of 2D photonic crystals with hexagonal air holes. LL: light line. PBG: photonic band gap. BZ: Brillouin zone.

where c is the light velocity and n is the refractive index of the cladding. Equation 1.1 is shown as a line, known as a *light line*, in a band diagram. Photons that are beyond this line satisfy the total reflection condition; hence they are vertically confined. The photonic crystal waveguide is theoretically lossless for a frequency component that is below the light line. According to Equation 1.1, the slope of the light line becomes steeper as the refractive

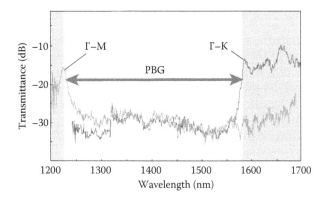

FIGURE 1.3
Transmittance spectrum of 2D photonic crystal in Γ–M and Γ–K directions.

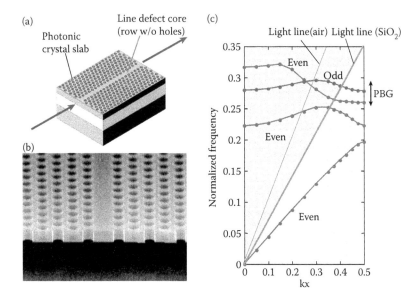

FIGURE 1.4
(a) Schematic image of a line defect silicon photonic crystal waveguide fabricated on a silicon-on-insulator wafer.
(b) Scanning electron microscope image of a fabricated air-bridge type silicon photonic crystal waveguide.
(c) Band diagram of a line defect photonic crystal waveguide.

index contrast between the slab and cladding increases. Indeed, Figure 1.4c shows light lines for air cladding ($n \sim 1$) and SiO_2 cladding ($n \sim 2.4$), where it can be seen that a larger light component will be confined in the air cladding case. Obviously, a steeper light line is preferred because it yields greater tolerance as regards the structural design. For this reason, an air-bridge structure, where the photonic crystal slab is sandwiched by air, is often preferred to a SiO_2 cladding structure.

The lowest reported propagation loss for an air-bridge photonic crystal waveguide is 2 dB/cm,[19,20] which is sufficiently small if we take the total size of the device into account. Since on-chip photonic devices will be much smaller than a centimeter in size, the total loss is less than 1 dB.

1.2.2 Ultrasmall Cavity: Photonic Crystal Nanocavity

Let us provide a simplified discussion to consider the smallest size optical cavity. Figure 1.5 is a schematic illustration of 1D Fabry–Pérot optical cavities of different lengths. The length of a cavity is multiples of a half wavelength, because it must satisfy the standing wave condition. Since, the wavelength of the light in a material with a refractive index of n is written as λ/n, the cavity length is given as $N(\lambda/2n)$, where N is a positive integer. Thus the shortest length possible for a 1D optical cavity is $\lambda/2n$, i.e. $N=1$, as shown in Figure 1.5c. Hence, the smallest volume of an optical cavity should be close to $(\lambda/2n)^3$. It should be noted that this discussion is just a simplified one, and the value is not accurately the theoretical limit of the small modal volume. However, this simplified picture helps us to understand intuitively the smallest size of an optical cavity.

There are various approaches that can be used to achieve small optical cavities. These include microrings,[21] microdiscs,[22] micro-pillars,[23] spherical,[24] and toroidal[25] cavities. However, in terms of size, photonic crystal nanocavities can achieve the smallest optical mode volumes. In fact, a cavity with a size of just $1.18(\lambda/2n)^3$ has recently been reported[26] using a point shifted cavity[27] on 2D photonic crystals. This value is close to the above discussed volume. Other types fabricated on 2D photonic crystals also have small volumes. For instance the mode volume is $1.6\,(\lambda/n)^3$ for a hexapole cavity and $1.1(\lambda/n)^3$ for a waveguide width modulated cavity, both of which are very close to an optical wavelength in size.

In addition to the small size, photonic crystal nanocavities are suitable for the integration on a chip. Microcavities such as microdiscs, micro-pillars, spherical, and toroidal cavities often couples light directly into space or into an optical fiber, which makes the integration

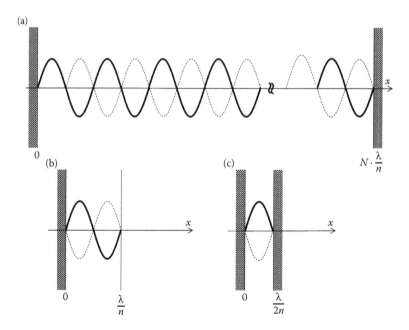

FIGURE 1.5
(a) Schematic illustration of an optical cavity with a length of $N\lambda/n$. (b) Optical cavity with a length of λ/n. (c) Smallest optical cavity with a length of $\lambda/(2n)$.

relatively difficult. On the other hand, photonic crystal nanocavities can couple light through photonic crystal waveguides, which are fabricated on the same chip. As a result, a large number of cavities can be connected in tandem or in parallel though photonic crystal waveguides. The surface of the 2D photonic crystal is flat, which makes the mechanical strength high. As a result, the optical coupling strength between the cavities and the waveguides is stable, once they are determined at the designing stage.

Small size, ultrahigh-Q and the ability of making integrated photonic circuit makes photonic crystal nanocavities attractive for various studies.

1.3 Designing High-Q Photonic Crystal Nanocavities

1.3.1 Design of High-Q Photonic Crystal Nanocavity

This section offers a quick review of the method used for designing a high-Q photonic crystal nanocavity system on a 2D photonic crystal slab. Q is determined by $Q = E/P_{out}$, where E is the energy stored in the cavity and P_{out} is the energy flux toward the outside of the cavity. Therefore, it is essential to reduce the optical loss (i.e. reduce the P_{out}) of the cavity to obtain a high Q.

Here, let us consider 2D photonic crystals with no structural errors. The theoretical Q of the nanocavity can be obtained by 3D finite difference time domain (FDTD) calculations. Since the nanocavity is surrounded by ideal photonic crystals on the horizontal plane, which has a perfect photonic bandgap, the light cannot leak towards the outside as regards the in-plane direction. Therefore, the sole loss is the out-of-slab radiation. Because the vertical confinement is achieved by total reflection, it is essential to look for a condition where as far as possible the light in the nanocavity satisfies the total reflection condition.

The idea of the light line provides a guideline for the design. To obtain stronger vertical confinement, it is essential to reduce the optical component that is above the light line. As discussed above, an air-bridge structure is preferred because the refractive index contrast between the semiconductor and the air is relatively large.

To reduce the number of optical components above the light line, Srinivasan and Painter proposed a momentum-space design,[28] where they considered the k-vector (the Fourier space of the real coordinate) distribution of the optical mode. To describe this briefly; if we divide the wavevector k into two components $k_{//}$ and $k_\perp$, it can be given as $k^2 = k_{//}^2 + k_\perp^2$, where $k_{//}$ and $k_\perp$ are the wavevectors of the in-plane and out-of-plane directions. Since the light line is given by Equation 1.1, $k_\perp^2 = (\omega/c)^2$ describes the cone in three dimensions for the air-cladding slab. Therefore, the strategy for achieving low vertical loss, hence for achieving an ultrahigh-Q in a 2D photonic crystal nanocavity system, is to find a structure where the Fourier transform of the optical mode yields very few components inside the light cone. Examples of the Fourier space distribution of the optical mode of the point defect photonic crystal are shown in Figure 1.6.

In terms of k-space design, the ideal optical mode profile is a sinc function, because it has a square shaped function in the k-space.[29] However it is not possible to find a structure that can exhibit a perfect sinc shaped optical mode, because sinc function requires infinite endpoints to define. Akahane et al. used a more convenient strategy to obtain a high-Q mode, namely they used a Gaussian function as a figure of merit.[30] Since the

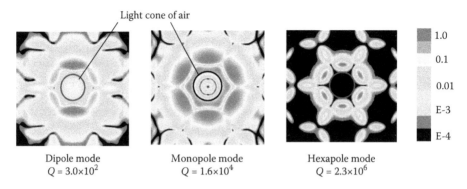

FIGURE 1.6
The *k*-space distributions of three different modes of a point defect photonic crystal nanocavity. The *Q* values calculated by 3D FDTD are shown in the panel.

Fourier function of a Gaussian curve is also Gaussian and its overlap with the light cone is relatively small, a Gaussian shape is a good practical shape for obtaining a high-*Q*. So far this strategy appears to work very well; in fact, a theoretical *Q* of about 2×10^7 has been obtained based on this strategy.[29]

1.3.2 Waveguide-Coupled High-*Q* Photonic Crystal Nanocavity

If we are to develop on-chip all-optical signal processing by using photonic crystal nanocavity systems, it is essential that we connect various nanocavities in tandem or in parallel by using in-plane photonic crystal waveguides. It should be noted that the *Q* values of waveguide-coupled cavities are different from that of an isolated cavity that has no input/output. The *Q* of an isolated cavity is called the unloaded *Q*. It is usually determined by the out-of-slab radiation loss, because the horizontal loss is negligible when the wavelength of the light is within the photonic bandgap. Hence the unloaded *Q* is written as,

$$Q_{\text{unloaded}} = Q_{\text{v}} \tag{1.2}$$

where Q_v is the *Q* determined by the out-of-slab radiation loss. On the other hand, the *Q* of the cavity coupled to the waveguides is called the loaded *Q*, and its value depends on the out-of-slab radiation loss and the coupling between the waveguides. The loaded *Q* can be written as

$$Q_{\text{loaded}}^{-1} = Q_{\text{v}}^{-1} + Q_{\text{h}}^{-1} \tag{1.3}$$

where Q_h is the *Q* value determined by the coupling strength with the waveguides. Figure 1.7 shows this schematically. Note that the unloaded *Q* is identical to Q_v for an isolated cavity because Q_h^{-1} is null. By employing this equation and simple coupled mode theory,[31,32] we can derive the relationship between *Q* and the transmittance. The transmittance *T* of a cavity/waveguide system is given as,[33]

$$T = \left(\frac{Q_{\text{loaded}}}{Q_{\text{h}}} \right)^2 = \left(\frac{Q_{\text{h}}^{-1}}{Q_{\text{v}}^{-1} + Q_{\text{h}}^{-1}} \right)^2 \tag{1.4}$$

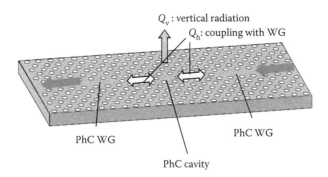

Q_v : vertical radiation

Q_h: coupling with WG

PhC WG

PhC WG

PhC cavity

FIGURE 1.7
Schematic illustration of photonic crystal nanocavity coupled to photonic crystal waveguides.

For a given Q_v, Equations 1.3 and 1.4 tell us that the loaded Q and the transmittance are in a tradeoff relationship. However, if we can design a high Q_v, it should be possible to achieve a high transmittance and a high loaded Q simultaneously. Therefore it is extremely important to find designs that can exhibit a high Q_v.

1.3.3 Various Types of High-Q Photonic Crystal Cavities

1.3.3.1 Line Defect Cavities with Modulated End-Holes

Of the various designs of photonic crystal nanocavities, line defect cavities with modulated end holes have been widely studied.[30,34,35] A schematic illustration of a shifted end-hole cavity is shown in Figure 1.8. To achieve a high Q, the position and the diameter of the end-holes are varied from those of the other holes. The position is slightly shifted towards the outside, and the hole diameter is smaller than that of the other holes. When the line defect consists of two missing holes, we call it an L2 cavity. When the line defect consists of x missing holes, it is called an Lx cavity. Figure 1.9a shows the k-space (in-plane wavevectors) distribution of an L2 cavity with shifted end holes. The component in the light cone is smaller for an end-hole shift of $0.199a$ than for a $0.064a$ shift. As a result, it exhibits a higher Q. The Q reaches its maximum value of 2.9×10^4 at a shift of ~$0.2a$ as shown in Figure 1.9b. Further optimization is possible by fine-tuning the end-holes, for example, by changing their diameter. Q exhibited a value of 3.8×10^5 when the innermost hole diameter was changed to $0.4a$.[36] Various sophisticated fine-tuning techniques have been reported in order to achieve higher Q values.[37]

Figure 1.10 shows an example of the fabricated L3 and L4 silicon photonic crystal nanocavities coupled to input and output photonic crystal waveguides. In addition to the modulation of the cavity end holes (r-holes), the end holes at both the input and output waveguides (c-holes) are modulated to control the coupling between the cavity and the waveguides. When the cavities and waveguides are placed in a straight line, this is referred to as in-line coupling. On the other hand, the L4 cavity shown in Figure 1.11 is coupled with the waveguides at an angle of 60 degrees, which we call a shoulder coupled configuration. Although the fabricated in-line coupled and shoulder coupled Lx cavities exhibit fairly similar optical properties, shoulder coupling has certain advantages in the design stage. As regards the design, shoulder coupling is more straightforward and easier to understand. This becomes clear when we consider the field distribution of the resonance for a Lx nanocavity. The electrical field of the cavity mode decays smoothly at an

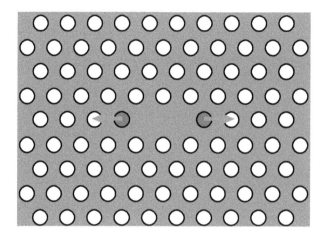

FIGURE 1.8
Schematic illustration of a line defect photonic crystal nanocavity with shifted end-holes.

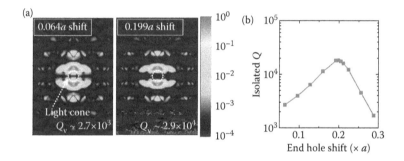

FIGURE 1.9
(a) Fourier space distribution (k-space) of the spatial mode of the L2 cavity with different end-hole shifts. The light cone is indicated by the dotted circle. a is the lattice constant. (b) Isolated Q (Q_v) of the L2 cavity with different end-hole shifts. Q reaches a maximum value of 2.9×10^4 at an end-hole shift of ~$0.2a$. The slab thickness is $0.5a$, and the hole diameter is $0.55a$.

angle of 60 degrees as shown in Figure 1.11, which makes it easier for the cavity field to overlap the waveguide ends when the waveguides are placed in a shoulder coupled direction. In addition, the optical field of the waveguide termination exhibits gradual decay in the 60-degree direction.[38] As a result the coupling between the cavity and the waveguides always becomes smaller as the cavity/waveguide distance is increased for shoulder coupling. This is more complicated with in-line coupling.

Figure 1.12 shows the calculated and measured spectra of the L3 cavity shown in Figure 1.10a and b. The resonance at 1550.36 nm yields a theoretical Q_v of ~3×10^4 compared with 1547.68 nm and $Q = 1.84 \times 10^4$ for the experiment. Since the transmittance for the fabricated sample was very small ($T = \sim 2.1\%$) the measured Q should be almost identical to the Q_v of the cavity. The mode volume calculated using 3D FDTD is 7.2×10^{-2} μm³.

Figure 1.13 is the measured spectrum of the L4 cavity shown in Figure 1.10b. It has two resonant modes, which we call mode C and mode S. The Q values are $Q_c = 1.15 \times 10^4$ for mode C and $Q_S = 2.3 \times 10^4$ for mode S. The corresponding photon lifetimes are $\tau_c = 9.3$ ps and $\tau_S = 19.1$ ps.

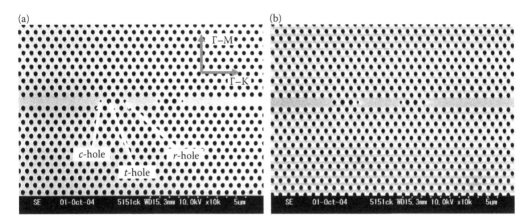

FIGURE 1.10
Scanning electron microscope image of fabricated silicon photonic crystal nanocavities. The lattice constant for both samples is $a = 420$ nm with a hole diameter of $d = 0.55a$. (a) An L3 nanocavity is fabricated with c-, t-, and r-hole diameters of 0.25a, 0.57a, and 0.3a, respectively. The r-holes are shifted 40 nm in the t-hole direction. (b) Four point defect nanocavity with c-, t-, and r-hole diameters of 0.45a, 0.57a, and 0.3a, respectively. The r-hole shift is 60 nm in the t-hole direction.

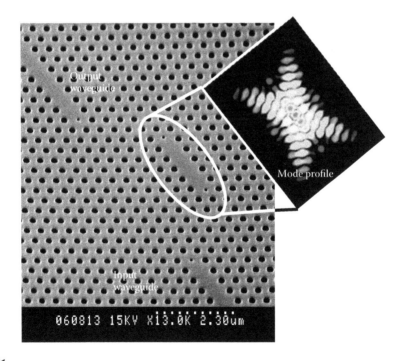

FIGURE 1.11
A silicon photonic crystal L4 cavity coupled to an input/output waveguide in the Γ–M direction. The inset is the mode profile shown on a log scale.

Historically, a line-defect cavity with shifted end holes has significantly increased the Q of 2D photonic crystal nanocavity systems and given them various possible applications.[30] Therefore, although there are now a number of other types of cavities that can exhibit a much higher Q, this type of cavity has been widely employed for various applications

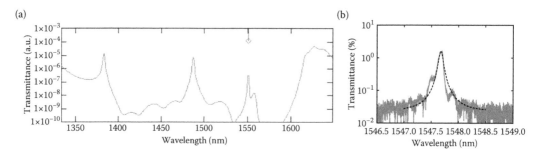

FIGURE 1.12
(a) Calculated transmission spectrum of an L3 cavity, which has the design parameters of the cavity shown in Figure 1.10. (b) The measured transmittance spectrum of the fabricated silicon photonic crystal L3 cavity shown in Figure 1.10. (Reprinted with permission from Tanabe, T. et al., *Appl. Phys. Lett.*, 87(15), 151112, 2005. © American Institute of Physics.)

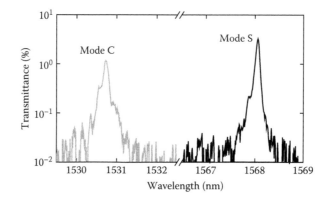

FIGURE 1.13
Transmittance spectrum of the silicon photonic crystal L4 cavity shown in Figure 1.10b. The Q values of mode C and mode S are $Q_C=1.15\times10^4$ and $Q_S=2.3\times10^4$, respectively. (Reprinted with permission from Tanabe, T. et al., *Appl. Phys. Lett.*, 87(15), 151112, 2005. © American Institute of Physics.)

such as quantum electro-dynamics studies,[39,40] ultrasmall wavelength add-drop filters,[41] all-optical switching[12] and optical bistable devices.[13]

1.3.3.2 Point Defect Hexapole Cavity with Rotational Symmetry Confinement

A schematic image of a point defect cavity is shown in Figure 1.14. The cavity consists of one missing hole, and it exhibits various optical modes, namely dipole, quadrupole, and hexapole modes. The optical field patterns for these cavity modes are shown in Figure 1.15. In particular, the hexapole mode exhibits an ultrahigh Q.[42] To obtain the highest Q, the parameters of the innermost holes are slightly different from those of the other holes. By setting the radius of the innermost holes at $0.23a$ (the radius of the other holes is $0.25a$) and shifting the holes slightly towards the outside ($c_k=1.18a$), a theoretical Q of 3.3×10^6 has been obtained.[43] Recently, it has been found that a high Q can be obtained without changing the hole diameter, which makes fabrication much easier.[44] The theoretical Q with respect to the hole shift is shown in Figure 1.16. A maximum Q of 1.6×10^6 was obtained at $c_k=1.26a$. The mode volume is $1.18(\lambda/n)^3\approx0.11\ \mu m^3$.

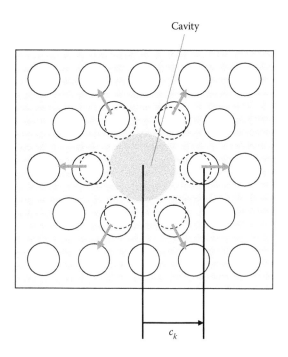

Cavity

c_k

FIGURE 1.14
Schematic image of a point defect photonic crystal nanocavity. The innermost holes are shifted slightly towards the outside to optimize the Q of the hexapole mode.

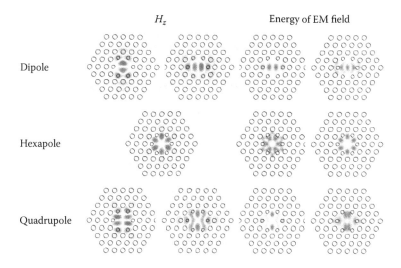

H_z Energy of EM field

Dipole

Hexapole

Quadrupole

FIGURE 1.15
Magnetic and electro-magnetic fields of the dipole, hexapole and quadrupole modes seen in a point defect photonic crystal nanocavity.

The explanation for the ultrahigh Q obtained in the hexapole mode is different from that for line-defect type photonic crystal nanocavities. Figure 1.17 shows the mode distribution and phase property of the magnetic field. The beauty of its spatial rotational symmetry originates in the matching between the symmetry of the optical mode and the pattern of the holes in the hexagonal photonic crystal lattice. In addition, because the phase of the

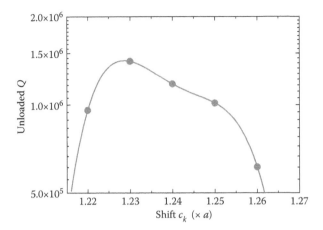

FIGURE 1.16
Calculated theoretical Q of the hexapole mode for different hole shifts c_k, where c_k is as shown in Figure 1.14. (Reprinted with permission from Tanabe, T. et al., *Appl. Phys. Lett.*, 91(2), 021110, 2007. © American Institute of Physics.)

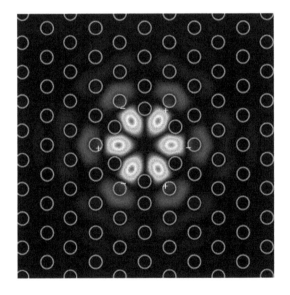

FIGURE 1.17
Profile of hexapole mode. The + and − represent the phase of the H_z magnetic field (z: perpendicular to the slab).

nearby H_z component is reversed (shifted by π), as indicated in Figure 1.17, the optical far field is cancelled out. As a result, the out-of-slab radiation is reduced by the destructive interference effect of the far-field pattern, which was initially discussed by Johnson et al.[45] Hence, the hexapole mode can exhibit an ultrahigh Q. We call this light confinement mechanism "rotational symmetric confinement".

The rotational symmetries yield different interesting characteristics of the hexapole mode when it is coupled with waveguides. Figure 1.18a and b, respectively, show the in-line coupling and side coupling of the hexapole mode with the waveguides. Owing

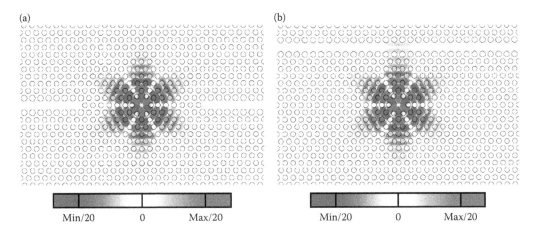

FIGURE 1.18
(a) Butt-coupled structure of separation=7 with H_z distribution. (b) Side-coupled structure of separation=7 with H_z distribution. (From Kim, G.-H. et al., *Opt. Express*, 12, 6624–6631, 2004. © Optical Society of America. With permission.)

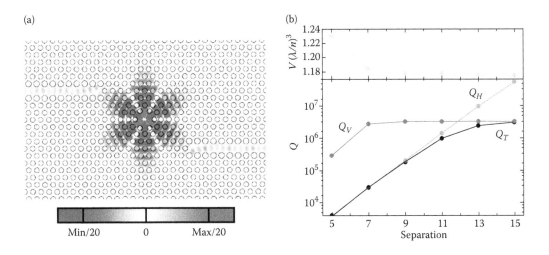

FIGURE 1.19
(a) Shoulder-coupled structure of separation=7 with H_z distribution. (b) Modal volume (V) and quality factors (Qs) of hexapole mode. (From Kim, G.-H. et al., *Opt. Express*, 12, 6624–6631, 2004. © Optical Society of America. With permission.)

to the symmetries of the optical mode, the light hardly couples with the waveguides. This characteristic enables the cavity to be extremely well isolated from closely positioned waveguides or cavities, which offers the possibility of the dense packing of cavity-waveguide systems. To couple the hexapole mode with the waveguides, the waveguides are placed in a shoulder-coupled configuration as shown in Figure 1.19. This configuration enables efficient coupling of the optical mode with the waveguides. When the separation is 7, Q_{loaded} is almost identical to Q_h. According to Equation 1.4, the transmittance is nearly 100%. For a separation of 9, Q_T is 1.9×10^5 and the transmittance is 88%. In fact, the transmittance of the fabricated sample was almost 100% when the separation was 9.[46]

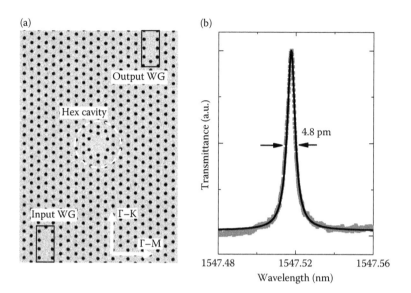

FIGURE 1.20
(a) Scanning electron microscope image of a hexapole cavity with photonic crystal waveguide ends fabricated on a silicon photonic crystal. WG: waveguide, Hex: hexapole. (b) Transmittance spectrum of the hexapole mode. The solid line is the fitted Lorenz curve. (Reprinted with permission from Tanabe, T. et al., *Appl. Phys. Lett.*, 91(2), 021110, 2007. © American Institute of Physics.)

Figure 1.20 shows a scanning electron microscope image and the transmittance spectrum of the fabricated sample. The obtained spectrum exhibits an extremely high Q of 3.2×10^5, which is the highest experimental Q yet reported for photonic crystal nanocavities, except for those with mode-gap confinement.[44] This cavity is a good candidate for achieving an ultrahigh Q in photonic crystals.

1.3.3.3 Width-Modulated Line Defect Cavity with Mode-Gap Confinement

Figure 1.21 shows a schematic diagram of the width-modulated line defect cavities.[10,47] The air holes are shifted slightly towards the outside to modulate the width of the line defect locally. It should be recalled that a line defect exhibits propagation modes as shown in Figure 1.4c. Since the propagation mode becomes flat as k_x increases, some frequency are not allowed to propagate the line defect, which is known as a mode gap. The frequency of the propagation mode can be shifted by changing the width of the line defect. Namely, a line defect with a larger width has a smaller mode frequency. This yields a shifting of mode gap when the width of a line defect is changed. As shown in Figure 1.21, the width-modulated cavities are composed of two line defects with different widths. The optical cavity consists of a line defect with a larger width and the termination is achieved by using line defects with smaller widths. Since the optical frequency mode of the line defect used for the cavity is lower than the line defects used for the termination, the light is localized at the cavity. In addition to the in-plane confinement, it is essential to reduce the out-of-slab radiation to obtain an ultrahigh Q. For this purpose, the original lossless propagation mode profile of the line defect must be kept in the k space. Therefore, the positional shifts of the holes at the cavity are kept small and they are gradually tapered along the line defect as shown in Figure 1.21. The typical value of the shift is a few nanometers. Other types of resonators that utilize the mode gap of photonic band gap waveguides have also been discussed by Inoshita et al.[48] and Song et al.[49]

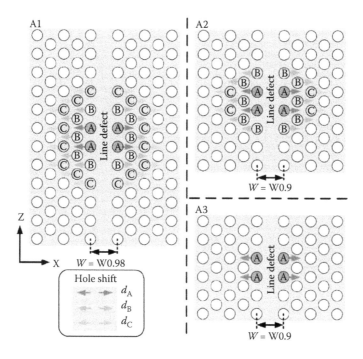

FIGURE 1.21
Different designs of width-modulated line defect cavity (A1, A2, A3). The width of the baseline defect is $0.98 \times a\sqrt{3}$ (W0.98) and $0.90 \times a\sqrt{3}$ (W0.90). Holes marked with A, B, and C are shifted slightly towards the outside of d_A, d_B, and d_C, respectively. Typical values for the A1 cavity are d_A=9 nm, d_B=6 nm, d_C=3 nm, a=420 nm, r=108 nm, t=204 nm, where r is the air hole radius and t is the slab thickness. (Reprinted with permission from Kuramochi, E. et al., *Appl. Phys. Lett.*, 88(4), 041112, 2006. © American Institute of Physics.)

Our first generation mode-gap confined silicon photonic crystal nanocavity is shown in Figure 1.22a. The obtained Q was not very high (Figure 1.22b and c), because the nanocavity was fabricated on SiO_2 cladding and the parameters were not optimized.[35]

Figure 1.21 shows the latest design, which exhibits a theoretical Q of 1.2×10^8 and a mode volume of $1.51(\lambda/n)^3$ for the cavity A1.[47] The calculated spatial mode profile is shown in Figure 1.23a. A scanning electron microscope image is shown in Figure 1.23b, where the circled region is the cavity. Although it is hard to distinguish, the holes at the cavity are slightly shifted towards the outside. The cavity exhibits an extremely high Q of 1.2×10^6, which is one of the highest Q values achieved by photonic crystal nanocavities.[10] This Q value corresponds to a photon lifetime of about 1 ns, which opens the possibility for various applications such as optical buffering and cavity quantum-electro dynamics. Note that there is an approximately two orders discrepancy between the theoretical and experimental values. This is caused by the material absorption and various types of scattering, such as absorption at the surface or scattering due to fabrication error. Of these we speculate that the variation in the hole diameter plays an important role in determining the experimental Q.

1.3.3.4 Other Photonic Crystal Nanocavities

It is difficult to provide a complete review of the various types of photonic crystal nanocavities, because there are a large number of different designs. Here, we introduce two cavities that are particularly important. One is the line defect type, which is called a heterostructure nanocavity. This was originally proposed by Song et al. in 2005.[49] It constituted

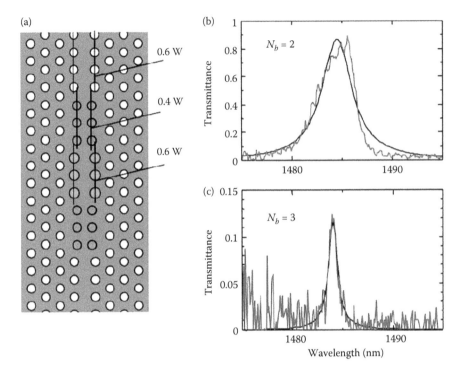

FIGURE 1.22
Resonant-tunneling filter using the mode-gap in the width-varied waveguides. The samples are fabricated on silicon-on-insulator photonic crystal slabs, where the undercladding is SiO_2. (a) Structural design. (b) Measured transmittance spectrum around the resonant wavelength for a barrier width of $N_b=2$. $Q=408$ and $T=86\%$. (c) Measured transmittance spectrum for $N_b=3$, $Q=1350$ and $T=12\%$. (From Notomi, M. et al., *Opt. Express*, 12, 1551–1561, 2004. © Optical Society of America. With permission.)

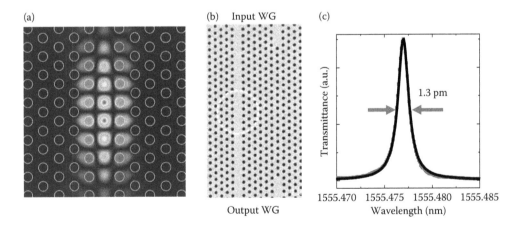

FIGURE 1.23
(a) Mode profile of width-modulated line defect cavity with mode-gap confinement shown in Figure 1.21 A1. (b) Scanning electron microscope image of the fabricated sample on silicon photonic crystal. The circle indicates the cavity region. (c) Transmittance spectrum of the cavity shown in (b).

a breakthrough in terms of achieving an ultrahigh Q with a value of ~6×10^5. The Q for this cavity is now ~2×10^6, which is the highest experimental Q value yet reported for a photonic crystal nanocavity.[11] It is a mode-gap confinement type cavity, where the lattice of the photonic crystal is changed to achieve the mode gap of the line defect.

Another important piece of work is the point-shift cavity proposed by Zhang and Qiu,[27] and recently fabricated by Nozaki et al.[26] It has a modal volume of only ~$1.18(\lambda/2n)^3$, which is the smallest value reported for any micro- or nanocavity, while maintaining a reasonably high Q of 2×10^4. Owing to the high Q, they achieved continuous-wave laser operation at room temperature at a very low power. It is also the smallest reported laser.

As shown by both examples above, photonic crystal nanocavities can exhibit an extremely high Q and also an ultrasmall mode volume. The optimal design has not yet been determined theoretically, and therefore there is still the possibility of finding an improved design that exhibits a higher Q and a smaller modal volume. In addition, it is also important to find a design that is as tolerant as possible in terms of such structural variations as hole diameter variation, position variation or sidewall roughness, because nanofabrication cannot produce an ideal structure.

1.3.4 Discussion of Structural Error and Q

The highest experimental Q reported in a photonic crystal nanocavity is about one order lower than the theoretical value. The cavity Q decreases if material absorption or optical scattering losses are present. To understand the mechanism behind the discrepancy between the theoretical and experimental Q values, 3D FDTD calculations have been performed that consider structure randomness.[20,44] Randomness was added with a Gaussian distribution to the radius of the air holes. Figure 1.24 shows the Q values for three different types of cavities with respect to the standard deviation of the Gaussian randomness. The Q value is 1.2×10^6 for a width-modulated line defect cavity at a randomness of 2.3 nm and 3.1×10^5 for a hexapole cavity at a randomness of 3 nm. These Q values are not far from

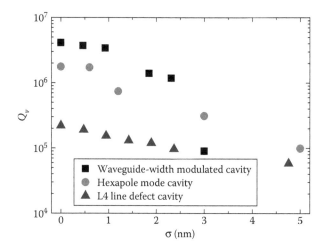

FIGURE 1.24
Calculated Q values for three different types of cavities (waveguide-width modulated cavity, hexapole mode cavity, and L4 line-defect cavity), with respect to radius randomness.

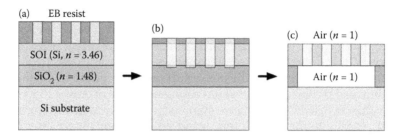

FIGURE 1.25
Schematic diagram of the air-bridged silicon photonic crystal slab. (a) Photonic crystals are patterned on resist by electron beam lithography. (b) Photonic crystal is formed on silicon by dry etching. (c) The underlying SiO_2 layer was removed by wet-etching.

those obtained experimentally. Hence, a standard deviation of 2–3 nm provides a good fit with the experimental results. Although such a small radius variation is very difficult to estimate with a scanning electron microscopic measurement, we believe that the value is not far from the actual fabrication accuracy. Therefore we speculate that the limitation of the experimental Q results from the randomness in the air-holes radius. Further detailed discussions can be found elsewhere.[20]

1.3.5 Fabrication of Photonic Crystal Slabs

We will briefly review the fabrication of the silicon photonic crystal slab. A schematic diagram of the fabrication process is shown in Figure 1.25. First we start with an SOI wafer spin coated with ZEP resist. Then, photonic crystals are patterned by 100-keV electron beam lithography. After the resist has been developed, the silicon is selectively etched by using inductively coupled plasma and fluorinated gas. To fabricate the air-bridge structure, the underlying SiO_2 layer was removed by selective wet etching using HF solution. Note that photonic crystals can also be fabricated using other materials such as GaAs or InP. Photonic crystal fabrication using other materials such as AlGaAs is detailed elsewhere.[50]

1.4 Characterization of Ultrahigh-Q Photonic Crystal Nanocavities

1.4.1 Spectral Domain Measurement

The Q of a cavity is given by

$$Q = \omega \times \frac{E_{cav}}{P_{out}} \tag{1.5}$$

where ω is the resonant angular frequency of the cavity, E_{cav} is the energy stored in the cavity, and P_{out} is the output power (i.e. optical energy loss per unit time) from the cavity. We can derive a different expression from Equation 1.1, where Q is described in the spectral domain.

$$Q = \frac{\omega}{\Delta\omega} \approx \frac{\lambda}{\Delta\lambda} \tag{1.6}$$

Here, $\Delta\omega$ and $\Delta\lambda$ are the linewidth of the resonant spectrum in terms of frequency and wavelength, respectively. λ is the resonant wavelength of the cavity. Obviously, Equation 1.6 suggests that the cavity Q can be determined by measuring the transmittance linewidth of the cavity when the cavity is coupled with the waveguides. Indeed the Q values of photonic crystal nanocavities have been characterized in the spectral domain. However, it should be noted that spectral domain measurement is now not an easy task because the Q of photonic crystal nanocavities has increased rapidly, and therefore, ultrahigh wavelength resolution is needed to measure the ultra-narrow transmittance linewidth of a nanocavity accurately. For a telecom wavelength, the linewidth for a cavity with a Q of 1.2×10^6 is about 1.3 pm as shown in Figure 1.22. Therefore a sub-pm wavelength resolution is required.

1.4.1.1 Spectrum Measurement with Frequency Tunable Laser

For a high-Q characterization measurement, a frequency tunable narrow-linewidth laser is often used because it generally provides better wavelength resolution than a spectrum analyzer composed of gratings. The transmittance at each wavelength is monitored with an optical power meter. Indeed the spectrum shown in Figure 1.22 was obtained by this method. Since the photon density in the cavity is extremely high owing to the large Q/V, the input power must be kept sufficiently low to prevent the cavity from exhibiting optical nonlinearity. Optical nonlinearity is observed at an input power of a few µW.[10]

1.4.1.2 Spectrum Measurement using Electro-Optic Frequency Shifter

Commercially available wavelength swept tunable laser diodes have a typical frequency resolution of a few 100 MHz. In contrast, the Q of a photonic crystal nanocavity is rapidly increasing. Therefore, it is important to develop a different measurement technique that allows higher frequency resolution. Frequency sweeping via fiber-coupled acousto-optic frequency modulators supports high frequency resolution, but the maximum achievable tuning range is limited to a few 100 MHz, and this range is often insufficient to obtain the complete transmission spectrum. A promising approach is to use an electro-optic single sideband modulator fabricated from lithium niobate as a frequency sweeper.[51] By applying radio-frequency waves to the device, we can shift the frequency of the laser light according to the frequency of the radio wave. As a result, the frequency resolution with this measurement is only limited by the linewidth of the input laser light, which is typically 100 kHz or less. Figure 1.26 shows the output spectrum from a single-sideband modulator when a 10-GHz ratio-wave frequency is applied. We can obtain a suppression ratio of more than 20 dB between the 1st and –1st order single sidebands. By adjusting the DC biases applied to the single-sideband frequency modulator, we can selectively maximize the 1st or –1st order single sideband.

The measured transmittance spectrum of the waveguide-modulated cavity in Figure 1.22b is shown in Figure 1.27. The data was fitted using a Lorenz curve, and we obtained a Q of 1.23×10^6 from the transmittance spectrum width. The laser frequency was set at 1564.85 nm, which is slightly detuned to a longer wavelength from the cavity resonance. The frequency of the laser light was shifted to a shorter wavelength with a single-sideband modulator. The plotting interval is 1 MHz, but it is possible to use a smaller value. The corresponding photon lifetime from the measured transmittance spectrum is 1.02 ns, which is in good agreement with the time domain measurement (1.09 ns) described in the

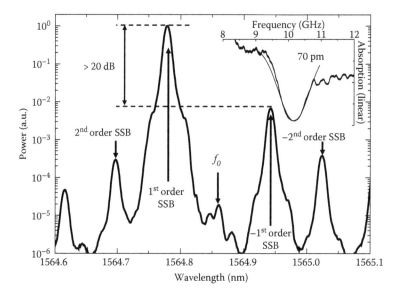

FIGURE 1.26
Optical output spectrum from single-sideband modulator at a radio frequency of 10 GHz. Inset is the spectroscope image of the absorption of acetylene ($^{13}C_2H_2$) gas at room temperature obtained with the single-sideband modulator setup to demonstrate the accurate spectrum measurement. (Reproduced by kind permission of the IET from Tanabe, T. et al., *Electron. Lett.*, 43, 187–188, 2007.)

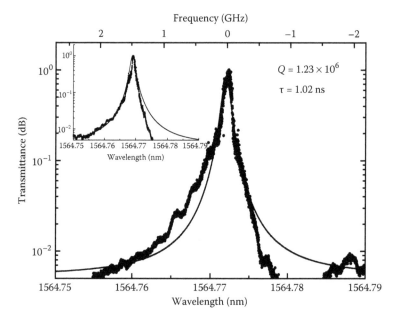

FIGURE 1.27
Measured spectrum obtained using first-order single-sideband light scan. Inset is the measured spectrum when a –1st-order single sideband is used for the scan. The dots are the measured plot and the solid line represents the fitted Lorenz function. (Reproduced by kind permission of the IET from Tanabe, T. et al., *Electron. Lett.*, 43, 187–188, 2007.)

next section. Measurement using a single-sideband modulator provides a high resolution (~100 kHz depending on the laser linewidth) and a wide measurement bandwidth (10 GHz or more), which will enable us to measure the Q value in the spectral domain even if the value continues to increase in the future.

1.4.2 Time Domain Measurement

Although the current spectral domain measurement technique is capable of measuring the highest Q yet achieved in photonic crystals, time domain measurement is an alternative and sophisticated way of characterizing the Q of a photonic crystal nanocavity. By using Equation 1.5, the Q of a cavity can be derived as,

$$Q = \omega\tau \tag{1.7}$$

where τ is the photon lifetime of the cavity. It is not very easy to measure τ when Q is small; but the τ measurement becomes more accurate as Q increases. Indeed a Q of 1.2×10^6 corresponds to a photon lifetime of ~1 ns at telecom wavelengths. The ring-down method is a direct way to characterize the τ and Q values of a cavity. First, the cavity is charged with an input CW light, and then the input is suddenly turned off. τ can be obtained by observing the decaying optical signal at the output waveguide. A schematic diagram of the measurement is shown in Figure 1.28. Note that the signal light must be kept sufficiently low to prevent the cavity exhibiting nonlinearity such as two-photon absorption or free-carrier absorption, which may modify the cavity Q. Therefore we employ time-correlated single photon counting for the ultrahigh-Q measurement, as this allows us to measure an extremely weak signal light with a time resolution of ~70 ps.[10]

The measured signal is shown in Figure 1.29. The decay is a smooth exponential curve, and the fitted decay is 1.01 ns, which is the photon lifetime of the cavity. This value corresponds to a Q of 1.2×10^6, which is in good agreement with the value obtained from a spectral-domain measurement. Time domain measurement provides a direct view of the photon trapping by the photonic crystal nanocavity.

Another aspect of the time domain measurement can be clarified by performing a ring-down measurement on a side-coupled cavity.[52] Figure 1.30a and b show a schematic diagram and the transmittance spectrum of a side-coupled cavity. The transmittance

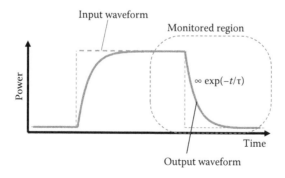

FIGURE 1.28
Schematic diagram of the ring-down measurement. The dotted line represents a rectangular input pulse and the solid line represents the output from a cavity. (From Tanabe, T. et al., *Opt. Express*, 15, 7826–7839, 2007. © Optical Society of America. With permission.)

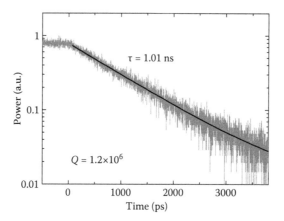

FIGURE 1.29
Output of a ring-down waveform from the waveguide-width modulated silicon photonic crystal nanocavity shown in Figure 1.22b.

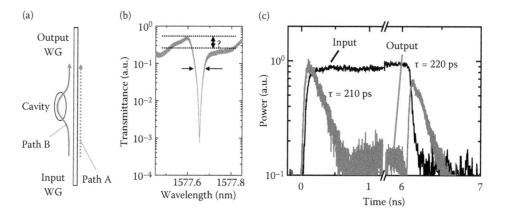

FIGURE 1.30
(a) Schematic diagram of a side-coupled cavity with a photonic crystal waveguide. (b) Transmission spectrum of the side-coupled cavity. Sometimes it is difficult to obtain the Q from the spectral bandwidth because of the background Fabry–Pérot oscillation. (c) The output waveform of a side-coupled cavity when a rectangular pulse is applied.

spectrum exhibits a dip at the resonance as shown in Figure 1.30b. Side-coupled cavity configurations are useful for constructing multi-channel add-drop filters or photonic DRAMs.[20,53] The transmitted waveform of a 6-ns square input pulse is shown in Figure 1.30c, along with the output pulse. The wavelength of the input light is adjusted to the cavity resonance; therefore, the light cannot propagate though the device in a steady state. However, the pulse transmits at the rising and falling edges of the square pulse. This can be explained as follows. The light propagates until the cavity is charged and forms interference. Interference occurs between the light that travels straight from the input waveguide toward the output waveguide (path A) and the output light from the side-coupled nanocavity (path B). When the light is turned off, path A is immediately cut off. As a result the light cannot interfere and the path B light is observed as a ring-down waveform. The discharging signal is observed after the input has been turned off, which constitutes an intuitive demonstration of light trapping in a photonic crystal nanocavity.

1.4.3 Technical Issues Related to Obtaining Accurate Q

With spectral domain measurement, it is not always easy to obtain an accurate quality factor when the transmittance spectrum is very narrow. The laser frequency must be precisely calibrated in order to obtain an accurate wavelength. In addition, we have to check the wavelength resolution of the measurement system in advance if we are to discuss the value. This wavelength resolution can be checked by measuring a reference sample, such as a high-finesse Fabry–Pérot etalon.[52]

Moreover, the presence of the temperature fluctuation makes measurement difficult. The thermo-optic coefficient of silicon is 3.9×10^{-5} K^{-1}. This results in a cavity wavelength shift of about 1 pm for a small temperature change of only 17 mK. Therefore it is essential to place the photonic crystal nanocavity sample in a temperature-stabilized environment, and complete the measurement as quickly as possible.

In practice, the stability and reproducibility (hence the accuracy) of the measurement can be confirmed by repeating the acquisition multiple times and obtaining the average value while also obtaining the standard deviation.

Most importantly, the Q values obtained with the spectral and temporal measurements should be identical. The measurement is reliable when the two results are the same. Figure 1.31 summarizes and compares the Q values for different cavities obtained with spectral and time domain measurements. Note that with the ring-down measurement for a lower Q cavity, it is possible to use a digital sampling oscilloscope that has a bandwidth of 28 GHz instead of using a time correlated single photon counter system, because we can input more light into the cavity without inducing any nonlinearity. Q_{time} and Q_{spec} show good

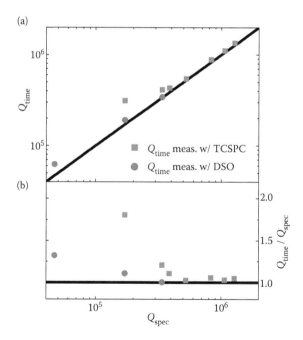

FIGURE 1.31
Q_{spec} is the Q value obtained from the transmittance spectrum bandwidth and Q_{time} is the Q value obtained from the decay of the ring-down waveform. (a) Square dots show the Q_{time} measured using time correlated single photon counting and round dots show the Q_{time} measured using a digital sampling oscilloscope. The dotted line indicates the ideal case. (b) Q_{time}/Q_{spec} with respect to Q_{spec}. (From Tanabe, T. et al., *Opt. Express.*, 15, 7826–7839, 2007. © Optical Society of America. With permission.)

agreement, which confirms the reliability of the measurements. Figure 1.31 also suggests that a detector speed of ~70 ps is sufficient to measure a cavity that has a Q of >10^5, but a faster response is required to obtain an accurate Q for those cavities with a value of <10^5. A detailed discussion of accurate measurement in the spectral and time domains is provided elsewhere.[52]

1.5 Applications of High-Q Photonic Crystal Nanocavities

1.5.1 Caging Light and Slow Light

1.5.1.1 Caging Light using Ultrahigh-Q Photonic Crystal Nanocavity

A high-Q nanocavity can store photons for a long time in a small area. Although the demonstrated photon lifetime for photonic crystal nanocavities fabricated on 2D silicon slabs is currently of the order of nanoseconds, the theoretical photon lifetime is more than 100 times longer. This offers the potential to develop optical buffers, which are the key to the exploitation of future all-optical signal processors. In addition, photonic crystal nanocavities can constitute a basic element of photonic DRAM, if their Q can be dynamically controlled.[20,53] For this purpose, the coupling between cavities and waveguides has to be dynamically tuned in order to store and release the photons.[20] Experimentally, the first step towards the dynamic tuning of the Q value is to control the intrinsic Q of the cavity. This can be achieved by injecting the light from the top of the slab and controlling the generation of free carriers and changes in the free carrier absorption within the photon lifetime of the cavity. As a result, the photons caged in the cavity can be eliminated faster than the original photon lifetime determined by the cavity Q. The demonstrated result is shown in Figure 1.32. The photon lifetime is 250 ps in the absence of the pump pulse. However, the decay immediately after a 2-ps wide, 800-nm pump pulse is 30 ps, which shows that the intrinsic Q of a cavity has been modified. The absorbed energy of the pump pulse is estimated roughly at 12 fJ. This demonstration shows that the trapped photons can be eliminated faster than the original photon lifetime by reducing the cavity Q dynamically, which enables the nanocavity to be discharged and reset. Although the operation described above is not a direct demonstration of read-write memory, coupled cavities will allow the sophisticated trapping and release of photons.[20,53,54]

1.5.1.2 Slow Light with Photonic Crystal Nanocavity

Slow light on a chip is attracting considerable attention, because it is expected to be used for compact optical delay lines. Slow light with a group velocity of ~$c/100$ has been demonstrated using photonic crystal waveguides, where c is the speed of light in a vacuum.[55,56] Recently, coupled resonator optical waveguides (CROW) have been studied extensively, due to their potential for achieving an extremely slow group velocity.[57] The half-bit shift of a 1 GHz signal has been demonstrated on a chip using a micro-ring based CROW.[58] The footprint of the device is only 4.5×10^4 μm^2 for $N=100$, where N is the number of cavities. Since the group velocity of CROWs relies on the Q of the cavity, it is essential to employ cavities with large Q/V values to achieve a large group delay and a small footprint simultaneously. From this point of view, the photonic crystal nanocavity based CROWs constitute promising candidates for achieving ultra-slow light in an ultra-compact device on a chip. Since the basic element of a CROW is a high-Q nanocavity, it is important to study the pulse delaying performance of a photonic crystal nanocavity to show the potential of the photonic crystal based CROW.

Although photon trapping and pulse delaying phenomena are closely connected, they are explained in different ways. Since the photon lifetime of a cavity is determined by the width of the cavity resonance spectrum, it can be regarded as the spectrum amplitude characteristic of the cavity (Figure 1.33a). On the other hand, the pulse delay is determined by the dispersion of the cavity. The delay is given by the tilt of the dispersion shown in Figure 1.33b. Thus the pulse delaying phenomenon can be regarded as the phase property of the cavity. Note that the amplitude and dispersion of a cavity are related by the Kramers–Krönig equation. As shown in Figure 1.33b, the slope of the phase curve gives the group delay and the maximum delay should be about two times greater than the photon lifetime for a relatively long input pulse.

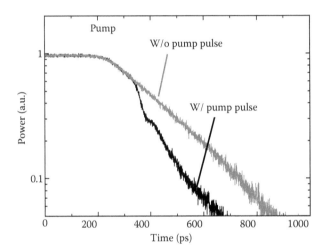

FIGURE 1.32
Example of the dynamic tuning of the photon lifetime of a silicon photonic crystal nanocavity. (Reprinted by permission from Macmillan Publishers Ltd from Tanabe, T. et al., *Nature Photon.*, 1, 49–52, 2007.)

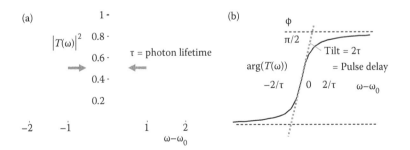

FIGURE 1.33
(a) Schematic illustration of the transmittance intensity of a Fabry–Pérot cavity. The photon lifetime is determined by the transmittance bandwidth. (b) Schematic illustration of the phase property of a Fabry–Pérot cavity. The pulse delay is given by the tilt of the dispersion curve.

Next, we discuss the relationship between the group velocities of a single cavity and a CROW. A schematic explanation is shown in Figure 1.34. Recall that the maximum delay is ~2τ. When we determine the device length L of a waveguide coupled single photonic crystal nanocavity as shown in Figure 1.34b, the smallest group velocity is given as $v_g \approx L/2\tau$. Now, let us consider a case where the cavities are tandem connected. If the horizontal coupling strength is the same as that for a single cavity; roughly speaking, the pulse delay is multiplied by the number of cavities. In other words, if a CROW consists of N cavities, the maximum delay will be N times larger than that of a single cavity as shown in Figure 1.34c. However, the resulting velocity is the same as that for a single cavity, because the device length is N times greater (Figure 1.34d). This means that the group velocity demonstrated in the single cavity should correspond to the minimum value that can be obtained in an ideal CROW system.

Figure 1.35 shows the experimental pulse transmittance results for three different input pulses using a single nanocavity with a Q of 1.2×10^6 and an L of 7.6 μm. The delay of a pulse at the nanocavity is measured by comparing the output from the photonic crystal nanocavity with that of a reference waveguide. A Gaussian-like input optical pulse is generated by using an electrical pulse shaper and a 40-GHz electro-optic modulator. A maximum delay of 1.4 ns was obtained. The obtained delay corresponds to a group velocity of 5.4 km/s, and this value is the smallest for any dielectric based slow-light material. Figure 1.35 shows that different input pulse widths yield different delays. Additionally, the pulse broadens when it propagates through a high-Q cavity, and this is due to the narrowing of the spectral bandwidth. Note that bandwidth narrowing does not explain the pulse delay.

To investigate the pulse delaying phenomenon in detail, a numerical model was studied using simple coupled mode equations.[32] The input and the output of the cavity are given as,

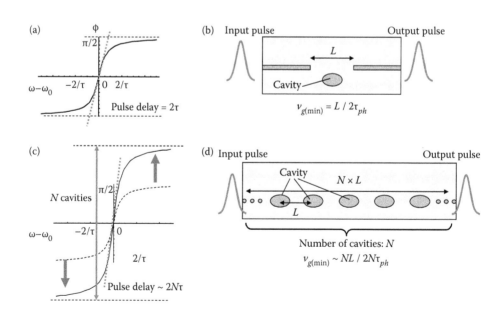

FIGURE 1.34
(a) Schematic illustration of the dispersion curve of a single cavity. (b) Schematic illustration of a single nanocavity coupled to waveguides. (c) Schematic illustration of the dispersion curve of a CROW consisting of N cavities. (d) Schematic illustration of a CROW consisting of N cavities.

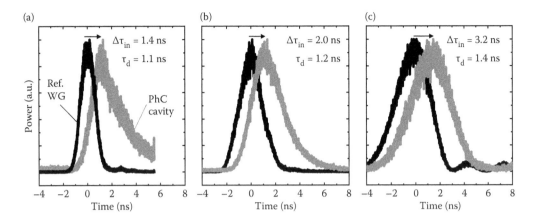

FIGURE 1.35
Output pulses from the reference waveguide (Ref. WG) and silicon photonic crystal nanocavity for different input pulse widths. $\Delta\tau_{in}$ is the input pulse width and the τ_d is the pulse delay.

$$\frac{da}{dt} = \left[j\omega_0 - (2\tau)^{-1} \right] a + (2\tau)^{-1/2} p_{in}(t) \tag{1.8}$$

$$p_{out}(t) = (2\tau)^{-1/2} a(t) \tag{1.9}$$

where,

$$P_{in}(t) = |p_{in}(t)|^2 \tag{1.10}$$

$$P_{out}(t) = |p_{out}(t)|^2 \tag{1.11}$$

P_{in}, P_{out}, and ω_0 are the input pulse power, output pulse power, and center angular frequency of the cavity resonance, respectively. We assumed a Gaussian shaped input pulse and normalized the time with the cavity photon lifetime τ. The calculated P_{in} and P_{out} values for different input pulse widths are shown in Figure 1.36. Different input pulse widths yield different delays. These calculations revealed that, for a practical input pulse width, the pulse exhibits a delay of about 1.6 times the photon lifetime. This is consistent with the experimental results.

The calculation results are summarized in Figure 1.37. As the width of the input pulse increases, the pulse delay gradually increases and then levels off at 2τ, which is consistent with the discussion in Figure 1.33b. However, the calculation also suggests that the delay is about 1.6τ for a practical pulse width, which agrees with the experiment. In Figure 1.33, the ratio of the pulse delay versus the output pulse width is given as the pulse shift, which reaches a maximum value of 0.43 when the input pulse width is ∼τ. Unfortunately, this result shows that single cavity is not very practical as an optical buffer. However, the delay can be significantly increased when a number of cavities are connected in tandem to form a CROW as shown in Figure 1.34. And regarding the group velocity, that demonstrated in a single cavity corresponds to the minimum value that can be obtained in an ideal CROW system. Indeed, photonic crystal based CROWs are currently being studied to obtain a large group delay with an extremely small footprint.[59]

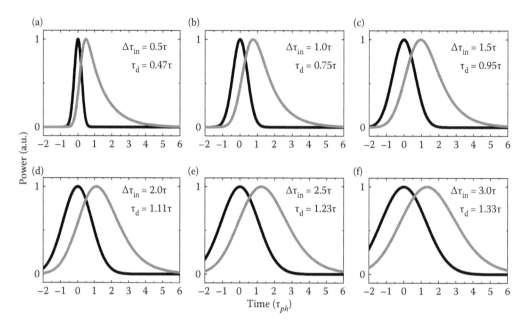

FIGURE 1.36
Calculated input (dotted line) and output (solid line) pulses from a photonic crystal nanocavity at different input widths. $\Delta\tau_{in}$ is the input pulse width and the τ_d is the pulse delay.

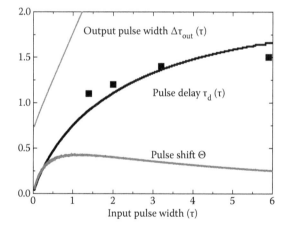

FIGURE 1.37
Three curves are shown with respect to the FWHM of the input pulse $\Delta\tau_{in}$. The FWHM of the output pulse $\Delta\tau_{out}$ (normalized with τ), the delay τ_d between the input and output pulse peaks (normalized with τ), and the pulse shift ($\Theta = \tau_d / \Delta\tau_{out}$). The square dots are the experimental pulse delay results.

1.5.2 Compact Optical Add-Drop Filter

By utilizing the photonic crystal nanocavity system as a wavelength selector, it is possible to construct a compact optical add-drop filter, which may be useful for wavelength division multiplexing applications. The basic idea behind the filter is to place a large number of photonic crystal nanocavities with different wavelengths, in parallel along a photonic crystal waveguide. A wavelength component that is in resonance

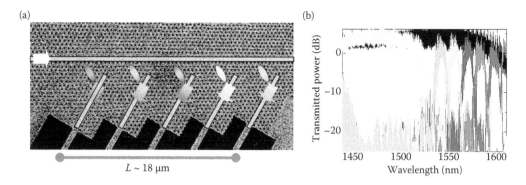

FIGURE 1.38

(a) Scanning electron microscope image and schematic diagram of a compact wavelength division multiplexer composed of L3 silicon photonic crystal nanocavities connected in parallel along a photonic crystal waveguide. The arrow indicates the flow of the input and output light, and the circles indicate the position and alignment of the cavities. The resonant wavelength of every cavity is shifted slightly in order to achieve wavelength multiplexing. (b) The output spectrum of the device.

with a photonic crystal nanocavity can transmit towards the drop port. The fabricated sample is shown in Figure 1.38a, where the system is composed of five L3 cavities with different resonant wavelengths. The device is only ~ 18 μm in length. In order to obtain the desired resonant wavelength, the shift at each cavity end is tuned. In addition, the width of the photonic crystal waveguide is fine tuned to control the transmittance bandwidth of the through port. The system design is discussed in detail elsewhere.[60] The transmittance spectrum of the device is shown in Figure 1.38b, where five wavelength components are separated in the through port spectrum. Since the device is ultra-small, this demonstration shows that photonic crystal based filters can be integrated in a photonic circuit on-chip.

1.5.3 All-Optical Switching

Network traffic is increasing rapidly owing to the rapid development of various information-oriented technologies, such as ubiquitous networking and computing. Although optical fiber can carry large amounts of data, the bandwidth of the network is limited by the speed of the electronics because network nodes rely on them. Hence, all-optical switching and signal processing are considered to be the keys to increasing network speed. However, the existing all-optical switches have had limited applications, because of their high switching energy and large device size, which are largely due to the small interaction between light and matter. These are general problems for photonics but are especially serious for silicon because of its small optical nonlinearity. Silicon has long been a key material for electronics, but not for photonics. However, silicon fabrication technologies have become highly sophisticated, and fusion with existing high-performance silicon CMOS technologies is now more promising. The use of mature fabrication techniques and the high refractive index of silicon means that photons can be confined in a small space and this has stimulated the current interest in silicon photonic research.

If we can achieve a high photon density by making use of artificial structures, it will constitute a promising approach to obtaining effective optical nonlinearity at a small input energy. Since the density of photons inside a cavity scales with Q/V for a given input optical power, a cavity with an ultrahigh Q and an ultrasmall V is a good candidate for achieving all-optical switching at a small operating energy. Indeed, all-optical switching at an extremely small input energy has been demonstrated on-chip using a silicon microring resonator.[61] Since photonic crystal nanocavities can achieve an even higher Q/V value, they are expected to switch at an even smaller input energy.

The $\chi^{(3)}$ nonlinearity in silicon results in two-photon absorption or the optical Kerr effect. In particular, two-photon absorption plays an important role in relation to 1550-nm wavelength light in silicon, because the carrier generation that results from the two-photon absorption modifies the refractive index and enables optical switching.

1.5.3.1 Switching by Thermo-Optic Effect

The thermo-optic effect is one of two effects caused by the two-photon absorption in silicon, and is a relatively slow effect. Since carriers recombine non-radiatively in silicon, carrier generation results in the accumulation of heat, which increases the refractive index.

Figure 1.11 shows a photonic crystal nanocavity that has been used for a switching experiment.[62] The cavity exhibits two resonant modes; one is the control mode (C-mode) and the other is the signal mode (S-mode), whose spectrum is similar to that shown in Figure 1.13. One of the resonant modes has a Q value of 7350 (mode C) and the other has a Q value of 33,400 (mode S). A schematic diagram of the operating principle is shown in Figure 1.39. Injecting a control light that is in resonance with mode C enables us to generate two-photon absorption carriers. As a result, heat is accumulated and the S-mode resonance shifts to a longer wavelength. When a signal light is in resonance with the mode-S (S_2), it exhibits ON to OFF switching. On the other hand, when the signal light is slightly detuned to a longer wavelength (S_1), the signal light is turned on when the control light is injected into the cavity. The results of a switching experiment based on the thermo-optic effect are shown in Figure 1.40. The switching was demonstrated at an input power of only 32 μW, which is one of the lowest operating powers reported for an all-optical switch. By performing a different experiment, we estimated that the corresponding input switching energy (i.e. the energy of the control light input) was about 11 pJ and the consumed energy was only about 280 fJ. It should be noted that the input and output light can be injected though the input/output photonic crystal waveguide, whose configuration presents the possibility of realizing all-optical switching devices integrated on a silicon chip.

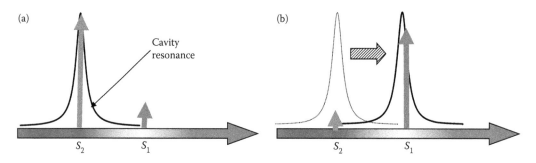

FIGURE 1.39
Schematic illustration of the mode-S spectrum of the cavity resonance and the signal light of a switching experiment. S_1 and S_2 represent the signal light with different detuning. The wavelength of S_1 is detuned slightly from the cavity resonance (to a longer wavelength) and S_2 is close to the initial cavity resonance. (a) When the control pulse is off. The transmittance of the S_1 light is low (off) and that of S_2 is high (on). (b) When the control pulse is injected and thermo-optic effect is present. The transmittance of S_1 is high (on) and that of S_2 is low (off).

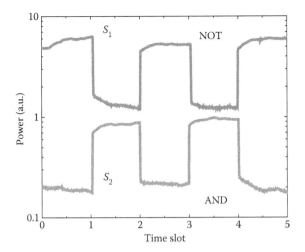

FIGURE 1.40
Switching demonstration based on thermo-optic effect using silicon photonic crystal nanocavity. Each time slot corresponds to approximately 1 ms.

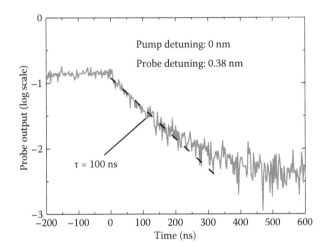

FIGURE 1.41
Temporal response of the probe output. The 64-μW pump light was suddenly turned off at 0 ns. The transmittance of the CW probe light exhibits a 100-ns decay.

Another important characteristic that must be considered for a switching device is the operating speed. The operating speed of thermo-optic effect based devices is considered to be relatively slow because it is determined by the effective heat relaxation time. However, this is not always the case with ultrasmall devices. To obtain an effective heat relaxation time for silicon photonic crystal nanocavity based switches, we performed a time-resolved measurement. We injected a 400-ns rectangular pulse with the mode-S resonance and measured the transmittance of a CW probe signal at a wavelength of 0.38-nm detuned from the mode-C resonance. The measured results are shown in Figure 1.41 where the probe light exhibits an exponential like decay. The exponential fitted decay is 100 ns, which we determined as the effective thermal relaxation time. This value is surprisingly small for a thermo-optic based device. The fast switch off is achieved because of the small size. Although the thermal diffusion constant is considered small in silicon, the value is sufficiently large for a nanocavity device, which is about 0.1 μm^3 in size. As a result, the heat diffuses efficiently outside the cavity, which makes the effective thermal relaxation time very fast.

1.5.3.2 Switching by Carrier Plasma Dispersion Effect

Although the switching energy and recovery time of a silicon photonic crystal nanocavity operating by employing the thermo-optic effect are very small, they can be further reduced by making use of different nonlinearities. Of these, the carrier-plasma effect is a good candidate. By injecting control pulses, two-photon absorbed carriers are created, which induces carrier-plasma dispersion. This effect reduces the refractive index of the silicon and hence switching becomes possible. Since, the carrier lifetime is much smaller than the thermal decay, it can yield fast switching. Moreover, it requires less energy because of the faster operation. The electron microscopic image of the cavity employed for this experiment is shown in Figure 1.10b and the transmittance spectrum is shown in Figure 1.13. The mode-S transmittance exhibits a nonlinear shift towards a shorter wavelength when a control pulse is injected. The wavelength of the control pulse is resonant with mode-C. We employed a control pulse with a width of 6.4 ps. Figure 1.42 shows the signal transmission where δ is the detuning wavelength from the mode-S resonance. The switching window for this device was 50 ps for off to on type switching. Using the slope of the switching recovery curve, we estimated the effective carrier lifetime to be 80 ps. This value is very fast for silicon optical devices. The required switching energy versus switching contrast is investigated in Figure 1.43. Although an input control pulse energy of ~ 100 fJ is required in this demonstration, it should be noted that the 6.4-ps control pulse that we employed has a much broader bandwidth than the mode-S of the cavity and the transmittance is

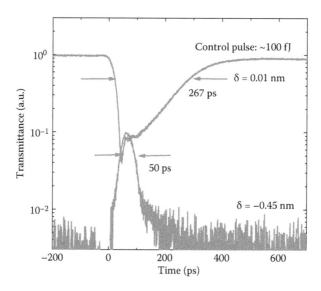

FIGURE 1.42

Transmittance of the signal light for different detuning values δ. A ~100-fJ control pulse is injected into the silicon photonic crystal nanocavity at $t=0$ ps.

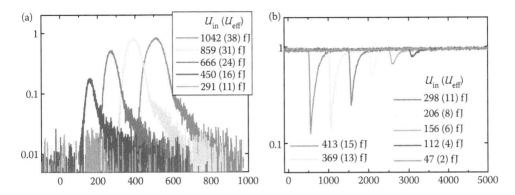

FIGURE 1.43
Switching contrast as a function of different input control pulse energies. U_{in} is the energy of the control pulse at the input photonic crystal waveguide and U_{eff} is the effective energy that couples with the cavity estimated from the spectrum shape and the transmittance. (a) is when $\delta = -0.3$ nm and (b) is when $\delta = 0$ nm.

only ~4%. Therefore, by optimizing the shape of the input pulse spectrum and the transmittance of the nanocavity with the waveguides, we expect to reduce the energy to a few 10 fJ. This shows that a control pulse of a few fJ is sufficient to operate this switching device. Indeed, the absorbed energy calculated from the cavity resonance shift is estimated to be only 2 fJ. This demonstration opens the possibility for the development of integrated all-optical switches and logic gates on a silicon chip.[20,63]

1.5.3.3 Numerical Study of Carrier Dynamics in Silicon Photonic Crystal

This section describes detailed numerical calculations that were performed in order to investigate the origin of the small effective carrier lifetime in silicon photonic crystal nanocavity switches.[64] 2D carrier transportation models were considered to describe the carrier dynamics in a silicon photonic crystal slab.

$$\frac{\partial n_e(\mathbf{r},t)}{\partial t} = \frac{1}{q}\nabla \cdot \mathbf{J}_n(\mathbf{r},t) - \frac{1}{\tau_{slab}}\left[n_e(\mathbf{r},t) - n_{e\_eq}(\mathbf{r},t)\right] \tag{1.12}$$

$$\mathbf{J}_n(\mathbf{r},t) = qn_e(\mathbf{r},t)\mu_e\nabla\left[\phi(\mathbf{r},t) + \phi_{ex}(\mathbf{r},t)\right] + qD_e\nabla n_e(\mathbf{r},t) \tag{1.13}$$

$$\frac{\partial n_h(\mathbf{r},t)}{\partial t} = \frac{1}{q}\nabla \cdot \mathbf{J}_p(\mathbf{r},t) - \frac{1}{\tau_{slab}}\left[n_h(\mathbf{r},t) - n_{h\_eq}(\mathbf{r},t)\right] \tag{1.14}$$

$$\mathbf{J}_p(\mathbf{r},t) = qn_h(\mathbf{r},t)\mu_h\nabla\left[\phi(\mathbf{r},t) + \phi_{ex}(\mathbf{r},t)\right] + qD_h\nabla n_h(\mathbf{r},t) \tag{1.15}$$

The subscript e represents electrons and h represents holes. n is the carrier density and n_{e_eq} and n_{h_eq} are the carrier densities under thermal equilibrium. μ and D are the carrier mobility and diffusion. τ_{slab}, q, ϕ, and ϕ_{ex} are the effective carrier lifetime for a 200-nm thick slab, electric charge, internal potential and external potential, respectively. These four equations describe the dynamics of the electrons and holes in the silicon substrate, where the diffusion and electron-hole recombination are taken into account. The interface between silicon and air is represented by the following equation.

$$D_{e,h}\nabla n_{e,h}(\mathbf{r}_{\text{surf}},t) = S\left[n_{e,h}(\mathbf{r}_{\text{surf}},t) - n_{e\_eq,h\_eq}(\mathbf{r}_{\text{surf}},t)\right] \tag{1.16}$$

where S is the surface recombination speed. Since the electrons and holes create internal potential, a Poisson equation is considered.

$$\nabla^2\phi(\mathbf{r},t) = \frac{1}{\varepsilon_0\varepsilon_r}[n_h(\mathbf{r},t) - n_e(\mathbf{r},t)] \tag{1.17}$$

The 2D electron and hole distribution is self-consistently calculated using Equations 1.12 through 1.15 and 1.17. Finally, the refractive index modulation is calculated using the following equation

$$\Delta n(t) = -\left[8.8\times10^{-22}n'_e(t) + 8.5\times10^{-18}n'_h(t)^{0.8}\right] \tag{1.18}$$

where $n'(t)$ are carrier densities normalized with the field profile of the cavity.

Before calculating the 2D photonic crystal nanocavity switches, we calculated the cross-sectional carrier dynamics of a silicon waveguide to determine the surface velocity S by comparing the obtained result with those obtained in the experiments. The cross-sectional optical mode profile of a 400×200-nm silicon waveguide was obtained with the beam propagation method and was employed to determine the initial electron and hole density distributions. The physical parameters of silicon were drawn from references, as $D_e=35$ cm²/s, $D_h=12$ cm²/s, $\mu_e=1450$ cm²/Vs and $\mu_h=500$ cm²/Vs.[65] And the bulk lifetime of silicon was set at 0.5 µs. Using these values, we calculated the effective carrier lifetimes for various S values, and the results are summarized in Figure 1.44. According to our experiment, the effective carrier lifetime of a silicon waveguide based switch is about

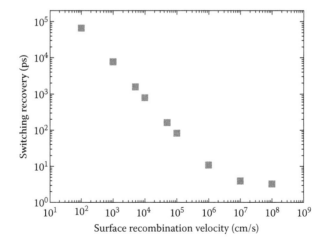

FIGURE 1.44
Calculated switching recovery time of a 200×400-nm silicon waveguide for different surface recombination velocities.

1.07 ns.[46] This value was obtained by injecting a CW signal light and short control pulses from the same input and monitoring the signal transmittance. Since reflection at the input and the output waveguide facets creates a Fabry–Pérot oscillator, the transmittance of the signal light is modified when the carrier-plasma dispersion effect is present. Different studies also show that the effective carrier lifetime of a silicon waveguide system is about 1 ns.[61,66] According to Figure 1.44, this corresponds to a surface velocity of $S=8\times10^3$ cm/s. This value is similar to previously reported silicon and air surface values.[67,68] Hence, we can now use this value to calculate the carrier dynamics in photonic crystal nanocavity switches. The initial distributions of electrons and holes are given by the power of the electrical field profile calculated by using the FDTD calculations. The calculated initial electron density distribution for an L4 cavity is shown in Figure 1.45a, when the initial electron density was 1×10^{17} cm^{-3}, which results in a wavelength shift of about −0.15 nm. The carrier densities after 8 and 24 ps are shown in Figure 1.45b and c, where rapid diffusion is present. Figure 1.46 shows the wavelength shift, which corresponds to the

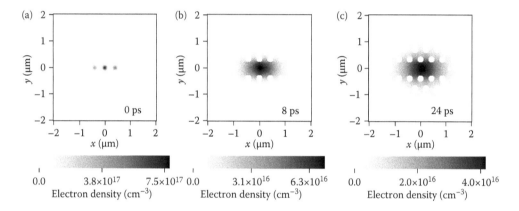

FIGURE 1.45
(a) Initial electron distribution of an L4 silicon photonic crystal nanocavity. (b) Electron density after 8 ps. (c) After 24 ps.

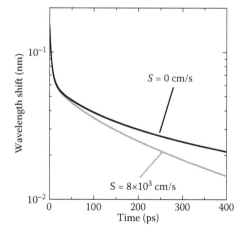

FIGURE 1.46
Calculated wavelength shift for different surface recombination velocities *S*.

switching recovery time, for two different S values of 8×10^3 and 0 cm/s. The difference between the two lines is not very significant, and this is particularly so at the initial stage ($t < 20$ ps). It suggests that surface recombination does not play a significant role in determining the effective lifetime in photonic crystal nanocavity switches, but that diffusion is an important factor. It should be noted that the exponential fitted decay was 0.26 ns, which is a much smaller than that of a silicon waveguide. The short lifetime is the result of the unique structure of the photonic crystal nanocavity switches, where the carriers can diffuse efficiently outside the cavity region. Although the diffusion coefficient of silicon is considered to be relatively small (i.e. slow diffusion), its value is noticeably large when it is compared with the ultrasmall volume of the nanocavity switches. In other words, it is confirmed by calculation that the very small effective carrier lifetime demonstrated using a photonic crystal nanocavity results from the small mode volume of the device.

1.5.3.4 5-GHz Return-to-Zero Pulse Train Modulation

Next we provide a practical demonstration in which a 5-GHz return-to-zero (RZ) optical pulse train is modulated by random control signals.[20,69] We employed a similar sample that has two resonant modes with Q values of 9,000 and 15,600. We injected optical pulse trains with different wavelengths; one with exactly the same wavelength as the S mode (S_{1in}) and the other slightly detuned to a shorter wavelength (S_{2in}). A schematic image of the operation and the experimental setup are shown in Figure 1.47. A random modulated 5-GHz

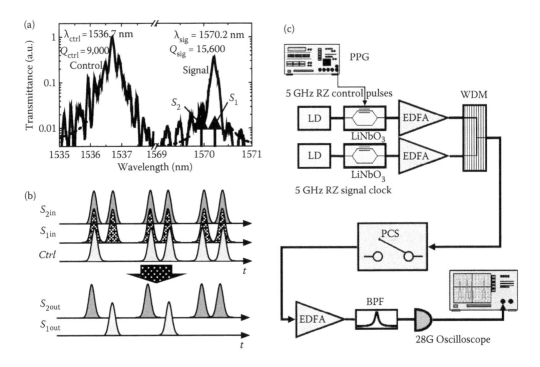

FIGURE 1.47
(a) Transmittance spectrum of the silicon photonic crystal nanocavity switch. (b) Schematic illustration of the pulse train modulation experiment. (c) Schematic diagram of the experimental setup. PPG: Pulse pattern generator, LD: Laser diode, WDM: Wavelength division multiplexer, PCS: Photonic crystal nanocavity switch, EDFA: Erbium doped fiber amplifier, BPF: Band pass filter, LiNbO$_3$: LiNbO$_3$ optical modulator.

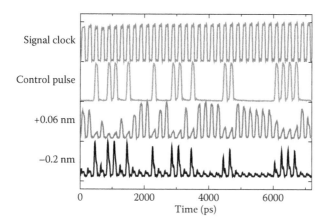

FIGURE 1.48
Signal clock (first row) is modulated with a random 5-GHz return-to-zero control pulse train (second row). The transmittances of the signal pulse trains (third and fourth rows) for two different wavelengths are shown in the panel.

return-to-zero optical pulse train is prepared and injected into the photonic crystal switch. Then the output was measured with a sampling oscilloscope. Injecting a control pulse into the cavity produces carriers, which cause Mode-S to shift toward a shorter wavelength. This results in a high transmittance for S_2. Logically, this is an AND operation. On the other hand, for S_1, pulses are transmitted only when no control pulses arrive at the switch. The experimental results are shown in Figure 1.48. Although the thermal accumulation results in a slightly smaller contrast after a sequence of identical control bits, the modulated signal recovers quickly before the next optical pulse arrives. Overall, the signal clock was successfully modulated. The launched energy of each control optical pulse was about 500 fJ.

1.5.3.5 Accelerating the Speed of All-Optical Switches using Ion-Implantation Technology

Although the operating speed of a photonic crystal nanocavity switch is extremely fast for a silicon based optical device, we tried to increase the speed further by employing ion-implantation technology.[46,69] It is known that the carrier lifetime decreases when ions are implanted as the result of the creation of defect modes. Larger ion doses result in shorter carrier lifetimes, which lead to a faster recovery time for carrier-based switches. According to the calculation shown in Figure 1.46, the longer tail of the switching recovery time is affected by the non-radiative recombination speed, for example the surface recombination velocity, which will limit the repetitive operation speed of the device. Therefore, if we can kill the carriers quickly by ion-implantation this may improve the repetitive operation characteristic of the device. With ion-implantation, we have to consider the optical loss, which is large for high ion-implanted devices. Thus, we first studied the relation between the propagation loss of a silicon waveguide and the carrier lifetime for different ion doses and different implantation conditions. Ar^+ ions were implanted at a target penetration depth of 100 nm at an ion beam energy of 100 keV. An as-fabricated single-mode silicon waveguide (600×200 nm) has a propagation loss of -1.7 dB/cm and a carrier lifetime of 1.07 ns. The carrier lifetime was determined by injecting pump pulses and a probe CW light into the same input, and measuring the transmittance of the probe. As shown in Figure 1.49, the carrier lifetime was 850 ps but the propagation loss increased significantly to -33.3 dB/cm

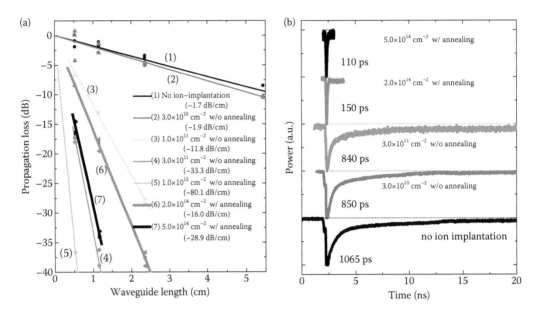

FIGURE 1.49
(a) Propagation loss of a silicon waveguide as a function of ion dose and processing conditions. (b) Measured carrier lifetime. (Reprinted with permission from Tanabe, T. et al., *Appl. Phys. Lett.*, 90(3), 031115, 2007. © American Institute of Physics.)

at an ion dose of 3.0×10^{11} cm^{-2}. At the cost of the propagation loss, this lifetime value is not small. In an attempt to reduce the propagation loss, we annealed the waveguide for 15 min at 900°C in a nitrogen atmosphere. This crystallizes the silicon and creates dislocation loops at the center of the slab.[66] As a result, the propagation loss was reduced to −16 dB/cm for an ion dose of 2×10^{14} cm^{-2}. Surprisingly, as shown in Figure 1.49, the carrier lifetime was only 150 ps, which proved that recombination via a dislocation loop is very effective. We employed this condition for the ion implantation of a photonic crystal nanocavity.

The fabrication process used for the ion implanted photonic crystal nanocavity is shown in Figure 1.50. First, photonic crystals are fabricated via electron-beam lithography and dry etching. Next, the wafer is covered with a resist mask and the photonic crystal nanocavity regions are selectively ion implanted. Finally, annealing is performed followed by the air-bridge process.

The photonic crystal nanocavity that we employed for this experiment was a point defect hexapole cavity. First, we compared the Q values of the photonic crystal nanocavity with and without ion implantation. The Q value was 3.6×10^4 for the sample without ion implantation and 2.1×10^4 for the sample with ion implantation. The Q value was smaller for the ion implanted sample but the difference is not significant for switching applications. Indeed the switching contrast with respect to the input pulse energy is shown in the inset of Figure 1.51 where the modulation depth is not significantly different. The recorded transmittance waveforms of the signal light when a control pulse was injected at 0 ps are shown in Figure 1.51a and b for the ion implanted sample and an as-fabricated sample. The switching speed of the implanted sample is only 1/3 that of the original device, while the required switching energy remains almost the same. This is one demonstration of our approach for increasing the switching speed of a silicon-based photonic crystal nanocavity switch.

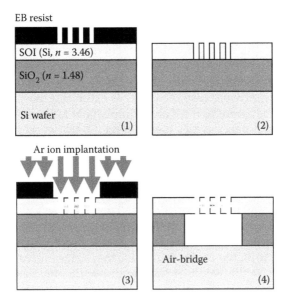

FIGURE 1.50
Fabrication process used for ion-implanted silicon photonic crystal nanocavity. (1) A photonic crystal is patterned using electron beam lithography. (2) A photonic crystal is fabricated by dry etching. (3) Ar ions are selectively implanted in the photonic crystal nanocavity region. (4) An air-bridge is formed by wet etching. (Reprinted with permission from Tanabe, T. et al., *Appl. Phys. Lett.*, 90(3), 031115, 2007. © American Institute of Physics.)

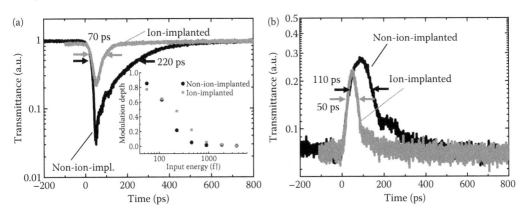

FIGURE 1.51
(a) Transmittance from ion-implanted and non-ion-implanted silicon photonic crystal nanocavity switch when the wavelength of the signal light is at the resonance of the nanocavity. The control pulse is injected at $t = 0$ ps. Inset is the switching contrast versus the input control pulse energy for ion-implanted and non-ion-implanted devices. (b) The transmittance waveform when the signal light is slightly detuned towards a shorter wavelength. (Reprinted with permission from Tanabe, T. et al., *Appl. Phys. Lett.*, 90(3), 031115, 2007. © American Institute of Physics.)

1.5.4 Ultra-Low Power Bistable Memory

Optical bistability was studied extensively during the 1980s and 1990s, with the expectation of replacing electrical transistors with optics, the operation of which is often based on optical nonlinearity in an optical cavity or etalon.[2,4] When the nonlinear medium changes the optical length of the cavity, the resonant wavelength of the cavity exhibits an input power dependent shift. If the wavelength of the input light is slightly detuned in the same direction as the shift of the cavity resonance, positive feedback occurs at a certain input power when it is increased. As a result, the input/output power characteristic of the cavity exhibits optical bistability. Optical bistability results in a sudden jump from the off (low transmittance) to the on (high transmittance) state. Once the cavity is on, it remains on until the input power is reduced significantly because the cavity is exhibiting positive feedback. Optical bistability is believed to be essential if we are to implement logic gates all-optically.

Because photonic crystal nanocavities have a large Q/V, they are expected to be capable of low power operation, which has long been considered difficult to achieve with conventional devices. Here, we employed a four point defect cavity with shifted end holes to demonstrate optical bistability. First, we studied bistable operation based on the thermo-optic effect. The input wavelength of the laser light is scanned from a shorter to a longer wavelength at different input powers. The recorded output power with respect to the input wavelength is shown in Figure 1.52a. At an input power greater than 25 µW, a sudden drop is observed at the long wavelength edge, which is one proof of optical bistability. Since the thermo-optic effect shifts the cavity resonance to a longer wavelength, we know that the measured bistability is the result of the thermo-optic effect caused by two photon absorption carriers. In addition, the input versus output power for a CW laser at different detuning values is shown in Figure 1.52b. When the input light is slightly detuned to a longer wavelength, the input/output curve exhibits hysteresis, which is additional evidence of optical bistability. The threshold power of optical bistability for this cavity was only ~40 µW, which is significantly smaller than previous reported values obtained using larger cavities. It is even possible to achieve a bistable threshold power of just a few µW by using a local-width-modulated line-defect photonic crystal nanocavity with an extremely high Q.[10] Since the operation is based on the thermal effect, the operating speed is limited by the thermal relaxation time, which is typically ~100 ns as discussed previously.

Since carrier-plasma dispersion is present in silicon, it is also possible to obtain optical bistability using this nonlinearity. We performed a time-resolved measurement to observe

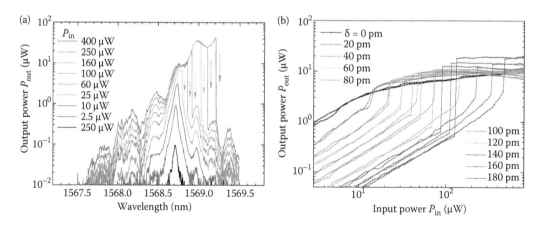

FIGURE 1.52
Optical bistability experiment based on the thermo-optic effect for an L4 silicon photonic crystal nanocavity with a Q value of 33,400 (i.e. spectrum FWHM of 47 pm). (a) Spectrum measured using a wavelength tunable laser diode at different input powers swept from a shorter to a longer wavelength. The displayed value is the power at the photonic crystal waveguide. (b) The input/output power characteristics as a function of wavelength detuning from the cavity resonance. (From Notomi, M. et al., *Opt. Express.*, 13, 2678–2687, 2005. © Optical Society of America. With permission.)

fast nonlinearity. We inject a step input pulse at a detuning of −0.15 nm in an L4 cavity that has a Q of 1.24×10^4 at a resonant wavelength of 1569.70 nm and recorded the transmittance. Note that the wavelength of the input light is set at a shorter wavelength than that of the cavity resonance because the carrier-plasma dispersion shifts the cavity resonance towards a shorter wavelength. The measured results are shown in Figure 1.53a. We observed a nonlinear jump at an input power between 0.40 and 0.63 mW. In addition, the outputs exhibit an overshoot at higher input powers, which is direct evidence of optical bistability.

By utilizing the nonlinear jump of a bistable device, we can observe transistor like behavior, often called transphasor behavior. The experimental results are summarized in Figure 1.54. Here, the output from the L4 cavity of a 200-ps wide input pulse was monitored. Region (2) is a linear regime, where the output increases linearly with respect to the input power. On the other hand, region (1) is a regime where the output exhibit nonlinearity.

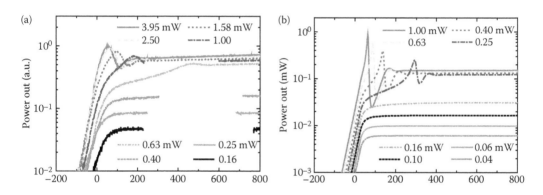

FIGURE 1.53
(a) Transmittance of a step input pulse at different input powers. (b) Calculated results. (From Tanabe, T. et al., *Opt. Lett.*, 30, 2575–2577, 2005. © Optical Society of America. With permission.)

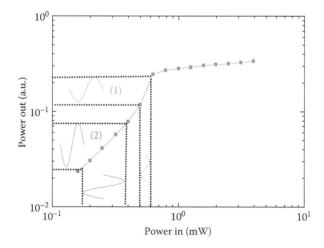

FIGURE 1.54
Experimental demonstration of an optical transistor (transphasor) by injecting 200-ps wide input pulses at different input powers.

To investigate the device in detail, we recorded the transmittance waveforms of a step input for different input wavelengths. A large number of temporal waveforms were collected and they are summarized in the 2D image shown in Figure 1.55. When the detuning is large, the overshoot peak is delayed. This is known as a critical slowing down phenomenon,[2] which is another proof of optical bistability. The slowing down phenomenon is also seen in Figure 1.53. Since this bistability is based on the carrier-plasma dispersion effect, this device can operate at a speed of a few 100 ps or less.

A numerical analysis was undertaken to study the optical bistability of a photonic crystal nanocavity in detail. Simple rate equations are employed to describe the dynamics of the cavity.

$$\frac{du(t)}{dt} = P(t)T\left[\delta + \Delta n(t)\lambda_0\right] - \frac{u(t)}{\tau_{ph}} \tag{1.19}$$

$$\frac{dN(t)}{dt} = \frac{2\lambda c\beta}{hn^2V^2}u(t)^2 - \frac{N(t)}{\tau_c} \tag{1.20}$$

$$\frac{dU_T(t)}{dt} = E_g\frac{N(t)}{\tau_c} - \frac{U_T(t)}{\tau_T} \tag{1.21}$$

$$\Delta n(t) = -\left[8.8\times10^{-22}N(t) + 8.5\times10^{-18}N(t)^{0.8}\right] + 1.12\times10^{-16}U_T(t) \tag{1.22}$$

where $u(t)$, $N(t)$, and $U_T(t)$ are the photon energy in the cavity, the two-photon absorbed carrier density, and the thermal energy density, respectively. n, β, and E_g are the refractive index, two photon absorption coefficient, and bandgap energy of silicon, respectively. c, h, and λ are the velocity of light, Planck's constant, and the wavelength of the pump light, respectively. τ_c and τ_T are the effective carrier lifetime and thermal relaxation time and they are set at 80 ps and 100 ns, respectively, where the values are estimated from the switching experiment. Equation 1.19 describes the photon energy stored in the cavity,

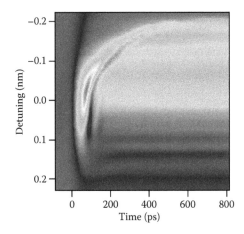

FIGURE 1.55
2D mapping of transmittance waveform for a step input light at different detuning values. (From Tanabe, T. et al., *Opt. Lett.*, 30, 2575–2577, 2005. © Optical Society of America. With permission.)

Equation 1.20 illustrates the generation and relaxation of two-photon absorption carriers, and Equation 1.21 is the generation and relaxation of heat. Equation 1.22 is the refractive index modulation induced by the carrier plasma dispersion and the thermo-optic effect. Equations 1.20 through 1.22 are solved in a self-consistent way, where the refractive index obtained using Equation 1.22 is fed back to the first term on the right hand side of Equation 1.20. The transmitted waveforms are calculated in Figure 1.53b for a step input at different input powers. The physical parameters are taken from references and the experiment; hence there are no freely adjustable parameters. We observed a nonlinear jump, overshoot and the slowing down effect and their behavior agrees well with the experiment.

Optical bistability is the key to achieving all-optical logic gates such as optical memory or optical transistors. It can constitute a basic element of all-optical random access memory (RAM)-based buffer.[70] A numerical and experimental demonstration of optical memory operation based on optical bistability achieved with the carrier-plasma effect is shown in Figure 1.56. A step input is injected into the cavity and the power is adjusted slightly below the bistable threshold power (off state). Injecting a set pulse enables the generation of two-photon absorption carriers, which turns the cavity on. The calculated transmittance is shown in Figure 1.56a. The calculation shows that the cavity can maintain the on state for more than 1 ns after the input pulse is injected. The experimental study shown in Figure 1.56b was performed based on the calculation. The transmittance of the step input is indicated with (1). When a set pulse is injected, it switches the bistable mode from off to on. As predicted by the numerical study, the cavity maintains the on state after the set pulse is injected, as shown on line (2). A negative pulse is applied to reset the cavity (3). Since the cavity stores on and off state information, it can be regarded as a bistable optical memory device.

1.5.5 Optical Logic-On Chip

1.5.5.1 Optical Flip-Flop

As discussed above, optical bistability is the key to developing all-optical logic gates. Of the various logic gate functional devices, the flip-flop device is the key to composing

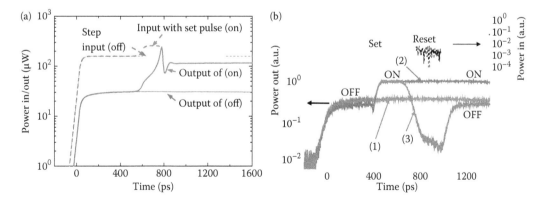

FIGURE 1.56
(a) Calculation of bistable memory operation with a silicon photonic crystal nanocavity. A 200-ps wide set pulse is added to the bias step input at 500 ps to switch the cavity to the ON state. (b) Experimental demonstration of bistable memory operation with an L4 photonic crystal nanocavity. The set pulse is added at 400 ps to set the cavity to the ON state. A negative reset pulse is added at 800 ps. (1) The output of a step input. (2) The output when a set pulse is added. (3) The output when a set pulse and a reset pulse are added. (From Tanabe, T. et al., *Opt. Lett.*, 30, 2575–2577, 2005. © Optical Society of America. With permission.)

highly sophisticated logic gate circuits, such as memory, retiming or regenerative circuits. In photonics, a flip-flop has been fabricated by combining two optical bistable devices.[4] However, the reported device is not suitable for on-chip integration owing to its bulky configuration and the large energy it requires for operation. The optical bistability demonstrated using silicon photonic crystal nanocavities in the previous section will provide a good platform for constructing an on-chip optical flip-flop device. Based on the idea of coupling two optical cavities to enable cross-feedback, which is the key to obtaining flip-flop operation, we propose the design shown in Figure 1.57a. This photonic crystal nanocavity system permits SR flip-flop operation, where the outputs can be switched by injecting set or reset pulses into the inputs. S, R, and B are the input waveguides for the set pulse, reset pulse and bias CW light, respectively. Q and $\bar{Q}$ are the output waveguides. A simplified conceptualized diagram of the system is shown in Figure 1.57b. If we compare this photonic system with that of the electrical SR flip-flop circuit, we can find various similarities. We performed a 2D FDTD calculation based on this structural design. The time sequence is shown in Figure 1.57d, where the set (CS) and reset (CR) pulses are applied with

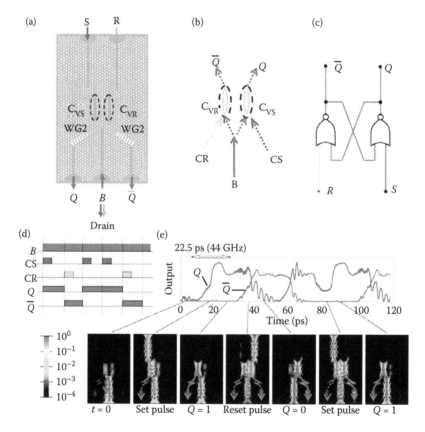

FIGURE 1.57
All optical SR flip-flop consisting of two bistable cavities coupled to waveguides. (a) Schematic illustration of the structural design based on hexagonal air-hole 2D photonic crystals. Two L7 cavities are connected through waveguides. The end holes of the L7 cavities are shifted by $0.30a$ with hole diameters of $0.24a$. (b) Simplified conceptualized diagram of the system. (c) Equivalent electronic circuit diagram of the SR flip-flop. (d) Time sequence of three inputs (bias, and set clock pulse, and reset clock pulse) and two outputs. (e) Calculated output waveform and optical mode profiles of the optical SR flip-flop by using 2D FDTD. (From Notomi, M., et al., *Opt. Express.*, 15, 17458–17481, 2007. © Optical Society of America. With permission.)

different timings. We assumed Kerr nonlinearity for the calculation. The cavities C_{VS} and C_{VR} have two resonant modes. The wavelength of one of the resonant modes has the same wavelength between the two cavities. This mode is employed for the bias input. When we inject a set pulse, the cavity C_{VS} is turned on, and this enables the bias light to travel toward port Q. On the other hand, when we inject a reset pulse, the cavity C_{VR} is turned on, which enables the bias light to flow toward port $\bar{Q}$. The calculated results are shown in Figure 1.57e along with the intensity profiles of the device. The calculated results show that a successive SR flip-flop operation at 44 Gb/s is possible by using this system. It should be noted that an optimized system will enable us to achieve a faster operating speed. We expect that this demonstration of an on-chip optical flip-flop by using a photonic crystal platform will constitute a significant step towards the development of all-optical integrated circuits.

1.5.5.2 Pulse Retiming Circuit

Another example of optical logic gates achieved by using a photonic crystal nanocavity system is shown in Figure 1.58. It is composed of two cavities (C1 and C2). This photonic circuit enables all-optical pulse retiming, which is important if we wish to fabricate high-speed optical circuits. The two cavities have a common resonant mode at a wavelength of λ_2, and the intensity profile extends to both cavities. In addition, each cavity has another resonant mode (λ_1 and λ_3), whose intensity profile is localized at each cavity. We consider that the data signal, which has a wavelength of λ_2, arrives at this photonic system with timing jitter. Optical clock signals at wavelengths of λ_1 and λ_3 are applied to ports P_A and P_B, respectively. The cavity C1 can be turned on (high transmittance mode) only when both the data and $\overline{\text{clock}}$ with respective wavelengths of λ_2 and λ_1 arrive simultaneously. When cavity C1 is on, the data (λ_2) can flow to C1 and λ_2 is fed to C2. Cavity C2 can be turned on only when both the clock (λ_3) and data (λ_2) are injected simultaneously. Turning cavity C2 on results in the high transmittance of the clock (λ_3) from P_B to P_D. As a result, the output from the P_D is synchronized with the clock and the data signals. Hence the jitter is corrected. Moreover, it is converted to a return-to-zero data stream from an arriving data signal. The operating principle is discussed in detail elsewhere.[20,63] The 2D FDTD calculation result is shown in Figure 1.56b. Kerr nonlinearity in AlGaAs is assumed for the

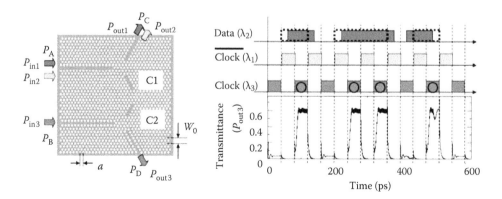

FIGURE 1.58
All-optical pulse retiming circuit based on two bistable cavities. (a) Schematic illustration of the photonic circuit. The lattice constant a is 400 nm and the hole diameter is $0.55a$. (b) Time sequence and calculated output waveform obtained with 2D FDTD. AlGaAs material is assumed for the calculation ($\chi^{(3)}/\varepsilon_0 = 4.1 \times 10^{-19}$ (m²/V²)). (From Notomi, M., et al., *Opt. Express.*, 15, 17458–17481, 2007. © Optical Society of America. With permission.)

calculation. The instantaneous driving power is 60 mW for all three inputs. The calculated result shows that the timing jitter of the data is completely removed and the input is converted to a return-to-zero output data stream with a speed of 50 Gb/s. It is surprising that such a complicated logic operation can be achieved by a photonic circuit composed solely of two nanocavities. Note that the structure can be further optimized to reduce the operating power or to enhance the operating speed.

1.6 Summary

The photonic crystal nanocavity system will enable us to confine photons in a very small space for an extremely long period. Since the linewidth of the transmittance spectrum and the footprint of the device are extremely small we are able to fabricate compact add-drop filters. If we focus on the temporal characteristics of the nanocavity system, we find that the system is capable of trapping photons and this constitutes an on-chip photonic buffer memory. Even though it was not discussed in this chapter, we can perform various operations to the trapped photons because the demonstrated photon lifetime is now of the order of nanoseconds, which enable novel manipulation of light.[20,71,72] An ultrahigh-Q nanocavity system will also result in the very slow propagation of pulses, and this will open the possibility of fabricating slow-light on-chip devices. In addition to the above mentioned applications that utilize the linear property of the cavity system, photonic crystal nanocavities are becoming more important for various light–matter interaction studies, owing to the high photon density achievable at a small input power. As a result of efficient light–matter interaction, active behavior such as all-optical switching or logic operation is becoming possible. In this chapter, we demonstrated all-optical switching, bistable memory operation, and various logic gate operations. It is extremely important that such operation is achieved at a significantly small input power. These demonstrations pave the way to the realization of on-chip photonic circuits.

References

1. M. Jinno, Y. Miyamoto, and Y. Hibino. 2007. Optical-transport networks in 2015. *Nature Photon.*, 1, 157–159.
2. H. M. Gibbs. 1985. *Optical bistability: Controlling light with light.* Academic Press, Orlando, FL.
3. S. D. Smith, A. C. Walker, F. A. P. Tooley, and B. S. Wherrett. 1987. The demonstration of restoring digital optical logic. *Nature*, 325, 27–31.
4. H. Tsuda and T. Kurokawa. 1990. Construction of an all-optical flip-flop by combination of two optical triodes. *Appl. Phys. Lett.*, 57, 1724–1726.
5. J. Takeda, K. Nakajima, S. Kurita, S. Tomimoto, S. Saito, and T. Suemoto. 2000. Time-resolved luminescence spectroscopy by the optical Kerr-gate method applicable to ultrafast relaxation processes. *Phys. Rev. B*, 62, 10083–10087.
6. E. Centeno and D. Felbacq. 2000. Optical bistability in finite-size nonlinear bidimensional photonic crystals doped by a microcavity. *Phys. Rev. B*, 62, 7683–7686.
7. M. Soljacic and J. D. Joannopoulos. 2004. Enhancement of nonlinear effects using photonic crystals. *Nature Mat.*, 3, 211–219.

•

8. E. Yablonovitch. 1987. Inhibited spontaneous emission in solid-state physics and electronics. *Phys. Rev. Lett.*, 58, 2059–2062.
9. S. John. 1987. Strong localization of photons in certain disordered dielectric superlattices. *Phys. Rev. Lett.*, 58, 2486–2489.
10. T. Tanabe, M. Notomi, E. Kuramochi, A. Shinya, and H. Taniyama. 2007. Trapping and delaying photons for one nanosecond in an ultra-small high-*Q* photonic-crystal nanocavity. *Nature Photon.*, 1, 49–52.
11. S. Noda, M. Fujita, and T. Asano. 2007. Spontaneous-emission control by photonic crystals and nanocavities. *Nature Photon.*, 1, 449–458.
12. T. Tanabe, M. Notomi, A. Shinya, S. Mitsugi, and E. Kuramochi. 2005. All-optical switches on a silicon chip realized using photonic crystal nanocavities. *Appl. Phys. Lett.*, 87, 151112.
13. T. Tanabe, M. Notomi, A. Shinya, S. Mitsugi, and E. Kuramochi. 2005. Fast bistable all-optical switch and memory on silicon photonic crystal on-chip. *Opt. Lett.*, 30, 2575–2577.
14. K. Ohtaka. 1979. Energy band of photons and low-energy photon diffraction. *Phys. Rev. B*, 19, 5057–5067.
15. E. Yablonovitch. 2007. Photonic crystals: What's in a name? *Optics & Photonics News*, 18, 12–13.
16. S. Ogawa, M. Imada, S. Yoshimoto, M. Okano, and S. Noda. 2004. Control of light emission by 3D photonic crystals. *Science*, 305, 227–229.
17. S. Takahashi, M. Okano, M. Imada, and S. Noda. 2006. Three-dimensional photonic crystals based on double-angled etching and wafer-fusion techniques. *Appl. Phys. Lett.*, 89, 123106, 1–3.
18. M. Fujita, S. Takahashi, Y. Tanaka, T. Asano, and S. Noda. 2005. Simultaneous inhibition and redistribution of spontaneous light emission in photonic crystals. *Science* 308, 1296–1298.
19. E. Kuramochi, M. Notomi, S. Hughes, L. Ramunno, G. Kira, S. Mitsugi, A. Shinya, and T. Watanabe. 2005. Scattering loss of photonic crystal waveguides and nanocavities induced by structural disorder. In *Proceeding of Pacific Rim Conference on Lasers and Electro-Optics, 2005.* Institute of Electrical and Electronics Engineers, New York, NY, pp. 10–11.
20. M. Notomi, T. Tanabe, A. Shinya, E. Kuramochi, H. Taniyama, S. Mitsugi, and M. Morita. 2007. Nonlinear and adiabatic control of high-*Q* photonic crystal nanocavities. *Opt. Express*, 15, 17458–17481.
21. M. Soltani, S. Yegnanarayanan, and A. Adibi. 2007. Ultra-high *Q* planar silicon microdisk resonators for chip-scale silicon photonics. *Opt. Express*, 15, 4694–4704.
22. B. Gayral, J. M. Gérard, A. Lemaître, C. Dupuis, L. Manin, and J. L. Pelouard. 1999. High-*Q* wet-etched GaAs microdisks containing InAs quantum boxes. *Appl. Phys. Lett.*, 75, 1908–1910.
23. J. M. Gérard, B. Sermage, B. Gayral, B. Legrand, E. Costard, and V. Thierry-Mieg. 1998. Enhanced spontaneous emission by quantum boxes in a monolithic optical microcavity. *Phys. Rev. Lett.*, 81, 1110–1113.
24. S. M. Spillane, T. J. Kippenberg, and K. J. Vahala. 2002. Ultralow-threshold Raman laser using a spherical dielectric microcavity. *Nature*, 415, 621–623.
25. D. K. Armani, T. J. Kippenberg, S. M. Spillane, and K. J. Vahala. 2003. Ultra-high-*Q* toroid microcavity on a chip. *Nature*, 421, 925–928.
26. K. Nozaki, T. Ide, J. Hashimoto, W.-H. Zheng, and T. Baba. 2005. Photonic crystal point shift nanolaser with ultimate small modal volume. *Electron. Lett.*, 41, 843–845.
27. Z. Zhang and M. Qiu. 2004. Small-volume waveguide-section high *Q* microcavities in 2D photonic crystal slabs. *Opt. Express*, 12, 3988–3995.
28. K. Srinivasan and O. Painter. 2002. Momentum space design of high-*Q* photonic crystal optical cavities. *Opt. Express*, 10, 670–684.
29. T. Asano, B.-S. Song, Y. Akahane, and S. Noda. 2006. Ultrahigh-*Q* nanocavities in two-dimensional photonic crystal slabs. *IEEE J. Sel. Top. Quantum Electron.*, 12, 1123–1134.
30. Y. Akahane, T. Asano, B. S. Song, and S. Noda. 2003. High-*Q* photonic nanocavity in a two-dimensional photonic crystal. *Nature*, 425, 944–947.
31. C. Manolatou, M. J. Khan, S. Fan, P. R. Villeneuve, H. A. Haus, and J. D. Joannopoulos. 1999. Coupling of modes analysis of resonant channel add-drop filters. *IEEE J. Quantum Electron.*, 35, 1322–1331.

32. H. Haus. 1984. *Waves and fields in optoelectronics.* Prentice-Hall, Englewood Cliffs, NJ.
33. M. Notomi, A. Shinya, E. Kuramochi, S. Mitsugi, H.-Y. Ryu, T. Kawabata, T. Tsuchizawa, T. Watanabe, T. Shoji, and K. Yamada. 2004. Photonic-band-gap waveguides and resonators in SOI Photonic crystal slabs. *IEICE Trans. Electron.,* E87-C, 398–408.
34. S. Mitsugi, A. Shinya, E. Kuramochi, M. Notomi, T. Tsuchizawa, and T. Watanabe. 2003. Resonant tunneling wavelength filters with high Q and high transmittance based on photonic crystal slabs. In *Proceedings of 2003 IEEE LEOS Annual Meeting.* Institute of Electrical and Electronics Engineers, New York, pp. 214–215.
35. M. Notomi, A. Shinya, S. Mitsugi, E. Kuramochi, and H. Ryu. 2004. Waveguides, resonators and their coupled elements in photonic crystal slabs. *Opt. Express,* 12, 1551–1561.
36. S. Mitsugi, A. Shinya, E. Kuramochi, M. Notomi, T. Tsuchizawa, and T. Watanabe. 2004. Wavelength adjustable photonic crystal wavelength filters with high-Q and high-transmittance. In *Technical Digest of PECS-V,* p. 153.
37. Y. Akahane, T. Asano, B.-S. Song, and S. Noda. 2003. Fine-tuned high-Q photonic-crystal nanocavity. *Opt. Express,* 13, 1202–1214.
38. A. Shinya, S. Mitsugi, E. Kuramochi, T. Tanabe, G. Kim, G. Kira, S. Kondo, K. Yamada, T. Watanabe, T. Tsuchizawa, and M. Notomi. 2005. Ultrasmall resonant tunneling/dropping devices in 2D photonic crystal slabs. *Proc. SPIE,* 5727, 72–85.
39. T. Yoshie, A. Scherer, J. Hendrickson, G. Khitrova, H. M. Gibbs, G. Rupper, C. Ell, O. B. Shchekin and D. G. Deppe. 2004. Vacuum Rabi splitting with a single quantum dot in a photonic crystal nanocavity. *Nature,* 432, 200–203.
40. A. Badolato, K. Hennessy, M. Atatüre, J. Dreiser, E. Hu, P. Petroff, and A. Imamoğlu. 2005. Deterministic coupling of single quantum dots to single nanocavity modes. *Science,* 308, 11580–1161.
41. A. Shinya, S. Mitsugi, E. Kuramochi, and M. Notomi. 2006. Ultrasmall multi-port channel drop filter in two-dimensional photonic crystal on silicon-on-insulator substrate. *Opt. Express,* 14, 12394–12400.
42. H.-Y. Ryu, M. Notomi, and Y.-H. Lee. 2003. High-quality-factor and small-mode-volume hexapole modes in photonic-crystal-slab nanocavities. *Appl. Phys. Lett.,* 83, 4294–4296.
43. G.-H. Kim, Y.-H. Lee, A. Shinya, and M. Notomi. 2004. Coupling of small, low-loss hexapole mode with photonic crystal slab waveguide mode. *Opt. Express,* 12, 6624–6631.
44. T. Tanabe, A. Shinya, E. Kuramochi, S. Kondo, H. Taniyama, and M. Notomi. 2007. Single point defect photonic crystal nanocavity with ultrahigh quality factor achieved by using hexapole mode. *Appl. Phys. Lett.,* 91, 021110.
45. S. G. Johnson, S. Fan, A. Mekis, and J. D. Joannopoulos. 2001. Multipole-cancellation mechanism for high-Q cavities in the absence of a complete photonic band gap. *Appl. Phys. Lett.,* 78, 3388–3391.
46. T. Tanabe, K. Yamada, K. Nishiguchi, A. Shinya, E. Kuramochi, H. Inokawa, M. Notomi, T. Tsuchizawa, T. Watanabe, H. Fukuda, H. Shinojima, and S. Itabashi. 2007. Fast all-optical switching using ion-implanted silicon photonic crystal nanocavities. *Appl. Phys. Lett.,* 90, 031115.
47. E. Kuramochi, M. Notomi, S. Mitsugi, A. Shinya, T. Tanabe, and T. Watanabe. 2006. Ultrahigh-Q photonic crystal nanocavities realized by the local width modulation of a line defect. *Appl. Phys. Lett.,* 88, 041112.
48. K. Inoshita and T. Baba. 2003. Lasing at bend, branch and intersection of photonic crystal waveguides. *Electron. Lett.,* 39, 844–846.
49. B.-S. Song, S. Noda, T. Asano, and Y. Akahane. 2005. Ultra-high-Q photonic double-heterostructure nanocavity. *Nature Mat.,* 4, 207–210.
50. T. Tawara, T. Ito, T. Tanabe, K. Tateno, E. Kuramochi, M. Notomi, and H. Nakano. 2006. Highly selective ZEP/AlGaAs etching for photonic crystal structures using Cl2/HI/Xe mixed plasma. *Jpn. J. Appl. Phys.,* 45, L917–L919.

51. T. Tanabe, M. Notomi, and E. Kuramochi. 2007. Measurement of an ultra-high-Q photonic crystal nanocavity using a single-side-band frequency modulator. *Electron. Lett.*, 43, 187–188.

52. T. Tanabe, M. Notomi, E. Kuramochi, and H. Taniyama. 2007. Large pulse delay and small group velocity achieved using ultrahigh-Q photonic crystal nanocavities. *Opt. Express,* 15, 7826–7839.

53. Y. Tanaka, J. Upham, T. Nagashima, T. Sugiya, T. Asano, and S. Noda. 2007. Dynamic control of the Q factor in a photonic crystal nanocavity. *Nature Mat.*, 6, 862–865.

54. Q. Xu, P. Dong, and M. Lipson. 2007. Breaking the delay-bandwidth limit in a photonic structure. *Nature Phys.*, 3, 406–410.

55. M. Notomi, K. Yamada, A. Shinya, J. Takahashi, C. Takahashi, and I. Yokohama. 2001. Extremely large group-velocity dispersion of line-defect waveguides in photonic crystal slabs. *Phys. Rev. Lett.*, 87, 253902.

56. S.-C. Huang, M. Kato, E. Kuramochi, C.-P. Lee, and M. Notomi. 2007. Time-domain and spectral-domain investigation of inflection-point slow-light modes in photonic crystal coupled waveguides. *Opt. Express*, 15, 3543–3549.

57. A. Yariv, Y. Xu, R. K. Lee, and A. Scherer. 1999. Coupled-resonator optical waveguide: a proposal and analysis. *Opt. Lett.*, 24, 711–713.

58. F. Xia, L. Sekaric, and Y. Vlasov. 2007. Ultracompact optical buffers on a silicon chip. *Nature Photon.*, 1, 65–71.

59. M. Notomi, E. Kuramochi, and T. Tanabe. 2008. Large-scale arrays of ultraigh-Q coupled nanocavities. *Nature Photon.*, 2, 741–747.

60. A. Shinya, S. Mitsugi, E. Kuramochi, and M. Notomi. 2006. Ultrasmall multi-port channel drop filter in two-dimensional photonic crystal on silicon-on-insulator substrate. *Opt. Express*, 14, 12394–12400.

61. V. Almeida, C. Barrios, R. Panepucci, and M. Lipson. 2004. All-optical control of light on a silicon chip. *Nature*, 431, 1081–1084.

62. M. Notomi, A. Shinya, S. Mitsugi, G. Kira, E. Kuramochi, and T. Tanabe. 2005. Optical bistable switching action of Si high-Q photonic-crystal nanocavities. *Opt. Express*, 13, 2678–2687.

63. A. Shinya, S. Mitsugi, T. Tanabe, M. Notomi, I. Yokohama, H. Takara, and S. Kawanishi. 2006. All-optical flip-flop circuit composed of coupled two-port resonant tunneling filter in two-dimensional photonic crystal slab. *Opt. Express*, 14, 1230–1235.

64. T. Tanabe, H. Taniyama, and M. Notomi. 2008. Carrier diffusion and recombination in photonic crystal nanocavity optical switches. *IEEE J. Lightwave Technol.*, 26, 1396–1403.

65. K.-H. Hellwege. 1982. *Landolt-Börnstein Numerical Data and Functional Relationships in Science and Technology Group III: Crystal and Solid State Physics.* Volume 17a. Springer-Verlag, Berlin.

66. K. Yamada, T. Tsuchizawa, T. Watanabe, H. Fukuda, H. Shinojima, T. Tanabe, and S.-I. Itabashi. 2006. All-optical signal processing using nonlinear effects in silicon photonic wire waveguides. In *The 19th Annual Meeting of the IEEE Lasers & Electro-Optics Society*. Institute of Electrical and Electronics Engineers, New York.

67. O. Palais and A. Arcari. 2003. Contactless measurement of bulk lifetime and surface recombination velocity in silicon wafers. *J. Appl. Phys.*, 93, 4686–4690.

68. H. Mäckel and A. Cuevas. 2003. Determination of the surface recombination velocity of unpassivated silicon from spectral photoconductance measurement. In *3rd World Conference on Photovoltaic Energy Conversion*. Institute of Electrical and Electronics Engineers, New York.

69. T. Tanabe, K. Yamada, K. Nishiguchi, E. Kuramochi, A. Shinya, H. Inokawa, S. Kawanishi, and M. Notomi. 2006. Fast all-optical pulse train modulation by silicon photonic crystal nanocavities. In *Proceedings of the 19th Annual Meeting of the IEEE Lasers and Electro-Optics Society*. Institute of Electrical and Electronics Engineers, New York.

70. K. Kitayama, S. Arakawa, S. Matsuo, M. Murata, M. Notomi, R. Takahashi, and Y. Itaya. 2007. All-optical RAM-based buffer for packet switch. In *Proceedings of Photonic in Switching, 2007*. Institute of Electrical and Electronics Engineers, New York.

71. M. Notomi and S. Mitsugi. 2006. Wavelength conversion via dynamic refractive index tuning of a cavity. *Phys. Rev. A*, 73, 051803.
72. M. Notomi, H. Taniyama, S. Mitsugi, and E. Kuramochi. 2006. Optomechanical wavelength and energy conversion in high-Q double-layer cavities of photonic crystal slabs. *Phys. Rev. Lett.*, 97, 023903.

2

Pillar Microcavities for Single-Photon Generation

Charles Santori and David Fattal
Hewlett-Packard Laboratories

Jelena Vučković
Stanford University

Matthew Pelton
Argonne National Laboratory

Glenn S. Solomon
National Institute of Standards and Technology and University of Maryland

Edo Waks
Institute for Research in Electronics and Applied Physics

David Press and Yoshihisa Yamamoto
Stanford University

CONTENTS

2.1 Introduction

This chapter describes devices designed to emit *single*-photons at controlled time intervals, as needed for various quantum information applications. Before describing how such a device might be constructed, let us define exactly what we mean by a source of single-photons. Such a light source is quite different from a classical source such as a laser or a light emitting diode, where the number of photons per pulse varies randomly, usually following a Poisson distribution. However, a realistic single-photon source will never be perfect, sometimes emitting zero or multiple photons, and thus the distinction between a classical photon source and a realistic single-photon source is subtle. We therefore need quantitative measures to distinguish between light sources based on photon statistics. From the viewpoint of system applications, other characteristics are also important. The most important parameters describing a single-photon source are:

1. The second-order coherence function $g_0^{(2)}$ [1,2] which places an upper bound on the two or multi-photon probability per pulse, $p_m = p(n \geq 2) \leq (1/2) g_0^{(2)} \langle n \rangle^2$, where $\langle n \rangle$ is the average photon number per pulse. A photon source with $\langle n \rangle = 1$ and with, $g_0^{(2)} = 1$, $g_0^{(2)} < 1$, or $g_0^{(2)} = 0$ is called a Poisson, sub-Poisson or ideal single-photon source, respectively. Since a realistic single-photon source has $0 < g_0^{(2)} < 1$, it is a sub-Poisson photon source operating in the regime of $\langle n \rangle \lesssim 1$. When a single-photon source is used in quantum key distribution systems, the robustness against a specific eavesdropping attack, called the photon number splitting attack, requires a small value of $g_0^{(2)}$.

2. An external quantum efficiency η_{total}, which defines an overall out-coupling efficiency of internally generated single-photons into a desired single waveguide mode.

3. Maximum repetition frequency f_{max} of single-photon generation, which determines an upper bound on the clock frequency of a system. These two parameters, η_{total} and f_{max}, must be reasonably high for practical applications. In the case of a laser, the stimulated emission of photons enhances these parameters. For a single-photon source, however, stimulated emission is not an option and we need a new trick to achieve high values of η_{total} and f_{max}. One of the most useful techniques

available for this purpose is to use cavity quantum electrodynamical effects [3] to enhance spontaneous emission.

4. The overlap $v(\Delta t=0)\equiv v(0)$ between two single-photon wavefunctions [4], which describes the degree of quantum indistinguishability of two single-photons when the time delay Δt between them is zero. This overlap can be computed either in the space-time or momentum-frequency domains, and an ensemble average is taken over all possible quantum states generated by the source. To use a single-photon source for such quantum information systems as a Bell state analyzer, quantum teleportation and linear optics quantum computation, single-photons must be identical quantum particles with $v(0)=1$. However, a realistic single-photon source has a finite timing jitter and dephasing for the generated electromagnetic field, which reduces $v(0)$ below one.

5. An emission wavelength uncertainty, $\Delta\lambda$, in different single-photon sources. When a single-photon source is used in a large scale photonic system, massively parallel generation of single-photons is needed. In such a case, the emission wavelength of single-photon sources must be identical.

In this chapter, we focus on semiconductor-based single-photon sources, especially a single InAs or InGaAs quantum dot (QD) embedded in a GaAs/AlAs pillar microcavity. The first proposal for a single-photon source in a semiconductor was based on the Coulomb blockade effect for a single electron and hole in a mesoscopic p–n junction [5,6]. A device based on this principle was demonstrated in 1999 [7]. This was an electrically pumped, compact p–n-junction device, but the operation temperature was limited to ~50 mK, and thus it could not be practical. Most semiconductor-based single-photon sources developed today employ optical excitation pulses to inject electron–hole pairs into a single QD, and use post-filtering to collect the last photon emitted. Due to strong Coulomb interactions, the emission wavelength depends on the number of charged particles inside a QD. Even if the laser pulse creates several electron–hole pairs, the last electron–hole pair emits a photon at a unique wavelength, which can be spectrally separated from other photons. Using this method, single-photon sources with $g_0^{(2)}<0.1$ were demonstrated [8,9].

For a high-fidelity and optically pumped single-photon source, an emissive system should have at least three discrete states. For incoherent generation of single-photons by a simple spontaneous emission process [8,9], an external pump source excites the atomic system from $|b\rangle$ to $|r\rangle$ states, which is followed by rapid relaxation to an $|a\rangle$ state, as shown in Figure 2.1a. The cavity field is resonantly coupled to the atomic transition between the $|a\rangle$ and $|b\rangle$ states. In the case of an electrically neutral QD, the $|b\rangle$, $|a\rangle$ and $|r\rangle$ states usually correspond to an empty state (or crystal ground state), one electron–hole pair at the 1e–1h transition energy, and one electron–hole pair at 2e–2h transition, respectively. In theory, better performance is possible in schemes for coherent generation of single-photons by a Raman scattering process [10,11]. Such coherent excitation schemes require two metastable ground states $|g\rangle$ and $|e\rangle$ which could be physically realized, for instance, by the Zeeman splitting of a single trapped electron inside a QD. The excited state $|r\rangle$ is the lowest exciton state with an unpaired electron. The two metastable ground states are optically coupled to a single excited state (Figure 2.1b). In such a Raman configuration, the excited state population can be kept small at all times to minimize decoherence due to spontaneous emission.

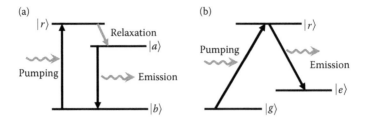

FIGURE 2.1
Three-level schemes states used for generating single-photons. (a) Configuration for incoherent generation of single-photons by spontaneous emission. (b) Configuration for coherent generation of single-photons by Raman scattering.

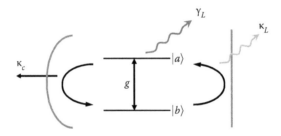

FIGURE 2.2
A single "atom"-cavity system and physical parameters.

To address the feasibility of single-photon generation in semiconductor microcavities, let us introduce a few physical parameters (Figure 2.2). The coupling rate between the cavity photon and the dipole of the atom or QD is [12],

$$g = \wp \sqrt{\frac{\omega_a}{2\hbar\varepsilon V}}, \tag{2.1}$$

where $\wp$ is the dipole moment, ω_a is the frequency of the dipole transition, ε is the permittivity of the semiconductor, and V is the cavity mode volume. For the present discussion we have assumed that the dipole is located at the cavity field maximum and that the dipole is perfectly aligned with the electric field. The spontaneous emission rate of the dipole in the bulk material (without the cavity) is [12],

$$\Gamma = \frac{n\omega_a^3\wp^2}{3\pi\varepsilon_0\hbar c^3}, \tag{2.2}$$

where n is the refractive index. It is convenient also to define a dimensionless oscillator strength,

$$f_{osc} = \frac{\Gamma}{3\Gamma_{ceo}} = \frac{1}{(4\pi K)^2}\left(\frac{\lambda}{n}\right)^3\omega_a\Gamma. \tag{2.3}$$

Here, $K = e/\sqrt{4\varepsilon m_e}$, where e and m_e are the electron charge and mass, λ is the wavelength in vacuum, and $\Gamma_{ceo} = n\omega_a^2 e^2 / (6\pi\varepsilon_0 m_e c^3)$ is the spontaneous emission rate from a classical electron oscillator [13]. We can then re-write the coupling rate as,

$$g = K\sqrt{3 f_{osc}/V}. \tag{2.4}$$

We note that g as defined here is half of the single-photon Rabi frequency.

The parameter V is the mode volume of the cavity, defined as,

$$V = \frac{\iiint \varepsilon(\vec{r}) |E(\vec{r})|^2 \, d^3\vec{r}}{\max(\varepsilon(\vec{r}) |E(\vec{r})|^2)}. \tag{2.5}$$

For the maximum possible interaction between a cavity photon and a QD dipole, the electric field of a single photon should be as large as possible at the location of the QD. This occurs when we confine the photon to the smallest possible space. Another important parameter is the rate at which photons leak out of the cavity. This is given by $\kappa = \kappa_C + \kappa_L$, where κ_C is the rate at which photons couple to a down-stream waveguide mode and κ_L is the rate at which photons are lost to other spurious modes or absorption in the cavity mirrors. These determine the length of time over which the atom and cavity photon can interact, and the coupling efficiency between a cavity and an external optical channel.

There are a number of figures of merit to be considered from these constants, which we now consider and discuss in the context of semiconductors. First is the quality factor of the cavity, $Q = \omega_{cav}/\kappa$ for cavity frequency ω_{cav}. The small size of semiconductor microcavities make them subject to surface defects, which prevents their Q from being as high as in larger cavities. For single-photon generation and quantum networking, the ratio κ_C/κ is also critical to ensure that all single photons that leak out of the cavity end up in the desired waveguide mode.

The Purcell factor is defined as the enhancement factor of the spontaneous emission decay rate in the weak-coupling limit ($g \ll \kappa$),

$$F(\Delta) = \frac{\kappa g^2/\Gamma}{\Delta^2 + \kappa^2/4}, \tag{2.6}$$

where $\Delta = \omega_a - \omega_{cav}$ is a detuning parameter between the QD or atomic transition frequency ω_a and cavity resonant frequency ω_{cav}, and Γ is the spontaneous emission rate without a cavity QED effect. The out-coupling efficiency into a desired waveguide mode is given by

$$\eta_{total} = \left(1 - \frac{\Gamma_L}{\Gamma F(0)}\right) \times \left(\frac{1}{1 + \kappa_L/\kappa_c}\right), \tag{2.7}$$

where Γ_L is the spontaneous emission rate into spurious modes.

The leakage rates κ and Γ may be compared to the coupling rate g in the context of the emission spectrum of the atom/cavity system. When $\omega_a = \omega_{cav}$, the frequencies of the two normal modes of the coupled system are,

$$\omega_{\pm} = \omega_a - i(\kappa + \Gamma)/4 \pm \sqrt{g^2 - (\kappa - \Gamma)^2/16}. \tag{2.8}$$

When $g > |\kappa-\Gamma|/4$, the system shows a splitting into two eigenmodes separated by a frequency called the vacuum Rabi splitting. This is the strong-coupling regime.

Strong coupling is not needed for the coherent generation and capture of single-photons, but it makes such schemes far simpler. In the weak coupling regime ($g < |\kappa-\Gamma|/4$), large detunings and very fast, critically shaped pulses must be used for the coherent generation and trapping of single-photons. The critical figure of merit for system applications of single-photon sources is not $g/4|\kappa-\Gamma|$, that is, whether we are in the strong or weak coupling regime, but rather the cooperativity parameter (or spontaneous emission coupling efficiency), $C = 4g^2/\Gamma\kappa$. An immediate physical indication that this parameter may take on a high value in existing semiconductor microcavity systems is the Purcell effect, in which the rate of radiation of an electric dipole is enhanced by the modified local density of photon states inside a cavity. This may be seen from Equation 2.4 by taking g, $\Gamma \ll \kappa$, in which case the $+$ solution is $\omega_a - i(\Gamma/2)(1 + C)$, indicating that the radiative damping of the atom is enhanced by the factor $1+C$. C may be quickly estimated as $C = 3/4\pi^2 \times (\lambda/n)^3 \times Q/V$. Semiconductor microcavity systems show a key advantage in achieving high values of C in comparison to larger cavities that might be used with trapped atoms or ions.

Pillar microcavities containing InAs quantum dots (QDs) combine small mode volumes with efficient coupling to free-space and thus can make excellent single-photon sources. The only major drawback in this system is the low operation temperature (<50 K) required to avoid excessive broadening and nonradiative losses in the QD transitions. We will begin by describing the design, crystal growth and lithography involved in making these devices at Stanford University. Next, we describe optical characterization of pillar microcavity structures, first demonstrating spontaneous emission modification in single InAs QDs, and then testing the degree to which these structures, excited by short optical pulses, can satisfy the requirements of reduced two-photon probability, high collection efficiency, and photon indistinguishability. We also describe quantum-optical characterization of pillar microcavities provided by the Forchel group at the Univerity of Würzburg which have high enough Q factors to reach the strong-coupling regime [14]. We then describe proof-of-principle demonstrations of simple applications for these devices, including quantum cryptography and experiments based on two-photon interference, and provide a detailed theory of coherent trapping and emission in a three-level Λ system coupled to a microcavity, which has implications for future applications in the area of quantum networking.

2.2 Design and Fabrication

Fabrication of pillar microcavity devices for single-photon applications involves many considerations in the areas of design, crystal growth and lithography. In this section we first describe the design issues involved in choosing the physical dimensions, number of mirror pairs, and refractive index contrast in pillar microcavities. The goal is to obtain the highest possible Q/V ratio for maximum coupling between a cavity photon and an InAs QD. We next describe the crystal growth methods used at Stanford University to create sparse InAs QDs with a relatively short emission wavelength (~900 nm), which is well suited for efficient detection by silicon avalanche-photodiode photon counters. Finally, we describe the lithographic processes used to create three "generations" of pillar structures, the goals being to obtain narrow pillars with straight sidewalls and minimal surface roughness.

2.2.1 Design of Pillar Microcavities

Pillar microcavities consist of a high-refractive-index region (spacer) sandwiched between two dielectric mirrors, as shown in Figure 2.3. Confinement of light in these structures is achieved by the combined action of distributed Bragg reflection (DBR) in the longitudinal direction (along the pillar axis), and total internal reflection (TIR) in the transverse direction (along the pillar cross-section). The pillars analyzed in this chapter are rotationally symmetric around the vertical axis. The DBR mirrors can be viewed as one dimensional (1D) photonic crystals generated by stacking high- and low-refractive-index disks on top of each other. The microcavity is formed by introducing a defect into this periodic structure. The periodicity of the photonic crystal is denoted as a, the thickness of the low-refractive-index disks is b, the diameter of the disks is D, and the refractive indices of the low- and high-refractive-index regions are n_l and n_h, respectively. The defect is formed by increasing the thickness of a single high-refractive-index disk from $(a-b)$ to s, as shown in Figure 2.3. The number of photonic crystal periods above and below the defect region (i.e., the number of DBR pairs) is labeled as MPT and MPB, respectively.

The mode of interest to us is the doubly degenerate fundamental (HE_{11}) mode, whose field pattern is also shown in Figure 2.3. The component of the electric field parallel to the DBR mirrors is dominant in this mode, and has an antinode in the center of the spacer. Furthermore, in this central plane, at the location of the vertical axis of the pillar, the electric field is polarized exactly parallel to the plane of the DBR layers. Because the cavity supports two degenerate modes with orthogonal polarizations, linear superpositions of these modes can produce any polarization in this plane, including linear, circular, or

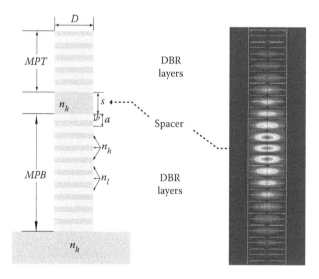

FIGURE 2.3
Left: Parameters for a pillar microcavity. The pillar is rotationally symmetric around the vertical axis. Right: Electric-field intensity for the fundamental (HE_{11}) mode in a pillar microcavity. The electric field is predominantly polarized in the plane of the distributed Bragg reflector layers. Since a pillar with perfect rotational symmetry supports two orthogonally polarized modes, any polarization in this plane (linear, circular, elliptical) is possible.

elliptical. Since epitaxial QDs have two electric dipole transitions polarized in this plane, the cavity and QD polarizations align perfectly, in theory. However, since asymmetry in the pillar shape can produce a polarization splitting of the cavity modes, and asymmetry in the QD can produce a polarization splitting in the dipole transitions, more complicated situations are possible, as will be mentioned in the next section. For the pillar modes, there is a small deviation from the in-plane polarization at larger distances from the pillar axis. All theoretical analyses presented in this section are performed by the Finite-Difference Time-Domain (FDTD) method; for the detailed description of this method, please refer to Pelton et al. [15].

The rule of thumb generally used for designing pillars is to make mirror layers one-quarter wavelength thick, and to choose the optical thickness of the spacer equal to the target wavelength. In the case of a planar DBR cavity (with diameter $D \rightarrow \infty$), this choice of parameters leads to the maximum reflectivities of the mirrors and the maximum Q-factor of the cavity mode: the cavity operates at the Bragg wavelength, for which the partial reflections from all high- and low-refractive-index interfaces add up exactly in phase. However, since the strength of the cavity QED phenomena is proportional to the ratio of the cavity Q-factor to the mode volume V, we will try to design pillars in such a way that this ratio is maximized. In order to do so, we will explore structures with small diameters D, and will attempt to improve their Q factors.

In general, Q reduces with a decrease in D, due to the combination of two loss mechanisms: *longitudinal loss* through DBR mirrors, and *transverse loss* due to imperfect TIR confinement in the transverse direction. Let us address the *longitudinal loss* first. The decrease in the post diameter D implies a change in the dispersion relation of the 1D photonic crystal, and the size and position of its bandgap, as illustrated in Figure 2.4. In this figure, it is

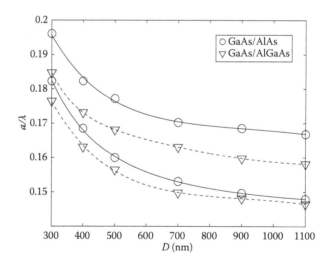

FIGURE 2.4

Bandgap edges, calculated using the FDTD method (points), of the fundamental (HE_{11}) mode in a cylindrical one-dimensional photonic crystal in the GaAs/AlAs or GaAs/Al_xGa_{1-x}As material systems. The lines are guides to the eye. The GaAs/AlAs photonic crystal has the following parameters: $n_h=3.57$, $n_l=2.94$, $b=85$ nm, and $a=155$ nm. The GaAs/Al_xGa_{1-x}As photonic crystal has the following parameters: $n_h=3.57$, $n_l=3.125$, $b=80$ nm, and $a=150$ nm. (See Figure 2.3 for definition of parameters.) The bandgap edges for $D \rightarrow \infty$ are positioned at a/λ equal to 0.1445 and 0.1634 for the GaAs/AlAs photonic crystal, and at a/λ equal to 0.1431 and 0.1565 for the GaAs/Al_xGa_{1-x}As photonic crystal. (From Vučković, J., Pelton, M., Scherer, A., and Yamamoto, Y., *Phys. Rev. A*, 66 (2002) 023808. © American Physical Society. With permission.)

assumed that the high- and low-refractive-index regions of the photonic crystal consist of GaAs and AlAs, with refractive indices of $n_h=3.57$ and $n_l=2.94$, and thicknesses of 70 and 85 nm, respectively, or that they consist of GaAs and $Al_x Ga_{1-x} As$, with refractive indices of $n_h=3.57$ and $n_l=3.125$, and thicknesses of 70 and 80 nm, respectively. The y-axis gives a/λ, where a is the mirror periodicity (155 or 150 nm; see figure caption) and λ is the vacuum wavelength corresponding to the bandgap edge. When the diameter D decreases, the frequencies of the bandgap edges increase, and the size of the bandgap decreases. For structure diameters larger than 2 μm, bandgap edges can be approximated by their values at $D \to \infty$. Therefore, as D decreases, the blue shift of the cavity mode wavelength λ increases relative to the target wavelength at which the 1D cavity operates [15]. Simultaneously, the size of the photonic bandgap decreases, implying that the cavity mode is less confined in the longitudinal direction than in the planar cavity case.

The cavity mode is strongly localized in real space, and consequently delocalized in Fourier space (k-space), meaning that it consists of a wide range of wave-vector components. Some of these components are not confined in the post by TIR; i.e., they are positioned above the light line, where they can couple to radiative modes, leading to *transverse loss*. A cavity mode which is strongly confined in the longitudinal direction by high-reflectivity mirrors is delocalized in Fourier space and suffers large transverse loss. Similarly, a mode that is delocalized in the longitudinal direction is more localized in Fourier space and suffers less transverse loss. Therefore, when optimizing the quality factor of three-dimensional (3D) pillars, there is a tradeoff between these two loss mechanisms.

In the middle of a large bandgap, the longitudinal confinement is strongest, but the Q-factor is limited by transverse loss. By shifting the resonant wavelength away from the mid-gap (e.g., by tuning the thickness of the cavity spacer), one can delocalize the mode in real space, localizing it more strongly in Fourier space, reducing the contribution of wave-vector components above the light line, and thereby decreasing the transverse radiation loss. Eventually, as the mode wavelength approaches the bandgap edges, the loss of longitudinal confinement starts to dominate and Q drops. Therefore, in the pillars with high reflectivity mirrors and finite diameter, it is expected that the maximum Q will be located away from the mid-gap position. Moreover, since the mode wavelength can be tuned from the mid-gap towards any of the two bandgap edges, two local maxima of Q (i.e., a double peak behavior in Q versus mode wavelength) are expected. Besides detuning the mode wavelength from the mid-gap, we can also suppress the transverse loss by relaxing the mode slightly in the longitudinal direction, i.e., by reducing the reflectivities of photonic crystal mirrors and decreasing the bandgap size. This can be achieved by shrinking the cavity diameter, or by changing the photonic crystal parameters (e.g., by reducing the refractive-index contrast).

In this section, we study both of these approaches to Q optimization: tuning the mode wavelength away from the mid-gap by changing the spacer thickness, and tuning the mirror reflectivities by changing photonic crystal parameters or cavity diameter [16]. We also show that the employment of very high reflectivity mirrors cannot lead to high-Q cavities with small diameters, as the transverse radiation loss is high, resulting from very strong mode localization in the longitudinal direction. Finally, we comment on the degradation of the pillar quality factor as a result of fabrication imperfections.

Let us first study the same set of pillar parameters as in our earlier work [15], chosen in such a way that the cavity would operate at the Bragg wavelength for $D \to \infty$: $a=155$ nm, $b=85$ nm, $s=280$ nm, $n_h=3.57$ and $n_l=2.94$, $MPT=15$ and $MPB=30$. Such a refractive index contrast corresponds to the GaAs/AlAs material system used in the experiments presented in this chapter. In an attempt to maximize the ratio of the cavity Q-factor to the mode volume, we here analyze only the behavior of the HE_{11} mode for the cavity diameter D below 0.5 μm.

In order to tune the mode frequency within the bandgap, we tune the spacer thickness s. Results for λ, Q, V, and Q/V are shown in Figures 2.5 and 2.6. From Figure 2.4, we see that the bandgaps in these structures extend from 875 to 969 nm, from 850 to 920 nm, and from 790 to 850 nm, for structure diameters of 0.5, 0.4, and 0.3 μm, respectively. As we have noted previously, when D decreases, the bandgap edges shift towards lower wavelengths, and the size of the bandgap decreases. The cavity mode wavelength is blue-shifted in this process, as can be seen in Figure 2.5.

The mode volume V is minimized when the mode wavelength is located near the middle of the bandgap, as shown in Figure 2.6. For the structures with D equal to 0.4 and 0.3 μm, the maximum Q-factor also occurs close to the mid-gap. Different behavior is seen for the

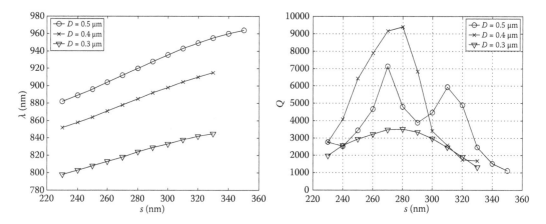

FIGURE 2.5
Wavelength λ and quality factor Q of the fundamental mode in a pillar with $a=155$ nm, $b=85$ nm, $n_h=3.57$, $n_l=2.94$, $MPT=15$ and $MPB=30$. The cavity diameter D and the spacer thickness s are tuned. (From Vučković, J., Pelton, M., Scherer, A., and Yamamoto, Y., *Phys. Rev. A*, 66 (2002) 023808. © American Physical Society. With permission.)

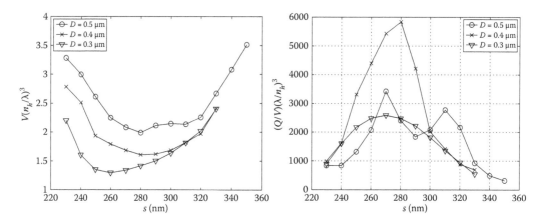

FIGURE 2.6
Mode volume V and ratio of quality factor Q to V for the HE_{11} mode in a pillar with $a=155$ nm, $b=85$ nm, $n_h=3.57$, $n_l=2.94$, $MPT=15$ and $MPB=30$. The cavity diameter D and the spacer thickness s are tuned. (From Vučković, J., Pelton, M., Scherer, A., and Yamamoto, Y., *Phys. Rev. A*, 66 (2002) 023808. © American Physical Society. With permission.)

structure with D equal to 0.5 µm, which has a local minimum of Q at mid-gap and exhibits a double-peak behavior.

The double-peak behavior is due to the mechanism described above. In the middle of the bandgap, where the longitudinal mode confinement is strongest and the mode volume is minimum, the radiation loss in the transverse direction is high, and the Q-factor is degraded. By shifting the resonant wavelength away from the mid-gap, the mode is delocalized in real space, leading to a reduction in the transverse radiation loss (e.g., at the positions of the two peaks in Q). Eventually, as the mode wavelength approaches the bandgap edges, the loss of longitudinal confinement starts to dominate, Q drops, and the mode volume increases. To support this explanation, we analyzed the same structure, with $D=0.5$ µm, but with the number of mirror pairs on top (MPT) increased from 15 to 25 (not shown). At mid-gap, Q did not increase significantly with MPT. The mode there is already strongly confined in the longitudinal direction, and the addition of extra pairs does not change the longitudinal loss. The modal Q-factor is determined by the radiation loss in the transverse direction, which is independent of MPT. On the other hand, the Q's at the two peaks increased with MPT. At these points, the mode is not confined as well in the longitudinal direction, and longitudinal loss can be reduced by adding more mirror pairs.

As an even stronger demonstration of our explanation for the double-peak behavior, we separate the radiation loss into the loss above the top pillar surface (L_a), and the loss below it (L_b). The total Q is a combination of two newly introduced quality factors, Q_a and Q_b, which are inversely proportional to L_a and L_b, respectively:

$$\frac{1}{Q} = \frac{1}{Q_a} + \frac{1}{Q_b}. \tag{2.9}$$

It follows from their definition that Q_a and Q_b are measures of the longitudinal and transverse loss, respectively. We analyze two sets of structure parameters, corresponding to the local maximum or minimum in Q. For $s=270$ nm and $D=0.5$ µm (local maximum), we calculate $Q_a \approx 14{,}500$ and $Q_b \approx 13{,}910$, while, for $s=290$ nm and $D=0.5$ µm (local minimum), we calculate $Q_a \approx 16{,}000$ and $Q_b \approx 5100$. These results show that the local minimum in Q is due to an increase in the transverse loss, manifested as a drop in Q_b.

Let us now address the single-peak behavior of Q as a function of cavity spacer thickness, when D is equal to 0.4 or 0.3 µm. Structures with smaller diameters have smaller bandgaps, as illustrated in Figure 2.4, and the cavity modes are more delocalized in the longitudinal direction, relative to the structure with $D=0.5$ µm. The defect modes must therefore be more localized in Fourier space, and will thus suffer less radiation loss in the transverse direction. This implies that the Q-factors are determined mostly by the longitudinal loss. They reach their maxima at the mid-gap, where the mode volume is minimum, and the longitudinal confinement is strongest.

The maximum Q/V ratio of almost 6000 (where V is measured in cubic wavelengths in the high-refractive index material) is achieved for the structure with $D=0.4$ µm, as shown in Figure 2.6. For this structure, the Q-factor is close to 9500, and the mode volume is $1.6(\lambda/n_h)^3$. For $D=0.4$ µm, a variation in the thicknesses of the mirror layers allows us to achieve a small increase in the Q-factor, to 10,500, and in the Q/V ratio, to 6500. This result is obtained for $a=155$ nm, $b=75$ nm, and $s=290$ nm. Further improvement of the Q/V ratio could possibly be achieved by employing more sophisticated methods such as genetic algorithms [17] and inverse design [18], in which more pillar parameters (e.g., individual mirror layers) are tuned.

In reality, imperfections in the growth and fabrication processes used to construct the pillars inevitably lead to reduced performance. Possible problems include deviations in the layer thickness, residual absorption, surface roughness, and nonvertical sidewalls. As discussed below, we find experimentally that the Q values of pillar structures are significantly reduced compared with the planar cavities before etching, and we attribute this reduction mainly to vertical undercut and insufficient etch depth through the lower DBR, which lead to an increase in the transverse losses and diffraction in the lower DBR, respectively, and cause a degradation of the quality factor of the fundamental mode, as shown in Figure 2.7 [15]. The problems of the insufficient etch depth and the vertical undercut have been gradually minimized in subsequent generations of our pillars, thereby leading to improved experimental Q-factors and better cavity QED results. Details on the microcavity fabrication and characterization are given in the following sections.

We noted above that a potential route to maximizing Q for small pillar diameters is the construction of a photonic crystal with a small refractive-index perturbation. As the perturbation gets smaller, the cavity mode becomes more delocalized in real space, and consequently more localized in Fourier space. This, in turns, leads to reduction in the transverse radiation loss. Furthermore, the cavity resonance can be located at lower frequencies,

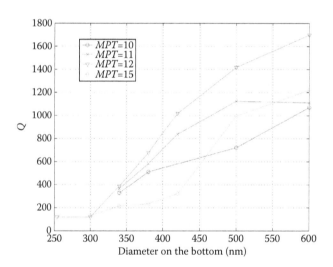
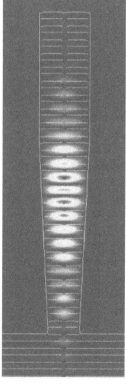

FIGURE 2.7

Left: Electric field intensity for the fundamental mode in a pillar with realistic wall profile. Right: The effect of vertical undercut on the Q-factor of the fundamental mode. For a pillar with a fixed etch depth (3535 nm, corresponding to etching through 21 mirror pairs and a spacer), with the diameter on top of 600 nm, and a straight etch through the top nine mirror pairs, the Q-factor drops dramatically with an increase in undercut (i.e., with the reduction of the pillar diameter on the bottom). Different curves correspond to different numbers of mirror pairs on top of the spacer.

where the density of free-space radiation modes is smaller. In order to compensate for the increased longitudinal loss, we need to put more MPT of these structures. We studied pillars in the GaAs/Al$_x$Ga$_{1-x}$As material system with lower index contrast (n_h=3.57 and n_l=3.125) and showed that even though Q larger than 14000 can be achieved for D=0.5 μm, V also increases, and the maximum Q/V ratio is similar to that calculated for the GaAs/AlAs system. Furthermore, this Q/V ratio requires more top mirror pairs than in the GaAs/AlAs system, thereby implying more challenging fabrication [16].

On the other hand, if we employ a material system with a high refractive-index contrast, such as GaAs/AlO$_x$ (n_h=3.57 and n_l=1.515), a larger bandgap (and thus a better longitudinal confinement of the cavity mode) could be achieved, but the mode suffers more radiation loss in the transverse direction, which limits its Q-factor to values below 1000, even with pillar diameters D of up to 1.3 μm [16].

2.2.2 Growth of QDs

There are several choices of optically active, discrete emitters in semiconductor systems, including deep-level states such as color centers in diamond, quantum well (QW) interface fluctuations, and colloidal and epitaxial QDs. It is possible, in principle, to integrate all of these emitters in microcavities. One of the most advanced planar microcavity systems is the epitaxial system utilizing groups III and V semiconductors such as GaAs and AlAs; this system is compatible with epitaxial QDs and QW fluctuations. Epitaxial QDs generally have deeper confining potentials and are well separated from the GaAs band edge. Even though they have size variations, their shape and polarization are more predictable than interface fluctuations.

Semiconductor crystal growth techniques, especially the molecular-beam epitaxy technique used here, can produce near monolayer control of the deposition thickness. This is used to make the high-quality planar microcavity structures described in this chapter, as well as QWs. A natural approach to fabricating atomic-like discrete emitters was to extend this control used to make QWs to 2D and 3D confinement with quantum wires and QDs through post-growth processing. These results have been encouraging, but not resounding, because the lithography techniques used to provide lateral confinement typically do not have the monolayer resolution that is available through epitaxial growth techniques. Furthermore, because of interfacial damage, it is difficult to directly pattern active regions using processing.

In homoepitaxial crystal growth (epitaxial layer and substrate are same material) of cubic materials, the (100) growth surface is the lowest energy facet plane [19]. If the growth temperature and flux rates are appropriate, adatom attachment on the growth surface is to kink and ledge sites or flat island nucleation regions. Crystal growth thus proceeds by the lateral growth of kinks and ledges, or by the expansion of flat, monolayer-scale high islands. Ideally, as one monolayer is filled, new, monolayer-high nucleation sites are created, and the 2D (100) growth surface propagates. In contrast, during heterogeneous crystal growth (epitaxial layer and substrate are different materials), as more adatoms are deposited onto the growing surface, this growth surface can go through structural changes that can be as dramatic as the complete loss of epitaxial coherency, or as small as changes in surface reconstruction, surface roughness, or the abruptness of a heterointerface [20–23]. In the case of InAs on GaAs, the lattice mismatch is 7.2%. The critical thickness for relaxation by dislocation generation is expected to be between a few monolayers [24] and 15 monolayers (4.5 nm) [25], depending on the theoretical model and the crystal growth conditions. Because of the similar InAs and GaAs crystal structures, at least one chemisorbed monolayer of InAs can be stable on the GaAs substrate. Below the critical thickness, or after a chemisorbed layer, a metastable phase can exist. This is phenomenologically known as the Stranski–Krastanow (SK) growth regime [26]. In this growth regime, the lattice-mismatched growth produces surface islands. The islands, which become the QDs, increase the total energy with respect to the bulk crystal, because they create additional interface surface area. However, the islanding surface allows for partial strain energy relaxation due to the surface curvature and distance from the hetero-interface. This growth regime is a transitional growth mode between the compliant, planar growth regime that characterizes ideal MBE growth, and plastically relaxed growth, since, as islands grow and merge, the surface area can no longer expand to accommodate the increasing strain energy.

Since the early 1990s, this technique has been used to make QDs in the InAs/GaAs and Ge/Si semiconductor systems by overgrowing the island layer with the host semiconductor [27]. The resulting QDs are narrow-bandgap regions commensurate with a host crystal and, for InAs, contain on the order of 20,000–50,000 atoms. The electron and hole potential confinement and small size are such that discrete electron and hole many-body states can be observed with transitions dominated by radiative processes.

This QD fabrication process produces random arrays of islands with slightly varying sizes, and ensembles which are either ordered or containing identical QDs have not been demonstrated to date. This is due to the secondary role of the energies that drive the QD formation process in epitaxial systems. The nonuniform QD size and shape lead to a nonuniform QD spectral distribution. By making dilute QD ensembles and accessing single QDs through lithographic patterning, the ensemble broadening can be an advantage, because single-QD states can be spectrally distinguished.

For MBE growth on planar or vicinal surfaces, the InAs QD size and density are controlled predominantly by the amount of InAs deposition and the surface temperature [28]. Other process controls that also contribute are the InAs growth rate and the V/III flux ratio [29]. All of these parameters control the kinetic limitations to the islanding process, including In adatom surface diffusion, In desorption, and alloy mixing. The InAs QDs used for the studies described in this chapter were deposited in a MBE system using an As_2 flux that is produced by thermally decomposing the dominant vapor species, As_4, present above heated solid arsenic. In addition, the As source has an adjustable valve, so that the V/III flux ratio can be precisely controlled. The V/III beam equivalent pressure ratio is kept between 8 and 20 for all layers. The GaAs growth rate was kept low, between 0.1 and 0.2 μm/hr, providing long In adatom surface diffusion lengths and, thus, a more equilibrium QD distribution.

The amount of material deposited has a significant impact on the QD distribution. Below approximately 1.8 monolayer (mls) of InAs, the surface is rough, but there is no island formation, since the lattice mismatch strain is accommodated by InAs lattice compression [28,30]. This region is called the wetting layer.

At a surface temperature of 480°C, with 2 mls of deposited InAs, the average island diameter is 15 nm. When the equivalent of three planar mls of InAs have been deposited, the additional material is accommodated by an increase in the island size and density. The island size increases because more material has been deposited and migrates to the islands. The increase in the island density increases as more InAs islands nucleate [28]. Between 1.8 and 3 mls the QD density changes from 0 to 1000 μm^{-2}. In order to incorporate single QDs into pillar microcavities, a density of less than 100 μm^{-2} is desired and less than 10 μm^{-2} is best.

When an additional 1 ml of InAs is deposited, so that the total InAs deposited is 4 mls, the islands begin to coalesce and this process no longer provides effective bulk strain relaxation through surface curvature. Plastic relaxation through dislocation generation begins to occur. Thus, there is a narrow range of InAs deposition between slightly larger than 1.8 mls (0.5 nm) to approximately 3 mls (0.9 nm) where useful QDs can be made. Furthermore, there is a correlation between the density of QDs and the average QD size: the greater the density, the larger the size.

A cross-sectional scanning tunneling microscopy (XSTM) image obtained by cleaving the QD sample in vacuum is shown in Figure 2.8. The cleaved surface is (110), and in this image tunneling occurs between the tip and empty conduction band states [31]. The QDs are covered with epitaxial GaAs. While the QDs nucleate on a planar wetting layer region, the image indicates that the wetting layer diffuses throughout the QD region, so that the QDs lie within a rough QW region formed by the wetting layer. This is most likely the result of strain-enhanced diffusion. Diffusion of In away from the QDs appears to be limited.

The surface temperature affects the QD size and density through the surface diffusion and desorption rates. Below 460°C, the In adatom surface diffusion is limited and island formation is inhibited. Conversely, with our low growth rates, by 540°C, the desorption rate of In is larger than the In input to the surface and no QDs are present. However, at temperatures above 500°C, alloying with the surrounding GaAs occurs and, coupled with desorption, can lead to a desired arrangement, where large QD sizes can be made with low density. This is the technique used here, where the amount of deposited InAs was varied from 1.9 to 2.2 mls, and the growth temperature was between 510 and 530°C.

By this method, we produced QDs with densities ranging from ~10 to 100 μm^{-2}, allowing isolation of a small number of QDs inside of pillar structures. At the same time, the photoluminescence wavelengths ranged from 850 to 950 nm, making quantum optics experiments much easier, since silicon avalanche-photodiode photon counters, with high efficiency and low dark counts, could be used. Measurements performed to determine other properties of these QDs are described below, but here we summarize their main properties. The radiative lifetimes usually varied from ~0.5 to 1.3 ns. If all of the electron–hole recombination is radiative, this corresponds to an oscillator strength f_{osc}~2.6–6.7 (see Equation 2.3). However, it is difficult experimentally to determine the radiative efficiency except when the photoluminescence lifetime is observed to decrease due to microcavity coupling. In this case a radiative efficiency into the cavity mode can be accurately estimated, and as discussed below, this efficiency can be quite high. The homogeneous linewidth (the linewidth of a single QD) is sample-dependent, and probably is affected by factors such as the etching process and the optical excitation method. Using an interferometric measurement (described below) we have observed coherence lengths up to 275 ps, corresponding to a homeogeneous linewidth of 1.16 GHz, or 0.003 nm. To observe such a linewidth, measurements must be performed at liquid-helium temperatures. There is little temperature dependence below 10 K, but by 50 K broadening and decreased photoluminescence efficiency can typically be observed with a grating spectrometer (0.02 nm resolution). The lowest-energy exciton level (one electron–hole pair) consists of four states due to the four possible spin combinations for one electron

FIGURE 2.8
Cross-sectional scanning-tunneling-microscope image of a vacuum-cleaved QD sample. The cleave plane is (110). Tunneling is to empty band states. The wetting layer is observed to diffuse in the growth direction and throughout the QD region. Diffusion from the disk-shaped QDs is limited. (From Wu, W., J. R., Tucker, G. S., Solomon, and Harris. J. S., *Appl. Phys. Lett.*, 71 (1997), 1083–1085.)

(spin$\pm1/2$) and one heavy hole (spin$\pm3/2$). Two of these states, the "bright" excitons, have allowed radiative recombination, and are degenerate in cubic symmetry. Their dipole moments for spontaneous emission to the empty QD state are polarized orthogonally to each other, and parallel to the growth plane. In actual QDs, asymmetry typically produces an energy splitting of these two bright-exciton states into components with orthogonal, linear polarizations along special axes determined by the asymmetry [32]. In the interferometric measurements we have observed splittings ranging from ~10 to 50 μeV, or 2.4 to 12 GHz. The inhomogeneous linewidth (ensemble linewidth determined by size variations among the QDs) is four orders of magnitude larger than the homogeneous linewidth and is ~30 nm.

2.2.3 Fabrication of Pillar Structures

The 3D microcavities are produced by etching the MBE-grown samples, producing pillars with diameters on the micrometer scale or smaller. Creating such structures requires a combination of precision patterning and dry-etching techniques. We developed the post fabrication over three successive "generations", with the improved processing in each stage leading to a corresponding improvement in the device performance.

2.2.3.1 First Generation

For the first-generation process, a layer of polymethyl methacrylate (PMMA) is spin-cast onto the MBE-grown sample, and a series of circular holes, with diameters from 0.5 to 6 μm, is exposed using an electron-beam lithography system (Leica StereoScan, running J. C. Nabity Nanometer Pattern Generation System). Following development of the PMMA, a thin layer of gold is evaporated onto the sample, and the PMMA is dissolved in acetone, so that the metal above it is lifted off. The resulting circular pads of gold serve as a mask for etching of the pillars.

The etch is performed in an electron-cyclotron-resonance plasma etcher (PlasmaQuest). In this system, permanent magnets set up a magnetic field, and a plasma of the reagent gases is formed above the sample by microwaves tuned to the cyclotron frequency of electrons in the applied magnetic field. A radio-frequency (rf) bias applied between the sample and the chamber wall sets up an effective bias that accelerates ions towards the sample. The plasma is formed from a dilute mixture of Cl_2 and BCl_3 in a background of Ar. Standard etch recipes, as developed for microelectronic fabrication, were used [33]. Figure 2.14a shows a scanning-electron microscope image of a pillar microcavity fabricated using this first-generation process [34]. The etch extends through the top DBR stack of 15 AlAs/GaAs layer pairs, the spacer layer containing the QDs, and a small portion of the lower DBR, which consists of 29.5 AlAs/GaAs layer pairs. This structure leads to three-dimensional confinement of light, and, as shown below, is sufficient to demonstrate strong enhancement of single-dot spontaneous emission rates. However, the limited etch depth and the tapered shape lead to degradation of the cavity quality factor Q with decreasing post diameter, as is illustrated in Figure 2.9.

In order to understand this loss of Q, we developed an approximate model to describe the microcavity modes [15]. Following Panzarini and Andreani [35], we assumed that the field distribution can be separated into independent transverse and longitudinal components. The longitudinal field dependence is calculated using a transfer-matrix method [36]. An effective refractive index n_{eff} can then be calculated by taking an average of the material indices in the longitudinal direction, weighted by the electric field intensity. The field

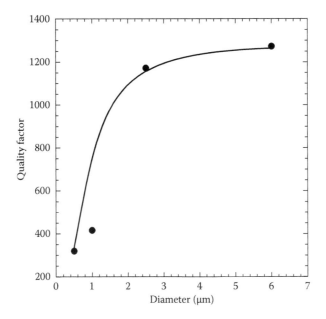

FIGURE 2.9
Quality factor of the fundamental modes of first-generation pillar microcavities. Points indicate experimentally measured values, and solid lines indicate calculated values.

dependence in the transverse direction is then determined by considering an infinitely long, cylindrical, dielectric waveguide with refractive index n_{eff}. Diffraction loss in the lower DBR is added heuristically, by allowing a wave with an initial transverse profile given by the cylindrical waveguide mode to propagate freely for twice the DBR penetration depth, as determined by the transfer-matrix calculation; this represents the spreading of the field as it penetrates into the unetched lower mirror and makes a round trip back to the post. The overlap integral between the propagated field and the waveguide mode is then calculated, giving the fraction of light that is recaptured by the post. Finally, nonideal DBR's are accounted for by introducing additional reflection losses, with the magnitude of the losses adjusted to reproduce the measured Q for the planar microcavity. The overall validity of this approximate theory was tested by verifying that predicted mode profiles, resonance frequencies, and quality factors agree well with the results of rigorous, finite-difference time-domain calculations, as described in the previous section.

Calculated Q values are shown in Figure 2.9, and agree well with experimental results. Since the only diameter-dependent loss considered is diffraction loss in the lower DBR, quality factors of small posts must be limited by this mechanism, and not by other phenomena such as sidewall scattering. The influence of diffraction is further illustrated in Figure 2.10, which shows the field distributions determined by FDTD calculations for an ideal post and for a realistic first-generation post, with a profile similar to that shown in Figure 2.14a. The effects of diffraction are apparent as a leakage of the fields outside of the lower portion of the realistic post.

2.2.3.2 Second Generation

In order to eliminate diffraction losses, the lithography and etching processes were improved. In particular, the etch depth was increased by reducing the undercut, visible in

(a)   (b)

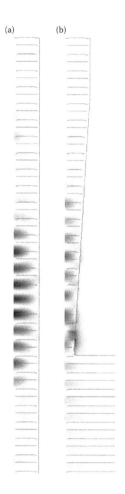

FIGURE 2.10
Electric field intensity for the fundamental mode of (a) an ideal and (b) a realistic first-generation pillar micro-cavity, each with a top diameter of 0.5 µm, calculated by the finite-difference time-domain method. The intensity is represented by a grey scale. Half a longitudinal cross-section of each post is shown. Lines represent interfaces between different materials. (From Pelton, M., Vučković, J., Solomon, G. S., Scherer, A., and Yamamoto, Y., *IEEE J. Quantum Electronics*, 38 (2002), 170–177.)

Figure 2.14a. This, in turn, was achieved by lowering the concentrations of Cl_2 and BCl_3, increasing the applied rf power, and decreasing the process pressure, thereby reducing the chemical component and increasing the physical component of the etch. As well, the etch process was divided into three stages; in each subsequent stage, the partial pressure of Cl_2 and the process pressure were decreased, while the applied RF power was increased to maintain a roughly constant DC bias. Finally, sample heating was reduced by improving heat transfer between the sample and its holder and by cooling the holder to approximately 3°C.

An effect of these changes in the etch process, though, is to make the etch less selective with respect to the mask material, so that the gold masks used in the first-generation process are completely eroded long before the etch is complete. The gold is therefore replaced with nickel, which is more resistant to plasma etching [37], and the metal thickness is increased. This, in turn, requires the use of a bilayer electron-beam resist, consisting of a top layer of high-molecular weight PMMA and a lower layer of a co-polymer of MMA and methacrylic acid (P(MMA-MAA)) [38]. Exposure and development of this resist leads to an undercut profile, allowing for the liftoff of Ni layers up to 100 nm thick and with feature sizes down to 150 nm.

Figure 2.11 shows a scanning-electron microscope image of a pillar microcavity produced with this second-generation process. The etch proceeds far into the lower DBR stack. Undercutting is still present, but is greatly reduced compared to the first-generation posts, and occurs mostly below the QD active layer. Light in the second-generation posts thus does not experience significant diffraction loss in the lower DBR, and measured quality factors for small pillar diameters are higher, as shown in Figure 2.12.

Despite the improvement, the quality factor still degrades as the post diameter decreases. This has previously been attributed to scattering by roughness at the post sidewalls [39], but FDTD calculations suggest that the loss is actually due to the tapered shape of the posts. The taper leads to coupling between the fundamental confined mode and other modes, including higher-order waveguide modes and radiation modes. The quality factors of the pillar microcavity modes could therefore still be increased by further improving the etch process.

FIGURE 2.11

Scanning-electron microscope image of a second-generation pillar microcavity with a top diameter of 0.6 μm and a height of 4.2 μm. (From Pelton, M., Santori, C., Vučković, J., Zhang, B., Solomon, G. S., Plant, J. and Yamamoto, Y., *Phys. Rev. Lett.*, 89 (2002), 233602. © American Physical Society. With permission.)

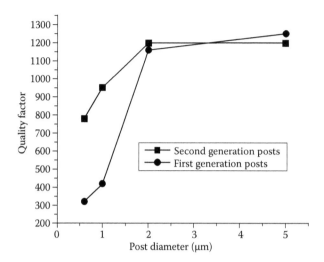

FIGURE 2.12
Measured quality factors for first-generation and second-generation pillar microcavities.

2.2.3.3 Third Generation

The third-generation pillars used in our experiments have an InAs QD layer embedded in the middle of an optical wavelength (274 nm) thick GaAs spacer sandwiched between 12 DBR mirror pairs on top, and 30 DBR mirror pairs on bottom. Each DBR pair consists of a 68.6 nm thick GaAs and a 81.4 nm thick AlAs layer (both layers are approximately a quarter-wavelength thick). The pillar structures (shown in Figure 2.13) were etched in a random distribution by chemically assisted ion beam etching (CAIBE), with Ar^+ ions and Cl_2 gas, and using sapphire (Al_2O_3) dust particles as etch masks. Sapphire was chosen as a mask because of its chemical stability and hardness, which enabled larger etch depths than other mask materials, such as the metal masks used in the previous generations. Unfortunately, these same properties impede its removal from the tops of the structures, without endangering the pillars [40]. The presence of sapphire on top of the structures decreases their quality factors and consequently reduces outcoupling efficiencies. This can be confirmed theoretically by the FDTD calculation of the Q-factor of the fundamental HE_{11} mode in a pillar with and without sapphire on top and with parameters as in the experimentally studied structures. Without sapphire, the Q-factor of the HE_{11} mode is 2600 and its wavelength is 882 nm; with a 0.5 μm thick sapphire disk on top of the post (with diameter equal to that of the post and with refractive index of 1.75), the Q-factor drops to 1400, while the mode wavelength remains unchanged. Due to the irregular shapes of the fabricated posts, the HE_{11} mode is typically polarization-nondegenerate, and many pillars have only one or two QDs on resonance with this fundamental mode.

2.3 Device Characterization

Next, we describe optical measurements on the pillar devices described above. We first show evidence for modification of spontaneous-emission lifetimes of single QDs. Then, we describe quantum-optical measurements that test the performance of optically excited

FIGURE 2.13
Scanning electron micrograph showing a fabricated array of third-generation pillar microcavities.

pillar microcavity devices as single-photon sources. These measurements include photon correlation measurements that place an upper bound on the multi-photon probability, measurements of the efficiency of collecting a single-photon, and measurements of quantum indistinguishability between two consecutively emitted photons. Finally, we describe quantum-optical measurements performed on structures exhibiting strong coupling between single QDs and cavity modes.

2.3.1 Modifying Single QD Spontaneous Emission

The radiative transition rate of an atom from an excited, initial state to a lower energy, final state depends on the availability of photon field states. This is expressed by Fermi's golden rule as

$$\Gamma_{if} = (2\pi/\hbar)\rho(\omega_c)|\langle b|H|a\rangle|^2,$$

where $\rho(\omega_c)$ is the photon field density of states at the transition frequency ω_c, $|b\rangle$ is the final state, $|a\rangle$ is the initial state, and H is the atom–vacuum field dipole interaction Hamiltonian.

Fermi's golden rule describes the weak-coupling regime, where the atomic excitation is irreversibly lost to the field. The spontaneous emission decay rate into a particular cavity mode is enhanced by increasing $\rho(\omega_c)$ at this cavity resonance. The enhancement factor is simply the ratio of the density of field states at the atomic frequency inside the cavity to that in free space [34]. For a Lorentzian emission line and a Lorentzian cavity mode resonance the enhancement is given by the Purcell factor, F, in Equation 2.6. This equation can be rewritten as

$$F = \frac{3Q\lambda_i^3}{4\pi^2 V} \frac{\Delta\lambda_0(\Delta\lambda_0 + \Delta\lambda_i)}{4(\lambda_0 - \lambda_i)^2 + (\Delta\lambda_0 + \Delta\lambda_i)^2} \cos^2\theta,$$

where Q is the cavity quality factor, and V is the cavity mode volume. The second term is the convolution of the Lorentizans of the cavity mode and emitter emission, with $\Delta\lambda_0$ and $\Delta\lambda_i$ the linewidths of the cavity and emitter, at center wavelengths λ_0 and λ_i, respectively. This term represents deturning of the emitter and cavity and accounts for varying linewidths. The third term accounts for the relative angle θ between the emitter dipole and cavity mode field. We assume no cavity mode degeneracy. For the ideal case where a nondegenerate emission state with a negligible linewidth is perfectly dipole-aligned and on resonance at the antinode of an ideal cavity, also with negligible linewidth, this reduces to $3Q\lambda^3/4\pi^2 V$. Thus, at a certain wavelength, one would like a cavity with a high Q and a small V, in order to achieve large enhancement. As the modeling described above has demonstrated, pillar microcavities are capable of achieving these conditions simultaneously, and are therefore well suited for spontaneous-emission enhancement.

2.3.1.1 First and Second Generation

Figure 2.14b shows the photoluminescence spectrum from a first-generation pillar microcavity, which has a cross-sectional area of approximately 0.04 µm³ at the location of the QD layer. Based on the spectrum, this pillar evidently contains three QDs, and the emission from a single QD is in resonance with the fundamental cavity mode. The cavity resonance wavelength and quality factor Q are determined under high pump power conditions, where the weak diffuse emission from the QD ensemble is filtered by the cavity. The independently measured modified spontaneous emission lifetime of the resonant QD is also shown, along with the theoretically predicted variation of lifetime with respect to the wavelength across the cavity mode. The spontaneous emission lifetime is measured with a streak camera. The theory is a first principles calculation, as described above.

The spontaneous emission from the QD ensemble without a microcavity structure is inhomogeneously broadened over a spectral range of approximately 30 nm, and is composed of sharp, instrumental resolution limited emission peaks associated with the discrete levels of individual QDs with slightly varying topologies [41]. The planar microcavity has its resonance centered at 932 nm, with a quality factor Q of 2300. When microposts are formed, the lateral confinement increases the fundamental cavity mode resonance to higher energies (blue shift) and increases the mode separation. In Figure 2.15, the fundamental cavity mode energy shift is plotted as a function of the effective post diameter, along with the theoretical results. The entire spectral distribution from a 5-µm pillar is shown in the insert, where arrows indicate calculated mode positions.

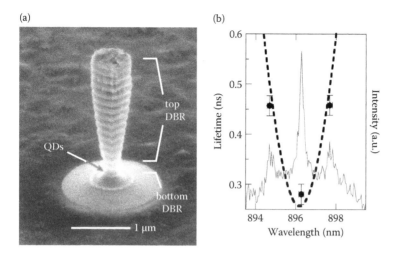

FIGURE 2.14
A single quantum dot (QD) post microcavity. (a) Scanning-electron microscope image of a first-generation pillar microcavity. The diameter at the top of the post is 0.5 μm. (b) A PL spectrum (solid) and the spectral dependence of the spontaneous emission lifetime (filled circles: experimental, dashed line: theory) at 4 K of a tapered post microcavity with a less than 0.5 μm cavity diameter. (From Solomon, G. S., Pelton, M., and Yamamoto, Y., *Phys. Rev. Lett.*, 86 (2001) 3903–3906. © American Physical Society. With permission.)

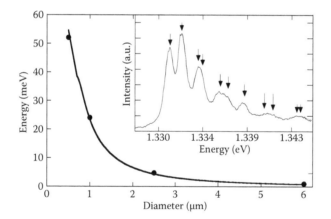

FIGURE 2.15
Measured and calculated energy shifts of mode centers as a function of post diameter for microcavity posts with QD active regions. Measured values determined from 4 K PL, calculations assuming the fields are separable into transverse and lateral components. The insert shows the 4 K luminescence from a 5 μm diameter post, where the confined modes are indicated by arrows. (From Solomon, G. S., Pelton, M., and Yamamoto, Y., *Phys. Rev. Lett.*, 86 (2001) 3903–3906. © American Physical Society. With permission.)

Without a cavity, the QDs have a spontaneous emission lifetime of 1.3 ns, and with the planar microcavity this is reduced to 1.1 ns, as shown in Figure 2.16a. In the planar microcavity case, there is a continuous distribution of cavity resonant modes from the cut-off wavelength, $\lambda_c = s \simeq 932$ nm to a wavelength corresponding to the stopband edge. Since the number of modes increases with decreasing wavelength from the cutoff wavelength, λ_c, the spontaneous emission lifetime is continually decreased from the cutoff wavelength to shorter wavelengths. For the 3D post microcavities, the modes are discrete, so that the minimum spontaneous emission lifetime occurs at the center of each mode. This

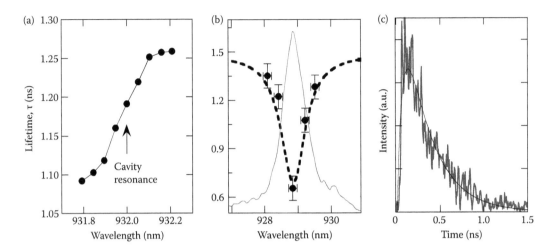

FIGURE 2.16
Spontaneous emission lifetime of QDs in microcavity posts. The average QD lifetime at spectral positions in the planar cavity is shown in (a), and for a post diameter of 2.5 μm in (b). The intensity is also shown in (b). In (c) the time-dependent spontaneous emission is shown for InAs QDs in bulk GaAs and a single InAs QD resonant with the fundamental mode of a 0.5-μm diameter microcavity post. (From Solomon, G. S., Pelton, M., and Yamamoto, Y., *Phys. Rev. Lett.*, 86 (2001) 3903–3906. © American Physical Society. With permission.)

unmistakable mark of the 3D confinement is seen in Figure 2.16b for the microcavity with a post diameter of 2.5 μm and also in Figure 2.14b for a post diameter of 0.5 μm. For the 2.5 μm diameter post microcavity, a continuous distribution of QDs emit photons into the single cavity mode. The reduction of spontaneous emission lifetime is largest when the QDs are on resonance, but the modification disappears off-resonance, near the mode edge. In Figure 2.16c the time-dependent spontaneous emission is shown for a single QD in resonance with a 0.5 μm diameter microcavity post (τ=0.28 ns), measured by a streak camera after the pulsed excitation from a 200 fs Ti:Al$_2$O$_3$ laser. The rise time (≈50 ps) is determined by the carrier capturing time by the QD and the decay time (≈0.30 ns) is the radiative lifetime. The spontaneous emission lifetime at the center of each fundamental mode is plotted against the post diameter in Figure 2.17, together with the theoretical prediction. The reduced spontaneous emission lifetime in a small post microcavity is primarily due to the reduced mode volume. This offsets the decrease in Q, which is approximately 300 in the 0.5 μm diameter post.

In the above analysis, we neglected the spatial dependence of the electric field intensity, $|E(r)|^2$, inside the micropost cavity. Because the QDs are not identical, it is likely there is variation in $\vec{\wp}\, E(r)$. One might suppose that the spectral QD emission distribution is coincidentally the same as the spatial QD distribution; that is, QDs detuned spectrally from the cavity are also not in the center of the post, and thus the QDs happen to be spatially detuned from the electric field maximum the same way they spectrally detuned. This would still be a cavity QED effect, but would be related to $\vec{\wp}\, E(r)$ and not Q/V. Because the QD ensemble is random, such ordering is unlikely. As well, the good agreement between theory assuming a constant $|\vec{\wp}\cdot\vec{E}|^2$ and experiment (Figure 2.14) justifies neglecting variations in $|\vec{\wp}\cdot\vec{E}|^2$. The theory contains no free parameters; the cavity Q used in the theory is obtained independently from decay rate measurements, using the filtered spontaneous emission under high pump conditions. Coincidental alignment of $|\vec{\wp}\, E(r)|^2$ and the spectral detuning for each QD will not satisfy our criterion.

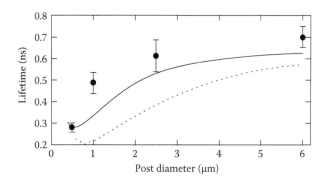

FIGURE 2.17

The spontaneous emission lifetime (•) at the center of the fundamental mode is plotted as a function of the post diameter. The lifetime for the 0.5 μm diameter post is for a single QD. Lines are the result of an approximate model for QDs at the center of the disk (dashed) or averaged radially over the disk (solid).

Nonetheless, in order to completely rule out the possibility of coincidental agreement between experiment and calculation, we observed the changes in lifetime of a particular QD as it is tuned into and out of resonance with a single cavity mode by changing the sample temperature [42]. This eliminates any other mechanisms, ensuring that the observed lifetime changes are due to coupling to the microcavity mode. As the temperature increases, the InAs bandgap decreases, leading to a redshift of the QD emission. The GaAs and AlAs refractive indices also change with temperature, but the resulting shift in the resonance energy of the cavity is smaller than the shift of QD emission energy. It is thus possible to bring a particular QD into and out of resonance with the cavity mode.

In Figure 2.18, the 4 K optical spectrum of a 0.6 μm diameter micropost is shown. The ground state (single exciton) emission from two QDs is shown, near 904.5 nm (called QD1) and near 902 nm (called QD2). The emission is known to be from the ground state, because there is a linear relationship between the pump power and the intensity. The fundamental cavity mode is also near 904.5 nm, so QD1 is initially close to resonance with the fundamental cavity mode. The fundamental cavity mode is determined under high pump power conditions, where the diffuse background from the QD layer is filtered by the cavity. It has a small temperature dependence, which is accounted for in the following temperature-dependent measurements.

A streak camera is used in conjunction with a spectrometer to capture both time and wavelength information on the QD emission. Streak camera data is taken from 4 to 70 K and the spontaneous emission decay rate is extracted by fitting a single exponential to the decay. In Figure 2.19, the temporal and spectral emission is shown. In the upper portion of Figure 2.19, emission is shown at a sample temperature of 4 K. The QD1 emission is brighter since it is nearly on resonance at 4 K, while the QD2 emission is dim since it is off resonance. In the lower portion of Figure 2.19, the image is shown for 50 K sample temperature.

The decay rates for the two QDs are plotted in Figure 2.20, first as a function of temperature, and then as a function of detuning from the cavity mode. As the temperature is increased, QD1 eventually becomes detuned from the cavity mode and the spontaneous emission decay rate decreases, while the opposite occurs for QD2. The different behavior of the two dots, and particularly the decrease in spontaneous emission rate with increasing temperature, rule out the possibility that a thermally activated process is responsible for

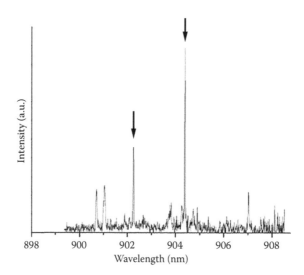

FIGURE 2.18
The 4 K photoluminescence spectrum of a micropost cavity, where the ground state excitonic emission from two QDs is indicated by arrows.

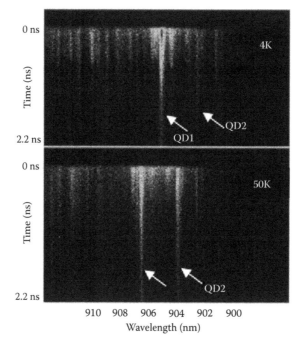

FIGURE 2.19
Streak camera image of temporal and spectral distribution from a 0.6 μm diameter micropost cavity. The ground state excitonic emission from a few QDs is observed. The QDs of interest are labeled. Upper: 4 K sample temperature; lower: 50 K sample temperature.

the observed lifetime changes. The detuning curve is found by measuring the detuning of each QD exciton state from the cavity mode. The enhancement for the two QDs are determined by Lorentzian fits, and are similar: QD1=4.1, and QD2=3.9. Several factors could account for the small difference. Since the spatial position of each QD is unknown, the field

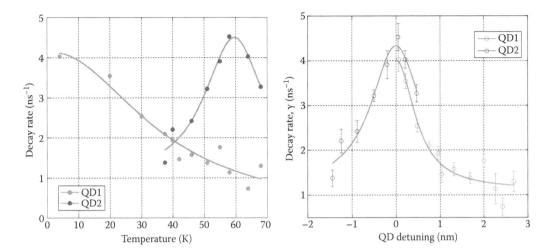

FIGURE 2.20
Spontaneous emission decay rate measured for two QDs as a function of sample temperature on the left and spectral detuning on the right. The detuning is the difference in wavelength from each QD state and the cavity mode.

amplitude at each QD is unknown, but could vary. The dipole strength and orientation of each QD could also vary. The fitted linewidths of the two Lorentzians, on the other hand, are the same within the errors, since they couple to the same cavity. The corresponding quality factor is 775 ± 160, in agreement with the results shown in Figure 2.12.

2.3.1.2 Third Generation

For the third-generation pillar microcavities, we performed experiments on a QD chosen for its bright emission under resonant excitation [43]. Tuning of the sample temperature was used to tune the emission wavelength relative to the cavity resonance [44]. The particular dot presented in Figure 2.21 is almost exactly on resonance with the cavity at low temperature, so by heating the sample and increasing the QD emission wavelength, one also increases the detuning from the cavity resonance and the radiative lifetime. Antibunching in the photon correlation measurements indicated that a single QD was probed in these experiments, as described below.

The data in Figure 2.21 show a good correspondence between the cavity resonance linewidth observed in the high-power photoluminescence spectrum, and the linewidth of the Lorentzian describing the variation of the spontaneous emission decay rate as a function of detuning from the cavity resonance. Figure 2.21 (bottom) shows the time-resolved photoluminescence of an emission line on and off resonance with the cavity mode. The decay lifetime differs by a factor of five for these two cases. The decay rate for this emission line as a function of the absolute value of its detuning from the cavity resonance ($|\lambda_i - \lambda_0|$) is plotted with circles in the top-right plot of Figure 2.21. The solid line corresponds to the Lorentzian fit to the experimental data [45], and a good match is observed between our experiment and the theoretically predicted behavior. The fitting parameters are the linewidth, maximum and minimum of the Lorentzian, while it is assumed that its central wavelength is equal to λ_0, the cavity resonance wavelength, λ_0 and the quality factor ($Q=1270$) are determined from the photoluminescence intensity taken at high pump powers (see the top-left plot of Figure 2.21). The cavity resonance wavelength red-shifts by roughly 0.3 nm with increasing temperature in the studied range, and this shift is included in plotting the data. Up to

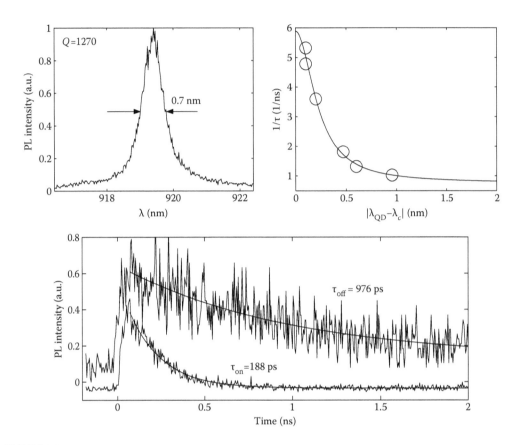

FIGURE 2.21

Top-left: Background emission filtered by a pillar cavity. Top-right: Decay rate of the emission line as a function of the absolute value of its detuning from the cavity resonance ($|\lambda_{QD} - \lambda_c|$). The dot emission wavelength was tuned by changing the sample temperature within the 6–40 K range. Bottom: time-dependent photoluminescence from the emission line on-resonance with the cavity, as opposed to this same emission line off-resonance. (Reprinted with permission from Vučković, J., Fattal, D., Santori, C., Solomon G., and Yamamoto, Y., *Appl. Phys. Lett.*, 82 (2003) 3597. © American Institute of Physics. With permission.)

fivefold spontaneous emission rate enhancement (Purcell factor F) is observed for the dot coupled to a cavity, as opposed to the dot off of resonance (top-right plot of Figure 2.21).

2.3.2 Photon Statistics

An ideal single-photon source emits exactly one photon each time it is triggered. For experimental reasons it is convenient to focus on two parameters of the photon number distribution: (1) the multi-photon probability, normalized to an equivalent Poisson distribution, and (2) the efficiency (probability to collect one or more photons). Here we consider the reduction of the multi-photon probability, which is the one feature of a single-photon source that cannot be obtained from attenuated laser pulses. For a Poisson distribution, the probability of a pulse to contain n photons is $P(n) = \mu^n e^{-\mu}/n!$, where $\mu = \langle n \rangle$ is the mean photon number, and the multi-photon probability is $P_m = 1 - (1+\mu)e^{-\mu}$. A good single-photon device should therefore have,

$$p_m \ll 1 - (1 + \langle n \rangle) e^{-\langle n \rangle}. \tag{2.10}$$

As noted in the Introduction, a convenient parameter describing this is $g_0^{(2)}$, which can be expressed as

$$g_0^{(2)} = \frac{\langle n(n-1)\rangle}{\langle n\rangle^2}. \tag{2.11}$$

It is straightforward to show from Equation 2.11 that for any possible photon number distribution, $p_m \leq (1/2) g_0^{(2)} \langle n\rangle^2$, as noted in the Introduction [2].

In this section, we first discuss the mechanism for multi-photon suppression in optically excited QDs. We then show experimental results for QDs in pillar microcavities, including a multi-photon suppression factor up to 50 compared with an equivalent Poisson source. We also describe some unexpected blinking and long-timescale antibunching effects we have observed, which are related to the optical excitation method.

2.3.2.1 Mechanism for Single-Photon Generation in QDs

Self-assembled InGaAs QDs can accommodate at least two excitons (electron–hole pairs), if the confinement potential is deep compared with the Coulomb charging energy (~10−20 meV), and more than two excitons if both the S and P shells are bound. With above-band optical excitation or pulsed electrical excitation, multiple electron–hole pairs can enter the QD, and these will recombine to emit multiple photons. In this case, the total number of photons emitted by the QD is likely to be nearly Poisson-distributed. However, the electrostatic interactions between excitons are typically strong enough that each photon produced in this radiative cascade will have a unique frequency.

This situation is observed in the time-resolved spectra shown in Figure 2.22 [46]. All of the spectral lines are from a single QD, with each line corresponding to a unique multiparticle state. Through a combination of power-dependent measurements and photon cross-correlation experiments we identified the lines labeled 1 and 2 as corresponding to exciton and biexciton emission, respectively. A third peak, 3, likely corresponds to a state with three electron–hole pairs. Many of the other peaks correspond to charged states such as two electrons and one hole (trion). In Figure 2.22, for small excitation powers, so that less than one exciton is injected on average ($\langle n_{ex}\rangle \approx 0.66$), single-exciton emission begins within tens of picoseconds of the excitation pulse and decays exponentially. For larger excitation power ($\langle n_{ex}\rangle \approx 1.34$), the biexciton emission appears first, and the single-exciton emission is delayed until the biexciton emission has decayed. For even larger excitation power ($\langle n_{ex}\rangle \approx 2.82$), even the biexciton emission is delayed. The quantitative time dependence of these lines can be described reasonably well by a simple rate-equation model in which the initial number of injected excitons follows a Poisson distribution, and the exciton and biexciton spontaneous emission lifetimes are 479 and 316 ps, respectively [46]. From the viewpoint of single-photon generation, the important result is that, despite the large number of spectral components involved in the radiative cascade, the single-exciton emission (line 1) is well separated from the other components. By using a spectral filter to select only this line, one can obtain a single-photon source [8,9,47,48]. This principle has also been used to demonstrate electrically driven single-photon devices [49], including devices incorporating DBR cavities [50]. For single-photon generation based on incoherent excitation, any spectral line can be used [51]. However, for applications requiring a particular energy-level structure, such as optically controlled quantum computation using electron spins in QDs [52], the charge state must be controlled.

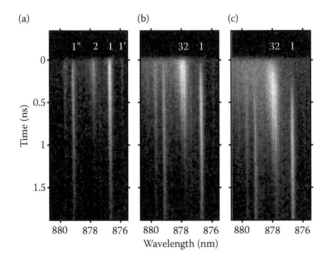

FIGURE 2.22
Streak camera images of emission from a single QD in a 0.2 μm mesa under pulsed, 708 μm laser excitation with powers (a) 27 μW, (b) 108 μW, and (c) 432 μW. For larger powers, multi-excitonic emission (lines 2, 3) occurs first, followed by single-exciton emission (line 1). (From Santori, C., Solomon, G. S., Pelton, M., and Yamamoto, Y., *Phys. Rev. B*, 65 (2002), 073310. © American Physical Society.)

The degree of two-photon suppression that can be obtained by this method also depends on the excitation method. The main issue is that, if carriers are injected too slowly, it is possible for a second electron–hole pair to be injected after a first exciton has recombined, and in this case two photons will be emitted at the single-exciton frequency. If we consider only a two-level system, a simple rate-equation model for incoherent excitation predicts that the degree of two-photon supression (the two-photon probability of this device divided by that for a Poisson source) is approximately Γt_{laser}, where Γ is the single-exciton spontaneous emission rate, and t_{laser} is the excitation pulse duration. If an excitation laser pulse is used to excite above the GaAs bandgap, then, even if the pulse is very short, the effective t_{laser} can be on the order of 50 ps, due to the carrier capture process. In this case, the degree of two-photon suppression is limited to a factor of approximately 10. This becomes worse when a microcavity is used to shorten the spontaneous emission lifetime. We have found that by tuning the excitation laser on resonance with a higher-energy transition within the QD, much better performance is possible. In this case, electron–hole pairs are created directly inside the QD. This excitation scheme is shown in Figure 2.23a. Initially, the laser excites the QD to an excited level X^*. Experimentally, such a resonance is found through a photoluminescence excitation (PLE) measurement, in which the intensity of the single-exciton ground-state emission is measured as a function of laser frequency [53]. In our samples, such resonances could typically be found at a wavelength approximately 20 nm shorter than that of the ground-state single-exciton transition. The transition from X^* to the lowest-energy exciton state X is apparently very fast, on the order to 10 ps or less, followed by emission of a single-photon from the X transition. It is possible sometimes for the laser to create a second electron–hole pair, raising the system to level XX^*; in this case, spectral filtering is required to reject the biexciton photon. For these experiments, we used a mod-elocked Ti–Sapphire laser with ~2 ps pulse duration and corresponding linewidth below 0.5 nm. If a biexcitonic energy shift of a few meV is also present in the excited transition (X^* to XX^*) [53], we have an additional mechanism for two-photon suppression, since the

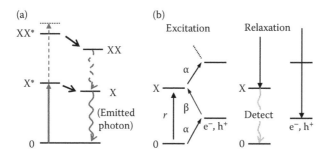

FIGURE 2.23
(a) Level diagram for resonant excitation scheme. (b) Diagram including single carrier injection pathways.

detuning of the laser from the XX* transition is substantially larger than the laser linewidth [9,54].

2.3.2.2 Experimental Results with Pillar DBR Devices

For a full characterization of photon statistics, we would ideally estimate each probability P_n of the photon number distribution, as would be possible if one could obtain a highly efficient photon-number-resolving detector. However, the best photon counters that are currently commercially available (silicon avalanche photodiodes) cannot differentiate between one photon or multiple photons, and also suffer from a dead time of ~50 ns following detection of a photon. In addition, losses will be present in the optical setup, and if the device itself has a low efficiency, detection events of three or more photons in the same pulse are extremely rare. The method most commonly used to characterize the multi-photon probability is therefore the "Hanbury Brown-Twiss" setup shown in Figure 2.24. Here, light collected from the device is first passed through a narrow spectral filter and optional polarizer before being split into two arms, each leading to a separate photon counter. Coincidence electronics (for example, a time-to-amplitude converter followed by a multi-channel analyzer) are used to obtain a histogram of the relative delay $\tau=t_2-t_1$ between single-photon detection events at detectors 1 and 2. This photon correlation measurement provides an estimate of the second-order coherence function $g^{(2)}(\tau)$ [1]. This measurement is insensitive to linear optical losses, aside from an overall normalization factor.

Figure 2.25 shows photon correlation data from one of the best devices from the third-generation sample described above (see also Figure 2.21). Peaks appear at integer multiples of the 13 ns laser repetition period. The m-th peak after $\tau=0$ corresponds to events in which one photon is detected from any pulse j on detector 1 and a second photon is detected from pulse $j+m$ on detector 2. The central peak at $\tau=0$ corresponds to events in which two photons are detected from the same pulse. One method of obtaining $g_0^{(2)}$ from this measurement is to divide the central peak area by the expected number of coincidences from a Poisson-distributed source, $r_1 r_2 t_{rep} t_{meas}$, where r_1 and r_2 are the average count rates at each detector, t_{rep} is the laser repetition period, and t_{meas} is the total measurement time. For a Poisson-distributed source, all peaks would have the same area, including the central peak at $\tau=0$. Thus, if the number of photons produced in each pulse is statistically independent, another way to estimate $g_0^{(2)}$ is to divide the central peak area by the area of any of the side peaks. However, in practice we find longer-timescale memory effects, and only the side peaks at large τ can be used for normalization. The extremely small central peak observed here indicates a two-photon probability only 2% of what is expected for a classical Poisson-

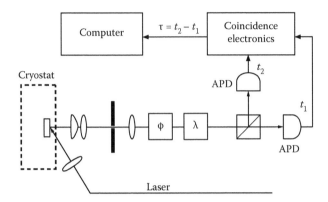

FIGURE 2.24
Schematic diagram of photon correlation setup. φ: adjustable half-wave plate followed by a polarizer. λ: spectral filter. (From Santori, C., Fattal, D., Vučković, J., Solomon, G. S., Waks, E., and Yamamoto, Y., *Phys. Rev. B*, 69 (2004), 205324. © American Physical Society.)

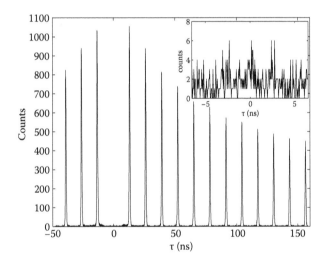

FIGURE 2.25
Histogram of the relative delay $\tau = t_2 - t_1$ for photons detected by counters 2 and 1, obtained from a QD coupled to a DBR pillar microcavity. The inset shows the small peak at $\tau = 0$, barely detectable above the background, indicating a large two-photon suppression factor. (Reprinted with permission from Vučković, J., Fattal, D., Santori, C., Solomon, G., and Yamamoto, Y., *Appl. Phys. Lett.*, 82 (2003) 3598. © American Institute of Physics. With permission.)

distributed source. This high degree of two-photon suppression is due partly to the narrow (0.1 nm) spectral filter used in this measurement. For other devices measured on this sample, the two-photon probability was typically suppressed by a factor of 10–20.

Another feature in the data which must be explained is the increase of the peak areas around $\tau = 0$. Such behavior suggests a "blinking" of the QD photon source: given that we detect a photon in pulse j, the probability of detecting another photon in the next pulse $j + 1$ is substantially higher than the probability of detecting a photon at a much later time. We observe this behavior for QDs under resonant excitation [9,4,55]. A variety of memory effects have been reported in the optical emission of single semiconductor QDs. Some occur on millisecond or longer timescales, including blinking [56–58], two-color blinking

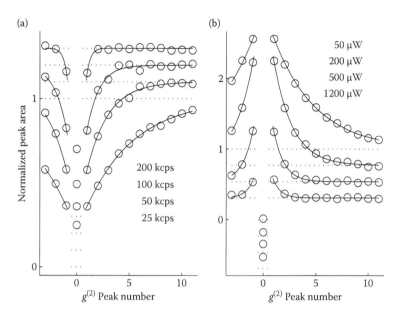

FIGURE 2.26
Normalized peak areas from photon correlation histograms plotted versus peak number (circles), and two-sided exponential fits (lines) using Equation 2.12. (a) Measurement of a single-photon device using above-band excitation, measured at four different excitation powers, resulting in the indicated count rates on one detector (shifted vertically for clarity). (b) Results for the device from Figure 2.25 under resonant excitation at four different excitation powers as indicated.

[59], and spectral diffusion [60,61]. However, the blinking observed here occurs on much faster timescales, from less than 10 ns to at least 800 ns. We have also observed a second type of memory effect using weak above-band excitation, in which the peaks in the photon correlation measurement increase slowly away from $\tau=0$ [55,62,63].

Several photon correlation measurements from single-photon devices under different excitation conditions are shown in Figure 2.26. In this case, only the normalized peak areas A_m are plotted versus peak number, with peak 0 corresponding to $\tau=0$. The normalization corresponds approximately to dividing the raw peak area by the expected peak area for a Poisson-distributed source, based on the measured singles count rates, with small corrections for detector dark counts. In all cases, the peak areas are well fitted by a simple two-sided exponential function:

$$A_{m\neq0}=1+g_1 \exp\left[-(|m|-1)t_{\mathrm{rep}}/\tau_b\right],\qquad(2.12)$$

where g_1 and τ_b are fitting parameters that characterize the amplitude and timescale of the memory effect, respectively. The experimentally observed sign of g_1 depends on whether we use resonant or above-band excitation.

These memory effects are most likely caused by charge fluctuation of the QD. Charge fluctuation can explain both the blinking (resonant excitation) and the long-timescale antibunching (above-band excitation) in terms of the level diagram in Figure 2.23b. These behaviors have simple qualitative explanations. When a first photon is detected from pulse *j*, the QD is empty immediately afterward. For single-carrier injection, which is possible under above-band excitation, two injections must occur before another photon can be emitted, and thus it is unlikely that another photon will be emitted from pulse

$j + 1$ if the injection rate is small. The opposite argument applies for the injection of entire electron–hole pairs (resonant excitation). If a photon is detected from pulse j, it is especially likely that another photon will be emitted from pulse $j + 1$, since only a single additional pair needs to be injected. At much later times, the QD will sometimes be charged, in which case it will appear dark at the particular wavelength being detected. It is possible to describe this behavior with a simple probabilistic model, which provides adequate fits to our data [55].

It may seem surprising that the charge of a QD should ever change under resonant excitation, for which the excitation energy is tuned below the GaAs bandgap and below the InAs wetting layer band edge. Perhaps it could be related to the "wettinglayer tail" observed in PLE spectra and attributed to continuum states associated with the combined wetting layer-QD system [64], or to Auger processes, which would allow an electron to escape from the QD. For a single-photon device, this charge fluctuation means a factor of approximately 2 reduction in total efficiency, which is a problem for some applications.

2.3.3 Efficiency

In the absence of a microcavity, few of the photons emitted by a single QD escape from the high-refractive-index semiconductor substrate containing the dot, and the fraction of photons that can be collected is very low. If the QD is in a pillar microcavity, on the other hand, the majority of the photons can be emitted into a single cavity mode, and the majority of those photons can subsequently escape from the cavity in a free-space mode with a nearly Gaussian spatial profile. A single QD in a pillar microcavity can thus serve as an efficient source of single-photons.

We demonstrated efficient single-photon generation using the second-generation microcavity structures [65]. We selected a particular post with a top diameter of 0.6 μm that exhibits a clear single-dot emission line, indicated by the arrow in Figure 2.27. The measured photon–photon correlation function for an average pump power of 10.9 μW is shown in Figure 2.28. The side peaks are separated by the laser repetition rate and all have nearly equal areas, while the central peak is nearly absent, reflecting strong suppression of the multi-photon probability.

Due to a low emission rate from QDs in this particular sample, adjacent peaks overlap, and it is necessary to fit the data in order to determine the area of the central peak. In order to reduce the number of free parameters in the fit, we measured the recombination rate of the QD as $1/\Gamma = 2.2 \pm 0.6$ ns, and the response time of the autocorrelation setup, using pump-laser light that is scattered off of the pillar, as $\tau_{\text{resp}} = 473 \pm 29$ ps. Each peak in the photon correlation data from the QD is thus described by a two-sided exponential with a decay constant $\tau_{\text{total}} \approx 1/\Gamma + \tau_{\text{resp}} = 2.7 \pm 0.7$ ns. Figure 2.28 shows the result of the fit, which involves only two free parameters: the area of the central peak and the area of all the other peaks. The ratio of these two areas provides an estimate of $g_0^{(2)}$ and gives an upper bound on the probability that two or more photons are present in the same pulse, as explained above.

Results for various pump powers are summarized in Figure 2.29. Normalized central peak areas as low as 0.10 were obtained, reflecting a tenfold reduction in the multi-photon probability as compared to a source with Poissonian statistics. The multi-photon probability increases with pump power, suggesting that other states, apart from the desired QD emission, are being excited by the pump laser and are contributing a background of unregulated photons. As explained above, resonant excitation and spectral filtering can greatly decrease this background.

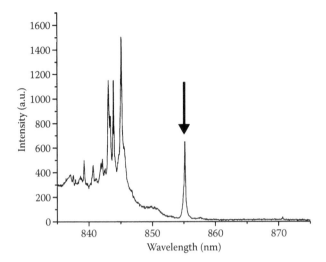

FIGURE 2.27
Photoluminescence from a single QD in a pillar microcavity, for excitation with photon energy above the GaAs bandgap.

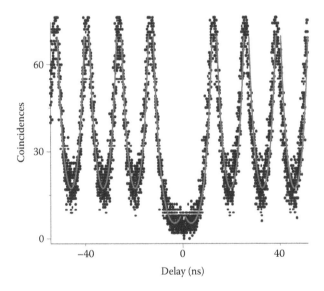

FIGURE 2.28
Measured autocorrelation for photons from a single QD in a pillar microcavity with an incident pump power of 10.9 μW (points), and corresponding fit (line). The measurement integration time is 600 sec. (From Pelton, M., Santori, C., Vučković, J., Zhang, B., Solomon, G.S., Plant, J., and Yamamoto, Y., *Phys. Rev. Lett.*, 89 (2002), 233602. © American Physical Society.)

In order to measure the efficiency of our single-photon source, we calibrate the detection efficiency of the measurement apparatus, by again scattering pump-laser light off the pillar. The scattered power is measured immediately after the collection lens, and is compared to the total photon count rate at the detectors, yielding an overall detection efficiency of 3.02±0.16%. The collection lens in front of the sample, in turn, captures only part of the light that escapes from the post; this capture fraction can be estimated by calculating the effective divergence angle of light emerging from the post. Both the approximate method

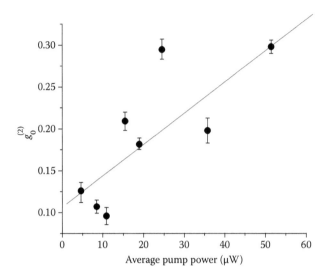

FIGURE 2.29
Normalized area of the central autocorrelation peak for photons from a QD in a pillar microcavity as a function of pump power (points), as well as a linear fit (line), as a guide for the eye. (From Pelton, M., Santori, C., Vučković, J., Zhang, B., Solomon, G.S., Plant, J., and Yamamoto, Y., *Phys. Rev. Lett.*, 89 (2002), 233602. © American Physical Society.)

outlined above and rigorous FDTD simulations indicate that the lens used, which has a numerical aperture of 0.5, collects approximately 22% of the light in the emitted beam. Using these values for the efficiencies of the collection lens and the measurement apparatus, then, it is possible to determine the mean photon number per pulse, $\langle n \rangle$, by comparing the measured photon count rate and the measured laser repetition rate.

The efficiency, η_{total}, of the source is then determined by assuming that the emitted light consists of a statistical mixture of perfectly antibunched single-photons and a background of photons with Poissonian statistics. The coupling of this state into the traveling-wave mode leaving the top of the pillar microcavity is modeled as an attenuation by a linear loss, giving

$$\eta_{\text{total}} = \langle n \rangle \sqrt{1 - g_0^{(2)}}. \tag{2.13}$$

Results are shown in Figure 2.30. The solid line is a fit according to the saturation equation

$$\eta_{\text{total}} = \eta_{\max}(1 - e^{-P/P_{\text{sat}}}), \tag{2.14}$$

where P is the pump power, P_{sat} is the saturation pump power, and $\eta_{\max}$ is the saturated efficiency. The maximum efficiency according to this fit is $37.6 \pm 1.1\%$. This number represents the external quantum efficiency of the device, and is approximately two orders of magnitude higher than the external efficiency for a QD in bulk GaAs.

The external efficiency is limited by two factors (see Equation 2.7). The first is the coupling coefficient, $\eta_{\text{qd,cav}}$, or the fraction of photons emitted by the QD that are captured by a single cavity mode. It depends on the cavity-enhanced emission rate Γ and the emission rate Γ_0 in

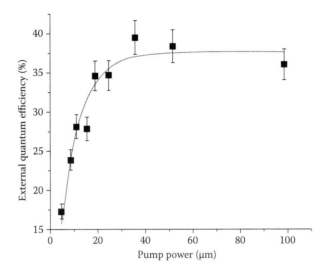

FIGURE 2.30
Measured efficiency of the single-photon source as a function of pump power (points), together with a saturation fit (line). (From Pelton, M., Santori, C., Vučković, J., Zhang, B., Solomon, G. S., Plant, P., and Yamamoto, Y., *Phys. Rev. Lett.*, 89 (2002), 233602. © American Physical Society.)

the absence of a cavity: $\eta_{qd,cav}=1-(\Gamma_0-\Gamma_c)/\Gamma$, where Γ_c/Γ_0 is the fraction of light that would be coupled into the cavity mode in the limit $Q \to 0$. In order to determine Γ_0, we measured the recombination lifetimes for excitons in other QDs on the same sample that are out of resonance with the cavity mode. The measured off-resonant lifetime is 25.4 ± 1.4 ns, corresponding to a coupling coefficient $\eta_{qd,cav}=83 \pm 23\%$.

The second factor in determining device efficiency is the light extraction efficiency, $\eta_{cav,out}$, or the fraction of light in the cavity that escapes into a single traveling-wave mode. This, in turn, can be determined by comparing the measured quality factor, $Q=628 \pm 69$, for the pillar microcavity and the measured quality factor, $Q_0=1718 \pm 13$, for the planar microcavity. Q_0 is almost entirely determined by the leakage rate through the top DBR; assuming that this rate is unchanged when the cavity is etched into posts, any difference between Q and Q_0 will be due to etch-induced loss. The extraction efficiency is then simply given by $\eta_{cav,out}=Q/Q_0=36.6 \pm 4.0\%$. Combining the estimated $\eta_{qd,cav}$ and $\eta_{cav,out}$ gives an expected external quantum efficiency of $\eta_{total}=30 \pm 9\%$. This agrees, within the error, with the measured efficiency at saturation.

2.3.4 Photon Indistinguishability

The two-photon suppression and efficient photon collection are sufficient for certain applications, such as the quantum-cryptography system described below. However, most potential applications also require at least that consecutive photons have identical wave packets. This is true for any schemes that use two-photon interference, including the quantum teleportation [66] and post-selective production of polarization-entangled photons [67] described below, as well as linear-optics quantum computation [68]. It is therefore important to demonstrate that consecutive photons emitted by a single-photon source are identical and exhibit mutual two-photon interference effects. For some applications, such as quantum networks [69], we have an even more difficult requirement that photons emitted from two different devices are indistinguishable. For a source based on a single quantum emitter, the emitter must therefore be excited in a deterministic way and interact little with its surrounding environment while the photon is emitted.

Here, we describe an experiment testing the indistinguishability of successive photons emitted by the same semiconductor QD in a DBR microcavity [4,70]. The experiment is based on "two-photon interference": when identical single-photons enter a 50–50 beam-splitter from opposite sides, quantum mechanics predicts that both must exit in the same direction if their wave packets overlap perfectly, an effect which can be attributed to the Bose–Einstein statistics of photons. This effect was first demonstrated experimentally in the Hong–Ou–Mandel experiment [71], using pairs of highly correlated photons produced by parametric downcoversion. Theoretically, this behavior is also expected for single, independently generated photons [72], but, until recently, sources of single-photons have not had the efficiency or spectral characteristics required to realize such an experiment. Besides the demonstration described here, two-photon interference has now also been demonstrated using photons from single atoms [73] and from QDs in photonic crystal cavities [74]. Recently two-photon interference has been observed using photons from two separate trapped ions [75], a major advance for quantum networks.

For this experiment, we used the third generation sample described above, cooled to 3–7 K. Many pillars with only one or two QDs on resonance with a fundamental cavity mode were found. To generate single-photons, we focused 3 ps Ti–Sapphire laser pulses every 13 ns onto these pillars from a steep angle. The laser was tuned to an excited-state absorption resonance of the QD. The QD emission was collected, and a single polarization was selected. The emission was then spectrally filtered with a resolution of approximately 0.1 nm using a diffraction grating, and coupled into a single-mode fiber. By this method, we obtained bright single-photon sources with excellent two-photon suppression and negligible background emission. We chose three QDs for this study, denoted as dots 1, 2, and 3, with emission wavelengths of 931, 932, and 937 nm, respectively. As explained above, the parameter $g_0^{(2)}$ quantifies two-photon suppression as the probability to generate two photons in the same pulse, normalized by an equally bright Poisson-distributed source. We estimate $g_0^{(2)}=0.053$, 0.067, and 0.071 for dots 1, 2, and 3, respectively. However, for the experiment described below, the important parameter is the probability to generate two photons in the same pulse, for either of two consecutive pulses, divided by the probability to generate one photon in each pulse. We estimate this quantity to be $g_0'=0.039$, 0.027, and 0.025 for QDs 1, 2, and 3, respectively. The difference between $g_0^{(2)}$ and g_0' is due to the blinking in our source, as discussed above.

The spectral properties of QDs can be highly variable, depending on the crystal growth and fabrication details. In many cases, quite broad lines are observed, which can be attributed to spectral diffusion of the exciton transitions [76]. However, some reports have indicated coherence times [77] or time-averaged linewidths [76,78,79] close to the radiative limit, showing that decoherence can be slow. Measurements of the temporal and spectral properties of devices 1, 2, and 3, averaged over many excitation pulses, are shown in Figure 2.31. In Figure 2.31a, the average emission intensity is plotted versus time following a resonant excitation pulse, measured using a streak camera. By fitting decaying exponential functions, we estimate the spontaneous emission lifetimes τ_s of devices 1, 2, and 3 to be 89, 166, and 351 ps, respectively. This variation is due largely to differences in how well each QD couples to its microcavity. In Figure 2.31b, a Michelson interferometer was used to measure the coherence length of the time-averaged emission. The curves show how the interference fringe contrast varies with path-length difference, and give the magnitude of the Fourier transform of the intensity spectrum. When we did not select a single polarization, we sometimes observed oscillatory behavior due to polarization splitting of the emission lines [32]. For devices 2 and 3 with splittings of 13 and 17 μeV, respectively, we were able to eliminate this effect by selecting a particular linear polarization. For device 1, the 45 μeV splitting could not easily be eliminated. One possible explanation is that the QD emission couples to just one cavity mode having a polarization rotated ~45° relative to the splitting axes of the QD. We estimate the 1/e coherence lengths τ_c (divided by c) for devices 1, 2, and 3 to be 48, 223, and 105 ps, respectively. Device 2 is closest to being Fourier transform-limited, with $2\tau_s/\tau_c=1.5$. When this ratio is equal to 1, perfect two-photon interference is expected.

The main elements of the two-photon interference experiment are shown in Figure 2.32. Now, the device is excited twice every 13 ns by a pair of equally intense pulses with 2 ns separation. Two pulses, each containing zero or one photon, emerge from the single-mode fiber. They are split into two arms by a beamsplitter, with one arm (2 ns+Δt) longer than the other. The beams recombine at a different place on the same beamsplitter. The two outputs are collected by photon counters, and a photon correlation histogram is generated of the relative delay time $\tau = t_2-t_1$ for two-photon coincidence events, where t_1 and t_2 are the measured detection times at photon counters 1 and 2, respectively. A series of histograms with varying Δt obtained in this way from device 2 are shown in Figure 2.33. Five peaks appear within the central cluster, corresponding to three types of coincidence events. For peaks "a" and "e" at $\tau=4$ ns, the first photon follows the short arm of the interferometer, the second photon follows the long arm, and one photon goes to each counter. For peaks "b" and "d" at $\tau=2$ ns, both photons follow the same arm. For the central peak "c" at $\tau=0$, the first photon follows the long arm, and the second photon follows the short arm, so that the two photons make their second pass through the beamsplitter at the same time. Only in this case can two-photon interference occur, and for perfect two-photon interference, peak "c" vanishes.

When the source successfully delivers a pair of photons, the two-photon state can be written as:

$$|\psi\rangle = \int ds\, x(s) \int dt\, y(t) a^\dagger(s) a^\dagger(t+2\,\text{ns})\,|0\rangle, \qquad (2.15)$$

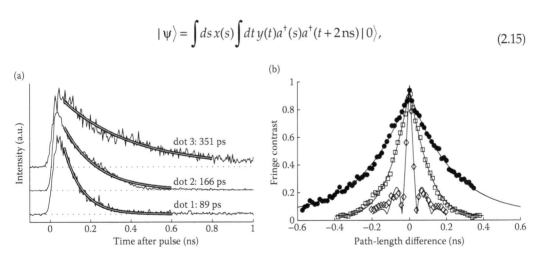

(a)

(b)

FIGURE 2.31

Time-resolved average spontaneous emission intensity following a laser pulse, measured by streak camera. (b) Results from a single-photon interference experiment showing fringe contrast versus path-length difference. Measurements were made using the setup in Figure 2.32 with the path-length difference varied about zero. The average output intensity was measured while a piezo-electric transducer modulated the path-length difference by a few wavelengths. The fringe contrast was estimated using the minimum and maximum vales from the resulting sinusoidal oscillations. (From Santori, C., Fattal, D., Vučković, J., Solomon, G. S., and Yamamoto, Y., *Nature* (London), 419 (2002) 594–597.)

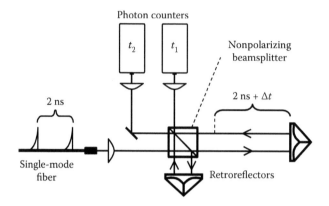

FIGURE 2.32

Two-photon interference experiment: Every 13 ns, two pulses, separated by 2 ns and each containing 0 or 1 photon, arrive through a single-mode fiber. The pulses pass through an interferometer with 2 ns path-length difference, and the outputs are detected by photon counters. When the first photon follows the long arm, and the second photon follows the short arm, two-photon interference is possible. Time interval analysis electronics allows these events to be distinguished from the other possibilities. (From Santori, C., Fattal, D., Vučković, J., Solomon, G. S., and Yamamoto, Y., *Nature* (London), 419 (2002) 594–597.)

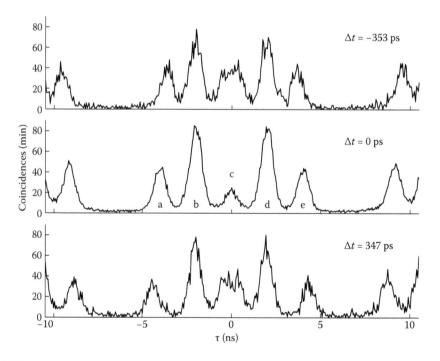

FIGURE 2.33

Time correlation histograms obtained by exciting device 2 with two pulses 2 ns apart and sending the resulting photon emission into the interferometer with path-length difference 2 ns+Δt. The central peak at $\tau = 0$ corresponds to events when photons from opposite arms arrive at the beamsplitter at the same time and exit in opposite directions.

where $a^\dagger(t)$ is the photon creation operator at time t, $x(s)$ and $y(t)$ define the photon wave packets, and $|0\rangle$ is the vacuum state (use of such a photon creation operator in the time domain requires a proper definition and certain assumptions [72]). We assume that the photon wave packets are much shorter than 2 ns. In the limit of low collection efficiency, the mean areas of peaks "a"–"e" are:

$$A_a = N\eta^{(2)}R_I^3 T_I ,\tag{2.16}$$

$$A_b = N\eta^{(2)}[R_I^3 T_I(1+2g_0') + R_I T_I^3],\tag{2.17}$$

$$A_c = N\eta^{(2)}[(R_I^3 T_I + R_I T_I^3)(1+2g_0') - 2v_1^2 R_I^2 T_I^2 v(\Delta t)],\tag{2.18}$$

$$A_d = N\eta^{(2)}[R_I^3 T_I + R_I T_I^3(1+2g_0')],\tag{2.19}$$

$$A_e = N\eta^{(2)}R_I T_I^3,\tag{2.20}$$

where N is the number of repetitions, $\eta^{(2)}$ is the combined two-photon generation and detection efficiency, and R_I and T_I are the beamsplitter intensity coefficients of reflection and transmission, respectively. As defined above, the parameter g_0' characterizes the two-photon emission probability (following a single pulse), with $g_0' = 0$ for an ideal single-photon source, and $g_0' = 1$ for a Poisson-distributed source. The parameter v_1 is the interference fringe contrast measured when an ideal monochromatic calibration source is sent into the interferometer, and accounts for optical surface imperfections. The parameter $v(\Delta t) = \langle|\,|\int dt x(t) y^*(t+\Delta t)|^2\,|\rangle$ in the expression for peak "c" is the mean overlap between the wave packets of the two photons for interferometer path-length difference (2 ns + Δt). An ensemble average is performed over all possible two-photon states generated by the source.

In Figure 2.33, the reduced area of peak "c" at $\Delta t = 0$ compared to peaks "b" and "d" can only be explained in terms of two-photon interference. As the delay Δt is changed, we observe that peak "c" becomes taller and also begins to split into two separate peaks as Δt becomes comparable to the detector time resolution (~200 ps). Let us define the quantity $M(\Delta t) = A_c/(A_b + A_d)$ in terms of the peak areas in Equations 2.17 through 2.19. This quantity is equal to the conditional probability, given that two photons collide at the beamsplitter, that the photons exit in opposite directions, in the limit $g_0' \to 0$. We measured $M(\Delta t)$ at many path length offsets Δt, and the results are shown in Figure 2.34. For all three devices, we observe reductions in the coincidence probability near $\Delta t = 0$, by factors of 0.61, 0.69, and 0.62 for devices 1, 2, and 3, respectively. The remaining coincidences are partly due to independently measured optical imperfections in our setup, $R_I/T_I = 1.1$ and $v_1 = 0.92$. Without these imperfections, the coincidence reduction factors would be $v(0) = 0.72$, 0.81, and 0.74 for devices 1, 2, and 3, respectively. We fitted the data in Figure 2.34 to the function $M(\Delta t) = 0.5[1 - a \exp(-|\Delta t|/\tau_m)]$, where the fitting parameters a and τ_m characterize the depth and the width of the coincidence dip, respectively. The fits, shown as solid lines, match the data well. For an ideal spontaneous-emission source and perfect interferometer we would expect $a = 1$ and τ_m equal to the spontaneous emission lifetime. The fitted values of τ_m we obtain are 80, 187, and 378 ps for devices 1, 2, and 3, respectively, in good agreement with the spontaneous emission lifetimes τ_s given above. For devices 1 and 3, this result is surprising given the short coherence lengths τ_c also given above. This suggests that for QDs

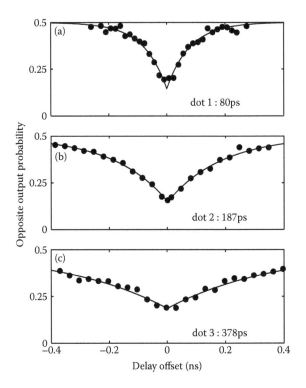

FIGURE 2.34
For devices 1 (a), 2 (b), and 3 (c), the probability that two photons that collide at the beamslitter exit in opposite directions, versus interferometer delay offset Δt. The data are fitted to a two-sided exponential function as described in the text. (From Santori, C., Fattal, D., Vučković, J., Solomon, G. S., Yamamoto, Y., *Nature (London)* 419 (2002) 594–597.)

1 and 3, the primary spectral broadening mechanism occurs on a timescale much longer than 2 ns. Such a "spectral diffusion" effect could occur, for example, due to charge fluctuations in the vicinity of the QD [76].

The best results are from device 2, with an estimated two-photon overlap of 0.81. We expect that the remaining imperfection is caused by two types of decoherence. The first is related to the excitation process. When the QD is first excited by a laser pulse, the generated electron–hole pair is initially in an excited state, and must relax to its lowest state through phonon emission before a photon can be emitted at the proper wavelength. Since this timing information is lost to the environment through the phonon, we can think of this process in terms of a classical random variable defining the start of the photon wavepacket. The second type of decoherence occurs after the QD is in the lowest single-exciton state, while the photon is being emitted. This includes decoherence caused by phonons [80,81], as well as spectral diffusion processes as discussed above. A simple model predicts that the two-photon overlap is equal to,

$$v(0) = \frac{\Gamma}{\Gamma + 2\gamma} \frac{r}{r + \Gamma'}, \tag{2.21}$$

where r is the relaxation rate of the upper exciton level into the lowest-energy single-exciton level, Γ is the excited-state population decay rate through spontaneous emission, and γ accounts for additional decoherence processes which create a broadened Lorentzian spectrum. This model predicts that, initially, the photon state purity can be improved by increasing Γ through the Purcell effect, so that $\Gamma \gg 2\gamma$. However, a maximum is reached when $\Gamma = \sqrt{2\gamma r}$, after which the excitation process becomes the dominant source of decoherence. For reasonable parameter values $1/\gamma = 1$ ns and $1/r = 10$ ps, this gives $\tau_r = 1/(\Gamma_{max}) \approx 70$ ps. This is not much shorter than the shortest lifetimes already observed with the present devices. Therefore, to see a large improvement over the present results we expect that a coherent excitation scheme is required. Schemes based on a vacuum-stimulated Raman process [10,11,82] have been explored for atomic single-photon sources, and could be applied to QDs, as well [83]. An improved coherent excitation scheme which can be applied to QDs coupled to microcavities is described in the Section 2.4.4. The degree of two-photon interference observed here is not yet high enough for realistic application to linear-optical quantum computing [68], but is high enough to demonstrate some simpler schemes such as generation of polarization entanglement through two-photon interference and post-selection, and single-mode teleportation. These demonstrations are also discussed below.

2.3.5 Strong Coupling

The devices discussed thus far have all operated in the weak coupling regime, meaning that the QD exciton and microcavity modes interact more strongly with the environment than with each other. However, recent advances in fabrication techniques have allowed the strong coupling regime to be realized in semiconductor systems [14,84–87]. In these strongly coupled devices, the QD exciton and microcavity modes coherently exchange energy back and forth, leading to a vacuum Rabi splitting of the two normal modes at resonance.

As described in the Introduction, the eigenfrequencies of the two modes on resonance are

$$\omega_{\pm} = \omega_0 - i\frac{\kappa + \Gamma}{4} \pm \sqrt{g^2 - \frac{(\kappa - \Gamma)^2}{16}}, \tag{2.22}$$

and strong coupling requires $g > |\kappa - \Gamma|/4$. For typical QDs and semiconductor cavities, $\kappa \gg \Gamma$, so the strong coupling condition reduces to $g > \kappa/4$. Since the coupling strength $g = \wp\sqrt{\omega/2\varepsilon_0\hbar V}$ and the cavity decay rate $\kappa = \omega_0/Q$, the figure of merit for a microcavity intended to achieve strong coupling is given by $Q/\sqrt{V}$ (rather than Q/V, which maximizes the Purcell enhancement). Strong coupling between a single (In,Ga)As QD and a semiconductor microcavity has been achieved using pillar microcavities [14], microdiscs [84], and photonic crystal resonators [85,86].

An electron microscope image of a pillar microcavity fabricated at the University of Würzburg, similar to those used to first demonstrate strong coupling [14], is shown in Figure 2.35a. The device contains Bragg mirrors consisting of 26 and 30 pairs of GaAs/AlAs layers above and below a GaAs cavity. At the central antinode of the cavity is a layer of InGaAs QDs with an indium content of about 40% and a density of 10^{10} cm^{-2}. A typical device contains on the order of 100 QDs, but their spectral density is low enough that single dots can be seen interacting with the cavity. Further details on device fabrication can be found in Löffler et al. [88]. Pillars with diameters from 1 to 4 µm were fabricated. The highest figure of merit $Q/\sqrt{V}$ was found in the 1.8 m pillars, which exhibited quality factors from 10,000 to 20,000 with a mode volume $V = 0.43$ µm^{-3}.

(a)

(b)

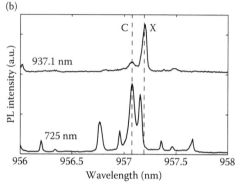

FIGURE 2.35

(a) SEM image of a 1.2 μm diameter pillar cavity. (b) Photoluminescence spectra of a 1.8 μm pillar with above-band pumping (725 nm) and resonant pumping (937.1 nm). Cavity emission is labeled C, QD exciton emission is labeled X. (From Press, D., Götzinger, S., Reitzenstein, S., Hofmann, C., Löffler, A., Kamp, M., Forchel, A., and Yamamoto, Y., *Phys. Rev. Lett.*, 98 (2007), 117402. © American Physical Society. With permission.)

Studies of high-Q microcavity-QD systems have shown that above-bandgap excitation leads to a surprisingly large amount of cavity emission, even when no QD is spectrally resonant with the cavity [14,84,85]. Part of this emission originates from the particular QD that is strongly coupled to the cavity, and part of it comes from background emitters, perhaps spectrally detuned QDs and wetting layer states [87]. In order to reduce this background contribution, we employ resonant excitation of QDs, as described above. The photoluminescence spectrum of a typical 1.8 μm pillar is shown in Figure 2.35b. When excited by a CW above-band pump, the cavity emission dominates the spectrum even though no QD is spectrally resonant with it. The background cavity emission is suppressed by nearly a factor of ten when the chosen QD's p-shell exciton is resonantly excited with 937.1 nm light.

Strong coupling can be identified in a particular pillar by an anticrossing of the QD exciton mode with the cavity photon mode as one mode is spectrally tuned relative to the other. As for the measurements of lifetime modification in the weak-coupling regime, we tune the QD emission energy through the cavity resonance by varying the temperature of the sample. The temperature-dependent photoluminescence of a 1.8 μm pillar exhibiting strong coupling is shown in Figure 2.36a. The QD is resonantly excited to suppress background emitters. A clear anticrossing of QD and cavity modes is observed, indicating the presence of strong coupling. The line centers and line widths of the two normal modes are shown as a function of temperature in Figure 2.36b and c. Below the resonance temperature of 10.5 K, the shorter wavelength line is narrower and tunes more quickly with temperature than the longer line. We thus identify the shorter wavelength line as a QD exciton-like mode, and the longer wavelength line as a cavity-like mode. As the temperature is increased, the exciton-like mode broadens and cavity-like mode narrows, until the two linewidths are equal at resonance. Above the resonance temperature, the shorter wavelength line continues to broaden into a cavity-like mode while the longer wavelength line narrows into an exciton-like mode.

At resonance, the half-cavity half-exciton modes are separated by a vacuum Rabi splitting of 56 μeV. The cavity linewidth is 85 μeV, corresponding to a Q of 15,200. Using Equation 2.22 we calculate $g=35$ μeV. The ratio of $g/\kappa=0.41$ exceeds 1/4, as required to satisfy the strong coupling condition.

It is necessary to prove that the observed anticrossing originates from a single QD, rather than a collection of quantum emitters, interacting with the cavity mode. Anticrossing

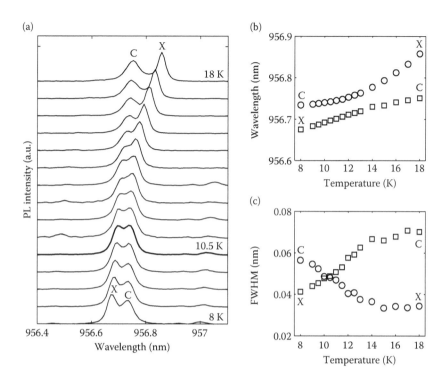

FIGURE 2.36

(a) Temperature dependent photoluminescence from a strongly coupled pillar under resonant excitation (pump wavelength 936.25–936.45 nm). Resonance is at 10.5 K. (b,c) Central wavelength and FWHM of longer wavelength (circles) and shorter wavelength (squares) lines as a function of temperature. (From Press, D., Götzinger, S., Reitzenstein, S., Hofmann, C., Löffler, A., Kamp, M., Forchel A., and Yamamoto, Y., *Phys. Rev. Lett.*, 98 (2007), 117402. © American Physical Society. With permission.)

behavior similar to that observed here has also been demonstrated using a QW embedded in a planar microcavity [89], in which many QW excitons contribute to the splitting at resonance. We may distinguish between the quantum case involving a single QD exciton and the semiclassical case involving a collection of emitters by measuring the autocorrelation function of the photons emitted from the resonantly coupled system $g_{r,r}^{(2)}(\tau)$. Due to the extremely short (~15 ps) lifetime of the coupled QD-cavity system, the system must be excited with short pulses to resolve antibunching behavior using much slower detectors.

Strong antibunching is observed in the photons emitted by the resonantly coupled QD-cavity system, as shown in Figure 2.37a. The measured value of the autocorrelation function $g_{r,r}^{(2)}(0) = 0.18 < 0.5$ proves that a single quantum emitter dominates photon emission.

When the QD was red detuned by 0.4 nm from the cavity, a surprisingly large amount of cavity emission persisted even though the resonant pump laser selectively excited only the chosen QD (see Figure 2.37e). Photon statistics were collected from the cavity and QD emission separately. The QD emission was antibunched as expected, with $g_{x,x}^{(2)}(0) = 0.19$ (Figure 2.37b). Interestingly, the cavity emission was also antibunched with $g_{c,c}^{(2)}(0) = 0.39$, demonstrating that most of the cavity emission originated from a single quantum emitter (Figure 2.37c). Finally the cross-correlation function between the cavity and QD emission was measured to be $g_{x,c}^{(2)}(0) = 0.22$ (Figure 2.37d), conclusively proving that both emission peaks originate from the single QD emitter.

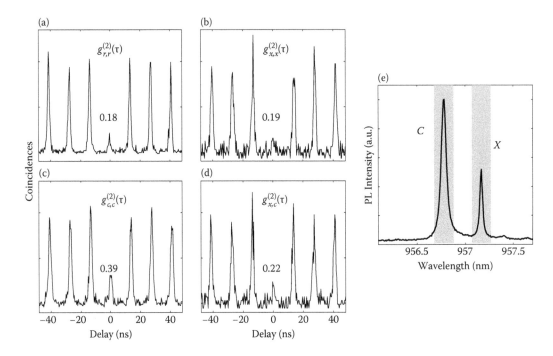

FIGURE 2.37
(a) Autocorrelation function of strong coupling system at resonance. (b–e) QD is detuned by 0.4 nm from cavity. (b) Autocorrelation function of QD emission. (c) Autocorrelation function of cavity emission. (d) Cross-correlation function of QD and cavity. (e) Photoluminescence spectrum. Shaded regions indicate pass bands of spectral filter used for correlation measurements.

The strong cavity emission cannot be attributed solely to radiative coupling to the QD at such a large detuning. Another mechanism, possibly mediated by the absorption or emission of acoustic phonons, must be responsible for efficiently channeling excitations from the QD to the cavity. Further studies are needed to fully understand this coupling.

2.4 Applications

Many applications have been proposed for single-photon devices in the area of quantum information, ranging from quantum key distribution, which uses the laws of quantum mechanics to enable fundamentally secure communication, to large-scale quantum computation, which uses "quantum parallelism" to allow for an exponential speed-up in certain problems over a classical computer. Here, we describe several applications relevant to single-photon devices based on pillar microcavities. For three of these applications— quantum key distribution [90], formation of polarization entanglement [91], and dual-rail teleportation [92]—we have already performed proof-of-principle demonstrations, as described below. The last part of this section presents a theory of coherent generation and trapping of photons in a three-level system coupled to a microcavity [93]. We expect this scheme to be the next step in improving the performance of QD-based single-photon devices, and we hope that such techniques will eventually enable construction of quantum networks for scalable quantum computation.

2.4.1 BB84 Quantum Key Distribution

The field of quantum information has opened up the potential to accomplish tasks previously considered impossible by classical information processing. Among these is quantum cryptography, a method to generate unbreakable cryptographic codes. Unlike classical cryptography, which often relies on the computational difficulty of performing numerical tasks, the security of quantum cryptography is guaranteed by the laws of quantum mechanics.

To date, there are a large number of protocols that have been proposed for secure quantum encryption. The oldest and most well studied of these is the protocol of Bennett and Brassard (BB84) [94]. In this protocol the sender, referred to as Alice, transmits a stream of single-photons to Bob. Each photon is prepared randomly in one of four polarization states: horizontal, vertical, right circularly polarized, or left circularly polarized. Bob randomly decides to measure in the linear or circular polarization basis. Later, Alice and Bob use a public channel to eliminate all instances where Bob measured in the wrong basis. The security of the BB84 protocol has been well established against the most general types of eavesdropping attacks allowed by the laws of quantum mechanics [95–97].

The security of the BB84 protocol critically relies on the assumption that information is encoded in single-photons. Nevertheless, the majority of cryptography systems that implement BB84 use attenuated lasers that emit a Poisson photon number distribution. When the source emits more than one photon, this opens the possibility for eavesdropping attacks based on photon-number splitting. In such an attack, the eavesdropper, referred to as Eve, splits off one of the photons while allowing the others to travel to Bob unperturbed. After the measurement basis is revealed, Eve can measure her photon and learn the bit value. A QD coupled to a pillar microcavity allows us to generate photon sources that perform much better due to the suppression of multi-photon states, which eliminates this potential security loophole.

To quantify the potential benefits of a single-photon source requires only two parameters: (1) the average photon flux, $\langle n \rangle$, given by

$$\langle n \rangle = \int_0^{t_{ph}} \langle a^\dagger(t) a(t) \rangle dt, \tag{2.23}$$

where t_{ph} is the time duration of the photon pulse, and $a^\dagger(t)$ and $a(t)$ are the photon creation and annihilation operators, respectively, at time t; and (2) the autocorrelation $g_0^{(2)}$, given by

$$g_0^{(2)} = \frac{\int_0^{t_{ph}} \int_0^{t_{ph}} \langle a^\dagger(t) a^\dagger(t') a(t') a(t) \rangle dt dt'}{\langle n \rangle^2}. \tag{2.24}$$

The communication rate of BB84 has been extensively studied, both for attenuated laser sources [98] and single-photon sources [2]. Figure 2.38 plots the calculated communication rate as a function of channel loss for several different values of $\langle n \rangle$ and $g_0^{(2)}$. The left panel plots the rate for several different values of $g_0^{(2)}$ when $\langle n \rangle = 1$. An attenuated laser would feature a communication rate characterized by $g_0^{(2)} = 1$, while the curve $g_0^{(2)} = 0$ is an ideal single-photon source. Note that the Poisson light bit rate decreases faster than

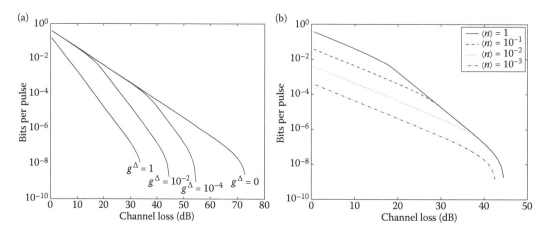

FIGURE 2.38
Security improvement for sources with supressed $g_0^{(2)}$ (indicated in the figure as g^Δ). (a) Communication rate as a function of channel loss for different values of $g_0^{(2)}$. (b) Communication rate for $g_0^{(2)} = 10^{-2}$ and different values of the average flux $\langle n \rangle$.

the ideal single-photon device. This is because the single-photon device does not suffer from photon splitting attacks. Thus, the rate decrease is only due to the increasing channel loss. For Poisson light, as the channel loss increases the effect of the multi-photon states is enhanced, forcing us to reduce the average number of photons. Intermediate devices with $0 < g_0^{(2)} < 1$ feature two types of behaviors. At low channel losses they behave similarly to the ideal device where the bit rate decreases in proportion to the channel transmission. At higher loss levels the multi-photon states start to make a significant contribution and the behavior gradually switches over to that of Poisson light. Note that each curve features a cutoff channel loss, beyond which secure communication is no longer possible. A smaller $g_0^{(2)}$ implies that more loss can be tolerated.

The right panel of Figure 2.38 plots the communication rate for different values of $\langle n \rangle$ with $g_0^{(2)} = 10^{-2}$. At low channel losses, $\langle n \rangle$ has a significant effect on the communication rate, but at higher loss levels most of the curves meet with the ideal curve. Only the extremely lossy device with efficiency of 10^{-3} fails to rejoin the ideal curve, and features a smaller cutoff loss. This implies that at sufficiently high loss levels the flux of the source does not have an impact on the communication rate, provided the flux exceeds a critical level. For the simulation parameters in Figure 2.38, this critical level occurs somewhere between 10^{-2} and 10^{-3}. Furthermore, the cutoff loss level, which defines the maximum communication distance, is unaffected.

We demonstrated the improvement in the performance of a quantum-cryptography system using our third-generation pillar microcavities. The performance of the particular structure used was characterized as described in the previous section. The left panel of Figure 2.39 shows the mean photon number per pulse for the micropillar device used. The emission rate initially increases linearly with pump power, but eventually saturates at a nominal rate of 0.07 photons per pulse. This average includes all losses in the measurement system. The autocorrelation of the QD is shown in the right panel of Figure 2.39, obtained for a pump intensity indicated by the point A in the left panel of the figure.

A full quantum-cryptography system requires active control and synchronization of the photon polarization state. Figure 2.40 illustrates a system used to achieve this control. The microcavity-QD structure is excited by a pulsed laser whose repetition rate defines the clock cycle of the experiment. An electro-optic modulator is used to prepare

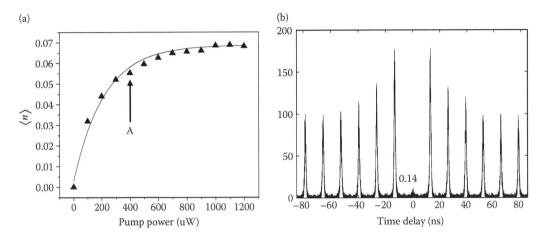

FIGURE 2.39
Characterization of QD. (a) Photon emission rate measurement as a function of excitation power.
(b) Autocorrelation taken at excitation power indicated by point A in left panel.

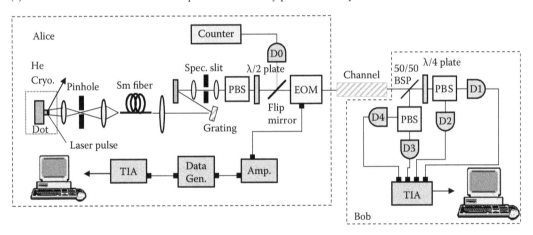

FIGURE 2.40
Apparatus for implementation of BB84 with a QD coupled to a micro-pillar cavity.

the polarization state of each photon before it enters the channel. A data generator, whose
signal is amplified by a high power amplifier, drives the modulator. The data generator is
synchronized to the laser pulses, and produces a random, four-level signal, corresponding
to the four different polarization states in the BB84 protocol. Bob's detection apparatus is
composed of a 50–50 beamsplitter, which partitions each photon randomly to one of two
polarization analyzers. Both Alice and Bob share a common clocking signal from the data
generator. Each of Bob's detection events is recorded by a time-interval analyzer (TIA),
together with a time stamp of the event relative to the common clock. A detection is also
used to generate a logic pulse (containing no information about the detection result) which
triggers a second TIA in Alice's apparatus. This TIA records the polarization state that was
prepared, along with a time stamp that can be used for later comparison with Bob's data.

The left panel of Figure 2.41 shows a correlation of the data recorded by Alice and Bob.
The central diagonal of the bar graph corresponds to error events, which are predominantly
due to effects such as phase drift and imperfections in the polarization optics. The overall
Bit Error Rate (BER) of the system is measured to be 2.5%. These errors can be handled in

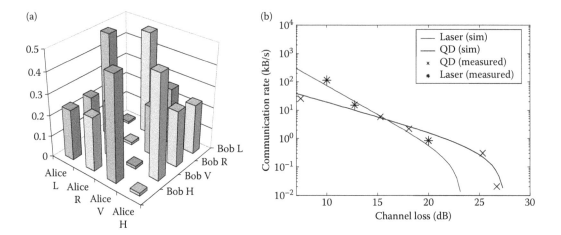

FIGURE 2.41
(a) Data correlation between Alice and Bob shows a 2.5% bit error rate. (b) Experimental comparison between a laser and QD coupled to a micro-pillar cavity. The QD achieves a 5dB improvement in channel loss over the attenuated laser.

the communication system using two way error correction methods [99]. Such error correction algorithms are very efficient and work within about 25% of the Shannon limit.

In the right panel of Figure 2.41 one can see a comparison between the performance of the microcavity QD to a standard attenuated laser in the presence of channel losses. At low loss levels, the communication rate with the attenuated laser is higher, because the laser starts out with a macroscopically large number of photons, which can be attenuated to any desired average. This is in contrast to the QD, which is limited by the device efficiency and losses in subsequent optics. However, at higher channel losses, the laser emits too many multi-photon states, causing a more rapid decrease in communication. At around 16 dB, the QD begins to outperform an attenuated laser. Above 23 dB of loss, secure communication is no longer possible with the laser, while the QD can withstand channel losses of about 28 dB. This demonstrates the security advantage of the source in the presence of channel losses.

In the final phase of the communication, the secret key is used as a one time pad to exchange the message. Figure 2.42 shows how this is done. A picture of Memorial Church at Stanford University serves as the message. A 20 kB secret key is exchanged over the communication system. Alice uses her copy of the key to do a bitwise exclusive OR operation with every bit of the message. The resulting encrypted message looks like white noise to anyone who does not posses a copy of the key, as shown in Figure 2.42b. This panel plots the pixel value histogram of the original and encrypted message. The original message has clear structure in the histogram, while the encrypted message features a flat pixel value histogram indicative of white noise. Bob decodes the encrypted message by performing a second bitwise exclusive OR using his copy of the key, faithfully reproducing the original message.

2.4.2 Entanglement Generation without a "True" Interaction

The ability to produce nearly identical single-photons in micro-pillar structures constitutes an enabling technology for quantum information processing, beyond QKD. The quantum interference between spatio-temporally identical photons of various joint polarization

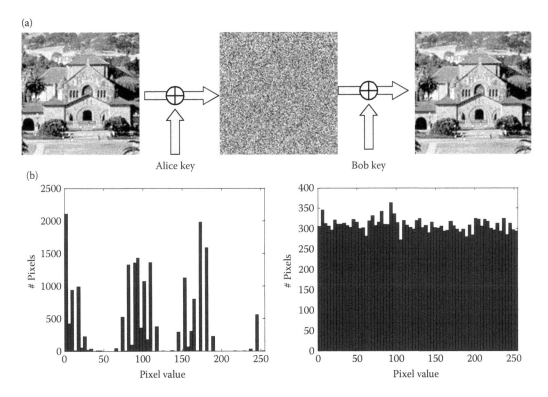

FIGURE 2.42
(a) One time pad encryption protocol using exchanged 20 kB key. (b) Pixel histogram of original and encrypted message. Encrypted message shows flat pixel distribution, as expected from white noise.

states, for instance, can be harnessed to produce interesting effects that would otherwise require the use of extremely (maybe prohibitively) large optical nonlinearities. In this section, we present the results of an experiment using quantum interference to produce *post-detected* polarization entangled photons, for which we obtained a violation of Bell's inequality. In the next section, we will present an experiment in which the quantum state of a single optical mode could be teleported to another optical mode via the same quantum interference effect.

Entanglement, the nonlocal correlations allowed by quantum mechanics between distinct systems, is a central concept in quantum information science [100]. These nonlocal correlations are often understood as the result of prior interactions between the quantum mechanical systems, something like a memory of those interactions. In light of relatively recent experiments in the field of quantum information [66]), this is too limited a view. Entanglement can be induced between noninteracting particles, provided they are quantum mechanically indistinguishable. Pioneering work by Shih and Alley [67], followed by Ou and Mandel [101], used a *post-selection* procedure to induce entanglement between two identical photons produced in a nonlinear crystal. More recently, entanglement swapping experiments [102,103] used two independent entangled photon pairs to induce entanglement between photons of different pairs which never interacted.

Our experiment makes use of the same underlying principle for entanglement generation. However, starting with two single-photons guarantees that at most a *single*-photon pair can be generated at a time. The entanglement formation relies on two crucial features of a micro-pillar QD single-photon source: (1) its ability to suppress multi-photon pulses

[9], and (2) its ability to generate consecutively two photons that are quantum mechanically indistinguishable [4].

When two spatio-temporally identical photons of orthogonal polarizations arrive together ("collide") at a nonpolarizing beam splitter (NPBS), a quantum interference effect ensures that, if detected simultaneously at *different* output ports of the NPBS, they should be entangled in polarization [101]. More precisely, when the two optical modes corresponding to the output ports, denoted "c" and "d," of the NPBS have a simultaneous single occupation, their joint polarization state is expected to be the EPR-Bell state:

$$|\Psi^-\rangle = \frac{1}{\sqrt{2}}(|H\rangle_c |V\rangle_d - |V\rangle_c |H\rangle_d).$$

The input port modes of the NPBS, denoted as "a" and "b", are related to the output modes "c" and "d" by the 50–50% NPBS unitary matrix according to:

$$a_{H/V} = \frac{1}{\sqrt{2}}(c_{H/V} + d_{H/V}),$$

$$b_{H/V} = \frac{1}{\sqrt{2}}(c_{H/V} - d_{H/V}),$$

where subscripts "H" and "V" specify the polarization (horizontal or vertical) of a given spatial mode. The quantum state corresponding to single-mode photons with orthogonal polarizations at port "a" and "b" can be written as:

$$a_H^\dagger b_V^\dagger |vac\rangle = \frac{1}{2}(c_H^\dagger c_V^\dagger - d_H^\dagger d_V^\dagger - c_H^\dagger d_V^\dagger + c_V^\dagger d_H^\dagger)|vac\rangle$$

As pointed out by Popescu et al. [104], this state already features nonlocal correlations and violates Bell's inequality without the need for post-selection, by using photo-detectors that can distinguish photon numbers 0, 1 and 2. If, however, we decide to discard the events when two photons go the same way (recording only coincidence events between modes "c" and "d"), we obtain the post-selected state:

$$\frac{1}{\sqrt{2}}(c_H^\dagger d_V^\dagger - c_V^\dagger d_H^\dagger)|vac\rangle = |\Psi^-\rangle$$

with a probability of 1/2.

The experimental setup is shown in Figure 2.43. Single-photon emission was triggered by resonant optical excitation of a single QD, with 3 ps Ti:Sa laser pulses, with a 76 MHz repetition rate. Each pulse was moreover split and one fraction delayed by 2 ns, so that the QD would emit two photons 2 ns apart, with a repetition period of 13 ns. The emitted photons were collected by a single mode fiber and sent to a Mach-Zehnder interferometer with 2 ns delay on the longer arm. A quarter wave plate (QWP) followed by a half wave plate (HWP) were used to set the polarization of the photons after the input fiber to linear and horizontal. An extra HWP was inserted in the longer arm of the interferometer to rotate the polarization to vertical. One time out of four, the first emitted photon takes the long path while the second photon takes the short path, in which case their

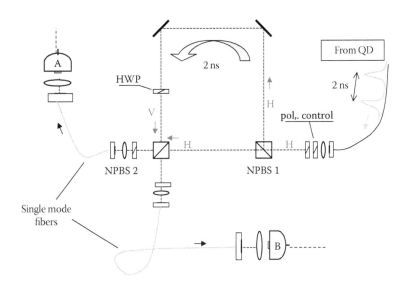

FIGURE 2.43
Experimental setup for entanglement generation. Single-photons from the QD microcavity device are sent through a single mode fiber, and have their polarization rotated to H. They are split by a first NPBS (1). The polarization is changed to V in the longer arm of the Mach-Zehnder configuration. The two path of the interferometer merge at a second NPBS (2). The output modes of NPBS 2 are matched to single mode fibers for subsequent detection. The detectors are linked to a time-to-amplitude converter for a record of coincidence counts.

wavefunctions overlap at the second nonpolarizing beam-splitter (NPBS 2). In all other cases (not of interest), the single-photon pulses "miss" each other by at least 2 ns which is greater than their width (100–200 ps). Two single-photon counter modules (SPCMs) in a start-stop configuration were used to record coincidence counts between the two output ports of NPBS 2, effectively implementing the post-selection (if photons exit NPBS 2 by the same port, then no coincidences are recorded by the detectors). Single-mode fibers were used prior to detection to facilitate the spatial mode-matching requirements. They were preceded by quarter-wave and polarizer plates to allow the analysis of all possible polarizations.

The detectors were linked to a time-to-amplitude converter and multi-channel analyzer to record histograms of coincidence events versus detection time delay τ. A typical histogram is shown on Figure 2.44, with the corresponding post-selected events. For given analyzer settings (θ_1, θ_2), we denote by $C(\theta_1, \theta_2)$ the number of post-selected events normalized by the total number of coincidences in a time window of 100 ns. This normalization is independent of (θ_1, θ_2), since the input of NPBS 2 are two modes with orthogonal polarizations. $C(\theta_1, \theta_2)$ measures the average rate of coincidences throughout the time of integration.

The single count rate at the output of the single-mode fiber was 9400 s^{-1}, from which we infer a total quantum efficiency of 0.13% (detection loss included). The total pair production rate for QD 1 was 12 s^{-1} after fiber, so that useful pairs were generated with a rate of 1.5 s^{-1} (we lose a factor 8 due to the post-selection and by excluding "bad-timing" events). High suppression of two-photon pulses and high mean overlap (indistinguishability) between consecutive photons were observed. The overlap was measured in a photon bunching experiment [4], as described above.

Several methods can be used to prove the success of entanglement generation. The traditional method is a Bell inequality test, that proves the nonlocal nature of the state of two quantum systems if they are sufficiently entangled. Another more complete way is to

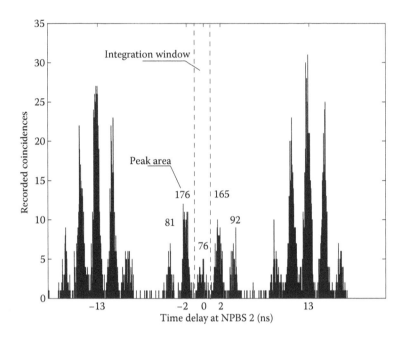

FIGURE 2.44

Zoom on a typical correlation histogram, taken on QD_1. Coincidences with delay τ between detectors A and B were actually recorded for $-50\ \text{ns} < \tau < 50\ \text{ns}$. The integration time was 2 min, short enough to guarantee that the QD is illuminated by a constant pump power. The central region $-1\ \text{ns} < \tau < 1\ \text{ns}$ corresponds to the post-selected events: the corresponding photons overlapped at NPBS 2 where they took different exit ports. (From Fattal, D., Inoue, K., Vučković, J., Santori, C., Solomon. G. S., and Yamamoto, Y., *Phys. Rev. Lett.*, 92 (2004) 37903. © American Physical Society.)

reconstruct the full density matrix describing the joint polarization state of two photons, a procedure known as *quantum state tomography*. The presence of entanglement can then be read from the density matrix by several mathematical methods.

A Bell-inequality test was performed for post-selected photon pairs from QD_1. Following Clauser et al. [105], if we define the correlation function $E(\theta_1, \theta_2)$ for analyzer settings θ_1 and θ_2 as

$$E(\theta_1,\theta_2) = \frac{C(\theta_1,\theta_2)+C(\theta_1^{\perp},\theta_2^{\perp})-C(\theta_1^{\perp},\theta_2)-C(\theta_1,\theta_2^{\perp})}{C(\theta_1,\theta_2)+C(\theta_1^{\perp},\theta_2^{\perp})+C(\theta_1^{\perp},\theta_2)+C(\theta_1,\theta_2^{\perp})},$$

then local realistic assumptions lead to the following inequality:

$$S = |E(\theta_1,\theta_2) - E(\theta_1',\theta_2)| + |E(\theta_1',\theta_2') + E(\theta_1,\theta_2')| \le 2.$$

Violation of this inequality is a signature of quantum-mechanical behavior.

Sixteen measurements were performed for all combinations of polarizer settings among $\theta_1 \in \{0°, 45°, 90°, 135°\}$ and $\theta_2 \in \{22.5°, 67.5°, 112.5°, 157.5°\}$. The corresponding values of the normalized coincidence counts $C(\theta_1, \theta_2)$ are reported in Table 2.1. The statistical error on S is quite large, due to the short integration time used to insure high stability of the QD device. Bell's inequality is still violated by two standard deviations: $S \approx 2.38 \pm 0.18$. Hence, nonlocal correlations were created between two single independent photons by linear optics and photon number post-selection.

TABLE 2.1

Normalized Coincidence Counts $C(\theta_1, \theta_2)$ 10^3 for Various
Polarizer Angles used in the Bell Inequality Test

$\theta_2 \backslash \theta_1$	0°	45°	90°	135°
22.5°	5.6	28.4	28.6	4.7
67.5°	9.0	8.3	25.2	25.1
112.5°	28.9	5.4	4.6	28.4
157.5°	26.0	24.9	8.6	8.8

Counts correspond to the coincidences in the integration window
(see Figure 2.44) divided by the total coincidences recorded for –50 ns
$< \tau < 50$ ns. Note that the quantity $C(\theta_1, \theta_2) + C(\theta_1^\perp, \theta_2^\perp) + C(\theta_1^\perp, \theta_2) + C(\theta_1, \theta_2^\perp)$ is constant for given settings θ_1 and θ_2.

Complete information about the two-photon polarization state is provided by a reduced density matrix, where only the polarization degrees of freedom are kept. This density matrix can be reconstructed from a set of 16 measurements with different analyzer settings, including circular [106]. This technique was applied to photon pairs emitted by our device. The reconstructed density matrix is shown in Figure 2.45. It can be shown to be nonseparable, i.e., entangled, by use of the Peres criterion [107]. The negativity is defined as the largest negative eigenvalue of the partial transpose of a bipartite density matrix, or zero if it has only positive eigenvalues. The density matrix in Figure 2.45 gives a negativity value of 0.43, where a value of 1 means maximum entanglement and a value of 0 means no entanglement.

We can account for the observed degree of entanglement by considering some parameters of the QD single-photon source. Due to residual two-photon pulses (reflected by a nonzero value of the intensity autocorrelation factor $g_0^{(2)}$), a recorded coincidence count can originate from two photons of the same polarization entering NPBS 2 from the same port. A multi-mode analysis also reveals that imperfect overlap $v(0) = |\int \psi_1(t)^* \psi_2(t)|^2$ between consecutive photon pulse amplitudes reduces the quantum interference responsible for the entanglement generation. Including those imperfections, we could derive a simple model for the joint polarization state of the post-selected photons. In the limit of low pump level, this model predicts the following density matrix in the $(H/V) \otimes (H/V)$ basis:

$$\rho_{\text{model}} = \frac{1}{\frac{R_I}{T_I} + \frac{T_I}{R_I} + 4g_0^{(2)}} \begin{pmatrix} 2g_0^{(2)} & & & \\ & \frac{R_I}{T_I} & -v(0) & \\ & -v(0) & \frac{T_I}{R_I} & \\ & & & 2g_0^{(2)} \end{pmatrix}$$

R_I and T_I are the intensity reflection and transmission coefficients of NPBS 2 ($R_I/T_I \sim 1.1$ in our case). Using the independently measured values for $g_0^{(2)}$ and $v(0)$, we obtain excellent quantitative agreement of our model to the experimental data, with a fidelity $Tr\left(\sqrt{\rho_{\text{exp}}^{1/2} \rho_{\text{model}} \rho_{\text{exp}}^{1/2}}\right)$ as high as 0.997.

The negativity of the state ρ_{model} is proportional to $(v(0) - 2g_0^{(2)})$, which means that entanglement exists as long as $v(0) > 2g_0^{(2)}$. This simple criterion can be applied to any single-photon source for which the intensity autocorrelation and photon overlap values are known. It indicates the extent to which that source will be able to generate entangled photons through two-photon interference and post-selection.

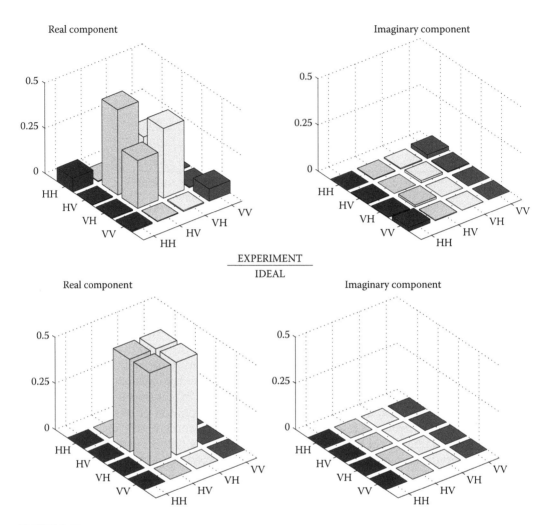

FIGURE 2.45
Reconstructed polarization density matrix for the post-selected photon pairs emitted by our device. The small diagonal HH and VV components are caused by finite two-photon pulses suppression ($g_0^{(2)} > 0$). Additional reduction of the off-diagonal elements originates from the imperfect indistinguishability between consecutively emitted photons. (From Fattal, D., Inoue, K., Vučković, J., Santori, C., Solomon, G. S., and Yamamoto, Y., *Phys. Rev. Lett.*, 92 (2004) 37903. © American Physical Society.)

2.4.3 Single-Mode Teleportation

Quantum teleportation is a general procedure by which a quantum state encoded in a physical system is transferred to another physical system in a possibly remote location. This procedure is useful for transportation of quantum states over long distances in a safe manner, and underlies many quantum-communication schemes (with the notable exception of BB84 QKD). Surprisingly, quantum teleportation can also be used to protect quantum information and turns out to be a critical element of the scalability of many quantum computation schemes. It will most likely become a major tool for the optical engineer of the future.

In this section, we show how to use quantum interference between spatio-temporally identical photons from a micro-pillar structure to realize quantum teleportation of a single mode of the optical field. A QD in a micro-pillar cavity was used to generate a stream of nearly identical photons, each defining a quantum bit of information in the so-called "dual-rail" representation. Upon interference and measurement of two such qubits, a target and an ancilla, the state of the target qubit is observed to change physical support (optical modes) with very high fidelity to the original state. In particular, the coherence between two different optical modes encoding the target qubit was transferred with high fidelity to the output modes. The observed fidelity is 80%, in agreement with the residual distinguishability between consecutive photons from the source. An improved version of this teleportation scheme using more ancillas constitutes a building block for scalable quantum computing using linear optics and photo-detectors.

Photons are almost ideal carriers of quantum information, since they have little interaction with their environment and are easy to manipulate individually with linear optics. The main challenge of optical quantum information processing is the design of controlled interactions between photons, necessary for the realization of nonlinear quantum gates. Photons do not naturally "feel" the presence of other photons, unless they propagate in a medium with high optical nonlinearity. The amount of optical nonlinearity required to perform controlled operations between single-photons is however not easily accessible. Large optical nonlinearities having noticeable effect at the single-photon level might become eventually available with the development of cavity QED or electromagnetically induced transparency (EIT) techniques, but at present these possibilities remain quite remote.

Probabilistic quantum-logic gates can be implemented with linear optics only [68,108,109], but as such, they are not suitable for scalable quantum computation. In a seminal paper, Gottesman and Chuang [110] suggested that quantum gates could be applied to photonic qubits through a generalization of quantum teleportation [111]. In such a scheme, the information about the gate is contained in the state of ancilla qubits. The implementation of a certain class of gates can then be reduced to the problem of preparing the ancilla qubits in some wisely chosen entangled state. Such a problem can be solved "off-line" with linear-optics elements alone, provided the photons used are quantum mechanically indistinguishable particles [66]. Following this idea, Knill, Laflamme and Milburn (KLM) [68] proposed a scheme for efficient linear-optics quantum computation (LOQC) based on the implementation of the controlled-sign gate (C-z gate) through teleportation. Since the C-z gate effectively acts on only one of the two modes composing the target qubit, a simplified procedure can be used where a single optical mode is teleported, instead of the two modes composing the qubit. This procedure will be referred to as *single mode teleportation*, to distinguish it from the usual teleportation scheme. In its basic version using one ancilla qubit, it succeeds half of the time. In its improved version using an arbitrarily high number of ancillas, it can succeed with a probability arbitrarily close to one [68,112].

Single-mode teleportation, in its simplest form, involves two qubits, a target and an ancilla, each defined by a single-photon occupying two optical modes (Figure 2.46). The target qubit can, *a priori*, be in an arbitrary state

$$| \psi_t \rangle = \zeta_0 | 0 \rangle_L + \xi_0 | 1 \rangle_L ,$$

where the logical $| 0 \rangle_L$ and $| 1 \rangle_L$ states correspond to the physical states $| 1 \rangle_1 | 0 \rangle_2$ and $| 0 \rangle_1 | 1 \rangle_2$, respectively, in a dual-rail representation. The ancilla qubit is prepared with a beam-splitter (BS a) in the coherent superposition

$$| \psi_a \rangle = \frac{1}{\sqrt{2}} (| 0 \rangle_L + | 1 \rangle_L) = \frac{1}{\sqrt{2}} (| 1 \rangle_3 | 0 \rangle_4 + | 0 \rangle_3 | 1 \rangle_4).$$

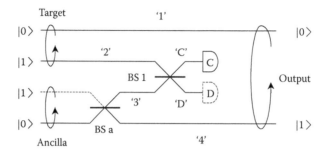

FIGURE 2.46

Schematic of single mode teleportation. Target and ancilla qubits are each defined by a single-photon occupying two optical modes (1–2 and 3–4). When detector C records a single-photon, the state in modes 1–4 reproduces the initial state of the target. In particular, the coherence between modes 1-2 of the target can be transferred to a coherence between modes 1–4. (From Fattal, D., Diamanti, E., Inoue, K., and Yamamoto, Y., *Phys. Rev. Lett.*, 92 (2004) 37904. © American Physical Society.)

One rail of the target is mixed with one rail of the ancilla with a beam-splitter (BS 1), for subsequent detection in photon counters C and D. The state after mixing can be written in terms of modes C, D, 1 and 4 as

$$| \psi_t \rangle_{12} | \psi_a \rangle_{34} = \frac{1}{2} | 10 \rangle_{CD} | \psi_t \rangle_{14} + \frac{1}{2} | 01 \rangle_{CD} (Z | \psi_t \rangle_{14}),$$

$$+ \frac{\zeta_0}{\sqrt{2}} | 00 \rangle_{CD} | 11 \rangle_{14} + \frac{\xi_0}{2} (| 02 \rangle + | 20 \rangle)_{CD} | 00 \rangle_{14}.$$

For a given realization of the procedure, if only one photon is detected at detector C, then the state of the output qubit (modes 1 and 4) is $|\psi_t\rangle$, so the teleportation was successful. Similarly, if only one photon is detected at detector D, then the output state is $Z|\psi_t\rangle$, so in this case we have to apply the Pauli operator Z (phase shift of π) to the output modes to retrieve the initial state $|\psi_t\rangle$ [113]. We did not implement this active feedforward here, in order to simplify the experiment. For this reason, our teleportation procedure succeeds with probability 1/4 (as compared to 1/2 had we used feedforward).

It is interesting and somewhat enlightening to describe the same procedure in the framework of single-rail logic. In this framework, each optical mode supports a whole qubit, encoded in the presence or absence of a photon, and the single mode teleportation can be viewed as entanglement swapping. Indeed, for the particular values $\zeta=\xi=1/\sqrt{2}$, modes 1 and 2 find themselves initially in the Bell state $|\psi^+\rangle_{12}$, while modes 3 and 4 are in a similar state $|\psi^+\rangle_{34}$. A partial Bell measurement takes place using BS 1 and counters C/D, which if it succeeds leaves the system in the entangled state $|\psi^+\rangle_{14}$, so that entanglement swapping occurs. In the rest of the chapter, we choose to consider the scheme in the dual-rail picture, since it is a more robust, and hence realistic, way of storing quantum information, albeit at the expense of using two modes per qubit.

The success of the teleportation depends mostly on the transfer of coherence between the pairs of modes (1–2) and (1–4). If the target qubit is initially in state $|0\rangle_L = |1\rangle_1 |0\rangle_2$, then the ancilla photon cannot end up in mode 4, because of the postselection condition, so that the output state is always $|1\rangle_1 |0\rangle_4$ as wanted. The same argument applies when the target qubit is in state $|1\rangle_L$. Hence, the success of the teleportation is granted when the target qubit is not in a superposition state. However, when the target qubit is in a coherent superposition of $|0\rangle_L$ and $|1\rangle_L$, the output state might not retrieve the full initial coherence. A good way to test the coherence transfer is by changing the optical path length Δ on mode (1). If the teleportation procedure does not randomize the phase between mode (1) and mode (4), then changing Δ in a controlled manner changes the well defined phase between modes (1–4), which can be observed by interfering modes (1) and (4) in an auxiliary setup. If, however, the teleportation randomizes the phase between modes (1)–(4), then changing the path length Δ will not have any effect on the interferometric signal.

The experimental setup is shown in Figure 2.47. Two photons emitted consecutively by the photon source are captured in a single-mode fiber. In the dual-rail representation, we refer to the first photon as the ancilla, and to the second photon as the target (see Figure 2.46). The ancilla qubit, initially in state $|0\rangle_L$, is delayed in free space to match the target qubit temporally at BS 1. The delay must be adjusted to within a fraction of the photon's temporal width (~200 ps or 6 cm in space). Note that the mode matching is significantly easier here than in similar experiments using photons from PDC, where the optical path length must be adjusted with a tolerance of only a few microns [114].

The ancilla is prepared in the superposition state $|\psi_{anc}\rangle=1/\sqrt{2}\,(|0\rangle_L+|1\rangle_L)$ with a beam-splitter "BS a". The target qubit is prepared in a similar maximum superposition state (with "BS t"). The path length Δ of mode (1) is changed in a controlled manner with a piezo-actuated mirror. The "partial Bell measurement" responsible for the teleportation is done at BS 1 by mixing the optical modes (2) of the target qubit and (3) of the ancilla qubit, with subsequent detection in counter C. A Mach-Zehnder type setup is used to measure the coherence between the two modes (1) and (4) of the output qubit. It is composed of a 50-50 beam-splitter BS 2 mixing modes (1) and (4), with subsequent detection

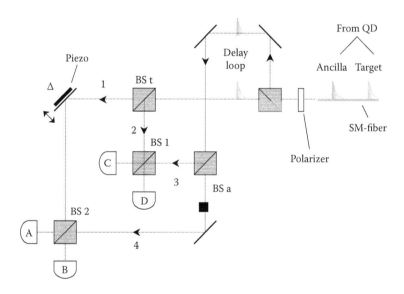

FIGURE 2.47
Experimental setup for single-mode teleportation. All the beam-splitters (BS) shown are 50–50 nonpolarizing BS. The teleportation procedure works when the ancilla photon is delayed, but the target is not. After preparation, the target photon occupies modes 1 and 2, and the ancilla occupies modes 3 and 4. Modes 2 and 3 are mixed at BS 1 and subsequently measured by detectors C and D, this step being the heart of the teleportation. When C records a single-photon, another single-photon occupies modes 1–4 (output qubit). The phase coherence between modes 1–4 in the output state is measured by mixing those modes at BS 2 and recording single counts at detector A or B. Note that since an event is recorded only if A *and* C or B *and* C clicked, more than one photon could not have reached detector C.

in counters A and/or B. The phase coherence of the teleportation is proven if modulating the path length on mode (1) results in the modulation of the count rate in detector A and B (conditioned on a click at detector C). Moreover, the degree of phase coherence between modes (1) and (4) can be quantified by the contrast (or visibility) of the count rate modulation.

Coincidences between counters A and C and between B and C were simultaneously recorded, using a start-stop configuration (each electronic "start" pulse generated by counter C was doubled for this purpose). This detection method naturally post-selects events where one photon went through BS 1, and the other went through BS 2, as required by the teleportation scheme. Since no more than one photon is emitted by the single-photon source, no more than one photon can reach detector C if detector A or B is to click. Typical correlation histograms are shown in Figure 2.48. The integration time was 2 min, short enough to keep the relative optical path length between different arms (1–4) of the interferometer stable. The whole setup was made compact for that purpose, and stability over time periods as long as 10 min was observed. A second post-selection was made, depending on the timing between target and ancilla photons, which is adequate only one time out of four. The resulting coincidence counts were recorded for different path lengths Δ of mode (1). The result of the experiment is shown in Figure 2.49. The number of counts recorded in

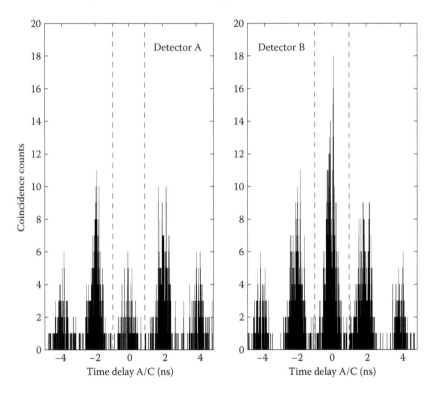

FIGURE 2.48
Typical correlation histograms taken simultaneously between detectors A/C and B/C. The central region indicated by the dashed lines correspond to the postselected events, when target and ancilla photons had such a timing that it is impossible to distinguish between them based on the time of detection. As the path length Δ varies, so does the relative size of the central peaks for detector A and B. The sum of count rates for the central peaks of detector A and B was 800/s, independently of φ as shown in Figure 2.49. (From Fattal, D., Diamanti, E., Inoue, K., and Yamamoto, Y., *Phys. Rev. Lett.*, 92 (2004) 37904. © American Physical Society.)

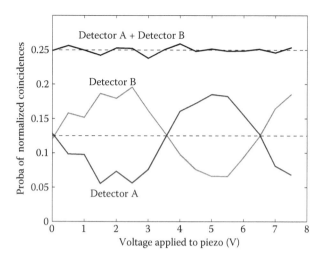

FIGURE 2.49

Verification of single mode teleportation. Coincidence counts between detector A/C and B/C are plotted for different voltages applied to the piezo transducer, i.e. for different path length Δ of mode (1). The observed modulation of the counts implies that the initial coherence contained in the target qubit was transferred to a large extent to the output qubit. The reduced contrast (~60%) is principally due to imperfect indistinguishability between the target and ancilla photons. (From Fattal, D., Diamanti, E., Inoue, K., and Yamamoto, Y., *Phys. Rev. Lett.*, 92 (2004) 37904. © American Physical Society.)

the post-selected window (−1 ns < τ < 11 ns) was normalized by the total number of counts recorded in detectors A and B in the broader window −5 ns < τ < 5 ns, corresponding to all events where one photon went through BS 1 and the other through BS 2 (but only one quarter of the time with right timing). Complementary oscillations are clearly observed at counter A and at counter B, indicating that the initial coherence was indeed transferred to the output qubit. In other words, mode (2) of the target qubit was "replaced" by mode (4) of the ancilla without a major loss of coherence.

Were the initial coherence fully conserved during the single mode transfer, the count rate at detectors A and B would be proportional to $\cos^2(\pi\Delta/\lambda)$ and $\sin^2(\pi\Delta/\lambda)$, respectively ($\lambda$ being the single-photon wavelength), giving a perfect contrast as the path length Δ is varied. More realistically, part of the coherence can be lost in the transfer, resulting in a degradation of the contrast. Such a degradation is visible in Figure 2.49. It arises mainly due to a residual distinguishability between ancilla and target photons. Slight misalignments and imperfections in the optics also result in an imperfect mode matching at BS 1 and BS 2, reducing the contrast further. Finally, the residual presence of two-photon pulses can reduce the contrast even more, although this effect is negligible here. The overlap $v(0) = |\langle \psi_t | \psi_{anc} \rangle|^2$ between target and ancilla wave-packets [4], the two-photon suppression factor $g_0^{(2)}$, as well as the nonideal mode matching at BS 1 and BS 2—characterized by the first-order interference visibilities v_1, v_2—were all measured independently, as described above. The results are $v(0) \sim 0.75$, $g_0^{(2)} \sim 2\%$, $v_1 \sim 0.92$ and $v_2 \sim 0.91$. The contrast C in counts at detector A or B when we vary the phase ϕ should then be

$$C = \frac{v(0) \cdot v_1 \cdot v_2}{1 + g_0^{(2)}/2} \sim 0.62.$$

This predicted value compares well with the experimental value of $C_{exp} \sim 0.60$.

The fidelity of teleportation is $F = (1 + C)/2 \sim 0.8$. This high value is still not enough to meet the requirements of efficient LOQC [68]. In particular, the quantum indistinguishability of the photons must be further increased to meet these requirements. As discussed above, the QD exciton from which the photon is generated dephases on a time scale of a few nanoseconds [115], which degrades the photon indistinguishability. Through further increases in the spontaneous emission rate, one can reduce the QD radiative lifetime well below this dephasing time. However, jitter in the photon emission time will eventually prevent any further reduction of the QD lifetime. Time jitter happens as a consequence of the incoherent character of our excitation method, and is currently on the order of 10 ps. Time jitter can, in principle, be completely suppressed using a coherent excitation technique described in the next section (see e.g., Kuhn et al. [11] for such a scheme with single atoms). It therefore seems vital to develop similar techniques with single QDs if they are to be used for optical QIP.

2.4.4 Coherent Single-Photon Emission and Trapping

In all the weak-coupling experiments described above, a micro-pillar cavity was used in a *passive* way, to increase the spontaneous emission rate of excited QDs, providing nearly identical photons of fixed spatio-temporal shape—a single-sided exponential pulse. Decreasing the photon lifetime helps, in this case, since it minimizes the effect of dephasing processes during the photo-emission. However this passive enhancement has its limit. There is still a 10-ps uncertainty in the precise "start" time of the photon pulse; hence, if one tries to further shorten the photon lifetime, one will make the photons distinguishable again by lack of temporal overlap. Clearly, in order to go further, another approach is needed, which does not rely on an incoherent process to excite the QDs.

In this section, we describe a theoretical scheme to produce single-photons by *coherent* optical excitation of a QD or, more generally, of an optically active quantum system. In addition to solving the time-jitter problem, this technique allows unprecedented capabilities, such as a total control over the single-photon waveform and the possibility to coherently *trap* a single-photon by running the procedure in reverse. The procedure does not require the strong-coupling regime (although strong coupling gives better performance) and works even for relatively large spectral detuning between emitters, cavities, and/or free photons. In the language of quantum information, this technique allows the coherent interconversion between photonic "flying" qubits and matter "stationary" qubits, a key element of most recent proposals for scalable quantum information processing.

We consider a general model for the coherent generation of a single-photon of arbitrary shape in an atomic-like three-level system, in a Λ configuration. We derive a single criterion for the time-varying amplitude of a single-photon pulse that determines whether a pulse of that amplitude can indeed be physically generated in a micro-pillar device of known characteristics. Usual restrictions made to understand the dynamics of these systems, such as adiabaticity, strong coupling, large or zero detuning, and symmetry between emitting and trapping sites are not necessary and reduce to special cases of the present theory. Given an arbitrary photon pulse that satisfies the criterion, the present theory explains how to drive the quantum system with a classical "control" light pulse so that it will emit a photon with a given pulse amplitude in a controlled, coherent fashion. Although the present theory makes no assumptions about the particular choice of light emitter, we specifically have in mind a semiconductor QD loaded with a single electron. In this particular system, the present theory offers the prospect of coherent emission of ultra-fast single-photons, with temporal width as short as a few picoseconds. We first present the core results of the model for ideal systems, then study the effect of losses in practical systems. The mathematical details of the model are presented last.

2.4.4.1 Coherent Photon Generation in Ideal Systems

Much like classical light pulses that are produced in lasers, single-photon pulses can have many different spatio-temporal profiles. These profiles are really photo-detection probability amplitudes [116] that are reminiscent of Schrodinger-like wavefunctions, and we will refer to them simply as "photon pulse amplitude. Photons emitted through the fluorescence of an excited two-level system [8,9,47–49], such as those used in the experiments of the previous sectionss have an amplitude which is a one-sided decaying exponential. Photon pulses with other shapes have been produced recently by driving single three-level atoms [11] or ions [117] through a stimulated Raman adiabatic passage (STIRAP) [82]. These first realizations of coherent single-photon production use quantum systems in a Λ configuration, with two ground states $|g\rangle$, $|e\rangle$ and an excited state $|r\rangle$. This will be the starting point of our analysis.

Consider the system shown in Figure 2.50. A quantum emitter with nondegenerate ground states $|g\rangle$ and $|e\rangle$ and excited state $|r\rangle$ is located in a one-sided optical microcavity. We will refer to this joint system as a "node". The $|r\rangle$ to $|g\rangle$ dipole transition has frequency ω_{rg} nearly resonant with the field of a single cavity mode (resonance frequency ω_0), with a vacuum Rabi frequency g. The cavity mode is coupled to an external radiation field, with energy decay rate denoted by κ. We control a classical laser pulse with center frequency ω_L nearly resonant with the $|e\rangle - |r\rangle$ dipole transition, and with a coherent Rabi frequency $\Omega(t)$. We assume that the control pulse and the cavity resonance have same detuning δ relative to their respective transition.

Suppose we want to *absorb* a single-photon pulse with amplitude $\psi(t)e^{-i\omega_{rg}t}$ incident on (and mode-matched to) the micro-pillar device, which is initialized in state $|g\rangle$. The complex photon pulse envelope $\psi(t)$ is known, and we want to design a control pulse $\Omega(t)$ so that the node completely absorbs the incident photon and ends up in state $|e, 0\rangle$ deterministically.

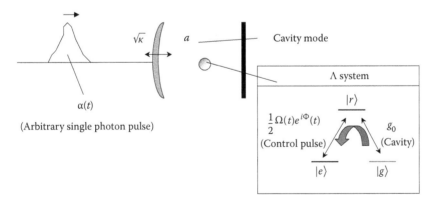

FIGURE 2.50

Composition of a "node": three-level atom or QD in a Λ configuration placed in a single mode optical microcavity. The $g-r$ transition frequency is close to the incident photon and cavity resonance, and has a vacuum Rabi frequency g with the cavity mode. The $e-r$ transition couples to a classical laser pulse called the "control pulse". This transition does not couple to the cavity, due, for instance, to polarization or frequency mismatch. The cavity mode, mathematically represented by annihilation operator a, is coupled to an external radiation mode, which gives it a finite decay rate κ. It is assumed that the cavity can exchange energy solely with that particular external mode.

The main result of this section is that this absorption process can be done perfectly if and only if the photon pulse satisfies the following condition for all times:

$$E(t) \equiv \int_{-\infty}^{t} |\psi(s)|^2 \, ds - \frac{|\psi(t)|^2}{\kappa} - \frac{1}{\kappa g^2} |\dot{\psi} - \frac{\kappa}{2} \psi(t)|^2 > 0 \qquad (2.25)$$

This relation tells us that the single-photon bandwidth cannot be much greater than κ, and that g cannot be much smaller than κ. However in general, it does not require an often-assumed adiabaticity condition ($|\dot{\psi}/\psi| \ll g$), nor does it require the regime of strong coupling ($g \gg \kappa$). We also learn that the detuning of the photon pulse relative to the cavity frequency can be as large as g. The proof of Equation 2.25 is given at the end of this section.

When this relation holds, we can always design a control pulse $\Omega(t)$ that enforces a destructive interference between two photon amplitudes: one reflected by the front mirror of the cavity and the other absorbed and re-emitted by the system (see Figure 2.51***). As a result, there is no net reflection of the incident photon. The absolute value of the required control pulse needs to be

$$\left|\frac{\Omega(t)}{2}\right| = \frac{|\ddot{\psi} - ((\kappa/2) - i\delta)\dot{\psi} + (g^2 - i\delta(\kappa/2))\psi(t)|}{\sqrt{\kappa g^2 E(t)}}. \qquad (2.26)$$

The control pulse must, in general, have a chirp to compensate both for a possible chirp in the incident photon pulse and for an eventual finite detuning δ. The exact expression for the chirp is rather complicated, and can be found in the mathematical details provided below. However if the photon envelope has no chirp, and if $\delta=0$, the control pulse does not need to have a chirp and is given by the following simple expression (also derived in Yao et al. [118] although not in closed form):

$$\frac{\Omega(t)}{2} = \frac{\ddot{\psi} - (\kappa/2)\dot{\psi} + g^2\psi(t)}{\sqrt{\kappa g^2 E(t)}}. \qquad (2.27)$$

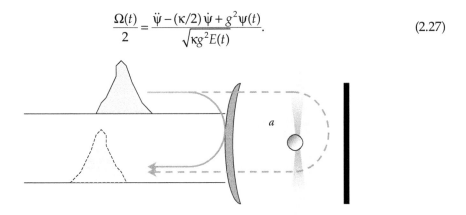

FIGURE 2.51
Physical picture of the photon trapping process. The incoming photon wave-packet can escape the node in two distinct ways: by direct reflection on the cavity front mirror (solid line) or by absorption and re-emission from the atom-cavity system (dashed line). If the control pulse applied to the Λ system is well designed, the probability amplitudes for these two events interfere destructively, and the photon stays trapped in the atom-cavity system with a probability of one.

Because of time-reversal symmetry, the solution to the photon trapping problem also gives a solution to the photon emission problem (Figure 2.52). This means that ultra-fast single-photon pulses can be produced deterministically, provided they obey the time-reversed relation (Equation 2.25). Relation 2.25 generally implies that arbitrary single-photon pulses can be perfectly emitted or trapped as long as they are not detuned from the cavity resonance ω_0 by more than g, and their bandwidth γ_p does not exceed min (κ, (g^2/κ)). This is true even in the weak coupling regime ($g \ll \kappa$). In the strong or intermediate coupling regimes, these processes can be made as fast as the cavity decay rate κ, in which case the control pulse becomes highly nonadiabatic and must be precisely designed to match the corresponding photon envelope (Figure 2.53). In the large detuning case, which was studied earlier by Cirac et al. [69], relation 2.25 can only be satisfied in the strong coupling regime, and the control pulse shape is given by

$$\left|\frac{g\Omega(t)}{2\delta}\right| = \frac{|\dot{\psi} - (\kappa/2)\psi(t)|}{\sqrt{\kappa\int_{-\infty}^{t} |\psi(s)|^2 \, ds - |\psi(t)|^2}}. \tag{2.28}$$

We write the result in this form, since $(g\Omega(t))/(2\delta)$ is the effective Raman coupling between the two ground states $|e\rangle$ and $|g\rangle$ after adiabatic elimination of level $|r\rangle$.

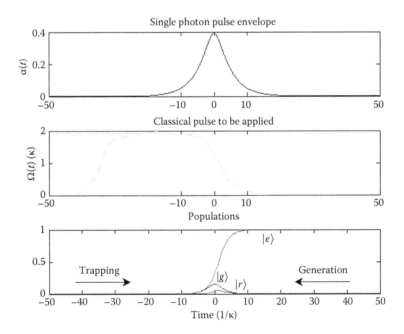

FIGURE 2.52
Control pulse $\Omega(t)$ to apply to trap a single-photon pulse of given amplitude $\psi(t)$. The bottom plot shows the simulated evolution of the state of the node with time. The parameters used in the simulation are $g = \kappa$, $\delta = 0$. Note that the control pulse is highly nonadiabatic ($\dot{\Omega}/\Omega \sim \Omega \sim g$). Looking at the plots from right to left (time-reversal) gives the solution to the photon emission problem. Modifying the control pulse in a time region where the photon amplitude is negligibly small does not significantly affect the dynamics. Hence, a control pulse that should be always "on" in the remote past can actually be turned on only a little while before the photon amplitude starts to rise (as indicated by the dashed line).

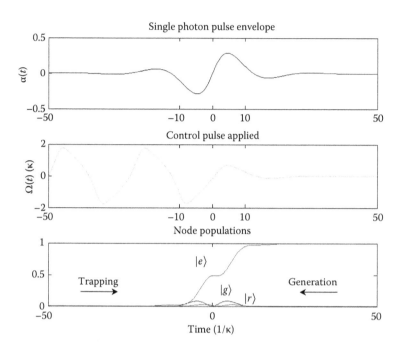

FIGURE 2.53
Generation and trapping of a single-photon pulse with oscillating amplitude in a nonadiabatic regime.

Additionally, if the incident photon has no chirp, the control pulse must have a necessary chirp $\dot{\Phi}(t)$ given by

$$\dot{\Phi}(t) = -\frac{|\Omega(t)|^2}{4\delta}. \tag{2.29}$$

Relations 2.28 and 2.29 agree with and extend the results of Cirac et al. [69], where identical nodes and time-reversal invariant photon shapes were assumed. The necessary chirp of the control pulse can be interpreted as a cancellation of the AC stark-shift induced by level $|r\rangle$ or level $|e\rangle$ via the classical control pulse.

Another limit of interest is the adiabatic regime, where the incident (emitted) photon pulse is slow enough so that $\gamma_p \ll \min[\kappa, (g^2/\kappa)]$. This is the regime in which existing coherent single-photon sources are operated. If we further assume $\delta = 0$, the control pulse has no chirp and takes the following simple form:

$$\frac{\Omega(t)}{2} = \frac{g}{\sqrt{\kappa}} \frac{\psi(t)}{\sqrt{\int_{-\infty}^{t} |\psi(s)|^2 \, ds}}. \tag{2.30}$$

Note that the denominator is the square root of the total energy carried by the photon up to time t and absorbed by the cavity. Therefore, the intensity $|\Omega(t)|^2$ of the control pulse is proportional to the instantaneous rate at which the single-photon energy enters the cavity. This result is intuitive: it means that any energy accumulated in level $|g\rangle$ from the radiation mode must be immediately transferred to level $|e\rangle$; otherwise, it will be re-radiated and lost. Equation 2.30 can be expressed equivalently as

$$\psi(t) = \frac{\sqrt{\kappa}}{g} \frac{\Omega(t)}{2} \exp\left[\frac{\kappa}{2g^2} \int_{-\infty}^{t} \left| \frac{\Omega(s)}{2} \right|^2 ds \right], \tag{2.31}$$

which gives the photon shape if we know the (adiabatic) control pulse. This result explains previous experimental observations and simulations of coherent photon emission via the STIRAP technique [11,117]. In the adiabatic regime, the emitted photon pulse follows the control pulse except for a slight correction given by the exponential term in Equation 2.31, causing the emitted photon pulse to rise more slowly and decay faster than the control pulse.

2.4.4.2 Performance of Practical Systems

Although many experimental systems implementing the present schemes can be found with stable ground states $|e\rangle$ and $|g\rangle$, they will usually suffer from decay of the cavity photon into spurious modes as well as from finite longitudinal decay and dephasing rates of the excited level $|r\rangle$. A proper study of these effects would require a master equation method, but here we provide estimates which assume that these additional processes cause only a small perturbation of the system dynamics.

Spurious cavity losses will affect the overall efficiency of the scheme. If κ_L denotes the total rate of such losses, assumed small compared to κ, both the trapping and emission schemes suffer from a reduction L_c of efficiency of approximately

$$L_c \sim \frac{\kappa_L}{\kappa}. \tag{2.32}$$

Spontaneous decay from level $|r\rangle$ at rate Γ_r further reduces the efficiency and can degrade the quantum-mechanical overlap between two photons emitted consecutively [4]. The reduction of efficiency L due to Γ_r can be estimated as

$$L \sim \frac{\Gamma_r}{g^2} \left[\frac{\kappa}{4} + \frac{1}{\kappa} \int_{-\infty}^{\infty} |\dot{\psi}(s)|^2\, ds \right], \tag{2.33}$$

and for most valid photon shapes does not exceed $(\Gamma_r \kappa)/(2g^2)$. In the photon-emission case, if level $|r\rangle$ decays spontaneously to level $|e\rangle$, emitted photons will have fluctuating shapes and their mean overlap $v(0)$ will be reduced. (If $\psi_1(t)$ and $\psi_2(t)$ are the amplitudes of two photons, their overlap $v(0)$ is defined as the quantity $\left| \int_{-\infty}^{\infty} \psi_1(t)\psi_2^*(t)dt \right|^2$). It can be shown, however, that

$$v(0) > 1 - \frac{\Gamma_r \kappa}{2g^2}, \tag{2.34}$$

and this bound is very loose. Dephasing of level $|r\rangle$ at rate Γ_d can also reduce the overlap between consecutive photons (by a factor as large as $2\Gamma_d/\gamma_p$) and, to a much lesser extent, the overall efficiency of the scheme. Note that the effect of losses and noise due to the transit through level $|r\rangle$ are greatly reduced in the strong coupling regime. Interestingly, even in the intermediate or weak coupling regimes, these effects can be small if the Purcell factor $(4g^2)/(\Gamma_r\kappa)$ is much greater than 1. This is the case of QD-cavity systems fabricated today by many groups [14,84,85,119].

Therefore, a large vacuum Rabi frequency g is not required for efficient, coherent photon generation or trapping, even in the presence of realistic experimental imperfections. This coherent scheme can reduce the impact of dephasing and finite lifetime of level $|r\rangle$, and achieves a high generation efficiency and quantum indistinguishability.

The success of this scheme also depends on the ability to initialize the system into states $|e\rangle$ or $|g\rangle$ and on the coherence lifetime of these states. In one possible implementation, using a neutral QD, states $|g\rangle$ and $|e\rangle$ are assigned to the bright exciton states, and $|r\rangle$ is the biexciton state. In this case initialization requires a π pulse to excite the QD from its empty (ground) state to one of the bright exciton states. The coherence lifetime of the bright exciton states is short (a few hundred picoseconds at best), causing serious degradation. A better implementation is to use a charged QD or a charged impurity such as a shallow donor. In singly charged QDs, a three-level Λ system (a system with an excited state coupled by optical transitions to two ground states) can be obtained by applying a magnetic field perpendicular to the growth direction of the QD [52]. In this case $|e\rangle$ and $|g\rangle$ are assigned to the two spin sublevels of the single electron, and $|r\rangle$ is one of the trion states having two electrons and one hole. State preparation with fidelity up to 99.8% has recently been reported in this system using optical pumping [120]. In III–V semiconductors, the electron spin coherence lifetime is expected to be ~1–10 ns, limited by the random effective magnetic field produced by nuclear spins in the lattice. Since the nuclear spins fluctuate on a long timescale, a much longer electron spin coherence lifetime of $T_2 \sim 1$ μs should be possible using spin echo techniques; see Yao et al. [121] for a recent discussion of electron spin coherence in QDs. These lifetimes are long enough to have little effect on the photon generation and trapping schemes described here. However, this electron spin T_2 sets the limit on how long quantum information can be stored.

2.4.4.3 Performance as a Single-photon Source

The coherent photon-trapping scheme presented here allows the deterministic generation of single-photon pulses with duration as short as κ^{-1}, even in the intermediate coupling regime (or κ/g^2 in the weak-coupling regime, Figure 2.54). With currently available QDs in micropillars, we expect theoretically that single-photon pulses with temporal width of a few picoseconds can be reliably produced, without time-jitter, with efficiency greater than 99%, and quantum-mechanical overlap of two consecutive photons as high as 98%. These figures could be further improved by increasing the vacuum Rabi frequency g and the cavity decay rate κ with respect to Γ_r. A new feature predicted by the present theory is the complete control of the photon pulse amplitude $\psi(t)$, as long as the condition (Equation 2.25) is satisfied. This feature gives access to a new degree of freedom to store and manipulate quantum information in single-photon pulses, never exploited in the past. An immediate application is a differential phase shift quantum key distribution (DPSQKD) protocol based on single-photon pulses [122]. This protocol requires single-photon pulses that are linear superpositions of elementary pulses well separated in time, with random 0 or π phase shifts between them. An example with two elementary pulses with relative phase shift of π is given in Figure 2.55.

2.4.4.4 Mathematical Details of the Theory

Next, we give a mathematical proof of the main result of the photon trapping theory, that, given an incident single-photon pulse with complex amplitude $\psi(t)$ satisfying Equation 2.25, there is a corresponding control pulse $\Omega(t) = |\Omega(t)e^{i\Phi(t)}|$ that can be applied to the $|r\rangle - |e\rangle$ transition to allow perfect trapping of the photon pulse.

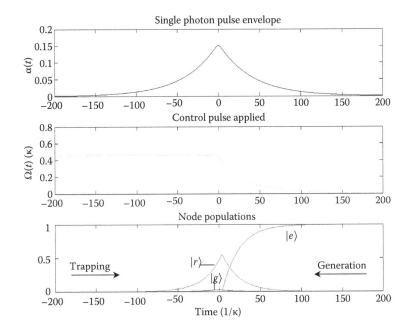

FIGURE 2.54
Deterministic photon trapping/generation in the weak-coupling regime. The parameters are $g/\kappa=0.1$ and $\delta=0$. The photon pulse cannot be as fast as in the intermediate or strong coupling cases, and the upper limit on the photon bandwidth is g^2/κ, instead of κ. Level $|r\rangle$ is also significantly more populated, causing increased loss by spontaneous emission ($\sim$22 Γ_r/κ here if the loss is small) and degradation of consecutive photon overlap. These undesired effects could still be small in a realistic QD-cavity system with $\kappa\sim(1\ \text{ps})^{-1}$, $g\sim(10\ \text{ps})^{-1}$ and $\Gamma_r\sim(1\ \text{ns})^{-1}$, for instance, where the quantum efficiency would be 98% and overlap greater than 95%.

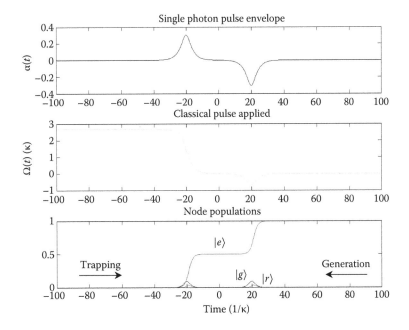

FIGURE 2.55
Generation and trapping of a composite single-photon pulse for differential phase shift QKD. The pulse is a superposition of two elementary pulses well separated in time and with a π phase shift between them.

The photon pulse $\psi(t)$ is incoming on a one-sided cavity with decay rate κ. We assume the photon is in a spatial traveling mode perfectly matched with a single confined mode of the cavity, described by annihilation operator a. Inside the cavity, there is a Λ-type quantum system with two metastable states $|g\rangle$ and $|e\rangle$ and an excited state $|r\rangle$. The transition $|g\rangle - |r\rangle$ is coupled to the vacuum field of the confined cavity mode, with vacuum Rabi frequency g. The $|r\rangle - |e\rangle$ transition is coupled to the control pulse $\Omega(t)$. We will assume an ideal system where none of the levels decay by spontaneous emission.

The output field of the cavity is related to the input field and the confined field by the following relation [123] in the Heisenberg picture :

$$a_{\text{in}}(t) + a_{\text{out}}(t) = \sqrt{\kappa}\, a(t). \tag{2.35}$$

This relation makes the implicit assumption that the coupling of the cavity mode to different longitudinal modes of the waveguide is the same in a frequency range κ around the cavity resonance. This condition might not be satisfied for photonic crystal structures unless designed carefully (e.g., one should avoid coupling near the waveguide band edge). The Heisenberg equation of motion for the cavity mode a is

$$\frac{da}{dt} = i[H_\Lambda, a] - \frac{\kappa}{2}a + \sqrt{\kappa}\, a_{\text{in}} \tag{2.36}$$

The first term represents coherent evolution of the confined field due to its coupling to the Λ system. The second and third terms describe damping and associated "noise" due to coupling of the cavity photon to a continuum of radiative modes outside of the cavity.

We treat the coupled atom-cavity system as an entity called a "node", interacting with the external radiation field. We define a nonnormalized wave-function $\psi(t)$ for the excited node, i.e., for a node that has absorbed a photon:

$$\Psi(t) = G(t)|g, 1\rangle + r(t)|r, 0\rangle + e(t)|e, 0\rangle \tag{2.37}$$

where $|X, n\rangle$ represents the state X of the Λ system and the number n of photon inside the cavity. The state $\psi(t)$ evolves according to:

$$\dot{G} = -ig\,r - \frac{\kappa}{2}G + \sqrt{\kappa}\,\psi(t), \tag{2.38}$$

$$\dot{r} = -i\delta r - ig\,G - i\frac{\Omega}{2}e, \tag{2.39}$$

$$\dot{e} = -i\frac{\Omega^*}{2}r, \tag{2.40}$$

where we used the rotating-wave approximation. The carrier frequency for the photon pulse is taken to be ω_c, the resonant frequency of the cavity. We assumed that the two-photon resonance condition is satisfied, and δ is the common detuning of the laser and cavity field from the excited level $|r\rangle$. From these equations one can show that

$$|G(t)|^2 + |r(t)|^2 + |e(t)|^2 = \int\limits_{-\infty}^{t} |\psi(s)|^2\,ds - \int\limits_{-\infty}^{t} |\sqrt{\kappa}G(s) - \psi(s)|^2\,ds. \tag{2.41}$$

This quantity represents the energy absorbed in the atom-cavity system from the incoming photon pulse up to time t. Note that it depends on the detuning implicitly through the amplitude $G(t)$. We therefore always have the intuitive result that

$$|G(t)|^2 + |r(t)|^2 + |e(t)|^2 \leq \int\limits_{-\infty}^{t} |\psi(s)|^2\,ds. \tag{2.42}$$

Moreover, the inequality is strict unless

$$\forall s, \sqrt{\kappa}\,G(s) = \psi(s). \tag{2.43}$$

If and only if Equation 2.43 is satisfied, the system absorbs the photon perfectly, and is left in state $|e, 0\rangle$. Indeed, as $t \to +\infty$, $\psi(t) \to 0$, therefore $G(t) \to 0$. Then, in virtue of relation 2.38, $r(t) \to 0$. Then, $e(t) \to \int_{-\infty}^{\infty} |\psi(s)|^2\,ds = 1$.

The condition (Equation 2.43) for perfect trapping can be seen as an impedance matching condition and has a straightforward physical interpretation. If we apply Equation 2.35 to the initial state $|\Psi_0\rangle$ of the system, and re-adopt the state-varying (Schrödinger) picture, we find that:

$$\langle \text{vac}|a_{\text{out}}(t)|\Psi_0\rangle = \left(\sqrt{\kappa}G(t) - \psi(t)\right). \tag{2.44}$$

If this impedance-matching condition is satisfied, a perfect destructive interference occurs at all times between the quantum amplitude of the photon pulse directly reflected off of the front cavity mirror and the amplitude of the photon absorbed and then re-emitted from the node. In this case, no light escapes from the cavity.

If Equation 2.43 is not satisfied at all times due to experimental imperfection (e.g., spurious fluctuation of control pulse) or to an incompatible photon pulse (e.g., with a sharp temporal feature violating relation 2.25 for some time), then this impedance mismatch causes losses L_{imp} equal to

$$L_{\text{imp}} = \int\limits_{-\infty}^{+\infty} |\sqrt{\kappa}G(s) - \psi(s)|^2\,ds. \tag{2.45}$$

In designing a control pulse, we are looking for a classical pulse $\Omega(t)$ that imposes the impedance matching condition 2.43 on the system. If such a classical pulse exists, it must satisfy the following equations:

$$\psi = \sqrt{\kappa}G, \tag{2.46}$$

$$\dot{G} = -igr + \frac{\kappa}{2}G, \tag{2.47}$$

$$\dot{r} = -i\delta r - igG - i\frac{\Omega}{2}e,$$ (2.48)

$$\dot{e} = -i\frac{\Omega^*}{2}r.$$ (2.49)

This leads to the existence of a solution given by:

$$|e|^2 = \int^t |\psi|^2 - \frac{\Gamma_r}{\kappa g^2}\left[\int^t |\psi|^2\right]\left[\int_t^\infty |\psi|^2\right]$$

$$- \frac{|\psi|^2}{\kappa}\left[1 + \frac{\Gamma_r\kappa}{2g^2}\right] - \frac{|\dot{\psi} - (\kappa/2)\psi|^2}{\kappa g^2},$$ (2.50)

$$\dot{\Phi}_e = \frac{|r|^2(\dot{\Phi}_r + \delta) - |G|^2 \dot{\Phi}_g}{|e|^2}.$$ (2.51)

This solution exists if and only if the right hand side of Equation 2.50 is positive at all times, which leads to the criterion (Equation 2.25) announced earlier. If this is true, then $|\Omega(t)|$ is bounded on any compact time interval and therefore is a maximal solution of the above system of differential equations for all times in virtue of the Cauchy–Lipschitz theorem [124].

Finally, we present solutions for the control pulse $\Omega(t)$ in two special cases. In the large-detuning limit, $\Delta \gg \kappa, g, |d\log(\psi)/dt|, |\Omega|, |d\log(\Omega)/dt|$. In this case level $|r\rangle$ has negligible population and can be adiabatically eliminated from the dynamics, reducing the three-level problem to a simpler two-level one, with an effective laser induced coupling $\Omega_{\text{eff}}(t) = g\Omega(t)/\Delta$. In this case, the trapping control pulse must be

$$\left|\frac{\Omega(t)}{2}\right| = \frac{|\delta||\dot{\psi} - (\kappa/2)\psi|}{\sqrt{\kappa g^2 \int_{-\infty}^t |\psi(s)|^2\, ds - g^2 |\psi(t)|^2 - |\dot{\psi} - \frac{\kappa}{2}\psi|^2}}.$$ (2.52)

A solution consistent with the condition $\delta \gg \Omega$ can be found only in the strong coupling regime where $g \gg \kappa$. In the strong coupling regime this solution simplifies to

$$\left|\frac{\Omega_{\text{eff}}(t)}{2}\right| = \frac{|\dot{\psi} - (\kappa/2)\psi|}{\sqrt{\kappa \int_{-\infty}^t |\psi(s)|^2\, ds - |\psi(t)|^2}},$$ (2.53)

with the phase of the control pulse given by

$$\dot{\Phi} = -\frac{|\Omega|^2}{4\delta},$$ (2.54)

if the single-photon pulse has no chirp (constant phase). This last relation can be simply interpreted as a compensation of the AC Stark shift created by the far detuned control pulse on the e–r transition, as was already noted by Cirac et al. [69]. In this large-detuning regime, fast photon pulses can still be trapped or emitted, the bandwidth being limited by the cavity decay κ only. However, strong coupling is necessary.

In the zero-detuning case the control pulse must have amplitude

$$\left|\frac{\Omega(t)}{2}\right|^2 = \frac{\left|\ddot{\psi} - (\kappa/2)\dot{\psi} + g^2\psi\right|^2}{\kappa g^2 \int_{-\infty}^{t} |\psi(s)|^2\,ds - g^2\,|\psi(t)|^2 - |\dot{\psi} - (\kappa/2)\psi|^2}. \tag{2.55}$$

If the single-photon pulse has no chirp, then neither does $\Omega(t)$. Also, we no longer need strong coupling for the denominator to be positive, and perfect trapping can occur even if $g \sim \kappa$.

A special case of interest is that of a slow photon pulse ($|\dot{\psi}/\psi| \ll \kappa$) in the strong-coupling regime $g \gg \kappa$. In this case we obtain a simple result:

$$\left|\frac{\Omega(t)}{2}\right|^2 \sim \frac{g^2\,|\psi(t)|^2}{\int_{-\infty}^{t} |\psi(s)|^2\,ds\,.\kappa}. \tag{2.56}$$

This can be differentiated to solve for the photon amplitude in terms of the control pulse:

$$|\psi(t)|^2 = \frac{\kappa}{g^2}\left|\frac{\Omega(t)}{2}\right|^2 \exp\left[\frac{\kappa}{2g^2}\int_{t}^{+\infty}\left|\frac{\Omega(s)}{2}\right|^2\,ds\right], \tag{2.57}$$

which is convenient for comparing the envelopes of the control and photon pulses. These results explain previous experimental observations and simulations of coherent photon emission via the STIRAP technique [11,117]. In the adiabatic setting used in these experiments, the emitted photon pulse "follows" the control pulse except for a slight correction given by the exponential term in Equation 2.57, causing the emitted photon pulse to rise more slowly and decay faster than the control pulse.

2.5 Conclusions

We have shown that single QDs in pillar microcavities can serve as high-quality sources of single-photons. Through advances in design and processing, we have been able to achieve strongly modified spontaneous-emission rates, efficient photon collection, and strong coupling between QD excitons and cavity photons. This high performance has enabled basic applications such as improved quantum cryptography, entanglement generation, and quantum teleportation. Further advances in such microcavity single-photon sources are expected to lead to a range of quantum technologies.

The entangled-photon source can be used, for example, in a BBM92 quantum-key-distribution system. If combined with a low-dark-count superconducting single-photon detector [125], channel loss of approximately 35 dB can be tolerated, so that a satellite-based, global QKD system can eventually be constructed.

Replacement of the self-assembled QD with a substitutional donor impurity may solve the problem of the large inhomogeneous linewidth of QDs. Indeed, donor-bound excitons in a high-purity GaAs bulk sample feature a linewidth of 1–10 GHz, which is close to the natural linewidth (radiative decay time) limit [126]. The control of emission wavelengths of many single-photon sources is an indispensible step toward the application to a scalable single-photon-based system. An electron spin is inherently provided by an unpaired bound electron of a neutral donor, and sometimes a single nuclear spin is naturally available as the donor atom itself. ^{19}F:ZnSe is such a promising material example, but stable isolation and precise positioning of a single impurity in a high-Q monolithic microcavity is yet to be demonstrated.

To produce massively parallel single-photons, a coupled array of microcavities with a single or few "atoms" might be used [127,128]. The performance of such a system has been modeled by the Bose–Hubbard Hamiltonian, and the quantum phase transition from a superfluid state to a Mott insulator state of cavity polaritons achieves one and only one photon generation from each cavity.

A single-photon source based on a wide-bandgap material may achieve a high operation temperature, possibly up to room temperature. A GaN-based single-photon source was successfully demonstrated at 200 K already [129]. Such a source can in principle be combined with an efficient single-photon frequency converter based on PPLN waveguides to obtain single-photons at any desired wavelength.

A single-photon source is also needed for efficient implementation of quantum repeaters. A quantum repeater based on coherent-state pulses enjoys a high success probability of generating an initial entangled state but has a limited fidelity due to photon loss in the channel [130]. A quantum repeater based on single-photon detection enjoys a high initial-state fidelity but has a low success probability [131]. If the coherent-state pulse in the first scheme is replaced by a single-photon pulse, we expect that high success probability and high fidelity can be simultaneously achieved.

References

1. R. J. Glauber. The quantum theory of optical coherence. *Physical Review*, 130: 2529–2539, 1963.
2. E. Waks, C. Santori, and Y. Yamamoto. Security aspects of quantum key distribution with sub-Poisson light. *Phys. Rev. A*, 66(4): 042315, 2002.
3. E. M. Purcell. Spontaneous emission probabilities at radio frequencies. *Phys. Rev.*, 69: 681, 1946.
4. C. Santori, D. Fattal, J. Vučković, G. S. Solomon, and Y. Yamamoto. Indistinguishable photons from a single-photon device. *Nature (London)*, 419(6907): 594–597, 2002.
5. A. Imamoğlu and Y. Yamamoto. Noise suppression in semiconductor p-i-n junctions: Transition from macroscopic squeezing to mesoscopic Coulomb blockade of electron emission processes. *Phys. Rev. Lett.*, 70(21): 3327–3330, 1993.
6. A. Imamoğlu and Y. Yamamoto. Turnstile device for heralded single-photons: Coulomb blockade of electron and hole tunneling in quantum confined p-i-n heterojunctions. *Phys. Rev. Lett.*, 72(2): 210–213, 1994.

7. J. Kim, O. Benson, H. Kan, and Y. Yamamoto. A single-photon turnstile device. *Nature (London)*, 397(6719): 500–503, 1999.

8. P. Michler, A. Kiraz, C. Becher, W. V. Schoenfeld, P. M. Petroff, L. Zhang, E. Hu, and A. Imamoğlu. A quantum dot single-photon turnstile device. *Science*, 290: 2282–2285, 2000.

9. C. Santori, M. Pelton, G. Solomon, Y. Dale, and Y. Yamamoto. Triggered single-photons from a quantum dot. *Phys. Rev. Lett.*, 86(8): 1502–1505, 2001.

10. M. Hennrich, T. Legero, A. Kuhn, and G. Rempe. Vacuum-stimulated Raman scattering based on adiabatic passage in a high finesse optical cavity. *Phys. Rev. Lett.*, 85: 4872–4875, 2000.

11. A. Kuhn, M. Heinrich, and G. Rempe. Deterministic single-photon source for distributed quantum networking. *Phys. Rev. Lett.*, 89: 067901, 2002.

12. M. O. Scully and M. S. Zubairy. *Quantum Optics*. Cambridge University Press, Cambridge, UK, 1997.

13. E. Siegman. *Lasers*. University Science Books, Mill Valley, CA, 1986.

14. J. P. Reithmaier, G. Sęk, A. Löffler, C. Hofmann, S. Kuhn, S. Reitzenstein, L. V. Keldysh, V. D. Kulakovskii, T. L. Reinecke, and A. Forchel. Strong coupling in a single quantum dot-semiconductor microcavity system. *Nature*, 432: 197–200, 2004.

15. M. Pelton, J. Vučković, G. S. Solomon, A. Scherer, and Y. Yamamoto. Three-dimensionally confined modes in micropost microcavities: Quality factors and Purcell factors. *IEEE J. Quantum Electron.*, 38(2): 170–177, 2002.

16. J. Vučković, M. Pelton, A. Scherer, and Y. Yamamoto. Optimization of three-dimensional micropost microcavities for cavity quantum electrodynamics. *Phys. Rev. A*, 66: 023808, 2002.

17. J. Goh, I. Fushman, D. Englund, and J. Vučković. Genetic optimization of photonic bandgap structures. *Optics Express*, 15: 8218–8230, 2007.

18. D. Englund, I. Fushman, and J. Vučković. General recipe for designing photonic crystal cavities. *Optics Express*, 13: 5961–5975, 2005.

19. W. A. Tiller. *The Science of Crystallization: Microscopic Interfacial Phenomena*. Cambridge University Press, Cambridge, UK, 1991.

20. H. Gao. A boundary perturbation analysis for elastic inclusions and interfaces. *Int. J. Solids Structures*, 28: 703–725, 1991.

21. L. B. Freund and F. Jonsdottir. Instability of a biaxially stressed thin film on a substrate due to material diffusion over its free surface. *J. Mech. Phys. Solid.*, 41(7): 1245–1264, 1993.

22. B. G. Orr, D. Kessler, C. W. Synder, and L. Sander. A model of strain-induced roughening and coherent island growth. *Europhys. Lett.*, 19(1): 33–38, 1992.

23. D. Vanderbilt and L. K. Wickham. Elastic energies of coherent germanium islands on silicon. *Mat. Res. Symp. Proc.*, 202: 555, 1991.

24. R. People and J. C. Bean. Calculation of critical layer thickness versus lattice mismatch for GeSi/Si strained-layer heterostructures. *Appl. Phys. Lett.*, 47: 322–324, 1985.

25. J. W. Matthews and A. E. Blakeslee. Defects in epitaxial multilayers. I. Misfit dislocations. *J. Cryst. Growth*, 27(1): 118–125, 1974.

26. N. Stranski and L. Krastanow. Akad. Wiss. Wien Math. Naturwiss. Kl. Abt. 2B Chemie, 146: 797, 1937.

27. G. W. Bryant and G. S. Solomon. *Optics of Quantum Dots and Wires*. Artech House Inc., Boston, MA, 2005.

28. G. S. Solomon, J. A. Trezza, and J. S. Harris Jr. Substrate temperature and monolayer coverage effects on epitaxial ordering of InAs and InGaAs islands on GaAs. *Appl. Phys. Lett.*, 66: 991–993, 1995.

29. G. S. Solomon, J. A. Trezza, and J. S. Harris Jr. Effects of monolayer coverage, flux ratio, and growth rate on the island density of InAs islands on GaAs. *Appl. Phys. Lett.*, 66: 3161–3163, 1995.

30. D. Leonard, K. Pond, and P. M. Petroff. Critical layer thickness for self-assembled InAs islands on GaAs. *Phys. Rev. B*, 50: 11687–11692, 1994.

31. W. Wu, J. R. Tucker, G. S. Solomon, and J. S. Harris. Jr. Atom-resolved scanning tunneling microscopy of vertically ordered InAs quantum dots. *Appl. Phys. Lett.*, 71: 1083–1085, 1997.

32. V. D. Kulakovskii, G. Bacher, R. Weigand, T. Kúmmell, A. Forchel, E. Borovitskaya, K. Leonardi, and D. Hommel. Fine structure of biexciton emission in symmetric and asymmetric CdSe/ZnSe single quantum dots. *Phys. Rev. Lett.*, 82: 1780–1783, 1999.

33. S. A. Campbell. *The Science and Engineering of Microelectronic Fabrication*. Oxford University Press, Oxford, UK, 1996.

34. S. Solomon, M. Pelton, and Y. Yamamoto. Single-mode spontaneous emission from a single quantum dot in a three-dimensional microcavity. *Phys. Rev. Lett.*, 86: 3903–3906, 2001.

35. G. Panzarini and L. C. Andreani. Quantum theory of exciton polaritons in cylindrical semiconductor microcavities. *Phys. Rev. B*, 60: 16799–16806, 1999.

36. G. Björk, H. Heitmann, and Y. Yamamoto. Spontaneous-emission coupling factor and mode characteristics of planar dielectric microcavity lasers. *Phys. Rev. A*, 47: 4451–4463, 1993.

37. A. Scherer, J. L. Jewell, Y. H. Lee, J. P. Harbison, and L. T. Florez. Fabrication of microlasers and microresonator optical switches. *Appl. Phys. Lett.*, 55: 2724–2726, 1989.

38. R. E. Howard. Multilevel resist for lithography below 100 nm. *IEEE Trans. Electron Dev.*, ED-28: 1378–1381, 1981.

39. T. Rivera, J.-P. Debray, J. M. Gérard, B. Legrand, L. Manin-Ferlazzo, and J. L. Oudar. Optical losses in plasma-etched AlGaAs microresonators using reflection spectroscopy. *Appl. Phys. Lett.*, 74: 911–913, 1999.

40. P. Levy, M. Bianconi, and L. Correra. Wet etching of Al_2O_3 for selective patterning of microstructures using Ar^+ ion implantation and H_3PO_4. *J. Electrochem. Soc.*, 145: 344–347, 1998.

41. J.Y. Marzin, J. M. Gérard, A. Izrael, D. Barrier, and G. Bastard. Photoluminescence of single InAs quantum dots obtained by self-organized growth on GaAs. *Phys. Rev. Lett.*, 73: 716, 1994.

42. G. S. Solomon, M. Pelton, and Y. Yamamoto. Response to comment on 'Single-mode spontaneous emission from a single quantum dot in a three-dimensional microcavity'. *Phys. Rev. Lett.*, 90: 229702, 2003.

43. J. Vučković, D. Fattal, C. Santori, G. Solomon, and Y. Yamamoto. Enhanced single-photon emission from a quantum dot in a micropost microcavity. *Appl. Phys. Lett.*, 82: 3596–3598, 2003.

44. A. Kiraz, P. Michler, C. Becher, B. Gayral, A. Imamoğlu, L. Zhang, E. Hu, W. V. Schoenfeld, and P. M. Petroff. Cavity-quantum electrodynamics using a single InAs quantum dot in a microdisk structure. *Appl. Phys. Lett.*, 78: 3932–3934, 2001.

45. M. Gérard, B. Sermage, B. Gayral, B. Legrand, E. Costard, and V. Thierry-Mieg. Enhanced spontaneous emission by quantum boxes in a monolithic optical microcavity. *Phys. Rev. Lett.*, 81: 1110–1113, 1998.

46. C. Santori, G. S. Solomon, M. Pelton, and Y. Yamamoto. Time-resolved spectroscopy of multiexcitonic decay in an InAs quantum dot. *Phys. Rev. B*, 65(7): 073310, 2002.

47. V. Zwiller, H. Blom, P. Jonsson, N. Panev, S. Jeppesen, T. Tsegaye, E. Goobar, M.-E. Pistol, L. Samuelson, and G. Björk. Single quantum dots emit single-photons at a time: Antibunching experiments. *Appl. Phys. Lett.*, 78: 2476–2478, 2001.

48. E. Moreau, I. Robert, J. M. Gérard, I. Abram, L. Manin, and V. Thierry-Mieg. Single-mode solid-state single-photon source based on isolated quantum dots in pillar microcavities. *Appl. Phys. Lett.*, 79: 2865–2867, 2001.

49. Z. Yuan, B. E. Kardynal, R. M. Stevenson, A. J. Shields, C. J. Lobo, K. Cooper, N. S. Beattie, D. A. Ritchie, and M. Pepper. Electrically driven single-photon source. *Science*, 295: 102–105, 2002.

50. A.J. Bennett, D. C. Unitt, P. See, A. J. Shields, P. Atkinson, K. Cooper, and D. A. Ritchie. Microcavity single-photon-emitting diode. *Appl. Phys. Lett.*, 86: 181102, 2005.

51. R.M. Thompson, R.M. Stevenson, A.J. Shields, I. Farrer, C.J. Lobo, D.A. Ritchie, M.L. Leadbeater, and M. Pepper. Single-photon emission from exciton complexes in individual quantum dots. *Phys. Rev. B*, 64(20): 201302, 2001.

52. A. Imamoğlu, G. Burkard D.D. Awschalom, D.P. DiVincenzo, D. Loss, M. Sherwin, and A. Small. Quantum information processing using quantum dot spins and cavity QED. *Phys. Rev. Lett.*, 83(20): 4204–4207, 1999.

53. H. Kamada, H. Ando, J. Temmyo, and T. Tamamura. Excited-state optical transitions of excitons and biexcitons in a single InGaAs quantum disk. *Phys. Rev. B*, 58: 16243, 1998.

54. C. Santori, D. Fattal, J. Vučković, G. S. Solomon, and Y. Yamamoto. Single-photon generation with InAs quantum dots. *New J. Phys.*, 6: 89, 2004.

55. C. Santori, D. Fattal, J. Vučković, G. S. Solomon, E. Waks, and Y. Yamamoto. Submicrosecond correlations in photoluminescence from InAs quantum dots. *Phys. Rev. B*, 69(20): 205324, 2004.

56. Al. L. Efros and M. Rosen. Random telegraph signal in the photoluminescence intensity of a single quantum dot. *Phys. Rev. Lett.*, 78: 1110, 1997.

57. M.-E. Pistol, P. Castrillo, D. Hessman, J. A. Prieto, and L. Samuelson. Random telegraph noise in photoluminescence from individual self-assembled quantum dots. *Phys. Rev. B*, 59: 10725, 1999.

58. M. Sugisaki, H. W. Ren, K. Nishi, and Y. Masumoto. Fluorescence intermittency in self-assembled InP quantum dots. *Phys. Rev. Lett.*, 86: 4883, 2001.

59. D. Bertram, M. C. Hanna, and A. J. Nozik. Two color blinking of single strain-induced GaAs quantum dots. *Appl. Phys. Lett.*, 74: 2666, 1999.

60. R. G. Neuhauser, K. T. Shimizu, W. K. Woo, S. A. Empedocles, and M. G. Bawendi. Correlation between fluorescence intermittency and spectral diffusion in single semiconductor quantum dots. *Phys. Rev. Lett.*, 85: 3301, 2000.

61. H. D. Robinson and B. B. Goldberg. Light-induced spectral diffusion in single self-assembled quantum dots. *Phys. Rev. B*, 61: R5086, 2000.

62. T. Aichele, V. Zwiller, and O. Benson. Visible single-photon generation from semiconductor quantum dots. *New J. Phys.*, 6: 90, 2004.

63. M.B. Ward, O. Z. Karimov, D. C. Unitt, Z. L. Yuan, P. See, D. G. Gevaux, A. J. Shields, P. Atkinson, and D. A. Ritchie. On-demand single-photon source for 1.3 μm telecom fiber. *Appl. Phys. Lett.*, 86: 201111, 2005.

64. Y. Toda, O. Moriwaki, M. Nishioka, and Y. Arakawa. Efficient carrier relaxation mechanism in InGaAs/GaAs self-assembled quantum dots based on the existence of continuum states. *Phys. Rev. Lett.*, 82: 4114, 1999.

65. M. Pelton, C. Santori, J. Vučković, B. Zhang, G. S. Solomon, J. Plant, and Y. Yamamoto. Efficient source of single-photons: A single quantum dot in a micropost microcavity. *Phys. Rev. Lett.*, 89: 233602, 2002.

66. D. Bouwmeester, J. W. Pan, K. Mattle, M. Eibl, H. Weinfurter, and A. Zeilinger. Experimental quantum teleportation. *Nature (London)*, 390: 575–579, 1997.

67. Y. H. Shih and C. O. Alley. New type of Einstein-Podolsky-Rosen-Bohm experiment using pairs of light quanta produced by optical parametric down conversion. *Phys. Rev. Lett.*, 61: 2921–2924, 1988.

68. E. Knill, R. Laflamme, and G. J. Milburn. A scheme for efficient quantum computation with linear optics. *Nature (London)*, 409: 46–52, 2001.

69. J. I. Cirac, P. Zoller, H. J. Kimble, and H. Mabuchi. Quantum state transfer and entanglement distribution among distant nodes in a quantum network. *Phys. Rev. Lett.*, 78: 3221–3224, 1997.

70. A. Bennett, D. Unitt, A. Shields, P. Atkinson, and D. Ritchie. Influence of exciton dynamics on the interference of two photons from a microcavity single-photon source. *Opt. Express*, 13(20): 7772–7778, 2005.

71. C. K. Hong, Z. Y. Ou, and L. Mandel. Measurement of subpicosecond time intervals between two photons by interference. *Phys. Rev. Lett*, 59: 2044–2046, 1987.

72. H. Fearn and R. Loudon. Theory of two-photon interference. *J. Opt. Soc. Am. B*, 6: 917–927, 1989.

73. T. Legero, T. Wilk, M. Hennrich, G. Rempe, and A. Kuhn. Quantum beat of two single-photons. *Phys. Rev. Lett.*, 93(7): 070503, 2004.

74. S. Laurent, S. Varoutsis, L. Le Gratiet, A. Lemaître, I. Sagnes, F. Raineri, A. Levenson, I. Robert-Philip, and I. Abram. Indistinguishable single-photons from a single-quantum dot in a two-dimensional photonic crystal cavity. *Appl. Phys. Lett.*, 87: 163107, 2005.

75. P. Maunz, D.L. Moehring, S. Olmschenk, K.C. Younge, D.N. Matsukevich, and C. Monroe, Quantum interference of photon pairs from two remote trapped atomic ions. *Nature Physics*, 3 (8): 538–541, 2007.

76. M. Bayer and A. Forchel. Temperature dependence of the exciton homogeneous linewidth in $In_{0.60}Ga_{0.40}As/GaAs$ self-assembled quantum dots. *Phys. Rev. B*, 65: 041308(R), 2002.

77. P. Borri, W. Langbein, S. Schneider, U. Woggon, R. L. Sellin, D. Ouyang, and D. Bimberg. Ultralong dephasing time in InGaAs quantum dots. *Phys. Rev. Lett.*, 87(15): 157401, 2001.

78. D. Gammon, E. S. Snow, B. V. Shanabrook, D. S. Katzer, and D. Park. Homogeneous linewidths in the optical spectrum of a single gallium arsenide quantum dot. *Science*, 273: 87–90, 1996.

79. C. Kammerer, G. Cassabois, C. Voisin, M. Perrin, C. Delalande, Ph. Roussignol, and J. M. Gérard. Interferometric correlation spectroscopy in single quantum dots. *Appl. Phys. Lett.*, 81: 2737, 2002.

80. X. Fan, T. Takagahara, J. E. Cunningham, and H. Wang. Pure dephasing induced by exciton-phonon interactions in narrow GaAs quantum wells. *Solid State Commun.*, 108(11): 857–861, 1998.

81. L. Besombes, K. Kheng, L. Marsal, and H. Mariette. Acoustic phonon broadening mechanism in single quantum dot emission. *Phys. Rev. B*, 63(15): 155307, 2001.

82. A. Kuhn, M. Hennrich, T. Bondo, and G. Rempe. Controlled generation of single-photons from a strongly coupled atom-cavity system. *Appl. Phys. B*, 69: 373–377, 1999.

83. A. Kiraz, M. Atature, and A. Imamoğlu. Quantum-dot single-photon sources: Prospects for applications in linear optics quantum-information processing. *Phys. Rev. A*, 69(3): 032305, 2004.

84. E. Peter, P. Senellart, D. Martrou, A. Lemaitre, J. Hours, J. M. Gerard, and J. Bloch. Exciton-photon strong-coupling regime for a single quantum dot embedded in a microcavity. *Phys. Rev. Lett.*, 95: 067401, 2005.

85. T. Yoshie, A. Scherer, J. Hendrickson, G. Khitrova, H. M. Gibbs, G. Rupper, C. Ell, O. B. Shchekin, and D. G. Deppe. Vacuum Rabi splitting with a single quantum dot in a photonic crystal nano-cavity. *Nature (London)*, 432: 200–203, 2004.

86. K. Hennessy, A. Badolato, M. Winger, D. Gerace, M. Atature, S. Gulde, S. Falt, E. L. Hu, and A. Imamoğlu. Quantum nature of a strongly coupled single quantum dot-cavity system. *Nature (London)*, 445(7130): 896–899, 2007.

87. D. Press, S. Götzinger, S. Reitzenstein, C. Hofmann, A. Löffler, M. Kamp, A. Forchel, and Y. Yamamoto. Photon antibunching from a single quantum dot-microcavity system in the strong coupling regime. *Phys. Rev. Lett.*, 98(23): 117402, 2007.

88. A. Löffler, J. P. Reithmaier, G. Sek, C. Hofmann, S. Reitzenstein, M. Kamp, and A. Forchel. Semiconductor quantum dot microcavity pillars with high-quality factors and enlarged dot dimensions. *Appl. Phys. Lett.*, 86: 111105, 2005.

89. C. Weisbuch, M. Nishioka, A. Ishikawa, and Y. Arakawa. Observation of the coupled exciton-photon mode splitting in a semiconductor quantum microcavity. *Phys. Rev. Lett.*, 69(23):3314–3317, 1992.

90. E. Waks, K. Inoue, C. Santori, D. Fattal, J. Vučković, G.S. Solomon, and Y. Yamamoto. Quantum cryptography with a photon turnstile. *Nature (London)*, 420(6917), 2002.

91. D. Fattal, K. Inoue, J. Vučković, C. Santori, G. S. Solomon, and Y. Yamamoto. Entanglement formation and violation of Bell's Inequality with a semiconductor single-photon source. *Phys. Rev. Lett.*, 92(3): 37903, 2004.

92. D. Fattal, E. Diamanti, K. Inoue, and Y. Yamamoto. Quantum teleportation with a quantum dot single-photon source. *Phys. Rev. Lett.*, 92(3): 37904, 2004.

93. D. Fattal, R. Beausoleil, and Y. Yamamoto. Coherent single-photon generation and trapping with imperfect cavity QED systems. *Arxiv preprint quant-ph/0606204*, 2006.

94. C. H. Bennett and G. Brassard. Quantum cryptography: Public key distribution and coin tossing. In *Proceedings of IEEE International Conference on Computers, Systems, and Signal Processing*, pp. 175–179, Bangalore, India, 1984.

95. H. Inamori, N. Lutkenhaus, and D. Mayers. Unconditional security of practical quantum key distribution. *Eur. Phys. J. D*, 41(3): 599–627, 2007.

96. E. Biham, M. Boyer, P. O. Boykin, T. Mor, and V. Roychowder. A proof of security of quantum key distribution against any attack. In *Proceedings of the Thirty-Second Annual ACS Symposium of Theory of Computing*, pp 715–724, Portland, Oregon, USA, 2000.

97. P. W. Shor and J. Preskill. Simple proof of security of the BB84 quantum key distribution protocol. *Phys. Rev. Lett.*, 85(2): 441–444, 2000.

98. N. Lütkenhaus. Security against individual attacks for realistic quantum key distribution. *Phys. Rev. A*, 61(5): 052304, 2000.

99. G. Brassard and L. Salvail. Lecture notes in computer science. In T. Hellseth, editor. *Advances in Cryptology–EUROCRYPT '93*. Vol. 765, 410–423. Springer, Berlin, Germany 1994.

100. H. Bennett and D. P. DiVicenzo. Quantum information and computation. *Nature (London)*, 404: 247–255, 2000.

101. Z. Y. Ou and L. Mandel. Violation of Bell's inequality and classical probability in a two-photon correlation experiment. *Phys. Rev. Lett.*, 61: 50–53, 1988.

102. J.-W. Pan, D. Bouwmeester, H. Weinfurter, and A. Zeilinger. Experimental entanglement swapping: Entangling photons that never interacted. *Phys. Rev. Lett.*, 80: 3891–3894, 1998.

103. J.-W. Pan, M. Daniell, S. Gasparoni, G. Weihs, and A. Zeilinger. Experimental demonstration of four-photon entanglement and high-fidelity teleportation. *Phys. Rev. Lett.*, 86: 4435–4438, 2001.

104. S. Popescu, L. Hardy, and M. Zukowski. Revisiting Bell's theorem for a class of down-conversion experiments. *Phys. Rev. A*, 56: R4353, 1997.

105. J. F. Clauser, M. A. Horne, A. Shimony, and R. A. Holt. Proposed experiment to test local hidden-variable theories. *Phys. Rev. Lett.*, 23: 880, 1969.

106. A. G. White, D. F. V. James, P. H. Eberhard, and P. G. Kwiat. Nonmaximally entangled states: Production, characterization, and utilization. *Phys. Rev. Lett.*, 83: 3103–3107, 1999.

107. A. Peres. Separability criterion for density matrices. *Phys. Rev. Lett.*, 77: 1413–1415, 1996.

108. T. C. Ralph, A. G. White, W. J. Munro, and G. J. Milburn. Simple scheme for efficient linear optics quantum gates. *Phys. Rev. A*, 65: 12314, 2002.

109. T. B. Pittman, M. J. Fitch, B. C. Jacobs, and J. D. Franson. Experimental controlled-not logic gate for single-photons in the coincidence basis. *Phys. Rev. A*, 68: 032316, 2003.

110. D. Gottesman and I. L. Chuang. Demonstrating the viability of universal quantum computation using teleportation and single-qubit operations. *Nature (London)*, 402: 390, 1999.

111. C. H. Bennett, G. Brassard, C. Crépeau, R. Jozsa, A. Peres, and W. K. Wootters. Teleporting an unknown quantum state via dual classical and Einstein-Podolsky-Rosen channels. *Phys. Rev. Lett.*, 70: 1895–1899, 1993.

112. J. D. Franson, M. M. Donegan, M. J. Fitch, B. C. Jacobs, and T. B. Pittman. High-fidelity quantum logic operations using linear optical elements. *Phys. Rev. Lett.*, 89: 137901, 2002.

113. S. Giacomini, F. Sciarrino, E. Lombardi, and F. De Martini. Active teleportation of a quantum bit. *Phys. Rev. A*, 66: 030302, 2002.

114. E. Lombardi, F. Sciarrino, S. Popescu, and F. DeMartini. Teleportation of a vacuum-photon qubit. *Phys. Rev. Lett.*, 88: 070402, 2002.

115. D. Fattal, C. Santori, J. Vučković, G. S. Solomon, and Y. Yamamoto. Indistinguishable single-photons from a quantum dot. *Phys. Stat. Sol (b)*, 238: 305, 2003.

116. H. J. Briegel, W. Dur, J. I. Cirac, and P. Zoller. Quantum repeater: The role of imperfect local operations in quantum communication. *Phys. Rev. Lett.*, 81: 5932–5935, 1998.

117. M. Keller, B. Lange, K. Hayasaka, W. Lange, and H. Walther. Continuous generation of single-photons with controlled waveform in an ion-trap cavity system. *Nature (London)*, 431: 1075–1078, 2004.

118. W. Yao, R.-B. Liu, and L. J. Sham. Theory of control of the spin-photon interface for quantum networks. *Phys. Rev. Lett.*, 92: 30504, 2005.

119. D. Englund, D. Fattal, E. Waks, G. Solomon, B. Zhang, T. Nakaoka, Y. Arakawa, Y. Yamamoto, and J. Vučković. Controlling the spontaneous emission rate of single quantum dots in a 2D photonic crystal. *Phys. Rev. Lett.*, 95: 013904, 2005.

120. M. Atatüre, J. Dreiser, A. Badolato, A. Högele, K. Karrai, and A. Imamoğlu. Quantum-dot spin-state preparation with near-unity fidelity. *Science*, 312: 551–553, 2006.

121. W. Yao, R.-B. Liu, and L. J. Sham. Theory of electron spin decoherence by interacting nuclear spins in a quantum dot. *Phys. Rev. B*, 74: 195301, 2006.

122. K. Inoue, E. Waks, and Y. Yamamoto. Differential phase shift quantum key distribution. *Phys. Rev. Lett.*, 89: 37902, 2002.

123. D. F. Walls and G. J. Milburn. *Quantum Optics*. Springer-Verlag, Berlin, 1994.

124. F. Laudenbach. *Calcul differentiel et integral*. Editions de l'Ecole Polytechnique, France, 2000.

125. H. Takesue, S. W. Nam, Q. Zhang, R. H. Hadfield, T. Honjo, K. Tamaki, and Y. Yamamoto. Quantum key distribution over 40 dB channel loss using superconducting single-photon detectors. *Nature Photonics*, 1: 343–348, 2007.

126. K. M. C. Fu, C. Santori, C. Stanley, M. C. Holland, and Y. Yamamoto. Coherent population trapping of electron spins in a high-purity n-type GaAs semiconductor. *Phys. Rev. Lett.*, 95(18): 187405, 2005.

127. M. J. Hartmann, F. G. S. L. Brandão, and M. B. Plenio. Strongly interacting polaritons in coupled arrays of cavities. *Nature Physics*, 2: 849–855, 2006.

128. D. G. Angelakis, M. F. Santos, and S. Bose. Photon blockade induced Mott transitions and XY spin models in coupled cavity arrays. *Arxiv preprint quant-ph/0606159*, 2006.

129. S. Kako, C. Santori, K. Hoshino, S. Gotzinger, Y. Yamamoto, and Y. Arakawa. A gallium nitride single-photon source operating at 200 K. *Nature Materials*, 5(11): 887–892, 2006.

130. P. van Loock, T. D. Ladd, K. Sanaka, F. Yamaguchi, K. Nemoto, W. J. Munro, and Y. Yamamoto. Hybrid quantum repeater using bright coherent light. *Phys. Rev. Lett.*, 96(24): 240501, 2006.

131. L. Childress, J. M. Taylor, A. S. Sørensen, and M. D. Lukin. Fault-tolerant quantum repeaters with minimal physical resources and implementations based on single-photon emitters. *Phys. Rev. A*, 72(5): 52330, 2005.

3

Crystalline Whispering Gallery Mode Resonators in Optics and Photonics

Lute Maleki, Vladimir S. Ilchenko, Anatoliy A. Savchenkov, and Andrey B. Matsko
OEwaves Inc.

CONTENTS

3.1 Introduction

Open dielectric resonators with whispering gallery modes (WGMs) are unique because of their ability to store light in microscopically small volumes for long periods of time. The small size of the resonator allows densely packaged optical elements, and the intrinsically high quality (Q) factors of WGMs result in reduced attenuation in photonic circuits.

Evidently, the first indirect observation of WGMs in optics was realized in crystalline WGM lasers when laser action was studied in Sm:CaF$_2$ crystalline resonators several millimeters in size [1]. Microsecond-long transient laser operation also was observed with a ruby ring of several millimeters in diameter at room temperature [2]. Transient oscillations were attributed to pulsed laser excitation of WGMs with Qs of 10^8–10^9.

WGMs were first directly observed by elastic light scattering from spherical dielectric particles in liquid resonators [3,4]. It was recognized that these modes could help in measurement of the size, shape, refractive index, and temperature of nearly spherical particles [5,6]. WGMs were also used to determine the diameter of an optical fiber [7], and the strong influence of WGMs on fluorescence [8–10] and Raman scattering [11–13] was recognized. Laser action in a free droplet was first studied by Tzeng et al. [14] and Lin et al. [15]. An extensive study of WGM resonators with a focus on their applications was started when high-Q optical modes were observed in solidified droplets of fused amorphous materials [16,17], and after an efficient scheme for coupling with the modes was developed [18–20].

The use of amorphous WGM resonators for practical applications is challenging because of several reasons. (i) The highest quality factor of WGMs in the resonator is determined by Rayleigh scattering occurring due to residual surface roughness and inhomogeneities of the resonator host material [21]. This is generally the case, even though a resonator formed by surface tension forces has nearly a defect-free surface characterized by molecular-scale inhomogeneities. (ii) Amorphous resonators are not environmentally stable. For instance, the quality factor of fused silica resonators degrades with time due to diffusion of atmospheric water into the material. (iii) It is difficult to produce both an optically transparent and a highly nonlinear amorphous dielectric material.

Despite multiple problems including those mentioned above, amorphous resonators are widely studied, and various approaches have been developed to mitigate problems associated with them. Many applications of these resonators have been proposed and some of them are discussed in other chapters of this book.

The primary goal of our research reviewed in this chapter is to return to the early studies of crystalline WGM resonators [1,2] and to develop technologies for controllable fabrication of the resonator, so that various useful properties of crystals that extend beyond the amorphous counterparts can be put to advantage. In what follows we show that this goal, for most parts, has been met. To meet these objectives, we have developed efficient methods of coupling to the high-Q modes of the resonators. Finally, we have used a crystalline WGM resonator for practical optical and photonic applications.

3.2 Fabrication Technique

Any residual roughness on the surface of the resonator results in serious limitation on the achievable WGM Q-factor. Hence, it was commonly believed that high-Q optical WGM resonators cannot be fabricated with transparent crystals. Melting is obviously not suitable for materials with crystalline structure, because it destroys the initial purity, crystalline structure, and stoichiometry of the material. Moreover, during solidification, the original spherical droplet of the melt turns into a rough body with multiple facets and crystal growth steps.

We have fabricated crystalline optical WGM resonators with very high Q-factors ($Q > 10^{11}$) limited in value only by the absorption of the material, and exceeding that of surface-tension-formed resonators. With simple polishing techniques we preserve the original crystalline structure and its composition. In these structures, the unique linear and nonlinear properties of the crystal are enhanced because of the small volume of the high-Q modes. Total internal reflection at the walls of the WGM resonators provides the effect of an ultra-broad band mirror, allowing very high Q-factors across the entire range of material transparency. This property makes crystalline WGM resonators a convenient tool for optical materials studies.

Crystalline WGM resonators, especially those fabricated with sapphire [22], were extensively studied in the microwave frequency range [23]. It was shown that sapphire resonators can have very high quality factors ultimately restricted by material impurities [24]. These resonators were also fabricated using mechanical polishing techniques. However, the requirements on the surfaces smoothness is significantly relaxed in the case of these structures that support microwave modes.

Earlier optical experiments indirectly suggested that mechanical polishing is very promising for the realization of ultra-high Q crystalline WGM resonators. Millimeter-sized calcium fluoride spheres were fabricated by mechanical polishing and WGM resonances were indirectly observed [1]. The fabrication technique described by Bond [25] is based on the mechanical polishing.

In another example, observation of laser action in ruby rings (e.g. 2.38 cm external diameter, 2.22 cm inner diameter with 0.08 cm width, and 0.32 cm height; the C axis was perpendicular to the plane of the ring) suggested the existence of WGMs with $Q \approx 10^9$ [2]. The surface of the ring (except for its inner part) was mechanically polished. A decade later CW laser operation in a Nd:YAG 5 mm sphere was demonstrated [27].

Mechanical polishing of crystalline resonators to obtain high-Q optical WGMs was reinvented at the Jet Propulsion Laboratory (JPL) (Quantum Sciences and Technology Group) 15 years later. A lithium niobate WGM with $Q = 5 \times 10^6$ (at 1.55 μm) resonator was first fabricated [28,29], and the result was repeated by the University of Southern California group

[26,30–32]. The high-Q obtained by mechanical polishing was not a miracle, though. Total internal reflection resonators (another type of open dielectric resonators resembling WGM resonators) made out of lithium niobate have $Q \approx 7 \times 10^8$ (optical path length 1.6 cm, index of refraction 2.19, finesse $F=6000$, at 2 μm) [33]. The JPL group explicitly demonstrated the capability of mechanical polishing, ultimately producing calcium fluoride WGM resonators with $Q > 10^{11}$ and showing that the material, not the polishing quality, determines the achievable quality factor [34].

Mechanical polishing, along with subsequent annealing, is routinely used to produce ultra-smooth crystalline surfaces. For instance, atomically smooth surface (0.25 nm) on $LiNbO_3$ and $LiTaO_3$ substrates were produced by thermal annealing of mechanically polished samples with 1–2 nm roughness [35]. Even better smoothness (0.1–0.2 nm) can be produced with quartz crystals using soft material polishers, fine abrasive powders, and suitable working environments, and by carefully reducing the mechanical action in the polishing process [36]. Mechanical polishing is not the only technique that can lead to achieving good surfaces. Single point diamond turning is also a very promising approach. For example, surfaces with 3 nm roughness have been achieved with diamond turning of CaF_2 [37].

We adopted various known techniques to fabricate WGM resonators out of different types of crystals. The basic procedure is the following. We start with a crystalline wafer and drill the wafer with a hollow drill to fabricate a cylindrical preform. The preform is then attached to a holder and is either polished or diamond turned. There is no unique recipe for production of a perfect surface. We have achieved ultra-high quality factors by selecting the polishing speed, and the size of the grain in a diamond slurry. As a rule, all the algorithms change from a crystal to a crystal, and from a crystal cut to a crystal cut. To date, the basic achievement is the absence of the Rayleigh scattering in our resonators, which is the basis of the claim that the surface roughness does not restrict the quality factor of crystalline resonators.

Hand polishing works well for fabrication of larger resonators. For resonators with diameters as small as tens of micrometer, diamond turning technique is more suitable. The diamond turning technique we used for fabrication of crystalline microresonators is based on a home-made apparatus [38], and allows precise engineering of geometrical features of the resonator such as surface curvature or profile. We have fabricated magnesium fluoride and calcium fluoride resonators to validate the technique. The quality of the surface of resonators produced with this apparatus is quite high after a final polishing, and is characterized with less than 0.2 nm surface roughness. The quality factor of the microresonators fabricated with CaF_2 exceeds 10^8.

The entire fabrication process of microresonators generally includes two steps. The first step is diamond turning, which employs computer control of a precision lathe. Structures obtained in this step are engineered to about 40 nm precision and have optical Q factors of up to 10^7 (surface quality limited). For instance, a chain of three CaF_2 resonators presented on Figure 3.1 was fabricated with this approach. Resonators are separated by approximately 70 μm and are coupled to each other for efficient energy transfer. If higher Q factors are needed, additional optical polishing has to be performed. This polishing step naturally modifies the structure that is initially obtained by the diamond turning process. Special diamond turning parameters must be employed to avoid brittle machining and to achieve the required surface smoothness. Parameters such as speed of rotation of the workpiece, feed and cutting angle, diamond cutter geometry and sharpness, as well as the type of lubricant are very important in achieving ductile regime of machining [39]. Special attention should also be paid to vibration isolation of the turning process. In our

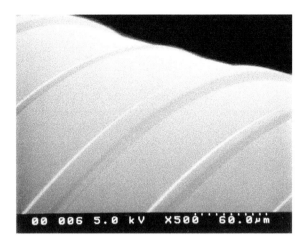

FIGURE 3.1
A stack of three WGM resonators fabricated by diamond turning technique on a cylindrical preform made out of calcium fluoride.

diamond turning setup we use a home-made air bearing to provide the required stiffness and repeatability of rotation of the workpiece. Optical polishing is performed by application of polycrystalline diamond abrasives. Polishing is performed in several steps with decreasing diamond grit size followed by a cleaning process. It was found that small particles are extremely difficult to remove from the surface of the material, and the cleaning process is most critical in achieving ultrahigh Q factor. Cleaning was performed in a clean room environment with the use of organic solvents.

3.3 Coupling Techniques

Efficient coupling of light to microresonators is one of the basic problems of photonics. It is not enough to build a low loss linear or nonlinear resonator; it is also important to be able to send/retrieve light into and out of the device. Thanks to the successful development of waveguide coupling techniques, the first commercial devices utilizing WGMs have now became available [40].

Fundamentally, coupling of light with the resonator is based on phase- and mode-matched power exchange between the mode and the guided wave in a specially engineered coupler (a waveguide or a prism). Parasitic coupling to unwanted modes is quantified by "ideality"—the ratio of power coupled to a desired mode to power coupled or lost to all modes (including the desired mode). Principles of any "ideal" coupling technique are based on (i) phase synchronism, (ii) optimal overlap of the selected WGM and the coupler mode, (iii) selectivity, and (iv) criticality. Phase-matched evanescent field couplers posses all these properties.

3.3.1 Critical Coupling

For the most efficient energy exchange, a resonator must be both ideally, as well as critically, coupled to a chosen waveguide or fiber. The notion of critical coupling, fundamental

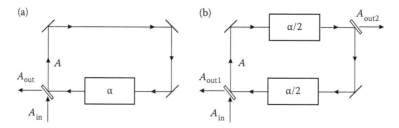

FIGURE 3.2
Models of WGM resonators.

in RF engineering (see e.g. [41,42]), has been recently applied to optical WGM resonators [43]. Indeed, unlike lossless Fabry–Perot (FP) resonators where intrinsic loss is identical to the coupling loss, the strength of evanescent wave coupling to WGM resonators is independent of their intrinsic loss. Criticality implies that the coupling strength between waveguide and resonator must match the loss of any nature to provide a 100% energy exchange at resonance.

Let us model a solid state WGM resonator as a ring cavity containing a linear absorber localized in space (Figure 3.2a). To find the intracavity field A we write the steady state condition

$$A = Ae^{i\omega Ln/c}e^{-\alpha}\sqrt{1-T_c} + \sqrt{T_c}A_{in}, \tag{3.1}$$

where $\alpha \ll 1$ is the (frequency independent) total amplitude absorption of the light in the resonator per round trip, $T_c \ll 1$ is the coupling factor, $L = 2\pi R$ is the circumference of the resonator, R is the radius of the resonator, A is the slow amplitude operator of the intracavity light, and n is the refractive index of the material. Equation 3.1 indicates that the steady state of the field amplitude is achieved by compensation of the round trip loss of the external optical pumping A_{in}.

The amplitude of the output pump light A_{out} can be found from equation

$$A_{out} = -Ae^{i\omega Ln/c}e^{-\alpha}\sqrt{T_c} + A_{in}\sqrt{1-T_c}, \tag{3.2}$$

and it is equal to

$$A_{out} = \frac{\sqrt{1-T_c} - \exp[i\omega Ln/c]\exp[-\alpha]}{1 - \exp[i\omega Ln/c]\exp[-\alpha]\sqrt{1-T_c}} A_{in}. \tag{3.3}$$

It is convenient to introduce notations

$$\gamma = \frac{\alpha}{L}\frac{c}{n}, \quad \gamma_c = \frac{T_c}{2L}\frac{c}{n}, \quad \omega\frac{Ln}{c} = (\omega - \omega_0)\frac{Ln}{c} + 2\pi l, \quad 2\pi l = \omega_0\frac{Ln}{c},$$

where ω_0 is the mode frequency, l is the mode number. In the vicinity of the resonance we have

$$A_{out} = -\frac{\gamma_c - \gamma - i(\omega_0 - \omega)}{\gamma_c + \gamma + i(\omega_0 - \omega)} A_{in}. \tag{3.4}$$

Equation 3.4 shows that all the power is absorbed in the resonator ($A_{out}=0$) if $\gamma_c=\gamma$. This is the condition of the criticality of coupling.

In the case of a WGM resonator with two couplers ("add-drop" configuration, Figure 3.2b) the criticality condition looses its meaning. In fact, the intracavity field in the case of two couplers is calculated from

$$A = Ae^{i\omega Ln/c}e^{-\alpha}\sqrt{1-T_{c1}}\sqrt{1-T_{c2}} + \sqrt{T_{c1}}A_{in}, \tag{3.5}$$

where $T_{c1} \ll 1$ and $T_{c2} \ll 1$ are the coupling factors, for two partially transparent mirrors. Output fields are described by equations

$$A_{out1} = -Ae^{i\omega Ln/c}e^{-\alpha}\sqrt{T_{c1}} + A_{in}\sqrt{1-T_{c1}}, \tag{3.6}$$

$$A_{out2} = Ae^{i\omega Ln/2c}e^{-\alpha/2}\sqrt{T_{c2}}, \tag{3.7}$$

solutions of which are approximated by expressions

$$A_{out1} = -\frac{\gamma_{c1}-\gamma_{c2}-\gamma-i(\omega_0-\omega)}{\gamma_{c1}+\gamma_{c2}+\gamma+i(\omega_0-\omega)}A_{in}, \tag{3.8}$$

$$A_{out2} = -\frac{2\sqrt{\gamma_{c1}\gamma_{c2}}}{\gamma_{c1}+\gamma_{c2}+\gamma+i(\omega_0-\omega)}A_{in}, \tag{3.9}$$

in the vicinity of the resonance. We have taken into account that $e^{i\omega Ln/2c} \approx -1$ in this case. It is easy to see now that the filter is nearly ideal if $\gamma_{c1}=\gamma_{c2}\gg\gamma$. The criticality of the coupling does not make sense here.

3.3.2 Prism

Prism coupling was initially developed to send light to optical waveguides [44–46]. Prism-waveguide coupling efficiency exceeding 90% has been demonstrated elsewhere [47,48]. Prism coupling to WGMs has been investigated both theoretically and experimentally [18–20,49–51]. The best efficiency of prism coupling to WGMs in microspheres reported to date is ~80% [49]. A coupling efficiency exceeding 97% was achieved in elliptical lithium niobate resonators [52]. In this chapter we report on achieving 99% overall nonlocal coupling with a toroidal WGM resonator having 4.2 mm in diameter.

The technique of prism coupling is based on three main principles. First, the input beam is focused inside the prism at an angle that provides phase matching between the evanescent wave of the total internal reflection spot, and the WGM, respectively. Second, the beam shape is tailored to maximize the modal overlap in the near field. And third, the gap between the resonator and the prism is optimized to achieve critical coupling. The loaded quality factor for the prism coupler in the case of the basic WGM sequence of microresonators is given by the expression (see [20]):

$$Q_c = \frac{\pi(n_r^2-1)m^{3/2}}{2n_r\sqrt{n_p^2-n_r^2}}\exp\left(4\pi\sqrt{n_r^2-1}\frac{d}{\lambda}\right), \tag{3.10}$$

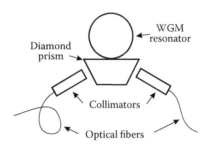

FIGURE 3.3
Schematic of prism coupling with a WGM resonator.

where n_r is the refractive index of the resonator, n_p is the refractive index of the prism, d is the shortest distance from the prism to the resonator, λ is the wavelength of the light in the vacuum, and m is the azimuthal number of the mode. It is easy to see that $n_p > n_r$ is expected.

The schematic diagram of a prism coupler is shown in Figure 3.3. The output of a single mode fiber is collimated and focused on the surface of the prism. The prism is located in the close proximity of the WGM resonator, and the collimated beam is focused on the contact spot between the resonator and the prism. The refractive index of the prism n_p is larger than the refractive index of the resonator n_r, and the phase velocity of the collimated beam along the surface of the prism is $c/n_p \sin\phi$, where c is the speed of light in vacuum, and ϕ is the angle of incidence. The phase velocity of the optical WGM is equal to c/n_r. This trivial approximation is good enough in practice. The modes are phase-matched when

$$\sin\phi \simeq n_r/n_p. \tag{3.11}$$

The phase matching of the modes is not the only factor required for efficient coupling. For example, when the focused beam is not adjusted with the prism–resonator contact point, then there is no coupling at all since there is no aperture matching. Aperture matching is the second important factor for achieving an efficient coupling. Aperture matching with a Gaussian beam occurs when

$$\cos\phi \simeq \sqrt{\frac{r}{R}}, \tag{3.12}$$

where r is the vertical radius of curvature of the resonator in the vicinity of the contact spot, and R is the horizontal radius of curvature of the resonator next to the contact spot. Equations 3.11 and 3.12 show that phase and aperture matching are generally impossible to obtain in the case of coupling with the basic WGM sequence in a microsphere. This is why the best achieved coupling with a microsphere is on the level of 80% [49].

It is possible though to improve the coupling by varying both the incidence angle and some other parameter. Two choices are available for this. One choice is to change the vertical radius of curvature of the resonator, r, i.e. make a toroidal resonator. The resonators rim must be shaped such that

$$r = R\left[1 - \left(\frac{n_r}{n_p}\right)^2\right]. \tag{3.13}$$

Another choice is to select a mode with the effective refractive index

$$n_r = n_p \sqrt{1 - \frac{r}{R}}. \tag{3.14}$$

This can be done by selecting modes located deeper in the resonator since

$$n_r > n_{reff} \simeq n_r \left[1 + \alpha_q \left(\frac{m}{2} \right)^{1/3} \right]^{-1}, \tag{3.15}$$

where α_q is the qth root of the Airy function, $Ai(-z)$, which is equal to 2.338, 4.088, and 5.521 for $q=1$, 2, and 3, respectively. Practically speaking, these approaches are both used together. At the onset, the resonator is made so that its curvature parameters satisfy the condition, Equation 3.13. Possible fabrication errors can be fixed by selection of a deeper mode.

It is possible to realize aperture matching with aspherical (astigmatic) optics. An axio-symmetric Gausian beam can be transformed to an elliptical beam that has perfect overlap with modes of an arbitrary WGM resonator. This method is now under development.

The third factor limiting the coupling efficiency is intermodal mixing. WGM modes in larger resonators overlap and interact with each other. The mode of interest emits into other modes limiting the maximum achievable optical coupling (in the sense of the critical coupling). To suppress this effect the number of modes should be reduced by, for example, shaping of the resonator [53]. However, it is possible to achieve excellent coupling even in a toroidal multimode resonator. For example, we have measured 99% nonlocal coupling in a toroidal WGM resonator having 4.2 mm in diameter and 120 μm in thickness Figure 3.4.

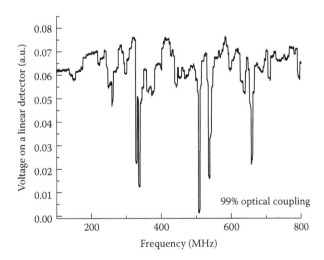

FIGURE 3.4
Spectrum of a WGM resonator taken using a prism coupler. The coupling efficiency approaching 99% is achieved.

It is important to note a common mistake made in the measurement of the coupling efficiency. It is possible to achieve a perfect contrast for the absorption resonance, even though the coupling is far from being critical, by simply changing the position of an output collimator that does not collect all the radiation exiting the resonator. Such an observation is in fact the signature of poor coupling, and the high contrast picture comes from the interference of the light exiting different WGMs. We are aware of claims of 100% coupling achieved with prisms having an index of refraction less than that of the resonator. One of the ways of avoiding such mistakes is to check the output beam profile, which will be Gaussian for the case of perfect coupling.

3.3.3 Angle-Cut Fiber

Another promising method for coupling light and WGM resonators is based on angle-cut fiber couplers. Initially side-polished fiber couplers [54–56] had limited efficiency owing to the residual phase mismatch between the guided wave and the WGMs. Subsequently, a phase-matched coupling technique was developed using an angle-polished fiber tip in which the core-guided wave undergoes total internal reflection [28,57]. Coupling the WGM resonators and semiconductor lasers was also studied by means of a similar technique [58,59].

The main idea of angle-polished fiber coupling is illustrated in Figure 3.5. The tip of a single-mode fiber is angle polished with a steep angle. The light propagating inside the core undergoes total internal reflection at the polished surface and escapes the fiber. Energy exchange between the waveguide mode of the single mode fiber and a WGM occurs at resonance when the resonator is positioned within the range of the evanescent field of the angle-cut fiber. The angle is chosen to fulfill the phasematching requirement (Equation 3.11). Generally, the system is equivalent to a prism coupler with the focusing optics eliminated. The applicability of this technique is limited by the availability of fibers with high refractive index.

3.3.4 Fiber Taper

Tapered fiber couplers provide the most efficient coupling scheme [60–70]. Fiber taper is a single mode bare waveguide of optimized diameter (typically a few microns for the near infrared wavelength, see Figure 3.6). The taper is placed along the resonator perimeter

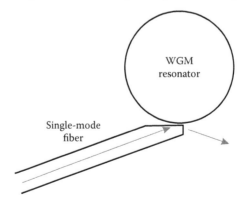

FIGURE 3.5
Angle-polished fiber coupler for WGMs.

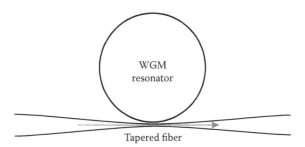

FIGURE 3.6
Tapered fiber coupler for WGMs.

allowing simple focusing and alignment of the input beam, as well as collecting the output beam. It saves the fundamental mode and filters other waveguide modes. The efficiency of tapered fiber couplers reaches 99.99% for coupling fused silica resonators [65]. Similar couplers were also used in photonic crystal resonators [66,68]. Unfortunately, fiber tapers are unsupported waveguides without cladding and are very fragile. They are only useful for resonators with refractive indices similar to silica (1.4–1.45), and cannot generally be used with higher index glass and crystalline resonators. A good coupling achieved with conventional fiber tapers, though, has been demonstrated with very small silicon resonators [71]. The coupling is possible in this case because phase matching is not very important if the size of the resonator is comparable with the wavelength.

The phase velocity depends on the taper's diameter and the refractive index. The diameter of the taper gradually decreases towards the tip, and so does the effective refractive index. By placing the resonator at a proper point of the taper, it is possible to achieve the ultimate phase matching and critical coupling. Practically speaking, this technique is limited by available fibers since the effective refractive index of the taper can be changed only between values of refractive index of the air/vacuum and the refractive index of the material the fiber is made of. When the resonator is made of lithium niobate or diamond, for example, it is not trivial to make a phase-matched taper, though the planarized silicon waveguide would likely do the trick.

3.3.5 Planar Coupling

For practical packaging of systems containing WGM resonators, the planar coupling is the most promising. Planar waveguides are generally used to couple to microring and microdisk WGM resonators [72–80]. Planar coupling to ultra-high Q crystalline WGM resonators is yet to be demonstrated.

The planar coupling technology to WGM resonators is under constant development. For example, strip-line pedestal antiresonant reflecting waveguides have been utilized for robust coupling to microsphere resonators and for microphotonic circuits [81–83]. The waveguides are highly wavelength dependent because they rely on the resonant nature of reflections from multiple dielectric layers in the cladding. A recently proposed gold-clad pedestal planar waveguide structure solves this problem and provides wide band coupling for WGM resonators [84].

An analytical method for calculation of coupling between a microdisk WGM resonator and a straight waveguide was presented elsewhere [61,85]. This problem has also been solved numerically [86,87].

3.4 Modal Structure and Spectrum Engineering

Properties of open dielectric resonators and conventional mirrored optical resonators have several significant differences. We discuss two such differences in this section. One is related to the density of the WGM spectra, and the other is related to the geometrical dispersion of the WGM resonators. Spherical WGM resonators as originally proposed are overmoded. Their spectra have a complex quasi-periodic structure with unequal mode spacings, which result from both material and geometrical dispersion.

The spectral properties of WGM resonators can be modified by changing their morphology. For example, a significant reduction in the modal spectral density is achieved in highly oblate spheroidal microresonators [88,89]. In the most extreme, but theoretically possible case, the resonator can be fabricated from a single mode fiber. Such a resonator would have only a single mode family.

The spacing of WGMs can also be modified. The problem of nonequidistant modes is rooted in the fact that radial distribution of WGMs is frequency dependent. Higher frequency modes propagate on paths that are slightly closer to the surface than those of lower frequency modes. Thus, higher frequency modes travel in trajectories of slightly larger radius and slightly longer optical path lengths. Within the current technology based on fabrication of microspherical resonators out of a uniform material, the smaller the resonator size, the more the geometrical dispersion is manifested in the unequal separation between adjacent modes. Again, a proper shaping allows fabrication of a WGM resonator with an equidistant spectrum. For example, prolate spheroidal microcavities have a nearly equidistant spectrum [90]. Fabrication of a resonator out of a single mode fiber would remove the geometrical dispersion because the paths of the modes would be restricted geometrically.

Geometrical dispersion can be mediated using properties of the refractive index of the resonator host material. The optical path length is a function of both the physical distance and the index of refraction. The dispersion can be reduced if resonators are made out of a cylindrically symmetric material whose index $n(r)$ decreases in the radial direction. With a proper choice of the gradient for the refractive index, circular trajectories corresponding to WGMs at different frequencies will have identical optical path lengths. This results in an equidistant WGM spectrum [91].

Except for the azimuthal dispersion compensation, a graded material resonator demonstrates radial dispersion compensation. This happens because modes do not encounter resonator boundaries for large refractive index gradients, but only the potential dip created due to the gradient. As a consequence, radial profiles of cavity modes are nearly symmetrical, much in the same way as with harmonic oscillator wave functions [91].

This conclusion follows from complex angular momentum theory [92], where an analogy between optics and mechanics is utilized and the resonator modes are described as eigenvalues of an effective potential U. For a WGM with index m this potential may be written as the sum of an attractive well of depth $(n^2(r)-1)k^2$ and the centrifugal potential $m(m+1)/r^2$.

The potential is asymmetric when n does not depend on radius r inside the sphere. For spheres possessing dielectric susceptibilities that increase with a decreasing radius, the shape of the potential function broadens with its apex shifting into the resonator, and its parabolic shape becoming more symmetrical; the minimum of the potential is still on the sphere surface. For some critical value of the susceptibility gradient, the potential resembles half of the oscillatory potential $U \sim (r-R)^2|_{r \to R-0}$. For gradients beyond the critical value the minimum the potential moves into the cavity [91,93]. The deeper the minimum, the better the potential can be described by the oscillatory function.

Both methods of spectral engineering mentioned above can be applied to crystalline WGM resonators. The index of refraction of resonators can be modified by doping of the host material. This, however, is a complex task, and the modification of the resonator shape to obtain a desired WGM spectrum appears much simpler. In what follows we provide several examples of how shaping the WGM resonators changes their spectra.

3.4.1 The Spectrum and the Shape of the Resonator

A modification of the shape of a WGM resonator allows modification of its spectrum. To explain this statement let us start with the usual wave equation for a plain wave

$$\nabla \times (\nabla \times \mathbf{E}) - k^2 \, \varepsilon(\mathbf{r})\mathbf{E} = 0, \tag{3.16}$$

where $k = \omega/c$ is the wave number, $\varepsilon(\mathbf{r})$ is the coordinate dependent dielectric susceptibility, E is the electric field of the mode, and $\mathbf{r} = (z, r, \phi)$ is the radius vector.

We assume that the resonator is axially symmetric and that its radius $R(z)$ weakly depends on z, the axis of symmetry. Using the adiabatic approximation along the z axis we present the scalar solution of Equation 3.16

$$\Psi(\mathbf{r}) \sim e^{\pm i\left[\int k_z dz + m\phi\right]} \Phi\left[\frac{r}{R(z)}\right], \tag{3.17}$$

where

$$k_z^2(z) = k^2 n^2 - k_{m,q}^2(z), \tag{3.18}$$

n is the index of refraction of the resonator host material,

$$k_{m,q} \simeq \frac{1}{nR(z)}\left[m + \alpha_q \left(\frac{m}{2}\right)^{1/3}\right] \tag{3.19}$$

is the radial wave number defined by the equation for the radial field distribution Φ

$$\frac{\partial^2 \Phi}{\partial^2 x} + \frac{1}{x}\frac{\partial \Phi}{\partial x} + \left[k_{m,q}^2 - \left(\frac{m}{x}\right)^2\right]\Phi = 0. \tag{3.20}$$

The light is confined along the z axis if the corresponding component of the wave vector (k_z) has two zeros, z_1 and z_2 (turning points), and is positive in the interval $z_2 > z > z_1$. Then the Wentzel–Kramers–Brillouin (WKB) quantization along the z axis leads to the equation that defines the eigenvalues $k_{m,q,p}$ (see e.g. [90])

$$\int_{z_1}^{z_2} \left[k_{m,q,p}^2 - k_{m,q}^2\right]^{1/2} = \pi\left[p + \frac{1}{2}\right]. \tag{3.21}$$

Such separation of variables as well as the expression for the resonator spectrum ($k_{m,q,p}$) is generally valid if [90,94]

$$k_z R \frac{dR}{dz} \ll 1, \quad z_2 > z > z_1. \tag{3.22}$$

Condition 3.22 restricts the class of resonators that can be analyzed with this technique. However, we can claim that for that class of resonators a selection of the shape of the surface $R(z)$ allows modification of the resonator spectrum. For instance, it helps to evaluate the shape of a WGM resonator with an equidistant spectrum [90].

The originally proposed spherical WGM resonators are "over-moded", with a complex quasi-periodic spectrum and unequal mode spacings. A significant reduction in the mode spectral density is achieved in highly oblate spheroidal microresonator (microtorus) [88,89]. Our experiments also show that changing the shape of a millimeter-sized crystalline dielectric resonator results in the rarefication of the observed WGM spectrum without destroying the quality factor.

As an example, we have measured [95] WGM spectra of disc resonators possessing spherically (Figure 3.7a), toroidally (Figure 3.7b), and conically shaped rims (Figure 3.7c). The toroidal geometry, with its transverse curvature diameter nearly equal to the thickness of the disk, resulted in a significant rarefaction of the spectrum compared to a spherical layer (separate modes are visible in Figure 3.7b as opposed to conglomerates in Figure 3.7a). However, with this geometry it was difficult to further increase the transverse curvature to eliminate all but one WGM per free spectral range (FSR): disks smaller than 100 μm in thickness were hard to polish into the toroidal shape. As an alternative to further improving the spectral regularity, we produced a conical resonator that was subsequently polished at the rim into a very sharp bend toroid with osculating curvature diameter of ~20 μm (Figure 3.8). The spectrum proved to contain only one major mode (distinct successive peaks) (Figure 3.7c). A high Q of 5×10^7 was obtained with this resonator.

It is useful to present here an analytical expression for the spectrum of an arbitrary spheroid found in [96]

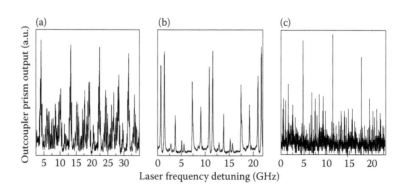

FIGURE 3.7
Spectrum of a disk LiNbO$_3$ resonator with (a) spherically shaped rim, (b) with torridally shaped rim, (c) with conically shaped rim.

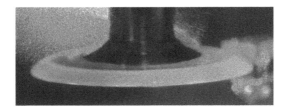

FIGURE 3.8
Conically shaped resonator spectrum of which is shown in Figure 3.7c. (From Savchenkov, A.A., Matsko, A.B. and Maleki, L., "White-light whispering gallery mode resonators", *Opt. Lett.* 31, 92–94 (2006)).

$$nk_{p,q,l}R \simeq l + \alpha_q \left(\frac{l}{2}\right)^{1/3} + \frac{2p(R-r)+R}{2r} - \frac{\chi n}{\sqrt{n^2-1}} + \frac{3\alpha_q^2}{20}\left(\frac{l}{2}\right)^{-1/3}$$

$$+ \frac{\alpha_q}{12}\left[\frac{2p(R^3-r^3)+R^3}{r^3} + \frac{2n\chi(2\chi^2-3n^2)}{(n^2-1)^{3/2}}\right]\left(\frac{l}{2}\right)^{-2/3}, \tag{3.23}$$

where R and r are equatorial and polar radiuses, $k_{p,q,l}$ is the wavenumber, $p = l - |m| = 0, 1, 2,...$ and $q = 1, 2, 3,...$ are integer mode indices, and $\chi = 1$ for quasi-TE and $\chi = 1/n^2$ for quasi-TM modes. The basic sequence of WGMs is given by $p = 0$ and $q = 1$.

Equation 3.23 does not show any actual rarefaction of the spectrum of a spheroidal resonator. It shows rather the removal of degeneracy by quantum numbers m and l. Hence, the results reported in Figure 3.7 come from the selectivity of coupling in a particular resonator, not from the real spectrum rarefaction. An additional reason for the clean spectrum of oblate spheroids is the reduction of the quality factors of the high-order modes. In what follows we discuss possible ways for obtaining real "single-mode" resonators beginning with a description of their opposite counterparts, "white light" resonators.

3.4.2 White Light Resonators

Not all applications of WGM resonators require a rare spectrum. Some applications are particularly burdened with rare and/or inhomogeneous spectra. These include coherent cavity ring-down spectroscopy [97] and electro-optical modulation with WGM resonators [98]. In the first case, narrow absorption features of a substance under study might fall between the resonator modes and would then be undetected unless the cavity modes are tunable. The spectrum also could be hard to reproduce in separate runs if the resonator is not stabilized. In the second case, the laser radiation that is modulated with a WGM-based electro-optic modulator requires a linewidth that is much narrower than the spectral width of a particular mode. The laser must also be locked to the mode of the resonator, or the resonator should be locked to the laser, which is not always feasible. A resonator with a continuum spectra, or a "white light" resonator, is extremely useful because it does not need to be locked to the laser, since there will always be modes that coincide with the laser frequency.

There are several methods for producing a continuum spectrum in a resonator and for keeping its ring-down time constant. It was suggested that a nonconfocal FP cavity would produce essentially a continuum excitation spectrum [97]. By exciting both the longitudinal and the transverse modes, one guarantees that there always are several modes at each frequency point and that light from an unstabilized narrowband laser will always be efficiently coupled into the resonator. This idea was widely used in coherent cavity ring-down spectroscopy. As another method, it was proposed to place a suitably prepared medium in between the resonator mirrors to cancel the variation of wavelength with frequency [99,100]. This would make the resonator simultaneously resonant for all frequencies, i.e. result in "whitening" the spectrum of the resonator. Such resonators were suggested for improving the sensitivity for gravitational wave detection.

The total achievable spectral width of the existing white light resonators is restricted by the spectral characteristics of the resonator mirrors. Usually the mirrors have good reflectivity within several nanometers of a selected optical frequency.

We theoretically proposed and experimentally demonstrated essentially "white light" WGM resonators in a frequency range exceeding a full octave [101]. The basic characteristics of the resonator are (i) power build-up at virtually any frequency inside the transparency window of the material it is made of; (ii) nearly the same Q-factor (ring-down time) at any frequency in the transparency window, and (iii) nearly critical coupling with cleaved fibers at any frequency in that window. These resonators can be used in cavity ring-down spectroscopy as an alternative for FP white-light cavities. The white light WGM resonators made of photorefractive materials also can be used in electro-optical modulators.

Let us explain the basic idea of the fabrication of the white light WGM resonator. The spectrum of WGMs in a finite cylinder of radius R can be approximated by Equation 3.21. The frequency distance between modes in any selected radial mode family is characterized with the FSR divided by the ratio of the resonator thickness and wavelength. The spectrum $k_{m,q,p}$ of a thick enough WGM resonator is very dense because changing the quantum number P does not change the mode frequency significantly. The Q-factors of various modes generally do not depend on the mode numbers, so all the modes have nearly identical Qs. Hence, a thick cylindrical or a prolate toroidal resonator can be considered as a white-light cavity.

We fabricated a high-Q CaF_2 WGM resonator ($Q > 10^9$) and loaded it heavily. The resonator had 0.5 mm thickness and 0.5 cm diameter. A proper coupling technique, e.g. a cleaved fiber (seen in Figure 3.9) readily results in the excitation of majority of the modes. The coupling efficiency was better than 40% for all frequencies at which the lasers could be tuned. The resonator was tested with 1320 nm (4 kHz linewidth), 780 nm (10 kHz linewidth) lasers, and 532 nm (incoherent) light. The spectrum of the resonator is essentially

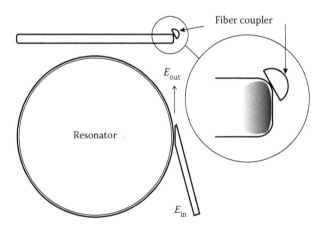

FIGURE 3.9
White light fiber coupled whispering gallery mode resonator. The resonator is thick (500 μm) which results in the coexistence of various mode families in the same geometrical volume in the vicinity of the rim of the resonator. The continuous spectrum of the resonator is achieved because the fiber coupler is slightly tilted with respect to the center of the rim of the resonator to maximize interaction with all the modes. (Reprinted from Savchenkov, A. A., Matsko, A. B., and Maleki, L., *Opt. Lett.* 31, 92–94, 2006. With permission from The Optical Society of America (OSA).)

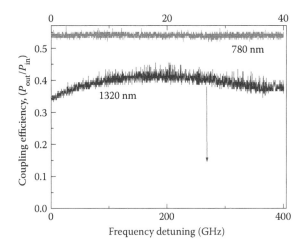

FIGURE 3.10
Spectra of the white light resonator at 1320 nm and 780 nm. Critical coupling corresponds to $P_{out}/P_{in}=0$, no coupling – to $P_{out}/P_{in}=1$. (Reprinted from Savchenkov, A. A., Matsko, A. B., and Maleki, L., *Opt. Lett.* 31, 92–94, 2006. With permission from OSA.)

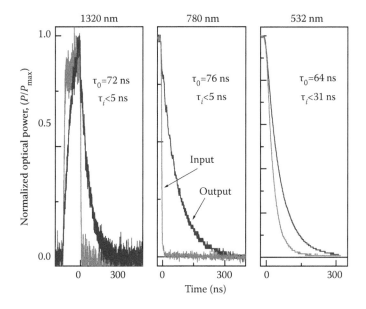

FIGURE 3.11
Normalized power of the input and output pulses of light taken at various carrier frequencies. (Reprinted from Savchenkov, A. A., Matsko, A. B., and Maleki, L., *Opt. Lett.* 31, 92–94, 2006. With permission from OSA.)

continuous if the frequency difference between modes is much smaller than their spectral width (Figure 3.10). The average distance between modes in a radial mode family was less than 20 MHz. We simultaneously excited several radial mode families.

To measure the quality factor at 1320 and 780 nm we used a ring-down technique with a $\tau_1 = 5$ ns shutter. For the green light we used an acousto-optic shutter with $\tau_1 = 31$ ns response time. Characteristic response times of the photodiode and circuits were less than 1 ns. The shape of the light pulse on the entrance and exit of the resonator are shown in Figure 3.11. The average ring-down time was $\tau_0 = 65$ ns, which corresponds to $Q \sim 10^8$.

Variations of the Q-factor with the carrier frequency of the lasers were less than 50%. Therefore, the resonator has all the necessary "white light" properties in a frequency interval larger than an octave.

Ring-down tails in Figure 3.11 have a small amplitude modulation. The modulation results from a slight loss of orthogonality between the modes of the resonator resulting from the coupling technique, and the beating between the modes. Changing the shape of the resonator allows suppression of the beats. Mode beats also disappear if the coupling with the resonator increases (Q-factor decreases).

Let us now discuss possible applications of this white light resonator. One application is a wide bandwidth electro-optic modulator. An approach to implement coupling between light and microwave fields in a WGM resonator was recently proposed. In that study, an efficient resonant interaction of several optical WGMs and a microwave mode was achieved by engineering the shape of a microwave resonator coupled to a micro-toroidal optical cavity. Based on this interaction a new kind of electro-optic modulator (as well as photonic microwave receiver) was suggested and realized [31,98]. The modulation frequency in the modulator is determined by the FSR of the resonator.

The problem of the application is in the necessity of lasers with very narrow linewidth, much smaller than the spectral width of the WGMs in the electro-optic resonator. The laser must be locked to a selected WGM to ensure continuous modulation. Application of white-light cavities would relax those requirements, as mentioned above. As an example, let us consider two lasers with different properties used in WGM electro-optical modulators: (i) a laser with a narrow linewidth but large frequency drift and (ii) a laser with a large linewidth.

Narrow linewidth lasers interact with essentially a single WGM mode at each instant of time. The modes naturally belonging to different mode families change with time, depending on the drift of the laser. The microwave field is applied to the WGM cavity using a proper strip-line microwave resonator placed along the rim of the cavity. The microwave field couples WGMs of the same mode family, and does not couple modes from different families. For instance, if the carrier frequency of the laser light coincides with a WGM characterized with wave vector $k_{m,q,p}$, the microwave radiation can excite modes with $k_{m-m1,q,p}$ and $k_{m+m_1,q,p}$ only, where m_1 is an integer number. Those modes, irrespective of which family, are nearly equidistant and their volumes and Q-factors are nearly equal, especially if the WGM resonator is large enough. Therefore, at each instant of time the WGM resonator can be considered as being "single mode", and the modulator operates unaffected by the frequency drift of the laser. There is no need to lock the laser to a mode of the resonator.

In the second case, a laser with a large linewidth simultaneously interacts with several modes belonging to different mode families in the white-light cavity EOM. Again, the modulation process "shifts" the photons from the carrier frequency to the frequency of the sidebands based on the same mode family interaction. If the different mode families were not equidistant, or they had different Q-factors, the noise in the sidebands would increase. However since the modes have identical properties, we expect no deterioration of the modulation.

Another application of the white light WGM resonators is in cavity ring-down spectroscopy, as mentioned above. WGM resonator-based chemical and biosensors were previously proposed and widely discussed [102–115]. Measurement of the material absorption as a function of frequency is one of the ways to detect specific chemical or biological substances. If the resonator surface is covered with a layer of the material to be studied, the quality factor decreases, and the ring-down time of the resonator τ changes as

$$\tau = \tau_0 \left(1 + \frac{V_{\text{ef}}}{V_{\text{r}}} \frac{\xi}{\xi_{\text{r}}} \right)^{-1} \tag{3.24}$$

where τ_0 is the ring-down time of the resonator in vacuum, V_{ef} is the effective volume in the added substance, V_r is the volume of the mode in the resonator, ξ is linear absorption per round trip of the material under study, and ξ_r is absorption of the material of the resonator per round trip. We neglected scattering in this expression.

White light WGM resonators, in the same way as white light FP resonators, can be useful for spectroscopy purposes. White light resonators allow tuning the laser interacting with the resonator instead of tuning the resonator, which can significantly simplify the experiments and the requirements for the laser characteristics could also be relaxed.

3.4.3 Single-Mode Resonators

Many practical applications of WGM resonators require their spectrum to be sparse, originating from a single mode family, and uncontaminated by the presence of other modes. The spectrum of a nonideal microsphere, the most common WGM resonator, is usually dense and special techniques are devised to clean it up [116]. Changing the resonator shape can reduce the density of high-Q modes which are effectively excited by the chosen coupler. Nearly single-mode operation of a microdisc WGM resonator was demonstrated in such a way (see the discussion at the beginning of this section and [88,89]).

Two general approaches exist for engineering optical modes. The most common approach is to use optical waveguides to confine and guide light. The waveguide is based on the refractive index contrast between its material and the surroundings to support light confinement and guiding. In the other approach, a photonic band-gap crystal provides confinement and guiding of photons using the morphology of the structure. This is achieved by placing defects in otherwise periodic arrays of a dielectric material. We have recently proposed a new approach that relies on generating an effective refractive index contrast produced by shaping the geometry of a WGM resonator [38,53]. With this technique single and multiply coupled WGM resonators can be designed for a wide variety of applications. We have found recently that a similar technique was proposed to clean the spectrum of FP resonators [117].

The utility of WGM-based devices and the efficiency of nonlinear optical interactions of WGMs depend on accurate engineering of the cavity mode structure. Specifically, this means producing cavities with extremely rarified (single-mode-family) spectra. Achieving the required level of control over the mode spectrum has eluded scientists working on WGM resonators. Thus, WGM resonators have as yet neither entered the mainstream of photonic device engineering nor have they been embraced by the mainstream of optical physics.

Researchers working on optical waveguides faced a similar set of problems in designing the single mode fiber. Single mode fiber has revolutionized the world of optics by enabling long-distance telecommunications and multichannel television broadcast systems, functions that would be impossible with multimode fibers. This is because the single mode fiber can retain the fidelity of a light signal over long distances. It exhibits no dispersion due to multiple modes and is characterized by lower attenuation than a multimode fiber.

A homogeneous dielectric waveguide becomes single-mode when the frequency of the propagating light is close to its cut-off frequency. This means that the thickness of the waveguide approaches the half-wavelength of light in the host material of the waveguide. It is impractical to fabricate a single-mode optical fiber by decreasing the fiber diameter. Instead, the single mode operation is ensured by the specially selected radial profile of the refractive index of the fiber material. The core of the fiber has a larger index of refraction than the cladding material that surrounds it. The difference of the indices is small, so only one mode propagates inside the core, while the others decay into the cladding (Figure 3.12A). For instance, the condition for the single mode operation of a planar waveguide is [118]

$$d_{co} < \frac{\lambda}{2\sqrt{\Delta\varepsilon}}, \tag{3.25}$$

where d_{co} is thickness of the core, λ is wavelength of light in vacuum, and $\Delta\varepsilon$ is the difference between the susceptibilities of the core and the cladding material, respectively. As a result, the core may have a reasonably large diameter. Note that the core of a single-mode fiber is a multi-mode fiber if the cladding is removed.

Let us consider the WGM resonator as a multi-mode gradient waveguide (Figure 3.12B) [119]. The resonator becomes an ideal single-mode-family resonator only if the waveguide is thin enough. Following this trivial approach the WGM resonator should be designed as an approximately half-wavelength-thick torus to support a single mode family. Recent experiments confirmed this conclusion and nearly single-mode resonators have been demonstrated [89].

There is another, nontrivial, approach to the problem based on the analogy with single-mode optical fiber. A WGM resonator made of any transparent material with any size can be transformed into a single-mode resonator if the appropriate geometrically defined "core and cladding" are developed (Figure 3.12C).

To show that WGM modes can be described using a waveguide formalism, we start with the usual wave equation (Equation 3.16). Higher-order WGMs (i.e. those with the wavelength much smaller than the resonator size) of both TE and TM modes are localized in

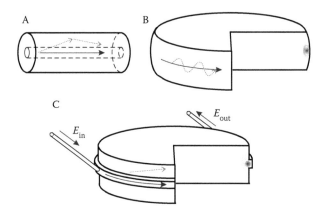

FIGURE 3.12
(A) A structure and mode localization in the optical fiber. Only one propagating mode, shown by the solid line, survives in the core of the fiber, while others (dotted line) penetrate into the cladding and dissipate. (B) Mode localization in the whispering gallery mode resonator. The resonator corresponds to a multimode gradient fiber where the index of refraction is set by the resonator shape, not by the change of the refractive index of the resonator host material, which is constant. Both the fundamental and auxiliary modes survive. (C) Mode localization in a low-contrast whispering gallery mode resonator. Only a single-mode family survives. The other modes penetrate into the cylinder rod and dissipate. The coupling to the single mode family is achieved with, e.g. a cleaved fiber coupler. (Reprinted from Savchenkov, A., Grudinin, I. S., Matsko, A. B., Strekalov, D., Mohageg, M., Ilchenko, V. S., and Maleki, L., *Opt. Lett.* 31, 1313–1315, 2006. With permission from OSA.)

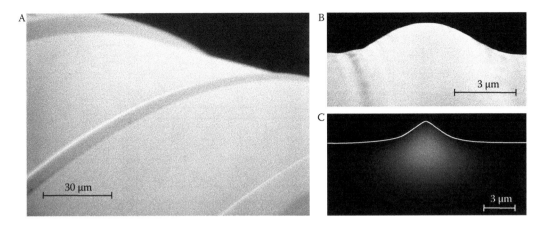

FIGURE 3.13
(A) Scanning electron microscope image of the resonator whose spectrum is shown in Figure 3.14. The resonator has nearly Gaussian shape with 2.5 μm height and 5 μm full width at the half maximum. (B) The image of the profile of the resonator shown in (A). (C) Intensity map of the field in the resonator shown in (A) simulated by numerical solution of Equation 3.26. (Reprinted from Savchenkov, A., Grudinin, I. S., Matsko, A. B., Strelkov, D., Mohageg, M., Ilchenko, V. S., and Maleki, L., *Opt. Lett.* 31, 1313–1315, 2006. With permission from OSA.)

the vicinity of the equator of the resonator. For this problem, cylindrical coordinates can be conveniently used. Applying the technique of separation of variables, and assuming that the resonator radius changes as $R = R_0 + L(z)$ ($R_0 \gg |L(z)|$) in the vicinity of the equator, we transform Equation 3.16, for TE mode family, to

$$\frac{\partial^2 E}{\partial r^2} + \frac{\partial^2 E}{\partial z^2} + \left[k^2 \varepsilon \left(1 + 2\frac{L(z)}{R_0} \right) - \frac{m^2}{r^2} \right] E = 0, \tag{3.26}$$

where m is the angular momentum number of the mode (we assumed that $m \gg 1$), $E(r,z)$ is the scalar field amplitude, ε is the susceptibility of the resonator host material. This equation is similar to the gradient waveguide equation. It is easy to see that, for instance, modes of a spherical WGM resonator coincide with modes of a gradient waveguide with parabolic distribution of refractive index in the z direction. Hence, it is the geometry of the surface that should be modified to produce an ideal single-mode WGM resonator. A "core" for the WGM "waveguide" can be realized by the proper design of the resonator surface in the vicinity of the equator. The rest of the resonator body plays the role of the "cladding" (Figure 3.1C).

Consider a resonator consisting of a cylindrical drum and a small, ring like protrusion, $L(z) = L_0$ for $d \geq z \geq 0$, on its surface. The drum's effective susceptibility does not depend on the z coordinate and is equal to ε. The effective susceptibility of the ring, $\varepsilon(1 + 2L_0/R_0)$ (see Equation 3.26), is slightly larger. Therefore, the ring is the core that confines the light in the z direction, while the drum is the cladding. The condition for "single mode" operation of the resonator using Equation 3.25 is:

$$1 > \frac{d}{\lambda} \sqrt{\frac{2L_0 \varepsilon}{R_0}} > \frac{1}{2}. \tag{3.27}$$

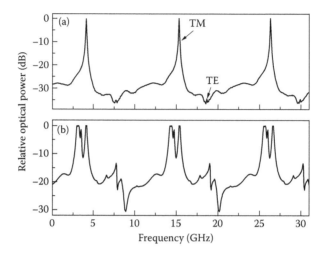

FIGURE 3.14
(A) The spectrum of a nearly ideal single-mode resonator obtained using a 980 nm laser as well as the cleaved fiber couplers (Figure 3.12C). The mode number is approximately $m = 3 \times 10^4$. The low-Q modes on the spectrum background belongs to the drum. (B) The spectrum of the multimode resonator made on the same drum (2.5 μm height and 15 μm width) and detected using the same coupling technique. The resonator is partially visible in the lower right corner of Figure 3.13A. (Reprinted from Savchenkov, A., Grudinin, I. S., Matsko, A. B., Strekalov, D., Mohageg, M., Ilchenko, V. S., and Maleki, L., *Opt. Lett.* 31, 1313–1315, 2006. With permission from OSA.)

This condition (Equation 3.27) stays valid for a resonator with an arbitrary large radius. Both the width and the height of the ring can be much larger than the wavelength of light. The ratio L_0/R_0 plays the same role as the ratio $\Delta\epsilon/\epsilon$ in an optical fiber.

To demonstrate the single-mode operation experimentally, we built such a resonator by using a two-step fabrication process that is described in one of the previous sections of this chapter. We fabricated a CaF$_2$ rod of 5 mm in diameter. WGMs in such a rod have extremely dense spectra. After that we fabricated a small ring with dimensions of the order of several microns on the surface (Figure 3.2).

We used an angle-polished fibers to couple light into the resonator. The spectrum shown was obtained with two angle-polished fiber couplers, one for the input and one for the output. The fiber couplers were made with standard single mode fiber with cladding diameter of about 150 μm and core of about 5 μm. The photodetector signal represents transmission of the resonator. The two families of modes shifted by a few GHz were observed, one for each polarization. One of such families is shown in Figure 3.14A. Logarithmic detector measurements have shown that there were no other modes down to −30 dB level, only the two mode families. It should be noted that, given the geometrical parameters of the resonator and the couplers, the single mode regime of this resonator is not a function of coupling but an intrinsic property. Moreover, the logarithmic measurements were performed in a loaded configuration, that is, when the couplers are in contact with the resonator. We also confirmed that this was a single-mode resonator by performing numerical simulation of its parameters which clearly demonstrated that only the fundamental modes survive. Intrinsic Q factors of both families of modes shown in Figure 3.14A are on the order of 10^7 (loaded Q is 2×10^6, Figure 3.15). The particular resonator's Q-factor was limited by residual surface roughness.

With this approach, one can make a single-mode optical WGM resonator of any size. A resonator the size of an apple requires a ring with dimension of tens of microns for single

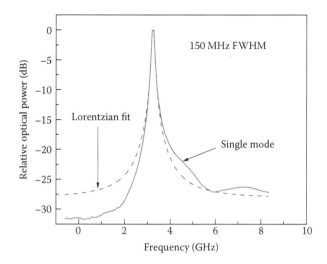

FIGURE 3.15
A Lorentzian fit of a WGM spectrum of which is shown in Figure 3.14A.

FIGURE 3.16
Elliptical LiNbO₃ resonators. (Reprinted from Mohageg, M., Savchenkov, A., and Maleki, L. *Opt. Express.* 15, 4869–4875, 2007. With permission from OSA.)

mode propagation of one micron wavelength light. This is counter-intuitive because the size of the single-mode channel is much larger than the wavelength. This type of experiment may be interesting as a fundamental research investigation. For practical applications small resonators are of more interest. In this context, our approach leads to a novel means for engineering of microcavity spectra.

The type of the dielectric media used for the proposed resonators is not important, as long as it is solid and transparent. Generally, since the behavior of the system is defined only by its geometry, it is not even important for what kind of waves the structures is a resonator. Microwave, acoustical or mechanical chains of resonators of this type have the same features and basis of design as these optical counterparts.

3.4.4 Elliptical Resonators

Mechanical fabrication that allows engineering the shape of the rim of a toroidal resonator, also enables producing elliptical and other kinds of geometrically shaped symmetric resonators. These resonators can retain high-Q WGMs, and also can possess unstable low-Q modes. Elliptical high-Q lithium niobate WGM resonators (see Figure 3.16) were studied elsewhere [52]. It was shown that the elliptical geometry of the resonator rim simplifies the coupling procedure. Coupling efficiency exceeding 97% was demonstrated.

Unlike modes in race track resonators, WGMs in elliptical resonators are stable. Their approximate spectrum can be found by means of the eikonal method [52]. Assuming a closed resonator we find for the high-order modes

$$2knRE(1, \varepsilon) = \pi m, \tag{3.28}$$

where $E(1,\varepsilon)$ is the complete elliptic integral of the second kind. Equation 3.28 has two parameters that define the geometry of the ellipse: eccentricity ε and semimajor axis length R.

Obtaining a rigorous expression for the modal spectra of elliptical resonators is rather involved. The determination of WGM spectrum in a spherical resonator is an easy task because the Helmholtz equation can be solved exactly by separation of variables. This is not the case for an elliptical resonator. A detailed mathematical analysis of the WGMs in elliptical resonators with large size compared to the resonant wavelength can be found in [121,122]. Waveguide coupling to elliptical WGM resonators were studied numerically [123]. It is worth noting that laser emission from ellipsoidal resonators was studied experimentally [124].

The fabrication procedure of the crystalline elliptical resonators is straightforward. A piece of crystal in the shape of a baseball diamond was cut from a Z-cut wafer of congruent $LiNbO_3$. The lengths of the axes of the diamond were selected to be slightly larger than the target dimensions of the desired ellipse. The edges of this structure were mechanically ground until the sharp edges of the diamond shape were smooth in a continuous curve resembling an ellipse. At this point, mechanical polishing techniques were applied to shape the edge curvature and polish the edge surfaces.

Three optical resonant cavities of elliptical cross section are shown in Figure 3.16. The eccentricity of the first cavity is 0.57. The lengths of the semimajor and semiminor axes are 4.4 and 3.6 mm, respectively. The second cavity has an eccentricity of 0.71, with semimajor and semiminor axis lengths of 4.6 and 6.5 mm, respectively. The thickness of both resonators is 0.5 mm. The third cavity has an eccentricity of 0.32, with semimajor and semiminor axis lengths of 3.4 and 3.6 mm, respectively.

Design of the resonator geometries is impacted by the coupling mechanism. Evanescent coupling of light into dielectric resonators requires phase matching between the wavenumber in the coupling medium and the wavenumber of the resonant mode. As is shown above, it means that the refractive index of the coupling medium should be larger than the refractive index of the resonator host material to couple to WGMs. Further, the coupling material should have a comparable optical transparency window to that of the host material to avoid additional round trip losses.

$LiNbO_3$ is a birefringent dielectric, and the coupling conditions for TE and TM modes in this material are slightly different. For example, to couple to TE modes of the resonator, the free space beam should have an incidence angle of 22° with respect to the plane tangent to the prism–resonator interface. A circular symmetric beam entering the prism at this angle will be elliptical in cross section at the interface plane. For TE modes, the spot size along the polar axis of the resonator is larger than the spot size measured along the azimuthal axis of the resonator by a factor of 2.7. There will be a drop in the overall coupling efficiency when the ratio achieved during fabrication is not equal to this value.

The ratio is fixed for resonators with circular symmetry, and is subject to the precision of the fabrication apparatus, knowledge of the refractive indices of the resonator and coupler materials, and knowledge of the wavelength of light being used. Realization of a critical curvature ratio at some point along the perimeter of the resonator is more probable for an elliptical resonator. The curvature ratio varies along the perimeter as the radius of curvature changes in those resonators.

Experiments with elliptical lithium niobate resonators were performed with 1550 nm light evanescently coupled to the resonators by means of a diamond prism [52]. The laser light was scanned over several FSRs of the resonators. Light reflected from the cavity-prism interface was collected by a 3 mm diameter light pipe and sent to an InGaAs photodetector. One of the assumptions of the approximation used to model WGMs in elliptical resonators is that the modes of the same modal indices travel the same optical path regardless of the input positions. That is, the FSR of the resonator is constant as the input position is varied. The position of the incident beam, governed by the location of the diamond prism along the cavity perimeter was varied between semimajor axes in 30° steps. It was found that throughout the various input angles, the same frequencies of light are resonant with the cavity and the same FSR was observed [52]. This observation confirms the theoretical assumption.

The FWHM of the mode was measured, and the value was used to calculate the Q-factor of this mode, yielding 1.5×10^8, which corresponds to the highest quality factor measured with congruent lithium niobate resonators (see next section). The measured FSR agrees well with the calculated values from Equation 3.28.

A coupling efficiency of 97.3% into the third resonator (Figure 3.16) was achieved. It was found that the coupling efficiency varied when the contact position between the prism and resonator was changed. The coupling efficiency refers to the fraction of light transmitted from within the coupling medium to a specific resonator mode and out through an output path symmetric with the input path. Light that couples from the mode to a different output channel, i.e. another WGM, is accounted for as additional loss. Fresnel losses at the air–prism interface are also present in the system. They could be mitigated by an appropriate antireflection coating of the prism. Taking Fresnel losses and loss associated with instrumentation into account we found that the best laser-to-detector transmission efficiency through the resonator remains in excess of 91%.

3.5 Quality Factor and Finesse of Crystalline Resonators

In this section, we discuss achievable values for quality factors that can be realized in crystalline WGM resonators [34]. Resonators are usually characterized with two partially dependent parameters—finesse (F) and quality factor (Q). The finesse of an empty FP resonator is defined solely by the reflectivity of its mirrors and is calculated as $F=\pi\sqrt{R}/(1-R)$. The maximum reported value of reflectivity $R \simeq 1-1.6\times 10^{-6}$ is achieved with dielectric mirrors [125,126]. A FP resonator made with these mirrors has finesse $F=1.9\times 10^6$. Further practical increase of the finesse of FP resonators is problematic because of the absorption and scattering of light in the mirror material [127,128], though the fundamental limit on the reflection losses given by the internal material losses and by thermodynamic density fluctuations is of the order of parts in 10^9 [127]. The quality factor of a resonator depends both on its finesse and its geometrical size. The one-dimensional FP resonator has $Q=2FL/\lambda$, where L is the distance between the mirrors, and λ is the wavelength. It is easy to see that the quality factor of the resonator is essentially unlimited because L is unlimited.

Finesse and Q are typically equally important in the majority of applications. In some cases, though, finesse is technically more valuable than the quality factor. For instance, the buildup of optical power inside the resonator and the Purcell factor [129] are proportional to finesse. Sometimes quality factor is more important. For example, the inverse threshold power of intracavity hyperparametric oscillation is proportional to Q^2 [130], and efficiency of parametric frequency mixing is proportional to Q^3 [131]. Therefore, it is important to know both the maximally achievable finesse and quality factor values of a resonator.

The knowledge of the resonator quality factor is important for the applications in material science. Not much data is available for the optical absorption coefficients, since there is no technique capable of measuring this coefficient with high precision when absorption becomes very weak. Interestingly, the WGM resonators may be used to measure internal absorption with good precision if the Q factor is only limited by the internal absorption. For example, the absorption coefficient reported by CaF_2 crystal producer is 2×10^{-5} cm^{-1}. According to the expression for the internal absorption limited Q factor, this corresponds to $Q = 4.2 \times 10^9$ at wavelength of 1064 nm. On the other hand, with this piece of crystal we have measured $Q = 5.3 \times 10^{10}$, which gives a more accurate upper estimate for the absorption coefficient 1.6×10^{-6} cm^{-1}.

WGM resonators can have larger finesse compared with FP resonators. For instance, fused silica resonators with finesse 2.3×10^6 [132] and 2.8×10^6 [133] have been demonstrated. Crystalline WGM resonators reveal even larger finesse values, $F = 6.3 \times 10^6$ [134], because of low attenuation of light in the transparent optical crystals. The large values of F and Q result in the enhancement of various nonlinear processes. Low threshold Raman lasing [135,136], opto-mechanical oscillations [137], frequency doubling [131], and hyperparametric oscillations [138,139] based on these resonators have been recently demonstrated. Theory predicts the possibility of achieving nearly 10^{14} for Q-factors of optical crystalline WGM resonators at room temperature [131,134], which correspond to a finesse level higher than 10^9.

Experiments, thus far, have shown numbers that are a thousand times lower. The difference between the theoretical prediction and the experimental values is due to material imperfections. To bridge this gap, a technique to substantially reduce the optical losses caused by the imperfections was developed [34]. A specific multi-step asymptotic process was utilized for removing imperfections in the host material of the resonator. This technique was previously used to reduce microwave absorption in dielectric resonators [24]. One step of the process consists of mechanical polishing performed after high temperature annealing. Several repeated subsequent steps lead to a significant reduction of the optical attenuation and, as a result, the increase of Q-factor and finesse of the resonator. With this approach we demonstrated a CaF_2 WGM resonator with $F > 10^7$ and $Q > 10^{11}$.

There are several practical issues related to the study of properties of ultra-high-Q resonators. The study is challenging because the spectral width of the corresponding WGM is expected to be less than a kHz. Direct observation of such a narrow line could be done with a narrowband tunable laser with a high level of frequency stability. Such lasers are not available in many laboratories. Another problem is related to the parasitic contributions from nonlinear effects. Because the threshold of stimulated Raman scattering (SRS) is very low in high-Q resonators [136] the direct observation of the intrinsic spectral width of a WGM with a continuous wave laser radiation is a problem. Thermal oscillations and drifts of the position of the optical resonance also make a time-averaged observation problematic.

A nonlinear ring-down technique that was recently demonstrated [140] helps avoiding these limitations. In this approach, the laser frequency is swept rapidly across the resonance line. A part of the light is accumulated in the mode during the sweep and is reemitted in the direction of the detector. The reemitted radiation interferes with the laser light, and beatnote signals can subsequently be observed. The power of the laser light is generally much larger than the power of the reemitted light. Hence, the envelope of the decaying oscillations follows the decay of the amplitude of the reemitted light. This decay is twice as long as the intensity decay. The optical Q-factor of the cavity is expressed as $Q = \omega\tau/2$.

We selected a fluorite WGM resonator with optical loss limited by the material attenuation, not by the surface scattering [141]. The resonator was 4.5 mm in diameter, had 0.5 mm in thickness, and 32 μm in diameter of sidewall curvature. The resonator was placed into the center of a 3-feet long air-filled transparent tube made of annealed fused silica. The tube was installed into a 20 cm long horizontal tube furnace. The heated furnace core had approximately one inch in diameter and three inches in length. We increased the temperature of the furnace core from room temperature to 650°C during a 3 hour period, and kept the temperature stabilized for 1 day. The core was then cooled back during 3 hours, and the fluorite resonator was subsequently repolished. We repeated this process three times keeping the same annealing duration but gradually decreasing the size of the grain of the diamond slurry used for polishing. This approach resulted in a significant increase of the ringdown time at the end of the process.

The measured ringdown spectrum did not change substantially after the first annealing stage. However, the ringdown time increased significantly after the third stage. The measured ringdown signals are shown in Figure 3.17. A five-fold increase of the optical ringdown time is clearly observed. It is also worth noting that the measured quality factor is several times larger when compared with the quality factor of calcium fluoride resonators observed previously at 1.55 μm (see [134,141]).

The power in the mode of the annealed resonator was clearly above the threshold of the nonlinear loss related to SRS [140]. The measurement method used in the experiment has a low dynamic range, so it was not possible to observe the nonlinear decay with the high-amplitude signals. For that reason the ringdown time was measured using a smaller optical power. In such a measurement two major precautions must be taken: (i) the laser wavelength must interact with only one mode, which means that no WGM doublets are

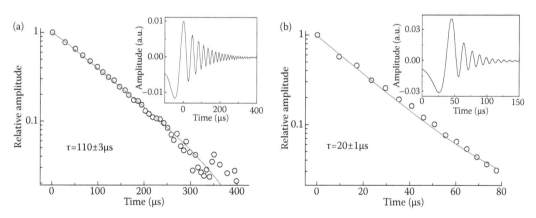

FIGURE 3.17
(a) Ring-down signal after one annealing step. (b) Ring-down signal after three annealing steps. The exponential fit (solid curve) is nonlinear in the logarithmic scale because the exponents have constant offsets. Keeping in mind that $Q = \omega\tau/2$, and $\omega = 2\pi c/\lambda$, $\lambda = 1.55\mu m$, we find that the values of the quality factors after the first and third annealing steps are $Q = 1.2 \times 10^{10}$ and $Q = 6.7 \times 10^{10}$, respectively. It is important to note that the initial value of the Q-factor corresponds to the earlier observations [134], while the final value is the apparent improvement at the given wavelength. (Reprinted from Savchenkov, A. A., Matsko, A. B., Ilchenko, V. S., and Maleki, L., *Opt. Express* 15, 6768–6773, 2007. With permission from OSA.)

allowed within the investigated frequency span. (ii) the sweep of the frequency of the laser must be controlled and the local frequency of the beatnote signal must never increase; such an increase can occur if the laser is swept back to the starting frequency rapidly. In this way the single excitation of a single WGM is ensured.

The best ringdown signal measured with all the precautions is shown in Figure 3.18, left. The oscillation period of the beatnote was evaluated in order to ensure that the measured line is not a result of several consecutive excitations of the optical mode caused by dithering of the laser frequency (Figure 3.18, right). This shows that the carrier frequency indeed moves gradually from the resonance and the WGM is not excited twice.

Let us discuss the fitting procedure to the theoretical curve in more detail. The positions of the maxima and minima of the beatnote peaks, as well as the zero crossings of the beatnote signal, were identified for the evaluation of the oscillator period. The time coordinate of each peak was subtracted from the coordinate of the adjacent peak and the averaged time for the two adjacent peaks was obtained. The same procedure was repeated for the zero-crossing points. In this way the time dependence of the period of the beat note signal, shown in Figure 3.18 right, was deduced. The dependence is linear at the tail of the curve. The initial period has a different time dependence because the frequency of the WGM changes much faster immediately after the frequency of the pump laser is tuned away from the WGM. The change is determined by multiple nonlinear processes, e.g. WGM frequency shift due to after-interaction cooling of the resonator.

Using linear approximation for the beatnote period we found the period of the waveform to fit the experimental data. We solved the equation $\phi(t + \text{Period}(t)/2) - \phi(t) = \pi$ approximating it by a linear differential equation $\dot{\phi}(t) = 2\pi/\text{Period}(t)$. Using the experimental results the $\text{Period}(t) = \xi - \zeta t$ is obtained, where $\xi = 71.1$ µs, and $\zeta = 0.11$ (time t is measured in microseconds). As a result, the expression

$$\phi(t) = \text{Const} - (2\pi/\zeta) \ln(1 - \zeta t/\xi) \tag{3.29}$$

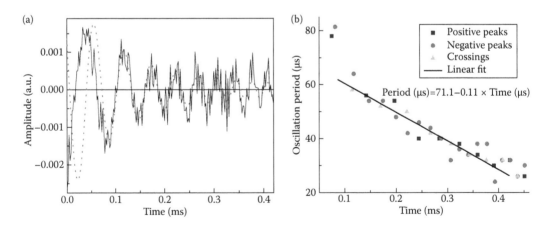

FIGURE 3.18
(a) Ringdown signal after the third annealing step taken with low-power laser radiation (solid line). The theoretical fit of the signal is shown by the dotted line. (b): Evaluated change of the period of the signal. The solid line stands for the linear fit of the time dependence of the period of the ringdown signal. (Reprinted from Savchenkov, A. A., Matsko, A. B., Ilchenko, V. S., and Maleki, L., *Opt. Express* 15, 6768–6773, 2007. With permission from OSA.)

follows for the phase. The final oscillation waveform presented in Figure 3.18, right is given by $\cos[\phi(t)]$.

The time dependence of the beatnote amplitude was extracted in the following way. The amplitude of a minimum of the oscillating beatnote signal was subtracted from the adjacent maximum. The value shows the relative oscillation amplitude. This value was related to the moment of time equally separated from the maximum and the minimum time coordinates. The result of the evaluation is shown in Figure 3.19. Using the first three points of the dependence the initial decay time $\tau_1 = 130$ μs was found. Using the part of the signal in the interval between 0.15 and 0.35 ms the final decay time, $\tau_2 = 510$ μs was found. Finally, we used the expression for the nonlinear decay rate discussed by Matsko et al. [142] to fit the amplitude decay rate. The resultant fit of the beatnote waveform is shown in Figure 3.18, left. In this way, we can conclude that the intrinsic linear quality factor of the CaF$_2$ WGM resonator approaches 3×10^{11} at 1.55 μm (20-fold improvement compared with the initial quality factor).

Let us calculate the finesse of such a resonator. The expression for the finesse, as well as the quality factor, of WGM resonators naturally includes index of refraction of the resonator host material, unlike the case of an empty FP resonator: $Q = 2\pi a F n_0/\lambda$, and $F = c/[2an_0(\gamma_0 + \gamma_c)]$. Here a is the radius of the resonator, n_0 is the index of refraction of the resonator host material, γ_0 and γ_c are the intrinsic and coupling amplitude decay rates, respectively. The values of light intensity inside I_{in} and outside I_0 the resonator are related as $I_{in}/I_0 = 2\gamma_c F/[\pi(\gamma_0 + \gamma_c)]$. Using the experimental data ($2a = 0.45$ cm, $n_0 = 1.42$) the finesse of the resonator is found to be $F = (2.1 \pm 0.6) \times 10^7$.

The annealing process discussed above improves the transparency of the material because an increased temperature results in the enhancement of the mobility of defects induced by the fabrication process, and also reduces any residual stress birefringence [143]. The increased mobility leads to the recombination of defects and their migration to the surface [24]. It is worth noting that this annealing technique is similar to the previously developed procedures [144,145].

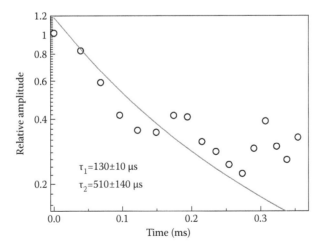

FIGURE 3.19
Change of the signal amplitude shown in Figure 3.18 left, with time. Initial and final quality factors are $Q = (7.9 \pm 0.5) \times 10^{10}$ and $Q = (3 \pm 1) \times 10^{11}$, respectively. Solid line stands for the theoretical fit of the ringdown signal found using formalism presented by Matsko et al. [142]. A small oscillation of the decaying signal may result from the residual scattering in the material becoming observable at the given value of quality factor.

The significant improvement of the Q-factor and finesse of a fluorite WGM resonator discussed above does not reach the fundamental limit. The straightforward annealing of a WGM resonator leads to $Q > 10^{11}$ at 1.55 μm while our earlier theoretical prediction gives $Q \simeq 10^{13}$ at this wavelength [141]. To improve the quality of annealing we suggest using a larger furnace with much lower thermal gradients. Thermal gradients within a sample play the leading role in the defect redistribution [143]. The amplitude of the gradient can be reduced not only by increasing the volume of the oven but also by increasing the thermal conductivity and optimizing the shape of the container the resonator is placed into during the annealing procedure. A fluorite container is the best choice for annealing fluorite WGM resonators.

In this section, we demonstrated that a proper thermal annealing procedure combined with mechanical polishing allows producing WGM resonators with vastly improved finesse and quality factors. We expect that the technique described here will help to improve quality factors and finesses of WGM resonators made of other transparent crystalline materials.

3.5.1 Fundamental Limits

Mechanical polishing techniques discussed above significantly suppress surface scattering and surface absorption as loss mechanisms [38] in crystalline resonators. The linear and nonlinear attenuation in the crystalline dielectric material is the remaining physical mechanisms limiting the ultimate achievable Q. We discuss such a fundamental attenuation in this section [146].

Optical transparency of ideal dielectric crystals is restricted by the fundamental blue and red wings of single photon absorption. The blue wing is defined by optically allowed transitions between valence and conduction electron bands. The red wing results from the interaction between light and phonons in the crystalline medium. Both absorption mechanisms have a strong frequency dependence, so the absorption of a dielectric within the transparency band can be extremely low. The question is how low the attenuation could be in reality, and is there any other mechanism that restricts the transparency of dielectrics. The question regarding the minimum attenuation of light propagating in a solid dielectric has been discussed for optical fibers by Lines [147]. It was shown that the scattering loss mechanisms such as Rayleigh, Brillouin and Raman scattering determine the extent of optical attenuation. These mechanisms also present in ideal optical crystals and limit their transparency.

We have analyzed [146] the minimum light attenuation achievable with ideal crystals at room and cryogenic temperatures. We used a real, nearly perfect, fluorite crystal as an example. It was shown previously that at room temperature the attenuation of UV light in artificially grown CaF_2 is nearly completely limited by Brillouin and thermodynamic Rayleigh scattering [148]. We have shown that the spontaneous Raman scattering is primarily responsible for attenuation in an ideal crystal at low temperatures. Applying this result to the crystalline WGM resonators we have found the limits to the quality factor. We also have shown that, in contrast to the low temperature regime, spontaneous Brillouin scattering as well as SRS determine the Q factor of a WGM resonator under realistic experimental conditions at room temperature. We have measured the influence of SRS on the Q factor of a millimeter-sized resonator using a ring-down technique and shown that this scattering becomes important at input power levels as low as a few microwatt.

Let us consider Rayleigh, Brillouin and Raman scattering in a perfect crystal. Thermodynamically limited Rayleigh scattering is small in an ideal crystal, and Brillouin scattering dominates. The quantitative ratio of Rayleigh and Brillouin scattering can be estimated using the Landau–Plachek relationship (see e.g. [149]):

$$\frac{\alpha_{Ri}}{\alpha_B} \simeq \frac{\beta_T}{\beta_S} - 1 \tag{3.30}$$

where β_T and β_s are the isothermal and isobaric compressibilities, respectively. This ratio is generally less than unity in crystals and is proportional to temperature in the low temperature limit [150]

$$\frac{\beta_T}{\beta_S} - 1 = \frac{C_P}{C_V} - 1 \sim T^4 \Big|_{T \to 0}, \tag{3.31}$$

where C_p and C_v are the specific heat capacities at constant pressure and constant volume, respectively. Rayleigh scattering can be neglected in a perfect crystal at low temperature, as a result of Equation 3.31).

Temperature dependence of spontaneous Raman and Brillouin scattering mechanisms is determined by the number of participating phonons, given by a Bose population factor [151]. The corresponding light attenuation coefficient accounting for both Stokes and anti Stokes components in a transparent solid may be estimated as follows (see also [149])

$$\alpha_{B,R} \approx \alpha_{B0,R0} \left(\frac{\lambda_0}{\lambda}\right)^4 \left[\left(\exp\frac{\hbar\Omega_{B,R}}{k_BT} - 1\right)^{-1} + \frac{1}{2}\right] \tag{3.32}$$

where λ is the wavelength of light, $\alpha_{B0,R0}$ are the scattering parameters given by the properties of a particular crystal corresponding to λ_0, Ω_B is the Brillouin frequency shift for $90°$ scattering and Ω_R is the Raman frequency shift.

We assume that the crystal has a single phonon branch. Using data evaluated for CaF_2 at $T=300$ K in [148]: $\hbar\Omega_B/k_BT \simeq 1/300$, $\hbar\Omega_R/k_BT \simeq 1/4$, $\lambda_0 = 0.532$ µm, $\alpha_B = 2.411 \times 10^{-7}$ cm^{-1}, $\alpha_R \simeq 4.34 \times 10^{-8}$ cm^{-1} we obtain $\alpha_{B0} \simeq 8 \times 10^{-10}$ cm^{-1}, $\alpha_{R0} \simeq 1.1 \times 10^{-8}$ cm^{-1}. It is easy to see that Brillouin scattering is significantly suppressed at low temperature and the attenuation of light in a perfect crystal is determined by the spontaneous Raman scattering. The room temperature attenuation, on the other hand, is given by spontaneous Brillouin scattering.

It is shown by Logunov and Kuchinsky [148] that the existing calcium fluoride crystals have transparency close to the fundamental limit (Equation 3.32) in the UV. However, such a low attenuation is yet to be demonstrated in experiments with visible and infrared light. The measured transparency is two to three orders of magnitude lower than the fundamental limit in those frequency bands [38,141,134] because of the extrinsic and intrinsic impurities of the material.

We now discuss the optical properties of fluorite WGM resonators. The maximum quality factor of the resonator is given by $Q_{max} = 2\pi n(\lambda)/[\lambda\alpha(\lambda)]$, where $n(\lambda)$ is the refractive index of the material, λ is the wavelength of light in vacuum, and $\alpha(\lambda)$ is the total loss coefficient of the bulk dielectric material. The corresponding wavelength dependence of the attenuation as well as Q factor for CaF_2 are shown in Figure 3.20. The approximation of the maximum quality factor of CaF_2 WGM resonators found elswhere [141,134] using existing experimental results nearly coincides with the fundamental limit presented in Figure 3.20.

The wavelength dependence of the index of refraction of the material at $T=300$ K was found using the four-term Sellmeier equation [152]. Blue and red absorption wings may

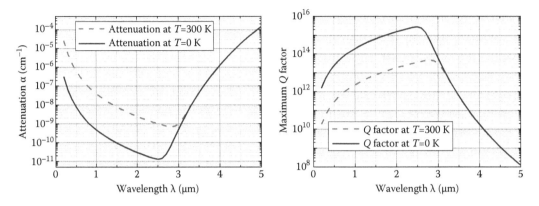

FIGURE 3.20

Attenuation in ideal CaF_2 and Q factor of ideal crystalline WGM resonators at room and nearly absolute zero temperature. Contributions from spontaneous Brillouin, Rayleigh and Raman scattering as well as blue and red wing absorption are added. (Reprinted from Grudinin, I. S., Matsko, A. B., and Maleki, L., *Opt. Express* 15, 3390–3395, 2007. With permission from OSA.)

generally be approximated by a simple exponential dependence [153,154]. Parameters for the blue wing were derived from Laufer et al. [155]; red wing parameters were obtained from Boyer et al. [156] and the experimental data publicly available for Corning CaF_2. For our analysis we assume both wings to be temperature independent, given that the actual temperature dependence is weak.

Not only spontaneous, but also stimulated scattering is important in macroscopic WGM resonators. Stimulated Rayleigh scattering is not experimentally observable in high-Q WGM resonators as it is suppressed by a mismatch of modal structure of a typical cavity and the gain profile for this process [149]. Stimulated Brillouin scattering is generally suppressed because of the absence of phase matching in the WGM resonators. The resonantly enhanced SRS is always possible if the frequency difference between subsequent modes of the same mode family (FSR) is less than the spectral width Γ of the optical phonons. In fluorite at room temperature $\Gamma=450$ GHz [157]. Hence, any resonator with radius $R \geq c/(2\pi n(\lambda)\Gamma) \simeq 75\,\mu m$ is not immune to SRS. The properties of the stimulated scattering processes are discussed in the other chapter available in this volume.

3.5.2 Technical Limits

The optical Q factor of a resonator after polishing is limited by contributions from two main technical sources: (i) absorption and scattering due to accumulation of dust on the resonator surface, (ii) absorption and scattering in the resonator material due to intrinsic and extrinsic impurities.

Absorption due to dust particles would be the main restriction on the quality factor if a clean room environment is not provided. Once inside the clean room environment or vacuum, the Q factor can be preserved on a very high level for an indefinite length of time. Aside from dust accumulation, optical polishing leaves some amount of nanoparticles embedded into the resonator surface. It is known that water is detrimental for Q factors of fused silica devices due to the formation of irregular adsorption layers. Fortunately, crystals such as CaF_2 and MgF_2 have chemical properties that reduce accumulation of water on the surface. Diffusion of water into crystalline lattice is inhibited, while water monolayer (a one molecule thick layer) is always present on the surface.

The intrinsic absorption of the resonator material is the second important effect restricting the value of the measured Q factor. The absorption is a function of crystal growing process and is defined by the optical properties of a crystal at a given wavelength. This parameter is particularly small in crystalline materials, since they can be produced with high purity (see previous section). Recent developments in UV lithography have stimulated the production of extremely pure CaF_2 monocrystals. However, even in those purest crystals the "reference-sheet" absorption rarely falls below 10 ppm/cm, which corresponds to quality factors on the order of 10^{10}.

The surface roughness is made small by the polishing technique, as confirmed by measurements with an AFM. In its contribution to surface scattering factor, the surface roughness is accompanied by the subsurface material damage. This damage is represented by the microfractures and embedded particles and is usually confined to within a shallow surface layer of the material. The thickness of this layer is determined by the polishing and grinding process. Studies with atomic force microscope (AFM) have shown the small amount of diamond particles embedded into the resonator surface. This could be useful in quantum optics, since quantum dots containing diamonds or other nanoparticles can be embedded in this way to create a coupled system without noticeable degradation of the Q factor.

For example, we have fabricated a calcium fluoride resonator with 100 μm in diameter and optical Q factor of 4×10^8. The surface roughness does not restrict the value of the quality factor. The measurements performed with an AFM showed that the surface roughness in small resonators is about the same as in the large disks with Q factor of about 10^9 and is practically negligible (σ=0.33 nm) [38]. Let us estimate the maximum Q factor that could be achieved in a resonator with measured surface roughness. We will use the expression derived by Gorodetsky [21]

$$Q \approx \frac{3\lambda^3 R}{8n\pi^2 B^2 \sigma^2} \tag{3.33}$$

where λ is the wavelength, R is the radius of the resonator, n is the refraction index of the resonator material, B is the correlation length, and σ is the roughness. The formula was obtained as an approximation which takes the correlation length to be much smaller than the wavelength of light ($\sigma B^2 \ll \lambda^3$). It was found that $Q=4 \times 10^{12}$ for $\lambda=1$ μm, $R=50$ μm, $n=1.43$, $B=2$ nm (was not calculated), and σ=0.33 nm [38]. The estimated value of the Rayleigh scattering limited Q factor scales proportionally to the radius for smaller resonators and is still much larger than any Q factor observed in WGM resonators.

More rigorous studies of the surface quality of ultra-high Q WGM resonators were performed in [134]. It was taken into account that the noise introduced by AFM is small (the AFM empty image standard deviation was 0.07 nm). The surface roughness was measured to be 0.19 nm and 0.15 nm in two different samples. Calculations were made with a standard expression for the unbiased variance. These values are significantly smaller than those typical of fused silica resonators, and demonstrate the capability of the fabrication technique. It is possible to achieve even better results if advanced polishing techniques are used.

In order to use the analysis, the correlation length B also has been carefully evaluated. The correlation length is defined as the decay length of the exponential autocorrelation function. We are interested in the correlation length along the direction of light propagation, which in our case is coincident with the direction of polishing traces. Computer algorithms of the image rotation generally alter the correlation function. In addition, the autocorrelation function depends on the direction of computation for the type of the surface discussed here. In order to overcome these problems, a two-dimensional correlation function may be computed with the use of a Wiener–Khinchin theorem and fast Fourier transform algorithms. It was found that $B \simeq 40$ nm, 20 times larger than the value assumed in [38]. The maximum corresponding Q is on the order of 7×10^{11} and is still higher that the quality factors measured experimentally.

It was also found that when an excimer grade CaF_2 is used, it is possible to achieve a Q factor higher than in any other open resonator and any other WGM resonator without annealing. The 5.5 mm in diameter CaF_2 WGM resonator was fabricated with Q factor of $(5.31 \pm 0.04) \times 10^{10}$ measured with a Nd:YAG laser at wavelength of 1064 nm. This is probably due to specific impurity content of the resonator host material.

To conclude this section, we note that a record high optical Q factor of a fluorite WGM resonators fabricated with a diamond grinding technique was demonstrated. Even higher Qs for WGM resonators made out of this and other materials were predicted after an analysis of possible loss mechanisms in crystals and the influence of nonlinear processes.

The extreme Q factors of crystalline resonators open new opportunities for optical filtering and spectroscopy. A single mode cavity made with lithium niobate could operate as a tunable fiber coupled etalon, more compact and with a higher finesse than a FP resonator. We discuss such filters in the following section.

3.6 Filters and Their Applications

Photonic filters based on optical WGM resonators are currently among the highest developed devices that utilise WGMs. For optical telecommunication purposes, the main task of the filters is to select channels, as in wavelength division multiplexing schemes. In this domain, where channel spacing is usually no less than 10 GHz, planar ring resonators with WGM and similar devices with $Q < 1 \times 10^5$ are adequate [163–168].

Ultra-high-Q WGM resonators with MHz range resonance bandwidths offer a unique opportunity for realization of photonic microwave filters in which optical-domain selection is used for separating the RF channels imprinted as sidebands on a stable optical carrier. Photonics filters based on optical resonators address the shortcomings of microwave filters. Multi-pole, high-Q filters based on cascaded WGM resonators fabricated with silica have been demonstrated allowing compact packages and robust performance. These filters provide passbands with a flat-top and sharp skirts, suitable for high performance applications. Since microwave signals in photonic systems are sidebands of an optical carrier, these filters can be used at any microwave frequency, providing the same characteristics throughout the band, from 1 to 100 GHz, and higher.

While delivering high optical Qs and desirable passband spectra, optical WGM filters based on silica ring resonators are limited in their microwave Q, and are not tunable. Crystalline WGM filters overcome this problem. Fabrication of optical WGM resonators with lithium niobate and tantalate [159] has led to the demonstration of high-Q tunable filters with linewidth of about 1 MHz and wider. A crystalline WGM resonators was used as a Lorentzian microwave filter with a tuning range in excess of tens of gigahertz [160]. A miniature resonant electro-optically tunable third-order Butterworth filter based on cascaded lithium niobate WGM resonators has also been demonstrated

[161]. Fabrication of a fifth-order narrow band tunable WGM filter was reported in Savchenkov et al. [162].

Existing WGM filters still do not solve all the problems of the microwave photonics. For example, a recently demonstrated lithium niobate WGM filter [161] has a third-order filter function, 30 MHz linewidth, and is tunable over a wider range of frequency (>20 GHz). Agile tunability accompanied by the third-order filter function, narrow linewidth, and low loss, make this filter exceptionally useful for various microwave photonics applications. However, the comparatively low-order filter function represents a limitation for the application of these filters in microwave systems that require efficient sidemode rejection, and high transmission/rejection contrast. While increasing the number of coupled WGRs theoretically yields the required high-order high-contrast filter function, the technical problems associated with the device fabrication pile up exponentially. Some of these problems were analyzed and solutions were proposed in Savchenkov et al. [162]. In what follows we review recent achievements in fabrication of filters using crystalline WGM resonators.

3.6.1 First-Order Filters

The simplest resonator-based filter includes a WGM resonator and an optical coupler, e.g. a prism or an angle-polished fiber coupler (Figure 3.21). Transmission of a monochromatic electromagnetic wave of frequency ω by an optical WGM resonator in a single prism configuration may be characterized by Equation 3.4.

The filter described by Equation 3.4 is a stop-band filter because it is characterized by the absorption resonance. A WGM resonator with two, input and output, couplers is characterized by a transmission resonance (Figure 3.21). This is a passband filter. An example of a packaged passband filter consisting of a single WGR is shown in Figure 3.22. The transmission and reflection coefficients through the resonator are given by Equations 3.8 and 3.9.

An example of the tunability of a stop-band LiNbO$_3$ WGM filter is shown in Figure 3.23. The maximum frequency shift of TE and TM modes may be found from [158]

$$\Delta v_{TE} = v_0 \frac{n_e^2}{2} r_{33} E_Z, \quad \Delta v_{TM} = v_0 \frac{n_0^2}{2} r_{13} E_Z, \tag{3.34}$$

where $v_0 = 2 \times 10^{14}$ Hz is the carrier frequency of the laser, $r_{33} = 31$ pm/V and $r_{13} = 10$ pm/V are the electro-optic constants, $n_e = 2.28$ and $n_0 = 2.2$ are the refractive indexes of LiNbO$_3$, E_Z is the amplitude of the electric field applied along the cavity axis. We worked with TM modes because they have larger quality factors than TE modes. If the quality factor is not

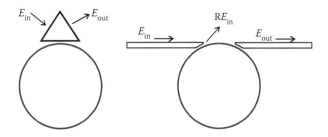

FIGURE 3.21
Scheme of single resonator filters with different coupling elements. The resonator with prism coupling element is a stop-band filter. The resonator with two angle-polished fiber couplers is a pass-band filter.

FIGURE 3.22
Filter setup with two prism couplers (passband configuration). The lithium niobate disc resonator is coated with metal.

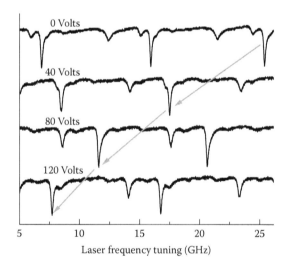

FIGURE 3.23
Tunability of the resonator spectrum, 14.7 GHz spectral shift per 100 V of applied voltage. Resonator FSR is 9.5 GHz.

very important, it is better to work with TE modes because their electro-optic shifts are three times as much as those of TM modes for the same value of the applied voltage.

Theoretically, Δv_{TE} and Δv_{TM} do not depend on the resonator properties and are related to the fundamental limitations of optical resonator-based high speed electro-optic modulators [169]. The results for different resonators measured in our experiment are not completely identical, a result that stems from the imperfections of the cavity metal coatings as well as a partial destruction of the coating during the polishing procedure.

The measured insertion loss for the pass band filter shown in Figure 3.22 was less than 5 dB (Figure 3.24). The insertion loss occurs primarily due to inefficient coupling to the

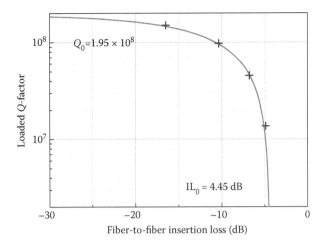

FIGURE 3.24
Insertion loss versus quality factor for a single resonator filter. The loading determined by the distance between the coupling prisms and the resonator was changed by temperature. The crosses stand for the experimental point and the line is the theoretical fit of the experimental results.

mode. We believe that antireflection coating of the coupling prisms or use of special gratings placed on high-index fibers may reduce the losses significantly.

For the pass band filter shown in Figure 3.22, we analyzed the operation of the filter in the switching regime. We sent microwave pulses to the filter and observed a change in light transmission as the microwave pulses entered the filter. An example of such a measurement is shown in Figure 3.25.

3.6.2 Periodical Poling and Reconfigurable Filters

Absolute tunability of optical resonator-based photonic filters is characterized by the ratio of their FSR and their linear tunability range. Tuning the filters does not change the FSR but only shifts the comb of the optical modes, making it to overlap with itself for each frequency shift proportional to the FSR. Hence, the filter can be tuned at any prescribed single frequency if the linear tunability exceeds the FSR.

Some photonic applications call for narrowband filters passing simultaneously both the carrier and the sidebands. For example, this is important for generation of spectrally pure microwave signals in opto-electronic oscillators, where beating of the optical sidebands and the carrier on a fast photodiode generates microwaves. Tunability of the microwave frequency of the oscillator requires a controlled change of the frequency difference between the filter passbands. This feature is lacking in existing tunable filters where the entire filter spectrum shifts as a whole as the tuning voltage is applied.

A filter with electro-optically reconfigurable spectrum was recently demonstrated [173]. The filter is based on a WGM resonator fabricated from a commercially available lithium niobate wafer. The crystalline resonator was fabricated with a special domain structure that results in controllable shift of the frequency of a single mode or a group of modes with respect to the other modes, as a DC bias voltage is applied across the resonator. This provides the possibility of tuning one resonance mode while keeping the rest of the spectra stationary. The filter may be characterized along the same lines as a FP filter with a tunable FSR. The proposed method has the potential for fabrication of resonant photonic filters with arbitrary passband spectrum.

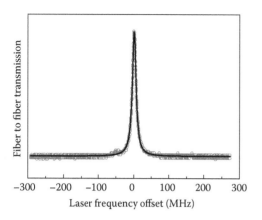 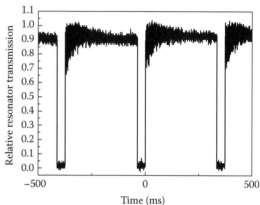

FIGURE 3.25
Left: the transmission resonance of the filter. The filter has 13.8 MHz FWHM and 4.3 dB insertion loss at the resonance. Right: 200 ns pulse front response in the switching regime. .

The filter operates at 1.55 μm wavelength, though the wavelength of operation is limited only by the absorption loss of lithium niobate and can be anywhere from about 1.0 to
1.7 μm. The reproducible value of the finesse of the filter (F) exceeds $F=300$, but in some
experiments finesse $F=1000$ was achieved.

The filter is a circular resonator made from electro-optic material (Figure 3.26). As discussed above, a homogeneous change of an applied electric field results in a homogeneous
change of the refractive index for a single-domain crystalline resonator and, as a consequence, in a frequency shift of the whole resonator spectrum. We propose to manipulate
the domain structure in the material to produce an inhomogeneous electro-optic effect in
the resonator in such a way that targetted mode families experience frequency shifts with
respect to other mode families.

The maximum frequency shift of the TE and TM mode can be found from

$$\Delta\omega_{TE} = \omega_0 \frac{n_e^2}{2}\left(\int_V r_{33}\,|E_{TE}|^2\,E_Z d\mathbf{r}\right)\left(\int_V |E_{TE}|^2\,d\mathbf{r}\right)^{-1}, \tag{3.35}$$

$$\Delta\omega_{TM} = \omega_0 \frac{n_o^2}{2}\left(\int_V r_{13}\,|E_{TM}|^2\,E_Z d\mathbf{r}\right)\left(\int_V |E_{TM}|^2\,d\mathbf{r}\right)^{-1}, \tag{3.36}$$

where ω_0 is the carrier frequency of the laser, r_{33} and r_{13} are the electro-optic coefficients,
sign of which is determined by the direction of the domain of the crystal, which could be
either $+z$ or $-z$, with respect to the homogeneous DC bias electric field E_z; n_o and n_e are the
refractive indices of lithium niobate, E_{TE} and E_{TM} are the amplitudes of the electric fields of
the modes. The integration is taken over the resonator volume V.

Various modes of a WGM resonator have various spatial dependences (Figure 3.26b).
The tunability of the modes can be changed in different ways by choosing spatial dependence of the electro-optic coefficients in a correct manner through poling the crystal.

A schematic diagram of the filter configuration is shown in Figure 3.26. Several 2.6 mm
diameter disk-shaped resonators of $LiNbO_3$ were fabricated, at 120 μm thickness. The rims

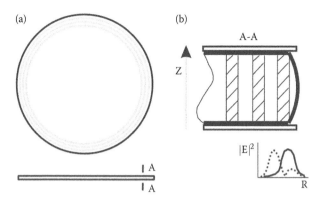

FIGURE 3.26
Schematic of the metalized disc resonator used in the reconfigurable filter. The resonator is made from a peri-odically poled lithium niobate wafer. (a) Picture of the disc resonator. The inverted domain structure is shown with gray circles. (b) Crossection of the resonator close to the disc rim. Inverted domain structure is shown by striped areas. Two geometrical structures of electro-magnetic modes of the resonator (radius dependence of the mode power distribution) are shown under the resonator crossection. The mode depicted by solid line is barely shifted by the applied voltage, while the mode shown by dotted line moves. (Reprinted from Mohageg, M., Savchenkov, A., Strekalov, D., Matsko, A., Ilchenko, V., and Maleki, L., *Electron. Lett.* 41, 91–92, 2005. With permission from Institute of Electrical Engineers (IEE).)

of the cylindrical disks were polished to resemble the surface of a sphere. A ferroelectric domain structure reminiscent of a set of rings concentric with the axis of the disk resonator was generated by dragging a 1 μm diameter electrode across the surface of the crystal while applying a 2.5 kV bias between the electrode and the bottom of the crystal, causing a permanent change in the structure of the material polarization. The poling process took place at room temperature and was visualized in situ by reflecting light from the bottom surface of the crystal. Domain walls, that is, barriers between polarizations parallel to +z and −z directions, are clearly visualized as dark bands in the reflected light.

The top and bottom surfaces of the polished and poled disk resonator were placed into contact with metal electrodes. These electrodes were connected to a 0–150 V regulated DC power supply. A probe beam of 1.55 μm scanned over 20 GHz was coupled into the resonator through a diamond prism. This allowed observing the absorption spectrum of the poled disk resonator and the motion of the modes as the voltage bias across the resonator axis was increased.

The first disk was poled with a ring-shaped domain pattern 5 μm edge-to-edge at the disk rim and a 35 μm thick ring-shaped domain 20 μm away. The probe was coupled into the WGMs of the resonator with quality factor of 10^7. When the probe beam was coupled into the high-Q modes of the first disk and voltage was applied, the radical mode was observed to change frequency with respect to the rest of the spectra at a rate of 21 MHz/V (Figure 3.27). The Q-factor of the resonance was maintained constant through the most of the motion. The mode motion was observed over 10 V of bias change.

3.6.3 Third-Order Filters

We made several miniature resonant electro-optically tunable third-order Butterworth filters. The filters are based on three metal coated WGM resonators fabricated from a commercially available lithium niobate wafer. The filters, operating at 1550 nm wavelength, have less than 30 MHz bandwidth and can be tuned in the range of ~±12 GHz by applying

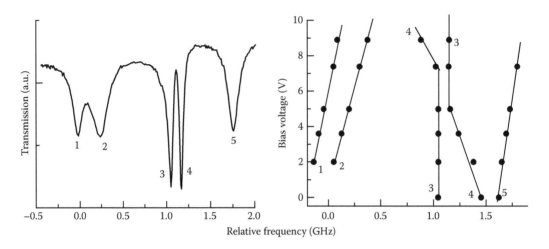

FIGURE 3.27

Spectrum of the WGM resonator and its tunability. The resonances 3 and 4 move through each other with bias voltage change. When the frequencies of the modes coincide, the mode interact due to residual light scattering in the disc resonator. Relative frequency shift between modes 3 and 4 versus applied bias voltage. (Reprinted from Mohageg, M., Savchenkov, A., Strekalov, D., Matsko, A., Ilchenko, V., and Maleki, L., *Electron. Lett.* 41, 91–92, 2005. With permission from IEE.)

a DC voltage of $\sim\pm 150$ V to an electrode. The basic operation principles of the filter are discussed in Savchenkov et al. [161].

The FSR of each resonator is approximately 10 GHz, and, therefore, the filter can be tuned practically at any optical frequency in the transparency window of lithium niobate. To clarify this statement, let us consider a single resonator first. If we are able to tune its spectrum by the FSR frequency, the resonator spectrum would overlap with itself, which means that it is always possible to find and to shift a resonator mode to coincide with an arbitrary external signal frequency. We do not need to use the same mode for filtering. Now, if we consider several interacting resonators, and tune the frequency of a mode of the smallest resonator (the largest FSR) by the FSR value, and hence the modes of the other resonators by values exceeding their FSR, then a high-order filter passband may be constructed at any desired frequency.

We note the following distinctive features/advantages of this type of filter over conventional filters: (i) agile tunability accompanied by a high-order filter function, (ii) narrow linewidth, and (iii) low fiber-to-fiber loss. Combination of these three features makes this a unique filter for a wide range of applications in optics and microwave photonics.

Tunable single-resonator filters are characterized by their finesse which is equal to the ratio of the filter's FSR and the bandwidth. The three-resonator filter has a significantly more sparse spectrum as compared with a stand alone WGM resonator. This feature is due to the so called Vernier effect [170] and is similar to the coupled fiber-ring resonators [171,172] which are noted for a rare spectrum due to a single sequence of modes. Finesse of the system is very large as the FSR exceeds 1 THz, though we were unable to measure it accurately because of the limited tuning range of the laser. The tuning speed of the filter is limited by the wiring layout and is approximately equal to 10 ns, while the actual shifting time of the spectrum is determined by the filter's bandwidth and does not exceed 30 ns.

The resonators were cut out from Z-cut metalized crystalline preforms. Each resonator perimeter edge was hand polished to a toroidal shape. The repeatable value of the loaded quality factor of the main sequence of the resonator modes was $Q \simeq 1.6 \times 10^7$, which corresponds to 15 MHz bandwidth of the mode. We studied several disks with nearly identical characteristics.

The resonators were arranged in the horizontal direction with home-made flex manipulators (Figure 3.28). There was no need in a vertical adjustment because the surface of the stage the resonators were placed on was optically polished and all resonators had the same thickness (with nanometer accuracy). The gaps between the resonators and between the prisms and the resonators were 50–100 nm, which corresponds to the evanescent field scale. We did not measure the gaps directly but adjusted them to have the appropriate system response.

Light was sent into, and retrieved out of, the first and last resonators in the array of the filter via coupling prisms. The repeatable value of fiber-to-fiber insertion loss, primarily due to inefficient coupling to the resonator modes, was less than 6 dB across the entire cascaded structure. The maximum transmission was obtained when light was resonant with the resonators modes. The antireflection coating of the coupling prisms or use of special gratings placed on high-index fibers may reduce the losses significantly. Tuning the filter was realized by applying a voltage to the top and bottom electrodes fabricated with gold. The gold coating is absent on the resonator perimeter edge where WGMs are localized.

The differences in the size of the resonators is important in the fabrication of the multi-resonator filter. Our aim was to produce spectral lines in all three resonators with a similar width to allow the realization of a complex line structure. If the interacting modes of resonators have different widths, then as they are tuned to approach each other, the height of the narrower resonance will simply track the shape of the wider ones, which is of no use for the filter application. The size of each resonator is important since cavities of similar size have similar optical coupling and similar FSRs. So similar optical modes with the same efficiency can be excited in each of them. The performance of the filter demonstrates that indeed similar resonators with approximately the same parameters were fabricated.

Figure 3.29 depicts the spectrum obtained in the experiment with three cascaded LiNbO$_3$ resonators. To highlight the filter performance the theoretical third-order Butterworth fit of the curve is also plotted. Obviously, the three-cavity filter has much faster roll off compared with the Lorentzian line of the same full width at half maximum. On the other

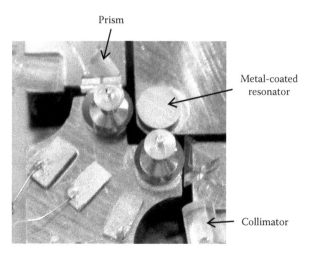

FIGURE 3.28
A packaged assembly of the three resonator filter.

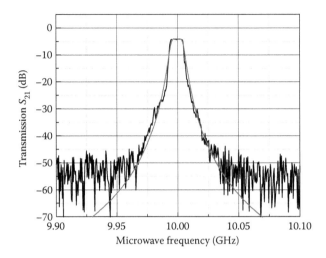

FIGURE 3.29
Transmission curve of the filter and its fit with Butterworth profile function $\gamma^6/[(v-v_0)^6+\gamma^6]$, where $\gamma=5.5$ MHz, v_0 determines the center of the filter function and primarily depends on the resonators' geometrical dimensions. Voltages applied to the resonators vary from near zero to 10 V to properly adjust modes of each individual resonator and construct the collective filter function as shown.

hand, the filter function does not look exactly like a third-order bandpass because of small differences between the cavity Q-factors and dimensions of the resonators.

The experimentally measured electro-optic tuning of the filter's spectral response, and tuning of its center frequency with an applied voltage, is shown in Figure 3.30. The filter exhibits a linear voltage dependence in the ±150 V tuning range, i.e. the total tuning span exceeds the FSR of the resonator. Changing the tuning voltage from 0 to 10 V shifted the spectrum of the filter by 1.3–0.8 GHz for TM mode polarization.

3.6.4 Fifth-Order Filters

We have also realized a five-WGR filter [162]. The filter (Figure 3.31), operating at the 1550 nm wavelength, has approximately 10 MHz bandwidth and can be tuned in the range of 10 GHz by applying a DC voltage of approximately 100 V to the graphite electrodes. The FSR of each resonator is approximately 10 GHz, and the filter may be tuned practically at any optical frequency in the transparency window of lithium niobate. This filter exhibits 12 dB of fiber-to-fiber insertion loss with.

For the five pole filter, the resonators were cut out from congruent Z-cut crystalline wafers and had approximately 4 mm in diameter and 100 μm thickness. Each resonator's rim was polished in the toroidal shape with a 50 μm curvature radius. The repeatable value of the loaded quality factor of the main sequence of the resonator modes was $Q=3\times10^7$ (the unloaded maximum was $Q=1.5\times10^8$). Resonators were arranged in the horizontal plane using flex manipulators. The gaps between the resonators, and between the coupling prisms and the resonators, were less than 100 nm, and were adjusted to maintain the appropriate system response.

3.6.5 Sixth-Order Filters

We also fabricated six-pole filters using metalized WGM resonators. Naturally, these filters had higher insertion loss compared with the lower order filters. It was also difficult to keep

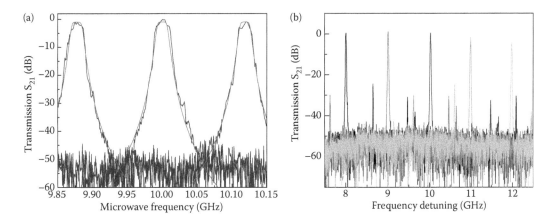

FIGURE 3.30
Demonstration of the tunability of the three-resonator filter. Left: Tuning of the filter and its Butterworth fit with 16.6 MHz pass band. Right: Broadband tuning of the filter. It is possible to see spurious pass bands not completely suppressed with the Vernier effect.

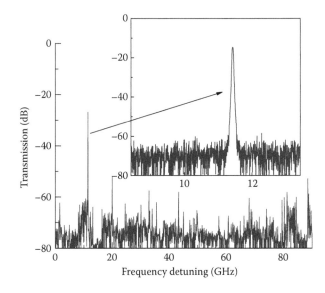

FIGURE 3.31
Demonstration of the five-resonator filter. (Reprinted from Savchenkov, A. A., Matsko, A. B., Ilchenko, V. S., Yu, N., Maleki, L., Millimeter and Submillimeter Waves and Workshop on Terahertz Technologies, 2007. The Sixth International Kharkov Symposium, 1, 79–84. With permission from Institute of Electrical and Electronics Engineers (IEEE).)

a stable the shape for the passband. An example of the six-pole filter function with 13 MHz pass band is shown in (Figure 3.32).

3.6.6 Tuning of the Multi-Resonator Filter

We now discuss in more detail considerations regarding tuning requirements of multi-resonator filters. A multiresonator filter has the required flat-top Butterworth shape of the order corresponding to the number of the resonators, if selected modes of the resonators have approximately the same frequency and loaded Q-factors, as discussed above. Generally it is not feasible to fabricate a pair of high-Q WGM resonators with coinciding

(a)   (b)

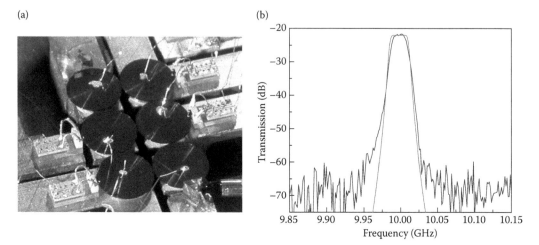

FIGURE 3.32
Demonstration of the sixth-order filter.

optical modes. For instance, the resonators have identical spectra if the error in their radius is much less than $\Delta R = R/Q$. This is approximately 0.01 nm for a $Q = 10^8$ resonator of 0.2 cm in diameter. Moreover, even if such resonators are fabricated, their relative temperature should be stabilized to better than $T = n/\kappa Q = 2$ mK to avoid relative drift of their spectra, where $\kappa \sim 10^{-5}$ K^{-1} and $n = 2.3$ are the combined thermal expansion/thermorefractive coefficient and the extraordinary index of refraction of the lithium niobate, respectively. Such a thermal stabilization is generally not convenient. Aside from these stringent requirements the Q-factors of the resonators must also be identical within a few percent margins to realize a proper filter function.

A partial solution to this is provided by the electro-optical properties of the material. It is possible to trim frequencies of each (nonidentical) resonator with a DC electric bias, or voltage, applied to the resonator. The value of the bias should follow the changes of the frequencies of the resonator and the laser. This scheme works well for a single WGM filter that requires only one bias [160]. The case of the three-WGR filter [161] is more challenging because it requires three independent biases; it is difficult to independently track frequencies of the three WGMs involved, especially the frequency of the central resonator. The growing complexity associated with filters having even a higher order makes their realization quite challenging.

We propose two schemes that simplify tuning the independent resonator frequencies. For the three-WGR filter the method involves optical probing the frequencies of two end resonators when the coupling between the resonators is small (tunable gaps are large, see Figure 3.33A). The frequency of the central resonator is selected by subsequently reducing the gaps between the resonators, and ramping the bias of the central resonator until the proper transmission function is realized. A similar technique works with the five-resonator filter (Figure 3.33B). In the first step a required second-order transmission function is realized with two uncoupled end pairs of the resonators. The central resonator is then tuned after the coupling is restored.

Nonetheless, this technique has several disadvantages: (i) it requires dark periods in the operation of the filter, (ii) it does not work with filters having higher order than five, and (iii) it requires a mechanical change of the shape of the filter (tuning the gaps between the WGM resonators). We have found a more efficient and robust technique to address these

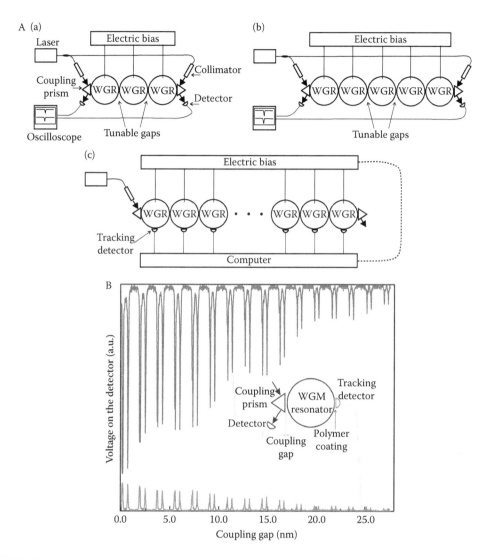

FIGURE 3.33

(A) Possible configurations for tuning the multiresonator WGM filters: (a) three resonator filters, (b) five resonator filters, and (c) multiple resonator filters. Scheme (c) involves tracking photodiodes. (B) Demonstration of the efficiency of the tracking photodiode method. The visibility of a WGM doublet as a function of the resonator load, regulated by the gap between the coupling prism and the resonator surface. This was observed by means of direct detection of light transmission through the resonator as well as by measurement with the tracking photodiode. (Reprinted from Savchenkov, A. A., Matsko, A. B., Ilchenko, V. S., Yu, N., Maleki, L., Millimeter and Submillimeter Waves and Workshop on Terahertz Technologies, 2007. The Sixth International Kharkov Symposium, 1, 79–84. With permission from IEEE.)

shortcomings. Fortunately, a slow photodiode coated with a hundred nanometer transparent polymer film and attached to the rim of a WGM resonator does not reduce its Q-factor significantly. The decrease is less than a couple of percents if the resonator's loaded quality factor is $Q = 10^8$. To demonstrate the operation of the "tracking detector" we recorded a segment of a WGM spectrum, monitoring the transmitted light and the photocurrent while decreasing the resonator loading. It is easy to see from Figure 3.33 that the detector operates satisfactorily. With this approach it is possible to design a computer controlled multipole filter (see, e.g. Figure 3.33C). The spectrum of each resonator in the chain is tracked by a photodiode, and is actively controlled by the microprocessor with feedback of the bias voltage applied to each resonator. There is practically no limitation on the number of WGRs in the filter with such a computer controlled scheme.

3.6.7 Resonator Coating Technique

Even when the problem of trimming of WGM frequencies is solved, the unequal Q-factors of the WGRs can hinder the desired response of the filter. Our mechanical polishing technique allows for fabricating resonator with quality factors determined exclusively by the intrinsic absorption of the material. Therefore, if all the resonators are made of the same material they will have the same Q-factors. The subsequent coating of the resonators with gold or any other conductor to form electrodes changes the Q-factor. The fabrication procedure is based either on repolishing the resonators after the coating process or on producing the resonators from gold-coated crystalline wafers. However, both solutions are difficult to realize. The resonator radius changes uncontrollably in the first case and the coating deteriorates in the second.

We have found that coating of the WGM resonators with colloidal graphite addresses these problems. Graphite has sufficient conductivity and is not expensive. It is easily removable from the resonator surface close to the rim where the electromagnetic field is localized. The graphite film can be literally wiped clean from the area without destruction of the quality of the polishing. These properties make the method advantageous for in-lab use.

We studied two possible ways of producing graphite coatings. One is more efficient from the point of view of tunability. The coating is removed only from the region where the WGM electromagnetic field is localized (Figure 3.34A). In this case the smallest distance between the electrodes (h), rather than the thickness of the resonator (H), determines the maximum tuning rate of the WGM spectrum per Volt. This allows fabrication of several-hundred micron thick resonators with tunability as if the resonator thickness is tens of microns. The method, nevertheless, has a disadvantage when applied to high-order filters. The coating in the vicinity of a WGM localization changes its Q-factor, which is undesirable. For this reason we analyzed a second approach for coating that does not influence the Q-factor (Figure 3.34B), though it has a lower efficiency compared with the first one. We found that it is not necessary to place the electrodes on the top and bottom of the WGM area. The DC electric field curves close to the resonator rim because of the large dielectric permittivity of the material, so the tuning efficiency does not change significantly if the electrodes are comparatively far from the rim. The Q-factor stays intact and the resonators can be placed in the chain.

3.6.8 Insertion Loss

Let us understand the nature of the insertion loss of the multiresonator filter. Using the formalism presented by Madsen and Zhao [174] we describe the field accumulated in each resonator with complex coefficients of transmission (T_j) and reflection (R_j). The coupling between resonators and between resonators and prism couplers is described by constants k_j (Figure 3. 35). The transmission and reflection coefficients are connected by a recurrent relation

$$\begin{bmatrix} T_{j+1}(37) \\ R_{j+1} \end{bmatrix} = \frac{1}{i\sqrt{k_j}\exp[-(i\delta_j + \alpha_j)]} \Phi_j \begin{bmatrix} T_j \\ R_j \end{bmatrix}, \tag{3.37}$$

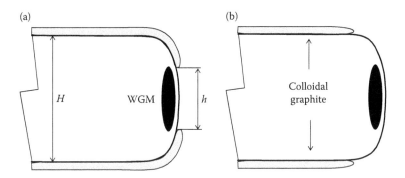

FIGURE 3.34
Application of the colloidal graphite for fabrication of DC electrodes mounted on crystalline WGM filters. (Reprinted from Savchenkov, A. A., Matsko, A. B., Ilchenko, V. S., Yu, N., Maleki, L., Millimeter and Submillimeter Waves and Workshop on Terahertz Technologies, 2007. The Sixth International Kharkov Symposium, 1, 79–84. With permission from IEEE.)

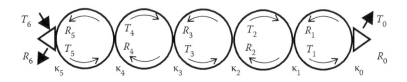

FIGURE 3.35
Parameters used to describe the resonator chain. (Reprinted from Savchenkov, A. A., Matsko, A. B., Ilchenko, V. S., Yu, N., Maleki, L., Millimeter and Submillimeter Waves and Workshop on Terahertz Technologies, 2007. The Sixth International Kharkov Symposium, 1, 79–84. With permission from IEEE.)

where δ_j and α_j represent the phase build-up and absorption per round trip in the resonator. For the sake of simplicity we assume that $\delta_j=\delta$ and $\alpha_j=0$. Then the transfer function for the five-resonator chain is

$$H_{11}=\frac{T_0}{T_6}e^{(i\delta+\alpha)/2}, \quad H_{21}=\frac{R_6}{T_6}e^{(i\delta+\alpha)}. \tag{3.38}$$

It is easy to find that $P_r/P_{in}=|H_{21}|^2$ and $P_{out}/P_{in}=|H_{11}|^2$. The relative power in resonator #1 is given by $P_1/P_{in}=|T_5/T_6|^2$, in resonator #2 is $P_2/P_{in}=|T_4/T_6|^2$, etc.

To further simplify the problem we assume that $k_0=k_5=k_a$, $k_1=k_4=k_b$, and $k_2=k_3=k_c$. The resonator chain becomes a fifth-order Butterworth filter with power transfer function

$$|H_{11}|^2\simeq\frac{1}{1+\left(2^{7/5}\delta/\kappa_a\right)^{10}}, \quad \text{if } \kappa_b\simeq\frac{\kappa_a^2}{8}, \kappa_c\simeq\frac{\kappa_a^2}{16}(1-\kappa_a), \text{and } \kappa_a \text{ is arbitrary.} \tag{3.39}$$

Finally, it is worth noting that

$$\gamma_a=\frac{\alpha}{2\pi a}\frac{c}{n}, \quad \gamma_c=\frac{\kappa_a}{2\pi a}\frac{c}{n}, \quad \delta=\Delta\omega\frac{2\pi an}{c},$$

where a is the radius of each resonator, and n is the refractive index of the material.

In reality $\alpha \neq 0$ and the filter is always absorptive ($|H_{11}|^2 < 1$ for $\delta=0$). We calculate $|H_{11}(\delta=0)|^2$ keeping the relative coupling between the resonators the same as when evaluated for the lossless case. The resultant absorption increases approximately exponentially with the increase of relative intrinsic absorption Figure 3.36. The resonators must be overcoupled to obtain small losses. In our experiment with the five-pole filter $\gamma_a/\gamma_c \sim 0.25$. The observed 12 dB fiber-to-fiber insertion loss results from the absorption in the chain as well as other miscellaneous losses (reflection from the prism surface, not perfect phase matching, etc).

3.6.9 Vertically Coupled Resonators

In this section we discuss a chain of one dimensional ring-like WGM resonators [175,176], as a configuration for coupled resonator optical waveguides (CROWs) [177] and other resonator-based waveguides [178]. We show that the basic features of the chain can be adjusted by changing the shape of the resonators as well as the distance between them. Unlike ordinary CROWs, where coupled resonators are considered as "lumped element" objects, our structure is amenable to tuning its photon density of states and thus , e.g. slowing light down, similar to such "distributed" systems as photonic crystals. This resonator chain is similar to the vertically stacked multi ring resonator (VMR) [179].

Let us consider a resonator waveguide shown in Figure 3.37. The linearly polarized narrow band laser light is coupled to a mode of one of the resonators using a prism or an optical fiber coupler. Light propagates due to the coupling between adjacent resonators, which occurs via the evanescent field inside the rod from that the resonators are made of, not through the air [38,53], as is the case with conventional CROWs. To understand the difference, note that the dimension of the evanescent field in the air outside a resonator scales as $l_e \sim \lambda/\pi(\varepsilon_0 - 1)^{1/2}$, where ε_0 is the electric susceptibility of the resonator material and λ is the wavelength of the light in vacuum. This value is usually equal to several hundreds of nanometers. In contrast, in this system coupling is efficiently realized between

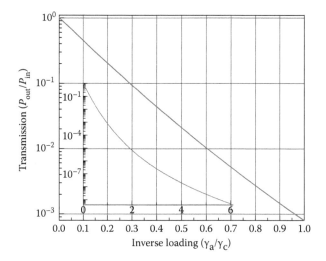

FIGURE 3.36
Dependence of the transmission coefficient of the filter on the resonator loading. (Reprinted from Savchenkov, A. A., Matsko, A. B., Ilchenko, V. S., Yu, N., Maleki, L., Millimeter and Submillimeter Waves and Workshop on Terahertz Technologies, 2007. The Sixth International Kharkov Symposium, 1, 79–84. With permission from IEEE.)

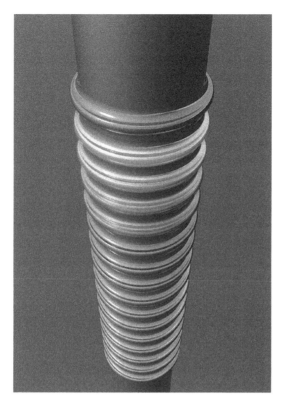

FIGURE 3.37
Schematic of the resonator waveguide. (Reprinted from Matsko, A. B., Savchenkov, A. A., and Maleki, L., *Opt. Lett.* 30, 3066–3068, 2005. With permission from OSA.)

the resonators that are within several microns of each other. The evanescent field of a resonator in the chain can be studied from the results of simulations made for the single-mode resonator (see Figure 3.13). The dimensions of the evanescent field inside the post exceed several micron. If several such resonators are placed close together on the same post, they will communicate through the "intrapost" evanescent field.

The propagation of light in the waveguide can be described with the usual wave Equation 3.16. We are interested in the case of high-order WGMs localized in the vicinity of the equator of the resonator. For the sake of simplicity we consider the TE mode family and change variables in Equation 3.16 as $E = \Psi e^{\pm im\phi}/\sqrt{r}$, m is the angular momentum number of the mode. We consider a low contrast structure, assuming that the resonator radius changes as $R = R_0 + L(z)$ and $R_0 \gg |L(z)|$. Then Equation 3.16 is transformed to

$$\frac{\partial^2 \Psi}{\partial r^2} + \frac{\partial^2 \Psi}{\partial z^2} + \left[k^2 \varepsilon \left(1 + 2 \frac{L(z)}{R_0} \right) - \frac{m^2}{r^2} \right] \Psi = 0, \tag{3.40}$$

where expressions $m \gg 1$ and $m^2 \simeq R_0^2 k^2 \varepsilon$ are used.

Considering an infinite waveguide, we separate variables and introduce $\Psi = \Psi_r \Psi_z$:

$$\frac{\partial^2 \Psi_z}{\partial z^2} + 2k^2 \varepsilon \frac{L(z)}{R_0} \Psi_z = -k_z^2 \Psi_z, \tag{3.41}$$

$$\frac{\partial^2 \Psi_r}{\partial r^2} + \left(k^2 \varepsilon - k_z^2 - \frac{m^2}{r^2} \right) \Psi_r = 0, \tag{3.42}$$

where k_z^2 is the separation parameter.

Equations 3.41 and 3.42 can be solved analytically if $L(z)$ is a periodic function with some period l. The solution is $\Psi = \Psi_0 e^{i\beta z} \varphi(k_z z) J_v(k_{v,q} r)$, where β is a propagation constant, $J_v(k_{v,q} r)$ is a Bessel function, $\Psi_z = \exp(i\beta z) \varphi(k_z z)$ is the Bloch function $[\Psi_z(z+1) = \exp(i\beta l)\Psi_z(z)]$, $k_{v,q}$ is the radial wave number, and q describes radial quantization of the WGMs.

Parameters β and k_z are to be derived from Equation 3.41. They stand for wave number and a counterpart of frequency of the wave propagating in the resonator chain (the frequency is determined by eigenvalues of both Equations 3.41 and 3.42). The parameters are connected by a dispersion relation which, in the simplest case of periodic grating of cylindrical WGM resonators with gaps between them characterized by depth L_0 and length $l_g (l_g < l)$, can be presented as

$$\cos(\beta l) = \cosh\left[l_g \sqrt{2k^2 \varepsilon L_0 / R_0 - k_z^2} \right] \cos\left[k_z (l - l_g) \right] \tag{3.43}$$

$$+ \frac{k^2 \varepsilon L_0 / R_0 k_z}{\sqrt{2k^2 \varepsilon L_0 / R_0 - k_z^2}} \sinh\left[l_g \sqrt{2k^2 \varepsilon L_0 / R_0 - k_z^2} \right] \sin\left[k_z (l - l_g) \right],$$

where we assumed that $k(2\varepsilon L_0 / R_0)^{1/2} > k_z$ (this assumption comes from the condition that a single localized WGM resonator has at least one confined mode). As follows from Equation 3.43 k_z is determined in allowed bands separated by forbidden band gaps (see in Figure 3.38).

Let us characterize the frequency spectrum of the entire system. We start from the expression for the frequency of WGMs derived from Equation 3.44.

$$\frac{\omega^2}{c^2} \varepsilon = k_{m,q}^2 + k_z^2. \tag{3.44}$$

In the case of small interaction between the resonators, when $k(2\varepsilon L_0 / R_0)^{1/2} \gg k_z$ and $k l_g (2\varepsilon L_0 / R_0)^{1/2} \gg 1$ we estimate from Equations 3.45 and 3.46 that each WGM mode in the resonator chain transforms to a frequency band with center.

$$\frac{\omega_c}{c} \sqrt{\varepsilon} \simeq k_{v,q} + \frac{1}{2k_{m,q}} \left[\frac{\pi p}{l - l_g} \right]^2 \tag{3.45}$$

and width

$$\frac{\Delta\omega}{c} \sqrt{\varepsilon} \simeq \frac{4\pi^2 p^2}{k_{m,q}^2 (l - l_g)^3} \sqrt{\frac{R_0}{2L_0}} \exp\left(-l_g k_{m,q} \sqrt{\frac{2L_0}{R_0}} \right). \tag{3.46}$$

The spectrum of the modes due to their radial confinement, being eigenvalues of Equation 3.42, is described by

$$k_{m,q} \simeq \frac{1}{R_0} \left[m + \alpha_q \left(\frac{m}{2} \right)^{1/3} - \sqrt{\frac{\varepsilon}{\varepsilon - 1}} \right], \tag{3.47}$$

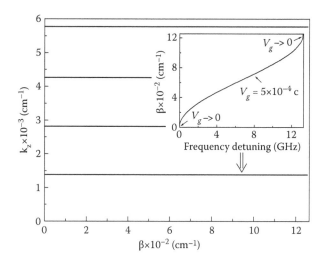

FIGURE 3.38
A solution of Equation 3.45 for period $l=25\ \mu m$ ($l_g=5\ \mu m$), wavelength $\lambda=1.5\ \mu m$, radius $R_0=0.3$ cm, bump height $L_0=30\ \mu m$, and $\varepsilon_0=2.2$. Inset: Structure of the first band evaluated using Equation 3.46. (Reprinted from Matsko, A. B., Savchenkov, A. A., and Maleki, L., *Opt. Lett.* 30, 3066–3068, 2005. With permission from OSA.)

where α_q is the qth root of the Airy function, $Ai(-z)$ (q is a natural number). This is, in fact, a spectrum of high-order WGMs of an infinite cylinder with radius R_0. We now have all the necessary elements to describe the waveguide.

One of its parameters is the group velocity, which can be evaluated using wave number β(1.43) and frequency ω(1.44). First of all, even without calculation and using the analogy with photonic band-gap structures, it is clear that the group velocity of light propagating in the system, determined as $V_g=d\omega/d\beta$, approaches zero if light is tuned to the band edge. However, at that point the dispersion of the group velocity is also large, so such a tuning is impractical. The dispersion is nearly linear in the center of the band-gap and group velocity there can be estimated as

$$\frac{V_g\sqrt{\varepsilon}}{c}\approx\frac{\Delta\omega\sqrt{\varepsilon}l}{2\pi c}=\frac{2\pi p^2 l}{k_{m,q}^2(l-l_g)^3}\sqrt{\frac{R_0}{2L_0}}\exp\left(-l_g k_{m,q}\sqrt{\frac{2M_0}{R_0}}\right).\tag{3.48}$$

It is easy to see that the group velocity decreases exponentially as the resonators are pulled apart, or the depth of trenches separating them increases. The result could be obtained even without analytically solving the problem. Indeed, because light propagates due to coupling between adjacent resonators, the smaller the coupling, the slower is the light propagation.

In reality, the resonator material possesses intrinsic losses due to scattering and surface inhomogeneities, as well as absorption. Because of these losses the modes have a finite ring-down time τ. The maximum group delay in the set of coupled resonators cannot exceed this ring-down time without a significant absorption of light. The minimum group velocity is then $V_{gmin}\approx l/2\pi\tau$, which corresponds to the propagation of light through a single uncoupled resonator. In the case of strong coupling between the resonators, when $\Delta\omega\pi\gg2\pi$, light interacts with many resonators and its propagation can be studied using the formalism presented above.

Finally, we made numerical simulations of light propagation in a finite delay line made of WGM resonators and shown in (Figure 3.39, left). We assumed that initially light is confined in the bottom resonator. After that, we switch the interaction between the resonators and observe pulse propagation toward the top of the resonator stack, a subsequent reflection, and backwards propagation (Figure 3.39, right). The ring-down time τ was chosen to be 1 ms, which is theoretically achievable in calcium fluoride WGM resonators [159].

Comparing group velocity of light in this waveguide with that of CROWs [177], we conclude that, under equivalent conditions, our waveguide always results in a group velocity lower than the group velocity in CROWs. The ratio of the group velocities, approximately equal to the ratio of thickness and diameter of the resonator, can easily give an order of magnitude of difference.

3.6.10 Microwave Photonics Applications

Optical WGM resonators have successfully been used in microwave photonics. The resonators exhibit bandwidth in the hundred kilohertz to gigahertz range. In microwave photonics applications, this bandwidth corresponds to a wide range of equivalent radio-frequency Q-factors (between 10 and 10^6 for the X-band). This unique flexibility, in combination with a proper choice of highly transparent and/or nonlinear resonator material, allows development of a number of high performance microwave photonic devices: tunable and multi-pole filters, resonant electro-optic modulators, photonic microwave receivers, and opto-electronic microwave oscillators. In this section we focus on the applications of linear WGM resonators in an opto-electronic oscillator as well as in a photonic microwave receiver.

3.6.10.1 Opto-Electronic Oscillator

The opto-electronic oscillator (OEO) produces microwave signals using optical techniques [180]. The generic scheme of the OEO is shown in Figure 3.40a. The conventional OEO consists of a laser as the source of light energy. The laser radiation propagates through an electro-optical modulator and an optical energy storage element, before it is converted to the electrical energy with a photodetector. The electrical signal at the output of the photodetector is amplified and filtered before it is fed back to the modulator, thereby completing a feedback loop with gain that generates sustained oscillation at a frequency determined by the filter. An OEO with a high-Q optical resonator in place of the electronic filter was

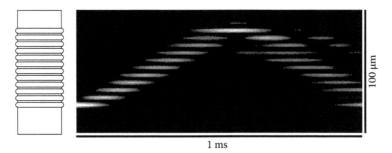

1 ms

FIGURE 3.39

Left: A set of coupled WGM rings placed on a single rod. Right: A map obtained with 2D numerical simulations describing ultra-slow propagation of an optical pulse. The group velocity is less than 20 cm/s. (Reprinted from Matsko, A. B., Savchenkov, A. A., and Maleki, L., *Opt. Lett.* 30, 3066–3068, 2005. With permission from OSA.)

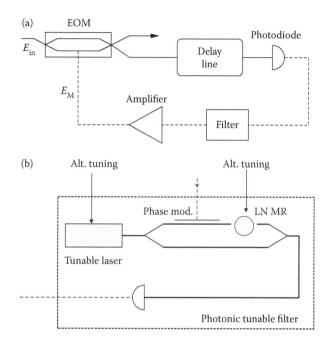

FIGURE 3.40

(a) Generalized scheme of an OEO. (b) A tunable microwave photonic filter that includes a lithium niobate WGM resonator.

studied theoretically by Strekalov et al. [181]. An OEO with a lithium niobate WGM modulator was demonstrated by Matsko et al. [182].

The fundamental noise performance of an OEO is determined by the energy storage time, or quality factor Q, of the optical storage element. Q-factor can be very large. A long fiber delay can easily have several microsecond long storage times, corresponding to the equivalent microwave Q of about a million at 10 GHz frequency. This is a high value compared to conventional dielectric microwave resonators used in oscillators. A high-Q optical resonator can play the role of the optical storage element, in addition to the role of the electronic filter. While an optical fiber is a true time delay and introduces group as well as phase delays. An optical resonator produces group delay only, and phase delay of the microwave signals due to these resonant elements is negligibly small.

We have made an OEO with a tunable microwave photonic filter as shown in Figure 3.40b. The OEO includes a DFB laser and 130 m delay line. The photonic filter has a tunable first-order pass band shown in Figure 3.41, left. As a result of the filter tunability we have achieved a smooth tuning of the oscillation frequency in the 8–9 GHz range, without degradation of the performance of the device Figure 3.41, right.

We also demonstrated an OEO with a lithium niobate WGM filter inserted into the optical loop Figure 3.42, left. The filter is used to suppress generation of spurious modes in the OEO loop. The noise power spectrum of the oscillator is shown in Figure 3.42, right.

3.6.10.2 Microwave Photonic Receivers

To demonstrate the filter performance in a photonic microwave receiver, we configured it for transmission of a video signal [160]. This type of narrowband signal link may be important for the development of portable navigation and communication devices

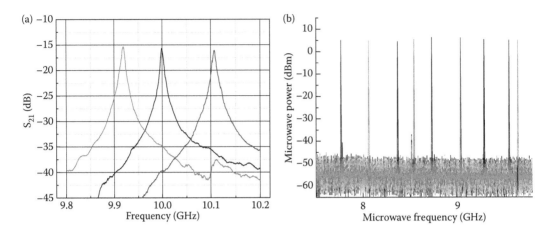

FIGURE 3.41
Left: Tunable 11 MHz photonic filter. Right: Demonstration of the frequency shift of the OEO.

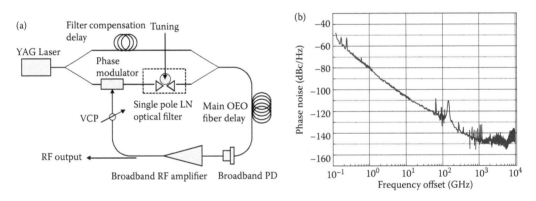

FIGURE 3.42
Left: Scheme of the OEO with optical WGM filter in the optical part of the oscillator. Right: Phase noise of the photonic oscillator with optical power −6 dBm and WGM resonator quality factor 2×10^8.

at microwave and millimeter-wave frequencies. For example, such a photonic architecture allows architecting a single platform for systems requiring operation in RF through millimeter-wave. These systems currently require multiple sets of hardware if implemented using all electronics schemes.

The scheme of the transmission experiment is shown in Figure 3.43. A video signal with approximately 20 MHz FWHM bandwidth and zero carrier frequency was sent from a CCD camera to a mixer, where it was mixed with a 10 GHz microwave carrier. The resultant modulated microwave signal was transmitted to a microwave receiver, filtered to suppress higher harmonics, amplified, and upconverted into light using a Mach-Zehnder electro-optic modulator. The modulated signal was then sent through the photonic filter, heterodyned and detected with a fast photodiode. The microwave signal from the photodiode output was mixed with a microwave carrier to restore the initial signal.

A high-Q WGM cavity may mix the amplitude and phase fluctuations of the light. The amplitude of the transmitted light through the filter may be presented as

$$E_{\text{out}}(t) \approx \frac{\tilde{E}_{\text{in}}(t - \tau_d)e^{-i\omega t}\gamma}{\gamma + i(\omega - \omega_0 + \dot{\phi}_{\text{in}}(t))}, \tag{3.49}$$

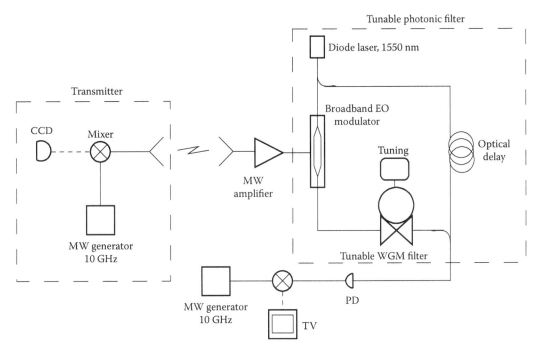

FIGURE 3.43
Schematic diagram of the video signal transmission experiment. The solid thin line corresponds to optical fibers, solid thick lines to microwave waveguides, and dashed lines to electric circuits. (Reprinted from Ilchenko, V. S., Savchenkov, A. A., Matsko, A. B., and Maleki, L., *Proc. SPIE* 4969, 195–206, 2003. With permission from Society of Optical Engineers (SPIE).)

where $\widetilde{E}_{in}(t-\tau_d)$ is the slow field amplitude, τ_d stands for the group delay, $\dot{\phi}_{in}(t)$ results from the phase diffusion of the pump laser. The filter works well when the phase diffusion of the pump laser is small $\gamma \gg \dot{\phi}_{in}(t)$; it transforms the phase fluctuations of the laser into the amplitude fluctuations for large phase diffusion.

It is important to note that to characterize and retrieve encoded information, the filter output should be mixed with a monochromatic light on a photodetector. The filter contains a high-Q cavity that introduces group delay τ_d into the signal. This group delay results in an additional source of frequency-to-amplitude noise conversion when the output signal from the filter is mixed with the light that did not pass through the filter. This happens unless the scheme is balanced. We inserted the filter into a Mach-Zehnder configuration with a fiber delay line L_f to compensate for the group delay. The delay line length was equal to $L_f = n_o dF/2n_f = 1.2$ m , where $n_f \simeq 1.5$ is the refractive index of the fiber material and $F = 300$ is the cavity finesse. Such a compensation is not required if the laser linewidth is much smaller than the width of the cavity resonance. In our case optical characterization of the filter was achieved using a semiconductor diode laser with a 30 MHz FWHM line, which is quite large. The laser power in the fiber was approximately 2.5 mW.

3.7 Frequency Stability of WGM Resonators

Ultrastable optical resonators play an important role in both classical and quantum optics, and generally are of FP type. They are fabricated from materials having ultra low thermal expansion coefficients, and are both thermally and mechanically isolated to achieve stability. The geometrical dimensions of this type of resonator is typically about 10 cm or larger. Lasers locked to these resonators demonstrate Hertz and even subHertz linewidth, with frequency stabilities as high as ~ 10^{-15} at 1 s.

WGM resonators can also be used as frequency etalons [183,184]. The basic advantages of WGM resonators compared with FP ones are: (i) the small size; (ii) a large wavelength range in which WGM resonators have high Q-factor; and (iii) the low sensitivity of WGM resonators to external mechanical noise. This latter property is the result of the unique orthogonality relations between optical and acoustical WGM modes, in addition to high-Q factors and low density of the WGM spectra. All these features make utilization of the WGM resonators attractive in place of FP resonators for a laser stabilization applications.

On the other hand, WGM resonators also have several disadvantages compared with FP resonators. (i) Reference FP resonators contain a specific mirror spacer material which has a low thermal expansion coefficient. WGM resonators cannot contain such a material. The polished rim surface of the WGM resonator plays the role of the resonator mirrors and the host material is the only resonator "mirror spacer" material possible. Such a material generally has a large thermal coefficient of expansion. This problem can be mediated by efficient thermal isolation of the resonator, which is simpler than the isolation of FP resonators because of the small size and structure rigidity of the WGM resonators. (ii) Reference FP resonators are empty, usually evacuated, and are comparably large. These properties reduce the fundamental thermodynamic fluctuations, which are important even if the resonator is kept at a constant temperature. WGM resonators are small and material-filled, and hence suffer more strongly from the thermodynamic fluctuations [185]. The fluctuations can be partially suppressed by passive and active frequency stabilization schemes. (iii) The optical nonlinearity of WGM resonators is much larger than the nonlinearity of FP resonators, because of the same reasons as indicated in the previous item. The circulating optical power level should be limited in a WGM resonator to mediate this problem. The power fluctuations of a laser interrogating the resonator should also be small.

The goal of this section, based on the materials published in [183,184], is elucidating the properties of WGM frequency fluctuations, resulting from the basic fundamental thermodynamic as well as quantum optic principles. We evaluate the frequency spectra of thermorefractive, thermal expansion, and thermoelastic fluctuations, together with the steady state WGM frequency uncertainty resulting from those fluctuations. We also study the photothermal and ponderomotive fluctuations originating from the measurement procedure and find their frequency spectra.

The stability of a passively stabilized millimeter-sized WGM resonator made of a certain class of crystalline materials is primarily determined by thermorefractive fluctuations [183]. Those fluctuations have been predicted and successfully observed in fused silica microspheres [185]. The frequency stability limit of a cylindrical WGM resonator having a thickness of one hundred microns and a diameter of several millimeters is on the order of one part in 10^{-12} at one second integration time [183]. Thermorefractive fluctuations increase inversely proportional to the mode volume, and the predicted stability is limited because of small volumes of the WGMs.

It was shown, however, that a proper selection of the resonator host material is essential for stabilization of the WGM frequency [184]. Photorefractive fluctuations can be suppressed in some materials like magnesium fluoride if the proper operation temperature is selected. But in those resonators thermal expansion fluctuations become dominant in the frequency stability limit. It is possible to design active schemes to further stabilize fluctuations that result from the residual thermal expansion, using specific thermal expansion properties of some crystals that are not homogeneous along different crystal axes. In this way, the achieved frequency stability could be higher than the stability determined by the fundamental thermodynamic limit.

We also discuss some methods for achieving the desired frequency stability. The simple passive temperature stabilization is not practical because it must sustain fluctuations smaller than submicroKelvin level. To solve this problem we make use of methods similar to technologies developed for the stabilization of quartz crystalline rf oscillators [186]. The scheme includes (i) compensation of the temperature drifts of the resonator by connecting it with special elements having appropriate linear or nonlinear thermal expansion (c.f. oven controlled crystal radio frequency oscillator) and (ii) usage of two WGM families having different thermo-optical constants for measurement and compensation of the resonator temperature fluctuations. These methods can be very efficient. It will be argued that the relative stability of two modes in the same resonator separated by an octave can be better than one part in 10^{-14}, at one second integration time.

3.7.1 Fundamental Thermodynamic Limits

In this section, we consider two types of resonators. We initially assume that a WGM resonator can be considered as a small part of a much larger object, so the thermal spectrum is continuous. This assumption is valid, for instance, for a low contrast resonator [53] or for a resonator being in good contact with the heat bath. The validity of the assumption is even more general. It was shown by Gorodetsky and Grudinin [185] that it is possible to neglect the discreteness of the spectrum of the thermal waves in a sufficiently large spherical resonator (radius exceeding several tens of microns) because the mode volume is small compared with the volume of the resonator. To confirm that conclusion, we also consider a thin resonator, so the discrete thermal spectrum would be studied. We show that the resonators with both discrete and continuous thermal spectra possess equivalent frequency fluctuations.

As a general rule, the relative uncertainty of the eigen frequency of WGMs can be found from the expression

$$\frac{\delta\omega}{\omega} = \frac{\Delta R}{R} + \frac{\Delta n}{n}, \tag{3.50}$$

where ω is the mean value of the frequency of a selected mode, $\delta\omega$ is the fluctuation of the frequency, R and ΔR are the values of the radius of the resonator and its fluctuation, n and Δn are the values of the index of refraction of the material of the resonator and its fluctuation. Equation 3.50 is valid for a WGM resonator of any relevant shape if the radius of the resonator determines the largest dimension of the geometrical localization of the mode, i.e. where $\omega_{WGM} \simeq cm/(Rn)$, where $m \gg 1$ is the azimuthal mode number.

The terms $\Delta R/R$ and $\Delta n/n$ in Equation 3.50 are responsible for thermoelastic/thermal expansion and thermorefractive noise, respectively. Both terms are of the same order and they both should be taken into account in crystalline WGM resonators.

3.7.1.1 Thermorefractive Fluctuations: Steady State

Thermodynamic fluctuation of temperature in the WGM volume results in fluctuations of the index of refraction in the WGM channel

$$\frac{\Delta n}{n}\bigg|_{(\Delta T)_m} = \alpha_n (\Delta T)_m, \tag{3.51}$$

where $\alpha_n = (1/n)(\partial n/\partial T)$ is the thermo-refractive coefficient of the resonator host material. The mean square value of the thermal fluctuation is

$$\langle (\Delta T)_m^2 \rangle = \frac{k_B T^2}{C_p V_m \rho}, \tag{3.52}$$

where k_B is the Boltzmann's constant, T is the absolute temperature, ρ is the density of the resonator host material, V_m is the mode volume (assumed to be much less than the volume of the resonator), C_p is the specific heat capacity at constant pressure of the resonator host material. In what follows we assume that $C_p = C_V = C$ for a crystalline material. The mode volume of a WGM belonging to the basic mode sequence of a spherical resonator is

$$V_m = 3.4 \pi^{3/2} \left(\frac{\lambda}{2\pi n} \right)^{7/6} R^{11/6}, \tag{3.53}$$

The value of the mode volume of a toroidal/cylindrical WGM resonator varies depending on the resonator thickness and can exceed the value for the spherical resonator.

3.7.1.2 Thermorefractive Fluctuations: Spectrum

The frequency spectrum of thermorefractive fluctuations can be found following the path described by Gorodetsky and Grudinin [185]. The complex temperature distribution u in the resonator is given by the heat transport equation containing a distributed external fluctuational thermal source [187]

$$\frac{\partial u}{\partial t} - D\Delta u = F(\mathbf{r}, t), \tag{3.54}$$

where $D = \kappa/(\rho C)$, κ is the thermal conductivity coefficient, and C is the specific heat capacity. The thermal source is normalized such that the quadratic deviation of the mode temperature coincides with Equation 3.52.

We are interested in the temperature fluctuations averaged over the mode volume

$$\bar{u}(t) = \int u(\mathbf{r}, t) \, |\Psi(\mathbf{r})|^2 d\mathbf{r}, \tag{3.55}$$

where $|\Psi(\mathbf{r})|^2$ is the normalized spatial power distribution for the light localized in a WGM, $\int |\Psi(\mathbf{r})|^2 d\mathbf{r} = 1$.

The spectral power density of the random process $\bar{u}(t)$ and the quadratic deviation of the average mode temperature $\langle \bar{u}(0)^2 \rangle$ are given by

$$S_{\bar{u}}(\Omega) = \int_{-\infty}^{\infty} \langle \bar{u}^*(t) \bar{u}(t + \tau) \rangle e^{-i\Omega\tau} d\tau, \tag{3.56}$$

$$\langle (\Delta T)_m^2 \rangle = \langle \bar{u}(0)^2 \rangle = \int_{-\infty}^{\infty} S_{\bar{u}}(\Omega) \frac{d\Omega}{2\pi}. \tag{3.57}$$

Let us consider a resonator formed on the surface of an infinitely long cylinder of radius R by a cylindrical protrusion of radius $R+\Delta R$ ($R \gg \Delta R$) and thickness L. We solve Equation 3.54 and numerically evaluate the spectral density for a calcium fluoride resonator with $R=0.3$ cm and $L=0.01$ cm (see Figure 3.44, left).

We also find the fluctuations for a thin cylindrical resonator of thickness L and radius R, $R \gg L$. A simple analytical approximation is possible in this case. We present the correlation of the fluctuation forces as

$$\langle F(\mathbf{r},t)F(\mathbf{r}',t')\rangle \simeq 16\pi \frac{k_B T^2 DL}{\rho C V_m}\delta(\mathbf{r}-\mathbf{r}')\delta(t-t'); \tag{3.58}$$

solving Equation 3.54, we derive an approximate expression for the spectral density of the thermorefractive noise

$$S_{\tilde{u}}(\Omega) = \frac{k_B T^2}{\rho C V_m}\frac{R^2}{12D}\left[1+\left(\frac{R^2}{D}\frac{|\Omega|}{9\sqrt{3}}\right)^{3/2}+\frac{1}{6}\left(\frac{R^2}{D}\frac{\Omega}{8m^{1/3}}\right)^2\right]^{-1}. \tag{3.59}$$

The spectral density of the frequency noise, given by $S_{\delta\omega/\omega}(\Omega)=\alpha_n^2 S_{\tilde{u}}(\Omega)$, is approximately the same for the case of a thin resonator and for a resonator being a part of an infinite cylinder (Figure 3.44 left).

We consider a calcium fluoride resonator of radius $R=0.3$ cm and thickness $L=0.01$ cm driven with $\lambda=1.55$ μm light. We find $m \simeq 2\pi Rn/\lambda \approx 2.7 \times 10^4$ and $R/m^{2/3} \simeq 1.1 \times 10^{-3}$. The thermal diffusivity for CaF_2 is equal to $D=3.6 \times 10^{-2}$ cm^2/s, hence characteristic frequencies for the process are $D/R^2=0.4$ s^{-1}, $Dm^{4/3}/R^2=3.2 \times 10^5$ s^{-1}, and $D/L^2=360$ s^{-1}. To find the factor $k_B \alpha_n^2 T^2/\rho C V_m \simeq 4 \times 10^{-24}$ we have taken $\alpha_n=-0.8 \times 10^{-5}$ K^{-1}, and $V_m=2\pi RL \times R/m^{2/3} \simeq 6 \times 10^{-6}$ cm^3.

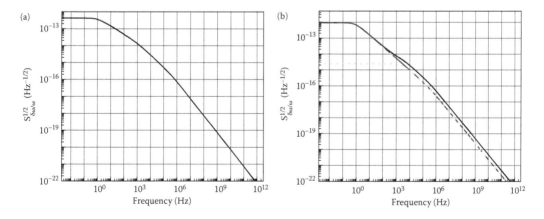

FIGURE 3.44

Left: Spectral density for a calcium fluoride resonator with $R = 0.3$ cm and $L=0.01$ cm. The resonator is formed on the surface of an infinitely long cylinder of radius R by a cylindrical protrusion of radius $R+\Delta R$ ($R \gg \Delta R$) and thickness L. Right: The fluctuations for a thin cylindrical resonator of thickness L and radius R, $R \gg L$. (Reprinted from Matsko, A. B., Savchenkov, A. A., Yu, N., and Maleki, L., *J. Opt. Soc. Am. B* 24, 1324–1335, 2007. With permission from OSA.)

We find the Allan variance of the WGM frequency by integrating the evaluated spectral density of fluctuations using the expression from Audoin [188]

$$\sigma^2(\tau) = \frac{2}{\pi} \int_0^\infty S_{\delta\omega/\omega}(\Omega) \frac{\sin^4(\Omega\tau/2)}{(\Omega\tau/2)^2} d\Omega, \tag{3.60}$$

where we took into account that $S_{\delta\omega/\omega}(\Omega)$ is a double sided spectral density. The evaluated Allan variance for the resonator is shown in Figure 3.45. To understand the change of slope of the dependence shown in Figure 3.45 we note that integration (averaging) in Equation 3.60 occurs in the vicinity of $\Omega=0$ when $\tau\to\infty$, where spectral density is approximately constant and $\sigma^2(\tau)\sim 1/\tau$. In the case of $\tau\to 0$ the integration in Equation 3.60 occurs in a wide band centered at frequency $\Omega\to\infty$, so that $S_{\delta\omega/\omega}(\Omega)\sim 1/\Omega^2$ and $\sigma^2(\tau)\sim 1/\tau$. The monotonic function $\sigma^2(\tau)$ naturally has a maximum at some specific value of τ. The increase of Allan variance with time for small τ is not counterintuitive because the thermorefractive fluctuations result in thermal drift of the WGM frequency. The maximum value of the drift is restricted and longer integration results in averaging down of the fluctuations.

An advantage of crystalline WGM resonators compared with other solid state resonators is that WGM resonators can be made out of various materials with various thermorefractive constants. For instance, it is important to mention the unique properties of MgF_2. This crystal has vanishing extraordinary (at $\sim 74°C$) and ordinary (at $\sim 176°C$) thermorefractive coefficients [189]. Tuning the temperature of a MgF_2 WGM resonator to the vicinity of zero thermorefractive coefficient $\alpha_n=0$ allows suppression of the fundamental thermorefractive noise $\langle(\Delta\omega_{TR})^2\rangle^{1/2}/\omega\to 0$. Technical thermorefractive noise is also compensated because temperature stability on the order of 2 mK required to reach $\Delta n_e/n_e=10^{-14}$ is feasible.

The basic conclusion of this section is that the thermorefractive noise does not limit the stability of WGM resonators made out of certain materials, even with moderate temperature stabilization and no sophisticated compensation. However, the other noise sources have to be dealt with to reach higher stability.

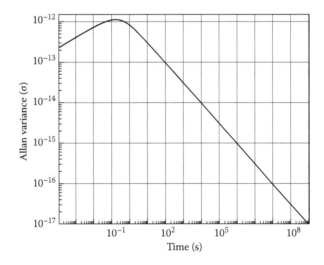

FIGURE 3.45
Thermorefractive Allan variance of the frequency of a mode of a cylindrical calcium fluoride WGM resonator with $R = 0.3$ cm and $L = 0.01$ cm. (Reprinted from Savchenkov, A. A., Matsko, A. B., Ilchenko, V. S., Yu, N., and Maleki, L., *J. Opt. Soc. Am. B* 24, 2988–2997, 2007. With permission from OSA.)

3.7.1.3 Thermoelastic and Thermal Expansion Fluctuations: Steady State

The variations of the mode volume of a WGM due to fundamental thermoelastic fluctuations do not change the WGM frequency. The frequency is given by the boundary conditions of the resonator and by the size of its radius. The fluctuation of the radius is determined by the fluctuation of the volume and the temperature of the entire resonator. To use the thermodynamic approach we assume that the resonator is a small part of a much larger sample. This is possible if the resonator is a thin disc cut on a solid state rod. Then the volume of the resonator V_r fluctuates as

$$\frac{\langle (\Delta V_r)^2 \rangle}{V_r^2} = k_B T \frac{\beta_T}{V_r}, \tag{3.61}$$

and $\beta_T = -[(1/V)\,(\partial V/\partial p)]_T$ is the isothermal compressibility of the resonator host material. Fluctuation of the radius of a spherical resonator due to fluctuations of its volume is

$$\frac{\Delta R}{R}\bigg|_{\Delta V} = \frac{\Delta V_r}{3 V_r}. \tag{3.62}$$

Fluctuation of the average resonator temperature

$$\langle (\Delta T)_r^2 \rangle = \frac{k_B T^2}{C_p V_r \rho}, \tag{3.63}$$

influences the radius as well:

$$\frac{\Delta R}{R}\bigg|_{\Delta T} = \alpha_l (\Delta T)_r, \tag{3.64}$$

where $\alpha_l = (1/l)\,(\partial l/\partial T)$ is the linear thermal expansion coefficient.

Using the fact that fluctuations $(\Delta V)_r$ and $(\Delta T)_r$ are statistically independent, and that the correlation of fluctuations of the temperature of the whole resonator and the average temperature in the WGM localization, $(\Delta T)_r$ and $(\Delta T)_m$, respectively, is small (mode volume is much smaller than the volume of the resonator) we arrive at

$$\frac{\langle (\Delta R)^2 \rangle}{R^2} = k_B T \frac{\beta_T}{9 V_r} + \alpha_l^2 \frac{k_B T^2}{C_p V_r \rho}. \tag{3.65}$$

It is important to note that the noise depends on the shape of the resonator. For instance, the thermoelastic fluctuations of a thin cylindrical resonator mounted on a thin stem should be considered in a different way compared with the above derivation. The above estimate is valid only if there is at least one continuous (quasi-continuous) dimension in the system, for instance, if the resonator is formed by a small protrusion on a long cylindric rod.

3.7.1.4 Thermoelastic Fluctuations: Spectrum

Let us find the spectral density of the frequency noise determined by the thermoelastic fluctuations. For the sake of simplicity we consider only one effective 1D elastic mode of the resonator of the lowest order. The contributions from other modes of higher order is comparably small. We assume that in the vicinity of the WGM localization the radius of the resonator changes in accordance with

$$\frac{\partial(\Delta R)}{\partial t} + (-i\Omega_0 + \Gamma_0)\Delta R = F_R(t), \tag{3.66}$$

where the oscillation frequency is taken to be equal to the eigenfrequency of the lowest order radial mode of a spherical liquid resonator of radius R

$$\Omega_0 = \frac{\pi v_s}{R}, \tag{3.67}$$

where v_s is the speed of sound. The decay rate of the acoustic mode can be very small. The minimal value of the rate is thermodynamically limited [190]

$$\Gamma_0 \geq \frac{\Omega_0^2 \kappa T \alpha_i^2 \rho}{9C^2}. \tag{3.68}$$

This is a very small value. The realistic value of the quality factor ($\Omega_0/2\Gamma_0$) of the acoustic mode is expected to exceed 5×10^4 [191].

We select the fluctuational force $F_R(t)$ such that it obeys

$$\langle F_R^*(t)F_R(t')\rangle = \Gamma_0 k_B T \frac{\beta_T R^2}{9V_r} \delta(t - t'); \tag{3.69}$$

and obtain an expression for the spectral density of the radius fluctuation:

$$S_{\Delta R/R} = k_B T \frac{\beta_T}{9V_r} \frac{\Gamma_0}{(\Omega - \Omega_0)^2 + \Gamma_0^2}. \tag{3.70}$$

The density is peaked at Ω_0 and is significantly suppressed at higher frequencies. For example, for a cylindrical resonator with $R=0.3$ cm, $L=0.01$ cm, and $v_s=5\times10^5$ cm/s we have $\Omega_0=5\times10^6$ rad/s. Assuming that $\Gamma_0=100$ rad/s we obtain Figure 3.46. It is easy to see that the low frequency branch of the spectral power density of the frequency noise given by the thermoelastic fluctuations ($S_{\delta\omega/\omega}=S_{\Delta R/R}$) is much smaller than the one due to thermorefractive fluctuations.

3.7.1.5 Thermal Expansion Fluctuations: Spectrum

Thermodynamic temperature fluctuations of the resonator result in the modification of the resonator radius and thickness, leading to frequency noise. To estimate this noise we assume that the basic contribution comes from the lowest order eigenfunction of the thermal diffusion equation. Using the reasoning of [183] for the noise spectral density we derive

$$S_{\Delta R/R} = \frac{\langle(\Delta\omega_{TE1})^2\rangle}{\omega^2} \frac{2R^2/\pi^2 D}{1 + (\Omega R^2/D\pi^2)^2}. \tag{3.71}$$

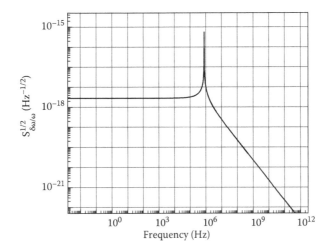

FIGURE 3.46
The spectral density of the frequency noise $S_{\delta\omega/\omega}(\Omega)=S_{\Delta R/R}(\Omega)$ due thermoelastic noise. (Reprinted from Matsko, A. B., Savchenkov, A. A., Yu, N., and Maleki, L., *J. Opt. Soc. Am. B* 24, 1324–1335, 2007. With permission from OSA.)

This expression basically tells us that the frequency dependence of the spectral density is determined by the slowest thermal diffusion time associated with the thermal diffusion along the radius of the resonator. Using Equation 3.60 we calculate the Allan variance of the frequency of the WGM resulting from the fundamental thermal expansion fluctuations of a z-cut magnesium fluoride resonator of radius $R=0.3$ cm and thickness $L=0.01$ cm (Figure 3.47). The thermal diffusivity for MgF_2 is equal to $D=7.2\times10^{-2}$ cm^2/s and the characteristic frequency for the process is $D/R^2=0.8$ s^{-1}. Equation 3.71 gives the top boundary of the low frequency spectral density.

In reality, the resonator can be placed on a metal plate possessing a high thermal conductivity, so eventually the time constant $R^2/\pi^2 D$ should be replaced with $L^2/\pi^2 D$; this would reduce the value of the low frequency spectral density significantly. For example, the thermal diffusivity of aluminium is $D=0.97$ cm^2/s at 300 K. Copper has a bit larger value: $D=1.15$ cm^2/s at 300 K. Placing the resonator on a polished copper plate (or squeezing the resonator between two copper plates) would result in more than an order of magnitude reduction of phase noise at zero frequency, leading to a reduction of the corresponding Allan variance below one part in 10^{-14} at one second integration time. It is important to note that placing a resonator onto a copper plate will reduce the quality factor of the mechanical modes of the resonator. This reduction will lead to an enhancement of the influence of thermoelastic fluctuations on the frequency stability.

We find the solution to the problem in the optimal shaping of the resonator allowing to increase the resonator volume without changing the characteristic time constant of the process. The resonator should have a nearly spherical shape, or the shape of a cylinder with equal radius and height. The light should travel in a small protrusion [53] that does not influence the thermal and mechanical modes of the resonator. For instance, a nearly spherical single mode MgF_2 WGM resonator of radius $R=0.3$ cm will have an Allan variance less than one part per 10^{-14} at one second integration time. This improvement will not increase the thermoelastic fluctuations.

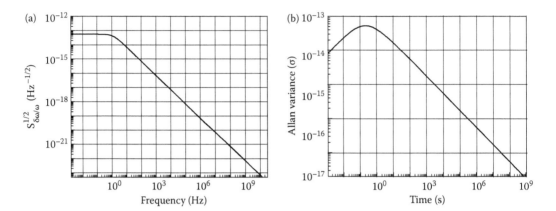

FIGURE 3.47
Power spectral density and Allan variance of the thermal expansion defined frequency fluctuations of a mode of a cylindrical magnesium fluoride WGM resonator with $R = 0.3$ cm and $L = 0.01$ cm. (Reprinted from Savchenkov, A. A., Matsko, A. B., Ilchenko, V. S., Yu, N., and Maleki, L., *J. Opt. Soc. Am. B* 24, 2988–2997, 2007. With permission from OSA.)

3.7.2 Fluctuations Originating from the Measurement Procedure

In this section, we consider fluctuations of the WGM frequency arising from the measurement procedure. The resonator is interrogated with laser radiation. Shot noise of the radiation causes two additional types of fluctuations: photothermal [187,192] and optoelastic [193].

3.7.2.1 Photothermal Fluctuations

These fluctuations appear as a result of the transfer of the shot noise of light absorbed in the resonator to the temperature fluctuations of the resonator host material, and to the subsequent fluctuations of the index of refraction of the resonator.

The temperature distribution in the resonator is described by the equation

$$\frac{\partial u}{\partial t} - D\Delta u = F_P(\mathbf{r}, t), \tag{3.72}$$

where $F_p(\mathbf{r}, t)$ is the fluctuational force describing noise due to the absorption of the photons in the resonator

$$\langle F_P(\mathbf{r},t)F_P(\mathbf{r}',t')\rangle = \frac{\hbar\omega_{m,q,p}}{\rho^2 C^2}\langle P_{\text{abs}}\rangle\,|\,\Psi(\mathbf{r})\,|^2\,\delta(\mathbf{r}-\mathbf{r}')\delta(t-t'), \tag{3.73}$$

$\omega_{m,q,p}$ is the angular frequency of the corresponding WGM, $\langle P_{\text{abs}}\rangle$ is the expectation value for the absorbed power, $|\Psi(\mathbf{r})|^2$ is the power distribution in a WGM ($\int |\Psi(\mathbf{r})|^2 d\mathbf{r} = 1$).

We find that the spectral density of the temperature fluctuations can be approximated by

$$S_{\bar{u}}(\Omega) \approx \frac{\hbar\omega_{m,q,p}}{\rho^2 C^2}\frac{\langle P_{\text{abs}}\rangle}{V_r^2}\frac{\pi^6 R^{3/2}/64L^{3/2}}{\Omega^2 + \pi^6 D^2/16L^3 R}. \tag{3.74}$$

A comparison of the exact numerical simulation and the approximation of the spectral density is presented in Figure 3.48.

3.7.2.2 Ponderomotive Fluctuations

Optoelastic ponderomotive fluctuations occur as a result of the fluctuations of the radiation pressure induced by light propagating inside the resonator [190,194]. The integral value of the pressure induced force is $F = 2\pi P/c$, where P is the power of the light inside the resonator. The force changes the resonator radius as well as the index of refraction. The index of refraction is involved due to the mechanical strain of the resonator host material:

$$\frac{\delta\omega}{\omega} = \left[1 + \frac{1}{2}K_\varepsilon\right]\frac{\Delta R}{R}, \tag{3.75}$$

where factor $K_\varepsilon = -E\varepsilon^{-1}\partial\varepsilon/\partial p$ ranges from 1 to 10 [190], E is Young's modulus of the material, and $\varepsilon = n^2$ is the electric susceptibility of the material.

For the sake of simplicity we take into account only one mechanical mode having the lowest mechanical frequency and assume that the probe light is resonant with the corresponding WGM. The last condition is required to avoid mechanical instability or additional rigidity added to the mechanical system by light. The mechanical oscillations of the resonator surface are described by the equation

$$\frac{\partial^2(\Delta R)}{\partial t^2} + \Gamma_0\frac{\partial(\Delta R)}{\partial t} + \Omega_0^2(\Delta R) = \frac{2\pi\delta P(t)}{m^*c}, \tag{3.76}$$

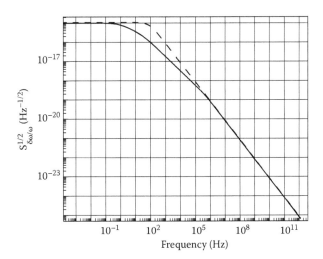

FIGURE 3.48
Spectral density of the photothermal fluctuations of a WGM frequency ($S_{\delta\omega/\omega}^{1/2} = \alpha_n S_\tau^{1/2}$) calculated for a fluorite resonator with $R=0.3$ cm and $L=0.01$ cm interrogated with coherent 3.55 μm light of 1 mW power assuming that the light is absorbed in the resonator. Solid (dashed) line stands for the simulation (analytical calculations). (Reprinted from Matsko, A. B., Savchenkov, A. A., Yu, N., and Maleki, L., *J. Opt. Soc. Am. B* 24, 1324–1335, 2007. With permission from OSA.)

where $m^* \approx \rho V_r$ is the effective mass of the oscillator, $\delta P(t)$ is the variation of the optical power in the corresponding WGM. Calculating $\delta P(t)$ using general Langevin formalism we derive an expression for the spectral density of the fluctuations:

$$S_{\Delta R/R} = \left(\frac{2\pi}{m^* cR}\right)^2 \frac{\hbar\omega_{m,q,p}\langle P\rangle}{\tau_0} \frac{2\gamma_R}{\gamma_R^2 + \Omega^2} \frac{1}{(\Omega_0^2 - \Omega^2)^2 + \Gamma_0^2\Omega^2}, \tag{3.77}$$

where γ_R is the spectral width of the mode. Because generally $\Omega_0 \gg \gamma_R, \Gamma_0$ we find an expression for the square deviation of the radius of the resonator resulting from the fluctuations of the radiation pressure:

$$\frac{\langle \Delta R^2(t)\rangle^{1/2}}{R} \simeq \frac{2\pi}{m^* cR}\left[\frac{\hbar\omega_{m,q,p}\langle P\rangle}{\tau_0\Omega_0^4}\left(1+\frac{\gamma_R}{\Gamma_0}\right)\right]^{1/2}. \tag{3.78}$$

Let us estimate the value for a fluorite resonator with $R = 0.3$ cm and $L = 0.01$ cm interrogated with coherent 1.55 μm light of 1 mW power. We also assume that $\gamma_R = 2\pi \times 10^4$ rad/s and $\Gamma_0 = 100$ rad/s. We find the averaged power inside the resonator $\langle P\rangle = 1$ mW $/\tau_0\gamma_R \simeq 250$ W, square deviation of the radius $(\langle\Delta R^2(t)\rangle / R^2)^{1/2} \simeq 5.5 \times 10^{-16}$, and low frequency spectral density $(S_{\Delta R/R}(0))^{1/2} = 3 \times 10^{-18}$ Hz$^{-1/2}$. The corresponding spectral density of frequency fluctuations for $K_\varepsilon = 4$ is shown in Figure 3.49.

Our calculations show that the radiation pressure fluctuations are comparably weak in a sufficiently large WGM resonator, and can be neglected in the majority of cases when the resonator is interrogated with low power light.

3.7.3 Stabilization Scheme: An Example

We have shown [184] that it is fundamentally possible to reach high frequency stability of WGMs by properly selecting the operating conditions along with the host material, as well as the morphology of the resonator. In what follows we discuss a triple-mode technique. This technique is purely photonic and it enables stabilization of the WGM frequency better than the fundamental thermodynamic limit.

We propose to use three WGMs to suppress both thermorefractive and thermal expansion noise. Let us consider a spherical magnesium fluoride resonator with crystalline axis corresponding to the Z axis of a coordinate frame. The resonator is kept at 176°C where modes polarized perpendicular to the Z axis have a negligible thermorefractive effect. We propose to excite the TM mode in the XY plane and the TE mode in the XZ plane. Both modes have identical vanishing thermorefraction. A comparison of the frequency difference between these modes ($\omega_{RF} + \Delta\tilde{\omega}_{RF2}$) with the frequency of an RF reference gives averaged resonator temperature because

$$\Delta\tilde{\omega}_{RF2} \sim \omega(\alpha_{lo} - \alpha_{le})\Delta T_R, \tag{3.79}$$

where $\Delta\tilde{\omega}_{RF2}$ is the variation of the frequency difference between the two modes determined by the temperature fluctuations of the resonator, ω is the optical frequency, and α_{lo} (α_{le}) is the thermal expansion coefficient for X and Y (Z) directions. The third mode, TE, is excited in the XY plane. The frequency difference between this mode and the TM mode in the same plane contains information about the temperature in the WGM channel. Both modes are influenced by the thermal expansion in the same way. Using results of the

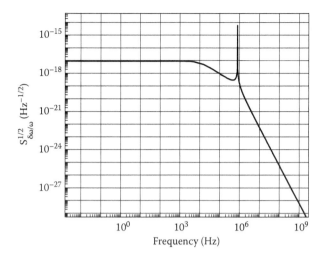

FIGURE 3.49
Spectral density of the ponderomotive backaction fluctuations of frequency ($S_{\delta\omega/\omega}^{1/2} = (1 + K_{\varepsilon}/2)S_{\Delta R/R}^{1/2}$) calculated for a fluorite resonator with $R = 0.3$ cm and $L = 0.01$ cm interrogated with 1 mW coherent light. (Reprinted from Matsko, A. B., Savchenkov, A. A., Yu, N., and Maleki, L., *J. Opt. Soc. Am. B* 24, 1324–1335, 2007. With permission from OSA.)

temperature measurements it is possible to design a proper feedback and/or compensation scheme that results in the suppression of both thermorefractive and thermal expansion fluctuations for the TM mode family in the XY plane. The relative stability of those modes is determined by expression

$$\frac{\Delta\omega_{TM}}{\omega} \sim \frac{\alpha_{lo}}{\alpha_{lo} - \alpha_{le}} \frac{\Delta\tilde{\omega}_{RF2}}{\omega}. \tag{3.80}$$

It is possible to achieve $\Delta\omega_{TM} \sim \tilde{\omega}_{RF2}$ if $\alpha_{lo} \neq \alpha_{le}$.

It is also possible to measure the resonator temperature with a sensitivity exceeding the fundamental thermodynamic limit. The measurement sensitivity is limited by $\Delta\tilde{\omega}_{RF2}/(\omega(\alpha_{lo} - \alpha_{le}))$, which can be very small if $\Delta\tilde{\omega}_{RF2}$ is small enough. Hence, the triple mode technique results in the possibility of compensation of the thermodynamic noise exceeding the fundamental thermodynamic limit. There is nothing special in the suppression of the fundamental thermorefractive frequency fluctuations of an optical mode, if one manages to lock this mode to an ultrastable optical frequency reference. The advantage of the proposed technique is in the possibility to stabilize an optical frequency beyond the thermodynamic limit using an RF reference. This feature will result in realization of stable UV, as well as FIR, lasers using crystalline WGM resonators.

3.7.4 Applications for Laser Stabilization

WGM resonators are naturally attractive for frequency stabilization of lasers as well as for fabrication of other kinds of frequency references. Preliminary experiments with stabilization of two diode lasers with a fused silica microsphere have been performed [58]. The result of that work was not very impressive compared with the best results of laser frequency stabilization using FP resonators since (i) the microsphere was not temperature stabilized, (ii) light coupling into and out of the microsphere was not stabilized, and

(iii) relative drift of the laser frequency resulting from the fluctuations of the driving currents was not stabilized.

Very recently, we have demonstrated injection locking of DFB lasers to crystalline WGM resonators. Initial measurements of the beat note of two lasers locked to calcium fluoride and lithium tantalate resonators show that the laser linewidth is less than 2 kHz (Figure 3.50). Further studies of the locking efficiency are required.

3.8 Conclusion

In this chapter, we have discussed the basic linear properties of crystalline optical WGM resonators. We have elucidated advantages of these resonators in practical application such as optical and microwave filtering. We have highlighted some important applications of the resonators in metrology for optical and microwave frequency stabilization, where a long photon storage time helps to suppress phase and frequency deviation of oscillators. The results of our R&D efforts clearly show the ample opportunities in applications of the resonators.

Because of the volume restrictions, we were unable to cover other applications of the microresonators, covered in the other chapters of the book. For instance, high-Q and long recirculation of light in compact WGM resonators offers interesting new capabilities in spectroscopy and sensing, where the change in Q or resonance frequency of WGMs can serve as a measure of absorption in the surrounding medium, or in a small (down to single molecule) quantity of deposited substance on resonator surface. The resonator can also be used for measurement of change in ambient parameters, such as temperature, pressure, motion, etc.

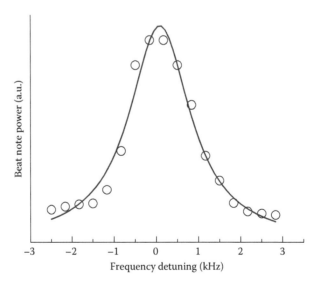

FIGURE 3.50
The beating signal of two lasers one of which is locked to a lithium tantalate resonator with loaded bandwidth approximately 3 MHz and calcium fluoride resonator with loaded bandwidth approximately 400 KHz. Laser power in the fiber at the exit of the resonator is 300 µW and 100 µW for $LiTaO_3$ and CaF_2 resonators, respectively. The resolution bandwidth of the measurement is 300 Hz and the video bandwidth is 30 kHz.

References

1. C. G. B. Garrett, W. Kaiser, and W. L. Bond. Stimulated emission into optical whispering gallery modes of spheres. *Phys. Rev.* 124, 1807–1809, 1961.
2. P. Walsh and G. Kemeny. Laser operation without spikes in a ruby ring. *J. Appl. Phys.* 34, 956–957, 1963.
3. A. Ashkin and J. M. Dziedzic. Observation of resonances in the radiation pressure on dielectric spheres. *Phys. Rev. Lett.* 38, 1351–1354, 1977.
4. P. Chylek, J. T. Kiehl, and M. K. W. Ko. Optical levitation and partial-wave resonances. *Phys. Rev. A* 18, 2229–2233, 1978.
5. A. Ashkin and J. M. Dziedzic. Observation of optical resonances of dielectric spheres by light scattering. *Appl. Opt.* 20, 1803–1814, 1981.
6. P. Chyek, V. Ramaswamy, A. Ashkin, and J. M. Dziedzic. Simultaneous determination of refractive index and size of spherical dielectric particles from light scattering data. *Appl. Opt.* 22, 2302–2307, 1983.
7. J. F. Owen, P. W. Barber, B. J. Messinger, and R. K. Chang. Determination of optical-fiber diameter from resonances in the elastic scattering spectrum. *Opt. Lett.* 6, 272–274, 1981.
8. R. E. Benner, P. W. Barber, J. F. Owen, and R. K. Chang. Observation of structure resonances in the fluorescence-spectra from microspheres. *Phys. Rev. Lett.* 44, 475–478, 1980.
9. S. C. Hill, R. E. Benner, C. K. Rushforth, and P. R. Conwell. Structural resonances observed in the fluorescence emission from small spheres on substrates. *Appl. Opt.* 23, 1680–1683, 1984.
10. S. C. Hill, R. E. Benner, C. K. Rushforth, and P. R. Conwell. Sizing dielectric spheres and cylinders by aligning measured and computed resonance locations: algorithm for multiple orders. *Appl. Opt.* 24, 2380–2390, 1985.
11. R. Thurn and W. Kiefer. Raman-microsampling technique applying optical levitation and radiation pressure. *Appl. Spectr.* 38, 78–83, 1984.
12. S.-X.Qian, J. B. Snow, and R. K. Chang. Coherent Raman mixing and coherent anti-Stokes Raman scattering from individual micrometer-size droplets. *Opt. Lett.* 10, 499–501, 1985.
13. J. B. Snow, S.-X.Qian, and R. K. Chang. Stimulated Raman scattering from individual water and ethanol droplets at morpholody-dependent resonances. *Opt. Lett.* 10, 37–39, 1985.
14. H. M. Tzeng, K. F. Wall, M. B. Long, and R. K. Chang. Laser emission from individual droplets at wavelengths corresponding to morphology-dependent resonances. *Opt. Lett.* 9, 499–501, 1984.
15. H. B. Lin, A. L. Huston, B. J. Justus, and A. J. Campillo. Some characteristics of a droplet whispering-gallery-mode laser. *Opt. Lett.* 11, 614–616, 1986.
16. L. Collot, V. Lefevre-Seguin, M. Brune, J.-M. Raimond, and S. Haroshe. Very high-Q whispering-gallery mode resonances observed in fused silica microspheres. *Europhys. Lett.* 23, 327–334, 1993.
17. M. L. Gorodetsky, A. A. Savchenkov, and V. S. Ilchenko. Ultimate Q of optical microsphere resonators. *Opt. Lett.* 21, 453–455, 1996.
18. S. Shiller and R. L. Byer. High-resolution spectroscopy of whispering gallery modes in large dielectric spheres. *Opt. Lett.* 16, 130–132, 1991.
19. D. R. Rowland, and J. D. Love. Evanescent wave coupling of whispering gallery modes of a dielectric cylinder, IEE Proceedings. *J. Optoelectronics* 140, 177–188, 1993.
20. M. L. Gorodetsky and V. S. Ilchenko. High-Q optical whispering gallery microresonators–precession approach for spherical mode analysis and emission patterns with prism couplers. *Opt. Comm.* 113, 133–143, 1994.
21. M. L. Gorodetsky, A. D. Pryamikov, and V. S. Ilchenko. Rayleigh scattering in high-Q microspheres. *J. Opt. Soc. Am. B* 17, 1051–1057, 2000.
22. D. G. Blair and I. N. Evans. High Q microwave properties of a sapphire ring resonator. *J. Phys. D* 15, 1651–1656, 1982.

23. V. B. Braginsky. V. P. Mitrofanov and V. I. Panov. *Systems with small dissipation*. Chicago University Press, Chicago, IL, 1985.

24. V. B. Braginsky, V. S. Ilchenko, and K. S. Bagdassarov. Experimental observation of fundamental microwave absorption in high quality dielectric crystals. *Phys. Lett. A* 120, 300–305, 1987.

25. W. L. Bond. Making crystal spheres. *Rev. Sci. Instr.* 25, 401–401, 1954.

26. D. A. Cohen and A. F. J. Levi. Microphotonic components for a mm-wave receiver. *Solid State Electron.* 45, 495–505, 2001.

27. T. Baer. Continuous-wave laser oscillation in a Nd:YAG sphere. *Opt. Lett.* 12, 392–394, 1987.

28. V. S. Ilchenko, X. S. Yao, and L. Maleki. Microsphere integration in active and passive photonics devices. *Proc. SPIE* 3930, 154–162, 2000.

29. V. S. Ilchenko and L. Maleki. Novel whispering-gallery resonators for lasers, modulators, and sensors. *Proc. SPIE* 4270, 120–130, 2001.

30. D. A. Cohen and A. F. J. Levi. Microphotonic millimetre-wave receiver architecture. *Electron. Lett.* 37, 37–39, 2001.

31. D. A. Cohen, M. Hossein-Zadeh, and A. F. J. Levi. Microphotonic modulator for microwave receiver. *Electron. Lett.* 37, 300–301, 2001.

32. D. A. Cohen, M. Hossein-Zadeh, and A. F. J. Levi. High-Q microphotonic electro-optic modulator. *Solid State Electron.* 45, 1577–1589, 2001.

33. D. K. Serkland, R. C. Eckardt, and R. L. Byer. Continuous-wave total-internal-reflection optical parametric oscillator pumped at 1064 nm. *Opt. Lett.* 19, 1046–1048, 1994.

34. A. A. Savchenkov, A. B. Matsko, V. S. Ilchenko, and L. Maleki. Optical resonators with ten million finesse. *Opt. Express* 15, 6768–6773, 2007.

35. G. Lee. Realization of ultrasmooth surface with atomic scale step structure on $LiNbO_3$ and $LiTaO_3$ substrates. *Opt. Express* 10, 556–560, 2002.

36. J. L. Yuan, P. Zhao, J. Ruan, Z. X. Cao, W. H. Zhao, and T. Xing. Lapping and polishing process for obtaining super-smooth surfaces of quartz crystal. *J. Mat. Process. Technol.* 138, 116–119, 2003.

37. J. W. Yan, K. Syoji, and J. Tamaki. Crystallographic effects in micro/nanomachining of single-crystal calcium fluoride. *J. Vac. Sci. Technol.* 22, 46–51, 2004.

38. I. S. Grudinin, A. B. Matsko, A. A. Savchenkov, D. Strekalov, V. S. Ilchenko, and L. Maleki. Ultra high Q crystalline microcavities. *Opt. Commun.* 265, 33–38, 2006.

39. J. W. Yan, J. Tamaki, K. Syoji, T. Kuriyagawa. Single-point diamond turning of CaF_2 for nanometric surface. *Int. J. Adv. Manuf. Technol.* 24, 640–646, 2004.

40. S. T. Chu, B. E. Little, V. Van, J. V. Hryniewicz, P. P. Absil, F. G. Johnson, D. Gill, O. King, F. Seiferth, M. Trakalo, and J. Shanton. Compact full C-band tunable filters for 50 GHz channel spacing based on high order micro-ring resonators. Optical Fiber Communication Conference 2004, presentation PDP9.

41. C. H. Townes and A. L. Shawlow. *Microwave spectroscopy*. McGraw-Hill, New York, 1955.

42. J. P. Gordon. Variable coupling reflection cavity for microwave spectroscopy. *Rev. Scientific Instruments* 32, 658–661, 1961.

43. A. Yariv. Critical coupling and its control in optical waveguide-ring resonator systems. *IEEE Photon. Tech. Lett.* 14, 483–485, 2002.

44. R. Ulrich. Theory of the prism-film coupler by plane wave analysis. *J. Opt. Soc. Am.* 60, 1337–1350, 1970.

45. P. K. Tien and R. Ulrich. Theory of prism-film coupler and thin film light guides. *J. Opt. Soc. Am.* 60, 1325–1336, 1970.

46. A. V. Chelnokov, and J.-M. Lourtioz. Optimized coupling into planar waveguides with cylindrical prisms. *Electron. Lett.* 31, 269–271, 1995.

47. D. Sarid, P. J. Cressman, and R. L. Holman. High-efficiency prism coupler for optical waveguides. *Appl. Phys. Lett.* 33, 514–515, 1978.

48. D. Sarid. High efficiency input-output prism waveguide coupler: an analysis. *Appl. Optics* 18, 2921–2926, 1979.

49. M. L. Gorodetsky and V. S. Ilchenko. Optical microsphere resonators: optimal coupling to high-Q whispering-gallery modes. *J. Opt. Soc. Am. B* 16, 147–154, 1999.

50. F. Treussart, V. S. Ilchenko, J. F. Roch, J. Hare, V. Lefevre-Seguin, J. M. Raimond, and S. Haroche. Evidence for intrinsic Kerr bistability of high-Q microsphere resonators in superfluid helium. *Eur. Phys. J. D* 1, 235–238, 1998.

51. Y. L. Pan and R. K. Chang. Highly efficient prism coupling to whispering gallery modes of a square μ cavity. *Appl. Phys. Lett.* 82, 487–489, 2003.

52. M. Mohageg, A. Savchenkov, and L. Maleki. High-Q optical whispering gallery modes in elliptical LiNbO$_3$ resonant cavities. *Opt. Express* 15, 4869–4875, 2007.

53. A. A. Savchenkov, I. S. Grudinin, A. B. Matsko, D. Strekalov, M. Mohageg, V. S. Ilchenko, and L. Maleki. Morphology-dependent photonic circuit elements. *Opt. Lett.* 31, 1313–1315, 2006.

54. A. Serpenguzel, S. Arnold, and G. Griffel. Excitation of resonances of microspheres on an optical fiber. *Opt. Lett.* 20, 654–656, 1995.

55. G. Griffel, S. Arnold, D. Taskent, A. Serpenguzel, J. Connoly, and N. Morris. Morphology-dependent resonances of a microsphere-optical fiber system. *Opt. Lett.* 21, 695–697, 1995.

56. N. Dubreuil, J. C. Knight, D. Leventhal, V. Sandoghdar, J. Hare, V. Lefevre-Seguin, J. M. Raimond, and S. Haroche. Eroded monomode optical fiber for whispering-gallery mode excitation in fused-silica microspheres. *Opt. Lett.* 20, 1515–1517, 1995.

57. V. S. Ilchenko, X. S. Yao, and L. Maleki. Pigtailing the high-Q microsphere cavity: a simple fiber coupler for optical whispering-gallery modes. *Opt. Lett.* 24, 723–725, 1999.

58. V. V. Vassiliev, V. L. Velichansky, V. S. Ilchenko, M. L. Gorodetsky, L. Hollberg, and A. V. Yarovitsky. Narrow-line-width diode laser with a high-Q microsphere resonator. *Opt. Comm.* 158, 305–312, 1998.

59. V. V. Vassiliev, S. M. Ilina, and V. L. Velichansky. Diode laser coupled to a high-Q microcavity via a GRIN lens. *Appl. Phys. B* 76, 521–523, 2003.

60. J. C. Knight, G. Cheung, F. Jacques, and T. A. Birks. Phase-matched excitation of whispering gallery mode resonances using a fiber taper. *Opt. Lett.* 22, 1129–1131, 1997.

61. M. K. Chin and S. T. Ho. Design and modeling of waveguide-coupled single-mode microring resonators. *J. Lightwave. Technol.* 16, 1433–1446, 1998.

62. B. E. Little, J. P. Laine, and H. A. Haus. Analytic theory of coupling from tapered fibers and half-blocks into microsphere resonators. *J. Lightwave. Technol.* 17, 704–715, 1999.

63. M. Cai, O. Painter, and K. J. Vahala. Observation of critical coupling in a fiber taper to a silica-microsphere whispering-gallery mode system. *Phys. Rev. Lett.* 85, 74–77, 2000.

64. M. Cai and K. J. Vahala. Highly efficient optical power transfer to whispering-gallery modes by use of a symmetrical dual-coupling configuration. *Opt. Lett.* 25, 260–262, 2000.

65. S. M. Spillane, T. J. Kippenberg, O. J. Painter, and K. J. Vahala. Ideality in a fiber-taper-coupled microresonator system for application to cavity quantum electrodynamics. *Phys. Rev. Lett.* 91, 043902, 2003.

66. P. E. Barclay, K. Srinivasan, and O. Painter. Design of photonic crystal waveguides for evanescent coupling to optical fiber tapers and integration with high-Q cavities. *J. Opt. Soc. Am. B* 20, 2274–2284, 2003.

67. A. Nurenberg and G. Schweiger. Excitation and recording of morphology-dependent resonances in spherical microresonators by hollow light guiding fibers. *Appl. Phys. Lett.* 84, 2043–2045, 2004.

68. K. Srinivasan, P. E. Barclay, M. Borselli, and O. Painter. Optical-fiber-based measurement of an ultrasmall volume high-Q photonic crystal microcavity. *Phys. Rev. B* 70, 081306, 2004.

69. H. Konishi, H. Fujiwara, S. Takeuchi, K. Sasaki. Polarization-discriminated spectra of a fiber-microsphere system. *Appl. Phys. Lett.* 89, 121107, 2006.

70. C. P. Michael, M. Borselli, T. J. Johnson, C. Chrystal, and O. Painter. An optical fiber-taper probe for wafer-scale microphotonic device characterization. *Opt. Express* 15, 4745–4752, 2007.

71. P. E. Barclay, K. Srinivasan, O. Painter, B. Lev, and H. Mabuchi. Integration of fiber coupled high-Q SiNx microdisks with atom chips. *Appl. Phys. Lett.* 89, 131108, 2006.

72. F. C. Blom, D. R. van Dijk, H. J. W. M. Hoekstra, A. Driessen, and Th. J. A. Popma. Experimental study of integrated-optics microcavity resonators: Toward an all-optical switching device. *Appl. Phys. Lett.* 71, 747–749, 1997.

73. D. Rafizadeh, J. P. Zhang, S. C. Hagness, A. Taflove, K. A. Stair, S. T. Ho, and R. C. Tiberio. Waveguide-coupled AlGaAs/GaAs microcavity ring and disk resonators with high finesse and 21.6-nm free spectral range. *Opt. Lett.* 22, 1244–1246, 1997.

74. B. E. Little, S. T. Chu, W. Pan, D. Ripin, T. Kaneko, Y. Kokubun, and E. Ippen. Vertically coupled glass microring resonator channel dropping filters. *IEEE Photon. Technol. Lett.* 11, 215–217, 1999.

75. P. Rabiei, W. H. Steier, C. Zhang, and L. R. Dalton. Polymer microring filters and modulators. *J. Lightwave Technol.* 20, 1968–1975, 2002.

76. D. V. Tishinin, P. D. Dapkus, A. E. Bond, I. Kim, C. K. Lin, and J. O'Brien. Vertical resonant couplers with precision coupling efficiency control fabricated by wafer bonding. *IEEE Photon. Technol. Lett.* 11, 1003–1005, 1999.

77. M. R. Poulsen, P. I. Borel, J. Fage-Pedersen, J. Hubner, M. Kristensen, J. H. Povlsen, K. Rottwitt, M. Svalgaard, and W. Svendsen. Advances in silica-based integrated optics. *Opt. Engin.* 42, 2821–2834, 2003.

78. S. J. Choi, K. Djordjev, S. J. Choi, P. D. Dapkus, W. Lin, G. Griffel, R. Menna, and J. Connolly. Microring resonators vertically coupled to buried heterostructure bus waveguides. *IEEE Photon. Technol. Lett.* 16, 828–830, 2004.

79. C. W. Tee, K. A. Williams, R. V. Penty, I. H. White. Fabrication-tolerant active-passive integration scheme for vertically coupled microring resonator. *IEEE J. Sel. Top. Quantum Electron.* 12, 108–116, 2006.

80. T. Le, A. A. Savchenkov, H. Tazawa, W. H. Steier, and L. Maleki, Polymer optical waveguide vertically coupled to high-Q whispering gallery resonators. *IEEE Photon. Technol. Lett.* 18, 859–861, 2006.

81. B. E. Little, J.-P. Laine, D. R. Lim, H. A. Haus, L. C. Kimerling, and S. T. Chu. Pedestal antiresonant reflecting waveguides for robust coupling to microsphere resonators and for microphotonic circuits. *Opt. Lett.* 25, 73–75, 2000.

82. J. P. Laine, B. E. Little, D. R. Lim, H. C. Tapalian, L. C. Kimerling, and H. A. Haus. Microsphere resonator mode characterization by pedestal anti-resonant reflecting waveguide coupler. *IEEE Photon. Technol. Lett.* 12, 1004–1006, 2000.

83. I. M. White, H. Oveys, X. Fan, T. L. Smith, and J. Zhang. Integrated multiplexed biosensors based on liquid core optical ring resonators and antiresonant reflecting optical waveguides. *Appl. Phys. Lett.* 89, 191106, 2006.

84. I. M. White, J. D. Suter, H. Oveys, X. Fan, T. L. Smith, J. Zhang, B. J. Koch, and M. A. Haase. Universal coupling between metal-clad waveguides and optical ring resonators. *Opt. Express* 15, 646–651, 2007.

85. A. Morand, K. Phan-Huy, Y. Desieres, and P. Benech. Analytical study of the microdisk's resonant modes coupling with a waveguide based on the perturbation theory. *J. Lightwave Technol.* 22, 827–832, 2004.

86. A. Belarouci, K. B. Hill, Y. Liu, Y. Xiong, T. Chang, and A. E. Craig. Design and modeling of waveguide-coupled microring resonator. *J. Luminescence* 94, 35–38, 2001.

87. C. Li, L. J. Zhou, S. M. Zheng, and A. W. Poon. Silicon polygonal microdisk resonators. *IEEE J. Sel. Top. Quantum Electron.* 12, 1438–1449, 2006.

88. V. S. Ilchenko, M. L. Gorodetsky, X. S. Yao, and L. Maleki. Microtorus: a high-finesse microcavity with whispering-gallery modes. *Opt. Lett.* 26, 256–258, 2001.

89. D. K. Armani, T. J. Kippenberg, S. M. Spillane, and K. J. Vahala. Ultra-high-Q toroid microcavity on a chip. *Nature* 421, 925–928, 2003.

90. M. Sumetsky. Whispering-gallery-bottle microcavities: the three-dimensional etalon. *Opt. Lett.* 29, 8–10, 2004.

91. V. S. Ilchenko, A. A. Savchenkov, A. B. Matsko, and L. Maleki. Dispersion compensation in whispering-gallery modes. *J. Opt. Soc. Am. A* 20, 157–162, 2003.

92. L. G. Guimaraes and H. M. Nussenzveig. Theory of Mie resonances and ripple fluctuations. *Opt. Commun.* 89, 363–369, 1992.

93. D. Q. Chowdhury, S. C. Hill, and P. W. Barber. Morphology-dependent resonances in radially inhomogeneous spheres. *J. Opt. Soc. Am. A* 8, 1702–1705, 1991.

94. B. Z. Katsenelenbaum, L. Mercader del Rio, M. Pereyaslavets, M. Sorolla Ayza, and M. Thumm. *Theory of nonuniform waveguides: the cross-section method.* Institute of Electrical Engineers, London, 1998.

95. V. S. Ilchenko, A. A. Savchenkov, A. B. Matsko, and L. Maleki. Tunability and synthetic line-shapes in high-Q optical whispering gallery modes. *Proc. SPIE* 4969, 195–206, 2003.

96. M. L. Gorodetsky and A. E. Fomin. Geometrical theory of whispering-gallery modes. *IEEE J. Sel. Top. Quantum Electron.* 12, 33–39, 2006.

97. G. Meijer, M. G. H. Boogaarts, R. T. Jongma, D. H. Parker, and A. M. Wodtke. Coherent cavity ring down spectroscopy. *Chem. Phys. Lett.* 217, 112–116, 1994.

98. V. S. Ilchenko, A. A. Savchenkov, A. B. Matsko, and L. Maleki. Whispering-gallery-mode electro-optic modulator and photonic microwave receiver. *J. Opt. Soc. Am. B* 20, 333–342, 2003.

99. A. Wicht, K. Danzmann, M. Fleischhauer, M. Scully, G. Müller, and R.-H. Rinkleff. White-light cavities, atomic phase coherence, and gravitational wave detectors. *Opt. Commun.* 134, 431–439, 1997.

100. G. S. Pati, M. Salit, K. Salit, and M. S. Shahriar. Demonstration of a tunable-bandwidth white-light interferometer using anomalous dispersion in atomic vapor. *Phys. Rev. Lett.* 99, 133601, 2007.

101. A. A. Savchenkov, A. B. Matsko, and L. Maleki. White-light whispering gallery mode resonators. *Opt. Lett.* 31, 92–94, 2006.

102. W. von Klitzing, R. Long, V. S. Ilchenko, J. Hare, and V. Lefevre-Seguin. Frequency tuning of the whispering-gallery modes of silica microspheres for cavity quantum electrodynamics and spectroscopy. *Opt. Lett.* 26, 166–168, 2001.

103. S. Blair and Y. Chen. Resonant-enhanced evanescent-wave fluorescence biosensing with cylindrical optical cavities. *Appl. Optics* 40, 570–582, 2001.

104. R. W. Boyd and J. E. Heebner. Sensitive disk resonator photonic biosensor. *Appl. Optics* 40, 5742–5747, 2001.

105. E. Krioukov, D. J. W. Klunder, A. Driessen, J. Greve, and C. Otto. Sensor based on an integrated optical microcavity. *Opt. Lett.* 27, 512–514, 2002.

106. S. Arnold, M. Khoshsima, I. Teraoka, S. Holler, and F. Vollmer. Shift of whispering-gallery modes in microspheres by protein adsorption. *Opt. Lett.* 28, 272–274, 2003.

107. F. Vollmer, S. Arnold, D. Braun, I. Teraoka, and A. Libchaber. Multiplexed DNA quanti. cation by spectroscopic shift of two microsphere cavities. *Biophys. J.* 85, 1974–1979, 2003.

108. C. Y. Chao and L. J. Guo. Design and optimization of microring resonators in biochemical sensing applications. *J. Lightwave Technol.* 24, 1395–1402, 2006.

109. S. Y. Cho and N. M. Jokerst. A polymer microdisk photonic sensor integrated onto silicon. *IEEE Photon. Technol. Lett.* 18, 2096–2098, 2006.

110. N. M. Hanumegowda, I. M. White, and X. D. Fan. Aqueous mercuric ion detection with microsphere optical ring resonator sensors. *Sensors Actuators B* 120, 207–212, 2006.

111. A. M. Armani, R. P. Kulkarni, S. E. Fraser, R. C. Flagan, and K. J. Vahala. Label-free, single-molecule detection with optical microcavities. *Science* 317, 783–787, 2007.

112. H. Zhu, I. M. White, J. D. Suter, P. S. Dale, and X. Fan. Analysis of biomolecule detection with optofluidic ring resonator sensors. *Opt. Express* 15, 9139–9146, 2007.

113. F. Xu, P. Horak, and G. Brambilla. Optical microfiber coil resonator refractometric sensor. *Opt. Express* 15, 7888–7893, 2007; *Opt. Express* 15, 9385–9385, 2007.

114. M. Sumetsky. Optimization of optical ring resonator devices for sensing applications. *Opt. Lett.* 32, 2577–2579, 2007.

115. A. T. Rosenberger. Analysis of whispering-gallery microcavity-enhanced chemical absorption sensors. *Opt. Express* 15, 12959–12964, 2007.

116. A. A. Savchenkov, A. B. Matsko, D. Strelalov, V. S. Ilchenko, and L. Maleki. Mode filtering in optical whispering gallery resonators. *Electron. Lett.* 41, 495–497, 2005.

117. M. Kuznetsov, M. Stern, and J. Coppeta. Single transverse mode optical resonators. *Opt. Express* 13, 171–181, 2005.

118. R. G. Hunsperger. *Integrated Optics: Theory and Technology.* Springer Verlag, Berlin, 2002.

119. S.-T. Ho, D. Y. Chu, J.-P. Zhang, S. Wu, and M.-K. Chin. In *Optical processes in microcavities* R. K. Chang and A. J. Campillo (Eds). Word Scientific, Singapore, 1996, p. 339.
120. M. V. Berry. Regularity and chaos mechanics, illustrated in classical by three deformations of a circular "billiard". *Eur. J. Phys.* 2 91–102, 1981.
121. A. Ancey, A. Folacci, and P. Gabrielli. Exponentially improved asymptotic expansions for resonances of an elliptic cylinder. *J. Phys. A: Math. Gen.* 33, 3179–3208, 2000.
122. S. Ancey, A. Folacci, and P. Gabrielli. Whispering-gallery modes and resonances of an elliptic cavity. *J. Phys. A: Math. Gen.* 34, 1341–1359, 2001; *J. Phys. A: Math. Gen.* 34, 2657–2657, 2001.
123. S. V. Boriskina, T. M. Benson, P. Sewell, and A. I. Nosich. Tuning of elliptic whispering-gallery-mode microdisk waveguide filters. *J. Lightwave Technol.* 21, 1987–1995, 2003.
124. L. Zhang, Y. -X. Wang, F. Zhang, and R. O. Claus. Observation of whispering-gallery and directional resonant laser emission in ellipsoidal microcavities. *J. Opt. Soc. Am. B* 23, 1793–1800, 2006.
125. G. Rempe, R. J. Thompson, H. J. Kimble, and R. Lalezari. Measurement of ultralow losses in an optical interferometer. *Opt. Lett.* 17, 363–365, 1992.
126. C. J. Hood, H. J. Kimble, and J. Ye. Characterization of high-finesse mirrors: Loss, phase shifts, and mode structure in an optical cavity. *Phys. Rev. A* 64, 033804, 2001.
127. H. R. Bilger, P. V. Wells, and G. E. Stedman. Origins of fundamental limits for reflection losses at multilayer dielectric mirrors. *Appl. Opt.* 33, 7390–7396, 1994.
128. H.-J. Cho, M.-J. Shin, and J.-C. Lee. Effects of substrate and deposition method onto the mirror scattering. *Appl. Opt.* 45, 1440–1446, 2006.
129. E. M. Purcell. Spontaneous emission probabilities at radio frequencies. *Phys. Rev.* 69, 681–681, 1946.
130. A. B. Matsko, A. A. Savchenkov, D. Strekalov, V. Ilchenko, and L. Maleki. Optical hyperparametric oscillations in a whispering-gallery-mode resonator: Threshold and phase diffusion. *Phys. Rev. A* 71, 033804, 2005.
131. V. S. Ilchenko, A. A. Savchenkov, A. B. Matsko, and Lute Maleki. Nonlinear optics and crystalline whispering gallery mode cavities. *Phys. Rev. Lett.* 92, 043903, 2004.
132. D. W. Vernooy, V. S. Ilchenko, H. Mabuchi, E. W. Streed, and H. J. Kimble. High-Q measurements of fused-silica microspheres in the near infrared. *Opt. Lett.* 23, 247–249, 1998.
133. T. J. Kippenberg, S. M. Spillane, and K. J. Vahala. Demonstration of ultra-high-Q small mode volume toroid microcavities on a chip. *Appl. Phys. Lett.* 85, 6113–6115, 2004.
134. I. S. Grudinin, V. S. Ilchenko, and L. Maleki. Ultrahigh optical Q factors of crystalline resonators in the linear regime. *Phys. Rev. A* 74, 063806, 2006.
135. S. M. Spillane, T. J. Kippenberg, and K. J. Vahala. Ultralow-threshold Raman laser using a spherical dielectric microcavity. *Nature* 415, 621–623, 2002.
136. I. S. Grudinin and L. Maleki. Ultralow-threshold Raman lasing with CaF_2 resonators. *Opt. Lett.* 32, 166–168, 2007.
137. T. J. Kippenberg, H. Rokhsari, T. Carmon, A. Scherer, and K. J. Vahala. Analysis of radiation-pressure induced mechanical oscillation of an optical microcavity. *Phys. Rev. Lett.* 95, 033901, 2005.
138. T. J. Kippenberg, S. M. Spillane, and K. J. Vahala. Kerr-nonlinearity optical parametric oscillation in an ultrahigh-Q toroid microcavity. *Phys. Rev. Lett.* 93, 083904, 2004.
139. A. A. Savchenkov, A. B. Matsko, D. Strekalov, M. Mohageg, V. S. Ilchenko, and L. Maleki. Low threshold optical oscillations in a whispering gallery mode CaF_2 resonator. *Phys. Rev. Lett.* 93, 243905, 2004.
140. A. A. Savchenkov, A. B. Matsko, M. Mohageg, and L. Maleki. Ringdown spectroscopy of stimulated Raman scattering in a whispering gallery mode resonator. *Opt. Lett.* 32, 497–499, 2007.
141. A. A. Savchenkov, V. S. Ilchenko, A. B. Matsko, and L. Maleki. Kilohertz optical resonances in dielectric crystal cavities. *Phys. Rev. A* 70, 051804R, 2004.

142. A. B. Matsko, A. A. Savchenkov, and L. Maleki. Ring-down spectroscopy for studying properties of CW Raman lasers. *Opt. Commun.* 260, 662–665, 2006.

143. B. Wang, R. R. Rockwell, and J. List. Linear birefringence in CaF_2 measured at deep ultraviolet and visible wavelengths. *J. Microlithography, Microfabrication, and Microsystems* 3, 115–121, 2004.

144. V. Deuster, M. Schick, Th. Kayser, H. Dabringhaus, H. Klapper, and K. Wandelt. Studies of the facetting of the polished (100) face of CaF_2. *J. Cryst. Growth* 250, 313–323, 2003.

145. Q. -Z. Zhao, J. -R. Qiu, X. -W. Jiang, C. -J. Zhao, and C. -S. Zhu. Fabrication of internal diffraction gratings in calcium fluoride crystals by a focused femtosecond laser. *Opt. Express* 12, 742–746, 2004.

146. I. S. Grudinin, A. B. Matsko, and L. Maleki. On the fundamental limits of Q factor of crystalline dielectric resonators. *Opt. Express* 15, 3390–3395, 2007.

147. M. E. Lines. Scattering losses in optic fiber materials. I. A new parametrization. *J. Appl. Phys.* 55, 4052–4057, 1984; Scattering losses in optic fiber materials. II. Numerical estimates. *J. Appl. Phys.* 55, 4058–4063, 1984.

148. S. Logunov and S. Kuchinsky. Experimental and theoretical study of bulk light scattering in CaF_2 monocrystals. *J. Appl. Phys.* 98, 053501, 2005.

149. I. L. Fabelinskii. *Molecular scattering of light*. Plenum Press, New York, 1968.

150. D. C. Wallace. *Thermodynamics of crystals*. Dover, New York, 1998.

151. R. Loudon. The Raman effect in crystals. *Adv. Phys.* 50, 813–864, 2001.

152. M. Daimon and A. Masumara. High-accuracy measurements of the refractive index and its temperature coefficient of calcium fluoride in a wide wavelength range from 138 to 2326 nm. *Appl. Opt.* 41 5275–5281, 2002.

153. E. Palik. *Handbook of optical constants of solids*. Academic, New York, 1998.

154. M. Lines. Ultralow-loss glasses, *Ann. Rev. Mater. Sci.* 16, 113–135, 1986.

155. A. Laufer, J. Pirog, and J. McNesby. Effect of temperature on the vacuum ultraviolet transmittance of lithium fluoride, calcium fluoride, barium fluoride, and sapphire. *J. Opt. Soc. Am.* 55, 64–66, 1965.

156. L. L. Boyer, J. A. Harrington, M. Hass, and H. B. Rosenstock. Multiphonon absorption in ionic crystals. *Phys. Rev. B* 11, 1665–1680, 1975.

157. S. Venugopalan and A. K. Ramdas. Effect of uniaxial stress on the raman spectra of cubic crystals: CaF_2, BaF_2, and $Bi_{12}GeO_{20}$. *Phys. Rev. B* 8, 717–734, 1973.

158. R. W. Boyd. *Nonlinear optics*. Academic Press, New York, 1992.

159. A. A. Savchenkov, V. S. Ilchenko, A. B. Matsko, and L. Maleki. KiloHertz optical resonances in dielectric crystal cavities. *Phys. Rev. A* 70, 051804(R), 2004.

160. A. A. Savchenkov, V. S. Ilchenko, A. B. Matsko, and L. Maleki. Tunable filter based on whispering gallery modes. *Electron. Lett.* 39, 389–391, 2003.

161. A. A. Savchenkov, V. S. Ilchenko, A. B. Matsko, and L. Maleki. High-order tunable filters based on a chain of coupled crystalline whispering gallery mode resonators. *IEEE Photonics Technol. Lett.* 17, 136–138, 2005.

162. A. A. Savchenkov, A. B. Matsko, V. S. Ilchenko, N. Yu, and L. Maleki. Microwave photonics applications of whispering gallery mode resonators physics and engineering of microwaves. Millimeter and Submillimeter Waves and Workshop on Terahertz Technologies, 2007. MSMW '07. The Sixth International Kharkov Symposium, 1, pp. 79–84, 2007.

163. B. E. Little, S. T. Chu, H. A. Haus, J. Foresi, and J. P. Laine. Microring resonator channel dropping filters. *J. Lightwave Tech.* 15, 998–1005, 1997.

164. J. V. Hryniewicz, P. P. Absil, B. E. Little, R. A. Wilson, and P. -T. Ho. Higher order filter response in coupled microring resonators. *IEEE Phot. Tech. Lett.* 12, 320–322, 2000.

165. S. T. Chu, B. E. Little, W. Pan, T. Kaneko, and Y. Kukubun. Cascaded microring resonators for crosstalk reduction and spectrum cleanup in add-drop filters. *IEEE Phot. Tech. Lett.* 11, 1423–1425, 1999.

166. S. T. Chu, B. E. Little, W. Pan, T. Kaneko, and Y. Kukubun. Second-order filter response from parallel coupled glass microring resonators. *IEEE Phot. Tech. Lett.* 11, 1426–1428, 1999.

167. K. Djordjev, S. J. Choi, S. J. Choi, and P. D. Dapkus. Microdisk tunable resonant filters and switches. *IEEE Phot. Tech. Lett*. 14, 828–830, 2002.

168. O. Schwelb and I. Frigyes. Vernier operation of series coupled optical microring resonator filters. *Microwave Opt. Technol. Lett*. 39, 258–261, 2003.

169. J.-L. Gheorma and R. M. Osgood. Fundamental limitations of optical resonator based high-speed EO modulators. *IEEE Photon. Tech. Lett*. 14, 795–797, 2002.

170. G. Griffel. Vernier effect in asymmetrical ring resonator arrays. *IEEE Phot. Technol. Lett*. 12, 1642–1644, 2000.

171. P. Urquhart. Compound optical-fiber-based resonators. *J. Opt. Soc. Am. A* 5, 803–812, 1988.

172. K. Oda, N. Takato, and H. Toba. Wide-FSR waveguide double-ring resonator for optical FDM transmission system. *J. Lightwave Technol*. 9, 728–736, 1991.

173. M. Mohageg, A. Savchenkov, D. Strekalov, A. Matsko, V. Ilchenko, and L. Maleki. Reconfigurable optical filter. *Electron. Lett*. 41, 91–92, 2005.

174. C. K. Madsen and J. H. Zhao. A general planar waveguide autoregressive optical filter. *J. Lightwave Technol*. 14, 437–447, 1996.

175. A. B. Matsko, A. A. Savchenkov, and L. Maleki. Vertically coupled whispering-gallery-mode resonator waveguide. *Opt. Lett*. 30, 3066–3068, 2005.

176. L. Maleki, A. B. Matsko, A. A. Savchenkov, and D. Strekalov. Slow light in vertically coupled whispering gallery mode resonators. *Proc. SPIE*, 6130, 61300R, 2006.

177. A. Yariv, Y. Xu, R. K. Lee, and A. Scherer. Coupled–resonator optical waveguide: a proposal and analysis. *Opt. Lett*. 24, 711–713, 1999.

178. J. E. Heebner, P. Chak, S. Pereira, J. E. Sipe, and R. W. Boyd. Distributed and localized feedback in microresonator sequences for linear and nonlinear optics. *J. Opt. Soc. Am. B* 21, 1818–1832, 2004.

179. M. Sumetsky. Vertically-stacked multi-ring resonator. *Opt. Express* 13, 6354–6375, 2005.

180. X. S. Yao and L. Maleki. Optoelectronic microwave oscillator. *J. Opt. Soc. Am. B* 13, 1725–1735, 1996.

181. D. Strekalov, D. Aveline, N. Yu, R. Thompson, A. B. Matsko, and L. Maleki. Stabilizing an optoelectronic microwave oscillator with photonic filters. *J. Lightwave Technol*. 21, 3052–3061, 2003.

182. A. B. Matsko, L. Maleki, A. A. Savchenkov, and V. S. Ilchenko. Whispering gallery mode based optoelectronic microwave oscillator. *J. Mod. Opt*. 50, 2523–2542, 2003.

183. A. B. Matsko, A. A. Savchenkov, N. Yu, and L. Maleki. Whispering-gallery-mode resonators as frequency references. I. Fundamental limitations. *J. Opt. Soc. Am. B* 24, 1324–1335, 2007.

184. A. A. Savchenkov, A. B. Matsko, V. S. Ilchenko, N. Yu, and L. Maleki. Whispering-gallery-mode resonators as frequency references. II. Stabilization. *J. Opt. Soc. Am. B* 24, 2988–2997, 2007.

185. M. L. Gorodetsky and I. S. Grudinin. Fundamental thermal fluctuations in microspheres. *J. Opt. Soc. Am. B* 21, 697–705, 2004.

186. F. L. Walls and J. R. Vig. Fundamental limits on the frequency stabilities of crystal oscillators. *IEEE Tran. Ferroelectrics and Freq. Control* 42, 576–589, 1995.

187. V. B. Braginsky, M. L. Gorodetsky, and S. P. Vyatchanin. Thermodynamical fluctuations and photo-thermal shot noise in gravitational wave antennae. *Phys. Lett. A* 264, 1–10, 1999.

188. C. Audoin. *Proceedings of the International School of Physics Enrico Fermi*, Course LXVIII, A. F. Milone and P. Giacomo (Eds). Amsterdam, North-Holland, p. 169, 1980.

189. A. Feldman, D. Horowitz, R. M. Waxler, and M. J. Dodge. *Optical materials characterization*. National Bureau of Standards Technical Note 993. U.S. Government Printing Office, Washington DC, 1979.

190. V. B. Braginsky, V. P. Mitrofanov, and V. I. Panov. *Systems with small dissipation*. The University of Chicago Press, Chicago, IL, 1985.

191. K. Numata, A. Kemery, and J. Camp. Thermal-noise limit in the frequency stabilization of lasers with rigid cavities. *Phys. Rev. Lett*. 93, 250602, 2004.

192. K. Goda, K. McKenzie, E. E. Mikhailov, P. K. Lam, D. E. McClelland, and N. Mavalvala. Photothermal fluctuations as a fundamental limit to low-frequency squeezing in a degenerate optical parametric oscillator. *Phys. Rev. A*, 72, 043819, 2005.
193. V. B. Braginsky and S. P. Vyatchanin. Frequency fluctuations of nonlinear origin in self-sustained optical oscillators. *Phys. Lett. A* 279, 154–162, 2001.
194. S. Chang, A. G. Mann, A. N. Luiten, and D. G. Blair. Measurements of radiation pressure effect in cryogenic sapphire dielectric resonators. *Phys. Rev. Lett.* 79, 2141–2144, 1997.

4

Microresonator-Based Devices on a Silicon Chip: Novel Shaped Cavities and Resonance Coherent Interference

---

Andrew W. Poon, Xianshu Luo, Linjie Zhou, Chao Li, Jonathan
Y. Lee, Fang Xu, Hui Chen, and Nick K. Hon
The Hong Kong University of Science and Technology

CONTENTS

4.1 Introduction

Optical microresonators [1–5] that partially confine light by total internal reflection (TIR) at the microresonator sidewalls are elegant micro-photonic structures. Such TIR-based resonators have demonstrated high-Q resonances and can often be realized by means of conventional fabrication techniques. Although early work on optical microresonators dates back more than six decades [6–8], many high-impact microresonator technologies from university and industry laboratories only came out in the 1990s. Recent progress in microresonator technology is in part fueled by the availability of advanced fabrication techniques for patterning fine structures on various material systems. Notably, we see microresonator-based lasers [9–16] in different material systems and in different cavity shapes. We also see microresonator-based filters [17–20], switches [21,22], and modulators [19,23,24] for next-generation wavelength-division multiplexing (WDM) optical communications. Moreover, refractive-index-based biochemical sensors using microresonators are also attracting considerable research interest [25–28].

Among all material systems, silicon, the most common semiconductor material for computer chips, has only recently attracted a surge of rekindled interest for photonics applications. Traditionally, silicon is not regarded as a photonic material other than for the manufacture of photodetectors and solar cells. It is well-understood that silicon is an indirect-bandgap material that does not efficiently emit light. Furthermore, bulk silicon is a centrosymmetric crystal that does not exhibit linear electro-optic effect. Nonetheless, there are also major merits that render silicon a good material platform for fabricating photonic devices. On the material front, silicon is a robust material that is widely available and can be purified to a high level. The conductivity of silicon can also be controlled by doping impurities. Moreover, the natural oxide of silicon, silicon dioxide (SiO_2), is a very stable and excellent insulator, offering a very high etching selectivity with respect to silicon. On the semiconductor optics front, silicon has a relatively wide bandgap of ~1.1 eV and thus is transparent to the telecommunication wavelengths (~1.3–~1.7 μm). The high material refractive index of ~3.45 also enables tight optical confinement. Hence, in hindsight it is not surprising that silicon emerges as a material platform for fabricating highly compact low-loss integrated waveguide devices for applications in telecommunication wavelengths. The prospect of integrating microelectronics and photonics on the same silicon platform using the complementary metal-oxide-semiconductor (CMOS) fabrication process is technologically appealing. Indeed, submicrometer-scale silicon wire waveguides have shown a propagation loss of only ~1–2.5 dB/cm [29,30]. The waveguide propagation loss mainly originates from sidewall roughness-induced scattering. Over recent years, considerable research work [31] has been devoted to minimize sidewall roughness by using CMOS-compatible deep-UV photolithography [29] or e-beam lithography [30] and optimized dry etching processes (e.g. double thermal oxidation [30]). Furthermore, these wire waveguides not only enable low-loss optical interconnection on a chip but also evanescent coupling to a microstructure that is positioned in close proximity. In this respect, silicon photonics offers a useful technology platform for fabricating and testing microresonator-based passive and active devices, which would otherwise be more difficult or costly to fabricate using other materials.

Over the past few years, we have witnessed unprecedented growth in research and development activities on exploiting silicon for photonics [32]. Excellent reviews [33–36] and textbooks [37,38] on silicon photonics have been written. One can perhaps argue that the surge of silicon photonics research is driven by the increasing need of ultra-short-reach broadband low-power interconnection on next-generation high-performance computer chips [39,40]. The potential impact of silicon photonics on the computer industry may also be derived from the scope of silicon photonics research at Intel [41,42] and IBM [30].

When it comes to the TIR-based optical microresonators, microring and microdisk are arguably the two most investigated structures on a silicon chip. The first silicon microring resonator-based channel dropping filter was demonstrated in 1998 [17], with the observed Q-factor of only ~250 in a 10-μm diameter microring resonator. Since then, microring and microdisk resonator-based wavelength filters (not necessarily silicon-based) have been attracting considerable research interest. On the silicon passive photonic devices front, research groups worldwide have demonstrated silicon microresonator-based passive wavelength filters [29,43–48]. With optimized device parameters and fabrication, silicon microresonators with Q exceeding 10^5 have been demonstrated [43,44,48,49]. In addition, more sophisticated photonic structures have been realized by coupling silicon microring resonators in serial and in parallel including coupled resonator optical waveguides (CROW) [46] and side-coupled integrated spaced sequences of resonators (SCISSOR) [50]. Such multiple-coupled microresonators enable tailoring of transmission spectra and dispersion properties, and also promise slow-light-based optical delay line [30].

On the silicon active photonic devices front, reconfigurable devices and fast modulators have been demonstrated in silicon microresonators using thermal-optics [51,52], MEMS [53,54], electro-optics [23,24,55,56], and all-optical methods [57,58]. Notably, in 2005 Xu et al. (M. Lipson's group) [23] pioneered the micro-scale GHz-speed carrier-injection-based silicon electro-optic modulator using a microring resonator integrated with a p-i-n diode. In 2006, Luxtera Inc. [55] announced the first 10 Gbit/s carrier-depletion-based silicon electro-optic modulators using microring resonators with integrated reverse-biased p-n diode. In addition, silicon microresonator-based light sources such as optically pumped Raman lasers [59], hybrid-integrated III-V-on-silicon lasers [60,61], and all-optical wavelength converters [62,63] have been demonstrated. Therefore, it is now feasible to make compact silicon photonic integrated circuits (PICs), based on optical microresonator technology, for photonic signal processing applications.

Nonetheless, it is also recognized that silicon microresonators using conventional circular-shaped microresonators impose shortcomings. The microresonator curved sidewall only enables a short interaction length with the side-coupled straight waveguide. In high-index-contrast (~3.45 to ~1.44 for silicon-to-silica) microresonators, such short interaction length imposes a submicrometer gap separation (typically ~0.1–~0.3 μm for a micro-sized resonator) for the lateral evanescent coupling. Fabrication-induced variations of the coupling gap distance thus significantly alter the coupling efficiency from the designed values. Repeatable fabrication of narrow gaps by lithographic and etching techniques is still technologically challenging, although deep-UV lithography can help ease the constraint [29].

In order to enhance the lateral waveguide-to-microresonator coupling, it has long been known that racetrack-shaped microring resonators with a straight waveguide section enable efficient evanescent coupling to a straight waveguide with a large gap tolerance [29,44,64]. However, racetrack-shaped microring has the major shortcoming of imposing modal transition losses between the curved waveguide and the straight waveguide. Besides, racetrack-shaped microring resonators impose the same limitations as circular-shaped microring resonators including waveguide-bending loss, inner-sidewall roughness-induced scattering loss, and polarization-mode conversion [65] due to the fabrication-induced ring-waveguide nonuniformities. Thus, it is desirable to exploit alternative microresonator designs for efficient waveguide coupling.

In this respect, our approaches to applications of optical microresonators on a silicon chip emphasize exploiting alternative non-circular-shaped microdisk resonators for enhanced waveguide-to-microresonator coupling. We focus on different cavity shape designs of microdisk resonators in order to eliminate complications from microring waveguides and inner-sidewall scattering. The noncircular-shaped microdisk resonators unveil modal characteristics that are distinct from the conventional circular-shaped microdisk resonators, giving us new insights to optical microresonators and inspiring new device applications. In order to make silicon microdisk resonators electro-optically switchable in high speed, we adopt the carrier-injection method using selectively integrated p-i-n diodes. The selective integration of the p-i-n diode intrinsic region along the microdisk rim enables the whispering gallery (WG) modes to spatially overlap with the injected carriers. Furthermore, we see that the microresonator resonance response can be tailored by means of interference of resonances with coherent background and coherent feedback. The microresonator-based devices with coupled waveguide interferometers thus open up novel device applications.

In this chapter, we review our approaches to microresonator-based devices on a silicon chip. We discuss device principles, design issues, numerical simulations, and experimental demonstrations of (i) alternative noncircular-shaped microdisk resonators that offer enhanced lateral waveguide-to-microresonator coupling, and (ii) coherent interference of resonances for switching and modulation functionalities. In Section 4.2, we discuss waveguide-coupled polygonal-shaped microdisk resonators for directional evanescent coupling. In Section 4.3, we introduce waveguide-coupled spiral-shaped microdisk resonators for nonevanescent coupling through the spiral notch. In Section 4.4, we discuss carrier-injection-based silicon electro-optic modulators using a microdisk resonator with selectively embedded p-i-n diodes. In Section 4.5, we present two schemes of coherent interference of microresonator resonances for applications as reconfigurable filters, modulators, and logic switches. In Section 4.6, we summarize the chapter and provide an outlook on exploiting silicon microresonators for promising photonic networks-on-chip applications on a silicon computer chip.

4.2 Polygonal-Shaped Microdisk Resonators with Directional Coupling

4.2.1 Overview of Polygonal-Shaped Microresonators

Based on the theoretical study on waveguide-coupled square-shaped microresonators from Manolatou et al. [66] (the late Prof. Herman Haus's group), and proof-of-principle experiments on square-shaped silica micropillar resonators using Gaussian-beam coupling by one of us (AWP) [67] and using prism coupling [68] (Prof. Richard Chang's group), we have studied waveguide-coupled polygonal-shaped microdisk resonators as alternative microresonator-based device structures for wavelength filtering [69–76] and also electro-optic switching [77]. Over the past few years, we have examined waveguide-coupled square- [69,70,76], hexagonal- [71–73,76], and octagonal- [74–76] shaped microdisk resonators on silicon nitride (SiN)-on-silica and silicon-on-insulator (SOI) substrates. Table 4.1 summarizes the key measurements of our silicon-based polygonal-shaped microdisk resonators.

Compared with circular microdisk- and microring-shaped resonators, polygonal-shaped microdisk resonators offer the key merit of flat cavity sidewalls for evanescent

TABLE 4.1

Summary of Our Experimental Demonstrated Polygonal-Shaped Microdisk Resonator-Based Add-Drop Filters

Year	Author	Materials	Cavity Shape	Modes	Maximum Q	Finesse	FSR (nm)	Extinction Ratio (dB)
2004	Ma et al. [71]	SiN-on-SiO$_2$	Hexagon	Single	1,300	7	7.7	20
2004	Li et al. [74]	SiN-on-SiO$_2$	Octagon	Single	1,150	5	7.1	3.6
2005	Li et al. [75]	SiN-on-SiO$_2$	Octagon	Single	6,400	30	7.4	14
		SOI	Octagon	Double	10,000	30	4.7	15
2006	Li et al. [76]	SOI	Hexagon	Double	7,600	20	4.2	20
		SOI	Square	Multi	2,600	8	4.8	25

input/output coupling with straight waveguides through a relatively wide gap, yet without the complications of using racetrack-shaped microring resonators. However, based on coupled-mode theory analysis, Manolatou et al. [66] also pointed out that resonances in square-shaped microresonators are standing-wave modes. This means that for a square-shaped microresonator-based add-drop filter with four symmetrically coupled waveguide ports, there is only a maximum of 25% transmission at the drop-port.

In order to apply waveguide-coupled polygonal-shaped microdisk cavities for device applications such as add-drop filters which impose traveling-wave resonances, our research group have addressed key issues regarding their (i) standing-wave or traveling-wave modal characteristics, (ii) waveguide-to-microresonator coupling characteristics, and (iii) cavity loss mechanisms.

On addressing (i), Fong et al. [69] numerically examined the standing-wave modal structures and transmissions of waveguide-coupled square-shaped microresonators (see Section 4.2.3), and proposed that by properly cutting the square cavity corners the traveling-wave components determined by the input-coupling direction can become dominant [70]. Ma et al. [71] experimentally demonstrated that traveling-wave resonances in waveguide-coupled hexagonal-shaped microdisk resonators can be preferentially coupled by optimizing the waveguide width for phase matching the waveguide mode with the cavity resonances (see Section 4.2.2). While, Li et al. [74] experimentally demonstrated traveling-wave resonances in waveguide-coupled octagonal-shaped microdisk cavity-based add-drop filters (see Section 4.2.6).

On addressing (ii), Li et al. [76] numerically simulated the effect of the cavity flat sidewall length on the waveguide evanescent coupling, and revealed that the coupling is directional (see Section 4.2.4). The coupled wavevector propagation direction essentially depends on the waveguide width and the coupling sidewall length. The directional evanescent coupling enables preferential coupling with only a few modes (and ideally singlemode) in albeit highly multimode large-sized polygonal-shaped microcavities. As a proof-of-concept, Li et al. [76] experimentally demonstrated only a few modes in waveguide-coupled 50-μm square, hexagonal, and octagonal-shaped microdisk cavities on SOI substrates.

On addressing (iii), Zheng et al. [73] and Li et al. [75] also experimentally demonstrated that the cavity loss due to the polygonal-shaped microcavity corners can be mitigated by rounding the sharp corners with a tailored radius of curvature (see Sections 4.2.5 and 4.2.6).

Meanwhile, Huang's group analytically investigated resonance modes in a discrete square-shaped microresonator. Huang's group systematically classified the modes in square-shaped microdisk resonators according to the symmetry point group C$_{4v}$ [78]. Based on the so-called mode match method [79], they showed that for modes with integer mode numbers in the *x*- and *y*-axes differing by two (see Section 4.2.3), the coupling between a standing-wave mode and its 90°-rotated mode results in mode splitting. One of the split modes displays WG-like standing-wave modal structure with zero-amplitude field at the cavity corners, and thereby favors high-Q resonance. The other split mode exhibits field extrema at the cavity corners and only supports low-Q resonance. Based on Fong et al. [70] Huang's group [80] further revealed that the cut-cornered square microresonator traveling-wave modes can be due to mode-coupling between two modes that are related by their spatial symmetry properties.

Other researchers also recognized the potentials of exploiting polygonal-shaped micro-cavities as microlaser cavities. In contrast to the conventional circular-shaped microdisk lasers that isotropically emit light along the cavity rim, polygonal-shaped microcavities have the advantage of attaining multiple directional emissions from the flat sidewalls [15]. Various polygonal-shaped microdisk or micropillar lasers have been demonstrated including hexagonal-shaped micropillar lasers in dye-doped zeolites [81], square-shaped microdisk lasers in dye-doped polymers [15], square-shaped micropillar lasers in polymer-coated silica [82], and triangular-shaped microdisk lasers in InP/GaInAsP systems [83].

4.2.2 N-Bounce Orbits in Polygonal-Shaped Microresonators

We can gain some physical insights to polygonal-shaped microresonator modes by using ray-optics. Polygonal-shaped microresonators support *N*-bounce orbits internally reflecting at the cavity flat sidewalls, as shown in Figure 4.1. Square-shaped microdisk cavities support four-bounce orbits, while hexagonal-shaped microdisk cavities support three- and six-bounce orbits, and octagonal-shaped microdisk cavities support four- and eight-bounce orbits. In order to excite resonances, the cavity round-trip wave and

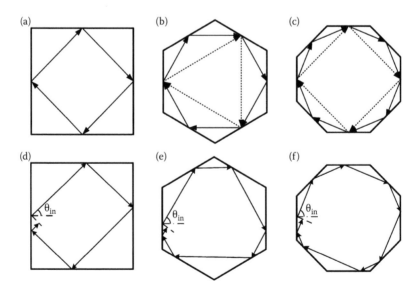

FIGURE 4.1
Ray-optics picture of resonance orbits in (a, d) square-shaped microcavities, (b, e) hexagonal-shaped microcavities, and (c, f) octagonal-shaped microcavities. (a–c) *N*-bounce closed round-trip orbits. The dotted arrows in (b) and (c) represent three- and four-bounce modes. (d–f) *N*-bounce open orbits that are wavefront-matched upon each round trip. The dashed lines represent the matched wavefronts. θ_{in}: incident angle.

the input-coupled wave should be wavefront-matched. For open round-trip orbits in polygonal-shaped microdisk cavities, as long as the round-trip ray travels co-parallel to the input-coupled ray, the cavity wavefront can spatially overlap and phase match with the input-coupled wave [67].

The open-orbit resonances concept has two implications: (1) Polygonal-shaped microcavities are multimode, as multiple open orbits have different wavefront-matched round-trip lengths. Early experiments by one of us (AWP) already confirmed multimode resonances in square-shaped microresonators [67]. (2) The cavity light can walk-off to the adjacent sidewall. For square-shaped microcavities, the walk-off light couples to the backward circulation of the same incident angle, resulting in standing-wave modes for all open-orbit resonances [67,70]. Whereas, for hexagonal- [72] and octagonal-shaped microcavities [74], the walk-off light constitutes cavity loss by refractive escape near the cavity corners without cross-coupling to the backward traveling-wave component, and thus the open-orbit resonances remain as traveling-wave resonances.

As different θ_{in}'s give different N-bounce orbits, we can selectively couple to N-bounce modes by choosing particular θ_{in}'s for the cavity light ray. In the case of waveguide side-coupling to a polygonal-shaped microcavity, θ_{in} can be determined by the waveguide propagation mode. Figure 4.2 depicts the N-bounce wavefront-matched open orbits with different θ_{in}'s in waveguide-coupled (a) square-, (b) hexagonal-, and (c) octagonal-shaped microcavities. Figure 4.2d shows the calculated waveguide dispersion (assuming symmetric slab waveguides) of three different effective refractive index contrasts for both the TE (electric field in plane) and TM (electric field normal to plane) polarization modes. We will detail the calculation parameters as needed in the following sections (Sections 4.2.3 through 4.2.5). We see that by properly designing the waveguide width w, we can choose waveguide input-coupling angles, ϕ_{in}, for selective wavefront-matching with cavity orbits. Furthermore, we note that the wavefront-matched round-trip path lengths are independent of the light input-coupling position along the cavity flat sidewall [67,71,74]. This means that the same resonance orbit can be coupled essentially along the entire sidewall, enabling a long lateral interaction length for polygonal-shaped microcavities.

4.2.3 Modes in Square-Shaped Microdisk Resonators

In order to derive further physical insights to resonance modes in polygonal-shaped microcavities, it is instructive to discuss wave-optics in square-shaped microcavities as an example. Rather than analyzing from the first principles which involves solving Maxwell's equations with careful consideration on the boundary conditions at the cavity corners, we heuristically model the four-bounce modes using **k**-space representation [67,69]. We assume that the cavity-field amplitude drops to zero at the high-index-contrast cavity sidewall (as in an infinite potential well). Thus, for a square-shaped microcavity of size a, we represent the optical resonance modes in terms of **k**-vector components k_x and k_y as [67,69], $k_x = m_x \pi/a$, $k_y = m_y \pi/a$, where m_x and m_y are integers. Each two-dimensional microcavity mode can be approximately described by a pair of integer numbers (m_x, m_y), representing orthogonal Fabry–Perot standing-wave modes with mode numbers m_x and m_y in the x- and y-directions. We relate the **k**-space picture to the ray-optics picture in terms of $\tan\theta_{in} = m_y/m_x$ [67,69]. We write the microcavity dispersion relation as $(n_{eff}k)^2 = k_x^2 + k_y^2$, where n_{eff} is the effective refractive index of the microcavity (accounting for the vertical dimension). Using the dispersion relation, we calculate the free-space resonant wavelength $\lambda = 2n_{eff}a/(m_x^2 + m_y^2)^{1/2}$. We also define total mode number $M = m_x + m_y$. Modes with identical M values have identical integer number of wavelengths in the wavefront-matched round-trip path lengths

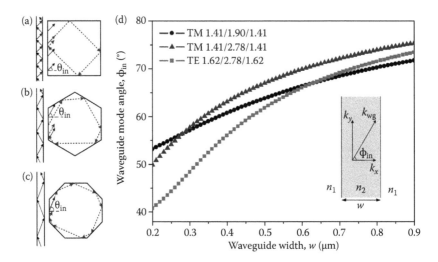

FIGURE 4.2
(a–c) Schematics of waveguide-coupled (a) square-, (b) hexagonal-, and (c) octagonal-shaped microdisk resonators. The cavity N-bounce ray orbits of incident angle θ_{in} are wavefront-matched with the waveguide propagation modes of different waveguide widths w. (d) Calculated waveguide mode angle ϕ_{in} as a function of waveguide width w, assuming symmetric slab structures with effective refractive index contrast $n_1/n_2/n_1$ relevant to our discussion. We choose $n_1/n_2/n_1$ of 1.41/1.90/1.41 for the TM mode (circle), 1.41/2.78/1.41 for the TM mode (triangle), and 1.62/2.78/1.62 for the TE mode (square). k_{wg}: waveguide wavevector, k_y: waveguide wavevector longitudinal component, k_x: waveguide wavevector transverse component.

[84]. Besides, odd M corresponds to opposite mode-field distribution parities in the x- and y-directions, whereas even M corresponds to the same mode-field distribution parity in the x- and y-directions.

We express the two-dimensional standing-wave scalar field pattern of mode (m_x, m_y) in a square-shaped microcavity as [69],

$$E_{mx,my}(x, y)\, e^{i\omega t} = A\, e^{i\omega t} \sin(m_x \pi x/a)\, \sin(m_y \pi y/a), \tag{4.1}$$

where ω is the mode angular frequency, and A is the field amplitude. We assume the x–y coordinates origin at the lower left-hand corner of the square-shaped microcavity (see Figure 4.3a). Figure 4.3a and b illustrates the calculated mode-field patterns of $(m_x, m_y) = (7, 9)$, and $(9, 7)$ using Equation 4.1.

We note that modes (m_x, m_y) and (m_y, m_x) in a square-shaped microcavity are spectrally degenerate yet spatially non-degenerate. The spatially non-degenerate modes are related by $90°$ rotation, as expected from a four-fold rotational symmetry. We also see from ray-optics that the two spatially nondegenerate modes have incident angles θ_{in} and $90°-\theta_{in}$ on the same cavity sidewall. Thus, when the square-shaped microcavity is waveguide-coupled, it is possible that only one of the spatially non-degenerate modes is wavefront-matched with the waveguide mode.

In the case of spatially nondegenerate (m_x, m_y) and (m_y, m_x) modes coherently coupled, the steady-state mode-field pattern is a linear superposition of the spectrally degenerate modes as follows [69]:

$$E_{mx,my}(x, y)\, e^{i\omega t} = A\, e^{i\omega t} \sin(m_x \pi x/a)\, \sin(m_y \pi y/a) + B\, e^{i(\omega t - \delta)} \sin(m_y \pi x/a)\, \sin(m_x \pi y/a), \tag{4.2}$$

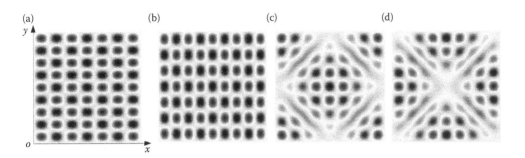

(a) (b) (c) (d)

FIGURE 4.3
Calculated standing-wave mode-field patterns of (a) $(m_x, m_y) = (7, 9)$, (b) $(m_x, m_y) = (9, 7)$, (c) superposition of modes (7, 9) and (9, 7) with relative phase $\delta = 0$, and (d) superposition of modes (7, 9) and (9, 7) with relative phase $\delta = \pi$.

where A and B are the normalized field amplitudes of the spectrally degenerate modes, and δ is the relative phase. Figure 4.3c and d illustrates the calculated mode-field patterns of the supposition of modes (7, 9) and (9, 7) with $t = 0$, $A = B$, $\delta = 0$ and π using Equation 4.2. The in-phase supposition with field maxima at the cavity corners may induce an excessive radiative loss, and thus suggest a low-Q resonance. Whereas, the π-out-of-phase superposition with zero-amplitude field along the square-shaped microcavity diagonals may favor a low radiative loss at the cavity corners, and thereby suggest a high-Q WG-like resonance. We remark that our heuristic modeling is consistent with the theoretical analysis by Huang's group [78,79].

Here, we numerically simulate the transmissions and mode-field distributions of silicon waveguide-coupled square-shaped microdisk resonators, using a commercially available finite-difference time-domain (FDTD) simulation tool [85]. We adopt an effective refractive index approach to approximately account for the vertical dimension in a two-dimensional (2-D) calculation. Figure 4.4 illustrates the silicon wire waveguide cross-section and the effective 2-D structure for the FDTD calculations. We consider a submicrometer singlemode silicon rib waveguide comprising a 0.34-μm thick silicon waveguide core, a 0.05-μm thick silicon slab layer, with an air upper-cladding and a silica under-cladding. We note that the thin slab layer structure adjacent to the waveguide sidewalls is instrumental for dopant implantations in electro-optic active devices (see Section 4.4.3). We simulate the transmission spectra for the TM polarization, to be consistent with the scalar-field approximation adopted in the **k**-space modeling. Both the waveguide core and the microdisk assume the same effective refractive index, $n_{\mathrm{eff}} = 2.78$, while the slab region assumes $n_{\mathrm{eff}} = 1.41$.

Figure 4.5a shows the simulated throughput-, drop- and add-port transmission spectra of the waveguide-coupled square-shaped microcavity-based add-drop filter of size $a = 5$ μm, waveguide width $w = 0.3$ μm, and gap separation $g = 0.2$ μm for the TM polarization. The transmission spectra are multimode. We identify the mode numbers using Fourier analysis of the simulated mode-field patterns. We label the resonance by the dominant mode numbers (m_x, m_y), indexed with the estimated relative phase δ (assuming 0 or π) based on the calculated temporal evolution of the mode-field pattern (e.g. $(11, 16)_0$, $(12, 16)_\pi$). We see that the drop-port and the add-port exhibit similar transmissions, with a maximum transmission of ~20%, as expected from standing-wave modes and in the presence of cavity radiative loss. Figure 4.5b shows the schematic of an open-orbit ray propagating in a waveguide-coupled square-shaped microcavity-based add-drop filter. The clockwise (CW) traveling-wave orbit (solid arrows) walk-offs and backward-couples to the counterclockwise (CCW)

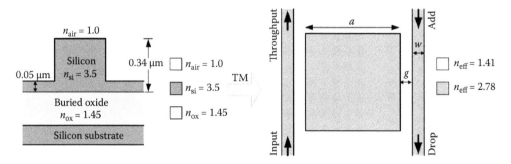

FIGURE 4.4

Cross-section of the silicon rib waveguide and the effective 2-D structure for the TM-polarization. n_{eff}=2.78 for the 0.34-μm thick silicon waveguide core and microdisk, and n_{eff}=1.41 for the 0.05-μm thick silicon slab region, assuming buried oxide and air cladding.

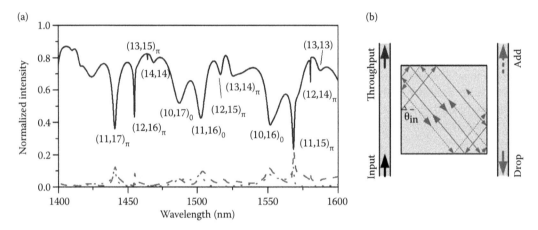

FIGURE 4.5

(a) 2-D FDTD-simulated TM-polarized transmission spectra of the waveguide-coupled square-shaped microcavity-based add-drop filter of a=5 μm, w=0.3 μm, and g=0.2 μm. Solid line: throughput. Dashed line: drop-port. Dotted line: add-port. The multimode resonances are labeled based on their simulated mode-field patterns. (b) Schematic of a standing-wave orbit in a waveguide-coupled square-shaped microcavity-based add-drop filter. Solid arrows: clockwise circulation; dotted arrows: counterclockwise circulation.

traveling-waves (dotted arrows) with the same incident angle, forming a standing-wave mode which yields undesirable transmissions in the add-port. Furthermore, our numerical simulations reveal that modes of odd and even M exhibit parity at different transmissions and mode-field patterns [69]. We find that the odd M modes display lower Q's than the even M modes. Figure 4.6a and b shows for instance the simulated mode-field pattern and the Fourier transform analysis for odd M mode $(11, 16)_0$. Figure 4.6c shows the calculated mode-field pattern of mode $(11, 16)_0$ in a discrete microcavity using Equation 4.2, with A=0.77 for mode (11, 16) and B=0.23 for mode (16, 11) following the Fourier analysis. In both the simulated and the calculated odd M mode-field patterns, we find field extrema at the cavity corners, suggesting high radiative loss at the cavity corners and thus a low-Q resonance. Whereas, Figure 4.6d and e shows the simulated mode-field pattern and the Fourier transform analysis of even M mode $(12, 16)_\pi$. Figure 4.6f shows the calculated mode-field pattern of mode $(12, 16)_\pi$, with A=0.62 for mode (12, 16) and B=0.38 for mode (16, 12). Here, we see zero-amplitude fields at the cavity corners, and zero-amplitude field patterns

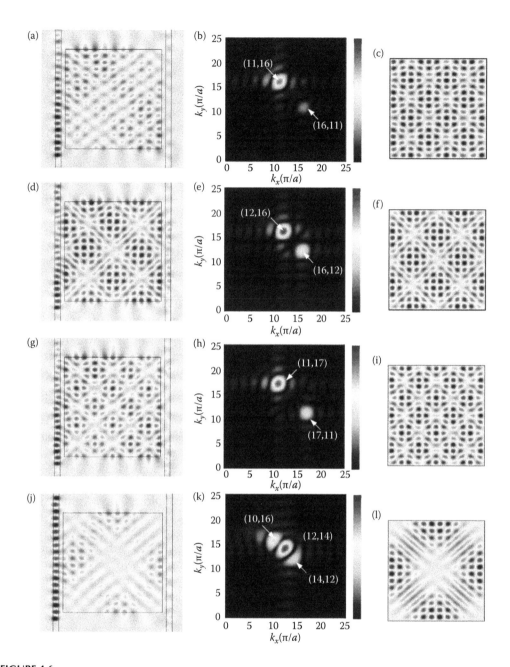

FIGURE 4.6
(a, d, g, j) FDTD-simulated mode-field patterns of a waveguide-coupled square-shaped microdisk resonator for modes $(11,16)_0$, $(12,16)_\pi$, $(11,17)_\pi$, and $(12,14)_\pi$. (b, e, h, k) Fourier analysis of the simulated cavity mode-field patterns. (c, f, i, l) Calculated mode-field patterns of a discrete square-shaped microcavity with the amplitude ratios of the superposed modes following (b, e, h, k).

surrounded by cross-like wavefronts in the cavity [66,86,87]. The zero-amplitude fields at the cavity corners favor low radiative loss and thus a high-Q resonance.

It is worth mentioning that all even M mode-field distributions exhibit periodic cross-like features in case $\delta=\pi$. The number of cross-like features scales with $|m_x - m_y|$ [69].

Figure 4.6g and h shows for instance the simulated mode-field pattern and the Fourier transform analysis of even M mode $(11, 17)_\pi$ ($|m_x - m_y| = 6$). Figure 4.6i shows the calculated mode-field pattern of mode $(11, 17)_\pi$, with $A = 0.69$ for mode $(11, 17)$ and $B = 0.31$ for mode $(17, 11)$. We see more cross-like features than mode $(12, 16)_\pi$ with $|m_x - m_y| = 4$ and mode $(7, 9)_\pi$ with $|m_x - m_y| = 2$ (see Figure 4.3d). Moreover, for the same even M, we note the resonance Q scales inversely as $|m_x - m_y|$. The maximum Q corresponds to $|m_x - m_y| = 2$, which gives a WG-like mode-field pattern [79] with a single cross-like feature at the cavity center. For instance, Figure 4.6j and k shows the simulated mode-field pattern and the Fourier transform analysis of mode $(12, 14)_\pi$, which displays high-Q. Yet, we note that the mode-field pattern is distorted due to the interference from mode $(10, 16)$. Figure 4.6l shows the calculated WG-like mode-field pattern of mode $(12, 14)_\pi$, taking into account the interference from mode $(10, 16)$.

4.2.4 Directional Coupling via Polygonal-Shaped Microdisk Flat Sidewalls

In order to develop some device design guidelines for waveguide-coupled polygonal-shaped microresonators, we further examine the waveguide coupling to microdisk structures with flat and curved sidewalls. Figure 4.7a and b illustrates the **k**-vectors in waveguide-coupled microstructures with (a) a long flat coupling sidewall and (b) a short curved coupling sidewall. In the case that the microstructure has a long flat coupling sidewall, it is intuitive that the evanescently coupled light has a plane wavefront. The coupled **k**-vectors can be directional with an incident angle θ_{in} following the waveguide mode angle ϕ_{in}. Whereas, in the case that the microstructure has a short curved coupling sidewall, we reason that the evanescently coupled light has a curved wavefront. The coupled **k**-vectors center angle can still be given by the waveguide mode angle, yet the **k**-vectors carry more diverse angular components.

Figure 4.7c and d shows the FDTD-simulated steady-state **H**-field patterns (TE-polarization) of silicon waveguide-coupled microstructures with a flat sidewall ($a = 10$ μm) and a curved sidewall (radius of curvature 5 μm) [76]. We only show the coupled fields in the microstructures. Our simulations confirm that waveguide coupling to a microstructure of identical n_{eff} with a flat sidewall occurs along the entire flat sidewall. The coupled plane wavefronts indicate a directional input-coupling. Whereas, we see that waveguide coupling to a microstructure with a curved sidewall occurs along a short arc length results in a divergent wavefront after initial focusing. We remark that the microstructures comprise only single sidewalls for coupling and perfectly matched layers (PMLs) for other sides. Our simulations adopt $n_{eff} = 2.78$ for the 0.2-μm thick silicon waveguide core and $n_{eff} = 1.62$ for the 0.05-μm thick silicon slab layer, assuming an air upper-cladding and a silica lower-cladding. We choose the waveguide width $w = 0.3$ μm and the air-gap spacing $g = 0.35$ μm.

In order to quantify the coupled **k**-vector directionality, we Fourier analyze the simulated field patterns in the microstructures within a spatial domain of size a. Figure 4.7e and f shows the spatial Fourier analysis in **k**-space of the simulated field patterns in the microstructures with a flat and curved coupling sidewall. In order to highlight the major **k**-vector angular components, the **k**-space contour plots only span full-width-half-maximum distributions over an angular width $\Delta\theta$. We use the ratio $\theta_{in}/\Delta\theta$ to measure the coupled **k**-vector directionality. We see that the coupled **k**-vector is only directional in the case that the coupling is via a flat sidewall.

Figure 4.8a and b shows the simulated coupled wavevector center angle θ_{in}'s and the directionality measured as $\theta_{in}/\Delta\theta$'s with various sidewall lengths a for different waveguide widths w's. For each w, θ_{in} rises sharply with short sidewall length, until saturates at the waveguide mode angle ϕ_{in}. The dashed lines indicate the analytically calculated

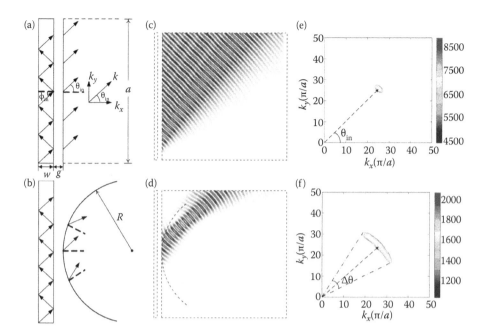

FIGURE 4.7

(a, b) Schematics of a waveguide-coupled microstructure with (a) a flat sidewall of length a, and (b) a curved sidewall of radius R. The coupled **k**-vectors along the flat sidewall are directional. The coupled **k**-vectors along the curved sidewall exhibit more angular components. ϕ_{in}: waveguide mode angle relative to the sidewall normal, θ_{in}: **k**-vector angle relative to the microdisk sidewall normal. (c, d) FDTD-simulated TE-polarized **H**-field patterns for a silicon waveguide-coupled microstructure with (c) a flat sidewall of length $a=10$ μm, and (d) a curved sidewall of radius 5 μm. The dashed-line square box indicates the area of size a under Fourier analysis. We adopt an effective refractive index contrast of 2.78–1.62, assuming a 0.2-μm thick silicon waveguide core and microdisk and a 0.05-μm thick silicon slab region in SOI for the TE mode. (e, f) Fourier analysis of the simulated field patterns in the microstructures. The contours represent the **k**-vectors spanning within half-maximum of the angular distributions. The dashed arrows indicate the coupled **k**-vectors center angles θ_{in}. The dashed lines indicate the **k**-space angular distribution $\Delta\theta$. (Reprinted from Li, C., Zhou, L., Zheng, S., and Poon, A. W., *IEEE J. Select. Topics Quantum Electron.*, 12, 1438–1449, 2006. With permission.)

waveguide mode angles ϕ_{in}'s (see Figure 4.2d). Whereas, $\theta_{in}/\Delta\theta$ linearly rises with a, and displays no saturation up to $\theta_{in}/\Delta\theta$ exceeding 30 for $a \approx 50$ μm, and is largely independent of the waveguide widths adopted here.

Furthermore, we examine the waveguide-microstructure field amplitude coupling coefficient κ under various configurations. We obtain κ from our simulations as the amplitude ratio of the total field along the entire upper and right boundaries of the dashed-line square (Figure 4.7c and d) to the input-coupled field amplitude at the waveguide. Figure 4.9 shows the extracted κ with flat sidewalls of various a's of different waveguide widths. We note that κ rises with a, yet drops with w. Thus, we see that with fixed w, we can obtain higher coupling coefficients for polygonal-shaped microdisks of a longer sidewall length.

Here, we assume the same waveguide coupling structure as above and numerically simulate a silicon waveguide-coupled octagonal-shaped microdisk resonator. In order to see the effect of the microcavity flat and curved coupling sidewall, we also numerically model a silicon waveguide-coupled circular-shaped microdisk resonator of the same cavity size. Figure 4.10a and b shows the FDTD-simulated TE-polarized throughput- and drop-port transmission spectra of the waveguide-coupled octagonal-shaped and circular-shaped microdisk resonator-based filters. The octagonal-shaped microdisk has a sidewall-to-sidewall

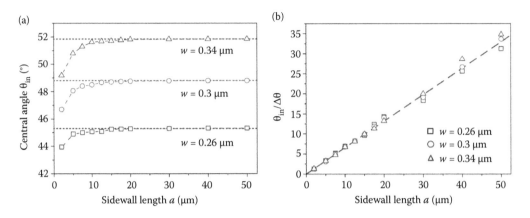

FIGURE 4.8
(a) FDTD-simulated coupled wavevectors center angle θ_{in} in the waveguide-coupled microstructures with various flat sidewall lengths a, for waveguide width w=0.34 μm, 0.3 μm, and 0.26 μm. θ_{in} saturates at angles determined by the waveguide mode (shown as horizontal dotted lines). (b) FDTD-simulated coupled wavevectors directionality measured as $\theta_{in}/\Delta\theta$ in the waveguide-coupled microstructures with various flat sidewall lengths a. The dashed line is visual aid. (Reprinted from Li, C., Zhou, L., Zheng, S., and Poon, A. W., *IEEE J. Select. Topics Quantum Electron.*, 12, 1438–1449, 2006. With permission.)

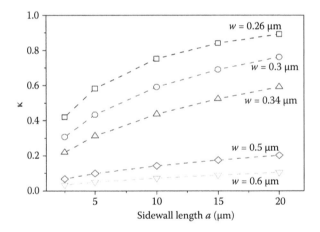

FIGURE 4.9
Simulated coupling coefficient κ for waveguide coupling to planar microstructures as a function of sidewall length with various waveguide widths. The dashed lines are visual aid. (Reproduced with permission from IEEE, Li, C., Zhou, L., Zheng, S., and Poon, A. W., *IEEE J. Select. Topics Quantum Electron.*, 12, 1438–1449, 2006. © IEEE. With permission.)

distance L=18 μm and a side-coupling length a≈7.5 μm. The circular-shaped microdisk has a radius of R=9 μm. Based on the waveguide-coupled microstructure analysis (Figure 4.8), we choose waveguide width w=0.3 μm with θ_{in}≈48.0° and $\Delta\theta$ exceeding 9.0°. Our simulation reveals only two traveling-wave modes (denoted as A and B) for the octagonal-shaped microdisk, whereas we discern five traveling-wave modes in the circular-shaped microdisk. This exemplifies that a waveguide-coupled polygonal-shaped microdisk resonator with a long flat sidewall is favorable for preferential coupling with a few and ideally single modes, whereas the waveguide-coupled circular-shaped microdisk resonator tends to couple with a large number of modes. Figure 4.10c shows the simulated steady-state **H**-field pattern at resonance

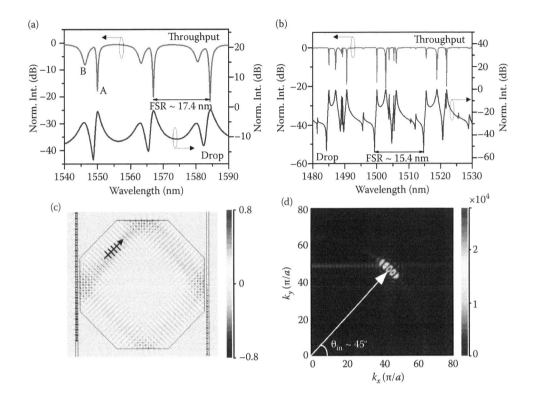

FIGURE 4.10
(a) FDTD-simulated TE-polarized throughput- and drop-port transmission spectra of a waveguide-coupled octagonal-shaped microdisk resonator-based filter with coupling length $a\approx7.5$ μm, $L=18$ μm, $w=0.3$ μm, and $g=0.3$ μm. (b) FDTD-simulated TE-polarized throughput- and drop-port transmission spectra of a waveguide-coupled circular-shaped microdisk resonator-based filter with $R=9$ μm, $w=0.3$ μm, $g=0.3$ μm. (c) Simulated steady-state **H**-field pattern of resonance A. (d) Fourier analysis of the mode-field pattern. The arrow indicates the center angle $\theta_{in}\approx45°$ and distribution $\Delta\theta\approx9.5°$ of the coupled **k**-vectors.

A of the octagonal-shaped microdisk. The mode-field pattern displays four-bounce traveling wavefronts with an incident angle near 45°, which is consistent with the free spectral range (FSR) of ~17.4 nm (Figure 4.10a). Figure 4.10d shows the Fourier analysis of the four-bounce mode-field pattern, revealing a $\theta_{in}\approx45°$ and a $\Delta\theta\approx9.5°$.

4.2.5 Sharp Corner Radiative Loss and Corner Rounding

One major drawback of polygonal-shaped microdisk resonators is that the sharp cavity corners impose considerable radiative loss [74]. Figure 4.11a depicts the waveguide-coupled sharp-cornered octagonal-shaped microdisk resonator-based filter. Intuitively, we see that the surface wave propagating along the flat sidewalls cannot follow the sharp transitions at the cavity corners, and thus results in radiative loss.

As a mitigation, we previously proposed [73,75] tailoring the polygonal-shaped microresonators with designed round corners. Figure 4.11b illustrates the round-cornered design. An arc of 45° with a radius of curvature R is applied to each cavity corner. We define the microresonator shape by the ratio of R to the sidewall-to-sidewall distance L. The round-cornered cavity sidewall length is $a=(\sqrt{2}-1)(L-2R)$. The tradeoff, however, is a reduced side-coupling length.

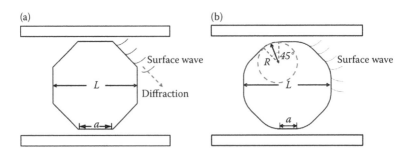

FIGURE 4.11
Schematics of waveguide-coupled octagonal-shaped microresonator-based filters with (a) sharp cavity corners that impose surface wave diffraction loss, and (b) rounded cavity corners that alleviate the surface wave diffraction loss.

4.2.6 Experimental Demonstrations

Here, we review representative measurements of waveguide-coupled octagonal-shaped microdisk resonators in a SiN-on-SiO$_2$ substrate [75]. Figure 4.12a through f shows the top-view scanning electron micrographs (SEMs) of the fabricated 50-μm microdisk resonator-based filters with R/L=0–0.5 (in steps of 0.1) [75]. The waveguide width is 0.6 μm. Figure 4.12g through l shows the measured TM-polarized throughput-port and drop-port transmission spectra. We observed that the sharp-cornered and round-cornered octagonal-shaped microdisk resonators are nearly singlemode. In contrast, the circular-shaped microdisk resonator (R/L=0.5) exhibits multimode. With R/L=0.4, we obtained an optimum extinction ratio of ~14.5 dB, a Q~6,400, and an on-off ratio of ~13.5 dB. The measured FSR for R/L=0.4 is ~7.35 nm, which suggests a WG-like mode grazing along the round-cornered octagonal-shaped microdisk circumference. Whereas, the measured FSR for R/L=0 and 0.1 is ~7.45 nm, which is consistent with relatively short eight-bounce round-trip lengths [74].

Figure 4.13a shows the rising trend of the measured Q with R/L ratio for w=0.6 μm, suggesting that the round-cornered microresonators favor high-Q resonances. Figure 4.13b shows the measured Q with various w for R/L=0.3, revealing that the resonance Q is optimized at w=0.8 μm. We attribute the optimized Q to the nearly **k**-vector matching between the microresonator mode and the waveguide mode. Based on the waveguide dispersion analysis shown in Figure 4.2d, we see that for an effective refractive index contrast $n_1/n_2/n_1$=1.41/1.90/1.41 (for our fabricated device with a 1.1-μm thick SiN device and a 0.15-μm thick SiN slab), the waveguide mode angle ϕ_{in}≈70° at w=0.8 μm is favorable for coupling to the octagonal-shaped microcavity eight-bounce or WG-like orbits [76].

However, we found a number of anomalies in our measurements that are worth detailed modeling. Specifically, we observed an exceptional case at R/L=0.2 (Figure 4.12i), which suggests weak coupling. Figure 4.13c shows the measured extinction ratio with R/L ratio (and a) for w=0.6 μm and w=0.5 μm. For both w's, the extinction ratio drops to a minimum at R/L=0.2 (a=12.43 μm). We attribute the appearance of weak coupling to possible destructive interference among multiple light paths that were coherently coupled along this particular side-coupling length. Interestingly, we also observed that with R/L=0.25 and w=0.5 μm, the measured extinction ratio in the throughput-port transmission spectrum displays a maximum of ~24 dB. We found a corresponding narrowband dip near the center of the drop-port peak transmission, as shown in Figure 4.13d. The drop-port sharp resonance dips resemble resonance line shapes expected in coupled microresonators [88].

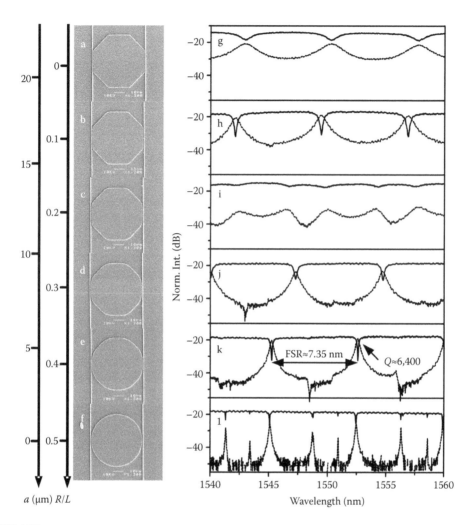

FIGURE 4.12
(a–f) Scanning electron micrographs of the fabricated filters with R/L ratios from 0 (sharp-cornered) to 0.5 (circular). (g–l) Measured TM-polarized throughput-port and drop-port transmission spectra. Throughput (solid line), drop (dashed line). (Reprinted from Optical Society of America. Li C., and Poon, A. W., *Opt. Lett.* 30, 546–548, 2005. With permission from the Optical Society of America.)

We attribute the sharp transmission dips to a destructive interference among coherent multiple orbits output-coupled to the drop-port.

4.3 Spiral-Shaped Microdisk Resonators with Nonevanescent Coupling

4.3.1 Overview of Spiral-Shaped Microresonators

Another scheme for enhancing the coupling between a waveguide and a TIR-based microresonator is to seek alternative non-evanescent coupling with the microresonator modes. To this end, researchers have long recognized that by globally deforming the

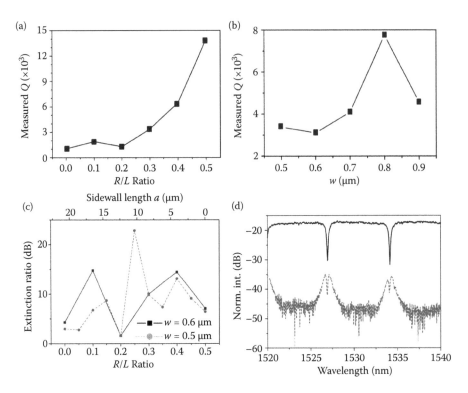

FIGURE 4.13

(a) Measured Q at various R/L ratios with $w=0.6$ μm. (b) Measured Q for $R/L=0.3$ at various waveguide width w. (c) Measured extinction ratio at various R/L ratios for $w=0.6$ μm (solid line) and $w=0.5$ μm (dashed line). (d) Measured TM-polarized throughput-port and drop-port transmission spectra with $R/L=0.25$ and $w=0.5$ μm. (Reprinted from Optical Society of America. Li C., and Poon, A. W., *Opt. Lett.* 30, 546–548, 2005. With permission form the Optical Society of America.)

symmetrical circular/spherical microresonators into asymmetrical resonant cavities (ARCs) [12,89], the TIR-confined cavity modes can be refractively coupled according to Snell's law. Such refractive coupling can be directional [12,89], with the coupling positions strongly dependent on the degree of deformation and the refractive index contrast [89]. However, there are drawbacks with the refractive coupling which are enabled by such deformed-shaped cavities: (i) the cavity mode can be refractively coupled at multiple locations in multiple directions, which is undesirable for on-chip light source applications requiring unidirectional light emission from a single output-coupling location, and (ii) the globally deformed-shaped cavities are not favorable for refractive coupling with integrated waveguides. In the case that the waveguide end-face is separated from the cavity sidewall, significant insertion losses due to Fresnel reflection and modal mismatch can compromise the advantage of the refractive coupling. In the case that the waveguide is butt-coupled to the deformed-shaped cavity sidewall, the cavity boundary condition can be severely perturbed.

In 2003, Chern et al. [16] (Prof. Richard Chang's group) in their pioneering work proposed and demonstrated spiral-shaped microcavities that enable refractive coupling at the spiral notch. The spiral shape is given in terms of a radius that linearly varies with azimuthal angle as: $r(\phi_r)=r_0(1-\varepsilon\phi_r/2\pi)$, where $r_0=r(\phi_r=0)$, and ε is the deformation parameter giving a spiral notch with width of $r_0\varepsilon$ at the maximum radius mismatch. Figure 4.14a illustrates the spiral-shaped microresonator. The spiral notch offers a unique window for refractive

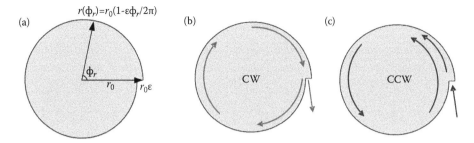

FIGURE 4.14
(a) Schematic of the spiral-shaped microdisk resonator. (b) Schematic CW traveling orbit with partial transmission at the spiral notch for each round trip. (c) Schematic CCW traveling orbit that allows input-coupling via the spiral notch yet the orbit bypasses the notch for each round trip.

coupling with the microresonator modes. Chern et al. [16] and various research groups [90–94] have experimentally demonstrated unidirectional lasing emission with refractive output-coupling from the spiral notch. Meanwhile, theoretical calculations of lasing modes revealed distinct modal structures in a spiral-shaped microcavity [95,96]. Furthermore, Chang's group [97,98] also demonstrated novel optical devices with high-Q resonances using refractive coupling via the spiral notch between a spiral-shaped microdisk laser and a semicircular-shaped microdisk amplifier. Thus, it is known that refractive coupling via the spiral notch preserves high-Q resonances.

Another uniqueness of spiral-shaped microresonators is that the cavity shape entirely lacks rotational symmetry. It is thus conceivable that the rotational asymmetry breaks the spatial degeneracy between CW and CCW traveling-wave modes, as depicted in Figure 4.14b and c. We see that CW traveling-wave component spatially overlaps with the spiral notch which enables partial transmission via the notch every round trip. Whereas, CCW traveling-wave component tends to bypass the spiral notch yet can be refractively input-coupled at the spiral notch.

Previously [99], in order to experimentally study CW and CCW traveling-wave modes in spiral-shaped microresonators, we devised passive waveguide-coupled spiral-shaped microdisk resonator-based filters, as shown in Figure 4.15. We seamlessly integrated a waveguide to the spiral notch in order to facilitate direct coupling to the spiral-shaped microdisk. Another waveguide was side-coupled to the microdisk, such that CW and CCW transmissions can be separately characterized by input-coupling light into the microdisk from each of the three waveguide ports. We remark that the waveguide notch-coupling is fundamentally different from butt-coupling a waveguide to a circular-shaped microdisk. In the spiral-shaped microdisk case, the cavity ray orbit has varying incident angle at each reflection (due to the linear spiral shape) and that potentially facilitates the cavity mode to be mode-matched with the notch-coupled waveguide mode. Whereas, in the circular-shaped microdisk case, the WG mode has a fixed incident angle at each reflection, and thus it is difficult to mode-match the butt-coupled waveguide with the WG mode without severely spoiling the phase-matching condition or the resonance Q. It is worth mentioning that Fujii et al. [93] also reported a spiral-shaped microdisk laser with a waveguide butt-coupled at the spiral notch.

The filter functionalities depend on to which waveguide port the light is input-coupled. Figure 4.15a shows the notch-filter functionality, which is based on CCW traveling-wave modes that are evanescently input-coupled from the side waveguide yet tend to by-pass the notch every round trip without direct output-coupling. Figure 4.15b shows the drop-filter

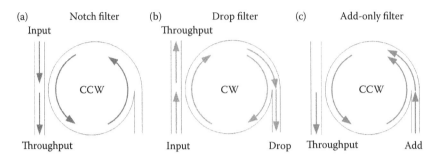

FIGURE 4.15

Schematics of the waveguide-coupled spiral-shaped microdisk resonator-based filters: (a) notch filter with side-coupled CCW traveling-wave modes, (b) drop filter with side-coupled CW traveling-wave modes, and (c) add-only filter with notch-coupled CCW traveling-wave modes. (Reprinted from Optical Society of America. Lee, J. Y., Luo, X., and Poon, A. W., *Opt. Express* 15, 14650–14666, 2007. With permission from The Optical Society of America.)

functionality, which is based on CW traveling-wave modes that are evanescently input-coupled from the side waveguide and can be directly output-coupled at the notch-coupled waveguide every round trip. Figure 4.15c shows the add-only filter functionality. Here, CCW traveling-wave modes that are directly input-coupled via the spiral notch tend to by-pass the notch every round trip without direct output-coupling, yet the cavity light can be evanescently output-coupled to the side waveguide.

4.3.2 Numerical Simulations

Here, we discuss 2-D FDTD simulations of a waveguide-coupled spiral-shaped microcavity. We choose radius $r_0 = 10\,\mu m$ and $\varepsilon = 0.04$, giving a singlemode notch-coupled waveguide of width $0.4\,\mu m$. The side-coupled waveguide width is $0.3\,\mu m$ with a gap separation of $0.3\,\mu m$. We choose an effective refractive index contrast of 1.9–1.4 for the TE mode in order to represent a 1.0-μm thick SiN-on-SiO$_2$ stack layered device region and a 0.1-μm thick SiN slab region (see Section 4.3.3).

Figure 4.16a shows the simulated TE-polarized throughput-port transmission spectra of notch and drop filters. Figure 4.16b shows the simulated TE-polarized drop-port transmission spectrum of drop filter and throughput-port transmission spectrum of add-only filter. Our simulations reveal reciprocal transmissions between side-coupled CCW and side-coupled CW traveling waves and between side-coupled CW and notch-coupled CCW traveling waves. The multimode resonances display a FSR of 21.6 nm, which is consistent with the spiral-shaped microdisk circumference.

The observation of reciprocal transmissions can be expected based on scattering theory [100]. In essence, reciprocity relations mean that light transmissions are preserved by interchanging positions of the source and the detector in a linear dielectric medium with symmetric permittivity tensor [100]. Thus, it is conceivable that the linear transmissions of the waveguide-coupled spiral-shaped microdisk resonator follow reciprocity relations, despite its geometric rotational asymmetry.

Nonetheless, the implications of reciprocal transmissions in a waveguide-coupled spiral-shaped microdisk resonator are still significant and not trivial. The observed identical resonances and Q's for CW and CCW transmissions imply that (1) CW and CCW waves preserve identical wavefront-matched round-trip path lengths for resonances, despite their spatial non-degeneracy, and (2) CW and CCW waves encounter identical total cavity losses.

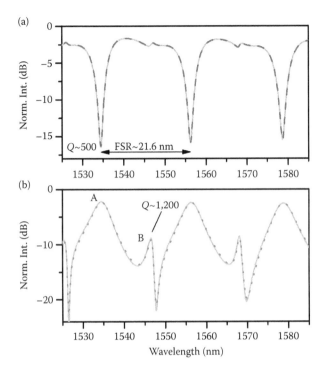

FIGURE 4.16
(a) FDTD-simulated TE-polarized throughput-port transmission spectra of notch filter (dashed line) and drop filter (solid line). (b) FDTD-simulated TE-polarized drop-filter drop-port transmission spectrum (solid line) and add-only filter throughput-port transmission spectrum (dotted line). We adopt $r_o = 10$ μm, $\varepsilon = 0.04$, and effective refractive index contrast of 1.9–1.4.

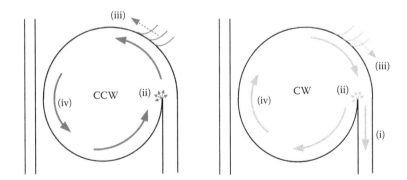

FIGURE 4.17
Loss mechanisms in a waveguide-coupled spiral-shaped microdisk resonator for CCW and CW traveling waves. (i) Direct output-coupling via the notch-waveguide. (ii) Diffraction and scattering at the notch junction. (iii) Distributed cavity loss along the cavity sidewall. (iv) Material absorption.

Figure 4.17 illustrates the various cavity loss mechanisms for CCW and CW waves. The total cavity loss comprises (i) direct output-coupling via the notch-waveguide, (ii) diffraction and scattering at the notch junction, (iii) distributed cavity loss along the cavity sidewall (including curved sidewall diffraction, refraction, and roughness-induced scattering), and (iv) material absorption. Given CW waves partially output-couple via the notch-waveguide, the losses from (ii) and (iii) should therefore be lower. Whereas, it is

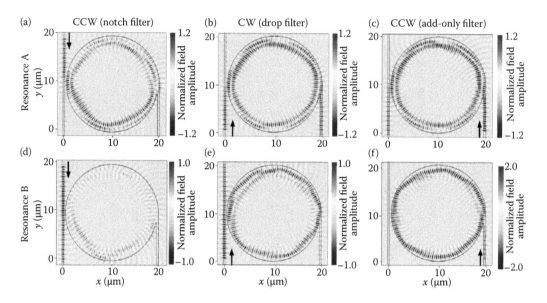

FIGURE 4.18
FDTD-simulated TE-polarized steady-state resonance mode **H**-field patterns for (a), (d) notch filter, (b, e) drop filter, and (c, f) add-only filter. (a–c) resonance A, (d–f) resonance B.

possible that CCW waves see relatively high losses from (ii) and/or (iii) in order to balance the relatively low-loss from (i). A large distributed cavity loss for CCW waves can be originated from a less confined traveling-wave orbit, also suggesting CCW and CW waves display different mode-field distributions. We remark that scattering at the notch junction and along the cavity sidewall may also result in cross-coupling between CCW and CW waves [16].

Figure 4.18a through f shows the FDTD-simulated TE-polarized mode-field (**H**-field) patterns at resonances A and B (see Figure 4.16) for notch filter (a and d), drop filter (b and e), and add-only filter (c and f). In both resonance wavelengths, our simulations reveal WG-like traveling-wave mode-field patterns that are rotational asymmetric. The simulated mode-field patterns are consistent with the theoretically calculated WG-like resonances in a spiral-shaped microdisk without waveguide coupling [95,96], suggesting that the perturbation from the notch-coupled waveguide to the spiral-shaped cavity modes is not significant. Both resonances display qualitatively similar field patterns, with weaker field amplitudes in notch filter than in drop filter or add-only filter. This shows that the side-coupled CCW mode and the side-coupled CW mode exhibit asymmetrical modal distributions despite their reciprocal transmissions. We attribute the asymmetrical modal distributions between the side-coupled CCW and CW modes to the rotational asymmetric cavity shape. This is also based on our earlier numerical work which suggested that the side-coupled CCW and CW traveling waves have essentially identical **k**-vector distributions at the input-coupling sidewall [99], and thus the input-coupling conditions do not differentiate the steady-state CCW and CW mode-field patterns.

Besides, the side-coupled CCW mode and the notch-coupled CCW mode display non-identical modal distributions, despite their same sense of circulation. We attribute the larger field amplitude in the notch-coupled CCW mode than the side-coupled CCW mode to a possible larger input-coupling at the notch. Whereas, we note that the side-coupled CW mode and the notch-coupled CCW mode, while both enable spatial overlap with the notch, exhibit field-patterns that are very similar to but not identical. Notably, for resonance B, the notch-coupled CCW mode field-amplitude is about twice as high as that of the side-coupled CW mode field-amplitude. This suggests that the notch input-coupling can be more efficient than the conventional evanescent input-coupling given particular resonance wavelengths.

4.3.3 Experimental Demonstrations

Here, we review our experimental demonstration of reciprocal transmissions and asymmetrical modal distributions in waveguide-coupled spiral-shaped microdisk resonators. We fabricated the resonators on a SiN-on-SiO$_2$ substrate using standard silicon microelectronics processes [99]. Spiral-shaped microdisks with fixed radius (r_0=25 μm) yet various deformations (ε=0.016, 0.032, 0.064, 0.08, and 0.16) were fabricated in the same chip. Figure 4.19a shows the SEM of the spiral-shaped microdisk filter with r_0=25 μm and ε=0.016. The ε value gives a notch size of 0.4 μm for singlemode waveguide notch-coupling. Figure 4.19b through d shows the zoomed SEMs of the notch coupling regions with ε=0.016, 0.032, and 0.064. The wider notch waveguide supports multimode. We note that the side-coupled waveguide width can be separately optimized for singlemode propagation.

We measured the transmission spectra of the notch/drop/add-only filter configurations by separately input-coupling light to the three waveguide ports [99]. Figure 4.20 schematically shows the experimental setup. The inset shows the optical micrograph of the filter configurations. For each input-coupling configuration, we also collected the out-of-plane light scattering near the notch coupling region and the side-coupling region by linearly scanning a vertical lensed-fiber at a height of a few microns above the cavity top surface. The light scattering spectra and spatial distributions indicated the cavity mode-field distributions.

Figure 4.21a shows the measured TE-polarized throughput-port transmission spectra of notch and drop filters [99] for the spiral-shaped microdisk with ε=0.016. The evanescently input-coupled CCW and CW traveling-wave transmissions are reciprocal related with essentially identical resonances and line shapes. The FSR of ~7.2 nm suggests a wavefront-matched round-trip path length that is consistent with the spiral-shaped

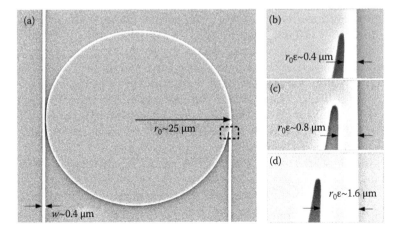

FIGURE 4.19
(a) SEM of the fabricated spiral-shaped microdisk resonator-based filter in SiN. (b–d) Zoom-in SEMs of the notch junctions with ε=0.016, 0.032, and 0.064.

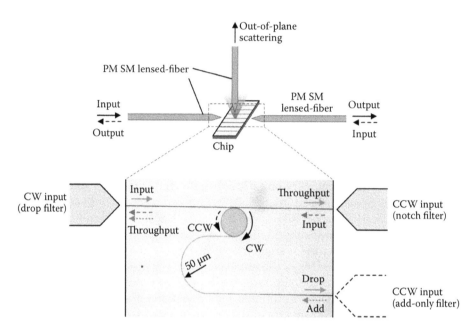

FIGURE 4.20

Schematic of the experimental setup for transmission and out-of-plane scattering measurements of the three filter configurations. Bottom inset: optical micrograph of the spiral-shaped microdisk filter with arrows denoting the three filter operations. PM: polarization maintaining, SM: singlemode. (Reprinted from Optical Society of America. Lee, J. Y., Luo, X., and Poon, A. W., *Opt. Express* 15, 14650–14666, 2007. With permissoin from The Optical Society of America.)

microdisk circumference. The observed highest Q is ~12,000, suggesting the spiral-shaped microdisk preserves high-Q resonances. Figure 4.21b shows the corresponding measured drop-port transmission spectrum of drop filter, and throughput-port transmission spectrum of add-only filter. The two transmission spectra are again nearly matched with each other, suggesting reciprocal transmissions between CW and CCW traveling waves. Similarly, Figure 4.21c through f shows the measured TE-polarized reciprocal transmission spectra of notch/drop filters and drop/add-only filters with $\varepsilon=0.032$ and $\varepsilon=0.064$. The reciprocity relation for transmissions holds in the larger ε devices. As ε increases, however, the measured maximum Q's drop significantly in both CW and CCW traveling waves, indicating the cavity becomes increasingly lossy.

In order to gain insights to the cavity modal structures that underlie the reciprocal transmissions, we collected the out-of-plane light scattering spectra from both CCW and CW traveling-wave modes. Figure 4.22a shows the measured light scattering spectra in the vicinity of the notch junction of $\varepsilon=0.016$ filter [99]. The estimated scanning fiber field of view and position are depicted in the SEM. The light scattering spectra exhibit the same resonance wavelengths as the transmission spectra (see Figure 4.21a and b). We find no evidence from the light scattering spectra that CW and CCW traveling-wave modes exhibit different Q's. However, the resonance line shapes in the scattering spectra differ among the three filters, possibly due to interference of the resonances with different coherent non-resonant light scattering. Noticeably, add-only filter displays a light scattering spectrum that suggests a large background scattering at the input-coupling notch junction. Figure 4.22b and c shows the light scattering intensity distributions spanning from the notch to ~11 μm in the x-direction at resonance wavelengths of 1552.06 nm and 1553.02 nm. In each filter configuration and for both resonance wavelengths, we see pronounced scattering

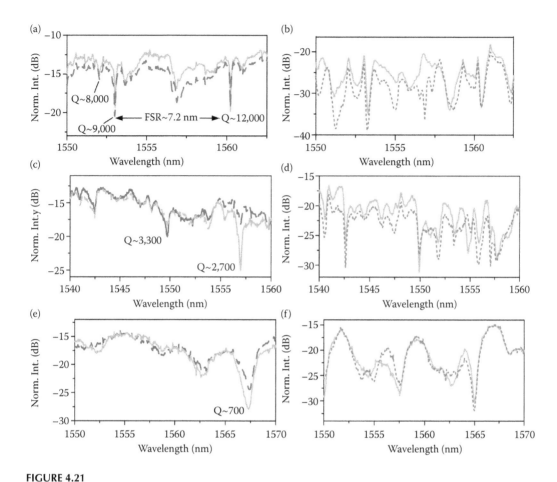

FIGURE 4.21

(a, c, e) Measured TE-polarized throughput-port transmission spectra of notch filter (dashed line) and drop filter (solid line). (b, d, f) Measured TE-polarized drop-port transmission spectrum of drop filter (solid line) and throughput-port transmission spectrum of add-only filter (dotted line). The spiral-shaped microdisk has $r_0 = 25$ μm. (a, b) ε = 0.016, (c, d) ε = 0.032, and (e, f) ε = 0.064. (Figure 4.21a and b are reproduced from Optical Society of America. Lee, J. Y., Luo, X., and Poon, A. W., *Opt. Express* 15, 14650–14666, 2007. With permission from The Optical Society of America.)

at the notch junction. Yet, the light scattering intensity profiles vary in detail among the three filter configurations, suggesting asymmetric modal distributions that depend on the input-coupling port. Each filter configuration displays light scattering intensity profiles that are consistent between the two resonance wavelengths.

Figure 4.22d depicts the measured light scattering spectra at the side-coupling region. The estimated Q values from the light scattering spectrum of the drop filter configuration are again consistent with those estimated from the transmission spectra. Figure 4.22e and f shows the light scattering profiles near the side-coupling region spanning ~13 μm in the x-direction at resonance wavelengths of 1552.06 nm and 1553.02 nm. At each resonance wavelength, the scattering intensity profiles again vary among the three filter configurations, suggesting cavity modal structures that depend on the input-port. Moreover, the light scattering profile distributions in the vicinity of the side-coupling region are significantly different from those collected in the vicinity of the notch-coupling region, indicating a nonsymmetric modal distribution for each filter configuration.

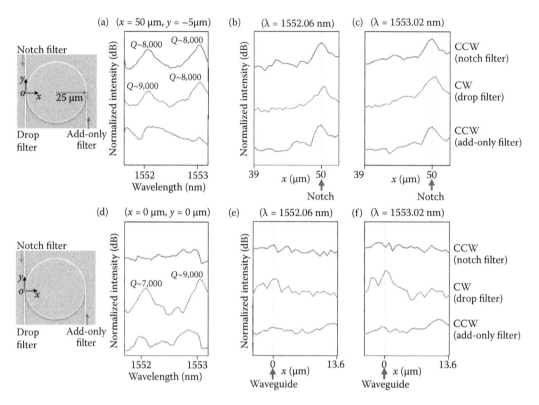

FIGURE 4.22

(a) Measured out-of-plane light scattering spectra near the notch junction of $\varepsilon = 0.016$ filter. The estimated Q values are labeled. (b–c) Light scattering intensity profiles over a ~11-μm span near the notch junction in the x-direction at resonances of (b) 1552.06 nm, and (c) 1553.02 nm. (d) Measured out-of-plane light scattering spectra near the side-coupling region. (e–f) Light scattering intensity profiles over a ~13-μm span near the side-coupling region in the x-direction at resonances of (e) 1552.06 nm, and (f) 1553.02 nm. The SEMs depict the scanning fiber field of view, positions, and the scanning directions. (Reprinted from Optical Society of America. Lee, J. Y., Luo, X., and Poon, A. W., *Opt. Express* 15, 14650–14666, 2007. With permission from the Optical Society of America.)

Based on both simulations and experiments, we therefore conclude that waveguide-coupled spiral-shaped microdisk resonators exhibit reciprocal CCW/CW traveling-wave transmissions with spatially nondegenerate CW and CCW traveling-wave modal distributions. Furthermore, we find that even the traveling-wave modal distributions in the same sense of (CCW) circulation depend on the side/notch in-coupling mechanisms. Only in hindsight we see that rotational asymmetric cavities in a linear dielectric medium should enable spatially nondegenerate yet spectrally degenerate CW and CCW traveling-wave modes, thereby preserving reciprocity in transmissions. We remark that the degeneracy breaking between CW and CCW traveling-wave modes have also been recently noted from the amplified spontaneous emission spectrum in a spiral-shaped quantum cascade laser [94].

4.3.4 Tilted Notch-Coupled Waveguide Design for Mode Matching

In order to attain efficient notch coupling, proper mode matching between the cavity mode and the waveguide mode is essential. To this end, detailed designs for the spiral shape, the notch shape, and the notch-coupled waveguide are imperative. Figure 4.23a shows

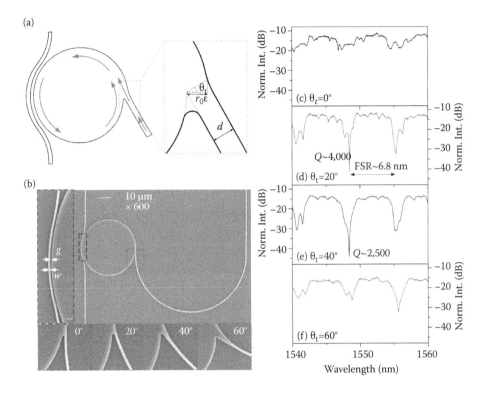

FIGURE 4.23
(a) Schematic of the waveguide-coupled spiral-shaped microdisk resonator-based filter with an angle-tilted notch-coupled waveguide. Arrows depict the light propagation. Dashed line: matched wavefront. Inset: Design parameters of the spiral notch-to-waveguide junction. θ_t: tilting angle; d: width of the tilted waveguide. (b) SEM of our fabricated device in SiN, with $r_0 \approx 25$ µm, $\varepsilon \approx 0.016$, $w \approx 0.5$ µm, and $g \approx 0.3$ µm. Left inset: zoom-in SEM of the side-coupling region. Bottom insets: zoom-in SEMs of the notch-coupling region with different tilting angles. (c–f) Measured TE-polarized throughput-port transmission spectra with different angle-tilted notch-coupled waveguides for $\varepsilon = 0.016$. Both Q and extinction ratio vary with the tilt angle.

the schematic of one design of a spiral-shaped microdisk resonator-based filter with an angle-tilted notch-coupled waveguide. The notch-coupled waveguide at a tilt angle θ_t has a width $d(\theta_t)$ that is slightly wider than the notch size (see inset). Furthermore, we impose two arcs on both sides of the notch-waveguide junction in order to mitigate the scattering loss. We wrap the side-coupled waveguide along the microdisk curved sidewall in order to enhance the side-coupling.

The mode matching scheme here is distinct from that for polygonal-shaped microdisk resonators (see Section 4.2). Figure 4.23a shows the schematic of light propagation in a spiral-shaped microdisk with a tilted notch-coupled waveguide. We design different notch-coupled waveguide tilting angles in order to match the waveguide mode with a particular spiral-shaped microdisk mode. The input-coupled **k**-vector angular distribution also varies with the tilting angle. Thus, with different angle-tilted notch-coupled wave-guide, we can selectively excite different cavity modes.

Figure 4.23b shows the SEM of the $\varepsilon = 0.016$ device in SiN. The fabricated microdisk radius is ~25 µm with a notch width of ~400 nm. The side-coupled waveguide width is ~500 nm, and the gap spacing between the cavity sidewall and the waveguide is ~300 nm (see left inset). The bottom insets show the notch-coupling region with θ_t of 0°, 20°, 40°, and 60°. Figure 4.23c through f shows the measured TE-polarized throughput-port transmission spectra of drop filter. The measured FSR of ~6.8 nm is consistent with the microcavity circumference. The observed inhomogeneously broadened resonance Q's and extinction ratios significantly vary with θ_t. The measured transmission spectra thus suggest that the notch coupling efficiency and/or the cavity loss can be tailored by varying the notch-waveguide tilting angle. We note that the detailed dependence on θ_t also depends on the spiral shape parameter ε.

4.4 Silicon Electro-Optic Modulators Using Microdisk Resonators

4.4.1 Overview of Silicon Electro-Optic Modulators

In the past 3–4 years, silicon electro-optic modulators based on carrier dispersion effect have been gaining major research interest from academia and industry [23,24,41,55, 56,101–105]. According to the seminal work of Soref and Bennet [106], the empirical relations for free-carrier induced refractive index change Δn and absorption change $\Delta\alpha$ in silicon at wavelength of 1550 nm are as follows,

$$\Delta n = \Delta n_e + \Delta n_h = -[8.8\times10^{-22}\Delta N_e + 8.5\times10^{-18}(\Delta N_h)^{0.8}] \tag{4.3}$$

$$\Delta\alpha = \Delta\alpha_e + \Delta\alpha_h = 8.5\times10^{-18}\Delta N_e + 6.0\times10^{-18}\Delta N_h \tag{4.4}$$

where Δn_e (Δn_h) is the change in refractive index resulting from a change in free electron (hole) carrier concentration (cm^{-3}), ΔN_e (ΔN_h), $\Delta\alpha_e$ ($\Delta\alpha_h$) is the change in absorption coefficient resulting from a change in free electron (hole) carrier concentration (cm^{-3}), ΔN_e (ΔN_h).

There are three common types of silicon electro-optic modulators depending on the microelectronic devices employed to induce carrier dispersion effect: (i) carrier-injection type using forward-biased p-i-n diodes [23,24,56], (ii) carrier-accumulation type using metal-oxide-semiconductor (MOS) capacitors [41,10], and (iii) carrier-depletion type using reverse-biased p-n diodes [55,102–105]. Figure 4.24a through c illustrates the three types of silicon electro-optic modulators.

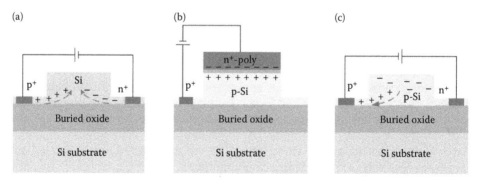

FIGURE 4.24
Schematics of three common mechanisms of silicon electro-optic modulators using carrier dispersion effect. (a) Carrier-injection-based using a forward-biased p-i-n diode. (b) Carrier-accumulation-based using a MOS capacitor. (c) Carrier-depletion-based using a reverse-biased p-n diode. The dashed arrows represent the carriers transport directions.

A carrier-injection-based modulator is a minority-carrier device enabling a Δn of up to ~10^{-3} with a carrier injection of ~3×10^{17} cm^{-3} according to Equation 4.3. This enables a microcavity resonance wavelength blueshift on the order of 0.5 nm. However, one major shortcoming is that the electrical switching time can be limited by the carrier recombination lifetime, which is in the order of ns for submicrometer-sized waveguide structures [23]. In order to speed up the electrical switching, a reverse-bias of a few-V can be applied in order to quickly sweep out the injected carriers [23].

In contrast, modulators using either MOS capacitors or reverse-biased p-n diodes are based on electric-field induced majority carrier dynamics. The modulation speed is therefore electrically limited by the resistance-capacitance (RC) time delay that is ideally in the order of ps [41]. However, one major drawback is that the effective refractive index change based on MOS capacitors and depletion-type p-n diodes are relatively small. For the MOS capacitor approach, by operating on a relatively small capacitance (in the order of pF) and a CMOS-compatible voltage (in the order of V), we expect the carrier-induced refractive index change to be in the order of 10^{-4} distributed in a thin layer ($\ll 0.1$ μm thick) in close proximity to the gate–oxide interfaces [41]. For the depletion-based p-n diode approach, the carrier concentration change is limited by the doped depletion region, which typically is designed to be lightly doped (~10^{17} cm^{-3} [103]) in order to avoid excessive dopant scattering and absorption losses (although moderately doped ~10^{18} cm^{-3} depletion region has also been used recently to demonstrate a GHz modulator [104]). Thus, despite the depletion-based p-n diode enabling a relatively large volume for spatial overlap with the optical mode (as in the injection-based modulator), the expected refractive index change is only in the order of 10^{-4} (assuming lightly doped ~10^{17} cm^{-3} depletion region). Although cm-long Mach-Zehnder interferometer (MZI) based modulators with either MOS-capacitors [41] or depletion-type p-n diodes [103,105] have been demonstrated, it is undesirable to have a large device footprint that also inevitably impose RF impedance matching.

In this respect, silicon microresonators can be instrumental to realizing high-speed compact modulators. The key is to enhance the sensitivity to small carrier-induced refractive index change via a sharp resonance [23,24,55,56].

4.4.2 Principle of Microresonator-Based Modulators

Here, we outline the principle of silicon microresonator-based electro-optic modulation using carrier-injection-based p-i-n diode as an example. Figure 4.25a illustrates the resonance modulation upon an ideal square-wave electrical waveform with forward and reverse biases. Upon a forward bias, the resonance mode that spatially overlaps with the injected carriers is spectrally blueshifted by carrier dispersion (see Equation 4.3). The resonance linewidth is also broadened due to the accompanied free-carrier absorption loss (see Equation 4.4). Upon a reverse bias, the injected carriers can be quickly swept out. Thus, optical intensity at a probe wavelength near the off-cycle resonance wavelength (λ_{off}) or the on-cycle blueshifted resonance wavelength (λ_{on}) can be modulated with a large extinction ratio. We caution that the carrier recombination and the ohmic heat can result in a slow thermo-optic resonance wavelength redshift of ($\lambda_{off} - \lambda_{on}$) for both λ_{off} and λ_{on}. Thus, the wavelength separation ($\lambda_{off} - \lambda_{on}$) is a proper metric to the carrier-induced resonance wavelength blueshift. Figure 4.25b schematically shows the free-carrier concentration modulation in the p-i-n diode intrinsic region, and the optical response at probe wavelengths λ_{off} and λ_{on}. A probe wavelength in the vicinity of λ_{off} enables marginally faster optical response and marginally larger extinction ratio than that near λ_{on}, due to the sharper and deeper resonance line shape at λ_{off}.

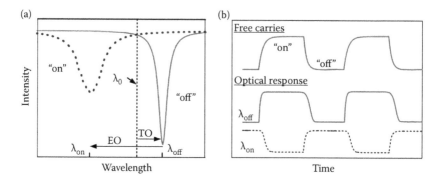

FIGURE 4.25

(a) Schematic resonance modulation upon a square-wave electrical signal with forward and reverse biases. Solid line: resonance during off-cycle of the electrical signal, dotted line: resonance during on-cycle. TO: Thermo-optic effect, EO: Electro-optic effect. (b) Schematic free-carrier concentration modulation and the resulting modulated optical signals at wavelengths λ_{off} and λ_{on}.

As the driving signal period becomes comparable to the carriers' transit times, the carrier injection into and extraction from the p-i-n diode intrinsic region cannot totally follow the driving signal. This results in congestion of carriers in the intrinsic region, and consequently narrows the dynamic range of the resonance wavelength shifts. The resonance wavelengths of the diode off- and on-cycles thus approach each other, resulting in an electrically-limited modulation depth.

The modulator bandwidth can therefore be electrically limited by the carrier dynamics or optically limited by the photon cavity lifetime τ_{ph} [56], depending on which sets the lower limit. For example, assuming a fast 10-ps carrier transit time in a carrier-injection or carrier-depletion-based modulator, and a relatively low-Q of 10^3, the modulator bandwidth is electrically limited as it only takes $\tau_{ph} \approx Q/\omega \sim 1$ ps to build up or decay the optical modes but ~10 ps to build up or decay the carrier concentrations. Whereas, for a relatively high-Q of 10^5 ($\tau_{ph} < 100$ ps), the modulator is optically limited as it now takes much longer time to build up or decay the optical mode than to build up or decay the carrier concentrations. We note that numerical modeling work suggested that silicon microring modulators using an integrated p-i-n-i-p diode structure can reach 40 Gb/s [107].

It is also important to attain a large extinction ratio for microresonator-based modulators by approaching the critical coupling condition [108]. In the case of a single-waveguide-coupled microresonator configuration (notch filter), the critical coupling imposes $\tau = A_c$, where τ is the bus waveguide field amplitude transmission coefficient, and A_c is the fraction of the electric field amplitude that remains upon a cavity round-trip ($A_c = 1$ means lossless). For instance, for modulator with an electrically limited bandwidth using a cavity of $Q \approx 10^3$, $A_c \sim 0.9$ and thus $\tau \sim 0.9$ in order for an extinction ratio larger than 20 dB. This suggests a relatively large waveguide coupling coefficient $\kappa \sim 0.4$ (assuming lossless waveguide-microresonator coupling $\tau^2 + \kappa^2 = 1$), imposing either a narrow coupling gap or a long interaction length for the conventional microdisk and microring resonators.

4.4.3 Microdisk Resonator-Based Modulators

Here, we review our work on silicon electro-optic modulators using circular-shaped microdisk resonators with selectively integrated lateral p-i-n diodes [24]. In contrast with microring resonators that suffer from scattering losses in both the inner and outer sidewalls and also from fabrication-induced waveguide nonuniformities, microdisk resonators have the key merit of having only the outer sidewall to worry about. As WG modes are localized near the microdisk rim, we can selectively inject carriers into the optical modes using an embedded p-i-n diode, which has the lossy heavily-doped regions only in the microdisk center and outside the microdisk rim. In addition, the heavily-doped region in the microdisk also serves to suppress high-order WG modes that extend deep into the cavity, enabling the microdisk modulator to be ideally singlemode [109].

Figure 4.26a shows the schematic of our silicon electro-optic modulator comprising a 10-μm diameter microdisk resonator selectively embedded with a lateral p-i-n diode on an SOI substrate. The microdisk rim region is left undoped, and thus enables low-loss propagation for the low-order WG modes. It is also critical that the doped region in the slab layer surrounds almost the entire microdisk [23], thus enabling a uniform carrier injection into the microdisk rim and a fast sweeping out of the injected carriers before carrier recombination. The inset shows the schematic cross-section of the lateral p-i-n diode in our microdisk modulator. Although the p-i-n diode intrinsic region width of 2.25 μm is sufficiently wide to house low-order WG modes, it is too wide for attaining a GHz-speed response.

Figure 4.26b shows the optical micrograph of our fabricated 10-μm diameter microdisk modulator. The dashed lines highlight the bus waveguide and the microdisk resonator underneath the oxide upper-cladding layer. Aluminum wires are connected to the p-i-n diode doped regions. We detailed the fabrication processes elsewhere [24].

4.4.3.1 *Experimental Demonstrations*

Figure 4.27a shows our measured TE-polarized throughput-port transmission spectrum of the modulator without biasing the p-i-n diode. Mode A has a Q value of ~16,900 and an extinction ratio of ~7 dB. Mode B has a Q value of ~10,300 and an extinction ratio of ~12 dB. Figure 4.27b shows the measured mode B wavelength blueshift of ~0.07 nm upon applying a DC forward bias of 0.85 V. We estimated the injection current to be ~24 μA. It is notable that the small blueshift (within a linewidth) is sufficient to result in a 9-dB transmission intensity modulation at a probe wavelength in the vicinity of the passive resonance wavelength.

Figure 4.27c shows the measured normalized modulation depth (relative to low-frequency modulation depth) with modulation frequency for modes A (open circles) and B (solid squares), using a square-wave radio-frequency drive signal with a forward-bias of 0.9 V and a reverse-bias of −6 V. For mode A, we measured the low-frequency modulation depth to be ~7 dB, consistent with the resonance extinction ratio, and the 3-dB modulation bandwidth to be ~510 MHz. For mode B, we measured the low-frequency modulation depth to be ~12 dB, consistent with the resonance extinction ratio, and the 3-dB modulation bandwidth to be ~410 MHz.

Our modulator speed is electrically limited by the ns-range fall times. Figure 4.27d shows the measured optical waveforms upon a single 5-ns electrical square waveform for modes A and B. We measured an optical rise time τ_{rise} of ~417 ps for mode A and ~850 ps for mode B, yet a relatively long optical fall time τ_{fall} of ~1.1 ns for mode A and ~2.1 ns for mode B. We attribute the relatively long transients to the relatively wide p-i-n diode intrinsic region in the microdisk rim. The ns-range fall times suggest that a significant fraction of the injected carriers end up recombining before they are swept out by the reverse-biased diode.

4.4.3.2 *Toward GHz-Speed Microdisk Resonator-Based Modulators*

Properly designed microdisk resonator-based modulators can have similar modulation characteristics as microring resonator-based modulators [23], yet without the complications

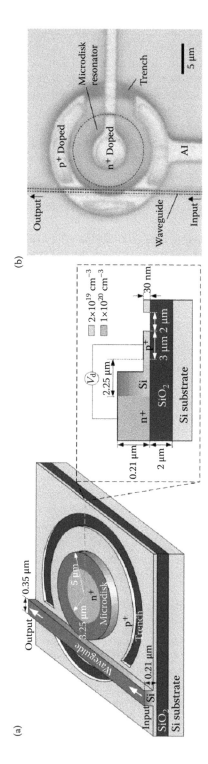

FIGURE 4.26

(a) Schematic of the silicon microdisk resonator-based modulator using a selectively embedded p-i-n diode. The p-i-n diode (and a 2-μm wide trench) surrounds almost the entire microdisk except in the vicinity of the coupled waveguide. Inset: cross-sectional view of the lateral p-i-n diode. The n⁺-doped region radius is 3.25 μm. The p⁺-doped region width is 3 μm. The p-i-n diode intrinsic region width is 2.25 μm. (b) Optical micrograph of our fabricated p-i-n diode integrated microdisk modulator. (Reprinted from Optical Society America. Zhou, L., and Poon, A. W., *Opt. Express* 14, 6851–6857, 2006. With permission form The Optical Society of America.)

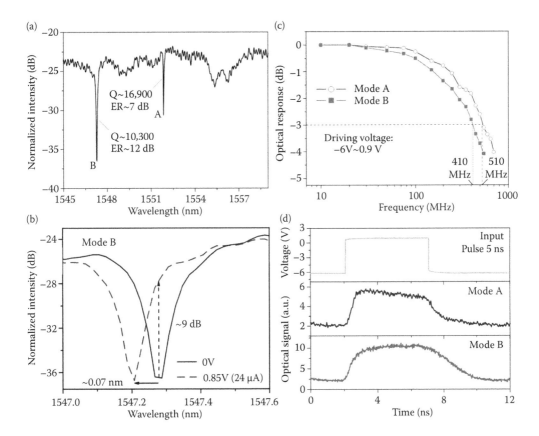

FIGURE 4.27

(a) Measured TE-polarized throughput-port transmission spectrum of the microdisk resonator-based modulator without biasing the embedded p-i-n diode. ER: extinction ratio. (b) DC tuning of mode B. Resonance wavelength blueshifts by ~0.07 nm upon a 0.85 V forward bias across the p-i-n diode. (c) Measured modulation bandwidths for mode A (open circles) and mode B (solid squares). The driving electrical signal is 0.9 V forward bias and −6.0 V reverse bias. (d) Optical transient responses to a single 5-ns input electrical pulse for modes A and B. Mode A: τ_{rise}~ 417 ps and τ_{fall}~1.1 ns; mode B: τ_{rise}~850 ps and τ_{fall}~2.1 ns. (Reprinted from Optical Society America. Zhou, L., and Poon, A. W., *Opt. Express* 14, 6851–6857, 2006. With permission from The Optical Society of America.)

from microring waveguides. Here, we outline a couple design improvements in order to reach GHz-speed modulation using carrier-injection-based microdisk modulators:

i. On the electrical front, our semiconductor device simulations (using MEDICI) suggest that with the p-i-n diode intrinsic region width of 1.5 μm and the microdisk etched height of 0.18 μm, we can attain a rise time of ~0.33 ns and a fall time of ~0.37 ns (i.e. GHz modulation bandwidth) using a forward bias of 1.25 V and a reverse-bias of −4 V. A sub-microampere injection current is sufficient to induce a refractive index change of ~10^{-3}, which enables tuning one resonance line width (0.36 nm) of a Q~4,500 mode.

ii. On the optical front, our 3D-FDTD simulations using a commercially available FDTD simulation tool [110] suggest that we can adopt a 5-μm diameter microdisk with the p-i-n diode as mentioned in (i) in order to confine the first-order WG mode within ~0.75 μm distance from the microdisk sidewall. Figure 4.28a and b shows that

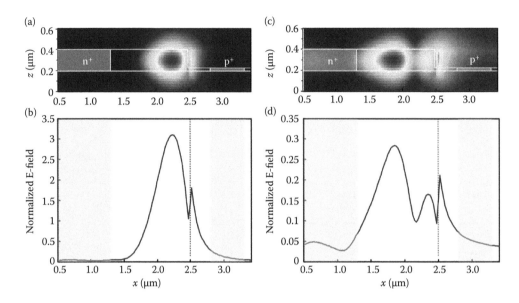

FIGURE 4.28

3-D FDTD-simulated TE-polarized (a), (b) first- and (c), (d) second-order WG mode-field distributions of a 5-μm diameter silicon microdisk resonator with the selectively embedded highly doped regions. (a, c) Cross-sectional view. (b, d) Radial distribution in the mid-height of the microdisk (z=0.3 μm). The cavity-field amplitude is normalized to the input E-field amplitude.

the WG mode can be properly confined without significant spatial overlaps with or attenuation by the highly doped regions inside and outside the microdisk. In contrast, Figure 4.28c and d shows that the mode field amplitude of the second-order WG-mode is about an order of magnitude smaller than the first-order mode.

4.5 Coherent Interference of Optical Resonances

4.5.1 Overview of Coherent Interference in Photonic Resonators

In order to enhance the functionalities of microresonator-based filters and switches, researchers have long employed coherent interference of optical resonances as a means to tailor the resonance response. One common structure is coupled microresonators. For instance, parallel [111] and serial [112] coupled multiple microresonators have been investigated for slow-light and fast-light propagation. Potential applications of coupled microresonators include wavelength-selective filtering [45,113], generating optical delays [112] and optical buffering [30,46] on a chip.

There are three common schemes for coherent interference of optical resonances: (i) a resonance pathway interferes with a coherent background pathway, resulting in an asymmetric resonance line shape (classical analogue of Fano resonances [114]), (ii) a resonance pathway interferes with a coherent feedback, resulting in resonance suppression or enhancement, and (iii) a resonance pathway interferes with another resonance pathway of identical or slightly detuned resonance wavelength, resulting in a narrowband transmission in the center of a transmission dip (classical analogue of electromagnetically-induced transparency

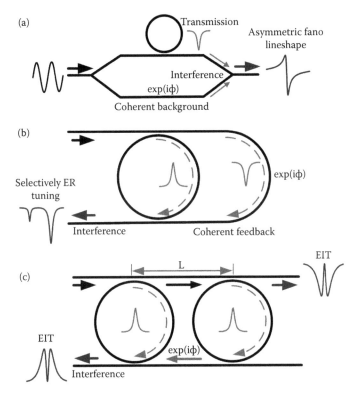

FIGURE 4.29
Schematics of three microresonator-based devices utilizing coherent interference of optical resonances. (a) Fano interference: a resonance pathway interferes with a coherent background pathway. (b) Coherent feedback: a resonance pathway interferes with a coherent feedback. (c) EIT: a resonance pathway interferes with another resonance pathway of the same or detuned wavelength. ER: extinction ratio.

(EIT) [88,115–117]). Figure 4.29a through c depicts three examples of microresonator-based devices utilizing the three schemes of coherent interference of optical resonances. Here, we review our work on (i) and (ii). For the purpose of demonstrating the concepts, we adopted singlemode microring resonators instead of multimode microdisk resonators.

4.5.2 Reconfigurable Microring Resonator-Based Add-Drop Filters Using Fano Resonances

Fano resonances [114] are ubiquitous, both quantum mechanically and classically, wherever a resonance pathway interferes with a coherent background pathway. Fano resonance line shapes are often asymmetric, with a characteristic dip adjacent to the resonance peak, although symmetric line shapes are also possible depending on the relative phase of the interference.

In order to tailor Fano resonance line shapes for device applications [27,118–121], it is of essence to access the relative phases and amplitudes between the resonance pathway and the coherent background pathway. Lu et al. [122] proposed to use the microresonator side-coupled MZI configuration (Figure 4.29a), which allows the resonance pathway (microresonator-coupled MZI arm) and the coherent background pathway (MZI reference arm) to be spatially separated in the interferometer such that a phase shifter in the MZI reference arm is used to vary the transmission Fano resonance line shape at the MZI throughput-port. Independently and almost concurrently, we experimentally demonstrated Fano resonance line shape tuning using a free-space MZI setup [123]. We adopted a prism-coupled silica micropillar resonator in one MZI arm acting as the resonance pathway, and a phase shifter along with a variable attenuator in the MZI reference arm acting as the coherent background pathway.

Here, we review our work using Fano resonances for applications as reconfigurable add-drop filters on a silicon chip [124]. Figure 4.30a schematically depicts our reconfigurable add-drop filter. The filter comprises a silicon racetrack-shaped microring resonator coupled to a MZI with two input-ports (input/add) and two output-ports (throughput/drop). The MZI comprises directional 3-dB input- and output-couplers. One interferometer arm, acting as the resonance pathway, is side-coupled to a racetrack-shaped microring resonator. The reference arm, acting as the coherent background pathway, is embedded with a lateral p-i-n diode for carrier-induced phase tuning. The inset shows the cross-sectional schematic of the reference arm wire waveguide with an embedded p-i-n diode.

We reconfigure the add-drop filter by interchanging the throughput- and drop-port functionality through tuning the relative phase of the coherent background pathway.

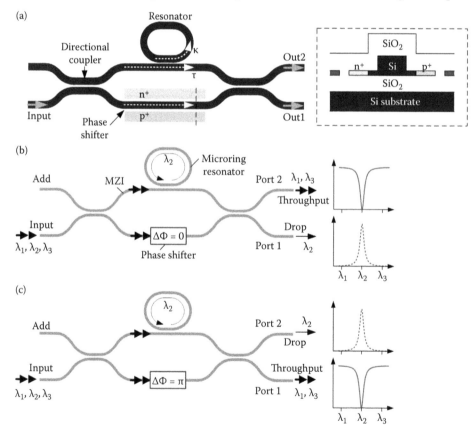

FIGURE 4.30
(a) Schematic of the MZI-coupled racetrack-shaped microring resonator-based reconfigurable add-drop filter. Inset: schematic cross-sectional view of the integrated lateral p-i-n diode in the phase shifter. (b, c) Schematics of the Fano resonance-based reconfigurable add-drop filter transmissions upon two relative phase conditions: (b) without phase shift ($\Delta\Phi=0$), and (c) with π phase shift ($\Delta\Phi=\pi$) of the nonresonance arm. (Reprinted from Optical Society of America. Zhou, L., and Poon, A. W., *Opt. Lett.* 32, 781–783, 2007. With permission from The Optical Society of America.)

Figure 4.30b and c illustrates the device principle. Without an induced phase shift at the coherent background pathway ($\Delta\Phi=0$), off-resonance wavelengths λ_1 and λ_3 are output-coupled at port-2 acting as a throughput-port, while resonance wavelength λ_2 is routed to port-1 acting as a drop-port. With a $\Delta\Phi=\pi$ phase shift imposed on the coherent background pathway, the output-port transmissions are swapped with off-resonance wavelengths λ_1 and λ_3 routed to port-1 and resonance transmission of λ_2 at port-2. As an add-function, in the case that resonance wavelength λ_2 is launched from the add-port, then λ_2 is output-coupled at port-2 with $\Delta\Phi=0$ and routed to port-1 with $\Delta\Phi=\pi$.

We model the MZI-microresonator output-ports transmission spectra using scattering-matrix approach, following Zhou and Poon [124]. We represent the output-ports electric fields, E_{o1} (port-1) and E_{o2} (port-2), and the input-port and add-port electric fields, E_{in} and E_{ad}, in matrix form as follows,

$$
\begin{bmatrix} E_{o1} \\ E_{o2} \end{bmatrix} = \begin{bmatrix} \cos\theta_2 & -i\sin\theta_2 \\ -i\sin\theta_2 & \cos\theta_2 \end{bmatrix} \begin{bmatrix} \left(\dfrac{\tau - A_c e^{-i\varphi_c}}{1-\tau A_c e^{-i\varphi_c}}\right)e^{-i\beta L} & 0 \\ 0 & te^{-i(\beta L - \Delta\Phi)} \end{bmatrix} \begin{bmatrix} \cos\theta_1 & -i\sin\theta_1 \\ -i\sin\theta_1 & \cos\theta_1 \end{bmatrix} \begin{bmatrix} E_{ad} \\ E_{in} \end{bmatrix} \quad (4.5)
$$

where θ_1 and θ_2 determine the directional couplers power coupling ratios, φ_c is the cavity round-trip phase, β is the propagation constant in the wire waveguide, L is the MZI arm length, and $\Delta\Phi=(\Delta\Phi_o+\Delta\Phi_c)$ is the phase difference between the resonance arm and the coherent background arm, comprising an initial phase mismatch $\Delta\Phi_o$ and a free-carrier induced phase shift $\Delta\Phi_c= -\Delta n_{eff}k_0 L$. We assume the waveguide effective refractive index is n_{eff} $(=\beta/k_0$, where k_0 is the free-space propagation constant) and its free-carrier induced dispersion and absorption are Δn_{eff} and $\Delta\alpha_{eff}$. The fraction of the electric field transmitted through the coherent background arm is given as $t=\exp(-\Delta\alpha_{eff}L/2)$.

Assuming ideal 3-dB couplers ($\theta_1=\theta_2=\pi/4$) and the light is launched from the input-port ($E_{ad}=0$), we express the input-normalized transmission intensities as follows [124],

$$
\frac{I_{o1}}{I_{in}} = \left|\frac{E_{o1}}{E_{in}}\right|^2 = \frac{1}{4}\left|-\left(\frac{\tau - A_c e^{-i\varphi_c}}{1-\tau A_c e^{-i\varphi_c}}\right)e^{-i\beta L} + te^{-i(\beta L - \Delta\Phi)}\right|^2 \quad (4.6)
$$

$$
\frac{I_{o2}}{I_{in}} = \left|\frac{E_{o2}}{E_{in}}\right|^2 = \frac{1}{4}\left|\left(\frac{\tau - A_c e^{-i\varphi_c}}{1-\tau A_c e^{-i\varphi_c}}\right)e^{-i\beta L} + te^{-i(\beta L - \Delta\Phi)}\right|^2 \quad (4.7)
$$

Equations 4.6 and 4.7 show that I_{o1} and I_{o2} are interchanged every π phase change of $\Delta\Phi$. Specifically, in the case that $\Delta\Phi$ equals even (odd) numbers of π, I_{o1} displays a symmetric resonance transmission peak (transmission dip) and I_{o2} exhibits a symmetric resonance transmission dip (transmission peak), suggesting drop/throughput functionalities. Likewise, we can deduce the add functionality assuming the light is launched from the add-port ($E_{in}=0$).

Figure 4.31a through c shows the top-view and zoom-in view optical micrographs of our Fano resonance-based device. The MZI arm length is 1.2 mm in order to accumulate 2π phase tuning. Two long aluminum pads are connected to the laterally embedded 1.2-mm long p-i-n diode. However, we caution that the long metal pads do not favor high-speed switching.

Our experiment revealed that the Fano resonance-based add-drop filter can be electrically reconfigured by a sub-0.1 V change in the driving voltage. Figure 4.32a through e

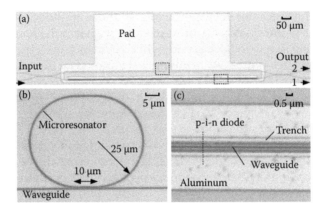

FIGURE 4.31
(a) Optical micrograph of the fabricated device. The MZI arm length is 1.2 mm. (b) Optical microscope zoom-in view of the racetrack-shaped microring coupled straight waveguide. The microring curved waveguide radius is 25 μm. The lateral interaction length is 10 μm. (c) Optical microscope zoom-in view of the p-i-n diode embedded phase shifter. (Reprinted from Optical Society of America. Zhou, L., and Poon, A. W., *Opt. Lett.* 32, 781–783, 2007. With permission from The Optical Society of America.)

shows the measured TE-polarized transmission spectra of the Fano resonance-based device as a function of the forward-bias voltage across the p-i-n diode. We discerned Fano resonances in various asymmetric line shapes. Specifically, upon driving voltage V_d=0.85 V, we observed a nearly symmetric resonance transmission peak at port-2 and a nearly symmetric resonance transmission dip at port-1. Upon increasing V_d by 0.07 V, we switched port-2 transmission to a nearly symmetric resonance dip, and switched port-1 transmission to a nearly symmetric resonance peak.

Figure 4.32f through j shows the modeled transmission spectra using Equations 4.6 and 4.7. We adopt A_c=0.965, τ=0.95, and $\Delta\Phi_o$=0.32π. The modeled spectra find good agreement with the measured spectra. We note that although the asymmetric resonance line shapes may not be directly utilized in the proposed reconfigurable add-drop filter application, it is conceivable that controllable sharp asymmetric line shapes should find niche applications in photonic signal processing (e.g. biochemical sensing [27]).

4.5.3 Coherent Interference between a Resonance Pathway and a Feedback Pathway

Several research groups have also investigated the coherent interference between an optical resonance pathway and an external feedback pathway [125–131] for applications including tailoring waveguide dispersion [125], resonance extinction ratio control [126], slow light propagation [127], CW and CCW modes mutual coupling [128], FSR expansion [129], and our previous work on reconfigurable filters, modulators, and electro-optic logic switches [130,131]. Here, we review our work on such coherent feedback-coupled silicon microring resonators.

4.5.3.1 Coherent Feedback-Coupled Filters

Figure 4.33 depicts our silicon waveguide cross-coupled microring resonator-based electrically reconfigurable filter. The resonance phase-matching condition strongly depends

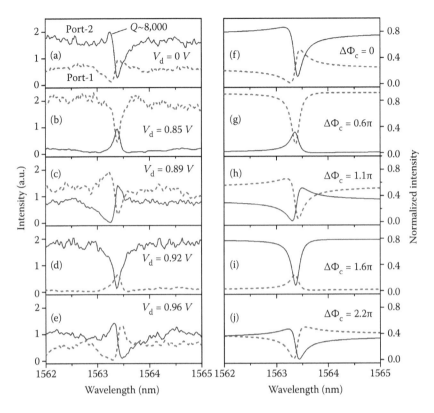

FIGURE 4.32
(a–e) Measured TE-polarized output-ports transmission spectra of the racetrack-shaped microring resonator-coupled MZI with bias voltages V_d=(a) 0 V, (b) 0.85 V, (c) 0.89 V, (d) 0.92 V, and (e) 0.96 V. (f–j) Scattering-matrix-modeled output-ports transmission spectra with the free-carrier induced phase shift $\Delta\Phi_c$=(f) 0, (g) 0.6π, (h) 1.1π, (i) 1.6π, and (j) 2.2π. Solid lines: port-2, dotted lines: port-1. (Reprinted from Optical Society of America. Zhou, L., and Poon, A. W., *Opt. Lett.* 32, 781–783, 2007. With permission from The Optical Society of America.)

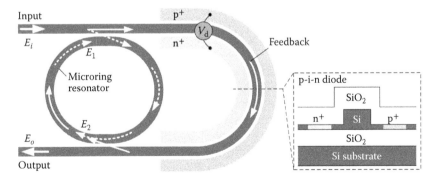

FIGURE 4.33
Schematic of a silicon waveguide cross-coupled microring resonator-based reconfigurable filter. Inset: cross-sectional schematic of the lateral p-i-n diode embedded in the external feedback-waveguide section. V_d: driving voltage. (Reprinted from Optical Society of America. Zhou, L., and Poon, A. W., *Opt. Express*, 15, 9194–9204, 2007. With permission from The Optical Society of America.)

on the coherent cross-coupled feedback phase and amplitude. Thus, by means of electrically tuning the feedback phase and amplitude using carrier dispersion effect, we can selectively enhance or suppress resonances in the waveguide transmission.

Using scattering-matrix approach [130], we express the steady-state relationship between the input- and output-coupled fields, E_i and E_o, as,

$$
\begin{bmatrix} E_o \\ E_2 \end{bmatrix} = \begin{bmatrix} \tau e^{-i\varphi} & \kappa e^{-i(\varphi+\pi/2)} \\ \kappa e^{-i(\varphi+\pi/2)} & \tau e^{-i\varphi} \end{bmatrix} \begin{bmatrix} b\gamma e^{-i\phi} & 0 \\ 0 & ae^{-i\theta} \end{bmatrix} \begin{bmatrix} \tau e^{-i\varphi} & \kappa e^{-i(\varphi+\pi/2)} \\ \kappa e^{-i(\varphi+\pi/2)} & \tau e^{-i\varphi} \end{bmatrix} \begin{bmatrix} E_i \\ E_1 \end{bmatrix}
\tag{4.8}
$$

We also express the steady-state relationship between the microring internal fields E_1 and E_2 as

$$
E_1 = ae^{-i\theta} E_2
\tag{4.9}
$$

where E_1 and E_2 are the fields inside the microring just prior to the input-coupler and just after the output-coupler, κ is the coupling coefficient of the symmetric input/output-couplers ($\tau^2+\kappa^2=1$ for lossless coupling), φ is the couplers transmission phase change ($\varphi=0$ for circular microring), a is the half-circular microring transmission factor, θ is the half-circular microring phase change, b is the feedback-waveguide amplitude transmission factor, γ is the amplitude transmission factor under free-carrier absorption ($\gamma=1$ means no carrier injection), and $\phi=\phi_o+\delta\phi$ is the feedback-waveguide phase change including a passive path-length phase change ϕ_o and a carrier-injection induced phase shift $\delta\phi$. We note that θ is related with the half-circular microring length L_a as $\theta = n_{eff}L_a(2\pi/\lambda)$. Whereas, ϕ_o is related with the feedback-waveguide length L_b as $\phi_o = n_{eff}L_b(2\pi/\lambda)$. We remark that $\delta\phi$ and γ are related by carrier dispersion effect.

From Equations 4.8 and 4.9, we obtain the filter transmission as,

$$
\frac{I_{out}}{I_{in}} = \left| \frac{E_o}{E_i} \right|^2 = \left| \frac{\tau^2 b\gamma e^{-i(\phi+2\varphi)} + \kappa^2 ae^{-i(\theta+2\varphi+\pi)} + a^2 b\gamma e^{-i(2\theta+4\varphi+\phi+\pi)}}{1 - \left[\tau^2 a^2 e^{-i(2\theta+2\varphi)} + \kappa^2 ab\gamma e^{-i(\theta+2\varphi+\phi+\pi)} \right]} \right|^2
\tag{4.10}
$$

We define the resonance phase-matching between the feedback-coupled round-trip field A and the input-coupled field at a point just prior to the input-coupler in terms of

$$
A = |A|e^{-i\Phi} = \tau^2 a^2 e^{-i(2\theta+2\varphi)} + \kappa^2 ab\gamma e^{-i(\theta+2\varphi+\phi+\pi)}
\tag{4.11}
$$

where Φ is the feedback-coupled round-trip phase change. Resonance occurs at $\Phi=2m\pi$ ($m=1, 2, 3\ldots$).

In order to gain physical insights to the phase-matching condition, it is instructive to consider approximations at two limiting cases. Equation 4.11 suggests that in the case of $\tau^2 a^2 \gg \kappa^2 ab\gamma$, the resonances can be approximated by the microring resonances given by $\Phi \approx (2\theta + 2\varphi) = 2m\pi$. We refer to this as microring approximation. Whereas, in the case of $\tau^2 a^2 \ll \kappa^2 ab\gamma$, the resonances can be given by $\Phi \approx (\theta+2\varphi+\phi+\pi) = 2m\pi$. We refer to this as feedback-loop approximation. Figure 4.34a and b illustrates these two round-trip phase change approximations.

The transmission according to Equation 4.10 yields oscillations of the resonance extinction ratio and line shape in the transmission spectrum. Figure 4.35a depicts the resonance-dependent coherent feedback cross-coupling. Resonances λ_{m+1} and λ_m that are spaced by an FSR see different feedback phase angles $\Delta\phi_{m+1}$ and $\Delta\phi_m$, resulting in different line shapes.

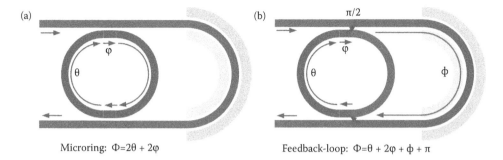

FIGURE 4.34
Schematics of two approximations of the round-trip phase change for the waveguide cross-coupled microring resonator. (a) Microring approximation. (b) Feedback-loop approximation. Arrows represent the light propagation directions.

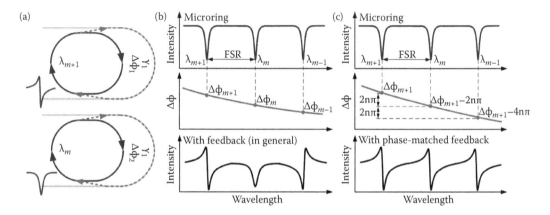

FIGURE 4.35
(a) Schematics of the resonance-dependent coherent feedback interference. Resonances λ_{m+1} and λ_m see different feedback phase angles $\Delta\phi_{m+1}$ and $\Delta\phi_m$, resulting in different line shapes. (b, c) Schematics of the microring resonance spectrum, the wavelength-dependent feedback phase $\Delta\phi$, and the coherent feedback-interfered transmission spectrum. (b) $\Delta\phi$ does not satisfy the phase-matched feedback condition. (c) $\Delta\phi$ nearly satisfies the phase-matched feedback condition. (Reprinted from Optical Society of America. Zhou, L., and Poon, A. W., *Opt. Express*, 15, 9194–9204, 2007. With permission from The Optical Society of America.)

Figure 4.35b illustrates the microring resonance spectrum, the wavelength-dependent feedback phase $\Delta\phi$, and the coherent feedback-interfered transmission spectrum. In general, $\Delta\phi$ values at resonance wavelengths λ_{m+1}, λ_m, and λ_{m-1} do not differ by integer number of 2π.

In order to obtain nearly uniform resonance extinction ratios and line shapes, we can design L_b such that every resonance wavelength sees nearly the same feedback phase and feedback amplitude. We refer to this as phase-matched feedback condition [130]:

$$L_b - L_a = \nu L_{res} \ (\nu = 1, 2, 3, ...) \tag{4.12}$$

where L_{res} is the resonance loop length. For devices close to the microring approximation, we see that $L_{res} \approx 2(L_a + L_c)$, where L_c is the coupler length. In the case that the phase-matched feedback condition is nearly satisfied by properly choosing L_b, $\Delta\phi$ values at resonance wavelengths differ by nearly integer number of 2π. Thus, our modeling suggests that the resonance line shapes are nearly uniform across multiple FSRs, as shown in Figure 4.35c.

However, we note that for devices close to the feedback-loop ring approximation, L_{res} exceeds L_b, and thus there is no L_b's satisfying Equation 4.12.

Figure 4.36a and b shows the optical micrographs of two typical feedback-coupled devices with different L_b design parameters. In both devices the microring is weakly coupled with the feedback-waveguide. Device I comprises a circular-shaped microring of radius 25 µm, an arbitrary chosen $L_b = 180$ µm, and an integrated p-i-n diode spanning 130 µm. Device II comprises a circular-shaped microring of radius 25 µm, $L_b = 230$ µm (with phase-matched feedback using $\nu = 1$), and an integrated p-i-n diode spanning 180 µm.

Figure 4.36c through e shows the measured and modeled TE-polarized singlemode transmission spectra of device I under forward biases of $V_d = 0$ V, 0.9 V, and 1.0 V. The FSR of 3.8 nm is consistent with the microring round-trip path length, suggesting that the device is in the microring approximation regime. While the resonance wavelengths are almost fixed upon the three V_d's, the resonance line shapes exhibit modulations over multiple FSRs. The extinction ratio values vary significantly (most pronounced for resonance C) yet with only slight variations in the Q values.

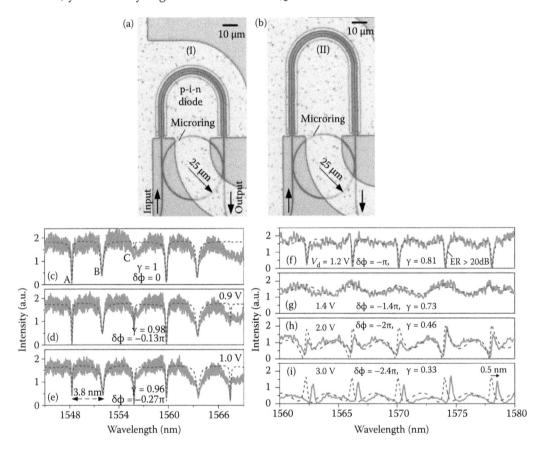

FIGURE 4.36
(a, b) Optical micrographs of two typical devices (I) and (II) with different feedback-coupled waveguide lengths on a SOI substrate. (c–e) Measured (solid lines) and modeled (dashed lines) TE-polarized transmission spectra of device (I) with bias voltages V_d of (c) 0 V, (d) 0.9 V, and (e) 1.0 V. (f–i) Measured (solid lines) and modeled (dashed lines) TE-polarized transmission spectra of device (II) under various bias voltages V_d of (f) 1.2 V, (g) 1.4 V, (h) 2.0 V, and (i) 3.0 V. ER: extinction ratio. (Reprinted from Optical Society of America. Zhou, L., and Poon, A. W., *Opt. Express*, 15, 9194–9204, 2007. With permission from The Optical Society of America.)

The modeled transmission spectra (using Equation 4.10) assume $a=0.96$ and $b=0.95$. We choose $\kappa=0.34$ for weak coupling. We determine the wavelength-dependent feedback phase ϕ_o by beam-propagation method (BPM) simulations and thereby numerically incorporating the waveguide dispersion. We fit the measured transmission spectra by choosing $\delta\phi=0$ and $\gamma=1$ for $V_d=0$ V, and γ remains ~1 for small V_d's. We assume $\delta\phi$ and γ are independent on wavelength within the spectral range of interest.

Whereas, device II displays nearly uniform electrically reconfigurable resonance line shapes over multiple FSRs as expected from the phase-matched feedback condition, and a FSR that is within the weakly-coupled microring approximation regime. Figure 4.36f through i shows the measured and modeled TE-polarized transmission spectra of device II upon $V_d=1.2$ V, 1.4 V, 2.0 V, and 3.0 V. We see uniformly distributed resonance dips with extinction ratio >20 dB ($V_d=1.2$ V), modulated all-pass transmission ($V_d=1.4$ V), asymmetric Fano-like line shapes ($V_d=2.0$ V), and resonance peaks within a broad resonance dip as in EIT-like transmission [88,115] ($V_d=3.0$ V).

4.5.3.2 Coherent Feedback-Coupled Modulators and Logic Devices

We can attain more device functionalities by embedding p-i-n diodes along both the cross-coupled waveguide and arcs of the microring. Our previous work [131] demonstrated a silicon microring resonator-based modulator with tunable operating wavelength and extinction ratio by biasing one of the embedded diodes while modulating the other. We also demonstrated electro-optic OR-logic switching in the optical transmission in the case that both diodes are simultaneously driven with different data inputs.

Figure 4.37a shows the optical micrograph of the coherent feedback-coupled microring modulator/switch with dual electrical inputs. The racetrack-shaped microring arc radius is 25 μm and the side interaction length is 10 μm. The feedback-waveguide length is ~280-μm long measured between the mid-points of the racetrack straight waveguide regions.

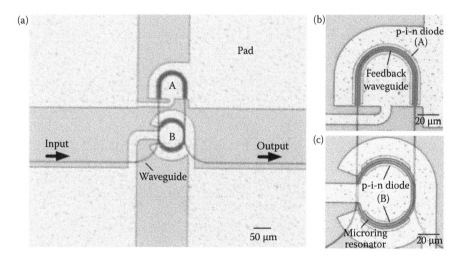

FIGURE 4.37
(a) Optical micrograph of the fabricated coherent feedback-coupled modulator/switch with dual electrical inputs (two pairs of metal pads) on a silicon chip. (b, c) Zoom-in view optical micrographs of the two embedded p-i-n diodes and their metal contacts along the feedback waveguide (node A) and the microring arcs (node B). (Reprinted from Optical Society of America. Li, C., Zhou, L., and Poon, A. W., *Opt. Express* 15, 5069–5076, 2007. With permission from The Optical Society of America.)

Figure 4.37b and c shows the zoom-in view optical micrographs of the diode-integrated feedback waveguide and microring. One p-i-n diode of ~135 µm in length (node A) is laterally integrated across the feedback waveguide, and two other p-i-n diodes of ~59 µm in length each (node B) with the same cross-section as node A are laterally integrated across the microring arcs.

4.5.3.2.1 Modulators with Tunable Operating Wavelengths and Extinction Ratios

Figure 4.38a through e shows the measured TE-polarized transmission spectra of the microring resonator with coherent feedback interference upon various DC forward-biases across nodes A and B of (a) $V_A=V_B=0$ V, (b) $V_A=1.2$ V, $V_B=0$ V, (c) $V_A=1.2$ V, $V_B=1.0$ V, (d) $V_A=0$ V, $V_B=1.1$ V, and (e) $V_A=1.0$ V, $V_B=1.1$ V. When the device is un-biased, we find modulated all-pass transmission with an extinction ratio of only ~4 dB (at ~1550.4 nm). In the case that we forward-bias node A only, the resonance wavelength blueshifts and reaches nearly critical coupling upon $V_A=1.2$ V (at ~1550.2 nm) with an enhanced extinction ratio of ~12 dB. In the case that we forward-bias both nodes A ($V_A=1.2$ V) and B ($V_B=1.0$ V), the resonance becomes moderately suppressed and blueshifts by ~0.24 nm from that upon $V_A=1.2$ V only. Thus, at a probe wavelength of 1550.2 nm (labeled by a dashed-dotted line) upon $V_A=1.2$ V, it is possible to attain ~11 dB extinction ratio by modulating V_B.

In the case that we forward-bias node B only with $V_B=1.1$ V, the resonance wavelength blueshifts by exceeding 1 nm (at ~1549.1 nm) from the unbiased case, with an enhanced extinction ratio of ~13 dB. We note that biasing node B phase-shifts the microring, and thus significantly blueshifts the resonance. In the case that we forward-bias both nodes B ($V_B=1.1$ V) and A ($V_A=1.0$ V), the resonance blueshifts by ~0.1 nm from that upon $V_B=1.1$ V only and is slightly suppressed. Thus, at a probe wavelength of 1549.1 nm (labeled by a dashed-dotted line) upon $V_B=1.1$ V, it is possible to obtain ~8.3 dB extinction ratio by modulating V_A. Hence, by biasing one of the nodes while modulating the other, we can reconfigure the modulator for different wavelengths with different modulation depths.

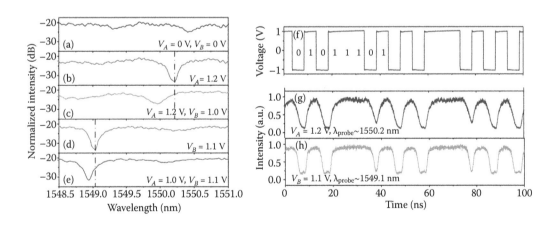

FIGURE 4.38
Measured TE-polarized transmission spectra of the coherent feedback-coupled microring resonator upon various DC-bias voltages across nodes A and B. (a) $V_A=V_B=0$ V, (b) $V_A=1.2$ V, $V_B=0$ V, (c) $V_A=1.2$ V, $V_B=1.0$ V, (d) $V_A=0$ V, $V_B=1.1$ V, and (e) $V_A=1.0$ V, $V_B=1.1$ V. (f) 200-Mbps NRZ electrical input data stream with ±1-V applied across node A or B. (g) Measured optical waveform upon node B modulation at a probe wavelength in the vicinity of the resonance upon $V_A=1.2$ V. (h) Measured optical waveform upon node A modulation at a probe wavelength in the vicinity of the resonance upon $V_B=1.1$ V. Reprinted from Optical Society of America. Li, C., Zhou, L., and Poon, A. W., *Opt. Express* 15, 5069–5076, 2007. With permission from The Optical Society of America.)

In order to examine the modulator response with a high data rate, we apply a 200-Mbps nonreturn-to-zero (NRZ) electrical driving signal of bit sequence "01011101" with ±1 V, as shown in Figure 4.38f. Figure 4.38g shows the modulated optical waveform at a probe wavelength of 1550.2 nm with the driving signal applied across node B upon a DC bias V_A=1.2 V. The modulation depth is consistent with the intensity modulation of ~11 dB upon DC biasing (see Figure 4.38b and c). Likewise, Figure 4.38h shows the modulated optical waveform at a probe wavelength of 1549.1 nm with the driving signal applied across node A upon a DC-bias V_B=1.1 V. The modulation depth is also consistent with the intensity modulation of ~8.3 dB upon DC biasing (see Figure 4.38d and e). We note that the modulator 3-dB bandwidth is only ~0.9 GHz, measured from modulating either node A or B. We attribute the bandwidth limitations to both the unoptimized device design (including the use of long metal pads) and the relatively low drive voltages.

4.5.3.2.2 Demonstration of OR-logic Functionality

Here, we discuss possible logic devices using the same feedback-coupled structure with dual electrical inputs. Figure 4.39a through d shows the measured TE-polarized transmission spectra for four electrical signal input states (V_A, V_B) of (0 V, 0 V), (1.0 V, 0 V), (0 V, 1.0 V), and (1.0 V, 1.0 V). We choose resonance "*" that displays an extinction ratio of ~12 dB when the device is unbiased. The resonance is partially suppressed (and blue-shifted) when only node A is switched on, and totally suppressed when only node B is switched on. When both nodes A and B are switched on, the resonance reappears with a spectral blueshift of ~0.25 nm. Thus, at a probe wavelength in the vicinity of resonance "*", the optical transmission varies between a high-level (Figure 4.39b through d) and a low-level (Figure 4.39a) upon different electrical signal input states.

Hence, in the case that two electrical data are simultaneously applied across nodes A and B, the electrical states can effectively impose logic functionality onto the optical transmission near a microring resonance. Figure 4.39e through g presents our measured electro-optic logic switching using the cross-coupled microring. We applied two different 200-Mbps NRZ electrical data with ±1 V across nodes A and B. The measured optical

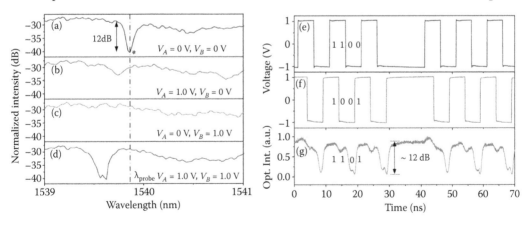

FIGURE 4.39
Measured TE-polarized transmission spectra of the coherent feedback-coupled microring resonator-based device with dual electrical inputs for four electrical signal input states (V_A, V_B) of (a) (0 V, 0 V), (b) (1.0 V, 0 V), (c) (0 V, 1.0 V), and (d) (1.0 V, 1.0 V). λ_{probe}: probe wavelength. (e, f) 200-Mbps NRZ electrical input data applied across nodes A and B. (g) Measured optical waveform at a probe wavelength near resonance "*". (Reprinted from Optical Society of America. Li, C., Zhou, L., and Poon, A. W., *Opt. Express* 15, 5069–5076, 2007. With permission from The Optical Society of America.)

transmission waveform exhibits OR-logic functionality that gives a low-transmission only when both nodes are negatively biased.

4.6 Summary and Outlook

In summary, we have devoted this chapter to elucidate some microresonator-based passive and active devices for integrated photonic circuits on a silicon chip. Specifically, we emphasized our previous and on-going work on exploiting novel shaped cavities and coherent interference of resonances for manipulating resonance response. We attempted to develop design rules or guidelines that facilitate alternative noncircular-shaped microresonators, injection microdisk modulators, and active devices that exploit coherent interference. In the case of waveguide-coupled polygonal-shaped microdisk resonators, we discussed the modal characteristics in different polygonal-shaped microdisk resonators, while highlighting the effects of waveguide dispersion and directional evanescent coupling along the cavity flat sidewall for phase matching to the N-bounce modes. In the case of waveguide-coupled spiral-shaped microresonators, we examined the observed reciprocal transmissions and input-port-dependent asymmetric modal distributions, and addressed the importance of mode matching between the microdisk mode and the waveguide mode for efficient notch coupling. For injection microdisk modulators, we selectively integrated the p-i-n diode with the intrinsic region in the microdisk rim to enable spatial overlap between the modes and the carriers, and to avoid excessive losses due to the heavily doped regions. We also proposed possible improved designs using narrow intrinsic width for shortening the electrical transition times. For tailoring resonance line shapes and wavelengths in active device applications, we devised ways to coherently interfere with the resonances. We wish that the device designs and experimental measurements reviewed here can collectively enrich the silicon photonic devices toolbox, and provide practical insights for microresonator applications. It should be emphasized that design rules for waveguide-coupled microresonators in high-index-contrast material systems are currently lacking. Therefore, it is very necessary to develop design rules that can be experimentally verified and theoretically understood in order to enable optical microresonators to truly deliver their promise as building blocks in nascent large-scale photonic integrated circuits for practical applications.

In light of our reviewed work, one potentially disruptive application of silicon microresonator-based devices and circuits is to act as wavelength-agile optical switches in next-generation multi-core computer chips [40]. We envision that by forming photonic cross-bar architectures with a microresonator-based switch side-coupled to a waveguide grid, we can route optical signals in a two-dimensional network on a silicon chip. Previously, our research group systematically studied microring-resonator-based cross-connect filters in silicon nitride-on-silica substrates [132]. However, the conventional waveguide crossing imposed severe insertion loss and crosstalk due to the light scattering at the waveguide intersection, limiting its application in photonic circuits. In order to alleviate the crossing induced losses and crosstalks, we recently introduced the multimode interference (MMI) principle into the crossing design in SOI substrates [133]. These MMI-based crossings exhibit low insertion loss and low crosstalk with broadband transmissions, enabling them as building blocks for compact photonic networks. Most recently, we took the vision one step further by demonstrating microring resonator-coupled MMI-based crossing array [134].

Hence, along with the carrier dispersion based resonance wavelength tuning techniques reviewed here, we see the potential of realizing nonblocking $N \times N$ microring resonator-coupled crossing-based switch nodes, for scalable photonic interconnect networks-on-chip in multi-core computing systems. It is conceivable that such photonic networks-on-chip technology offers an optical solution for broadband high-throughput short-distance data communications with low latency and low-power consumption. Nonetheless, many critical issues are still debatable such as whether to put light sources on- or off-chip, and how logic devices can be embedded for intelligent microresonator switching, while keeping low-power consumption for the entire microsystem. Besides, in order to overcome the inherent limitations of photonics in networks-on-chip applications, namely buffering and signal processing, we believe the integration of microelectronics and photonics is inevitable and will bring up many new major design/fabrication challenges and re-thinking [40] for computer architectures, optoelectronics integrated systems, and silicon photonics.

Acknowledgments

Andrew W. Poon would like to dedicate this Chapter to his thesis advisor, Professor Richard K. Chang of Yale University, for his inspiration and foresight on optical microresonators. Many of the concepts reviewed here including Fano resonances, square-shaped microresonators, and spiral-shaped microresonators are based on early work done and ideas discussed in Chang's lab while Andrew W. Poon was still a graduate student. Andrew W. Poon would also like to thank all his former and present students at the Photonic Device Laboratory. Their collective effort over the past 6 years has certainly put PDL at HKUST on the world map of silicon photonics and optical microresonators research. In particular, Andrew W. Poon would also like to acknowledge former students N. Ma, C. Y. Fong, H. T. Lee, K. K. Tsia, and S. Zheng for their early contributions. The research on microresonators and silicon photonics have been substantially supported by the Research Grants Council and the University Grants Committee of The Hong Kong Special Administrative Region, China, under Project HIA01/02.EG05, Project 6166/02E, Project 6254/04E, Project 618505, and Project 618506. Andrew W. Poon would also like to acknowledge support by the Institute of Integrated Micro Systems of HKUST, under Project I2MS01/02.EG07. X. Luo would like to acknowledge support from NANO Fellowship of HKUST.

References

1. R. K. Chang and A. J. Campillo, Eds., *Optical Processes in Microcavities*. Singapore: World Scientific, 1996.
2. K. J. Vahala, Ed. *Optical Microcavities*. Singapore: World Scientific, 2004.
3. R. K. Chang and Y. Pan. Linear and non-linear spectroscopy of microparticles: basic principles, new techniques and promising applications. *Faraday Discussions* 137, 9–36, 2008.
4. A. B. Matsko and V. S. Ilchenko. Optical resonators with whispering-gallery modes—part I: basics. *IEEE J. Select. Topics Quantum Electron.* 12, 3–14, 2006.
5. V. S. Ilchenko and A. B. Matsko. Optical resonators with whispering-gallery modes—part II: applications. *IEEE J. Select. Topics Quantum Electron.* 12, 15–32, 2006.

6. R. D. Richtmyer. Dielectric resonators. *J. Appl. Phys.* 10, 391–398, 1939.
7. E. A. J. Marcatilli. Bends in optical dielectric waveguides. *Bell Syst. Tech. J.* 48, 2103–2132, 1969.
8. A. Ashkin and J. M. Dziedzic. Observation of resonances in the radiation pressure on dielectric spheres. *Phys. Rev. Lett.* 38, 1351–1354, 1977.
9. S. L. McCall, A. F. J. Levi, R. E. Slusher, S. J. Pearton, and R. A. Logan. Whispering-gallery mode microdisk lasers. *Appl. Phys. Lett.* 60, 289–291, 1992.
10. Y. Yamamoto and R. E. Slusher. Optical processes in microcavities. *Phys. Today* 46, 66–73, 1993.
11. Z. Zhang, D. Chu, S. Wu, S. Ho, W. Bi, C. Tu, and R. Tiberio. Photonic-wire laser. *Phys. Rev. Lett.* 75, 2678–2681, 1995.
12. J. U. Nöckel, A. D. Stone, G. Chen, H. L. Grossman, and R. K. Chang. Directional emission from asymmetric resonant cavities. *Opt. Lett.* 21, 1609–1611, 1996.
13. C. Gmachl, F. Capasso, E. E. Narimanov, J. U. Nöckel, A. D. Stone, J. Faist, D. L. Sivco, and A. Y. Cho. High-power directional emission from microlasers with chaotic resonators. *Science* 280, 1556–1564, 1998.
14. S. M. Spillane, T. J. Kippenberg, and K. J. Vahala. Ultralow-threshold Raman laser using a spherical dielectric microcavity. *Nature* 415, 621–623, 2002.
15. G. D. Chern, A. W. Poon, R. K. Chang, T. Ben-Messaoud, O. Alloschery, E. Toussaere, J. Zyss, and S.-Y. Kuo. Direct evidence of open ray orbits in a square two-dimensional resonator of dye-doped polymers. *Opt. Lett.* 29, 1674–1676, 2004.
16. G. D. Chern, H. E. Tureci, A. D. Stone, R. K. Chang, M. Kneissl, and N. M. Johnson. Unidirectional lasing from InGaN multiple-quantum-well spiral-shaped micropillar. *Appl. Phys. Lett.* 83, 1710–1712, 2003.
17. B. E. Little, J. S. Foresi, G. Steinmeyer, E. R. Thoen, S. T. Chu, H. A. Haus, E. P. Ippen, L. C. Kimerling, and W. Greene. Ultra-compact $Si-SiO_2$ microring resonator optical channel dropping filters. *IEEE Photon. Technol. Lett.* 10, 549–551, 1998.
18. S. T. Chu, B. E. Little, W. Pan, T. Kaneko, S. Sato, and Y. Kokubun. An eight-channel add–drop filter using vertically coupled microring resonators over a cross grid. *IEEE Photon. Technol. Lett.* 11, 691–693, 1999.
19. P. Rabiei, W. H. Steier, C. Zhang, and L. R. Dalton. Polymer micro-ring filters and modulators. *J. Lightwave Technol.* 20, 1968–1975, 2002.
20. A. Kazmierczak, M. Brière, E. Drouard, P. Bontoux, P. Rojo-Romeo, I. O'Connor, X. Letartre, F. Gaffiot, R. Orobtchouk, and T. Benyattou. Design, simulation, and characterization of a passive optical add-drop filter in silicon-on-insulator technology. *IEEE Photon. Technol. Lett.* 17, 1447–1449, 2005.
21. R. A. Soref and B. E. Little. Proposed N-wavelength M-fiber WDM crossconnect switch using active microring resonator. *IEEE Photon. Technol. Lett.* 10, 1121–1123, 1998.
22. K. Djordjev, S. J. Choi, S. J. Choi, and P. D. Dapkus. Microdisk tunable resonant filters and switches. *IEEE Photon. Technol. Lett.* 14, 828–830, 2002.
23. Q. Xu, B. Schmidt, S. Pradhan, and M. Lipson. Micrometre-scale silicon electro-optic modulator. *Nature* 435, 325–327, 2005.
24. L. Zhou and A. W. Poon. Silicon electro-optic modulators using p-i-n diodes embedded 10-micron-diameter microdisk resonators. *Opt. Express* 14, 6851–6857, 2006.
25. R. W. Boyd and J. E. Heebner. Sensitive disk resonator photonic biosensor. *Appl. Opt.* 40, 5742–5747, 2001.
26. E. Krioukov, D. J. W. Klunder, A. Driessen, J. Greve, and C. Otto. Sensor based on an integrated optical microcavity. *Opt. Lett.* 27, 512–514, 2002.
27. C. Y. Chao and L. J. Guo. Biochemical sensors based on polymer microrings with sharp asymmetrical resonance. *Appl. Phys. Lett.* 83, 1527–1529, 2003.
28. K. D. Vos, I. Bartolozzi, E. Schacht, P. Bienstman, and R. Baets. Silicon-on-insulator microring resonator for sensitive and label-free biosensing. *Opt. Express* 15, 7610–7615, 2007.
29. W. Bogaerts, D. Taillaert, B. Luyssaert, P. Dumon, J. Van Campenhout, P. Bienstman, D. Van Thourhout, and R. Baets. Basic structures for photonic integrated circuits in silicon-on-insulator. *Opt. Express* 12, 1583–1591, 2004.

30. F. Xia, L. Sekaric, and Y. Vlasov. Ultracompact optical buffers on a silicon chip. *Nature Photonics* 1, 65–71, 2007.
31. D. Van Thourhout, P. Dumon, W. Bogaerts, G. Roelken, D. Taillaert, G. Priem, and R. Baets. Recent progress in SOI nanophotonic waveguides. In *Proceedings of the 31st European Conference on Optical Communication (ECOC '05)*, Glasgow, Scotland, September 25–29, 2005, 241–244.
32. See for example *Proceedings of the 1st–4th IEEE International Conferences on Group IV Photonics*, 2004–2007.
33. M. Lipson. Guiding, modulating, and emitting light on silicon - challenges and opportunities. *J. Lightwave Technol.* 23, 4222–4238, 2005.
34. B. Jalali and S. Fathpour. Silicon photonics. *J. Lightwave Technol.* 24, 4600–4765, 2006.
35. C. Gunn, CMOS photonics technology platform. *Proceedings of SPIE Photon. West*, San Jose, CA, January 25, 2006, Invited Paper 6125-01.
36. R. Soref. The past, present, and future of silicon photonics. *IEEE J. Select. Topics Quantum Electron.* 12, 1678–1687, 2006.
37. G.T. Reed and A.P. Knights. *Silicon Photonics: An Introduction.* John Wiley, Chichester, UK, 2004.
38. L. Pavesi and D. J. Lockwood, Eds. *Silicon Photonics.* New York: Springer, 2004.
39. B. A. Small, B. G. Lee, K. Bergman, Q. Xu, and M. Lipson. Multiple-wavelength integrated photonic networks based on microring resonator devices. *J. Opt. Netw.* 6, 112–120, 2007.
40. A. Shacham, K. Bergman, and L. P. Carloni. On the design of a photonic network-on-chip. *Proceedings of the First International Symposium on Networks-on-Chip*, Nocs'07, 53–64, 2007.
41. A. Liu, R. Jones, L. Liao, D. Samara-Rubio, D. Rubin, O. Cohen, R. Nicolaescu, and M. Paniccia. A high-speed silicon optical modulator based on a metal-oxide-semiconductor capacitor. *Nature* 427, 615–618, 2004.
42. H. Rong, A. Liu, R. Jones, O. Cohen, D. Hak, R. Nicolaescu, A. Fang, and M. Paniccia. An all-silicon Raman laser. *Nature* 433, 292–294, 2005.
43. J. Niehusmann, A. Vörckel, P. H. Bolivar, T. Wahlbrink, W. Henschel, and H. Kurz. Ultrahigh-quality-factor silicon-on-insulator microring resonator. *Opt. Lett.* 29, 2861–2863, 2004.
44. I. Kiyat, A. Aydinli, and N. Dagli. High-Q silicon-on-insulator optical rib waveguide racetrack resonators. *Opt. Express* 13, 1900–1905, 2005.
45. M. A. Popović, T. Barwicz, M. R. Watts, P. T. Rakich, L. Socci, E. P. Ippen, F. X. Kärtner, and H. I. Smith. Multistage high-order microring-resonator add-drop filters. *Opt. Lett.* 31, 2571–2573, 2006.
46. F. Xia, L. Sekaric, M. O'Boyle, and Y. Vlasov. Coupled resonator optical waveguides based on silicon-on-insulator photonic wires. *Appl. Phy. Lett.* 89, 041122, 2006.
47. P. Koonath, T. Indukuri, and B. Jalali. Monolithic 3-D silicon photonics. *J. Lightwave Technol.* 24, 1796–1804, 2006.
48. M. Soltani, S. Yenagnarayanan, and A. Adibi. Ultra-high Q planar silicon microdisk resonators for chip-scale silicon photonics. *Opt. Express* 15, 4694–4704, 2007.
49. M. Borselli, T. J. Johnson, and O. Painter. Accurate measurement of scattering and absorption loss in microphotonic devices. *Opt. Lett.* 32, 2954–2956, 2007.
50. S. Blair and K. Zheng. Intensity-tunable group delay using stimulated Raman scattering in silicon slow-light waveguides. *Opt. Express* 14, 1064–1069, 2006.
51. M. S. Nawrocka, T. Liu, X. Wang, and R. R. Panepucci. Tunable silicon microring resonator with wide free spectral range. *Appl. Phys. Lett.* 89, 071110, 2006.
52. I. Kiyat, A. Aydinli, and N. Dagli. Low-power thermooptical tuning of SOI resonator switch. *IEEE Photon. Technol. Lett.* 18, 364–366, 2006.
53. G. N. Nielson, D. Seneviratne, F. Lopez-Royo, P. T. Rakich, Y. Avrahami, M. R. Watts, H. A. Haus, H. L. Tuller, and G. Barbastathis. Integrated wavelength-selective optical MEMS switching using ring resonator filters. *IEEE Photon. Technol. Lett.* 17, 1190–1192, 2005.
54. M. M. Lee and M. C. Wu. Variable bandwidth of dynamic add-drop filters based on coupling-controlled microdisk resonators. *Opt. Lett.* 31, 2444–2446, 2006.
55. A. Huang, C. Gunn, G. Li, Y. Liang, S. Mirsaidi, A. Narasimha, and T. Pinguet. A 10Gb/s photonic modulator and WDM MUX/DEMUX integrated with electronics in 0.13 µm SOI CMOS. *2006 IEEE International Solid-State Circuits Conference*, San Francisco, CA, 2006.

56. Q. Xu, S. Manipatruni, B. Schmidt, J. Shakya, and M. Lipson. 12.5 Gbit/s carrier-injection-based microring silicon modulators. *Opt. Express* 15, 430–436, 2007.

57. S. F. Preble, Q. Xu, B. S. Schmidt, and M. Lipson. Ultrafast all-optical modulation on a silicon chip. *Opt. Lett.* 30, 2891–2893, 2005.

58. V. R. Almeida, C. A. Barrios, R. R. Panepucci, and M. Lipson. All-optical control of light on a silicon chip. *Nature* 431, 1081–1084, 2004.

59. H. Rong, S. Xu, Y. Kuo, V. Sih, O. Cohen, O. Raday, and M. Paniccia. Low-threshold continuous-wave Raman silicon laser. *Nature Photonics* 1, 232–237, 2007.

60. A. W. Fang, R. Jones, H. Park, O. Cohen, O. Raday, M. J. Paniccia, and J. E. Bowers. Integrated AlGaInAs-silicon evanescent race track laser and photodetector. *Opt. Express* 15, 2315–2322, 2007.

61. J. Van Campenhout, P. Rojo-Romeo, P. Regreny, C. Seassal, D. Van Thourhout, S. Verstuyft, L. Di Cioccio, J.-M. Fedeli, C. Lagahe, and R. Baets. Electrically pumped InP-based microdisk lasers integrated with a nanophotonic silicon-on-insulator waveguide circuit. *Opt. Express* 15, 6744–6749, 2007.

62. Q. Xu, V. R. Almeida, and M. Lipson. Micrometer-scale all-optical wavelength converter on silicon. *Opt. Lett.* 30, 2733–2735, 2005.

63. S. F. Preble, Q. Xu, and M. Lipson. Changing the colour of light in a silicon resonator. *Nature Photonics* 1, 293–296, 2007.

64. A. Vörckel, M. Mönster, W. Henschel, P. H. Bolivar, and H. Kurz. Asymmetrically coupled silicon-on-insulator microring resonators for compact add–drop multiplexers. *IEEE Photon. Technol. Lett.* 15, 921–923, 2003.

65. A. Melloni, F. Morichetti, and M. Martinelli. Polarization conversion in ring resonator phase shifters. *Opt. Lett.* 29, 2785–2787, 2004.

66. C. Manolatou, M. J. Khan, S. Fan, P. R. Villeneuve, H. A. Haus, and J. D. Joannopoulos. Coupling of modes analysis of resonant channel add–drop filters. *IEEE J. Quantum Electron.* 35, 1322–1331, 1999.

67. A. W. Poon, F. Courvoisier, and R. K. Chang. Multimode resonances in square-shaped optical microcavities. *Opt. Lett.* 26, 632–634, 2001.

68. Y. Pan and R. K. Chang. Highly efficient prism coupling to whispering gallery modes of a square μ cavity. *Appl. Phys. Lett.* 82, 487–489, 2003.

69. C. Y. Fong and A. W. Poon. Mode field patterns and preferential mode coupling in planar waveguide-coupled square microcavities. *Opt. Express* 11, 2897–2904, 2003.

70. C. Y. Fong and A. W. Poon. Planar corner-cut square microcavities: ray optics and FDTD analysis. *Opt. Express* 12, 4864–4874, 2004.

71. N. Ma, C. Li, and A. W. Poon. Laterally coupled hexagonal micropillar resonator add-drop filters in silicon nitride. *IEEE Photon. Technol. Lett.* 16, 2487–2489, 2004.

72. N. Ma, F. K. Tung, S. Lui, and A. W. Poon. Hexagonal micro-pillar cavities: multimode resonances and open-loop resonance linewidth broadening. *Proceedings of SPIE* 4986, 153–160, 2003.

73. S. Zheng, N. Ma, and A. W. Poon. Experimental demonstration of waveguide-coupled hexagonal micropillar resonators with round-corners in silicon nitride. *Proceedings of Conference on Lasers Electro-Optics.* IEEE and Optical Society of America, 443–445, 2005.

74. C. Li, N. Ma, and A. W. Poon. Waveguide-coupled octagonal microdisk channel add-drop filters. *Opt. Lett.* 29, 471–473, 2004.

75. C. Li and A. W. Poon. Experimental demonstration of waveguide-coupled round-cornered octagonal microresonators in silicon nitride. *Opt. Lett.* 30, 546–548, 2005.

76. C. Li, L. Zhou, S. Zheng, and A. W. Poon, Silicon polygonal microdisk resonators. *IEEE J. Select. Topics Quantum Electron.* 12, 1438–1449, 2006.

77. C. Li, L. Zhou, and A. W. Poon. Waveguide-coupled square micropillar resonator-based devices: channel filters and electro-optic switches with embedded p-i-n diodes. *IEEE Proceedings of the 8th International Conference on Transparent Optical Networks*, 96–99, 2006.

78. W. H. Guo, Y. Z. Huang, Q. Y. Lu, and L. J. Yu. Whispering-gallery-like modes in square resonators. *IEEE, J. Quantum Electron.* 39, 1106–1110, 2003.

79. W. H. Guo, Y. Z. Huang, Q. Y. Lu, and L. J. Yu. Modes in square resonators. *IEEE, J. Quantum Electron.* 39, 1563–1566, 2003.

80. Q. Chen, Y. D. Yang, and Y. Z. Huang. Finite-difference time-domain analysis of deformed square cavity filters with a traveling-wave-like filtering response by mode coupling. *Opt. Lett.* 32, 967–969, 2007.

81. U. Vietze, O. Krauß, and F. Laeri, G. Ihlein and F. Schüth, B. Limburg and M. Abraham. Zeolite-dye microlasers. *Phys. Rev. Lett.* 81, 4628–4631, 1998.

82. H. J. Moon, K. An, and J. H. Lee. Single spatial mode selection in a layered square microcavity laser. *Appl. Phys. Lett.* 82, 2963–2965, 2003.

83. Q. Chen, Y. H. Hu, Y. Z. Huang, Y. Du, and Z. C. Fan. Equilateral-triangle-resonator injection lasers with directional emission. *IEEE J. Quantum Electron.* 43, 440–444, 2007.

84. A. W. Poon. Optical resonances of two-dimensional microcavities with circular and non-circular shapes. PhD thesis, Yale University, 2001.

85. FullWAVE, Rsoft Inc. Research Software, http://www.rsoftinc.com.

86. M. Lohmeyer. Mode expansion modeling of rectangular integrated optical microresonators. *Opt. Quantum Electron.* 34, 541–557, 2002.

87. M. Hammer. Resonant coupling of dielectric optical waveguides via rectangular microcavities: the coupled guided mode perspective. *Opt. Commun.* 214, 155–170, 2002.

88. Q. Xu, S. Sandhu, M. L. Povinelli, J. Shakya, S. Fan, and M. Lipson. Experimental realization of an on-chip all-optical analogue to electromagnetically induced transparency. *Phy. Rev. Lett.* 96, 123901, 2006.

89. J. U. Nöckel and A. D. Stone. Ray and wave chaos in asymmetric resonant optical cavities. *Nature* 385, 45–47, 1997.

90. M. Kneissl, M. Teepe, N. Miyashita, N. M. Johnson, G. D. Chern, and R. K. Chang. Current-injection spiral-shaped microcavity disk laser diodes with unidirectional emission. *Appl. Phys. Lett.* 84, 2485–2487, 2004.

91. T. Ben-Messaoud and J. Zyss. Unidirectional laser emission from polymer–based spiral microdisks. *Appl. Phys. Lett.* 86, 241110, 2005.

92. A. Fujii, T. Nishimura, Y. Yoshida, K. Yoshino, and M. Ozaki. Unidirectional laser emission from spiral microcavity utilizing conducting polymer. *Jap. J. Appl. Phys.* 44, L1091–L1093, 2005.

93. A. Fujii, T. Takashima, N. Tsujimoto, T. Nakao, Y. Yoshida, and M. Ozaki. Fabrication and unidirectional laser emission properties of asymmetric microdisks based on poly (p-phenylenevinylene) derivative. *Jap. J. Appl. Phys.* 45, L833–L836, 2006.

94. R. Audet, M. A. Belkin, J. A. Fan, B. G. Lee, K. Lin, F. Capasso, E. E. Narimanov, D. Bour, S. Corzine, J. Zhu, and G. Höflerf. Single-mode laser action in quantum cascade lasers with spiral-shaped chaotic resonators. *Appl. Phys. Lett.* 91, 131106, 2007.

95. T. Y. Kwon, S. Y. Lee, M. S. Kurdoglyan, S. Rim, C. M. Kim, and Y. J. Park. Lasing modes in a spiral shaped dielectric microcavity. *Opt. Lett.* 31, 1250–1252, 2006.

96. S. Y. Lee, S. Rim, J. W. Ryu, T. Y. Kwon, M. Choi, and C. M. Kim. Quasiscarred resonances in a spiral shaped microcavity. *Phys. Rev. Lett.* 93, 164102, 2004.

97. R. K. Chang, G. E. Fernandes, and M. Kneissl. The quest for uni-directionality with WGMs in μ-Lasers: coupled oscillators and amplifiers. *Proceedings of the 8th International Conference on Transparent Optical Networks*, 1, 47–51, 2006.

98. G. D. Chern, G. E. Fernandes, R. K. Chang, Q. Song, L. Xu, M. Kneissl, and N. M. Johnson. High-Q-preserving coupling between a spiral and a semicircle μ-cavity. *Opt. Lett.* 32, 1093–1095, 2007.

99. J. Y. Lee, X. Luo, and A. W. Poon. Reciprocal transmissions and asymmetric modal distributions in waveguide-coupled spiral-shaped microdisk resonators. *Opt. Express* 15, 14650–14666, 2007.

100. M. Born and E. Wolf. *Principles of Optics*, 7th Ed. Cambridge: Cambridge University Press, 1999, 724–726.

101. L. Liao, D. Samara-Rubio, M. Morse, A. Liu, D. Hodge, D. Rubin, U. Keil, and T. Franck. High speed silicon Mach-Zehnder modulator. *Opt. Express* 13, 3129–3135, 2005.

102. F. Y. Gardes, G. T. Reed, and N. G. Emerson. A sub-micron depletion-type photonic modulator in silicon on insulator. *Opt. Express* 22, 8845–8854, 2005.

103. A. Liu, L. Liao, D. Rubin, H. Nguyen, B. Ciftcioglu, Y. Chetrit, N. Izhaky, and M. Paniccia. High-speed optical modulation based on carrier depletion in a silicon waveguide. *Opt. Express* 15, 660–668, 2007.

104. D. Marris-Morini, L. Vivien, S. Maine, E. Cassan, S. Laval, P. Lyan, and J-M. Fédéli. Low loss optical modulator in a silicon waveguide based on a carrier depletion horizontal structure. *Proceedings of the IEEE 4th International Conference on Group IV Photonics*, 195–197, 2007.

105. L. Liao, A. Liu, D. Rubin, J. Basak, Y. Chetrit, H. Nguyen, R. Cohen, N. Izhaky, and M. Paniccia. 40 Gbit/s silicon optical modulator for high-speed applications. *Electron. Lett.* 43, 20072253, 2007.

106. R. A. Soref and B. R. Bennett. Electrooptical effects in silicon. *IEEE J. Quantum Electron.* QE-23, 123–129, 1987.

107. S. Manipatruni, Q. Xu, and M. Lipson. PINIP based high-speed high-extinction ratio micron-size silicon electro-optic modulator. *Opt. Express* 15, 13035–13042, 2007.

108. A. Yariv. Universal relations for coupling of optical power between microresonators and dielectric waveguides. *Electron. Lett.* 36, 321–322, 2000.

109. L. Zhou and A. W. Poon. Silicon-on-insulator tunable waveguide-coupled microdisk resonators with selectively integrated p-i-n diodes. *Proceedings of Conference on Lasers Electro-Optics*. IEEE and Optical Society of America, 116–118, 2005.

110. OmniSim, Photon Design Ltd., http://www.photond.com.

111. J. E. Heebner and R. W. Boyd. "Slow" and "fast" light in resonator-coupled waveguides. *J. Mod. Opt.* 49, 2629–2636, 2002.

112. J. K. S. Poon, L. Zhu, G. A. DeRose, and A. Yariv. Polymer microring coupled-resonator optical waveguides. *J. Lightwave Technol.* 24, 1843–1849, 2006.

113. B. E. Little, S. T. Chu, J. V. Hryniewicz, and P. P. Absil. Filter synthesis for periodically coupled microring resonators. *Opt. Lett.* 25, 344–346, 2000.

114. U. Fano. Effect of configuration interaction on intensities and phase shifts. *Phys. Rev.* 124, 1866–1878, 1961.

115. D. D. Smith, H. Chang, K. A. Fuller, A. T. Rosenberger, and R. W. Boyd. Coupled-resonator-induced transparency. *Phys. Rev. A* 69, 063804, 2004.

116. L. Maleki, A. B. Matsko, A. A. Savchenkov, and V. S. Ilchenko. Tunable delay line with interacting whispering-gallery-mode resonators. *Opt. Lett.* 29, 626–628, 2004.

117. A. B. Matsko, A. A. Savchenkov, D. Strekalov, V. S. Ilchenko, and L. Maleki. Interference effects in lossy resonator chains. *J. Mod. Opt.* 51, 2515–2522, 2004.

118. S. Fan. Sharp asymmetric line shapes in side-coupled waveguide-cavity systems. *Appl. Phys. Lett.* 80, 908–1000, 2002.

119. A. Chiba, H. Fujiwara, J. Hotta, S. Takeuchi, and K. Sasaki. Fano resonance in a multimode tapered fiber coupled with a microspherical cavity. *Appl. Phy. Lett.* 86, 261106, 2005.

120. H. T. Lee and A. W. Poon. Fano resonances in prism-coupled square micropillars. *Opt. Lett.* 29. 5–7, 2004.

121. H. T. Lee, L. Zhou and A. W. Poon. Fano resonances in prism-coupled multimode square micropillar resonators. *Opt. Lett.* 30. 1527–1529, 2005.

122. Y. Lu, J. Yao, X. Li, and P. Wang. Tunable asymmetrical Fano resonance and bistability in a microcavity-resonator-coupled Mach-Zehnder interferometer. *Opt. Lett.* 30, 3069–3071, 2005.

123. N. K. Hon and A. W. Poon. Silica polygonal micropillar resonators: Fano lineshapes tuning by using a Mach-Zehnder interferometer. *Proceedings of SPIE* 6101, 176–182, 2006.

124. L. Zhou and A. W. Poon. Fano resonance-based electrically reconfigurable add-drop filters in silicon microring resonator-coupled Mach-Zehnder interferometers. *Opt. Lett.* 32, 781–783, 2007.

125. G. Lenz and C. K. Madsen. General optical all-pass filter structures for dispersion control in WDM systems. *J. Lightwave Technol.* 17, 1248–1254, 1999.

126. W. Green, R. Lee, G. DeRose, A. Scherer, and A. Yariv. Hybrid InGaAsP-InP Mach-Zehnder racetrack resonator for thermooptic switching and coupling control. *Opt. Express* 13, 1651–1659, 2005.
127. G. T. Paloczi, Y. Huang, A. Yariv, and S. Mookherjea. Polymeric Mach-Zehnder interferometer using serially coupled microring resonators. *Opt. Express* 11, 2666–2671, 2003.
128. S. Mookherjea, Mode cycling in microring optical resonators. *Opt. Lett.* 30, 2751–2753, 2005.
129. M. R. Watts, T. Barwicz, M. Popovic, P. T. Rakich, L. Socci, E. P. Ippen, H. I. Smith, and F. Kaertner. Microring-resonator filter with doubled free-spectral-range by two-point coupling. *Proceedings of Conference on Lasers and Electro-Optics* 1, 273–275, 2005.
130. L. Zhou and A. W. Poon, Electrically reconfigurable silicon microring resonator-based filter with waveguide-coupled feedback. *Opt. Express* 15, 9194–9204, 2007.
131. C. Li, L. Zhou, and A. W. Poon. Silicon microring carrier-injection-based modulators/switches with tunable extinction ratios and OR-logic switching by using waveguide cross-coupling. *Opt. Express* 15, 5069–5076, 2007.
132. S. Zheng, H. Chen, and A. W. Poon. Microring-resonator cross-connect filters in silicon nitride: rib waveguide dimensions dependence. *IEEE J. Select. Topics Quantum Electron.* 12, 1380–1387, 2006.
133. H. Chen and A. W. Poon. Low-loss multimode-interference-based crossings for silicon wire waveguides. *IEEE Photon. Technol. Lett.* 18, 2260–2262, 2006.
134. F. Xu and A. W. Poon. Microring resonator-coupled multimode-interference-based crossings in 2x2 cross-grid array filters. *Proceedings of the IEEE 4th International Conference on Group IV Photonics*, 16–18, 2007.

5

Electro-Optic Polymer Ring Resonators for Millimeter-Wave Modulation and Optical Signal Processing

William H. Steier
University of Southern California

Byoung-Joon Seo and Bart Bortnik
University of California Los Angeles

Hidehisa Tazawa
University of Southern California

Yu-Chueh Hung, Seongku Kim, and Harold R. Fetterman
University of California Los Angeles

CONTENTS

5.1 Introduction

Optical waveguide ring resonators, shown in Figure 5.1, are the simplest form of an integrated optical resonant structure. They were first proposed by Marcatili[1] and in the past several years photonic ring resonators have been investigated and demonstrated in several material systems including SiO2,[2] Si,[3,4] III-V semiconductors,[5] and polymers.[6] They have been demonstrated as passive wavelength filters,[7,8] slow wave devices,[9] and complex multipole filters.[2] If the rings are made of an electro-optic material the resonant wavelength can be voltage or current tuned and the devices become optical switches or modulators. Tuning based on charge injection, depletion layer modulation, and the Pockel's effect[10] is in the literature. The modulation bandwidth or the switching speed is limited by the bandwidth of the resonator or the inherent speed limitation of the electro-optic effect. The highest speed devices are based on the Pockel's effect.

In first section of this chapter we review the basics of micro-resonators, Pockel's effect modulators, and the bandwidth limitations. In the second section, our work on

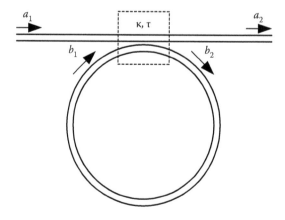

FIGURE 5.1
The generic geometry of ring resonator (one bus waveguide).

operating polymer ring resonant modulators at modulation frequencies over 100 GHz will be reviewed. This very high speed operation is based on the modulation frequency matching the free spectral range (FSR) of the resonator. The modulation frequency matches the FSR but the band width is still limited by the Q of the resonator. This high frequency band-limited modulation has been demonstrated in LiNbO$_3$ disc resonators and in polymer ring waveguide resonators.

In the third section, the great potential of using tunable ring resonators for very high speed optical signal processing will be reviewed. One of the most promising technologies for the general optical signal processing is the *optical delay line circuits*.[11–17] The ring resonator is a crucial building block for the optical delay line circuits enabling the feedback control of the optical signal. Together with other optical building blocks such as optical coupler, we have proposed this general optical structure as *optical signal processor* (OSP).[18] In addition to generating arbitrary and complex functionality by employing optical delay line structures, use of the electro-optic polymer OSP permits operation at high speed.

5.2 Ring Resonator Basics and EO Modulators and Switches

5.2.1 Ring Resonator Basics

In this section, the basic transmission characteristics of ring resonators are briefly described for later discussions. The generic geometry of a ring resonator coupled to a bus waveguide is illustrated in Figure 5.1. Under the condition that a single mode of the resonator is excited and that the coupling is lossless, the interaction can be described by means of two real coupling constants κ and τ and a unitary scattering matrix

$$\begin{pmatrix} a_2 \\ b_2 \end{pmatrix} = \begin{pmatrix} \tau & i\kappa \\ i\kappa & \tau \end{pmatrix} \begin{pmatrix} a_1 \\ b_1 \end{pmatrix}, \quad \tau^2 + \kappa^2 = 1 \tag{5.1}$$

The transmission around the ring is given by

$$b_1 = \alpha e^{-i\theta} b_2, \quad \alpha = e^{-(\rho L/2)}, \quad \theta = \beta L \tag{5.2}$$

where β, ρ, and L are propagation constant, intensity attenuation coefficient, and perimeter of a ring, respectively. Thus α and θ represent the loss factor and phase shift after one round trip. From Equations 5.1 and 5.2, one obtains

$$a_2 = \frac{\tau - \alpha e^{-i\theta}}{1 - \tau\alpha e^{-i\theta}} a_1. \tag{5.3}$$

The same result can be obtained by the summation of multiple round trips

$$a_2 = \tau a_1 - \kappa^2 \alpha e^{i\theta} a_1 - \kappa^2 \tau (\alpha e^{-i\theta})^2 a_1 - \dots$$

$$= \left[\tau - \kappa^2 \sum_{n=1}^{\infty} \tau^{n-1} (\alpha e^{-i\theta})^n \right] a_1 \tag{5.4}$$

$$= \frac{\tau - \alpha e^{-i\theta}}{1 - \tau\alpha e^{-i\theta}} a_1$$

This expression is useful in analyzing the properties of EO modulation later. The transmission of a ring resonator is

$$T(\theta) = \left|\frac{a_2}{a_1}\right|^2 = 1 - \frac{(1-\alpha^2)(1-\tau^2)}{(1-\alpha\tau)^2 + 4\alpha\tau\sin^2(\theta/2)}. \tag{5.5}$$

Optical resonances occur at the points $\theta = 2m\pi$, where m is some integer. When the internal losses in ring, α, are equal to the coupling losses, τ, the transmitted power vanishes. This condition is called critical coupling. For $\alpha < \tau$, the resonator is said to be undercoupling and for $\alpha > \tau$, the resonator is said to be overcoupled.

The critically coupled ring resonator can be used as a notch filter. In case of low loss ($\alpha \approx 1$) and over-coupling ($\alpha > \tau$), the relatively flat intensity response and the rapid phase change at resonances can be used as a phase filter or all-pass filter. Figure 5.2 shows the optical

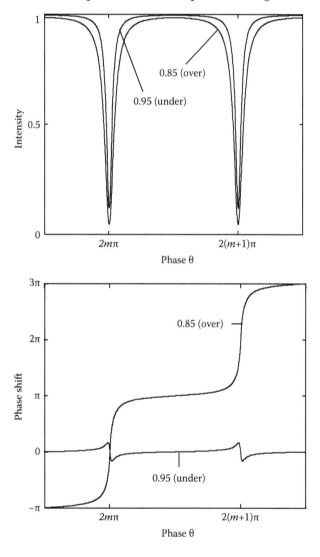

FIGURE 5.2
The intensity transfer function and phase shift of a ring resonator ($\alpha = 0.9$, $\tau = 0.95$ and 0.85).

intensity transfer function $T(\theta)$ and the phase shift of slightly over and under coupled ring resonators.

The full phase width at half maximum (FWHM), $\Delta\theta$, and the finesse of a transfer function are obtained from Equation 5.5

$$
\Delta\theta = 4\sin^{-1}\left(\frac{1-\alpha\tau}{\sqrt{2+2\alpha^2\tau^2}}\right) \approx \frac{2(1-\alpha\tau)}{\sqrt{\alpha\tau}}
$$

$$
F = \frac{2\pi}{\Delta\theta} \approx \frac{\pi\sqrt{\alpha\tau}}{1-\alpha\tau}.
$$

(5.6)

The separation of each resonance peaks is the free spectral range, FSR, which is given by

$$
\text{FSR}_\lambda = \frac{\lambda^2}{n_g L}
$$

(5.7)

where n_g is the waveguide effective index of refraction. Using the relation, $F = \text{FSR}/\Delta f$, the full frequency and wavelength width at half maximum (FWHM) are

$$
\Delta f = \frac{c}{F n_g L}, \quad \Delta\lambda = \frac{\lambda^2}{F n_g L}.
$$

(5.8)

The quality factor Q of a resonator is given by

$$
Q = \frac{f}{\Delta f} = \frac{\lambda}{\Delta\lambda} = \frac{F n_g L}{\lambda}.
$$

(5.9)

To derive the expression of intrinsic or unloaded Q_U, setting $\tau = 1$ and $\alpha \approx 1$,

$$
Q_U \approx \frac{\pi n_g L}{(1-\alpha)\lambda} \approx \frac{2\pi n_g}{\rho\lambda} \approx \frac{2.73 \times 10^5 n_g}{\rho(\text{dB/cm})\lambda(\mu\text{m})}.
$$

(5.10)

The final portion of Equation 5.10 is useful when r is given in dB/cm and λ in μm. Thus, Q_U is dominated only by the attenuation coefficient, which will be a combination of material absorption, bending loss, and scattering loss. The intrinsic finesse is also obtained as

$$
F_U \approx \frac{2\pi}{\rho L}.
$$

(5.11)

It is sometimes convenient to write the transmission as an inverted Lorentzian. If θ_0 is the detuning from resonance, critical coupling is assumed, and the finesse is relatively large

$$
T = 1 - \frac{1}{1 + \left[\frac{F}{\pi}\right]^2 \theta_o^2}
$$

(5.12)

5.2.2 Electro-Optic Ring Modulators

When a voltage V_0 is applied to a ring made with EO polymer, the phase shift θ changes

$$\theta = \theta_0 + \Delta\beta L, \quad \Delta\beta = \frac{\pi n_g^3 r \Gamma V_0}{\lambda g},$$ (5.13)

where θ_0 is the bias phase, L is the perimeter of the ring, n_g is the effective index of the ring waveguide, λ is the free space optical wavelength, r is the EO coefficient, g is the electrode gap, and Γ is the electrical-optical overlap integral. If the wavelength of the incident light is set to the high slope point on the transmission, the output intensity will be strongly modulated with a small modulating voltage. This situation is illustrated in Figure 5.3. The high modulation depth results from the sharp slope of the transmission spectrum and higher F provides higher sensitivity.

From Equation 5.12, the slope of the transmission, $dT/d\theta_0$, is maximum at

$$\theta_0 = \frac{\pi}{\sqrt{3}F}$$ (5.14)

At this point $T = 0.25$ and

$$\Delta T = \frac{9F}{8\sqrt{3}\pi}\Delta\theta_0 = \frac{9F}{8\sqrt{3}\pi}\frac{\pi n_0^3 r \Gamma L}{\lambda g}\Delta V$$ (5.15)

Equation 5.15 gives the change in transmission for a change in voltage when the transmission is optically biased to $T = 0.25$, the maximum slope point.

By comparing $|\Delta T/\Delta V|_{max}$ of a ring modulator with that of a Mach-Zehnder (MZ) modulator, an equivalent V_π^{eq}, can be defined as

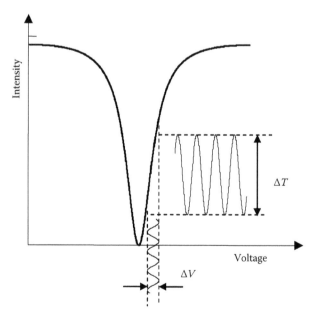

FIGURE 5.3
Conceptual drawing of the modulation by a ring modulator.

$$V_\pi^{eq} = \frac{\pi}{2}\left(\left|\frac{\Delta T}{\Delta V}\right|_{max}\right)^{-1} = \frac{\pi}{2}\left(\left|\frac{\Delta T}{\Delta\theta}\frac{\Delta\theta}{\Delta V}\right|_{max}\right)^{-1} \tag{5.16}$$

Since the V_π of a single arm driven MZ modulator is expressed by

$$V_\pi^{MZ} = \pi\left(\left|\frac{\Delta\theta}{\Delta V}\right|\right)^{-1}, \tag{5.17}$$

the enhancement of the modulation sensitivity by optical resonance is the factor of

$$\frac{V_\pi^{MZ}}{V_\pi^{eq}} = 2\times|\Delta T\Delta\theta| = \frac{9F}{4\sqrt{3}\pi} \tag{5.18}$$

if the device parameters are same. For example, for a finesse of 10 and $|dT/d\theta|_{max}=2.07$ at $\theta_0=0.058\pi$. An MZ modulator must have an electrode length of $4.14\times L$ to obtain the same modulation sensitivity as the ring modulator with a finesse of 10. The resonator with $F=10$ increased the EO interaction length by a factor of 4.14.

5.2.3 Bandwidth of Ring Resonant Modulators

The bandwidth of a ring modulator is limited either by the intrinsic speed of the electro-optic effect, the resonant nature of the modulator, or by the electrode design. The modulation bandwidth of semiconductor ring or disk modulators based on the EO effect of free carrier injection mechanism or depletion width modulation can be limited by the speed of these two types of EO effect. The Pockels effect in nonlinear inorganic crystals and EO polymers is a very fast process, enabling the material limitation to be negligible. When the electrode of a modulator is a lumped element, there are two types of bandwidth limitations: the capacitance limited bandwidth and the transit-time limited bandwidth. When the capacitance between electrodes, C, is parallel to termination load R, the capacitance limited bandwidth is given by $BW_{cap}\approx 1/\pi RC$. When the transit-time of light passes through the electrode region, $t=n_gL/c$, is comparable to the modulation period $1/f_m$, the light does not see a constant voltage but an effective voltage given by the integral over transit, t, of the applied voltage. This bandwidth is conventionally given by $BW_{opt}\approx c/2n_gL$. Optical resonator based modulators impose another bandwidth limitation, that is based on the optical resonator line-width, the FWHM of optical resonance, Δf. This bandwidth is given by $BW_{opt}\approx\Delta f$ and occurs around the frequencies of $n\times$FSR, where n is an integer. This resonator imposed bandwidth is the trade-off for an increase in sensitivity by optical resonance structures. In EO polymer microring modulators, $BW_{opt}<BW_{transit}=FSR/2<BW_{cap}$, since the capacitance C is small. Therefore, the modulation bandwidth of lumped element microring modulators is BW_{opt} and it can be only operated in baseband ($0\sim BW_{opt}$) because of the $BW_{transit}$ limitation. However for a traveling wave electrode the transit-time and the capacitance bandwidth limitation is removed and the modulator can be operated at multiple of the FSR with a bandwidth of BW_{opt}.

The general expression to analyze modulation bandwidth properties of ring modulators is derived using the multiple round trip approach. When a modulation RF signal $V_0\sin\omega_m t$ is applied to an EO ring, the output amplitude $E_{out}(t)$, using Equation 5.4 is given by

$$E_{out}(t) = \left[\tau - \kappa^2\sum_{n=1}^{\infty}\tau^{n-1}\alpha^n\exp[-i(n\theta+\delta_n\sin(\omega_m t - n\phi))]\right]E_{in}(t) \tag{5.19}$$

where $\phi = \omega_m / FSR$. It is assumed that the group index and effective index of a waveguide are the same. The dispersions of propagation constants, coupling coefficients, and loss are neglected.

5.2.3.1 Lumped Circuit Electrode

The modulation index δ_n depends on the electrode structure. In case of a lumped electrode, it is given by

$$\delta_n \sin(\omega_m t - n\phi) = \int_0^{nL} \Delta\beta \sin\left[\omega_m\left(t + \frac{n_0}{c}z\right) - n\phi\right]dz$$

$$= \Delta\beta L \frac{\sin(n\phi/2)}{\phi/2}\sin\left(\omega_m t - \frac{n}{2}\phi\right). \tag{5.20}$$

Equation 5.20 takes into account the transit-time. Now the optical power of the transmitted signal can be given by substituting Equation 5.20 into Equation 5.19 and calculating the first component of the Fourier expansion of the output intensity

$$I_{\omega_m} = \frac{2}{T}\int_0^T |E_{out}(t)|^2 e^{i\omega_m t}dt \tag{5.21}$$

where $T = 2\pi/\omega_m$. The electrical signal power in a photo detector is proportional to the square of the optical signal power and this must be taken into account in calculating the bandwidth of the electrical signal power. Figure 5.4 shows the small signal frequency responses in dBe of the modulated output from lumped ring modulators, optically biased to the maximum slope with a finesse of 10. The response is normalized by low modulation frequency response. The modulating frequency is normalized by the FSR. The modulation is clearly limited by optical resonator bandwidth. The 3 dBe bandwidth BW_{opt} is $0.0845 \times FSR$. Notice that this bandwidth is somewhat larger than the half of the resonator line-width ($0.05 \times FSR$ for $F = 10$).

5.2.3.2 Traveling Wave Electrode

For operation at multiples of the FSR, the circumference of the electrode is a multiple of the RF wavelength and the bandwidth will be limited by $BW_{transit}$ unless the electrode is a traveling wave structure or a resonant electrode. For the resonant electrode, the RF resonance must coincide with a FSR modulation frequency as demonstrated elsewhere.[19–21] In work by Hossein-Zadeh and Levi,[22] the electrode was a traveling wave RF resonator whose resonant frequency equaled the FSR of the disc resonator. In this section, the performance of a ring modulator with a traveling wave electrode is analyzed, and is compared to that of a MZ modulator. The effects of an optical/microwave velocity mismatch and the loss of the microwave transmission line on the performance comparison are included.

The traveling wave ring modulator is illustrated in Figure 5.5. A microstrip line electrode, that is impedance matched to the driving cable and termination, covers the EO ring resonator. Assuming no microwave loss, one considers the drive signal

$$V(z,t) = V_0 \sin\omega_m\left(t - \frac{n_m}{c}z\right) \tag{5.22}$$

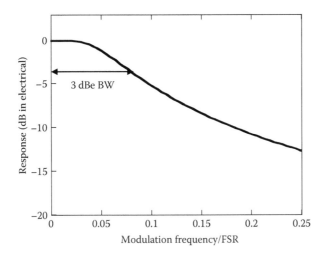

FIGURE 5.4
The small signal frequency responses (in dBe) of the modulated output from a ring modulator with a lumped electrode, optically biased to the maximum slope with a finesse of 10. The responses are normalized by low modulation frequency response.

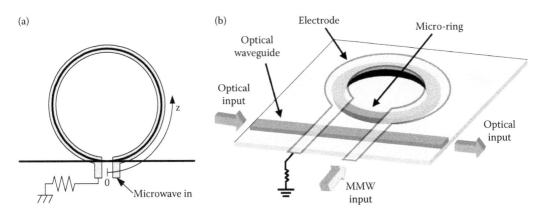

FIGURE 5.5
Traveling wave ring modulator.

where n_m is the microwave effective index. The voltage seen at position z along the ring resonator waveguide by photons that enter the $z=0$ at $t=t$ is given by

$$V(z,t) = V_0 \sin \omega_m \left(t - \frac{\Delta n}{c} z \right) \tag{5.23}$$

where $\Delta n = n_m - n_g$ and n_g is the optical index. The modulation index δ_n is given by

$$\delta_n \sin(\omega_m t - n\phi) = \sum_{k=0}^{n-1} \int_0^L \Delta\beta \sin\left[\omega_m \left((t + kt_r) - \frac{\Delta n}{c} z \right) - n\phi \right] dz$$

$$= \Delta\beta L \frac{\sin(\psi/2)}{(\psi/2)} \frac{\sin(n\phi/2)}{\sin(\phi/2)} \sin\left(\omega_m t - \frac{\psi}{2} - \frac{n+1}{2}\phi \right) \tag{5.24}$$

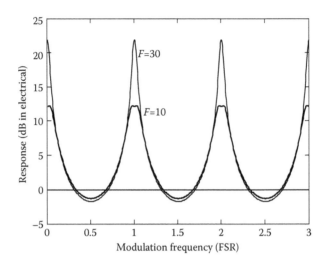

FIGURE 5.6
Modulation frequency responses of velocity matched traveling wave ring modulators with the finesse of 10 and 30. The responses are normalized by the response by an MZ modulator with same electrode length.

where $t_r = n_0 L/c$ is the optical round trip time and $\psi = \omega_m \Delta n L/c$ is the velocity matching factor. The optical power of the transmitted signal can be obtained by substituting Equation 5.24 into Equation 5.19, and then using Equation 5.21.

The small signal frequency responses in dBe of the modulated output from velocity matched ($\Delta n = 0$) ring modulators, optically biased to the maximum slope, with a finesse of 10 and 30 are shown in Figure 5.6. The responses are normalized by the small signal response of a MZ modulator biased at quadrature with same electrode length. The modulating frequency is normalized by the FSR. The velocity matched traveling wave ring modulators provide high modulation efficiency at frequencies around all multiples of FSR. The 3 dBe bandwidth for $F=10$ is 0.174 and for $F=30$ is 0.0516. Notice that this bandwidth is somewhat larger than the resonator line-width (0.1 for $F=10$ and 0.033 for $F=30$). Comparing Figures 5.4 and 5.6 for the $F=10$ and baseband resonance curves, the 3 dBe bandwidth of the traveling wave electrode, 0.087, is slightly wider than that for the lumped electrode, 0.0845. This means there is a small transit-time effect on the bandwidth of the lumped electrode.

In addition to the velocity mismatch, another bandwidth limitation in traveling wave modulators is the loss of the microwave transmission line. For a given electrode dimension, the high frequency microwave loss is determined by the skin depth and one expects a loss in dB/cm of $a = a_0 f^{1/2}$ where a_0 depends on electrode conductivity and geometry. Assuming no velocity mismatch, the effect of loss on the modulation index δ_n is[10]

$$\delta_n \sin(\omega_m t - n\phi) = \sum_{k=0}^{n-1} \int_0^L \Delta\beta e^{-\alpha_m z} \sin[\omega_m(t + kt_r) - n\phi] dz$$

$$= \Delta\beta L \frac{1 - e^{-\alpha_m L}}{\alpha_m L} \frac{\sin(n\phi/2)}{\sin(\phi/2)} \sin\left(\omega_m t - \frac{n+1}{2}\phi\right) \qquad (5.25)$$

where $a_m = a/8.7$ converts from power loss in dB/cm to an exponential amplitude loss coefficient.

TABLE 5.1

Numerical Example Values of an EO Polymer Device

Index of polymer no	1.6
EO coefficient r	50 pm/V
Overlap integral Γ	1
Wavelength of light λ	1.31 μm
Electrode gap g	10 μm
Tuning sensitivity df/dV	1.47 GHz/V
$V_\pi^{MZ}L$	6.4 Vcm
$V_\pi^{eq}L$ (Finesse 10)	1.58 Vcm

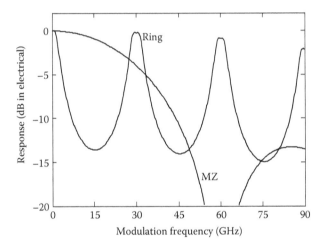

FIGURE 5.7

Effect of velocity mismatch ($\Delta n=0.2$, no microwave loss) in modulation frequency response of the ring modulator with the FSR of 30 GHz ($L=6.25$ mm) and $F=10$ compared to the equivalent broadband MZ modulator ($L=4.04\times6.25$ mm). The responses are normalized by the low-frequency response.

To evaluate the effect of velocity mismatch and microwave loss in traveling wave ring modulators, the small signal frequency responses of the output with a MZ modulator with equal low-frequency V_π is compared. As a numerical example, the parameters in Table 5.1 are used. Assuming a FSR of 30 GHz, a finesse of 10, and critical coupling, the perimeter of the ring resonator is set to be 6.25 mm and the V_π^{eq} is 2.53 V. The electrode length of the equivalent MZ modulator is 2.53 cm.

Figure 5.7 shows the effect of velocity mismatch in the modulation frequency response of the ring modulator and the equivalent MZ modulator for $\Delta n=0.2$. The mismatch in LiNbO$_3$ is much larger ($\Delta n\sim2$) but can be reduced by design while the mismatch is negligible in EO polymer modulators. $\Delta n=0.2$ was chosen as a compromise which clearly shows the effect. Figure 5.8 shows the effect of microwave loss in the modulation frequency response of the ring modulator and the MZ modulator with equal low-frequency V_π. The microwave power loss coefficient of the microstrip line is 0.7 dB/cm $\sqrt{\text{GHz}}$ which is typical for modulators. No velocity mismatch is assumed. In the both cases, the ring modulator shows higher modulation efficiency at frequencies around multiples of FSR than the equivalent MZ modulator because of the shorter electrode length.

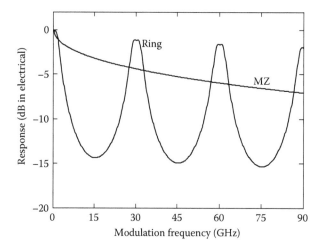

FIGURE 5.8
Effect of microwave loss (a=0.7 dB/cm$\sqrt{\text{GHz}}$, Δn=0) in modulation frequency response of the ring modulator with the FSR of 30 GHz (L=6.25 mm) and F=10 compared to the equivalent broadband MZ modulator (L=4.04×6.25 mm). The responses are normalized by the low-frequency response.

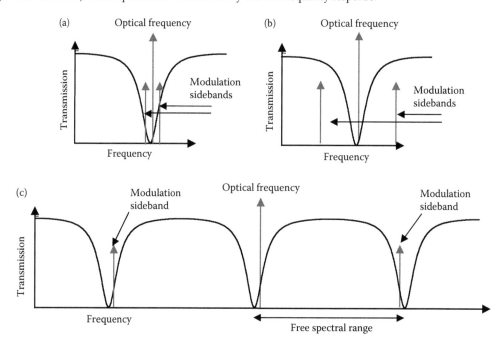

FIGURE 5.9
This sideband picture of the modulation provides another explanation for the peaks in the response of the ring resonant modulator at a modulation frequency which matches the FRS.

The high response of the ring resonant modulator at the FSR modulation frequency can be understood by considering the position of the optical sidebands due to the modulation as shown in Figure 5.9. In part A the modulation frequency is low enough to fit the sidebands in the first resonant response of the ring. In part B, the modulation frequency has increased and the modulation sidebands fall outside the ring response and the modulator has poor efficiency. In part C the modulation frequency has increased to near the FSR and the sidebands again fall under an optical response on the ring.

In summary, ring resonator based traveling wave modulators show high modulation efficiency at frequencies around all multiples of the FSR and have better tolerance for the velocity mismatch and microwave loss of the electrode than traveling wave MZ modulators, thus making them useful in microwave photonics applications. An example other than communication applications is metrology systems.[23] Efficient generation of modulation sidebands improves the precision of measurements. A ring modulator is mathematically identical to a Fabry–Perot modulator.[24] In the traveling wave Fabry–Perot modulator,[25] however, the microwave can interact only with one direction of propagated light. In traveling wave ring modulators, the microwave can interact with light in the full length of a round trip due to unidirectional propagation. Hence, they have a great potential not only for modulators but also for other applications such as comb generation and pulse generation previously demonstrated by Fabry–Perot modulators.[25,26]

5.2.4 Electro-Optic Polymer Traveling-Wave Ring Modulator

5.2.4.1 Fabrication

The top view optical microscope image and the schematic cross section of the fabricated modulator are shown in Figure 5.10. The microstrip line electrode covers the optical waveguide ring resonator as a traveling wave electrode. The modulator consisted of three polymer layers: the lower cladding, the core, and the upper cladding. The 5 μm thick lower cladding was made from UV15LV (Master Bond Co.) on a silicon substrate with the 200 nm thick Au bottom electrode. The core was made from EO polymer AJL8/APC.[27] The racetrack shape ring resonator with the bending radius of 1 mm and the straight coupling region of 150 μm is laterally coupled to a bus waveguide with 2 μm gap. Both the ring and bus waveguide are a rib waveguides with 2 μm width, 1 μm high rib, and 1 μm high slab. The waveguides were formed using an oxygen plasma to etch a trench in the lower cladding. The 4 μm thick upper cladding was made from UFC170A (Uray Co.). The refractive indices of UV15LV, AJL8/APC, and UFC170A were 1.51, 1.61, and 1.50, respectively. The commercial mode solver, Olympios (C2V), confirmed that the waveguides were single mode and that the mode in 1 mm bending radius had negligible bending loss. The device was corona-poled μm to align AJL8 chromophore before forming the top electrode. The top Au electrode was formed by vacuum evaporation and electroplating with 2 μm thick and 17 μm wide. The gap between the top and bottom electrodes was 10 μm. The characteristic impedance of the microstrip was designed to be 57 Ω. Figure 5.11 shows the fabrication steps.

5.2.4.2 Optical and Electro-Optical Properties

Using a tunable laser (New Focus 6200) at the wavelength of 1.31 μm, the basic characteristics of the modulator were measured. The input and output light were coupled through the small core fibers (Nufern UHNA3). The fiber-to-fiber insertion loss is –12 dB due mainly to fiber/waveguide coupling loss. Figure 5.12 shows the transmission spectrum of the TM mode of the ring modulator. The data was measured by the frequency modulation of the laser source. The full width at half maximum (FWHM) is 5.1 GHz and the FSR is 28 GHz. The experimental or loaded Q is 4.5×10^4 and the finesse is 5.5. From the extinction ratio of –14 dB, the intrinsic Q and the waveguide loss are estimated to be 7.1×10^4 and 4.8 dB/cm,

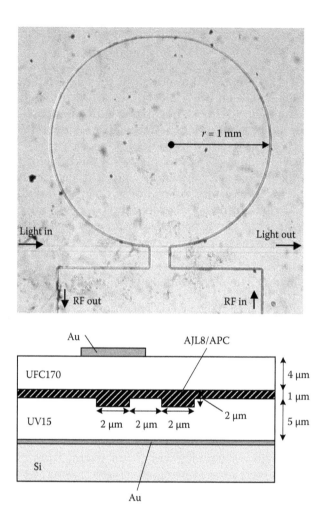

FIGURE 5.10
The top view optical microscope image and schematic cross section of the fabricated modulator.

respectively. Since the material loss of AJL8/APC is approximately 2 dB/cm, the excess 2.8 dB/cm is the scattering loss from the sidewall roughness due to fabrication. From this data, the parameters in the theoretical transfer function (Equation 5.5) are $\alpha=0.696$ and $\tau=0.783$. The EO tuning sensitivity, measured by applying a 100 Hz triangular signal, (Figure 5.13) was found to be 1 GHz/V, which corresponds to an effective EO coefficient $r=33$ pm/V of the core layer. The V_π^{IM} of the modulator at a low modulation frequency are estimated to be 7.5 V.

5.2.4.3 Traveling Wave Electrode Properties

The 50 Ω microstrip line electrode consists of the microstrip section, feeding pad, and termination pad. The feeding and termination pad provide a transition between an RF probe and the microstrip. The detailed dimensions of the feeding pad are shown in Figure 5.14. To maintain the 50 Ω input impedance in the feeding pad, the ground plane was cut out underneath the pad so that the contact pad and the cutout ground plane form a quasi-CPW

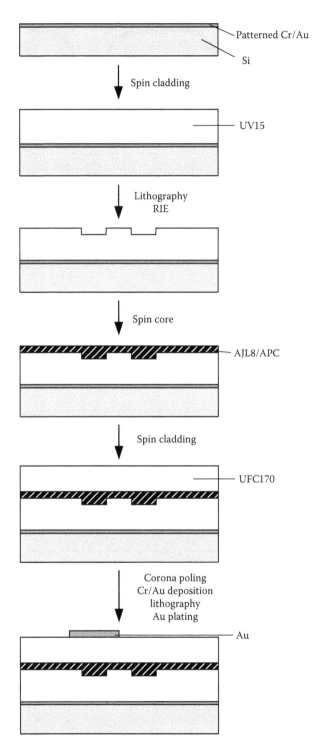

FIGURE 5.11
Fabrication of traveling wave ring resonator.

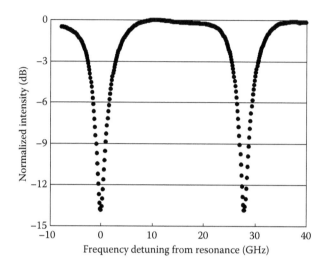

FIGURE 5.12
The transmission spectrum of the TM mode of the ring modulator.

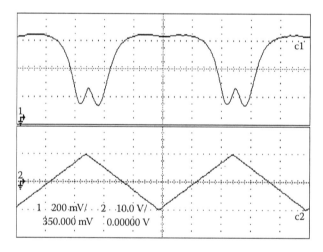

FIGURE 5.13
The modulated TM light from the ring modulator (V_{pp}=20 V).

transmission line. At the other end of the line, a RF terminator is required to minimize the reflection. A 50 Ω chip resistor (State of the Arts S0202AF) is put on the side of the pad, and then the contact on resistor is connected to the termination pad on the device by Au ribbon. The ground plane is also bonded to common ground by Au ribbons. The situation is illustrated in Figure 5.15. Figure 5.16 shows the reflection S-parameter, S11, from 1 to 40 GHz, as measured by a network analyzer. S11 is less than −10 dB from 0 to 30 GHz, which indicates the RF impedance match is reasonable.

5.2.4.4 Modulation at the First FSR Spacing of 28 GHz

The frequency modulation response around the free spectral range at 28 GHz was measured by observing the modulation sideband power in an optical spectrum analyzer (Ando

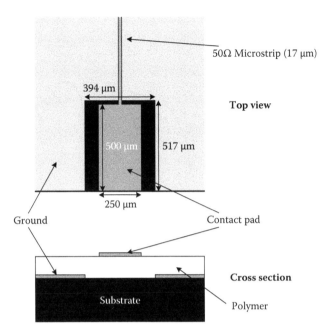

FIGURE 5.14
The detailed dimensions of the feeding pad.

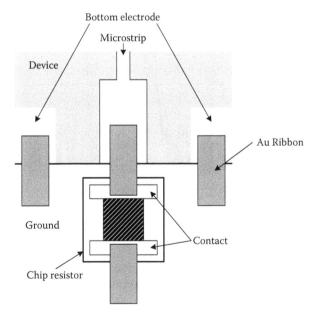

FIGURE 5.15
Schematic drawing of the microstrip termination.

AQ6317B). Sinusoidal modulation signals up to 40 GHz are generated by a signal generator. The microwave power from the signal generator (Agilent 8244A) was 10 dBm over the frequency range used. Optical spectra of the modulated light at 22 GHz (0.125 nm), 28 GHz (0.16 nm), and 34 GHz (0.194 nm) are shown in Figure 5.17. In order to reliably repeat

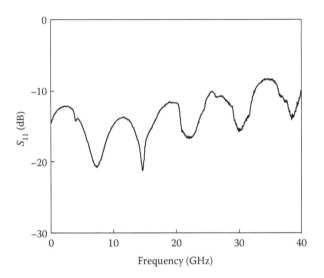

FIGURE 5.16
S_{11} (reflection) of the traveling wave electrode.

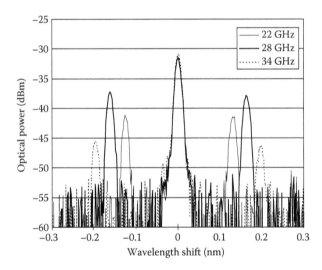

FIGURE 5.17
Optical spectra of the modulated light at 22 GHz, 28 GHz, and 34 GHz.

the same bias point in each modulation frequency, the input light was tuned to minimum transmission which is at resonance of the ring modulator. The observed sidebands were therefore due to phase modulation and the carrier is suppressed. The data shows that the sideband power peaks at the modulation frequency of 28 GHz, which is the FSR of the ring, and the sideband power decreases as the RF frequency is tuned away from the FSR.

Figure 5.18 shows the modulation frequency dependence of optical sideband powers. The side band power shows the minimum at the modulation frequency of 14 GHz, which is off-resonance of the resonator, and the maximum at 28 GHz. The 3 dBe bandwidth of the detected signal power is the same as the 3 dB bandwidth of the sideband power. From Figure 5.18, the signal 3 dBe bandwidth is ~7 GHz. This modulator would therefore find application in analog optical links with the sub-carrier frequency of around 28 GHz with a useable bandwidth of ~7 GHz.

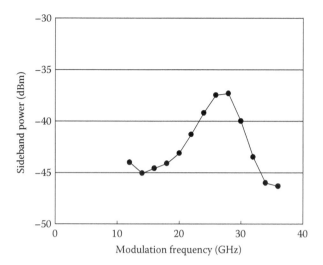

FIGURE 5.18
The modulation frequency dependence of optical sideband powers.

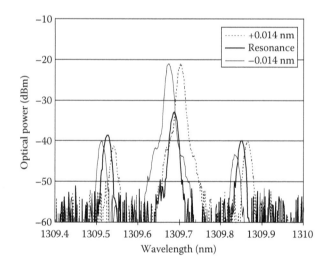

FIGURE 5.19
The bias dependence of 28 GHz modulation spectra.

The bias dependence of 28 GHz modulation spectra is shown in Figure 5.19. When the input light is tuned to a resonance, the carrier is suppressed by 15 dB by the optical resonance and the 28 GHz modulation sidebands are observed. At this point there is only a small intensity modulation at 28 GHz. When the input light is tuned at near a half transmission point, ±2.5 GHz (±0.014 nm) from a resonance, the carrier suppression is approximately 3 dB and the 28 GHz modulation sidebands are also observed. At this point there is significant intensity modulation at 28 GHz.

5.2.4.5 Modulation at Multiples of the FSR

The diagram of the experimental setup for modulation at high microwave frequencies is shown in Figure 5.20. The light source was a tunable laser (New Focus 6200) at the

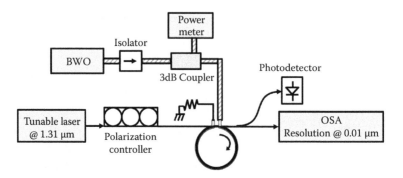

FIGURE 5.20

Millimeter-wave ring modulator test setup. The optical coupler after the ring modulator allowed continuous power monitoring at the photodetector to ensure proper wavelength bias for pure phase modulation. To obtain the optical modulation response, the modulated sideband power to optical carrier ratio was measured on the optical spectrum analyzer (OSA) and divided by the monitored input modulation power from the backward wave oscillator (BWO) millimeter-wave source, as measured at the millimeter-wave 3 dB coupler.

wavelength of 1.31 µm. After passing through a polarization controller, the light was butt-coupled to the device. The tunable laser was tuned to a minimum of the ring modulator's transfer function, ensuring that pure optical phase modulation with suppressed carrier occurs. The modulated optical output was divided into two paths by a 3 dB optical coupler. One arm of the optical output was connected to an optical power meter to ensure phase modulation of the ring modulator throughout the experiment by holding the transmitted optical power to a minimum and performing fine tuning of the wavelength of the tunable laser as needed. The other arm was connected to an optical spectrum analyzer (OSA), the Ando AQ6319, which has a spectral resolution of 0.01 nm, yielding a measurement uncertainty of about 1 GHz.

A backward wave oscillator (BWO), micro-now model 705C with model 728 BWO head, was used as a millimeter wave source, producing up to 17 dBm in millimeter-wave power within the W-band (75–110 GHz). The millimeter-wave setup consisted of entirely W-band waveguide components to minimize resonances before the device. An isolator was placed between the BWO and the waveguide to reduce back-reflected waves and a 3 dB coupler was used after the isolator. One of the coupler's outputs was used to drive the ring modulator and the other output was coupled to an Anritsu ML83A power meter with W-band MP81B head. The power transfer function from both outputs of the coupler was fully characterized across the W-band measuring a power difference between both outputs of less than 2 dB. In addition, this frequency scan showed a smooth spectral profile of the BWO output, indicating that resonances within the BWO and the external components were negligible. The millimeter-wave signal was coupled to the device via a co-planar (G-S-G) W-band probe (Picoprobe). The 10 µm EO polymer stack was scratched on either side of the microstrip to expose the ground plane. Direct contact was then established between the signal tips and ground tips of the probe with the microstrip and exposed ground plane areas, respectively. The microstrip line was terminated by an on-chip commercial 50 Ω resistor that has been rated to 50 GHz.

We observed the modulation sideband power on the optical spectrum analyzer and divided by the millimeter-wave power input to the probe, as measured at the coupler before the probe. The largest power input into the probe was 8 dBm. The experimentally obtained optical modulation response data across the W-band is plotted in Figure 5.21. A simulated optical modulation response for phase modulation is also plotted in the same

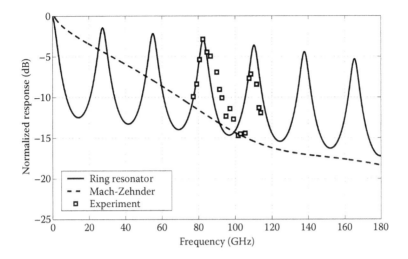

FIGURE 5.21
Experimental optical modulation response plotted along with simulated optical modulation response of a ring modulator and an MZ modulator. Both velocity mismatch (Δn=0.1) and microwave loss (0.8 dB/cm$\sqrt{\text{GHz}}$) are considered in the simulation. Both simulated devices have the same effective low-frequency and are biased at the nulls of their transfer function to produce pure phase modulation. The agreement between the experimental data and simulation validates the proposed analysis.

figure. Pure phase modulation occurred since the ring modulator was wavelength biased at the null of its transfer function. For comparison purposes, the simulation results plotted in Figure 5.21 give the optical modulation response for both the ring resonator and an equivalent MZ modulator when both are biased for phase modulation. The experimental data was normalized to the simulation results by matching the responses at the 84 GHz resonance. The simulation included a velocity mismatch of Δn=0.1 and microwave loss of the electrodes of 0.8 dB/cm$\sqrt{\text{GHz}}$. There is a good agreement between the analysis/simulation above and the experimental results. The third and fourth modulation resonances at 84 GHz and 111 GHz are evident in the experimental data, rising approximately 10 dB above the experimental off-resonance response. Figure 5.22a and b show the optical carrier (center) and modulation sidebands that were seen on the OSA when obtaining the modulation response data at the third and fourth order resonances. If the FSR is 28 GHZ, the fourth resonance should occur at 112 GHz. The difference could be error in the microwave frequency measurement or that the FSR is slightly less than 28 GHz.

In addition, the modulation response in the D-band (110–180 GHz) was also obtained. The microwave module, including the BWO source (Model 731 BWO head), the coupler and the waveguide, were replaced by components in the D-band. The D-band coupler was also characterized by the same procedure as above. The probe accepts a W-band waveguide and therefore a D-band to W-band waveguide transition was used in this setup. The probe used to apply the millimeter wavelength driving signal is rated to 120 GHz and has higher millimeter-wave loss and potential oscillations in its transmission spectrum above 120 GHz. This makes full characterization of the ring modulator throughout the D-band difficult. Nonetheless, the device showed substantial modulation at the fifth and sixth order resonances of 139 GHz and 165 GHz, while remaining mostly ineffective at off-resonance frequencies. Figure 5.23a and b are the experimentally obtained OSA traces showing the optical carrier and modulated sidebands at 139 GHz and 165 GHz, illustrating the observed modulation response at resonance.

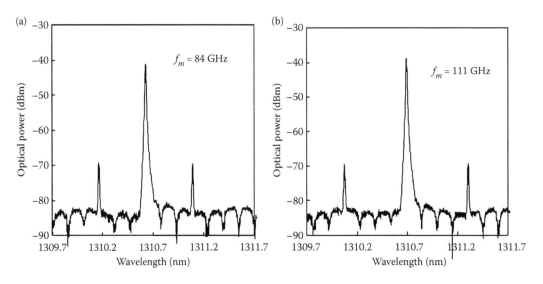

FIGURE 5.22
(a) and (b). Phase modulation sidebands observed on the OSA at 84 GHz (3×FSR) and 111 GHz (4×FSR).

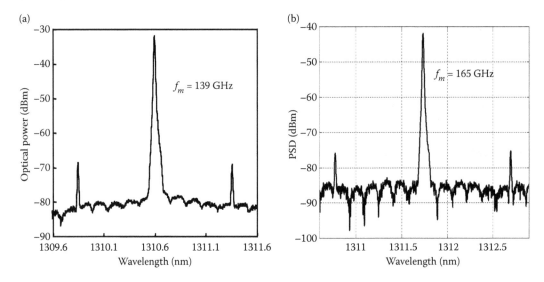

FIGURE 5.23
(a) and (b). The measured output optical spectrum of the ring modulator upon modulation by the BWO at the resonant millimeter-wave frequencies of 139 GHz (5×FSR), and 165 GHz (6×FRS).

5.3 Optical Signal Processing Using Ring Resonator

5.3.1 Theory

5.3.1.1 Fundamentals of OSP

The fundamental enabling concept of the OSP we have proposed stems from the multiple and temporal interference of delayed optical signals. By controlling the amplitude and the phase of the interfered signals, the optical signal can be processed in any arbitrary way.

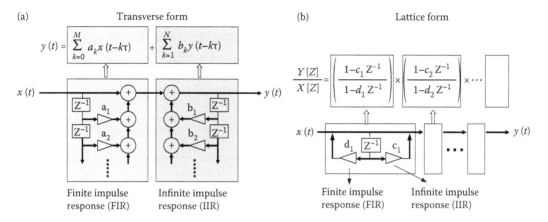

FIGURE 5.24
Example block diagrams of OSP circuits. The input is split into multiple waveguides. The individual optical signal experiences equally different time delays represented by Z^{-1}, and amplitude and phase changes represented by the a_k, b_k, c_k, and d_k coefficients. At the output port, they are combined again to generate an output signal.

Example block diagrams of OSP circuits are shown in Figure 5.24. The input optical signal is launched at the left side and split into multiple waveguides. The individual optical signal experiences equally different time delays represented by τ and Z^{-1} in the diagrams, and amplitude and phase changes represented by the a_k, b_k, c_k, and d_k coefficients. At the output port located at the right side of the diagram, they are combined again to generate an output signal.

Two typical configuration structures are distinguished; *transversal form*[28] and *lattice form*[13,14] as seen in Figure 5.24a and b, respectively. The transversal form is a parallel structure since the signal is processed in a parallel way while the lattice form is a serial structure where signals are processed in series. If we define a waveguide branch *arm*, an arm can be classified into two types; forward feeding arm and backward feeding arm. Since the response of forward feeding arms is finite in time, it is called finite impulse response (FIR). In a similar way, the response of backward feeding arms is called infinite impulse response (IIR) since it is infinite. The number of the forward feeding arms and of backward feeding arms represent the order of FIR and IIR respectively. The origin of this terminology is from the digital filter theory in electronics.[29] According to this theory, it is well known that any arbitrary response can be obtained when FIR and IIR are combined together. The order of FIR and IIR is associated with how arbitrary an OSP can describe the response.

Several physical values are important for characterizing an OSP. They are the unit delay time, τ, and the number of arms that split or combine lights (N and M in Figure 5.24a). The unit delay time of an OSP is analogous to the sampling time in discrete-time signal processing. It represents a time resolution for the OSP to process a signal. Due to this time resolution, the frequency response of an OSP is periodic and the period is proportional to $1/\tau$. This period is called the free spectral range (FSR).

The number of optical splitting and combining arms determines the spectral resolution of the OSP within a FSR. As the number of arms increases, a sharper frequency response can be obtained and hence a more arbitrary response is possible. In that sense, it is analogous to the number of eigen functions in the frequency domain because the overall frequency response of an OSP is a response combination of the individual arms.

5.3.1.2 Representations of OSP

If $(x)t$ and $(y)t$ denote the input and output signal respectively, an OSP performs as an operator to transform $(x)t$ into $(y)t$. Several methods can be used for describing an OSP. The first method is to use a characteristic equation, already shown in Figure 5.24a. If an OSP has M number of forward feeding arms and N number of backward feeding arms, the characteristic equation of the processor can be generally written as Equation 5.26.

$$y(t) - \sum_{n=1}^{N} b_n y(t - n\tau) = \sum_{m=0}^{M} a_m x(t - m\tau) \tag{5.26}$$

where the coefficients, a_m and b_n, stand for the amplitude and phase changes of the m'th forward feeding arm and the n'th backward feeding arm respectively. The coefficients are complex values in general because the signals, $(x)t$ and $(x)t$, stand for coherent optical fields, and are thus considered to have both amplitude and phase. The amplitude and the phase of a coefficient corresponds to the amplitude and phase change of an individual signal.

The second method for describing this system is the transfer function representation. In general, the coefficients, a_m and b_n, and the delay time, τ in Equation 5.26 can be functions of time as well if we change (or modulate) those parameters in time. If their varying rate is comparable to that of the optical signal such as in electro-optic polymer waveguides, their time-dependent effect must be considered. We sometimes apply time-varying signals to the time-delay lines to effectively change τ in our experiments. However, we assume that the time-varying rate is small enough or the *delay control mode* (see Section 5.3.1.3 for its definition) is only considered in the OSP operation. In this case, the OSP becomes a linear system and is similar to digital filter in many ways. Indeed, the OSP has the same transmission characteristics and features as those of a corresponding digital filter. Therefore, the conventional Z transform analysis, which is frequently used with digital filters, can be applied to analysis of the OSP. In Z transformation, the transfer function is represented as the rational functions of Z.

If we calculate the Fourier transformation of Equation 5.26 with respect to the eigen functions of $e^{jp\omega t}$ where p is any integer and ω is the optical angular frequency and replace the $e^{-j\omega t}$ with Z^{-1}, we obtain the transfer function, $H(Z)$, in Z domain.

$$H(Z) = \frac{\displaystyle\sum_{m=0}^{M} a_m Z^{-m}}{1 - \displaystyle\sum_{n=1}^{N} b_n Z^{-n}} \tag{5.27}$$

where $Z^{-1} (= e^{-j\omega t})$ stands for one unit delay in Z domain.

Another convenient and more visual way to describe an OSP is to use a pole/zero diagram.[30] Because the transfer function, $H(Z)$, is a complex function of the complex variable, Z, the value of $H(Z)$ can be plotted on the complex Z space. The zeros are the Z values which make $H(Z)=0$ and the poles are the ones that make $|H(Z)|=\infty$. Then, another form of $H(Z)$ can be written as,

$$H(Z) = A \frac{\displaystyle\prod_{i=1}^{M} (1 - \rho_i Z^{-1})}{\displaystyle\prod_{j=1}^{N} (1 - \gamma_j Z^{-1})} \tag{5.28}$$

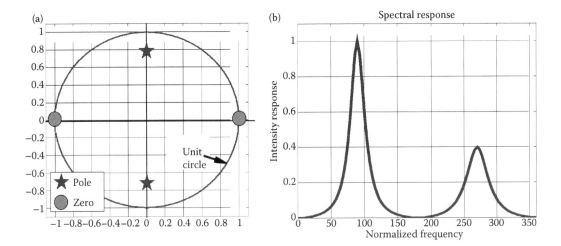

FIGURE 5.25
Pole/zero diagram examples. (a) Pole/zero diagram. Suppose that an OSP is configured such that it has poles (star dots) and zeros (circle dots) located in the Z space (b) Corresponding transfer function in frequency domain (linear scale). Around zeros, the magnitude of the transfer function is small while it becomes large near poles.

where ρ_i and γ_j are the ith zero and jth pole and A is the amplitude factor. Figure 5.25a and b shows an example of pole/zero diagram and its corresponding transfer function, respectively. As the optical frequency increases, the variable Z in Equation 5.28 follows the unit circle counter-clockwise in the Z space. Around the zeros, the magnitude of the transfer function becomes small while it increases near the poles.

Once we find the transfer function, $H(z)$, the functionality of the OSP is exactly defined. This can be done by defining the coefficients, a_m and b_n. Therefore, operation of an OSP depends on how flexible and stable these coefficients can be tuned.

5.3.1.3 Operations of OSP

In general, the coefficients, a_m and b_n, and the delay time, τ in Equation 5.26 can be functions of time as well as the input light signal. Three different methods are possible to operate the OSP depending on which parameters are used for optical processing. They are the *coefficient control mode, delay control mode*, and *frequency control mode*.

The coefficient control mode performs optical processing via controlling the coefficients, a_m and b_n. When a typical continuous wavelength light is input, the output light can be processed by modulating or controlling these coefficients. The delay control mode is via controlling the delay time, τ. For a continuous or varying wavelength light input, the delay time can be controlled or modulated to generated processed output light. The frequency control mode utilizes frequency dependence of the OSP due to its delay lines.

Most PLC structures using silica waveguides regard the coefficients and the delay time as constant variables since their controlling method (mostly thermal tuning) is much slower than the optical light signal. Therefore, the silica waveguides mostly use the frequency control mode for their optical signal processing, such as in optical filters.[11,16,28] However, if the time-varying or modulating rate of the coefficients or the delay time is comparable to the propagation of the optical signal through the OSP, such as in electro-optic polymer waveguides, their time-dependent effect will be useful and should be considered. Indeed, as will be shown in Sections 5.3.4 and 5.3.5, we sometimes apply time-varying signals to the delay lines or power splitting control elements to effectively change the delay time or the coefficients, respectively.

We investigate the OSP applications using the delay control mode rather than the frequency control mode or the coefficients control mode when we discuss its applications in Section 5.3.5. If the frequency control mode is used in an OSP with the electro-optic polymer waveguides, the OSP can tune such optical filters much faster. However, the filter performance will be limited if electro-optic polymers are used in place of silica waveguides due to its intrinsic optical loss. Furthermore, operating our polymer OSP in the coefficients control mode is not particularly interesting, either. The OSP we investigate is classified into the lattice form structure. In the lattice form structure, the c_k and d_k coefficients in Figure 5.24b are complex functions of actual electrical biases, which makes it difficult to find the relationship between its output function and the electrical biases.[13,14] In addition, we have not found any useful and unique applications using this mode partly because the optical loss limits the performance of the OSP building blocks.

One can find that the frequency control mode is based on the same principle as the delay control mode. In Equation 5.27, the delay time τ and the optical frequency ω are multiplied together. Therefore, either affects the transfer function in the same way. We can change the delay time using the electro-optic effect with a fixed input optical frequency (wavelength) or the optical frequency can be changed using a fixed delay time. Either approach will generate the same output response. This is the basic concept for the *arbitrary waveform generator*, we investigate in Section 5.3.5.1 and it becomes more clear with the driving formula of Equation 5.34.

5.3.2 Structure

Since multiple power splitting is hard to obtain and not effective in the optical waveguide, the lattice form structure is considered for implementation of our OSP. The lattice form is a series structure with a certain type of a unit block. The unit block which we have chosen consists of a symmetric MZ interferometer and a racetrack waveguide as shown in Figure 5.26. "Symmetric" means that the lengths of the two waveguide arms of the MZ interferometer are the same. A racetrack structure is used so that the straight waveguide section has an extended coupling region for coupling inside and outside of the ring.

This unit block, originally proposed by Jinguji,[14] has two input ports and two output ports. Any input port can be used for operation while the two output ports are related to

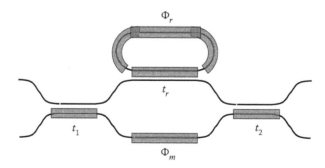

FIGURE 5.26
Unit block of OSP. It is a two-port input and two-port output system consisting of a symmetric MZ interferometer and a racetrack structure. Four electrodes control the locations of a zero and a pole.

each other by a conjugate power relation. A conjugate power relation means that the total sum of two output powers is the same as the input power if the system is lossless.

This structure generates one zero and one pole simultaneously. The degree of freedom to locate a zero or a pole in the Z space is two because a zero or a pole is a complex value, and hence has both amplitude and phase. The locations of the zero and the pole in the Z space are controlled by the four different electrodes. Note that t_1 and t_2 are functionally redundant.

A multiple-block OSP consists of cascaded multiple unit blocks as shown in Figure 5.27. With N cascades of unit blocks, our OSP can generate N zeros and pole pairs. Therefore, the degree of freedom of the N cascaded OSP is $N \times 4$, which are controlled by $N \times 4$ electrodes on top of the waveguides. In the strict sense, the definition of the unit block in Figure 5.26 is wrong if the multiple-block is considered. It should be one of the divided sections in Figure 5.27 except the first block, which is just a configurable coupler. To avoid confusion, we name the structure in Figure 5.26 as a "one-block OSP" and one of the divided sections in Figure 5.27 as a "unit block".

The unit block contains two configurable couplers and two phase shifters. They are the "splitting coupler", the "racetrack coupler", the "MZ phase shifter", and the "racetrack phase shifter", labeled as t_1, t_r, ϕ_m and ϕ_r, respectively, shown in Figure 5.26. They are used for configuring one zero and one pole generated by a unit block and controlled by microstrip electrodes located on top of the waveguides via the electro-optic effect.

In order to have a useful phase shift at the racetrack phase shifter, the perimeter of the racetrack must be large enough. On the other hand, the round trip loss of the racetrack will be too large if the perimeter is large. Therefore, we design the radius of the racetrack to be 1.2 mm. The interaction lengths of the racetrack coupler and the splitting coupler are designed as 2 mm and 6 mm, respectively. The detail design parameters of the individual building blocks or components are discussed in the following sections.

5.3.3 Analysis

As discussed earlier, the one-block OSP generates a single zero and a single pole simultaneously. The degrees of freedom to locate a zero or a pole in the Z space is two since a zero or a pole is a complex value, and thus has both amplitude and phase. Their locations in the Z space are controlled by the four different electrodes shown by the red shaded bars in Figure 5.26. With N cascades of the unit block, an OSP can generate N number of zero and pole pairs as shown Figure 5.27. The poles of the whole OSP system are the same as those of the individual unit blocks. On the other hand, the zeros of the whole system are not the

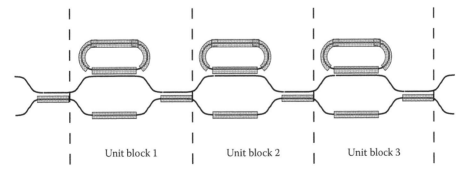

FIGURE 5.27
Multiple structure of OSP. It consists of N cascaded unit blocks.

same as those of the unit blocks since both the output power and the conjugated power of a unit block are cascaded to the next unit block by coupling each other. Because of this problem, it is not trivial to identify the zeros of a multiply cascaded structure. Jinguji et al. demonstrated the synthesis method for analyzing these structures.[14] However, their technique cannot be applied generally to a lossy system since their technique assumed that the unit structure is lossless.

However, dealing with just a one-block OSP is relatively easy and straight forward. By using the scattering matrices, we can derive the scattering parameters, which are useful to understand the operation of the one-block OSP. First, before deriving the scattering matrices, we define the scattering matrices of the individual components. And then, we cascade their matrices to find the scattering matrices of the entire one-block OSP.

5.3.3.1 Configurable Couplers

The matrix form of optical couplers can be derived from coupled mode theory. The form can be summarized as,

$$S = \alpha \begin{pmatrix} t & -j\kappa \\ -j\kappa^* & t^* \end{pmatrix} \tag{5.29}$$

where t and κ ($|t|^2+|\kappa|^2=1$) are the transmission and coupling ratio of each coupler and α represents the optical losses inside the coupler. Note that the matrix should be an unitary matrix if a lossless case is assumed ($\alpha=1$).[31] In a real situation, all these numbers are complex. If we change or tune the coupling ratio, the phases as well as the amplitudes of all values change. Due to the electro-optic effect, they are all functions of the applied voltage.

5.3.3.2 Racetrack

The schematic diagram for a racetrack structure is shown in Figure 5.28a. In order to find the transfer function for the racetrack, we use the scattering matrix of the racetrack coupler, which is Equation 5.29. In the steady state condition, the input electric field, E_{in} and the output electric field, E_{out}, satisfy Equation 5.30.

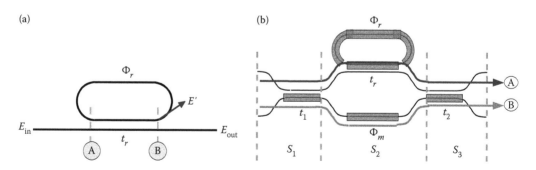

FIGURE 5.28
(a) Racetrack structure with appropriate symbols for mathematical analysis. (b) Unit block of OSP with appropriate symbols for mathematical analysis. The individual scattering matrix transfer function consists of two different terms. One, labeled as "Path A", is associated with the light beam which propagates through the racetrack and the other, labeled as "Path B", is with the light beam which propagates through the other MZ arm.

$$\alpha_r e^{j\tau_c\omega}\begin{pmatrix} t_r & -j\kappa_r \\ -j\kappa_r^* & t_r^* \end{pmatrix}\begin{pmatrix} E'e^{j\tau_r\omega}\alpha_{RT1}e^{j\phi_r} \\ E_{in} \end{pmatrix} = \begin{pmatrix} E' \\ E_{out} \end{pmatrix} \qquad (5.30)$$

where E' is a steady state solution at the starting point of the racetrack as indicated in Figure 5.28a, τ_c and τ_r are the transition time in the coupler (from A to B counter-clockwise) and in the racetrack (from B to A counter-clockwise), and α_r and α_{RT1} is the optical loss factor from A to B and from B to A, respectively. The optical loss factors represent the electric field attenuation on a linear scale and become unity in lossless waveguides. The parameters t_r and k_r denote the transmission and coupling coefficients for the racetrack coupler. The optical loss factor is a measure of electric field intensity attenuation on a linear scale.

We solve Equation 5.30 to find out the transfer function of the racetrack, H_{RT}, which is the complex ratio of E_{out} with respect to E_{in}.

$$H_{RT}(t_r,\phi_r) = \alpha_r e^{j\tau_c\omega}\frac{t_r^* - \alpha_{RT}e^{j\tau\omega}e^{j\phi_r}}{1 - t_r\alpha_{RT}e^{j\tau\omega}e^{j\phi_r}} \qquad (5.31)$$

where $\tau(=\tau_c+\tau_r)$ is the total round trip time, and $\alpha_{RT}(=\alpha_{RT1}\alpha_r)$ is the total round trip loss.

5.3.3.3 MZ Section

In the symmetric MZ section, two light beams propagate independently. Hence, the scattering matrix for the MZ section, S_2 becomes,

$$S_2 = \begin{pmatrix} H_{RT} & 0 \\ 0 & \alpha_m e^{j\phi_m} \end{pmatrix} \qquad (5.32)$$

where α_m and ϕ_m are the optical loss factor and the phase shift in the MZ phase shifter.

5.3.3.4 One-Block OSP

Having defined the individual scattering matrices for the one-block OSP, the scattering matrix, S, for a one-block OSP shown again in Figure 5.28b is the multiplication of the individual scattering matrices.

$$S = S_1 S_2 S_3 = \begin{pmatrix} S_{11} & S_{21} \\ S_{12} & S_{22} \end{pmatrix} \qquad (5.33)$$

where S_1 and S_3 are the scattering matrices for the two splitting couplers and S_2 is the MZ section as shown in Figure 5.28b. The calculation is straight forward and the scattering parameters can be summarized as,

$$S_{lm} = \alpha_{lm}\left(\underbrace{\frac{t_r^* - \alpha_{RT}e^{j\tau\omega}e^{j\phi_r}}{1 - t_r\alpha_{RT}e^{j\tau\omega}e^{j\phi_r}}}_{\text{Path A}} + \underbrace{R_{lm}e^{j\phi_m}}_{\text{Path B}} \right) \qquad (5.34)$$

where,

$$\alpha_{11} = \alpha_1\alpha_2\alpha_r(+1)t_1t_2$$

$$\alpha_{12} = \alpha_1\alpha_2\alpha_r(-j)t_1c_2^*$$

$$\alpha_{21} = \alpha_1\alpha_2\alpha_r(-j)c_1t_2$$

$$\alpha_{22} = \alpha_1\alpha_2\alpha_r(-1)c_1c_2^*$$

$$R_{11} = (-1)\frac{c_1^*}{t_1}\frac{c_2}{t_2}\frac{\alpha_m}{\alpha_r}e^{-j\tau_c\omega}e^{j\phi_m}$$

(5.35)

$$R_{12} = (+1)\frac{c_1^*}{t_1}\frac{t_2^*}{c_2^*}\frac{\alpha_m}{\alpha_r}e^{-j\tau_c\omega}e^{j\phi_m}$$

$$R_{21} = (+1)\frac{t_1^*}{c_1}\frac{c_2}{t_2}\frac{\alpha_m}{\alpha_r}e^{-j\tau_c\omega}e^{j\phi_m}$$

$$R_{22} = (-1)\frac{t_1^*}{c_1}\frac{t_2^*}{c_2^*}\frac{\alpha_m}{\alpha_r}e^{-j\tau_c\omega}e^{j\phi_m}$$

As seen in the under brackets in Equation 5.34, the individual scattering matrix transfer function consists of two different terms. One, labeled as "Path A", is associated with the light beam which propagates through the racetrack and the other, labeled as "Path B", is for the light beam which propagates through the other MZ arm. The Path A and B terms correspond to two paths indicated in Figure 5.28b. Therefore, the Path A term contains the transfer function of the racetrack while the Path B term is independent of the optical delay line formed by the racetrack.

The two terms are summed at the output port depending on the coupling ratio of the two splitting couplers. The amplitude of R_{lm} stands for the normalized intensity of Path B with respect to the maximum intensity of Path A. Note that the amplitude and the phase of Path B beam of light are controlled by the two splitting couplers and the ϕ_m electrode, respectively and that the transfer function and the resonant wavelength of the racetrack (Path A light beam) are controlled by the t_r and the ϕ_r electrodes, respectively. The mathematical representation in Equation 5.34 is useful for an intuitive and physical understanding of the one-block OSP structure and it is used when we verify the OSP experimentally in Section 5.3.4.6.

Another useful and more mathematical way to represent the scattering matrices is using the concept of poles and zeros. Further simplification of Equation 5.34 results in:

$$S_{lm} = A_{lm}\left(\frac{1-\rho_{lm}e^{j\tau\omega}e^{j\phi_r}}{1-\gamma e^{j\tau\omega}e^{j\phi_r}}\right)$$

(5.36)

where

$$A_{lm} = \alpha_{lm}(t_r^* + R_{lm})$$

$$\gamma = \alpha_{RT}t_r$$

(5.37)

$$\rho_{lm} = \gamma\frac{1+t_rR_{lm}}{|t_r|^2+t_rR_{lm}}$$

All transfer functions have the same pole, γ, while ρ_{lm} is the zero obtained from the S_{lm} scattering matrix element. As we discuss before in Section 5.3.1.2, using poles and zeros is convenient to understand the output characteristics of the OSP.

Note that the phase of γ defined in Equation 5.36 is not configurable. The definition of the pole should include γ as well as either $e^{j\theta_r}$ or $e^{j\omega\tau}$ in Equation 5.36 because the two terms contribute to the pole phase. However, we define our pole as shown in Equation 5.36 since it is convenient to understand two similar operation modes; frequency control mode or delay control mode. In this way, the similarity between two operation modes of the OSP, as discussed in Section 5.3.1.3, becomes clear. In the frequency control mode, the $e^{j\tau\omega}$ term in Equation 5.36 is varying and represents the unit delay Z^{-1} in the Z space. In this case, the actual pole becomes $\gamma e^{j\theta_r}$, thus, the pole phase is controlled by the ϕ_r electrode. From a practical point of view, the frequency response of the OSP is periodic with respect to the FSR and the response shifts in the frequency domain with respect to the ϕ_r value. On the other hand, in the delay control mode, the delay is controlled by $e^{j\theta_r}$ in Equation 5.36, which represents the unit delay Z^{-1} in the Z space. In this case, the actual pole becomes $\gamma e^{j\omega\tau}$, thus, the pole phase is controlled by the input optical frequency (ω) With a similar analogy from a practical point of view, the amplitude and phase response of the OSP is periodic with respect to ϕ_r and the response shifts in the ϕ_r (or voltage) domain with respect to the input optical frequency. Note also that the phase of the zero also depends on both parameters in the same manner. By changing any of two parameters, the phases of both pole and zero are changing in the same amount. However, the zero has additional configurable degrees of freedom such as t_1, t_2 and ϕ_m as seen in Equation 5.36.

5.3.3.5 Generality of One-Block OSP

As we discussed earlier, the one-block OSP has one zero and one pole. If the zero can be located in entire space and the pole can be located in any region within the unit circle in the Z space, the generality condition is satisfied.

The amplitude of a pole depends on the total round trip loss factor (α_{RT}) and transmission coefficient of the racetrack coupler (t_r) as seen in Equation 5.36. In the lossless case when α_{RT} becomes unity, the pole can be located at any point within the unit circle if t_r can be adjusted from 0 to 1. However, polymer waveguides intrinsically has certain amount of intrinsic optical loss and the configurable amount of t_r is difficult to design to have complete configurability. In this case, the actual implementable amplitude of the pole locations becomes limited. The phase of the pole depends on the optical phase change controlled by the ϕ_r phase shifter. Since the interaction length of the ϕ_m is designed 912 mm, 360° of the optical phase change can be obtain with applied voltage, hence any arbitrary phase of the pole is possible.

Once we determine the pole location, the zero depends on the transmission coefficients of the splitting couplers and optical phase shift in the MZ phase shifter as seen in Equation 5.36. Since the splitting couplers are designed long enough to have as large tuning ratio as possible and the interaction length for the ϕ_m electrode is relatively long (5 mm), the phase and amplitude of the zero can cover almost entire range of the Z space.

We discuss the generality of the fabricated OSP with experimental data in Section 5.3.4.5.

5.3.4 Verification and Operation

There are several individual components or building blocks of the OSP. They are the race-tracks, configurable couplers, and phase shifters. The OSP will function correctly once all these components are working correctly. Therefore, it is important to characterize and verify the individual components before integrating them. In order to verify the components, we also fabricate individual components. In this section, we discuss the experimental results for individual components" verification.

5.3.4.1 Racetrack

First, we characterize the racetrack, which is one of the crucial components in an OSP. Figure 5.29 shows the spectral response of a racetrack and its experimental measurement setup. The racetrack measured has the same design as the OSP, where the bending radius of the racetrack is 1.2 mm and the interaction length of the racetrack coupler is 2 mm. An Ando AQ4321D is used for the tunable laser source at 1.5 μm with TM mode polarization control. As seen in Figure 5.29b, the measurement is done in a 0.6 nm wavelength span and the FSR, extinction ratio, finesse and Q-factor (loaded) are measured 0.12 nm (15.5 GHZ),18 dB, 3.36 and 4.34×10^4, respectively. Using these values, the effective group refractive index, α_{RT} value, t_r value, total round trip optical loss and the optical loss inside the ring are calculated as 1.66, 0.608, 0.535, 4.4 dB and 3.85 dB/cm, respectively. Figure 5.29b also shows the simulated spectral response using these values. Since the propagation loss of the straight waveguide is measured around 2 dB/cm, the excess loss inside the racetrack is expected to be from scattering loss due to the roughness at the interface between core and cladding materials. A similar propagation loss inside the racetrack has been obtained using 1 mm bending radius and similar electro-optic polymer material.[32]

We drive voltages of ±20 V on the t_r electrode to verify the operation of the racetrack coupler as seen in Figure 5.30a. Figure 5.30b shows the spectral response while varying the driving voltages. The spectral response when no voltage is applied is shown as well for purpose of a comparison. As seen in Figure 5.30b, the response (extinction ratio) is changed depending on the voltage. By fitting the measured response, we find that t_r becomes 0.535 ± 0.25 with an applied voltage to the racetrack coupler of ±20 V. We also find that the local minima shift with applied voltage.

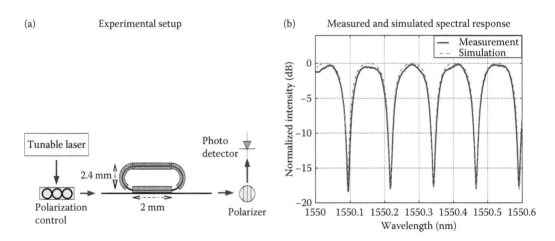

FIGURE 5.29
Experimental setup and spectral response of a racetrack which has the same design as the OSP used. The free spectral range (FSR), extinction ratio, finesse and Q-factor are measured 0.12 nm (15.5 GHz), 18 dB, 3.36 and 4.34×10^4, respectively.

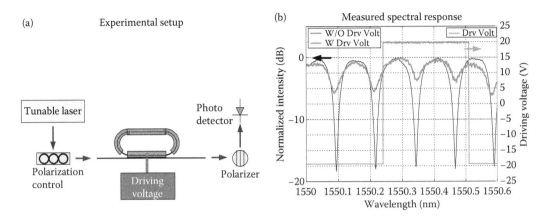

FIGURE 5.30
Experimental setup and spectral response of a racetrack when ±20 V is applied to the racetrack coupler. By fitting the measured response, we find that t_r becomes 0.535 ± 0.25 with an applied voltage to the racetrack coupler of ±20 V.

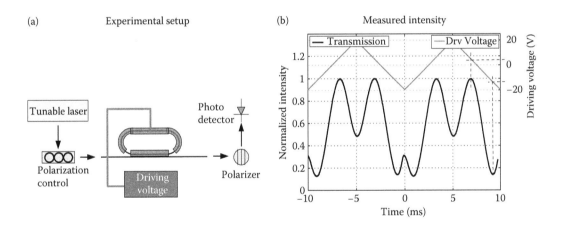

FIGURE 5.31
Intensity response of a racetrack when ±20 V peak-to-peak triangular signal is applied to the racetrack phase shifter. The measurement shows that the half-wave voltage (V_π) of the ϕ_r phase shifter is 18.3 V. The corresponding r_{33} coefficient is also calculated as 23 pm/V.

For verification of the racetrack phase shifter, we fix the input optical wavelength at 1.5 μm and apply ±20 V peak-to-peak triangular signal to the ϕ_r electrode. Its experimental setup and response are shown in Figure 5.31a and b, respectively. The measurement shows that the half-wave voltage (v_π) of the racetrack phase shifter is 18.3 V. The corresponding r_{33} coefficient is also calculated as 23 pm/V.

5.3.4.2 Configurable Couplers

Two different types of couplers are considered in the OSP. They are the racetrack couplers and the splitting couplers. For the racetrack couplers, the separation widths of 4.5 μm, 4.6 μm and, 4.5 μm are considered between two waveguides and their interaction length is 2 mm. The splitting coupler has a separation of 5.1 μm and interaction length of 6 mm.

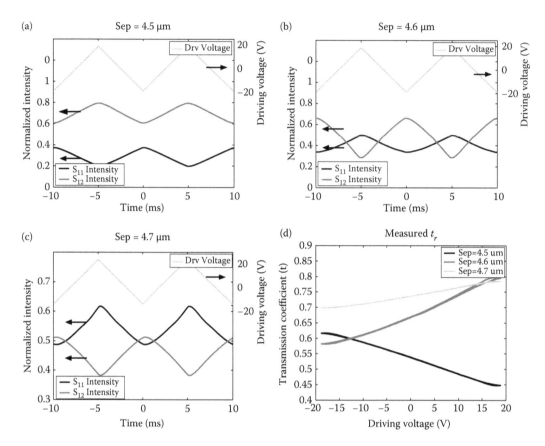

FIGURE 5.32
Experimental measurements of S_{11} and S_{12} scattering parameters for the different racetrack couplers.

Figure 5.32 shows the measured S_{11} and S_{12} responses of the racetrack couplers. From the measured response, we calculate the transmission coefficients for the couplers, which are also shown in Figure 5.32d. Comparing to the computer simulations, the measurements show that the measured transmission coefficients are smaller than the simulated values. Furthermore, the different polarity of voltages leads to different output response even though output responses should be even functions with respect to applied voltages since the waveguides are symmetric. This implies that two waveguides inside the couplers are mismatched already due to imperfect fabrication, which has also been found in conventional electro-optic MZ devices.[33]

Figure 5.33 shows measured S_{11} and S_{12} responses of the splitting couplers. Since the interaction length is designed large enough (6 mm), both outputs reach their maximum and minimum intensities during the application of the triangular voltage waveform on the electrodes. The turn on/off voltage of the splitting coupler is 30 V and its extinction ratio is approximately 10 dB, implying that its transmission coefficient, t_1 or t_2, can be configurable between 0.4 and 0.9 with an applied voltage of 30 V.

5.3.4.3 MZ Phase Shifter

The MZ phase shifter (or ϕ_m phase shifter) is for tuning the phase of Path B in Figure 5.28b. For verification of the MZ phase shifter, we also apply a ±20 V peak-to-peak triangular

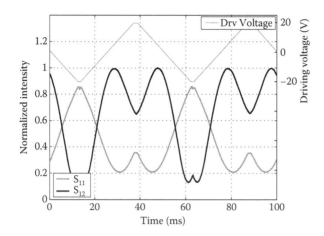

FIGURE 5.33
Experimental measurements of S_{11} and S_{12} scattering parameters for the splitting coupler.

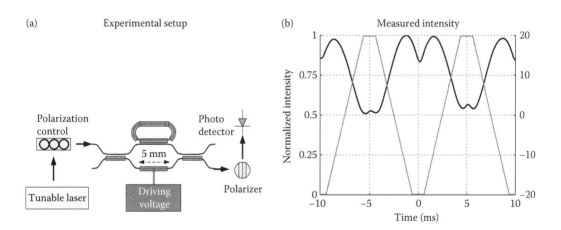

FIGURE 5.34
Intensity response of a racetrack when a ±20V peak-to-peak triangular signal is applied to the ϕ_m electrode. The measurement shows that the half-wave voltage (V_π) of the MZ phase shifter is 33 V. The corresponding r_{33} coefficient is calculated as 20 pm/V.

signal to the ϕ_m electrode. Its experimental setup and response are shown in Figure 5.34a and b, respectively. The measurement shows that the half-wave voltage (V_π) of the MZ phase shifter is 33 V. The corresponding r_{33} coefficient is calculated as 20 pm/V.

5.3.4.4 Summary of One-Block OSP

Based on the discussion in Section 5.3.4, we summarize the parameters and their configurable range with applied voltages in Figure 5.35.

- The FSR of the racetrack is 0.12 nm (15.5 GH$_Z$) and the α_{RT} value is constant 0.608. The transmission coefficient, t_r, varies between 0.535−0.25 and 0.535+0.25 with ±20 V applied voltage to the t_r electrode.

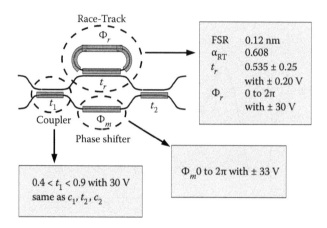

FIGURE 5.35
Summary of one-block OSP. Based on the measurement of individual components, configurable parameters are summarized with their configurable amount and driving voltages.

- The phase, ϕ_r, can be fully configurable (0–2π) with $\pm$30 V applied voltage to the ϕ_r electrode.
- The phase, ϕ_m, of the MZ phase shifter can be fully configurable (0–2π) with $\pm$33 V applied voltage to the ϕ_m electrode.
- The splitting coupler, t_1 and t_2, can be configurable between 0.4 and 0.9 with 30 V.

5.3.4.5 Pole/Zero Locations Using One-Block OSP

As we discussed earlier, a one-block OSP generates one pole and one zero. By applying voltages on the electrodes, and hence, reconfiguring and changing the biases of the OSP components, we change the locations of the pole and the zero. Based on the discussion in Section 5.3.4, we can find possible locations of the pole and the zero of the fabricated one-block OSP.

The location of the pole is determined by $\alpha_{RT}t_r$ as derived in Equation 5.36. α_{RT} is a constant parameter while t_r is configurable using the racetrack coupler. From the components' summary in Figure 5.35, we find the amplitude of the pole $|\gamma|$ can be between 0.17 and 0.48. The phase of the pole can be fully configured with either ϕ_r or ω depending on the operation mode. (The detail of operation mode is discussed in Section 5.3.1.3 and the pole phase in Section 5.3.3). Therefore, the pole can be located and configured inside the shaded area in Figure 5.36.

On the other hand, location of the zero depends on that of the pole. According to Equation 5.36, the zero has an offset from the pole; $(1+t_rR_{lm})/(|t_r|^2+t_rR_{lm})$. Since this offset is a complex number as well, it is a vector in the Z space. Location of the zero is determined by this offset vector. The offset vector depends on two parameters; R_{lm} and t_r. First, assume that $|R_{lm}|$ is infinity implying all the input light goes through Path B and there is no power flow in Path A. Path A and Path B are shown in Figure 5.28b. In this case, the offset vector becomes 1 and the location of the zero is same as that of the pole, thus the zero cancels the pole and there will be no zero and no pole. If $|R_{lm}|$ is 0, then, the offset vector becomes real number $1/|t_r|^2$ and the zero will be α_{RT}/t_r^*, which is the same zero which the transfer function of the racetrack structure represents as in Equation 5.31.

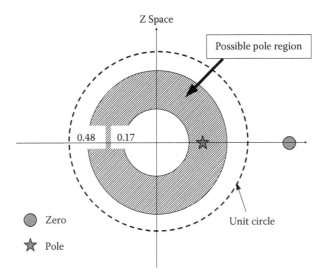

FIGURE 5.36
Pole locations in Z space which can be implemented by one-block OSP. The amplitude of poles is bounded from 0.17 to 0.48 while the pole phase is configured from 0 to 2π as indicated in the shaded region.

However $|R_{lm}|$ is bounded due to the bounded transmission coefficients of the splitting couplers. As in Equation 5.34, $|R_{lm}|$ is the electric field amplitude ratio between Path A and Path B, which is configured by the splitting couplers. From the bounded transmission coefficients of the splitting couplers, we find the maximum and minimum values of $|R_{lm}|$.

$$0.2 < |R_{lm}| < 5 \tag{5.38}$$

On the other hand, the phase of R_{lm} depends on ϕ_m. Since ϕ_m can be configured completely from 0 to 2π by ± 33 V as shown in Figure 5.35, the phase of R_{lm} is also configured completely from 0 to 2π. Therefore, the zero can be located in the entire Z space except near two points, which are the pole and the zero of the racetrack (α_{RT}/t_r^*). The shaded area in Figure 5.37a shows conceptually possible zero locations when the pole and the zero of the racetrack are given as shown. We use computer simulations to find the possible locations of the zeros that can be implemented by the fabricated one-block OSP. Figure 5.37b, c, and d shows their results in the Z space when the pole has the minimum (0.17), middle (0.325), and maximum (0.48) possible amplitude, respectively. For three plots, a dot closer to the origin and the other dot are indicating the pole and the zero of the racetrack, respectively. The zero cannot be located inside the two circles, which indicate boundaries set by the bounded $|R_{lm}|$. As the pole locates near to the unit circle, the zero can be located throughout the entire Z space.

5.3.4.6 DC Operation

The multiple-block OSP is complicated and it is difficult to understand its operation intuitively. However, the one-block OSP is somewhat simple and it is possible to understand its operation with physical insights. The output of a one-block OSP consists of two different light paths as seen in Figure 5.28b. Path A is through the racetrack structure while Path B is through the other straight MZ branch. The amplitude and phase responses of Path A

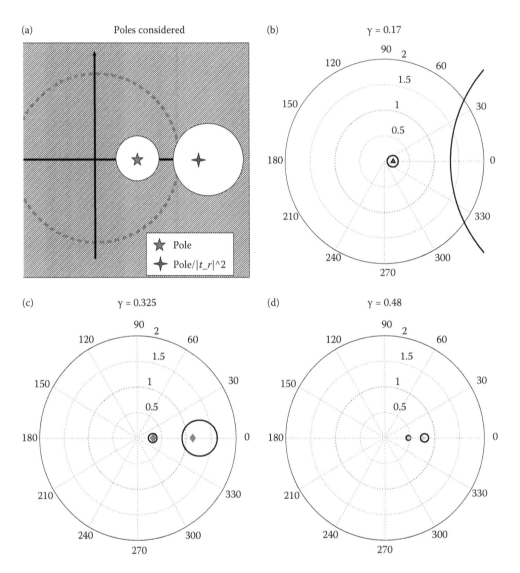

FIGURE 5.37
The shaded area in (a) shows conceptually possible zero locations when the pole and the zero of the racetrack (pole/|t_r|²) are given. (b), (c), and (d) show simulation results of the possible zero locations in the Z space when the pole has the minimum (0.17), middle (0.325), and maximum (0.48) amplitude, respectively. For three plots, a dot closer to origin and the other dot are indicating the pole and the zero of the racetrack, respectively. The zero cannot be located inside the two circles, which indicate boundaries set by the bounded |R_{lm}|.

are same as those of ring resonators. On the other hand, the amplitude response of Path B is only a function of splitting couplers (t_1 or t_2) and its phase response is independent of ϕ_r. Figure 5.38 summarizes an intuitive understanding of the roles of bias change on the operation of the one-block OSP. The splitting couplers (t_1 or t_2) change mostly the extinction ratio of the intensity response of a one-block OSP. The change of ϕ_m changes the shape of the intensity response of a one-block OSP. We also find that the racetrack coupler (t_r) changes mostly the sharpness of the amplitude and phase response of a one-block OSP.

Figure 5.39 shows various amplitude responses of a one-block OSP with different bias values and its experimental setup. We apply a ±20 V peak-to-peak triangular voltage signal

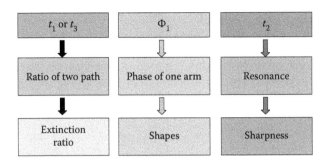

FIGURE 5.38
Summary of biases' role.

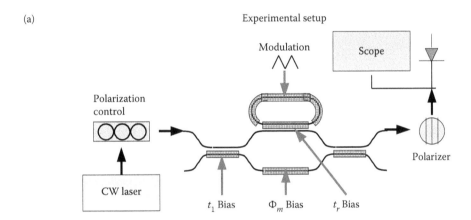

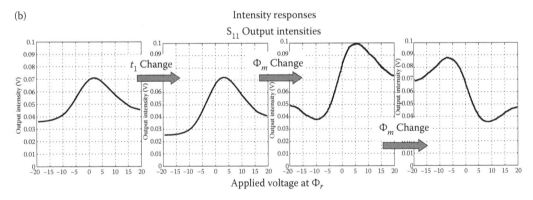

FIGURE 5.39
Various intensity responses of one-block OSP and its experimental setup when different biases are applied. The splitting couplers (t_1 or t_2) mostly change their extinction ratios while the MZ phase shifter (ϕ_m) mostly changes their shapes.

to the ϕ_r electrode and different bias voltages to the t_1, ϕ_m, and t_r electrodes. A continuous laser source at 1.55 μm is used with a polarization control to launch TM mode light into the input of the OSP and its S_{11} output port is measured with an oscilloscope, which is connected to a photodetector. Using the X–Y display feature of the oscilloscope, we see the amplitude response of the OSP in Figure 5.39b.

The first plot in Figure 5.39b shows the amplitude response when no biases are applied to the OSP. It is one of the typical response shapes of the one-block OSP. The second plot in Figure 5.39b shows the amplitude response when a different voltage is applied to the t_1 electrode. Due to the change of the t_1 value, the second plot shows an extinction ratio of the response changes as expected. The third and fourth plots in Figure 5.39b shows another amplitude response when different voltages are applied to the ϕ_m electrode. Due to the change of the ϕ_m value, the plots show changes of the response shapes as we expected.

5.3.5 Applications

Due to its intrinsic arbitrariness, the OSP can be used for many applications. Applications presented here are for high speed situations, for which a conventional PLC cannot be used. As for the potential high speed analog applications, we investigate *arbitrary waveform generators* and *true-time delay elements* using our one-block OSP. For discrete-time applications, *optical pulse code generator* and *optical discrete-time signal processor* will be introduced.

5.3.5.1 Arbitrary Waveform Generator

It is known to be possible to implement arbitrary optical filters using PLC structures such as a notch filter, a linear dispersion filter, and a band-pass filter.[11–14] Such filters have been experimentally investigated in silica waveguides, where thermal tuning was used to change the index of refraction. Since the OSP employs the PLC for its structure and it is based on the fast electro-optic effect, much higher data rates (more than tens of gigahertz rate) can be accessible using the OSP. Therefore, not only is the OSP useful in fast reconfigurable optical filters, but also the OSP can be used for high speed arbitrary waveform generators. High speed arbitrary waveform generators are useful for modulator linearization and correction of amplifier distortion.

Assume that we apply a sawtooth signal to a one-block OSP at the ϕ_r electrode and its peak-to-peak voltage is two times the half-wave voltage of the racetrack phase shifter. Then, during one cycle of the voltage signal, the output amplitude response of the OSP takes on the optical filter shape as a function of time. By changing other biases, the output response will also change, and hence the OSP generates arbitrary waveforms. The degree of arbitrariness of the generated signal depends on the number of unit blocks if a multiple-block OSP is used and the generality of the OSP.

The detailed concept and theory of the arbitrary waveform generators have been investigated by Fetterman and Fetterman.[34] If we summarize,[34] its principle of operation is based on the similarity of two operation modes of the OSP as we discussed in Section 5.3.1.3; frequency control mode and delay control mode. According to Equation 5.34, the transfer function, S_{lm}, depends on both optical frequency and ϕ_r. If the OSP is operated with a fixed ϕ_r value and varying optical frequency (frequency control mode), the OSP performs optical filter as described elsewhere.[11–14] On the other hand, if the OSP is operated with a fixed optical frequency and varying ϕ_r value (delay control mode), the response of the OSP with respect to ϕ_r should be the same as an optical filter shape.

Since the OSP is based on high speed electro-optic polymer, it can generate high speed arbitrary waveforms. However, a sawtooth signal is difficult to obtain at high frequencies with high power.[35] Instead, we examine a simple sinusoidal signal input. Using a sinusoidal input, the OSP generates the desired output shape with proper adjustment of t_1, t_r, and ϕ_m.

Figure 5.40 shows a measured rectangular signal generated by the one-block OSP and its experimental setup. The continuous laser source at 1.55 μm is applied to the one-block OSP

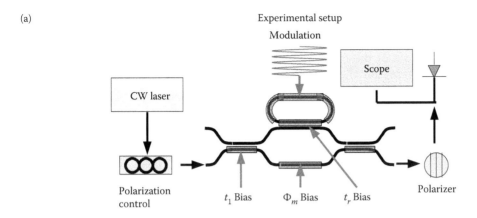

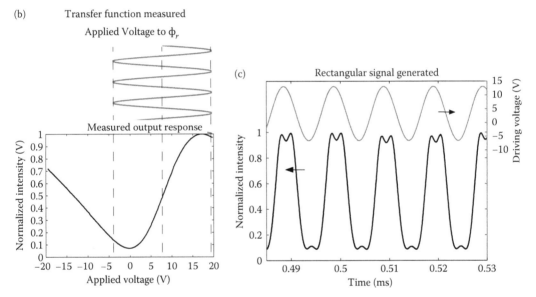

FIGURE 5.40
Rectangular signal generation using a one-block OSP and its experimental setup. (a) A ±10 V peak-to-peak sinusoidal voltage input with a proper offset bias is applied to the ϕ_r electrode and other biases are properly adjusted by voltage supplies. (b) We obtain a proper transfer function for the rectangular signal generation. (c) We utilize the sharp transition region in the transfer function to generate a rectangular output signal.

with polarization control and the TM mode output of S_{22} is measured by the photodetector and the oscilloscope. First, we apply a ±20 V peak-to-peak sinusoidal voltage input to the ϕ_r electrode. When the other biases are properly adjusted by voltage supplies as shown in Figure 5.40a, we obtain the OSPs transfer function as a function of the applied voltage as shown in Figure 5.40b. We then apply a ±10 V peak-to-peak sinusoidal voltage input with a proper offset bias to the ϕ_r electrode while the other biases are properly adjusted. We utilize the sharp transition region in the transfer function in Figure 5.40b to generate the sharper rectangular output signal. As shown in Figure 5.40c, a rectangular voltage signal is obtained. As the number of unit blocks increases, a more rectangular shape is possible. Furthermore, this waveform can be quickly changed to another desired shape with different sets of parameters due to the fast electro-optic effect. In Section 5.3.5.2, we use a similar transfer function from Figure 5.40b to generate another linear signal.

5.3.5.2 Linearized Modulator

Another useful application of the OSP is a linearized electro-optic modulator. Electro-optic modulators are one of the most important devices of lightwave communications. The most common scheme for this device is the use of a MZ interferometer. However, the inherent disadvantage of this technique is a large nonlinear distortion that limits the dynamic range in analog applications.[36] Several efforts have been performed to increase the dynamic range of the optical modulator including dual-polarization techniques,[37] parallel or series cascaded configurations,[38,39] and electronic predistortion schemes.[40] A dual-section directional coupler modulator using electro-optic polymer waveguides has also been developed.[41]

The one-block OSP can also perform as a linearized amplitude modulator if the applied electric field modulates the optical phase inside the racetrack (delay control mode). When multiple coherent lights interfere, the overall intensity response is nonlinear (sinusoidal) as the optical phase changes in one of the interfered light beams. Therefore, the intensity response in a simple MZ structure is always nonlinear since the applied electric field changes the optical phase of the interfered lights in a "linear" way. The fundamental concept in the linearized modulator using the one-block OSP lies in the OSP's "nonlinear" response to the optical phase in the applied electric field. The optical phase change inside the racetrack recursively changes the overall optical phase leading to a nonlinear response, which compensates the nonlinear response of the phase modulation under proper conditions, thus, generating a "linear" intensity response overall. Note that this principle is analogous to that of the arbitrary waveform generator as discussed in Section 5.3.5.1. The linear amplitude response is a specific kind of arbitrary waveform generation.

A ring resonator assisted MZ (RAMZ) structure, which is similar to the one-block OSP, has been proposed to function as a linearized modulator[42] and has been studied in detail including the influence of optical loss.[43] Studies have found that the higher-order nonlinear harmonic terms can vanish (up to 5th order) with proper design of the waveguide structures. As seen in Equation 5.36, the transfer function with respect to the ϕ_r phase change depends on the pole and the zero. Therefore, its linearity can be calculated by a common Taylor expansion technique using various poles and zeros. The first and higher-order harmonic terms are calculated by:

$$P_n = \frac{d^n \, |S_{lm}|^2}{d\phi_r^n} \tag{5.39}$$

where P_n is the nth high-order harmonic term and n is an integer larger than 0. Instead of using the analytical Taylor expansion technique, we utilize a numerical method to calculate the high-order nonlinear harmonic terms with various poles, zeros and biases. As a result, we find the most linear region of the response (in terms of smallest higher-order terms[43]) when the pole, the zero, and the bias are $0.27\angle -1.037$ rad, $1\angle 0$ and 2.2019 rad, respectively. In this case, the third, fourth, and fifth harmonic terms vanish while the first and second harmonic terms are calculated as $0.287/\mathrm{rad}^1$, $0.001/\mathrm{rad}^2$, respectively.

The simulated normalized transfer function using these values are shown in Figure 5.41a, where the linear bias is indicated with a red thick line. The normalized transfer function of a conventional MZ modulator is also shown for the purpose of comparison. The corresponding pole/zero diagram is also shown in Figure 5.41b. The calculation shows

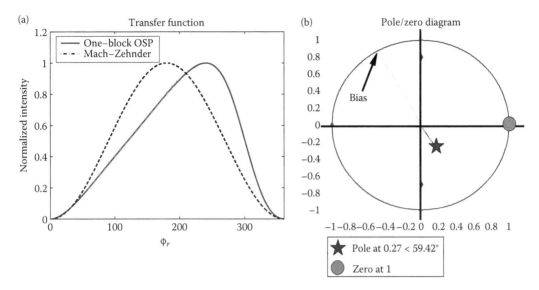

FIGURE 5.41
(a) Simulated linearized transfer function using a one-block OSP. It is also compared with that of a conventional MZ modulator. (b) Pole/zero and bias location for the linearized modulation. The zero, the pole, and the bias are located in 1, 0.27 < −1.037 rad, and 2.2019 rad, respectively.

that the most linearized modulation occurs away from the pole location, implying that a small resonance is required for linearization. The relatively small correction of the nonlinear response from phase change is sufficient. Therefore, the optical loss issue is somewhat mitigated in the linearized modulator. In order to demonstrate a linearized modulator, we use the same experimental setup as shown in Figure 5.40 except that we utilize the linear slope of the transfer function and we apply a triangular signal rather than a sinusoidal signal since a triangular signal allows easier verification of linearity. As shown in Figure 5.42c, the output response linearly follows the input signal.

However, more rigorous methods should be applied to check the linearity of a modulator such as two-tone test measurement method.[44]

5.3.5.3 True-Time-Delay Element

During the last few years, photonic true-time-delay (TTD) units have been widely investigated.[16,45–48] Several different techniques have been applied to implement TTD including fiber grating prisms,[45] fiber Bragg gratings,[46] single sideband modulation approaches,[47] hybrid techniques[48] and ring resonators.[16] Using resonance in the racetrack of the one-block OSP, we can implement a TTD element.[49] The TTD element using an one-block OSP is classified into ring resonators and optical all-pass filters (APF, see Lenz et al.[16] for its definition).

We perform a computer simulation to estimate the operation of our one-block OSP for the TTD application. For the simulation, we consider a one-block OSP, which has all parameters matched to the experimentally obtained parameters summarized in Figure 5.35. The transmission coefficient, t_r, is considered unchanged at 0.535. We use the FTT function in Matlab with the sampling rate of 1 MHz for the Fourier transformation. First, a Gaussian optical pulse with 1.06 ns FWHM pulse width (assumed transform-limited,

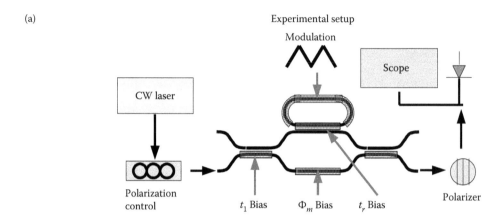

(a)   Experimental setup

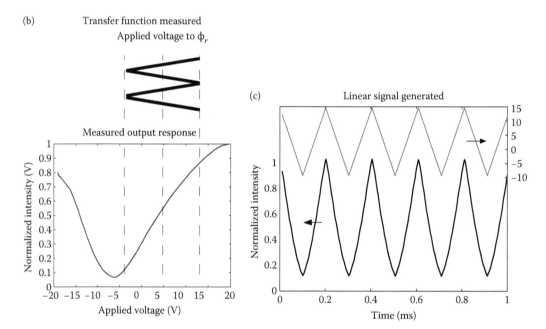

FIGURE 5.42
Linear signal generation using a one-block OSP and its experimental setup. (a) A ±12 V peak-to-peak triangular voltage input with a proper offset bias is applied to the ϕ_r electrode and the other biases are properly adjusted by voltage supplies. (b) We obtain a proper transfer function for linear signal generation. (c) We utilize the linear transition region in the transfer function to generate a linear output signal.

FWHM bandwidth of 0.2 GHz) is launched into the one-block OSP and then the voltage applied to the electrode on top of the ring waveguides is set to 0 and V_π. Figure 5.43 shows the intensities of the output pulses with different delay times of 123 ps and 33 ps, respectively. Hence, 90 ps difference between maximum and minimum time delay is obtained. The insertion loss is approximately 10 dB and the shape and the insertion loss of the output pulses are almost the same regardless of the delay time.

Figure 5.44a shows the experimental setup for the TTD measurement using the one-block OSP. In order to generate an input optical pulse for the OSP, we use a mode-lock pulse laser (Clark-MXR Inc.). The optical pulse from the mode-lock laser is set to generate

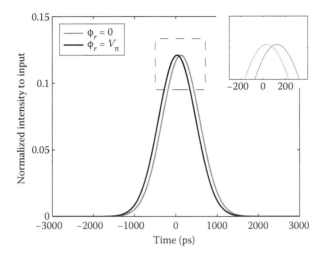

FIGURE 5.43
Simulated intensities of the output optical pulses when a Gaussian optical pulse with 1.06 ns pulse width (FWHM) is launched into the one-block OSP. The difference between maximum and minimum time delay of 90 ps is obtained by applying a different voltage to the ϕ_r electrode.

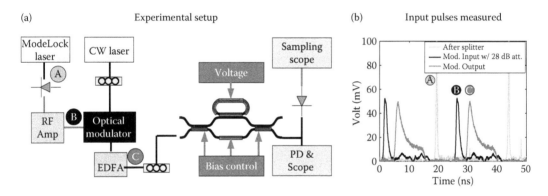

FIGURE 5.44
TTD measurement experimental setup. (a) Experimental setup. (b) Photograph of laboratory setup. (c) Input optical pulses corresponding to the points labeled in (a).

a short electrical pulse, which is shown and labeled as A in Figure 5.44a and b. We then amplify the electrical pulse with an RF amplifier, which has a bandwidth from 1 MHz to 1 GHz. The amplified pulse is shown and labeled as B in Figure 5.44a and b after 28 dB attenuation because the maximum peak voltage of the amplified pulse is about 2 V, which is too high for detection with our sampling scope. We apply the amplified electrical pulse signal as the driving input of an optical LiNbO$_3$ amplitude modulator (Lucent) and obtain a few nanosecond optical pulses at the modulator output, which is used for the input of the OSP after optical amplification from an erbium-doped fiber amplifier (EDFA). The input pulse for the OSP is also shown and labeled as C in Figure 5.44a and b.

By using multiple voltage supplies, a function generator and oscilloscope, we find a bias condition which satisfies the APF condition. Once the APF condition is met, the intensity response seen on the oscilloscope is constant as a function of voltage at the ϕ_r electrode. Next, the output optical pulses from the OSP are measured by a sampling oscilloscope,

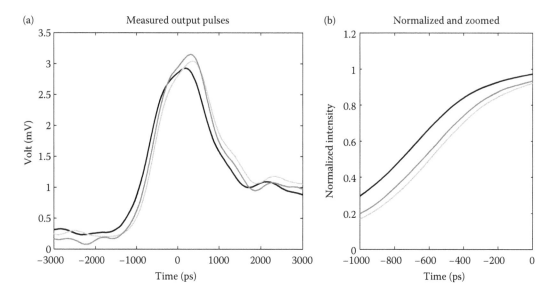

FIGURE 5.45
Output pulses from the TTD measurement setup with different voltages. (a) The output pulses detected at the sampling oscilloscope and (b) their normalized and zoomed plots. The outputs have similar output shape and insertion losses, and they are time-delayed each other. Maximum delayed amount measured is 135 ps.

triggered by the mode-lock laser. Figure 5.45a shows the output signals measured by the sampling oscilloscope with three different voltages applied to the ϕ_r electrode. As seen in Figure 5.45a, outputs have similar shapes and are time-delayed with respect to each other. Figure 5.45b also shows a close-up of the normalized output signals. The maximum delayed amount measured is 135 ps corresponding to a free space delay of 4 cm, which is close to our simulation result.

5.3.5.4 Discrete-Time Applications of OSP

The processing and handling of discrete-time signals is already popular in electronic circuits due to their well-known flexibility and has also already been adapted to optical fiber communications. If the optical pulse is used for the input of an OSP and its pulse width is the same as the optical round trip time, the OSP processes a discrete-time signal. Since the OSP is directly based on the concept of the discrete-time signal processing,[29] this theory can also applied to the OSP, which generates many interesting applications. In this section, we discuss the pulse code generator and optical discrete-time signal processor.

If an optical pulse is used for the input of an OSP and its temporal pulse width is much shorter than the optical round trip time, the OSP can generate arbitrary optical pulse codes. In the previous section, we find that an OSP can be represented as a Z-transformed signal of $H(Z)$. We can expand this as an infinitely numerated form:

$$H(Z) = a_0 + a_1 Z^{-1} + a_2 Z^{-2} \dots \tag{5.40}$$

Since Z^{-1} stands for one unit delay as we discussed before, this OSP generates pulse codes with coefficients of a_0, a_1, a_2, a_3 and so forth. These coefficients or sequence numbers are complex in general, having amplitudes and phases, which are interpreted as the amplitudes and phases of the pulses. As in the case of analog applications, the individual parameters

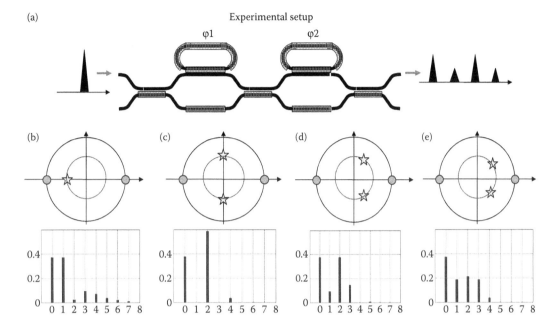

FIGURE 5.46
Pulse codes are generated using the OSP. For example, four different pulse codes are shown using a two-block OSP in this figure with four different pole phases and same zeros as in the pole/zero diagrams (stars for poles and dots for zeros) from (a) to (d). Intensities of the generated pulses are normalized to the input pulse intensity. In (a), the parameters are set to generate zeros and poles at (−1, 1) and (−0.5, −0.5) and the generated intensities are (0.37, 0.37, 0.02, 0.09, 0.07, 0.03, 0.02, …).

of the OSP determine these coefficients (pulse sequence) all together and the complexity of the generated pulse sequence depends on the number of OSP blocks.

Figure 5.46 shows a simulation result of the generated pulse codes from a two-block OSP. In this simulation, four different pulse codes are generated with four different pole phases while maintaining the zeros as in the pole/zero diagrams from Figure 5.46b through e. Since a two-block OSP is assumed, two zeros and two poles are configurable. The individual subfigures show the pole/zero diagrams denoting their locations in the complex Z domain and also show the intensity of the generated pulses sequentially, which are normalized to the input pulse intensity. In Figure 5.46b, the parameters are set to generate zeros and poles at (−1,1) and (−0.5, −0.5) and the generated intensities are (0.37, 0.37, 0.02, 0.09, 0.07, 0.03, 0.02, …). In Figure 5.46c, the poles are changed to (−0.5*j*, −0.5*j*) while maintaining the zeros and the generated intensities are (0.38, 0, 0.59, 0, 0.04,0, …) (likewise in Figures 5.40 and 5.46d and e).

Since pulse code generators can be operated at GHz rates, it will be useful in generating security codes of optical pulses and in general optical communications.

If *x(n)* and *y(n)* are used to denote an input and output discrete-time signal respectively, the OSP behaves like an operator to transform *x(n)* into *y(n)*. Consider a system where the input and output satisfy an Nth-order linear constant coefficient differential equation of the form:

$$y(n) - \sum_{k=1}^{N} b_k y(n-k) = \sum_{k=0}^{M} a_k x(n-k) \tag{5.41}$$

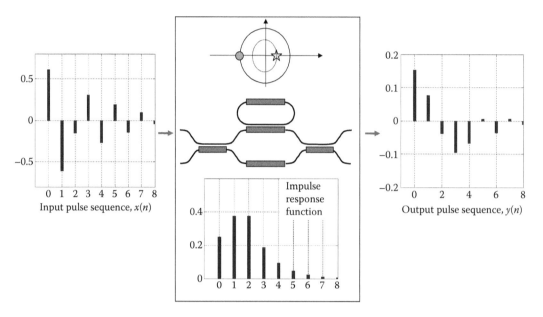

FIGURE 5.47

Equation 5.42 is computed in real time using the one-block OSP for the input sequence of $x(n)$, which was generated in Figure 5.46b. The input sequence is $(0.61, -0.61, -0.15, 0.31, -0.27, 0.19, -0.14, \ldots)$ and the output sequence is $(0.15, 0.08, -0.04, -0.10, -0.07, 0, -0.04, 0.01, -0.01, \ldots)$. Pole/Zero diagram for the configured OSP and its impulse response function are also shown.

This is similar to the equation introduced in Section 5.3.1.2. Here we replace the round trip time with a symbolic sequence. Therefore, if an OSP is configured properly to have these a_m and b_n coefficients, the OSP can compute this differential equation to generate a real time solution for output $y(n)$ with given input $x(n)$. For example, consider a system which has a differential equation of:

$$y(n) - \frac{1}{2}y(n-1) = \frac{1}{4}x(n) + \frac{1}{4}x(n-1) \tag{5.42}$$

Since this system has one zero (one delayed) and one pole system (one recursive), only one unit block is needed. By using the Z-transform, we also find out that this system has a zero of -1 and a pole of $1/2$. Figure 5.47 shows the computed sequence of $y(n)$ for the input sequence of $x(n)$, which was generated in Figure 5.46b. For the input sequence of $(0.61, -0.61, -0.15, 0.31, -0.27, 0.19, -0.14, \ldots)$, the one-block OSP computes Equation 5.42 to generate an output code as $(0.15, 0.08, -0.04, -0.10, -0.07, 0, -0.04, 0.01, -0.01, \ldots)$ in real time. Note that pulse codes are plotted in units of electric fields instead of the intensity since electric field is the actual signal for the OSP. Because the electric field has phase in general, the pulse codes are all complex numbers. In Figure 5.47, we set the parameters in such a way that the pulses are all real numbers for simplicity.

The ability to compute and process optical signals at very high speeds will be truly useful for optical packet communication networks and therefore are a very exciting prospect for communication providers.

References

1. E. Marcatili. Bends in optical dielectric guides. *The Bell System Technical Journal*, 48, 2103–2132, 1969.
2. B. E. Little, S. T. Chu, W. Pan, and Y. Kokubun. An eight-channel add-drop filter using vertically coupled microring resonators over a cross grid. *IEEE Photonics Technology Letters*, 11, 691–693, 1999.
3. B. E. Little, J. S. Foresi, G. Steinmeyer, E. R. Thoen, S. T. Chu, H. A. Haus, E. P. Ippen, L. Kimerling, and W. Greene. Ultra-compact si-sio2 microring resonator optical channel dropping filters. *IEEE Photonics Technology Letters*, 10, 529–551, 1998.
4. Q. Xu, B. Schmidt, S. Pradhan, and M. Lipson. Micrometre-scale silicon electro-optic modulator. *Nature*, 435, 325–327, 2005.
5. K. D. Djordjev, S. J. Choi, S. J. Choi, and P. D. Dapkus. Active semiconductor microdisk device. *IEEE Journal of Lightwave Technology*, 20(1), 105–113, 2002.
6. D. Rafizadeh, J. P. Zhang, S. C. Hagness, A. Taflove, K. A. Stair, S. Ho, and R. C. Tiberio. Waveguide-coupled algaas/gaas microcavity ring and disk resonators with high finesse and 21.6-nm free spectral range. *Optics Letters*, 22(16), 1244–1226, 1997.
7. P. Rabiei, W. H. Steier, C. Zhang, and L. R. Dalton. Polymer micro-ring filters and modulators. *IEEE Journal of Lightwave Technology*, 20(11), 1968–1975, 2002.
8. B. E. Little, S. T. Chu, H. A. Haus, J. Foresi, and J.-P. Laine. Microring resonator channel dropping filters. *IEEE Journal of Lightwave Technology*, 15, 998–1005, 1997.
9. S. O. Transmission, group delay, and dispersion in single-ring optical resonators and add/drop filters-a tutorial overview. *IEEE Journal of Lightwave Technology*, 22, 1380–1394, 2004.
10. R. C. Alferness. Waveguide electooptic modulators. *IEEE Transactions on Microwave Theory and Techniques*, MTT-30, 1121–1137, 1982.
11. C. K. Madsen and G. Lenz. Optical all-pass filters for phase response design with applications for dispersion compensation. *IEEE Photonics Technology Letters*, 10(7), 994–996, 1998.
12. C. K. Madsen. General iir optical lter design for wdm applications using all-pass filters. *IEEE Journal of Lightwave Technology*, 18(6), 860–868, 2000.
13. K. Jinguji and M. Kawachi. Synthesis of coherent two-port lattice-form optical delay-line circuit. *IEEE Journal of Lightwave Technology*, 13(1), 73–82, 1995.
14. K. Jinguji. Synthesis of coherent two-port optical delay-line circuit with ring waveguides. *IEEE Journal of Lightwave Technology*, 14(8), 1882–1898, 1996.
15. N. Takato, T. Kominato, A. Sugita, K. Jinguji, H. Toba, and M. Kawachi. Silica-based integrated optic mach-zehnder multi/ demultiplexer family with channel spacing of 0.01–250 nm. *IEEE Journal of Selected Areas in Communications*, 8(6), 1120–1127, 1990.
16. G. Lenz, B. J. Eggleton, C. K. Madsen, and R. E. Slusher. Optical delay lines based on optical filters. *IEEE Journal of Quantum Electronics*, 37(4), 525–532, 2001.
17. K. P. Jackson, S. A. Newton, B. Moslehi, M. Tur, C. C. Cutler, J. W. Goodman, and H. J. Shaw. Optical fiber delay-line signal processing. *IEEE Transactions on Microwave Theory and Techniques*, MTT-33(3), 193–210, 1985.
18. B.-J. Seo, S. Kim, B. Bortnik, H. Fetterman, D. Jin, and R. Dinu. Optical signal processor using electro-optic polymer waveguides. *IEEE Journal of Lightwave Technology*, submitted.
19. V. S. Ilchenko, A. A. Savchenkov, A. B. Matsko, and L. Maleki. Sub-micro watt photonic microwave receiver. *IEEE Photonics Technology Letters*, 14, 1602–1604, 2002.
20. D. A. Cohen, M. Hossein-Zadeh, and A. F. J. Levi. High-q microphotonic electro-optic modulator. *Solid-State Electronics*, 45, 1577–1589, 2001.
21. V. S. Ilchenko, A. A. Savchenkov, A. B. Matsko, and L. Maleki. Whispering-gallery-mode electro-optic modulator and photonic microwave receiver. *Journal of the Optical Society of America B*, 20, 322–342, 2003.

22. M. Hossein-Zadeh and A. Levi. A new electrode design for microdisk electro-optic rf modulator. The *Conference on Lasers and Electro-Optics, proceedings are published by IEEE*, 863–865, 2003.

23. O. P. Lay, S. Dubovitsky, R. D. Peters, J. P. Burger, S.-W. Ahn, W. H. Steier, H. R. Fetterman, and Y. Chang. Mstar: a submicrometer absolute metrology system. *Optics Letters*, 28, 890–892, 2003.

24. I.-L. Gheorma and R. M. Osgood Jr. Fundamental limitations of optical resonator based on high-speed eo modulators. *IEEE Photonics Technology Letters*, 14, 795–797, 2002.

25. T. Saitoh, S. Mattori, S. Kinugawa, K. Miyagi, A. Taniguchi, M. Kourogi, and M. Ohtsu. Modulation characteristic of waveguide-type optical frequency comb generator. *IEEE Journal of Lightwave Technology*, 16, 824–832, 1998.

26. M. Kato, K. Fujiura, and T. Kurihara. Generation of super-stable 40 ghz pulses from fabry-perot resonator integration with electro-optic phase modulator. *Electronics Letters*, 40, 299–301, 2004.

27. T. Sadagopan, S. J. Choi, P. D. Dapkus, and A. E. Bond. Optical modulators based on depletion width translation in semiconductor microdisk resonators. *IEEE Photonics Technology Letters*, 17(3), 567–569, 2005.

28. K. Sasayama, M. Okuno, and K. Habara. Coherent optical transversal filter using silica-based waveguides for high-speed signal processing. *IEEE Journal of Lightwave Technology*, 9(10), 1225–1230, 1991.

29. A. V. Oppenheim and R. W. Schafer. *Discrete-Time Signal Processing*. Prentice Hall, New Jersey, 1989.

30. C. J. Kaalund and G.-D. Peng. Pole-zero diagram approach to the design of ring resonator-based filters for photonic applications. *IEEE Journal of Lightwave Technology*, 22(6), 1548–1559, 2004.

31. Yariv. Critical coupling and its control in optical waveguide-ring resonator systems. *IEEE Photonics Technology Letters*, 14(4), 483–485, 2002.

32. H. Tazawa, Y.-H. Kuo, I. Dunayevskiy, J. Luo, A. K.-Y. Jen, H. R. Fetterman, and W. H. Steier. Ring resonator-based electrooptic polymer traveling-wave modulator. *IEEE Journal of Lightwave Technology*, 24(9), 3514–3519, 2006.

33. K. Geary, S.-K. Kim, B.-J. Seo, and H. R. Fetterman. Mach-zehnder modulator arm length mismatch measurement technique. *IEEE Journal of Lightwave Technology*, 23, 1273–1277, 2005.

34. M. R. Fetterman and H. R. Fetterman. Optical device design with arbitrary output intensity as a function of input voltage. *IEEE Photonics Technology Letters*, 17(1), 97–99, 2005.

35. I. Y. Poberezhskiy, B. Bortnik, S.-K. Kim, and H. R. Fetterman. Electro-optic polymer frequency shifter activated by input optical pulses. *Optics Letters*, 28(17), 1570–1572, 2003.

36. T. R. Halemane and S. K. Korotky. Distortion characteristics of optical directional coupler modulators. *IEEE Transactions on Microwave Theory and Techniques*, 38(5), 669–673, 1990.

37. L. M. Johnson and H. V. Roussell. Reduction of intermodulation distortion in interferometric optical modulators. *Optics Letters*, 13(10), 928–930, 1998.

38. D. J. M. Sabido, M. Tabara, T. K. Fong, C.-L. Lu, and L. G. Kazovsky. Improving the dynamic range of a coherent am analog optical link using a cascaded linearized modulator. *IEEE Photonics Technology Letters*, 7(7), 813–815, 1995.

39. J. L. Brooks, G. S. Maurer, and R. A. Becker. Implementation and evaluation of a dual parallel linearization system for am-scm video transmission. *IEEE Journal of Lightwave Technology*, 11(1), 34–41, 1993.

40. R. B. Childs and V. A. O'Byrne. Multichannel am video transmission using a high-power nd:yag laser and linearized external modulator. *IEEE Journal on Selected Areas in Communications*, 8, 1369–1376, 1990.

41. Y.-C. Hung and H. R. Fetterman. Polymer-based directional coupler modulator with high linearity, *IEEE Photonics Technology Letters*, 17, no. 12, pp. 2565–2567, Dec 2005.

42. X. Xie, J. Khurgin, J. Kang, and F.-S. Chow. Linearized mach–zehnder intensity modulator. *IEEE Photonics Technology Letters*, 15(4), 531–533, 2003.

43. J. Yang, F. Wang, X. Jiang, H. Qu, M. Wang, and Y. Wang. Influence of loss on linearity of microring-assisted mach–zehnder modulator, *Optics Express*, 12, no. 18, pp. 4178–4188, Sep 2004.

44. P.-L. Liu, B. J. Li, and Y. S. Trisno. In search of a linear electrooptic amplitude modulator. *IEEE Photonics Technology Letters*, 3(2), 144–146, 1991.
45. H. Zmuda, R. A. Soref, P. Payson, S. Johns, and E. N. Toughlian. Photonic beamformer for phased array antennas using a fiber grating prism. *IEEE Photonics Technology Letters*, 9(2), 241–243, 1997.
46. J. L. Corral, J. Marti, S. Regidor, J. M. Fuster, R. Laming, and M. J. Cole. Continuously variable true time-delay optical feeder for phased-array antenna employing chirped fiber gratings. *IEEE Transactions on Microwave Theory and Techniques*, 45(8), 1531–1536, 1997.
47. T. Kawanishi, S. Oikawa, K. Higuma, and M. Izutsu. Electrically tunable delay line using an optical single-side-band modulator. *IEEE Photonics Technology Letters*, 14(10), 1454–1456, 2002.
48. N. A. Riza, M. A. Arain, and S. A. Khan. Hybrid analog–digital variable fiber-optic delay line. *IEEE Journal of Lightwave Technology*, 22(2), 619–624, 2004.
49. B.-J. Seo and H. R. Fetterman. True time delay element in lossy environment using eo waveguides. *IEEE Photonics Technology Letters*, 18(1), 10–12, 2006.

6

Organic Micro-Lasers: A New Avenue onto Wave Chaos Physics

Melanie Lebental

Ecole Normale Superieure of Cachan
Université Paris-Sud

Eugene Bogomolny

Université Paris-Sud

Joseph Zyss

Ecole Normale Superieure of Cachan

CONTENTS

6.1 Introduction

This chapter is mainly devoted to the basic concepts that underlie the dynamic features of light confined in quasi two-dimensional micro-cavities. Moreover, in accordance with the general orientation of this book, there is special emphasis on practical applications, or the onset of such in this relatively new and emerging field at the border of science and technology. The implementation of physical phenomena in organic materials is justified by the appealing assets of this broad class of materials, both in terms of their functional versatility, that lends them to a molecular engineering approach and the availability and maturity of a relatively easy enabling technology. Nevertheless, the foundations as well as major applications of organic materials outreach any specific material technology, similar possibilities in glass and semiconductor materials bear specific pros and cons. In the first section, we will review the main motivation which have lead us to select organic materials and the candidate technology in the context of solid state micro-cavity lasers. Their structural and functional flexibility, compliance to etching, abiding to relatively easy and unconstrained design rules and basic technological fabrication steps will be considered in the second part. In the final sections, different experimental lay-outs and methods to extract relevant information will be considered and subsequently challenged at different levels for specific generic cavity shapes, from nonchaotic and integrable to chaotic.

6.2 Introduction to Flat Organic Micro-Cavities

6.2.1 Context

Organic materials play an increasing part in various aspects of fundamental as well as applied photonics. This is due to a unique list of built-in assets, reflected in higher figures of merits as well as in the development of new and appealing technological processing avenues. The former aspects are related to the optimization of a broad range of optical properties, an area sometimes referred to as molecular engineering [1–3]. Noteworthy successes are due to the well-known structural flexibility and synthetic know-how offered by organic synthesis. At the more downstream end of technological developments, attractive features pertain to mechanical and other processing properties that allow for example low temperature fabrication techniques of large-area flexible foils, embedding many connected or independent components [4,5]. Such technologies point the way towards new functionalities and manufacturing methods that are more suited for large scale and low cost production than for example, with usual inorganic semiconductors.

Among the optical properties of interest, the control of the index of refraction via polymer processing has enabled the design and fabrication of a variety of wave confining structures towards integrated optics applications. Organic materials endowed with higher nonlinear optical properties have benefited from such investigations and are being considered for implementation in telecommunication systems [3,6–9].

Among the most promising applications of organic materials for photonics, luminescent materials made of low molecular weight dyes or conjugated polymers have made their way to the market in the form of highly efficient and compact organic light emitting diodes (OLEDs) with applications ranging from display to biomedicine [10]. The success of this approach derives from a sequence of optimization steps starting from the tailoring of upstream material properties all the way to the fabrication of confining devices allowing for the concentration and steering of light in structures of adequate shape.

This area of investigation has been active for more than three decades, experiencing periodic revivals each time some significant progress has emerged in physical concepts, material engineering methods or enabling technologies. In that respect, silica has always been a starting material of choice, hard to compete with in view of its excellent optical quality, but depending on rather demanding and rigid processing conditions at higher flame temperatures which have limited its application in integrated optics. Nevertheless, optical microspheres have been investigated for their record high figures of merit and ability to sustain long lived whispering gallery modes (WGM) with high confinement features along microtore volumes close to the equator [11,12]. Such configurations, while offering the advantage of high-Q factors, are inherently deprived of the planarity that is demanded for easy coupling and connecting with usual optical integrated inputs and outputs. Moreover, the rather rigid structure of silica makes it difficult to vary at will its composition and the outer shape of micro-spheres, which calls for other types of materials and technologies, providing more flexibility both in terms of internal composition and boundary definition.

It was therefore natural to consider the possibilities of polymers, and more generally of organic materials, to further investigate and exploit the properties of WGMs in a more accessible and flexible two-dimensional format, i.e. a thickness comparable to the wavelength (~1 μm) and planar dimensions about one or two orders of magnitude higher. Indeed earlier technological processes have evidenced a quasi-unlimited freedom of design by tailoring the outer boundary of a limited film portion by adequate masking techniques (see Section 6.3.2).

Such planar cavity structures can also be achieved with inorganic semiconductors however with less ease of fabrication and a significantly higher refractive index (typically from 2.5 to 3.5 in contrast with 1.5 for usual organic materials). As the refractive index governs the total internal reflection angle, the higher its value, the more confining the cavity will be. Referring to a ray picture in the limits of geometric optics (e.g. a cavity size much larger than the wavelength, typically hundred times more), the number of confined modes abiding to the total internal reflection condition will then be significantly smaller for lower index materials. This leads to "cleaner" spectral features for the out-coupled emission, making it amenable to simpler physical interpretation. Moreover, this phenomenon implies direct practical applications, for instance if light must be emitted in a planar geometry along a preferential direction. It is therefore of primary interest to distort the circular boundary of a micro-disk onto a more adequate shape so as to favor such a desired emission pattern. But this distortion is sometimes not sufficient to achieve the directional emission since a large amount of modes can coexist in the cavity. Then the low refractive index brings an additional way to select the required modes [13,14].

As specified in the title, we are working on micro-lasers and not on passive resonators. However the laws governing an active cavity in contrast to a passive one are still understudied, the materials being either organic or inorganic. In fact the standard principles of laser can be hardly extended to these open resonators due to their small size and their relatively low quality factor which usually implies strong spatial hole-burning and no orthogonality between resonances [15]. The successful analysis with fundamental equations coupling field, polarization and inversion density shows that the active modes are not so different from that of passive cavities [16,17]. Only passive properties are thus involved[*] while, from an experimental point of view, the use of dye-doped polymers reveal a double asset. First it opens the way to well-behaved micro-lasers [18] with potential applications as light emitters or sensors; and secondly the laser effect appears as an ideal probe located in the resonator itself, thereby allowing to select (or not) the modes to be excited by an efficient and relatively straightforward optical pumping scheme.

Our investigations aim at answering a number of fundamental questions, some of which are driven by practical considerations. Among the central issues that we wish to address are the behaviors of a so-called chaotic cavity (for example stadium or cut disk) whereby periodic (closed) orbits are an exception somehow buried within a sea of dominant chaotic trajectories where rays experience erratic bouncing [19]. Indeed at the onset of the research reported here, little was known about the spectral and angular emission properties of such cavities, where the erratic behavior of rays could be expected to lead to featureless spectral and angular properties. We will show that, in general, this is not the case. Moreover, these results can be accounted for by calling on the more elaborate wave chaos tool-box [19] and the design of the cavity shape can then be optimized for applications where specific properties of the spectrum or the directional emission are involved.

In summary, we will proceed to show that polymer-based micro-cavities are appealing structures in the investigation of a variety of both technologically and physically meaningful issues due to their external boundary shapes. We will confront experimental results based on the emissive properties of a series of cavities with generic shapes so as to test various theoretical modeling approaches ranging from fully numerical (solution of Helmholtz equations) to more analytical methods. A most interesting outcome of this work, is the definition of a general methodology by which the nature of the underlying orbits can be inferred from the Fourier transform of the emission spectrum, which is applicable to a broad variety of shapes (chaotic or not) and class of materials.

6.2.2 Basic Elements

The main purpose of this chapter is to provide a comprehensive basic understanding of what happens when light is confined in a dielectric cavity[†]. In general, one can state that light escapes, decays, and finally disappears. At intermediate time-scales, it is spatially organized in resonances, which differ from what is usually called "eigen-functions" because they are not stationary. Among the broad diversity of resonances, we are interested in the better confined modes, referred to as quasi-stationary or high-Q modes. In a general way, resonances are arranged in symmetry classes and characterized by wave-numbers, which are complex numbers contrary to the wave-numbers of true eigen-functions: their real part retains the same meaning, while their imaginary part represents the loss. Resonances are therefore *a priori* neither orthogonal nor real. This usual and convenient formalism originates from the S-matrix theory of diffusion. An introduction to this approach in direct connection with our subject is proposed by Cvitanović et al. [20] and Lebental [21].

To be more specific, we study the resonant solutions of Maxwell equations for homogeneous dielectric cylindrical cavities such as in Figure 6.1a: the cavity with n refractive index leans on a layer of smaller refractive index (buffer layer) and is surrounded by air. In

---

[*] For specific properties where the influence of lasing has been proven, comments will specify how to take it into account.

[†] We restrict our study to passive cavities, the influence of lasing being then deduced from general passive properties as described in Section 6.3.

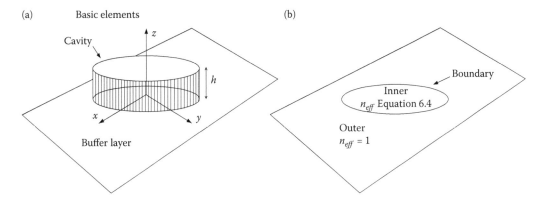

FIGURE 6.1
(a) Schematic view of a cylindrical cavity with h thickness. (b) Schematic view of the system after reduction to a two-dimensional problem.

this frame, we assume that the electromagnetic fields, far away from the cavity, behave like outgoing waves with an e^{ikr} propagative phase factor, without any ingoing waves and the related e^{-ikr} phase factor. The difficulty of dealing with this problem in general comes from its three-dimensional and vectorial nature. To our knowledge, the only case where a quasi analytical solution could be found is that of an infinite cylinder with a circular basis [22].

To overcome this difficulty, the electromagnetic fields are commonly separated into two different polarizations: the electric (resp. magnetic) field lies in the (x,y) plane, while the magnetic (resp. electric) field has a nonzero component in the z direction, referred to as TE (resp. TM) polarization[‡]. This separation is based on a Fourier transform and assumes that the H_z and E_z components are uncoupled. This is therefore exact for a cavity infinite either in the z or in the xy directions. But this assumption fails due to the boundaries where these components are coupled by Maxwell equations (see e.g. Dubertrand et al. [23]). Far from the boundaries, in the cavity bulk, the separation into TE and TM polarizations seems usually a good approximation. This condition is more or less verified for our micro-lasers, since the h thickness is less than 1 μm for a typical size of 100 μm in the xy plane. It will thus be used in this chapter as in most papers dedicated to this subject (see e.g. [24,25]).

The problem then reduces to the scalar Helmholtz equation for the E_z or B_z component (called Ψ hereafter), the other one being directly inferred from Maxwell equations:

$$(\Delta_{\vec{x}} + k^2 n^2(\vec{x}))\Psi(\vec{x}) = 0 \tag{6.1}$$

where $k = \omega/c$ is the vacuum wave-number and $n(\vec{x})$ the local refractive index. The Ψ wavefunction depends on the three space variables and, due to the assumption of infinite layers in the xy plane, it can be expressed as $\Psi(x,y,z) = \phi(z)\psi(x,y)$. This separation of the xy and z variables is called *effective refractive index approximation* and leads to

$$(\Delta_{xy} + k^2 n_{eff}^2)\psi(x,y) = 0 \tag{6.2}$$

It assumes a well-confined wave-function in the cavity layer with evanescence in the adjacent layers (air and buffer layer, see notations in Figure 6.2a). The n_{eff} effective refractive index takes into account the wave profile in the vertical direction and can be directly

---

[‡] This definition is consistent in this chapter. In the literature, these terms are sometimes interchanged.

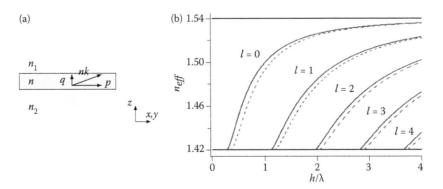

FIGURE 6.2
(a) Cross-section of the layers and notations for refractive indexes and propagation wavenumbers. The propagative layer is a polymer (PMMA) doped with a laser dye (DCM) ($n = 1.54$) between confinement layers assumed to be infinite in the vertical direction: the air ($n_1 = 1$) and the buffer layer (another polymer (SOG) ($n_2 = 1.42$) or silica ($n_2 = 1.45$)). (b) Effective refractive index versus the thickness over wavelength variable calculated from Equation 6.4. The refractive indices are assumed to be constant: 1 for air, 1.42 for the buffer layer, and 1.54 for the cavity layer (horizontal lines). The TE polarization is plotted with solid lines and TM polarization with dotted lines. Integer l (see Equation 6.4) increases from left to right starting from $l = 0$.

inferred from $\phi(z)$. In fact, from Equation 6.1, it follows that $\phi(z) \propto e^{iqz}$ in the cavity layer, and so $k^2 n_{eff}^2 = k^2 n^2 - q^2$. The q value is determined from the phase coherence condition after one round trip (length $2h$) and one reflection on both interfaces (cavity layer/air and cavity layer/buffer layer):

$$r_1 r_2 e^{2iqh} = 1 \qquad (6.3)$$

where r_1 (resp. r_2) is the reflection coefficient at the interface between air (resp. the buffer layer) and the cavity layer. It reduces to a pure dephasing due to the assumption of an evanescent wave in the vertical direction outside the cavity layer (total internal reflection at both interfaces). From Equation 6.3 the expression of q follows and then the effective refractive index (here, for instance, for TM polarization):

$$kh\sqrt{n^2 - n_{eff}^2} = \arctan\left(\frac{\sqrt{n_{eff}^2 - n_1^2}}{\sqrt{n^2 - n_{eff}^2}}\right) + \arctan\left(\frac{\sqrt{n_{eff}^2 - n_2^2}}{\sqrt{n^2 - n_{eff}^2}}\right) + l\pi \quad l \in N \qquad (6.4)$$

where n_1 and n_2 stand for the refractive indexes of the adjacent layers. The refractive indexes are inferred from ellipsometric measurements and the plots of Figure 6.2b are directly calculated from Equation 6.4. This graph shows that under a certain thickness cut-off (about 0.3 μm) no mode propagates in the cavity, and that between this inferior cut-off and the next threshold (typically $h \sim \lambda$) there exists only one vertical excitation for each polarization. Our cavities are designed according to this latter configuration, in general close to the upper boundary ($h/\lambda \sim 1$) to maximize the volume and then the gain. This refractive index approximation is not specific to organic material and is commonly used also for inorganic devices.

At this latter step, the problem reduces to the scalar and two-dimensional Helmholtz equation (Equation 6.2) and the boundary shape is defined from the cylinder section (see Figure 6.1b). Then the n_{eff} effective refractive index outside the cavity is $n_{eff} = 1$ and is calculated from Equation 6.4 inside the cavity. The boundary shape has a prominent influence on spectra and resonance patterns while the effective refractive index approximation fails close to the boundary. This paradoxical approach could be only decided with a full three-dimensional treatment. For in-plane scale much larger than the wave-length (typically 100 times larger for our cavities), effective refractive index approximation has proved to be relevant. For consistence, the boundary conditions are inferred from Maxwell equations of the three-dimensional system (see Dubertrand et al. [23]):

$$\psi_{in} = \psi_{out} \tag{6.5}$$

$$\frac{\partial \psi_{in}}{\partial \nu} = \frac{\partial \psi_{out}}{\partial \nu} \text{ for TM polarization} \tag{6.6}$$

$$\frac{1}{n^2}\frac{\partial \psi_{in}}{\partial \nu} = \frac{\partial \psi_{out}}{\partial \nu} \text{ for TE polarization} \tag{6.7}$$

where the label *in* (resp. out) stands for inside (resp. outside) and ν is the coordinate normal to the one-dimensional boundary.

This problem is significantly simpler than the original one, since it reduces to a two-dimensional one. It is however still hardly solvable due to two separate difficulties: the diffraction at the corners and the boundary shape.

Actually the diffraction coefficient of a plane wave incident on a dielectric corner is not yet known, but it was analytically calculated for a metallic one by Sommerfeld at the end of the nineteenth century [26]. As a consequence, even a system as simple as a square shaped dielectric cavity currently remains an unsolved challenge. This problem is directly connected to more general ones such as the propagation of electromagnetic waves in urban environments, where some solutions derived from micro-cavity studies could also prove useful.

To explain the additional difficulty due to the boundary shape, we resort to geometrical optics, i.e. when the wavelength is much smaller than the typical size, and therefore rays follow the trajectories of point-like particles. In fact the dynamics of rays depends strongly on the boundary shape. It can be integrable for instance inside a square or a disk, chaotic inside a stadium or a cut disk (see Figure 6.4), mixed, pseudo-integrable, etc.[§] [19]. The route from a geometrical optics system to the equivalent wave optics one (usually referred to as quantization) is straightforward for an integrable system because, by definition, there exists as many constants of motion as degrees of freedom.[¶] For instance in a circle, the constants of motion are the energy and the angular momentum (or the incident angle). But for a chaotic system, where, by definition, there are less constants of motion than degrees of freedom, the standard quantization method fails and other ways must be explored. This is the broad domain of research called wave chaos.[**] Earlier works were based mainly on metallized micro-cavities in the millimeter wave regime, whereby electromagnetic modes are fully confined within the metallic boundaries, as a result of strict cancelation of the field outside of the cavity and therefore referred to as "closed system" as in the following.

[§] This is true for closed cavities such as metallic ones, from which light cannot escape, while for dielectric ones, refraction selects some rays and modifies the dynamics.

[¶] Moreover the constants of motion must commute with each other [27].

[**] In general it is called "quantum chaos" in reference to the original studies on quantum systems. But the concepts and results easily extend to any physics domain where waves are concerned.

In our case, a dielectric interface does not impose cancelation of the field out of the cavity and allows indeed, not only for refractive out-coupling but also for evanescent wave generation, even in the case where total internal reflection conditions are fully satisfied. Connection of the inside and outside parts of the cavities provides new physical properties that cannot occur in strictly confining cavities and that is further enhanced by the use of low refractive index materials such as polymers. As the topic of this book is "practical applications", we will not emphasize this aspect and restrict our chapter to potentially applicable results derived from these studies, in particular in Section 6.5. However, a wealth of literature has been devoted to this field [19,20,27].

6.3 Polymer-Based Technology and Process

Organic materials feature a wealth of attractive properties allowing for new approaches in fabrication technologies of photonic devices at the required micron and nanometer scales, with applications ranging from information technologies to life sciences. Polymers are an important class of organic materials with potential for passive or active optoelectronic properties (e.g. wave-guiding, nonlinear optics, fluorescence and lasing, electron transport) in bulk form (e.g. conjugated polymers) or when doped with active molecules in a guest-host configuration.

Whereas inorganic materials, especially III–V semiconductor alloys are well known for their robustness and superior track record of performances in similar areas of physical properties and applications, the distinctive case of organics is built on specific inherent advantages to be exploited for easier and more flexible device fabrication. Indeed the structure and functionality of III–V semiconductor multi-layer stacks is inherently limited by the stringent bottleneck of epitaxial matching conditions which governs the compatibility between adjacent crystalline layers. In contrast, the soft amorphous nature of polymers, possibly at the expense of their diminished resilience (lower melt temperature for example, typically down to 200–300°C, however still very much within the demands of workable devices in normal conditions, eases the fabrication techniques to conditions that will not degrade sophisticated guest molecules), permits the relaxation of deposition conditions over any kind of substrates (polymer over inorganic substrates or polymer over polymer). They are now governed at the more manageable statistical level, by physical chemistry parameters, such as solubility, evaporation rate, hydrophilic versus hydrophobic nature, or chemical affinity and interactions (from looser Van Der Waals forces to hydrogen bonding or even locking by chemical bonds). Deposition techniques, be it by evaporation, centrifugation or dipping, are compatible with large area and eventually flexible substrates which allow for new application perspectives with considerable and still largely unexplored potential in the realm of printing ("e-paper" and "e-book") and large panel display or even new types of functional textiles.

Moreover, the less robust nature of organics and polymers turns out to be an asset as compared to semiconductors, in view of their relevance in the emerging realm of soft technologies, whereby patterning by molding, stamping or ink-jet printing can be readily achieved by simple techniques allowing for nano-meter scale resolution [28]. The inherently accommodating nature of polymer host matrices which can be engineered to solubilise a diversity of guest compounds, allows for the incorporation of guest molecules as well as nano-particles of eventual inorganic nature [29]. Such potential for hybrid organic-inorganic blending where the polymer host provides an ideal scaffold for a diversity of guests and properties (such as would allow for simultaneous lasing and electro-optic properties) which cannot be readily achieved in purely crystalline inorganic materials, opens the way to multi-functional structures.

Finally, keeping in mind the perspective of life science applications and the need for a new generation of bio-chips with sensing and diagnostic potential, there is clearly a need for a patterned and eventually active organic or polymer intermediate functional layer, capable of connecting a biological recognition layer (e.g. targeted to specific cells, tissues or macromolecular assemblies of proteins) with an inorganic substrate (e.g. a silicon chip to drive or retrieve and process input or output electronic signals, such as based on an array of field effect transistors). A physically functional polymer layer bears more potential than a purely inorganic substrate to allow for additional biological functions, for example by addition of a biopolymer over-layer, or by featuring intrinsic biopolymer character, so as to provide adequate and sensitive linkage between the biological and physical parts of a biosensor [30].

In the context of this chapter, we illustrate the implementation of such advantages in a more modest, however highly rewarding context, that of patterning flexibility, achieving with simple means over the same substrate, a quasi-unlimited variety of micro-cavity contours. The advantage of ease and speed of fabrication, by way of easily accessible technological tools, has been used to explore a wealth of cavity shapes over a broad range of dimensions, which would hardly have been possible under the more rigid conditions imposed by inorganic semiconductor fabrication. We have been mainly using, as will be detailed in the following sections, relatively basic equipment for micro-photonics device fabrication processes, such as micro-photo-lithography, reactive ion etching or metal evaporation, excluding the use of ultra-vacuum deposition techniques. When higher quality factors than those achieved by such well established and "elementary" techniques are needed, direct patterning of polymers by electron or focused ion beam is feasible and provides the required level of resolution, down to the nano-scale.

6.3.1 Materials

One of the main advantages of organic compounds is the broad diversity of materials with very specific properties. Various arrangements lead to efficient organic micro-lasers, but each one must fulfill the following requirements.

- The polymer *matrix*, which makes up the cavity, must be as transparent as possible in the pumping and emission ranges of the active material to avoid additional losses. More specifically, the choice depends on the etching technique (e.g. UV sensitivity) and on the active material (e.g. solvent compatibility and glass transition temperature). In our case, polymethyl methacrylate (PMMA) is well-adapted and some studies even report that it makes the photo-degradation of the laser dye reversible [31,32]. It may be responsible for the very long sample life-time, up to several years of storage at regular room conditions.

- The *active material* can be organic or inorganic [29]. Our micro-lasers are based on [2-[2-[4-(dimethylamino)phenyl]ethenyl]-6-methyl-4H-pyran-4-ylidene]-propanedinitrile (DCM) but other laser dyes can also be used: nile-red, rhodamine [33], etc. The typical doping amounts to a few percents in weight and cannot be increased by much due to the dye emission quenching.

- A *buffer layer* is usually needed to ensure the vertical confinement since the refractive index of the gain layer ($n \sim 1.5$) is in general lower than that of the wafer ($n = 3.9$ for silicon at 600 nm). Two microns of silica ($n = 1.45$) or Spin-On Glass ($n = 1.42$) fulfill this purpose perfectly well.
- The *wafer* can be either opaque (silicon) or transparent (glass) depending on the application. Micro-lasers on flexible wafers have also been realized [34].

These requirements allow for a broad choice of emission wavelengths and optical properties (e.g. addition of nonlinear materials [3]).

6.3.2 Etching Methods

Thin film deposition by spin-coating is one of the great advantages of organic materials as opposed to their inorganic counterparts which generally require much more elaborate high vacuum techniques. In fact this method to make a layer is cheap, fast, easy, reproducible and allows for a fine control of the thickness (typical uniformity ±10 nm). Once the necessary layers have been laid, standard photolithography and reactive ion etching (RIE) steps can be performed as described in Figure 6.3. The use of a commercial Cr/quartz mask ensures a good etching quality (roughness scale less than 500 nm) and a broad diversity of shapes (see Figure 6.4). The simplicity of the device (the single PMMA-DCM layer) reduces the processing time: in 2 hours, several hundred of micro-cavities of different sizes and shapes can be simultaneously made on a standard wafer.

Usually we use the photolithography and RIE process, but different etching processes are available, each one with various grading in etching and ease-of-use:

- The *direct photolithography* process takes advantage of the photo-sensitive nature of the polymer used for the matrix (as e.g. PMMA). It can thus be exposed through a mask and subsequently developed according to its specific recipe [33]. While this method is fast and does not require equipment (as RIE), the chemical attack does not ensure a good edge quality and the preservation of the gain layer properties.
- The *photolithography+RIE* process is a spin-off from inorganic semi-conductor technology. Very small details such as corners or cusps are not well reproduced (see Figure 6.4), but it is relatively fast and the roughness is small enough to ensure a good compatibility with optical applications [18].
- Currently, *electron-beam* etching provides a more efficient way to define small details: the cavity layer is directly exposed to an electron beam and then developed as a photo-resist. The pictures of Figure 6.5 show a standard resolution better than 100 nm, which is compatible with the requirements for micro-lasers emitting in the visible range. However this technology is quite expensive (due to the sophistication of the e-beam machine) and time-consuming (more or less proportional to the removed area). Nevertheless this one-step process fits perfectly with specific and limited applications requiring a high-quality design [23].
- Recently, *nanoembossing* techniques, based on the cavity layer molding, have demonstrated their efficiency [35,36]. It is only the mold fabrication step which requires sophisticated technologies. Then the replication is fast and easy, achieving a resolution up to a few dozen nanometers [37].

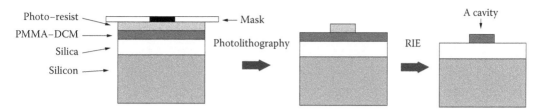

FIGURE 6.3
Cross-sectional view for the "photolithography+RIE process" steps, following the spin-coating of the PMMA-DCM and photoresist layers.

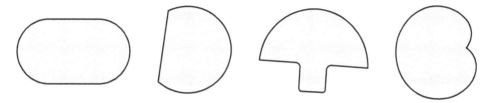

FIGURE 6.4
Top view *photographs* of micro-cavities by an optical microscope. From left to right: stadium (a rectangle between two half-circles), cut disk, mushroom, and cardioid. Their typical size is of the order of 100 μm with a thickness (direction perpendicular to the sheet) less than 1 μm.

FIGURE 6.5
Scanning electron microscope images of PMMA cavities etched with an electron beam (Courtesy of C. Ulysse, LPN, France). The specific scale is shown in the bottom right corner of each picture.

The fabrication of standard organic micro-lasers only requires a relatively cheap technology, such as available in standard clean room conditions. Furthermore, the use of organic materials gives access to a broad variety of optical properties.

6.4 Optical Tests

6.4.1 Background

The main features of the experimental set-up are shown in Figure 6.6: a single[††] cavity is pumped from the top while it emits light laterally (i.e. parallel to the substrate plane) at room temperature and atmosphere. The light is further collected by a detector (e.g. a

---

[††] Each wafer contains a few hundred cavities. The cavity to be pumped is selected with a microscope.

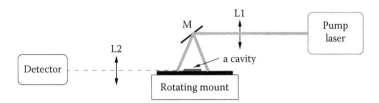

FIGURE 6.6
Schematic view of a standard experimental set-up. M stands for mirror and L for lens.

spectrometer) via a lens (L2). Concerning the pumping, the size of the beam cross-sectional area can be adjusted with a lens (L1) and its shape eventually controlled to select specific lasing modes of the cavity [38]. In the following, we will consider uniform pumping only, in a so-called "flooded" geometry.

This set-up is similar to optical pumping experiments with inorganic micro-lasers. The only difference is the constraint on the pulse duration of the pump laser. In fact laser dyes show dark states, referred to as triplet states, with typical transition times from the excited state of about 100 ns. The pump pulse duration must thus be less than a few nano-seconds [39,40] (30 ps in our case).

The laser threshold depends on the cavity shape and the organic compounds. For a typical stadium made of 5% DCM doped PMMA, it amounts to about 100 pJ/μm^2 [18] and can be decreased using a better etching process and an integrable shape such as square or disk because of the better confinement of the modes therein. Besides, as the available gain decreases with the size of the cavity, the minimal size to get a laser effect is imposed by the damage threshold of the layer.

One of the advantages (or disadvantages, depending on the point of view) of laser dyes is their homogeneous gain broadening. We will take advantage of it in Section 6.4. Concerning the lasing dynamics, it induces a competition between modes, probably controlled by the fluorescence statistics, and then implies variability from pulse to pulse: peaks stay at the same position, but their relative intensities change. This variation disappears after averaging over a few dozen pulses. In the following, we will be considering spectra after integration over 30 pump pulses.

6.4.2 Various Set-Up

Many versions of the standard experimental set-up can be developed depending on the study. In general they can be used for inorganic micro-cavities as well, either active or passive.

- *Spectra* and *direction of emission*: these properties can be studied at the same time with the standard set-up described in Figure 6.6. In this configuration, the detector is a spectrometer coupled to a cooled CCD camera with a rotating mount to turn the cavity in front of it. The intensity emitted in each direction is then deduced from integration over all the pixels of the spectrum. The resolution depends mainly on the angular aperture of the detection system. In Figure 6.6, it is connected to the size of the lens and the distance to the sample. In our experiments, it is about ±5° and can be easily decreased.
- *Lateral far-field imaging*. The side of the cavity can be imaged, changing the detector system of Figure 6.6 into a microscope objective and a cooled CCD camera. Figure 6.7 shows two examples of stadium images collected for different positions of the cavity

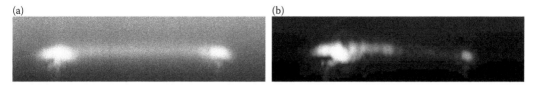

FIGURE 6.7
Images of stadium sides collected in the far-field (a) in the direction of the small axis, (b) in the direction of maximal emission. The area of highest intensity are indicated in white. The maximal intensity in (b) is about 15 times higher than in (a).

with respect to the detector. This set-up provides additional information about the distribution of light around the cavity [41]. But emission mechanisms at the boundary can not be directly inferred because of the far-field collection: near-field imaging is then required.

- *Top far-field imaging.* At low pump intensity, it is possible to look trough a microscope and observe directly the cavity from the top when it is emitting. The use of a CCD camera provides quantitative information about vertical emission [42,43]. But this set-up is not adapted to the study of low loss resonances, since these modes are confined in the plane of the cavity and thus present an evanescent amplitude profile in the vertical direction.

- *Near-field imaging* is the most suitable approach to test the spatial pattern of low loss modes. A few reports are devoted to this approach in the case of quasi two-dimensional cavities [44–46]. But these studies have been further developed in micro-spheres (see e.g. [47]) and in photonic-crystal micro-cavities (see e.g. [48]).

- To *track a wave-packet*, two different methods have been demonstrated. The first one [49,50] is based on the near-field optical set-up described elsewhere [44,45]. The time resolution is obtained by detecting the femto-second pulse injected inside the resonator and using interferences with a referent beam. The second method [51] takes advantage of the nonlinear properties of the laser dye. The femto-second pulse injected inside the cavity acts as a pump beam while a delayed probe beam lights the cavity uniformly. The wave-packet is then tracked via the two-photon triggered fluorescence. In both cases, the cavity is used as a passive resonator.

6.4.3 Spectra

The spectrum of an organic micro-laser (or a standard laser) is made of two main elements: its envelope and the peaks. The former one results from the gain property of the material. For instance, Figure 6.8a shows the fluorescence and amplified spontaneous emission (ASE) spectra of a PMMA-DCM layer before etching. The width of the fluorescence spectrum is of the order 50–70 nm, which is typical for a laser dye, while the ASE spectrum is narrower and shifted to the red, as a result of competition between gain and absorption losses [52]. Besides, the peak properties are imposed by the resonator itself, both for their width and position, at least in a pumping range close to the laser threshold[‡‡]. And their amplitudes depend on the envelope as shown in Figure 6.8b.

[‡‡] The peaks could become narrower due to lasing and be shifted by frequency pulling but we have not observed so far such a phenomenon in our experiments. To our knowledge, there does not exist any report on mode-locking or, more generally, interaction between modes in polymer micro-lasers, but these nonlinear effects could occur.

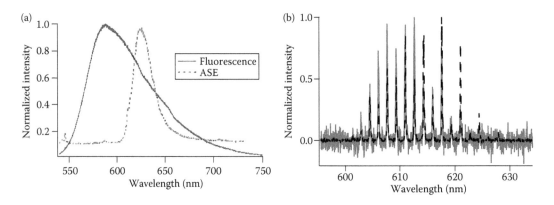

FIGURE 6.8
(a) Spectra collected with the set-up of Figure 6.6. Continuous line: fluorescence spectrum of a PMMA-DCM layer on a glass substrate excited with a continuous pumping. The refractive index of the glass is higher than the refractive index of the active layer, preventing any propagation in the latter one. Dotted line: ASE spectrum of a PMMA-DCM layer on SOG on Si layer excited with a pulse pumping. (b) Laser spectra of the same cavity exposed to a pulse pump energy of 8 a.u. (in continuous) and 20 a.u. (in dotted).

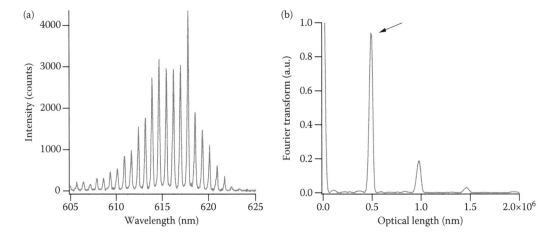

FIGURE 6.9
(a) Spectrum of a Fabry–Perot micro-laser of width $w = 150\mu m$. (b) Normalized Fourier transform of the spectrum (a) expressed as intensity versus wavenumber. The arrow indicates the "first peak".

The positions of the peaks hold a wealth of information about the light pattern inside the cavity. In fact they are connected to geometric features of periodic orbits. For instance, let us consider a Fabry–Perot resonator[§§] of w width (distance between mirrors). The experimental spectrum in Figure 6.9a is made of almost regularly spaced peaks. The corresponding wavenumbers must verify the phase coherence condition along the single periodic orbit of $L = 2w$ length

$$r^2 e^{iLkn_{eff}} = 1$$

where r stands for the reflection coefficient and n_{eff} for the effective refractive index (see Section 6.3). The solutions of this equation are complex numbers: the imaginary part

---

[§§] The corresponding polymer micro-cavity is a very long stripe [53].

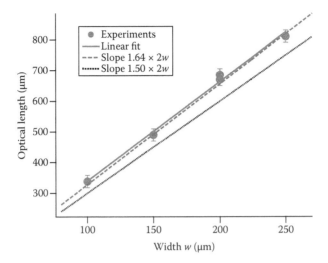

FIGURE 6.10
Optical length versus Fabry–Perot width w. The experiments (circle points) are linearly fitted by the solid line. The dashed line corresponds to the theoretical prediction for a refractive index $n = 1.64$ without any adjusted parameter. For comparison, the dark dotted line is similar to the dashed line, but for a refractive index $n = 1.50$, i.e. the effective refractive index in the absence of group velocity corrections.

corresponds to the width of the resonance and the real part (called k_m) gives the position of a peak in the spectrum and verifies

$$Lk_m n_{eff} = 2\pi m, \; m \in N \quad (6.8)$$

Some groups use the resonance widths, for instance towards sensor applications [54], while we focus here on their positions through Equation 6.8 as a way to identify the lasing periodic orbits via their length L. From Equation 6.8, the peaks are regularly spaced by a wave-number interval expressed as

$$\delta k = k_{m+1} - k_m = \frac{2\pi}{L n_{eff}} \quad (6.9)$$

This relation suggests that the periodic orbit length can be directly inferred from an experimental spectrum, by way of measuring the distance between adjacent peaks[¶¶]. Nevertheless the use of Equation 6.9 for organic micro-lasers leads to a systematic deviation from the expected length, as illustrated in Figure 6.10. This graph shows the optical length versus the width w of Fabry–Perot resonators. The circle points correspond to the quantity $2\pi/\delta k$ inferred from the experimental spectra[***] and are linearly fitted by the solid line. For comparison the straight line $2wn$ has been plotted with a dotted line with $n = 1.50$ standing for the effective refractive index, showing some disagreement with the error bars[†††].

[¶¶] The effective refractive index n_{eff} is inferred independently from the cavity thickness and bulk refractive indexes (see Section 6.3).

[***] In fact this quantity can be directly measured via the Fourier transform of the experimental spectrum when expressed as intensity versus wavenumber. The optical length corresponds to the position of the first peak which is indicated by an arrow in Figure 6.9b.

[†††] In our experiments, the measure of the optical length is limited by the precision of the discrete Fourier transform. So the error bars correspond to the width of the first peak (see Figure 6.9b).

This discrepancy comes from the dispersion of the refractive index, whereas Equation 6.9 is correct if the refractive index does not depend on the wavenumber. Obviously dispersion can not be discarded and a more complete expression is derived from Equation 6.8 after differentiating with respect to k, namely:

$$\delta k_m [n_{eff}(k_m) + k_m \frac{\partial n_{eff}}{\partial k}(k_m)]\, L = 2\pi. \tag{6.10}$$

Thus a more comprehensive expression of the refractive index is given by the well-known group velocity formula:

$$n_{full} = n_{eff} + k_m \frac{\partial n_{eff}}{\partial k} \tag{6.11}$$

The term n_{eff} corresponds to the phase velocity and $k_m\,(\partial n_{eff}/\partial k)$ to the group velocity. The latter contribution accounts for about 10% of the former one, which explains the discrepancy in Figure 6.10. The group velocity correction is made of two different contributions resp. accounting for bulk and effective dispersions as summarized in Figure 6.11. The bulk refractive index of the active layer is inferred from ellipsometric measurements, evidencing a nonnegligible slope due to the proximity of the absorption band. The dispersion originating from the effective refractive index must also be taken into account, as the thickness is of the same order of magnitude as the wavelength. Both contributions can be calculated without any micro-laser spectrum. Thus the group velocity term is inferred independently. For our structures, the full-index value accounts for $n_{full} = 1.64 \pm 0.01$ and does not depend significantly on thickness and polarization, allowing its use as such in the following. Taking into account other error sources (spectrometer calibration, discretization, etc.), the global uncertainty on the geometrical length of the periodic orbit is estimated to be less than 3%. The excellent agreement between this theoretical prediction and experiments as shown in Figure 6.10 confirms our approach.

Having demonstrated the efficiency of our orbit length measurement method, we can proceed using it to study different resonators with more complex shapes. For instance, let us

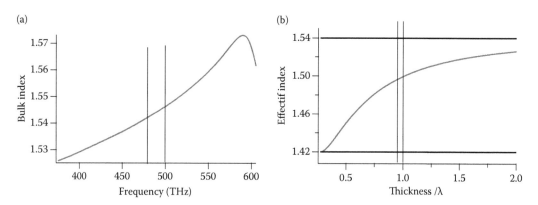

FIGURE 6.11
(a) Refractive index versus the frequency for a 5% DCM doped PMMA layer, inferred from ellipsometric measurements. (b) Effective refractive index versus the parameter "thickness over wavelength" for the TE fundamental mode calculated for constant bulk refractive indexes shown by the horizontal lines. For both graphs, the vertical lines indicate the experimental range: from 600 nm to 630 nm for a 0.6 µm thickness.

consider the square and try to find out its dominant periodic orbit. Such an orbit is known to minimize the refractive losses, so Fabry–Perot[‡‡‡] orbits should be excluded whereas the so-called diamond, such as drawn in Figure 6.12a, is to be expected. Figure 6.12b shows the comparison between the length inferred from experiments and the expected diamond orbit length. Such good agreement confirms our expectation that the dominant periodic orbit is indeed the diamond. Based on such encouraging results, we propose in Section 6.5.2 a general method to predict the dominant periodic orbit in more complex situations of increasing complexity.

6.5 Theoretical Approaches

The main physical properties of polymers that are being emphasized in this chapter are mainly their low refractive index on the one hand, and their availability towards relatively easy photonic functions via adequate modifications on the other hand. In the present case where emphasis is put more on physical properties by way of a simple and routine fabrication techniques, we chose to resort to the simpler approach of blending a polymer host matrix with a low concentration (a few % in mass) of a robust and easily accessible low molecular weight dye molecule endowed of efficient (high quantum yield) fluorescence properties.

A dominating feature is the lower value of the refractive index which appears to play a major role in enhancing the dissipative nature of the cavity, thus favoring the emergence of more resistant periodic orbits, eventually down to a single dominating one, at the expense of the more lossy or unstable ones which are being out-coupled by refraction. In addition

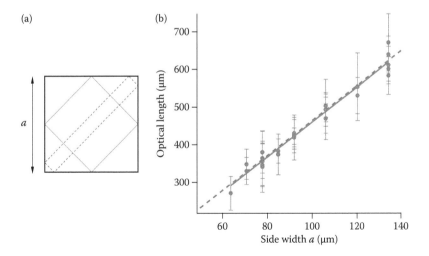

FIGURE 6.12

(a) Two representations accounting for the diamond-like periodic orbit. They have the same fundamental features: length and incident angles. (b) Optical length versus square side width a. The experiments (circle points) are linearly fitted by the solid line. The dashed line corresponds to the theoretical prediction (diamond periodic orbit) without any adjusted parameter, namely $2\sqrt{2}an_{full}$.

[‡‡‡] In general, we call Fabry–Perot a periodic orbit connecting two opposite plane sides.

to easier technological fabrication conditions allowing to test many cavity shapes, such less stringent confinement conditions entailing stronger dissipation is a central cornerstone of the use of polymer in the context of this chapter.

The aim of this section is to provide practical tools to predict the main features of the electromagnetic field both inside and outside micro-cavities. After a paragraph dealing with the application range and numerical simulations, three topics will be more specifically explored: spectral features, directions of emission and intra-cavity light pattern. These will encompass most of the necessary elements to optimize a device or account for an already existing one, for instance in order to combine ultra-low loss and directional light output in a same resonator [55].

6.5.1 General Methodology

As noted in Section 6.2, we focus hereafter on passive resonators. Actually the influence of the additional nonlinearity due to lasing seems to be very limited [16,17] and can be accounted for from simple arguments to be developed at the end of Section 6.5.2. Therefore tools proposed in the following sections can be directly used for passive devices and readily extended to active ones.

As demonstrated in Section 6.2, the underlying physics of micro-cavities is based on solutions of the passive two-dimensional stationary Helmholtz equation (see Equation 6.2):

$$(\Delta + n_{i,o}^2 \, k^2)\psi_{i,o} = 0$$

where the label i (resp. o) stands for *inside* (resp. *outside*), and ψ corresponds either to the vertical component of the electric or the magnetic field, depending on the polarization. In this modeling, the effective refractive index (so far noted as $n_{i,o}$) takes into account the vertical direction (i.e. perpendicular to the substrate). The values $n_i = 1.5$ and $n_0 = 1$ are used in the following as being typical for organic materials, but the methods developed in this section can be readily extended to higher values such as $n_i \sim 3$ for inorganic materials. However a low refractive index is sometimes a great advantage because refraction leads to a strong selection and the specificities of nonescaping light are then emphasized.

Our purpose here is to provide "user-friendly" recipes to account for the main features of the devices, with predictions from geometrical optics used whenever possible. Such treatment makes sense considering the much larger typical size of cavities ($a \sim 100\ \mu m$) as compared to the wavelength ($\lambda \sim 0.6\ \mu m$). Full wave solutions are used only if no other way is known so as to account for the resonance patterns as shown at the end of Section 6.5.

To reinforce the relevance and validity of our approach, it is compared with numerical simulations based on ray propagation or Maxwell equations, for which the precision is consistent with our experimental studies and has been tested apart [56,57].

The ray simulations consist in statistics over a large number of rays (about 10^6) with random starting points. Each ray propagates inside the cavity and is reflected or refracted at the boundary depending on its incident angle.

More elaborate solutions of Maxwell equations become necessary if wave properties are involved. Different methods have been proposed, that can be roughly separated into two classes: time-dependent and stationary. The former one, finite-difference time-domain (FDTD), is broadly developed and available as a routine package but requires a time consuming discretization procedure both inside and outside the cavity. However it has already been performed close to the experimental range (see e.g. [33]). In the following section, we use a stationary algorithm, known as boundary element method [56,57], which provides spectra and resonance patterns from a discretized mesh localized only on the cavity boundary.

6.5.2 Spectrum

The signature of a resonator is its spectrum. It is related to the scattering matrix (see e.g. [20]), which is useful for numerical simulations and fundamental approaches. But in the context of practical applications, we will consider the spectrum as a sequence of peaks such as in Figure 6.9a to be further accounted for and predicted. Their relative intensities depend on the gain media and pumping conditions. We will therefore focus on the more robust quantities of position and periodicity.

In the case of a Fabry–Perot cavity (width w), the light is organized along the single straight line periodic orbit of length $L = 2w$. The spectrum reveals this structure from the phase coherence condition (see Section 6.4)

$$r^2 e^{iLkn} = 1$$

resulting in peaks spaced by

$$\delta k = \frac{1}{nL}$$

for a medium without any refractive index dispersion. In general the peaks of an arbitrary resonator belong to different combs, more or less in relation with the different periodic orbits. For a Fabry–Perot resonator, there is a single periodic orbit and then a single comb.

At this stage, two questions can be raised:

1. Can the spectrum of any resonator be described with periodic orbits?
2. When assertion (1) is true, which is the dominating comb? In other terms, on which periodic orbit does the light preferentially concentrate?

A rigorous and quantitative answer to issue (1) is given by the *trace formula*. This approach connects the density of states (i.e. the spectrum) to a sum over periodic orbits in the limit of geometrical optics. To be more precise, for a cavity without loss the spectrum is a sequence of peaks located at k_m that can be individually represented as Dirac functions:

$$d(k) = \sum_m \delta(k - k_m) \tag{6.12}$$

This expression underlies the wave point of view. In the geometrical optics approximation, it has been analytically proved[§§§] to be equivalent to

$$d(k) \propto \bar{d}(k) + \tilde{d}(k) = \bar{d}(k) + \sum_p C_p \cos(kL_p + \varphi_p) \tag{6.13}$$

[§§§] The demonstration has been carried out for integrable and chaotic shapes [27,59]. The terms "integrable" and "chaotic" refer to the formally equivalent dynamical system where a particle stands for the light and is reflected at the boundary.

Here $\bar{d}(k)$ represents a mean of the density of states [58,59], which does not contain orbit contributions leading us to concentrate on the more relevant $\tilde{d}(k)$ oscillating part. The latter is expressed as a sum over the periodic orbits labeled by p, L_p corresponding to the length of the p periodic orbit and C_p being a factor depending only on the geometrical properties of the p periodic orbit. The additional phase φ_p will be of no use in the following and can be omitted. The transition from Equations 6.12 to 6.13 is exact in the limit of geometrical optics, that means when $ka \to \infty$ where a is the typical size of the system.

The trace formula then provides a positive answer to issue (1): the spectrum of any resonator is made of a sum over periodic orbits. But with some reservations. First the transition from Equations 6.12 to 6.13 has been demonstrated for a broad diversity of systems, but not for all of them. Secondly, this demonstration applies for closed systems (i.e. systems where light can not escape the cavity) and the extension to resonators has only been conjectured and not proved [20]. Thus the expression of the C_p coefficients in Equation 6.13 is not precisely known in the case of dielectric cavities, requiring further investigations.

With the answer to issue (1) being positive, in general, question (2) becomes relevant to predict the spectrum of an arbitrary shape. For instance it has been experimentally demonstrated in Section 6.4 that the dominating periodic orbit in a square is the diamond. But the diamond apart, there exists a lot of different periodic orbits in a square. Their weighing depends on the attached C_p coefficients, which appear directly in the Fourier transform of the spectrum (Equation 6.13). This point clarifies the data processing proposed in Section 6.4 for experimental spectra, while a similar process can also be applied with spectra inferred from numerical simulations [60,61].

The analytical expression of the C_p coefficients is well-known for closed systems. For dielectric resonators, we have shown in some particular cases [21,53] that they seem identical, however additionally weighted by refractive losses:

$$\tilde{d}(k) = \sum_p r_p \, C_p \, \cos(knL_p + \varphi_p) \tag{6.14}$$

For instance, the r_p coefficient for the Fabry–Perot periodic orbit corresponds to

$$r_p = r^2$$

where

$$r = \frac{n-1}{n+1}$$

leading to $r_p = 0.04$ for $n = 1.5$. On the other hand, the diamond periodic orbit in the square is confined by total internal reflection for $n > \sqrt{2} \simeq 1.41$, then $r_p = 1$. So we notice that the refractive losses have a significant influence. Thus if there exist periodic orbits confined by total internal reflection, they should dominate the spectrum.

In a general way, it is necessary to evaluate the C_p coefficients. For instance in a circular cavities, all N-polygons with $N \geq 4$ (i.e. square, pentagon, hexagon, etc.) are confined by total internal reflection. But their C_p coefficients are different and we have demonstrated that the square periodic orbit is dominating [53].

The C_p coefficients are known for two kinds of periodic orbits, namely those *isolated* and *in family*. These notions are illustrated in Figure 6.13 on the example of the pentagonal cavity.

One of the most obvious periodic orbits is the inscribed pentagon plotted in Figure 6.13a. Hereafter it is referred to as a simple pentagon periodic orbit so as to avoid possible confusion. It belongs to the class of isolated orbits. Actually if one of the bounces is slightly moved, it leads to a very different trajectory. This is illustrated in Figure 6.13b and c. On the other hand, the double pentagon periodic orbit plotted in Figure 6.13d, is *in family* since it can be continuously deformed without losing its fundamental features: length and incident angles. For instance, the arrows in Figure 6.13d indicate how to deform the double pentagon all the way to the five-pointed star periodic orbit of Figure 6.13e, which is a boundary case of the double pentagon periodic orbit.

In a circular cavity, the polygons are in family. For illustration, Figure 6.14a shows two representations of the square periodic orbit family. When all these representations are plotted, we notice that a portion of the cavity is not crossed by any ray. Its boundary, called

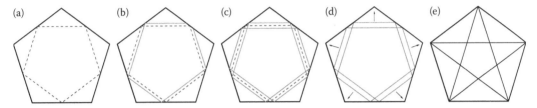

FIGURE 6.13
Periodic orbits in a pentagonal cavity. (a) Simple pentagon periodic orbit. (b) In continuous line, slight displacement of the simple pentagon periodic orbit. (c) Dotted line: simple pentagon periodic orbit. Continuous line: double pentagon periodic orbit. (d) Arrows indicate possible deformations of the double pentagon periodic orbits. (e) Five-pointed star periodic orbit obtained from continuous deformation of the double pentagon periodic orbit.

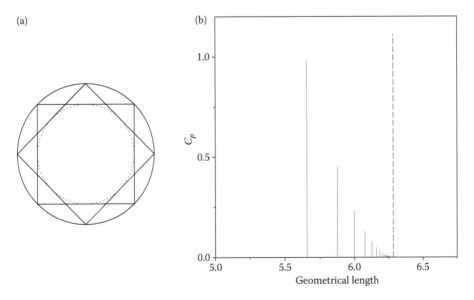

FIGURE 6.14
Circular cavity. (a) Two representations of the square periodic orbit family. The caustic is indicated in dotted line. (b) C_p coefficients versus the length of the periodic orbits for the disk from Equation 6.15. From left to right, the vertical solid lines correspond to square, pentagon, hexagon, etc., and the vertical dotted line indicates the position of the perimeter. The C_p coefficient for the square periodic orbit has been normalized to unity. Triangle and diameter are not plotted since they are not confined by total internal reflection for $n = 1.5$.

caustic, is indicated in dotted in Figure 6.14a. In a general way, for a periodic orbit in family, the C_p coefficients are

$$C_p = \frac{A_p}{\sqrt{L_p}} \tag{6.15}$$

where A_p is the area covered by the family. For a circular cavity of radius R and the family of a polygon with N sides, A_N corresponds to the area between the boundary and the caustic and is written as $A_N = \pi R^2 \sin^2(\pi/N)$. As the length of the orbit is $L_N = 2NR \sin(\pi/N)$, the C_p coefficients are

$$C_N \propto \frac{1}{\sqrt{N}}\left(R \sin \frac{\pi}{N}\right)^{3/2}$$

These coefficients are plotted in Figure 6.14b. The coefficient corresponding to the square periodic orbit is higher and is thus expected to dominate the spectrum, which has been experimentally demonstrated [53]. As the C_p coefficients of the other periodic orbits are not so different, they also contribute to the spectrum but with a smaller weight.

If both kinds of periodic orbits coexist, i.e. isolated and in family, the trace formula predicts that in family periodic orbits dominate the spectrum. For instance, in the case of the pentagonal cavity where the simple pentagon periodic orbit is isolated and the double pentagon is in family, it has been demonstrated both experimentally and via numerical simulations, that the dominant periodic orbit is indeed the double pentagon [53].

There exists a last configuration where it is possible to predict the dominating periodic orbits: if all the periodic orbits are unstable, and thus isolated. Figure 6.15b explains how the stability of a periodic orbit can be characterized. The continuous curve corresponds to the periodic orbit. For a typical starting point slightly shifted by an infinitesimal displacement $|d\vec{r}|$, one considers the end point of the resulting trajectory after one round trip $d\vec{r}'$. If $|d\vec{r}'|$ is smaller than $|d\vec{r}|$, the orbit is stable, whereas it is unstable if it is larger. The λ_p Lyapounov coefficient quantifies this instability. Appendix A specifies how it can be calculated in a practical way for optical cavities. In the case of an unstable periodic orbit, the C_p coefficient is

$$C_p = \frac{1}{\sinh \frac{\lambda_p L_p}{2}} \tag{6.16}$$

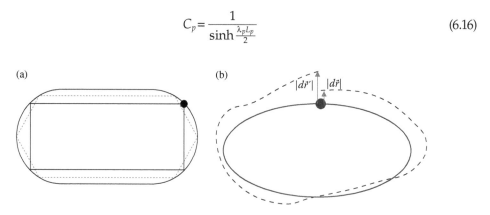

(a)   (b)

FIGURE 6.15
(a) Periodic orbits in a stadium-shaped cavity: rectangle in solid line and elongated hexagon in dotted line. (b) Diagram to explain how to quantify the stability of a periodic orbit.

This means that the dominating periodic orbit minimizes the $\lambda_p L_p$ product, so as to make it both shorter and less unstable (small L_p and λ_p).

As an example, let us consider a stadium-shaped cavity (see Figure 6.15a). It has been proved to be fully chaotic [62], leading to instability for all the periodic orbits[¶¶¶]. The *a priori* dominating periodic orbit is confined by total internal reflection, so the competing orbits are shaped as elongated polygons[****] as plotted in Figure 6.15a. They exhibit comparable lengths, corresponding approximately to the perimeter. But their Lyapounov coefficients are significantly different. It is shown in Appendix A, that they do increase with the number of bounces on circular boundaries. Thus the dominating periodic orbit is expected to have the minimum number of bounces and so to be the rectangle. A more refined approach is proposed in [61].

To summarize, in the limit of geometrical optics, resonator spectra are made of peak combs in relation with the periodic orbits. The competing periodic orbits are weighted by a coefficient $r_p C_p$, taking into account their refractive (r_p) and geometrical (C_p) features. At this time, it is possible to predict the dominating periodic orbits in three configurations

- The periodic orbits are living in family with their weighing given by formula (Equation 6.15).
- The periodic orbits are unstable (and so isolated) with their weighing given by formula (Equation 6.16).
- Isolated and in family periodic orbits coexist. Then the in family confined periodic orbits dominate.

To conclude this section devoted to spectra, we must come back to the influence of the detection set-up and the laser effect. Actually, predictions from trace formula stand for what happens *inside* a *passive* resonator, and thus favor well-confined periodic orbits. But in practice, we detect what is emitted and so the hierarchy can be seen as different from that inferred from trace formula predictions. For instance the diamond is the dominant periodic orbit in a square-shaped micro-cavity, as appears in Figure 6.12b. These experimental spectra have been collected in a direction perpendicular to the square sides for a low pump energy. If the pump energy is increased high enough, then the Fabry–Perot periodic orbit can start to lase. As its outside coupling is very efficient, it dominates the spectrum at the expense of the diamond periodic orbit, which is confined by total internal reflection [21]. Moreover the energy injection scheme can deeply modify the hierarchy (e.g. to pump a ring from an adjacent waveguide). Depending on the application, this selection can turn out to be detrimental or advantageous. It can be prevented with a uniformly pumped active medium to ensure a homogeneous distribution of light or, on the other hand, enhanced by local pumping.

In this section, a concrete method based on periodic orbits has been presented to predict the spectrum of an arbitrary resonator. This approach is exact for passive cavities and can be easily extended depending on the pumping and the emission process.

[¶¶¶] Actually this definition also allows for a countable set of periodic orbits with marginal stability living in family. In a stadium-shaped cavity, there is one such periodic orbit: the Fabry–Perot orbit which connects the straight lines. But in practice, it is neglected because of its important refractive losses.

[****] The elongated polygons are confined by total internal reflection only in the strong limit of geometrical optics, i.e. $|kr| \gg 1$. For $|kr| \simeq 1$, these orbits have refracted losses due to the curvature of the boundary.

6.5.3 Directions of Emission

Information pertaining to the emission angular pattern of a micro-cavity is no less important than that on the spectrum. For applications, the far-field emission is needed to optimize a sensor or the coupling between cavities. In short, outgoing emission appears through three process: tunneling, diffraction, and refraction. The knowledge of the dominating periodic orbit is of great advantage when predicting the kind of process and then the direction of emission. For instance, a Fabry–Perot periodic orbit escapes by refraction in a direction perpendicular to the sides. In this section, the three processes will be studied one by one, stressing their respective range of application and their consequences onto far-field emission.

Tunneling is a purely wave phenomenon. It corresponds to energy transfer when, from a geometrical optics point of view, light is confined by total internal reflection. This kind of emission is often used[++++] since the great sensitivity of the evanescent wave with respect to the distance allows for fine control of the energy transfer between structures (e.g. a waveguide and a ring) [64,65]. For a single cavity, it is not really useful, apart from bearing the hallmarks of well-confined modes. A typical example is the circular cavity, where the so called WGM are localized along the boundary. The losses of these well-confined modes decrease rapidly when $|kR|$ is increasing: the theoretical[++++] quality factor already amounts to $Q \simeq 3.10^7$ for $|kR| = 40$. The geometrical optics picture corresponds to a plane wave totally reflected at the boundary. If the interface were plane and infinite, the evanescent wave generated outside would not propagate any energy. But with a curved interface, the evanescent wave is transformed into a propagative wave at a certain distance from the boundary. This explains why light confined in a circular cavity can be detected in the far-field.

The second emission process, *diffraction*, happens if a wave comes across a singularity. For the cavities studied here, the singularities take place at the boundary, as the polygon corners or the cardioid cusp. If the boundary is parameterized by an application $\rho = f(\theta)$ in polar coordinates, then the singularity is referred to an *n*-order singularity where $d^n f/d\theta^n$ is not continuous. For $n = 1$, it is usually called a corner. For instance, the stadium presents four similar second-order singularities at the connections between the straight and circular parts.

In this section, we are interested in the diffraction coefficient, i.e. the angular distribution of the emitted light in the far-field, also called the far-field pattern. For an *n*-order singularity with $n \geq 2$, an analytical method has been developed in [21], based on an idea originating from Kaminetzky and Keller [66]. This diffraction exists but it seems to be usually overcome by another emission process, and no experimental evidence has been reported. We do not therefore develop it here, but the detailed calculations for a second-order singularity can be found in [21].

However the diffraction of a plane wave by a dielectric corner is still an unsolved problem, although it had been solved by Sommerfeld for a metallic corner at the end of the nineteenth century [26]. The knowledge of this diffraction coefficient would be of great use for micro-cavity studies and in a wide range of applications (e.g. propagation of electromagnetic waves in city areas). As no analytical answer is known, only a simplified approach can be proposed [57] as illustrated in Figure 6.16a with the example of a square. In a square-shaped cavity and as demonstrated in the previous section, the diamond is the dominant

---

[++++] For a review, see [63].
[++++] The imaginary part of a wave-number is numerically calculated from e.g. [23]. The corresponding quality factor is then directly inferred.

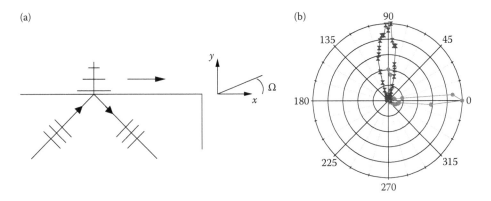

FIGURE 6.16
(a) Proposed diffraction process at a polygon corner. (b) Normalized far-field patterns for micro-squares etched by different process. The triangle hourglass correspond to a square shaped micro-laser etched by a "photolithography+RIE" process and then selected for the quality of its corners, while the circled points refer to a cavity etched by e-beam.

periodic orbit. As it is confined by total internal reflection, it generates an evanescent wave outside, for instance in the y direction, which propagates along the side in the x direction. When it reaches the corner, it is diffracted according to the usual principles of wave optics, and leads to an emission mainly parallel to the side. The polar graph in Figure 6.16b confirms that this prediction is easily extended to other polygons. We have noticed a high sensitivity of the directions of emission with the etching quality of the corners. Thus a precise etching technology is recommended for the study of far-field patterns if diffraction is the dominating emission process.

The third and last emission process, *refraction*, appears via two different ways, depending on the dominant periodic orbit. If this orbit is not confined by total internal reflection, the light is emitted in the direction of the refracted ray. This case is well illustrated by the Fabry–Perot resonator where the single periodic orbit bounces on the boundary at normal incidence and thus the far-field pattern indicates an emission perpendicular to the sides (see Figure 6.4b in [53]). On the other hand, if the dominant periodic orbit is confined by total internal reflection, no emission is expected via refraction. Sometimes the dominant emission process is tunneling as for the circle or scattering as for the square. But if the cavity is chaotic such as a stadium or a cut disk, the high output coupling points to a refraction process. Moreover the good agreement between experiments and geometrical optics simulations is an additional evidence that refraction is indeed the dominant process.

As chaotic cavities can be involved in many applications, in particular for their directional emission, they are developed hereafter. First of all, experiments show in Figure 6.17 that these cavities present a highly directional emission, which respects the obvious symmetries of the shapes [18]. Repeating the same experiments under the laser threshold, the directions of emission do not change, demonstrating that they are connected to passive properties of the cavities. Another interesting feature is related to the wave approach. In fact the far-field pattern can be directly deduced from wave numerical simulations [13]. Then it appears that each high-Q wave-function presents a peak in a precise direction, as illustrated in Figure 6.18a, and this direction agrees with predictions from ray simulations although the wave numerical simulations have been performed in the wave domain (see Figure 6.19). This is a great advantage, since we can limit our study to geometrical optics without missing some decisive elements.

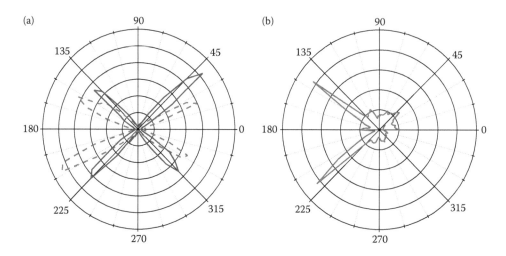

FIGURE 6.17
Experimental far-field patterns in polar coordinates for chaotic cavities. (a) Stadium-shaped cavities with $l/r=0.5$ (continuous line) and $l/r=1$ (dashed line). See notations in Figure 6.18b. (b) Cut-disk with $d=0.9$. The parameter d is defined as the radius proportion which has not been removed from the circle (see [21]).

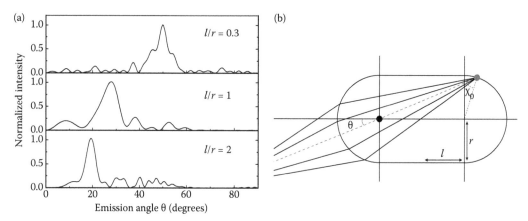

FIGURE 6.18
(a) Far-field pattern of individual well-confined wave-functions for micro-stadiums with different l/r shape ratios. From top to bottom: (l/r=0.3, $Re(kr)$=39.31); (l/r=1, $Re(kr)$=13.88); (l/r=2, $Re(kr)$=23.29). The θ angle is defined according to the main axis of the stadium as in (b). (b) Outline of the lens model and some notations.

Experiments and ray simulations, which agree with each other, show a great dependence on the direction of emission with the l/r shape parameter (see Figures 6.17a and 6.19). This feature is very useful as well for practical applications as for testing our model. This model, which is referred to as "lens model" in the following, is based on three assumptions, illustrated on the example of the stadium:

- The confined light is localized close to the boundary of the cavity to limit the refraction losses, like the so-called WGM. This assumption is confirmed by wave and ray simulations. Thus, due to the stadium geometry, the emission occurs at a half-circle, originating from the other half-circle.

- A kind of ergodicity appears on the half-circles, which means that the source points (dark spot on the right of Figure 6.18b) are more or less uniformly distributed on the

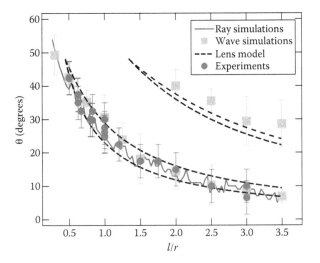

FIGURE 6.19
Direction of maximal emission versus the l/r shape for a stadium-shaped cavity. Comparison between experiments, ray simulations, wave simulations, and predictions of the lens model. The wave simulation points correspond to an averaging over a representative set of high-Q wave-functions, while their error bars represent the standard deviation.

half-circles according to the S curvilinear coordinate. The same assumption applies for the χ starting directions, provided that the ray corresponds to a confined one (i.e. χ is greater than the critical angle). This assumption is confirmed for stadium-shaped cavity by ray simulations [21], and can probably be used for other shapes if they are sufficiently chaotic to lead to ergodicity despite their refraction losses.

- The third and last assumption is not absolutely necessary, but allows for simple and analytical calculations. For rays starting from one of the half-circle (for instance the right one, as in Figure 6.18b), it states that they are subsequently focused by the other half-circle, this half-circle acting then as a lens. To go a further step towards simplification, we consider only the ray crossing the center of the lens half-circle (χ_0) and represented by the dotted line in Figure 6.18b. So aberrations are neglected, which is consistent with our experimental resolution.

Then within this model, we just need to calculate the average over the θ angles for a starting point covering the right half-circle. Due to the equivalent averaging over S or χ_0, the direction of emission is defined by

$$<\theta> = \frac{1}{\chi_2 - \chi_1} \int_{\chi_1}^{\chi_2} \theta(\chi_0)\, d\chi_0 \tag{6.17}$$

It seems that the choice of an integration density such as $d\chi_0$, $d\sin(\chi_0)$ or $d\varphi(\chi_0)$ is not relevant compared to the experimental precision. The limiting angles, χ_1 and χ_2, are directly connected to refraction or geometrical constraints. In the stadium example, χ_1 is the critical angle, and χ_2 the boundary angle, χ_0 corresponding to the incident ray at the connection between the circular and straight part.

This $<\theta>$ value can be analytically calculated versus the l/r stadium shape and is plotted as the inferior dashed line in Figure 6.19. The other dashed lines correspond to equivalent quantities but considering reflections on the straight boundaries. The detailed formulae are developed elsewhere [13,21] and their predictions agree with ray simulations (second branch not represented in Figure 6.19) and wave simulations. Probably due to the limited resolution, the second peak has not yet been shown experimentally.

The good agreement between this lens model, experiments and simulations, confirms in some way the above assumptions: a kind of ergodicity and focusing due to circular boundaries. Even without calculating Equation 6.17, these tools allow to qualitatively predict and interpret the directions of emission of a wide range of chaotic polymer[§§§§] cavities.

6.5.4 Light Patterns Inside the Cavities

The structures of resonances inside the cavity are of great importance for practical purposes, for instance to optimize the positions of electrodes [67], coupling wave-guides [64], or localized gain media. *A priori*, the Helmholtz equation has to be solved either analytically or numerically. The circular cavity being the only integrable system, numerical simulations must be performed for each boundary shape and cavity size, which is time consuming and not always appealing to intuition. To by-pass these drawbacks, we propose two different approaches: one is a very efficient *ad hoc* model, the other originates from a perturbation treatment leading to exact equations.

6.5.4.1 Benefit of "Scarring"

The first approach is based on a specific advantage of low index cavities: the high-Q modes are in general localized along the dominant periodic orbit. This property, called "scarring", appears only sparingly in closed cavities, while it has been widely observed in numerical simulations of low index cavities. This fundamental phenomenon is still under study but it is agreed that this effect is a consequence of the selecting losses.

Polygonal cavities are considered as an example because their specific geometry allow for a more complete understanding. In particular, the pentagon offers a nontrivial configuration to test this approach. We want to connect the structure of the high-Q resonances to that of the dominant periodic orbit, which is the double pentagon, according to Section 6.5.2. Figure 6.20b illustrates the proposed connecting scheme. A representative of the double pentagon periodic orbit is plotted in dotted line with a conventional drawing explained in Figure 6.20a: the ray propagates in a straight line whereas the cavity is reflected. In this representation a family of periodic orbits is clearly identified as a continuous set of straight lines. In Figure 6.20b, it appears that the cavity corners limit this family. In agreement with this observation, the ray connecting the corners corresponds to the five-pointed star periodic orbit described in Figure 6.13e. In Figure 6.20b, the family extends over the hatched area, referred to as a channel.

Our method, called "superscar", is based on two assumptions.

- The wave-functions propagate along the periodic orbit and imitate its periodicity:

$$\Psi(x,y) = \psi(\kappa_y y)\, e^{i\kappa_x x}$$

---

§§§§ The extension to high refractive index (n~3–4) is far from obvious because of the increasing number of rays confined in the cavity, which leads to a more complicated dynamics [14].

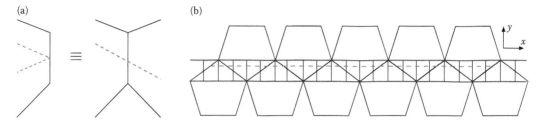

FIGURE 6.20
(a) To clarify the plotting of ray trajectories, the rays are not reflected when bouncing at the boundary, but propagate along a straight line, while a symmetrical cavity is being plotted across the boundary. The cavity boundary is indicated in continuous line and the ray in dotted line. (b) The channel of the double pentagon periodic orbit covers the hatched area. Their lateral boundaries are plotted in continuous horizontal line and one representative of this orbit in dotted line.

with

$$\kappa_x L_p + \delta = 2m\pi \quad m \in N \tag{6.18}$$

where δ takes into account the dephasing at each reflection on the boundary, and L_p is the length of the periodic orbit. In the limit of geometrical optics, the transversal excitation is negligible compared to the longitudinal one, i.e. $\kappa_y \ll \kappa_x$, then Equation 6.18 is similar to the basic elements of Section 6.2.

- The transversal profile, $\psi(\kappa_y y)$, is defined by geometrical constraints. For the pentagonal cavity, the corners force the high-Q wave-functions to vanish in their vicinity so as to minimize their diffraction losses. In the case of closed cavities [68], it leads to a vanishing along all lateral boundaries of the channel, as indicated with the horizontal continuous lines in Figure 6.20b. As the diffraction coefficient on a dielectric corner is still unknown, no similar demonstration can be invoked for dielectric cavities, but it seems reasonable to propose a similar assumption, and so

$$\psi(\kappa_y y) = \sin(\kappa_y y)$$

with

$$\kappa_y l_p = q\pi \quad q \in N^*$$

At this step, the wave-functions can be folded back to the original cavity. Once symmetry class and polarization have been taken into account, the agreement with wave numerical simulations is very satisfactory [53]. Beyond its quantitative efficiency, the main interest of this method lies in predictions of resonance structures. In fact, it appears in Figure 6.20 that the channel never cross the cavity center as the rays passing through the corners delineate the limit of the double pentagon periodic orbit. Thus it draws a pentagonal caustic which is clearly seen in the wave-function plotted in Figure 6.21a. This scheme can be generalized to other shapes such as the hexagonal cavity presented in Figure 6.21b.

The efficiency of this approach encourages an extension to a broad diversity of shapes, based on the clearly dominating periodic orbit which often underlies the high-Q modes. However the prediction of resonance structures requires transversal boundary conditions. In the case of the polygonal cavities reported above, they can be deduced from the

(a)   (b)

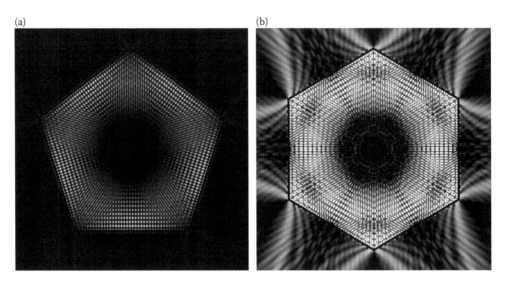

FIGURE 6.21
Numerical simulations: modulus squared of wave-functions for the TM polarization and a particular symmetry class. In this grey scale, black area corresponds to minimal intensity zones. (a) Pentagonal cavity, ak = 123.0−0.02i. (b) Hexagonal cavity, ak = 90.99−0.12i. a is the side length.

corresponding metallic cavities where a robust theory has been developed [68]. For other cavity shapes, such as chaotic, strong evidence inclines to possible extensions and studies are still in progress.

6.5.4.2 Perturbation Approach

Another approach to predict the resonance structures both inside and outside the cavity lies in a perturbation treatment of an analytic solution. As the circular shape is the only integrable system for a two-dimensional dielectric cavity, we focus on small variations from the circular cavity and the boundary defined as

$$r = R + \lambda f(\theta)$$

in polar coordinates (r,θ). The f function can be either positive or negative as in Figure 6.22 with λ as the small perturbation parameter. The expansion with respect to λ converges if

$$\delta a \, k^2 \ll 1$$

where δa is the area modified from the circular cavity, represented by dashed regions in Figure 6.22. From this criterion, the wavelength must be larger than the deformation scale.

Different expansion methods are well-documented, mainly in the context of quantum physics. Hereafter, we focus on the Green function method since it avoids definition problems in the modified area [23]. It is directly based on the two-dimensional Helmholtz equation

$$\left(\Delta + k^2 n^2(\vec{x})\right)\psi(\vec{x}) = 0$$

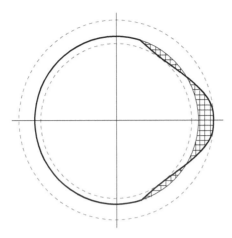

FIGURE 6.22
Example of a cavity shape which can be studied by the perturbation method.

with position dependent "potential" $n^2(\vec{x})$. For perturbed cavity

$$n^2(\vec{x}) = n_0^2(\vec{x}) + \delta n^2(\vec{x})$$

where $n_0^2(\vec{x})$ is the "potential" for the pure circular cavity

$$n_0^2(\vec{x}) = \begin{cases} n^2 & \text{when} \quad |\vec{x}| < R \\ 1 & \text{when} \quad |\vec{x}| > R \end{cases}$$

and $\delta n^2(\vec{x})$ is the perturbation to the regular potential. Then the Helmholtz equation can be rewritten in the form

$$\left(\Delta_{\vec{x}} + k^2 n_0^2(\vec{x})\right)\psi(\vec{x}) = -k^2\delta n^2(\vec{x})\psi(\vec{x})$$

and its formal solution is given by the following integral equation

$$\psi(\vec{x}) = -k^2 \int G(\vec{x},\vec{y})\delta n^2(\vec{y})\psi(\vec{y})d\vec{y} \tag{6.19}$$

Here $G(\vec{x},\vec{y})$ is the Green function for the dielectric circular cavity and describes the field produced at point $\vec{x}$ by a delta-function source situated at point $\vec{y}$. The explicit expressions of this function are presented in Appendix C of [23]. "Expression" is written in the plural since the plane—and then the study—is divided into three circular regions, $r<R_1$, $R_1<r<R_2$, and $r>R_2$ defined by

$$R_1 = \min_\theta(R, R + \lambda f(\theta))$$
$$R_2 = \max_\theta(R, R + \lambda f(\theta)). \tag{6.20}$$

Their boundaries are indicated by dashed circles in Figure 6.22, the deformation "potential" $\delta n^2(\vec{x})$ being nonzero only in the second region $R_1<r<R_2$. Due to singular character

of the Green function, wave functions inside each region are represented by different expressions, leading to different perturbation expansions. But continuity of the perturbation wave-functions are maintained thanks to the Green function.

The main steps of the calculations are detailed in [23], where they are performed up to second order λ. In [23], this method is applied to the example of the cut disk and leads to very good agreement with numerical simulations both for the wave-functions[¶¶¶¶] and the spectrum, while this system presents a double difficulty due to its chaoticity and dielectric corners. In a more general way, this approach seems very promising for cavity shapes where wave numerical simulations reveal a low efficiency such as cusp or concave boundaries. All the more so as there are no restrictions due to the refractive index, and it can also be directly applied to inorganic cavities.

To summarize this section, we have proposed two different approaches to predict and explain the resonance structures. The former one is an *ad hoc* "scar" model based on the dominant periodic orbit and therefore useful in the geometrical optics limit. The other approach is a perturbation expansion from the circular cavity and leads to analytical formulae valid for wavelengths of the same order of magnitude or larger than the small perturbation parameter.

6.6 Conclusion

The engineering and physical properties of polymer based micro-cavities have to be considered in the dual scope of light–matter interactions, where physical and chemical viewpoints are inter-twinned and are considered on a par. A supposedly "purely" physical approach, is traditionally more attached to photons *per se*, in the legitimate search of general physical laws governing their behavior, the validity of which should extend above and beyond any specific class of materials. However, the legitimate search for generality does not imply ignorance of the material point of view, as photonic effects obviously require material interactions to manifest themselves. In our opinion, this quest has been amply illustrated above in the case of polymer based micro-lasers.

Polymers provide, both in terms of their specific technological assets as well as physical properties, a unique enabling arena allowing progress towards more general (e.g. ultimately material independent) mathematical physics rules governing the confinement and out-coupling of light in dissipative billiard-like micro-cavities. As far as technological assets are concerned, it is the versatility of polymer technologies allowing to fabricate and screen a wealth of micro-cavity shapes (in fact much beyond those that could be discussed within the limited space of this chapter). This versatility has been taken advantage of here, so as to experimentally probe and theoretically analyze previously unexplored and hardly accessible configurations with the potential to challenge existing models.

As far as the physical properties of polymers are concerned, it is their low index of refraction and the ensuing dissipative nature of micro-cavities (as opposed to higher index of inorganic semiconductor based cavities), that has been specifically targeted in this study. This leads to the dominance of easily identified "scar" orbits which provide a simple and

---

[¶¶¶¶] It must be stressed that this approach leads to the wave-functions inside *and* outside the cavity, and thus predicts the far-field as well.

illustrative geometric support to the field distribution. Such a situation is in contrast with that of micro-cavities based on higher index materials where the more messy electromagnetic speckle-like wave-functions do not lend themselves to the extraction of clearly distinctive geometric optics orbits. As shown in Section 6.5, this approach has lead to conjecture the extension of the trace formula to the case of chaotic dissipative micro-cavities, thus building-up on the work by Gutzwiller et al. [53,61] and extending its validity range, based on the interpretation of our polymer based micro-cavity experiments.

Far from an embarrassment or a degrading factor as opposed to the universality of physics, respectful consideration of materials in their diversity and specificity entails a benefit. It would be a shame to ignore the opportunities of exploiting different material properties and functions, if only as a useful help and handle towards reaching an increased level of "immaterial" generality. But material science and, more generally, chemistry deserve better than being considered as a transient and utilitarian technological playground towards physics. Indeed, the application and engineering potential of micro-cavity lasers will critically and specifically depend on and benefit from specific material choices, making progress in this important and promising area of research critically dependant on joint efforts in physics, both theoretical and experimental, as well as material science and technology

Acknowledgments

The authors are grateful to their collaborators at Laboratoire de Photonique Quantique et Moléculaire, N. Djellali, R. Hierle, J.-S. Lauret, and J. Lautru, at Laboratoire de Physique Théorique et Modèles Statistiques, R. Dubertrand and C. Schmit, and at Laboratoire de Photonique et de Nanostructures, C. Ulysse and Y. Chen.

Appendix A: Lyapounov Coefficient for Unstable Periodic Orbits

The stability of an orbit is determined by its attached Lyapounov coefficient λ deduced from the maximal eigenvalue Λ of the monodromy matrix by:

$$|\Lambda| = e^{\lambda L}$$

where L is the length of the periodic orbit. The monodromy matrix is calculated by using the basic matrix $\Pi(\mathcal{L})$ (propagation over a distance $\mathcal{L}$) and $R(\chi)$ (reflection with incident angle χ and curvature radius ρ).

$$\Pi(\mathcal{L}) = \begin{pmatrix} 1 & \mathcal{L} \\ 0 & 1 \end{pmatrix} \quad R(\chi) = \begin{pmatrix} -1 & 0 \\ \dfrac{2}{\rho\cos(\chi)} & -1 \end{pmatrix}$$

For instance, in a stadium-shaped cavity, the monodromy matrix of the rectangle periodic orbit (Figure 6.15a) can be written $[\Pi(\mathcal{L}_1)R(\pi/4)\Pi(\mathcal{L}_2)R(\pi/4)]^2$ with $\mathcal{L}_1 = 2l/r + \sqrt{2}$ and $\mathcal{L}_2 = \sqrt{2}$, assuming $r=1$. For a periodic orbit in a shape of an elongated polygon with m bounce on each half-circle, the trace of the monodromy matrix can be written as

$$trM = 2 + \frac{16\,m}{\sin\left(\frac{\pi}{2m}\right)}\frac{l}{r} + \frac{16\,m^2}{\sin^2\left(\frac{\pi}{2m}\right)}\left(\frac{l}{r}\right)^2$$

from which the Lyapounov coefficient follows since Det $M = 1$.

References

1. H. S. Nalwa and S. Miyata editors. *Nonlinear optics of organic molecules and polymers*. CRC Press, 1997.
2. B. Valeur. *Molecular fluorescence. Principle and applications*. Wiley-VCH, Weinheim, 2002.
3. J. Zyss. Molecular nonlinear optics: materials, physics, and devices. Academic Press, Boston, MA, 1994.
4. H. C. Song, M. C. Oh, S. W. Ahn, W. H. Steier, H. R. Fetterman, C. Zhang. Flexible low voltage electro-optic polymer modulators. *Applied Physics Letters*, 82, 4432, 2003.
5. Y. Huang, G. T. Paloczi, J. K. S. Poon, and A. Yariv. Demonstration of flexible free standing all polymer integrated optical ring resonator devices. *Advance Materials*, 16, 44, 2004.
6. R. Levenson and J. Zyss. Polymer based optoelectronics: from molecular nonlinear optics to device technology. In *Material for optoelectronics*, M. Quillec editor. Kluwer, Dordrecht, 1996.
7. P. Labbe, A. Donval, R. Hierle, E. Toussaere, and J. Zyss. Electro-optic polymer based device technology for optical telecommunication. In *Molecular photonics: material, physics, and devices*, special issue (in English) of the *Compte-rendu de l'Académie des Sciences*, 3, 2002.
8. S. Kim, W. Yuan, K. Geary, Y.-C. Hung, H. R. Fetterman, D.-G. Lee, and C. Zhang. Electro-optic phase modulator using metal-defined polymer optical waveguide. *Applied Physics Letters*, 87, 011107, 2005.
9. R. Song and W. Steier. Overlap integral factor enhancement using buried electrode structure in polymer Mach-Zehnder modulator. *Applied Physics Letters*, 92, 031103, 2008.
10. A. Moliton. *Optoelectronics of molecules and polymers*. Springer, 2006.
11. V. Sandoghar, F. Treussart, J. Hare, V. Lefèvre-Seguin, J.-M. Raimond, and S. Haroche. Very low threshold whispering-gallery-mode microsphere laser. *Physical Review A*, 54, R1777, 1996.
12. D. W. Vernooy, V. S. Ilchenko, H. Mabuchi, E. W. Streed, H. J. Kimble. High-Q measurements of fused-silica microspheres in the near infrared. *Optics Letters*, 23, 247, 1998.
13. M. Lebental, J. S. Lauret, J. Zyss, C. Schmit, and E. Bogomolny. Directional emission of stadium-shaped microlasers. *Physical Review A*, 75, 033806, 2007.
14. S. Shinohara, T. Fukushima, and T. Harayama. Light emission patterns from stadium-shaped semiconductor microcavity lasers. arxiv:0706.0106, 2007.
15. H. Türeci, D. Stone, and L. Ge. Theory of the spatial structure of nonlinear lasing modes. *Physical Review A*, 76, 013813, 2007.
16. T. Harayama, T. Fukushima, P. Davis, P. O Vaccaro, T. Miyasaka, T. Nishimura, and T. Aida. Lasing on scar modes in fully chaotic microcavities. *Physical Review E*, 67, 015207(R), 2003.
17. S.-Y. Lee, M. S. Kurdoglyan, S. Rim, and C.-M. Kim. Resonance patterns in a stadium-shaped microcavity. *Physical Review A*, 70, 023809, 2004.
18. M. Lebental, J.-S. Lauret, R. Hierle, and J. Zyss. Highly directional emission of stadium-shaped polymer microlasers. *Applied Physics Letters*, 88, 031108, 2006.

19. H.-J. Stöckmann. *Quantum chaos, an introduction*. Cambridge University Press, Cambridge, 1999.

20. P. Cvitanović, R. Artuso, R. Mainieri, G. Tanner, and G. Vattay. *Chaos: classical and quantum*. chaosBook.org, Niels Bohr Institute, Copenhagen, Cambridge, 2005.

21. M. Lebental. Quantum chaos and organic micro-lasers. PhD dissertation, http://tel.archives-ouvertes.fr/2007.

22. H. G. L. Schwefel, A. D. Stone, and H. E. Tureci. Polarization properties and dispersion relations for spiral resonances of a dielectric rod. *Journal of the Optical Society of America B*, vol. 22, 2295, 2005.

23. R. Dubertrand, E. Bogomolny, N. Djellali, M. Lebental, and C. Schmit. Circular dielectric cavity and its deformations. *Physical Review A*, 77, 013805, 2008.

24. P. K. Tien. *Light waves in thin films and integrated optics*, 10, 2395, 1971.

25. C. Vassallo. *Optical waveguide concepts*. Elsevier, Amsterdam-Oxford-New York-Tokyo, 1991.

26. Sommerfeld. *Optics. Lectures on theoretical physics volume IV*. Academic Press, 1964.

27. M. C. Gutzwiller. *Chaos in classical and quantum mechanics*. Springer, 1991.

28. B. Gates, Q. Xu, M. Stewart, D. Ryan, C. Willson, and G. Whitesides. New approaches to nanofabrication: Molding, printing, and other techniques. *Chemical Review*, 105, 1171, 2005.

29. V. Klimov, A. Mikhailovsky, S. Xu, A. Malko, J. Hollingsworth, C. Leatherdale, H.-J. Eisler, and M. Bawendi., Optical gain and stimulated emission in nanocrystal quantum dots., *Science*, 290, 314, 2000.

30. C.-Y. Chao, W. Fung, and L. J. Guo. Polymer microring resonators for biochemical sensing applications. *IEEE Journal of Selected Topics in Quantum Electronics*, 12, 134, 2006.

31. B. F. Howell and M. G. Kuzyk. Amplified spontaneous emission and recoverable photodegradation in polymer doped with dispersed orange 11. *Journal of the Optical Society of America*, 19, 1790, 2002.

32. B. F. Howell and M. G. Kuzyk. Lasing action and photodegradation of disperse orange 11 dye in liquid solution. *Applied Physics Letters*, 85, 1901, 2004.

33. W. Fang and H. Cao. Wave interference effect on polymer microstadium laser. *Applied Physics Letters*, 91, 041108, 2007.

34. Kallinger, M. Hilmer, A. Haugeneder, M. Perner, W. Spirkl, U. Lemmer, J. Feldmann, U. Scherf, K. Müllen, A. Gombert, and V. Wittwer. A flexible conjugated polymer laser. *Advanced Materials*, 10, 920, 1998.

35. Y. Huang, G. Paloczi, J. Scheuer, and A. Yariv. Soft lithography replication of polymeric microring optical resonators. *Optics Express*, 11, 2452, 2003.

36. M. Armani, A. Srinivasan, and K. J. Vahala. Soft lithographic fabrication of high Q polymer microcavity arrays. *Nano Letters*, 7, 1823, 2007.

37. Y. Chen. Nanoembossing techniques. In *Encyclopedia of Nanoscience and Nanotechnology*, H.S. Nalwa, editor. American Scientific Publisher, 2004.

38. T. Ben Messaoud and J. Zyss. Unidirectional laser emission from polymer-based spiral microdisks. *Applied Physics Letters*, 86, 241110, 2005.

39. O. G. Peterson, J. P. Webb, W. C. McColgin, and J. H. Eberly. Organic dye laser threshold. *Journal of Applied Physics*, 42, 1917, 1971.

40. P. P. Sorokin, J. R. Lankard, V. L. Moruzzi, and E. C. Hammond. Flashlamp-pumped organic-dye lasers. *Journal of Chemical Physics*, 48, 4726, 1968.

41. H. G. L. Schwefel, N. B. Rex, H. E. Tureci, R. K. Chang, A. D. Stone, T. Ben Messaoud, and J. Zyss. Dramatic shape sensitivity of directional emission patterns from similarly deformed cylindrical polymer lasers. *Journal of the Optical Society of America*, 21, 923, 2004.

42. W. Fang, H. Cao, and G. S. Solomon. Control of lasing in fully chaotic open microcavities by tailoring the shape factor. *Applied Physics Letters*, 90, 081108, 2007.

43. J. Gao, P. Heider, C. J. Chen, X. Yang, C. A. Husko, C. W. Wong. Observations of interior whispering gallery modes in asymmetric optical resonators with rational caustics. arxiv:0707.2552, 2007.

44. M. L. M. Balistreri, D. J. W. Klunder, F. C. Blom, A. Driessen, H. Hoekstra, J. P. Korterik, L. Kuipers, and N. F. van Hulst, Visualizing the whispering gallery modes in a cylindrical optical microcavity, *Optics Letters*, vol. 24, 1829, 1999.

45. M. L. M. Balistreri, D. J. W. Klunder, F. C. Blom, A. Driessen, J. P. Korterik, L. Kuipers, and N. F. van Hulst. Experimental analysis of the whispering-gallery modes in a cylindrical optical microcavity. *Journal of the Optical Society of America*, 18, 465, 2001.

46. R. Quidant, J.-C. Weeber, A. Dereux, G. Leveque, J. Weiner, and C. Girard. Addressing and imaging microring resonators with optical evanescent light. *Physical Review B*, 69, 081402(R), 2004.

47. J. Knight, N. Dubreuil, V. Sandoghdar, J. Hare, V. Lefevre-Seguin, J.-M. Raimond, and S. Haroche. Mapping whispering-gallery modes in microspheres with a near-field probe. *Optics Letters*, 20, 1515, 1995.

48. N. Louvion, D. Gérard, J. Mouette, F. de Fornel, C. Seassal, X. Letartre, A. Rahmani, and S. Callard. Local observation and spectroscopy of optical modes in an active photonic-crystal microcavity. *Physical Review Letters*, 94, 113907, 2005.

49. M. L. M. Balistreri, H. Gersen, J. P. Korterik, L. Kuipers, and N. F. van Hulst. Tracking femtosecond laser pulses in space and time. *Science*, 294, 1080, 2001.

50. H. Gersen, D. J. W. Klunder, J. P. Korterik, A. Driessen, N. F. van Hulst, and L. Kuipers. Propagation of a femtosecond pulse in a microresonator visualized in time. *Optics Letters*, 29, 1291, 2004.

51. Courvoisier, V. Boutou, J. P. Wolf, R. K. Chang, and J. Zyss. Deciphering output coupling mechanisms in spiral microcavities with femtosecond light bullets. *Optics Letters*, 30, 738, 2005.

52. K. Geetha, M. Rajesh, V. P. N. Nampoori, C. P. G. Vallabhan, and P. Radhakrishnan. Propagation characteristics and wavelength tuning of amplified spontaneous emission from dye-doped polymer. *Journal of Applied Optics*, 45, 764, 2006.

53. M. Lebental, N. Djellali, C. Arnaud, J.-S. Lauret, J. Zyss, R. Dubertrand, C. Schmit, and E. Bogomolny. Inferring periodic orbits from spectra of simply shaped micro-lasers. *Physical Review A*, 76, 023830, 2007.

54. M. Armani and K. J. Vahala. Heavy water detection using ultra-high-Q microcavities. *Optics Letters*, 31, 1896, 2006.

55. J. Wiersig and M. Hentschel. Combining directional light output and ultralow loss in deformed microdisks. *Physical Review Letters*, 100, 013910, 2008.

56. J. Wiersig. Boundary element method for resonances in dielectric microcavities. *Journal of Applied Optics: Pure and Applied Optics*, 5, 53, 2003.

57. J. Wiersig. Hexagonal dielectric resonators and microcrystal lasers. *Physical Review A*, 67, 023807, 2003.

58. R. Balian and C. Bloch. Distribution of eigenfrequencies for the wave equation in a finite domain. *Annals of Physics*, 60, 401, 1970.

59. E. Bogomolny. Quantum and Arithmetical Chaos, pre-print online/0312061 (2003), Les Houches summer school: Frontiers in Number Theory, Physics and Geometry, 2003.

60. H. Alt, C. Dembowski, H.-D. Graef, R. Hofferbert, H. Rehfeld, A. Richter, and C. Schmit. Experimental vs. numerical eigenvalues of a bunimovich stadium billiard – a comparison. *Physical Review E*, 60, 2851, 1999.

61. R. Dubertrand, E. Bogomolny, N. Djellali, M. Lebental, and C. Schmit. Trace formula in open dielectric billiards, in preparation.

62. L. A. Bunimovich. On the ergodic properties of nowhere dispersing billiards. *Communications in Mathematical Physics*, 65, 295, 1979.

63. Matsko and V. Ilchenko. Optical resonators with whispering-gallery modes - Part I: basics. *IEEE Journal of Selected Topics in Quantum Electronics*, 12, 3, 2006.

64. M. T. Hill, H. J. S. Dorren, T. de Vries, X. J. M. Leijtens, J. H. den Besten, B. Smalbrugge, Y.-S. Oei, H. Binsma, G.-D. Khoe, and M. K. Smit. A fast low-power optical memory based on coupled micro-ring lasers. *Nature*, 432, 206, 2004.

65. T. Aoki, B. Dayan, E. Wilcut, W. P. Bowen, A. S. Parkins, T. J. Kippenberg, K. J. Vahala, and H. J. Kimble. Observation of strong coupling between one atom and a monolithic microresonator. *Nature*, 443, 671, 2006.

66. L. Kaminetzky and J. Keller. Diffraction coefficients for higher order edges and vertices. *SIAM Journal of Applied Mathematics*, 22, 109, 1972.

67. M. Choi, T. Tanaka, T. Fukushima, and T. Harayama. Control of directional emission in quasistadium microcavity laser diodes with two electrodes. *Applied Physics Letters*, 88, 211110, 2006.

68. E. Bogomolny and C. Schmit. Structure of wave functions of pseudointegrable billiards. *Physical Review Letters*, 92, 244102, 2004.

7

Optical Microfiber Loop and Coil Resonators

Misha Sumetsky

OFS Laboratories

CONTENTS

7.1 Introduction

This Chapter reviews recent progress in theory, fabrication, and application of optical microfiber (MF) resonators, i.e., a microfiber loop resonator (MLR) and a microfiber coil resonator (MCR). Generally, MF devices and circuits can be created by manipulation (bending, looping, coiling, twisting, crossing, etc.) of uniform and tapered MFs. An optical MF is usually fabricated from a standard telecom optical fiber by drawing and has a diameter of ~1 μm. The interest in photonic devices fabricated of MFs is caused by small losses and potential ability of micro-assemblage in 3D. Eventually, these properties could make possible the creation of MF devices, which are more compact and perform better than those fabricated lithographically. Furthermore, some MF-based devices possess functionalities, which are not possible or much harder to achieve by other means. MLR and MCR resonators together with MF waveguide are the basic elements of MF photonics, which is introduced in Section 7.2. Section 7.3 reviews the theory, experimental demonstration, and application of an MLR. Section 7.4 reviews the theory, experimental demonstration, and application of an MCR.

7.2 Microfiber Photonics

Microfiber photonics is a field of optics exploring devices fabricated from MFs. The history of MF fabrication is not easy to follow. It may be as long as the history of the glassmaking art: whenever a piece of glass is melted in a flame and then pulled out, the melted part becomes thinner and thinner. The thickness of the melted glass segment may achieve micron and submicron dimensions. In applications of optical MF, it is suggested that the MF diameter is comparable to or less than the wavelength of propagating light, i.e. of the order of 1 µm. For this reason, an MF supports a single or a few propagating modes. These modes are strongly guided and can survive fairly strong bending and tapering. To the best of the author's knowledge, the first optical application of MFs was demonstrated in 1959 by Kapany, who created a fiber bundle consisting of numerous micron and submicron diameter MFs and used it for transmission of images [1,2]. Thirty years later, Caspar and Bachus fabricated the first MLR of a 8-µm diameter fiber immersed in a polymer [3]. In 1990, Mackenzie and Payne implemented an MF immersed in a dye solution for optical amplification [4]. The large fraction of the evanescent field propagating in a dye solution allowed a significant reduction in the pump power [4–6]. In 1997, Knight and co-authors applied an MF for low-loss excitation of the whispering gallery mode resonances in a microsphere [7]. In 1999, Bures and Ghosh experimentally investigated transmission of silica MF down to ~400 nm diameter [8]. In 2000, Birks and co-authors demonstrated strong nonlinear effects and supercontinuum generation in an MF [9]. In 2003, Tong and co-authors fabricated uniform MFs with diameters as small as 50 nm [10]. This paper has renewed attention to MFs as potential building blocks for miniature photonic devices. A method for the fabrication of MF photonic circuits by wrapping MFs onto central optical rods has been suggested in Ref. [11]. Very low transmission losses of an MF, ~0.001 dB/mm, had been demonstrated experimentally [12,13]. Several methods of fabrication of MFs, whose diameter can be as small as a few tens of nanometers, were implemented. These methods include conventional flame drawing [4,7], indirect flame drawing [10], indirect laser drawing [14], and electric microheater drawing [15]. A comprehensive method for MF diameter characterization with sub-nanometer accuracy has been developed [16]. A MF with a diameter much smaller than a micron is often called a nanofiber. An optical nanofiber has the diameter of the order of 100 nanometers. Very thin nanofibers cannot transmit light due to fundamental limitations [17–20].

Generally, a photonic circuit fabricated from MFs is a step back from a photonic circuit based on lithographic technology (similar to wired circuits as compared to printed-in circuits in electronics). However, two advantages of MFs compared to lithographically-fabricated planar waveguides should be mentioned: significantly smaller losses for a given index contrast and ability of micro-assemblage in 3D [11]. Figure 7.1 illustrates an example of how an optical MCR can be replaced by a planar photonic circuit consisting of several ring resonators coupled to a bus waveguide. From this Figure, the following advantages of an MCR device become clear: (a) smaller dimensions due to 3D assemblage; (b) simple low-loss input and output connections; (c) creation of a single continuous fiber; (d) possibility of fabrication with feedback; and (e) smaller transmission loss than that of a planar wave-guide, whose surface roughness is introduced by etching. Possible functional elements, which can be created by wrapping an MF around a cylindrical surface, are a delay line, a sequence of uniform and chirped MLR, and a uniform MCR illustrated in Figure 7.2a, b, c and d, respectively. Eventually, the MF coils can be assembled into more complex functional MF circuits as shown in Figure 7.3.

In order to create a photonic circuit composed of MFs, one has to learn how to manipulate them. The simplest elements of an MF circuit are straight segments, bends, couples, and loops. These elements can be composed in a plane and are similar to the corresponding elements of lithographically-fabricated photonic circuits. Using MFs, one can create more compact photonic circuits compared to the conventional planar circuits by assembling them in 3D, e.g., by wrapping on an optical rod. Figure 7.4a shows an example of the MF macro-manipulation, which was used in earlier studies [14,21,22] for creating of an MF loop

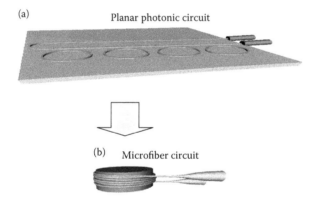

FIGURE 7.1
Example of shrinking a planar ring resonator circuit into an MCR. (a) Planar circuit consisting of ring resonators coupled to a bus waveguide; (b) MCR with the same functionality. (From Sumetsky, M., *IEEE J. Lightwave Technol.*, 26, 21–27, 2008.)

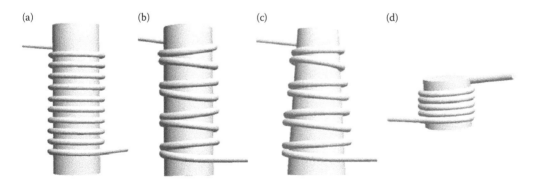

FIGURE 7.2
Simplest MF devices, which can be fabricated by coiling an MF around optical rod: (a) a delay line; (b) and (c) a uniform and a chirped sequence of loop resonators; (d) a coil resonator.

in free space illustrated in Figure 7.5a and described in Section 7.3. Macro-manipulation of an MF in free space does not allow the creation of MF bends with small radii of curvature. However, this problem can be solved with optical rods [23–26] as described in Section 7.4. Figure 7.4b illustrates the micro-manipulation of MFs using the tips of a scanning tunneling microscope. With this technique, fabrication of short MF segments, an MF coupler, and an MF knot illustrated in Figure 7.5b were demonstrated [10,27–31].

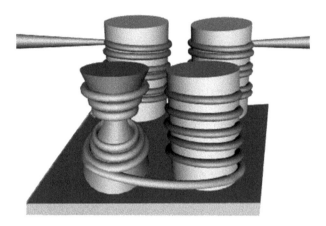

FIGURE 7.3
Illustration of a MF photonic circuit.

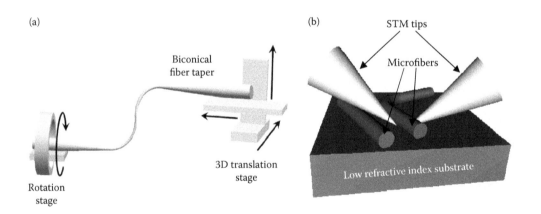

FIGURE 7.4
(a) MF macro-manipulation with translation and rotation stages; (b) MF micro-manipulation with two STM tips, which hold and bend an MF.

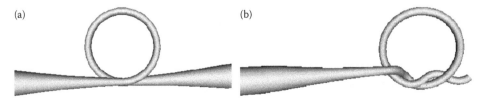

FIGURE 7.5
(a) an MF loop resonator; and (b) an MF knot resonator.

7.3 Microfiber Loop Resonator (MLR)

The MLR is a miniature version of a fiber loop resonator, which was created for the first time back in 1982 from a single-mode fiber and a directional coupler [32]. In 1989, Caspar and Bachus [3] demonstrated a 2-mm diameter MLR fabricated from a 8.5-μm diameter silica MF. The MF thickness was too large to ensure sufficient coupling in air. Therefore, in order to enhance the coupling efficiency, the MLR was immersed into a silicone-rubber with the refractive index close to the refractive index of silica. The Q-factor of the MLR fabricated by Caspar and Bachus [3] was Q~27,000. Recently [10,14,21,22,27–31], MLR fabrication of a micrometer-order diameter microfiber were investigated. In Tong et al. [10], a knot MLR was fabricated of a 0.95 μm diameter MF with a Q-factor of ~1500. In Sumetsky et al. [14], a microloop fabricated of a 0.7 μm diameter microfiber exhibited strong periodic interferometric transmission power spectral oscillations with extinction ratio ~18 dB at wavelengths of 1.5 μm. In Sumetsky et al. [21], the MLR fabricated of a 0.66 μm diameter MF had a Q-factor ~15,000. In the same paper an MLR was tuned to the regime of critical coupling with transmission spectral oscillations of ~34 dB. In Jiang et al. [28], a knot MLR with Q~31,000 was demonstrated. MLRs fabricated by Sumetsky et al. [22] exhibited smooth and uniform spectral characteristics in the full C-band having a loaded Q-factor up to 120000 and an intrinsic Q-factor of up to 630,000. In Sumetsky et al. [22] the MLR resonator was demonstrated as a sensitive and fast sensor of temperature variations. Recently, an MF knot resonator was used to demonstrate an MF knot laser [29,30]. In addition, a knot MLR embedded in a polymer matrix was demonstrated by Vienne et al. [31]. This section reviews the theory, the progress in experimental realization, and applications of an MLR.

7.3.1 Theory of an MLR

It is assumed below that the MLR depicted in Figure 7.6 is comprised of a single-mode MF. For simplicity, we ignore coupling between polarization states. This assumption usually

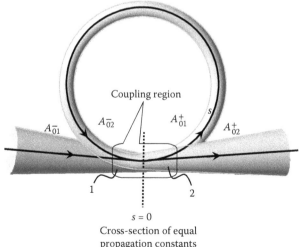

FIGURE 7.6
Illustration of an MLR. The coupling region is outlined. The numbers 1 and 2 correspond to the coupled microfiber segments 1 and 2. (From Sumetsky, M., Dulashko, Y., Fini, J. M., Hale, A., and DiGiovanni, D. J., *IEEE J. Lightwave Technol.*, 24, 242–250, 2006.)

holds for an MLR shown in Figure 7.5a and may be wrong for a knot MLR shown in Figure 7.5b where the twisting effects are significant. In the latter case, the results presented in this subsection should be modified by taking into account coupling between polarization states [33].

7.3.1.1 Transmission Amplitude

The electromagnetic field propagating along the MF is characterized by the propagation constant $\beta(s)$, which can slowly vary along the MF length s, e.g., due to MF diameter variation. Generally, function $\beta(s)$ is polarization dependent. In the region of coupling, near point $s=0$ (Figure 7.6), we define $\beta_1(s)$ and $\beta_2(s)$ as the propagation constants of the adjacent MF segments 1 and 2, respectively. The electromagnetic field amplitudes in segments 1 and 2, $A_1(s)$ and $A_2(s)$, are determined by the coupled wave equations [11,33]:

$$\frac{d}{ds}\begin{pmatrix} A_1 \\ A_2 \end{pmatrix} = i \begin{pmatrix} 0 & \kappa(s)\exp\left(i\int_{s_1}^{s}\Delta\beta(s)ds \right) \\ \kappa(s)\exp\left(-i\int_{s_1}^{s}\Delta\beta(s)ds \right) & 0 \end{pmatrix} \tag{7.1}$$

$$\Delta\beta(s) = \beta_1(s) - \beta_2(s)$$

where $\kappa(s)$ is the inter-fiber coupling coefficient. From this equation, the transmission amplitude of the MLR, T, can be found in the form, which is similar to that of the ring resonator [34]:

$$T = \frac{e^{-\alpha/2}e^{i\Psi} - \sin(K)}{1 - \sin(K)e^{-\alpha/2}e^{i\Psi}},$$

$$\Psi = \text{Re}(\Theta), \quad \alpha = 2\,\text{Im}(\Theta) \ll \Psi, \tag{7.2}$$

$$\Theta = \frac{1}{2}\int_{0}^{S}(\beta_1(s) + \beta_2(s))ds + \varphi, \quad \varphi \ll \Psi.$$

Here, the parameter K and phase φ characterizes the coupling between the adjacent MF segments, parameter α characterizes the roundtrip attenuation, and S is the circumference of the MLR. The MLR is nonreflecting device and, for lossless propagation, $|T|=1$. The losses in the MF loop are taken into account by the complex-valued propagation constant β, and, consequently, by a complex valued phase Θ. If losses are uniformly distributed along the MF length, we have $\text{Im}(\Delta\beta)=0$ and the imaginary component of propagation constant does not enter the coupling wave equation, Equation 7.2. If the coupling parameter $K=0$ then the loop is decoupled and, from Equation 7.2, the MLR transmission behavior is similar to that of a straight MF. In the opposite case of strong coupling, the resonances in transmission amplitude occur only if the coupling parameter K is close to the values

$$K_m = \frac{\pi}{2}(2m+1), \quad m = 1,2,\ldots, \tag{7.3}$$

which correspond to the full transmission of electromagnetic field from one of the adjacent MF segments to another. Then the resonances in transmission amplitude correspond to the condition:

$$\Psi = \Psi_{mn} = \pi(m+2n) \tag{7.4}$$

where n is an integer.

If the propagation losses are ignored, then the propagation constant is real and $|T|=1$. In this case, the MLR performs as an all pass filter [35] and the resonances appear in the group delay, t_d, only. The group delay is determined through the derivative of the phase of T with respect to radiation wavelength λ:

$$t_d = \frac{n_f \lambda^2}{2\pi c} \frac{d\ln(T)}{d\lambda}, \tag{7.5}$$

where c is the speed of light and n_f is the effective refractive index of the MF.

For relatively low losses, $\alpha \ll 1$, the resonances are well separated. In this case, in the neighborhood of the resonance wavelength λ_0 calculated from Equations 7.3 and 7.4, the expression for the transmission amplitude is simplified:

$$T = \frac{(\gamma_a - \gamma_c) + i(\lambda - \lambda_0)}{(\gamma_a + \gamma_c) - i(\lambda - \lambda_0)},$$

$$\gamma_a = \frac{1}{2}\left(\frac{\partial \Psi}{\partial \lambda}\right)^{-1}\Bigg|_{\lambda=\lambda_0} \alpha, \quad \gamma_c = \frac{1}{2}\left(\frac{\partial \Psi}{\partial \lambda}\right)^{-1}\Bigg|_{\lambda=\lambda_0} (K - K_m)^2 \tag{7.6}$$

Here the parameters γ_c and γ_a are the coupling and attenuation parameters, respectively.

From Equation 7.6, the transmission power of an MLR is:

$$P = \frac{(\gamma_a - \gamma_c)^2 + (\lambda - \lambda_0)^2}{(\gamma_a + \gamma_c)^2 + (\lambda - \lambda_0)^2} \tag{7.7}$$

7.3.1.2 Q-Factor, Extinction Ratio, and Finesse

The performance of a resonator is characterized by its Q-factor, which is determined as the ratio of the radiation wavelength in free space, λ, to the FWHM of the resonance at this wavelength, $\Delta\lambda$: $Q = \lambda/\Delta\lambda$. The Q-factor of the MLR is determined from Equation 7.7 as:

$$Q = \frac{\lambda_0}{2(\gamma_a + \gamma_c)}. \tag{7.8}$$

The Q-factor defined by Equation 7.8 is often called the loaded Q-factor. Sometimes it is also useful to introduce the intrinsic Q-factor,

$$Q_{\text{int}} = \frac{\lambda_0}{2\gamma_a}. \tag{7.9}$$

This Q-factor takes into account only internal losses of the MLR. From Equation 7.7, the extinction ratio of the resonance is:

$$R = 10 \cdot \log\left[\left(\frac{\gamma_c + \gamma_a}{\gamma_c - \gamma_a}\right)^2\right] \tag{7.10}$$

The finesse of a resonator is the ratio of its free spectral range (FSR) to the resonance FWHM: $F = \text{FSR}/\Delta\lambda$. From Equation 7.2

$$F = \frac{2\pi}{(K - K_m)^2 + \alpha}. \tag{7.11}$$

Equations 7.8 through 7.11 show that the MLR parameters contributing to its Q-factor, extinction ratio, and finesse can be divided into geometric and material parameters. The quantity $\gamma_c \sim (K - K_m)^2$ is a geometrical parameter. It characterizes the deviation from the condition of full transmission of electromagnetic field from one of the adjacent MF segments to another. This parameter is determined by the shape of the coupling MF segments and their mutual position. The quantity $\gamma_a \sim \alpha$ is a material parameter because it characterizes losses in the MF due to microscopic elastic and inelastic scattering of light. Currently, the propagation losses achieved for the silica microfiber with diameter $d \sim 1$ μm is $\sim 10^{-3}$ mm^{-1} [13]. From Equation 7.9 the corresponding estimate of the intrinsic Q-factor is $Q_{int} \sim 10^7$.

7.3.1.3 Models of Directional Coupling

The coupling parameter K in Equation 7.2 is calculated from the coupling wave equations, Equation 7.1. In Sumetsky et al. [22], two situations were considered: (a) a case when the propagation constants of the adjacent microfiber segments are close, $\Delta\beta(s) \ll \kappa(s)$, and (b) a case when the propagation constants of the adjacent microfiber segments are equal at a point $s = 0$ (Figure 7.6) and their difference linearly changes near this point. The latter case can be treated with the so called Landau-Zener model [36].

For coupling of the MF segments with close and slow varying diameters, the solution of Equation 7.1 can be found by perturbation theory. In the neighborhood of a resonance, $|K_0 - K_m| \ll 1$, the expression for K calculated along the coupling region $s_1 < s < s_2$ has the form [11,22]:

$$K = K_m + \sqrt{(K_0 - K_m)^2 + \varepsilon^2},$$

$$\varepsilon = \int_{s_1}^{s_2} \Delta\beta(s)\sin\left(2\int_{s}^{s_2} ds'\kappa(s')\right)ds, \quad K_0 = \int_{s_1}^{s_2} \kappa(s)ds. \tag{7.12}$$

Equation 7.12 allows to estimate the reduction of the MLR Q-factor due to microfiber non-uniformity. From this equation, the minimum possible deviation of the coupling parameter, K, from its resonance value K_m is ε. Assuming $\kappa(s_2 - s_1) \sim 1$, $\varepsilon \sim \Delta\beta/\kappa \sim \Delta d\beta/\kappa d$, $\partial\beta/\partial\lambda \sim \beta/\lambda$, (where d is the MF diameter and Δd is the characteristic diameter variation), and neglecting losses, $\gamma = 0$, we have from Equation 7.9 [22]:

$$Q_{int} \sim \frac{\kappa^2 d^2 S}{(\Delta d)^2 \beta}. \tag{7.13}$$

Setting the coupling length $s_2 - s_1 \sim \kappa^{-1} \sim 10$ μm, the MF diameter $d \sim 1$ μm, the loop length $S \sim 10^3$ μm, the propagation constant $\beta \sim 5$ μm, and the relative diameter variation $\Delta d/d \sim 10^{-3}$ over 10 μm, we have $Q_{int} \sim 10^6$. From this estimate, the MF uniformity is critical for achieving the high Q-factor. In certain situations, Equation 7.13 can significantly underestimate the Q-factor. For example, if the coupling region is symmetric with respect to the plane

$s=0$ then $\kappa(s)$ is a symmetric and $\Delta\beta(s)$ is an antisymmetric function. Then, at resonance, $K_0 = (\pi/2)(2m+1)$, Equation 7.12 gives $\varepsilon=0$. This means vanishing of γ_c in the considered first-order approximation and possibility of much greater Q-factors.

Assume now that the diameters of adjacent MF segments and, hence, their propagation constants are equal at point $s=0$ (see Figure 7.6), i.e., $\Delta\beta(0)=0$, and diverge away from this point. In this case, coupling can be studied using the Landau-Zener approximation developed in quantum mechanics [36]. If the coupling region is symmetric with respect to the plane $s=0$ then $\Delta\beta(0)=0$ and $d\kappa/ds\,|_{s=0}=0$. In the immediate neighborhood of the point $s=0$ we can set $\Delta\beta(s)=vs$ and $\kappa(s)=\kappa(0)$. Then the coupling wave equation, Equation 7.1 can be reduced to the parabolic cylinder differential equation. As a result, the coupling parameter is determined by the following Landau-Zener formula [36,37]:

$$K = \arcsin\left[\sqrt{1-\exp\left(-\frac{\pi\kappa^2(0)}{v}\right)}\right]. \tag{7.14}$$

From this equation, if the propagation constant is an adiabatically slow varying function, so that $v \ll \pi\kappa^2(0)$, then K is exponentially close to the resonant coupling parameter $K=\pi/2$. In the opposite case, $v \gg \pi\kappa^2(0)$, the coupling coefficient K is close to zero and the adjacent MF segments are decoupled. Equation 7.14 is important for modeling of the MLR. It defines the rate of MF tapering at which one can expect the proximity of the coupling parameter to its resonance value.

7.3.2 Experimental Demonstration and Applications of MLR

Fabrication of the MLR consists of drawing a MF and bending it into a self-coupling loop. Two types of MLR have been demonstrated experimentally: a regular MLR illustrated in Figure 7.5a [3,14,21,22] and a knot MLR illustrated in Figure 7.5b [10,27–31]. Methods of fabrication, transmission properties, and application of a regular MLR are reviewed in Section 7.3.2.1. Methods of fabrication, transmission properties, and application of a knot MLR are reviewed in Section 7.3.2.2.

7.3.2.1 MLR Fabricated by Macro-Manipulation

The setup for the MRL fabrication is illustrated in Figure 7.4a. It consists of three translation stages, which can move the MF ends with respect to each other, and a rotation stage, which can twist the MF. In papers [14,21,22], a biconical taper with a MF waist was fabricated with the CO_2 laser indirect drawing technique [14]. A standard single-mode fiber was placed into a sapphire capillary, which was heated by the radiation of a CO_2 laser and played a role of a microfurnace. The CO_2 laser drawing technique may have certain advantages compared to MF drawing in a flame [3–7] or in close proximity to a flame [10] due to its cleanness and strong suppression of air flow near the MF. The latter allows fabricating of very thin MFs having diameters below 100 nm [14]. In experiment [21], an MLR was fabricated of a MF with diameter 0.66 μm. The achieved Q-factor was 15,000. In Sumetsky et al. [22], an MLR was fabricated of 0.9 μm diameter MF. The circumference length of MLR was around 2 mm and the total length of the biconical taper was 25 mm. The MF loop was created by manipulating both the translation stages and rotation stage. Controlling the shape of the MF with an optical microscope, the authors coiled the taper waist into a self-touching loop. Figure 7.7 shows an optical microscope image of this loop with the input

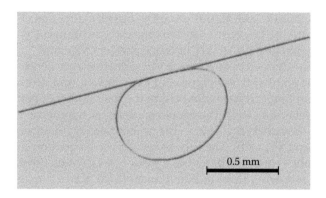

0.5 mm

FIGURE 7.7
Optical microscope image of an MLR. (From Sumetsky, M., Dulashko, Y., Fini, J. M., Hale, A., and DiGiovanni, D. J., *IEEE J. Lightwave Technol.*, 24, 242–250, 2006.)

and output ends aligned parallel to each other. The parallel alignment of the adjacent MF segments was possible due to surface attraction forces (Van der Waals and electrostatic), which kept the ends together, apparently overcoming the elastic forces that would tend to straighten out the microfiber. In Sumetsky et al. [22], the characteristic diameter of a MF used for fabrication of a MLR was ~0.9 μm. For this MLR, the MF diameter variation in the coupling region was slower than for the MLR of Ref. [21], which resulted in smaller losses, higher coupling efficiency, and, eventually, in higher Q-factor of the MLR.

7.3.2.1.1 Transmission Spectrum

The spectrum of the unpolarized light transmitted through the MLR was measured [21,22]. Figure 7.8 shows an example of very uniform and smooth transmission spectra in the C-band (from 1520 to 1570 nm), which were obtained by translation of the MLR ends to different positions with respect to each other [22]. It was found that the loaded/intrinsic Q-factor and finesse of the resonances in Figure 7.8a achieve the values of 22,000/60,000 and 9, in Figure 7.8b these values are 95,000/630,000 and 42, and in Figure 7.8c these values are 120,000/155,000 and 35, respectively. The achieved maximum loaded Q-factor is close to the largest Q-factors demonstrated for the planar ring resonators [38–41]. A small asymmetry in each of resonances in Figure 7.5a is due to the superposition of TE and TM modes, which are shifted in phase and have different Q-factors. In Figure 7.8b and c, the resonances of different polarization states are well separated. The amplitudes of resonances are slowly changing with wavelength. The change is caused by the weak dependence of coupling and attenuation parameters in Equation 7.7 on wavelength. In order to calculate the Q-factor of an MLR the experimental data has been analyzed using the theory presented in Section 7.3.1. The total power transmitted through the MLR, P, is a sum of partial transmission powers for TE and TM polarizations:

$$P = P_1 + P_2 \tag{7.15}$$

where each of the transmission powers, P_1 and P_2 corresponds to one of the polarization states, TM or TE. The transmission powers P_j in Equation 7.15 are defined by Equation 7.7. The experimental spectrum was fitted with Equations 7.6 and 7.7 and averaged over the interval of instrumental averaging equal to 0.01 nm. For the case of Figure 7.8c, an example

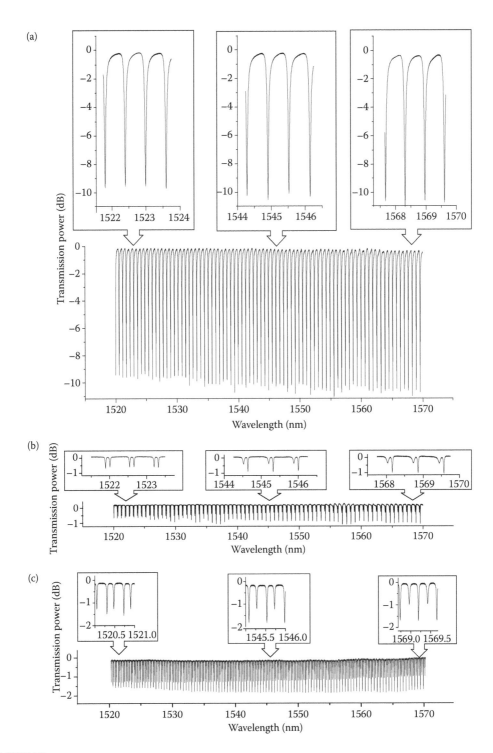

FIGURE 7.8
Transmission spectra of MCR tuned to loaded Q-factor of (a) 22,000, (b) 98,000, and (c) 120,000. (From Sumetsky, M., Dulashko, Y., Fini, J. M., Hale, A., and DiGiovanni, D. J., *IEEE J. Lightwave Technol.*, 24, 242–250, 2006.)

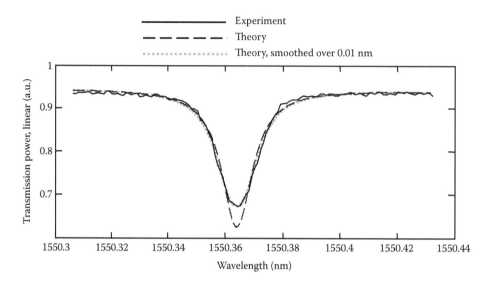

FIGURE 7.9
Fitting the experimental data of Figure 7.8c near a resonance (solid) with the theoretical calculation (dashed) averaged over the interval of instrumental averaging equal to 0.01 nm (dotted). (From Sumetsky, M., Dulashko, Y., Fini, J. M., Hale, A., and DiGiovanni, D. J., *IEEE J. Lightwave Technol.*, 24, 242–250, 2006.)

of fitting the theoretical model to the experimental resonance peak is shown in Figure 7.9. The best fit is achieved for the following parameters: $\alpha_1=0.14$, $\alpha_2=0.19$, $\sin(K_1)=0.981$, and $\sin(K_2)=0.985$. From the image of the MLR shown in Figure 7.7, the width of the coupling region is ~10–100 μm; therefore the coupling coefficient $\kappa \sim 10^{-1}-10^{-2}$ μm^{-1}. It can be shown that for the propagation constant $\beta \sim 5$ μm^{-1} and the characteristic length of its variation $S_c \sim 10^3$ μm, the coupling parameter defined by Equation 7.14 can be away from the resonance value as well as exponentially close to it. For the known MF diameter variation (measured, e.g. with an SEM or with the more accurate method [16]), the accurate values for the parameters entering Equation 7.14 can be calculated numerically.

7.3.2.1.2 Ultra-Fast Direct Contact Gas Temperature Sensor

Microring, microsphere, microdisk, microcylinder, and microcapillary resonators have been suggested as sensors of physical, chemical, and biological changes of the environment [42–49]. Narrow spectral resonances of these devices are very sensitive to variations in the effective refractive index and dimensions of the resonator and the refractive index of the ambient medium. These parameters are changing with temperature, pressure, and applied radiation. A MLR positioned in free space has a much larger interfacial contact area with the environment than a microsphere and a microring resonator mounted on a substrate. Therefore, the MLR can perform as a more sensitive detector of environmental changes. In this Section, the performance of an MLR as an ambient gas local temperature sensor is reviewed following Ref. [22].

The characteristic intrinsic response time of an MF to temperature variations is $t_{int}=r^2\rho c/2k$, where r is the microfiber radius, ρ is its density, c is the specific heat, and k is the heat conductivity of silica. For a 1 μm diameter microfiber, $t_{int}\approx 1$μs. However, the equilibration time of an MF responding to change of ambient air temperature is greater: $t_{amb}=r\rho c/2h$, where $h\sim 400$ Wm^{-2}K^{-1} is the convection heat-transfer coefficient. For the same MF, $t_{amb}\approx 2$ms. Figure 7.10 illustrates the experiment performed in Ref. [22], which models the

MLR temporal response. An MLR, whose transmission spectrum is shown in Figure 7.11a, was placed into a 7.5 Hz frequency on/off modulated CO_2 laser field. First, the OSA was tuned to 1550.4 nm corresponding to a flat region of the spectrum in Figure 7.11a and recorded. The result for the time dependence of transmitted power shown in Figure 7.11b indicated no visible time response. However, at 1550.12 nm, which corresponded to a steep region near a resonance, the authors of Ref. [22] observed oscillations of transmission power

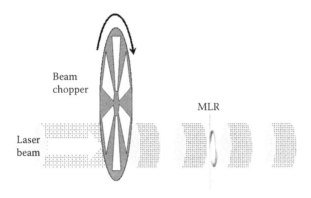

FIGURE 7.10
Periodic on/off heating of a MLR using a CO_2 laser beam and a beam chopper. (From Sumetsky, M., Dulashko, Y., Fini, J. M., Hale, A., and DiGiovanni, D. J., *IEEE J. Lightwave Technol.*, 24, 242–250, 2006.)

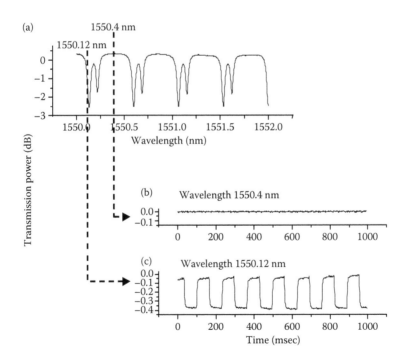

FIGURE 7.11
(a) Transmission spectrum of MLR used as a temperature sensor. Transmission power of MLR periodically heated by CO_2 laser as a function of time at wavelength 1550.4 nm (b) and 1550.12 nm (c). (From Sumetsky, M., Dulashko, Y., Fini, J. M., Hale, A., and DiGiovanni, D. J., *IEEE J. Lightwave Technol.*, 24, 242–250, 2006.)

shown in Figure 7.11c. From the transient oscillations in Figure 7.11c, the magnitude of temperature variation was estimated as 0.4 K, and the thermal equilibration time constant as 3 ms, in agreement with the theoretical prediction. For a regular optical fiber, which has a diameter 100 times greater, this constant is, proportionally, 0.3 s. The measured temperature range in this experiment was ~1 K. The range can be increased by orders of magnitude with decreasing Q and/or choosing a spectral region having smaller steepness. The temperature resolution of the demonstrated MLR thermometer is determined by resolution of the used optical spectrum analyzer and can be as small as ~0.1 mK.

7.3.2.2 Knot MLR Fabricated by Micro-Manipulation

A knot MLR is illustrated in Figure 7.5b. It was fabricated [10,27–31] with the micro-manipulation technique described in Section 7.2. As opposed to the regular MLR considered in Section 7.3.2.1, an MF knot MLR cannot be fabricated from a biconical fiber taper. Actually, in order to fabricate an MF knot, an MF open end should be available. This end is bent into a relatively large loop about a few millimeters in diameter, which is then tightened into a smaller knot by pulling the free end of the MF. Figure 7.12a shows an SEM image of a 290 μm diameter MF knot from Ref. [28]. A knot makes the loop more stable. This allows the fabrication of more robust MLR devices and, in particular, to immerse them in a polymer matrix [30]. The spectrum of a knot MLR may be complicated by coupling of polarization states due to the twisted geometry of a knot. Recently, an MF knot resonator was used to demonstrate an MF knot laser [29,30].

7.3.2.2.1 Transmission Spectrum

For optical characterization of the knot MLR, its open end is coupled to a second MF as illustrated in Figure 7.12b [28]. Figure 7.13 shows typical transmission spectra of two microfiber knots with diameters of 396 μm and 850 μm [28]. The 396 μm diameter MLR (Figure 7.13a), created from a 2.66-μm diameter MF, had a Q factor and a finesse ~10,000 and 9.2, respectively. The 850 μm diameter knot (Figure 7.13b), created from a 1.73-μm diameter microfiber, has a Q factor and a finesse ~ 57,000 and 22, respectively. These Q-factors [28] were of the same order of magnitude but smaller than the largest Q-factor obtained for a

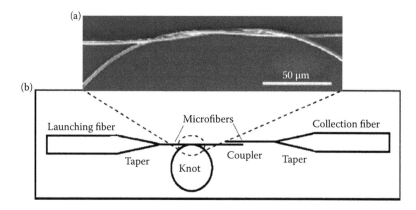

FIGURE 7.12
(a) SEM image of the twisted overlap region of a knot MLR with diameter of 290 μm. The MLR is fabricated from a 2.66-μm diameter MF. (b) Schematic diagram of a microfiber knot with two ends connected to standard fibers for light launching and collection through fiber tapers. (From Jiang, X., Tong, L., Vienne, G., Guo, X., Tsao, A., Yang, Q., and Yang, D., *Appl. Phys. Lett.*, 88, 223501, 2006.)

regular MLR in Ref. [22] (see Section 7.3.2.1.1). In addition, the transmission spectrum of a regular MLR shown in Figure 7.8 is noticeably smoother and more uniform than that of a knot MLR shown in Figure 7.13. Presumably, this is the result of twisting and polarization mode coupling in the knot MLR. Nevertheless, a knot MLR is a much more robust device than a regular MLR. This allows use of a knot MRL in contact with liquid and solid surrounding medium. In particular, the authors of Ref. [28] demonstrate a knot MLR in water with a Q-factor of 31,000. This indicates on opportunity of using a knot MLR as a microfluidic chemical and biological sensor. In addition, a knot MLR can be safely imbedded into a UV curable polymer. After curing, the device is secured in a solid polymer matrix and exhibits the resonant spectrum [31]. An optical microscope image of a knot MLR embedded in fluoroacrylate from Ref. [31] is shown in Figure 7.14.

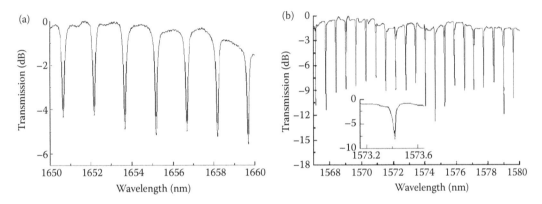

FIGURE 7.13
Transmission spectra of (a) a 396-µm diameter knot MLR assembled using a 2.66-µm diameter MF and (b) a 850-µm diameter knot MLR assembled using a 1.73-µm diameter MF. The inset shows a single resonance peak. (From Jiang, X., Tong, L., Vienne, G., Guo, X., Tsao, A., Yang, Q., and Yang, D., *Appl. Phys. Lett.*, 88, 223501, 2006.)

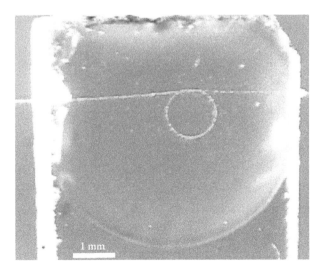

FIGURE 7.14
Optical microscope image of a knot MLR embedded in fluoroacrylate. The MF diameter is 5 µm. (From Vienne, G., Li, Y., and Tong, L., *IEICE Trans. Electron.* E90, 415–421, 2007.)

7.3.2.2.2 MF knot laser

Lasing of a knot MLR was demonstrated and reported in Refs. [29,30]. With an MLR, the length of optical light propagation in the active medium and the optical amplification can be resonantly increased. In Ref. [29], a knot MLR laser was formed from an Er:Yb doped 3.8 MF. The authors draw a 3.8 μm diameter MF from a standard single mode fiber and tied it into a knot MLR with diameter 2 mm as illustrated in Figure 7.12. The idea of Jiang et al.'s paper [30] was similar to Refs. [4–6], where an MF immersed in a dye solution was applied for optical amplification. To form a knot dye laser, the authors of Ref. [30] tied a 3.9-μm diameter MF into a knot MLR with a 350-μm diameter. The MLR was immersed into a rhodamine 6 G dye doped ethylene glycol solution. The pump light was sent into the MLR through the taper. Because of the small diameter of the MF and relatively low index contrast between the MF and the solution (1.46 vs 1.43), a noticeable fraction of the pump light was guided outside the fiber core as evanescent waves. When the pump power exceeded the threshold for lasing oscillation, a laser signal was clearly observed. Figure 7.15 shows three lasing groups A, B, and C, located around 567-, 570- and 580-nm wavelength, respectively.

7.4 Microfiber Coil Resonator (MCR)

The microfiber coil resonator (MCR) [11] is a 3D generalization of the loop and ring resonators [34]. This resonator can be created by wrapping an optical MF on a dielectric rod with smaller or the same refractive index as illustrated in Figure 7.2. Unique properties of an MCR offer a potential of using this device as a basic functional element for the MF photonics. Similarly to the 2D resonant microring structures [34,35] MCRs can perform complex filtering, time delay, switching, and lasing functions. An illustration of possible similarity between an MCR and a planar photonic circuit is given by Figure 7.1. In addition, MCR-based optical devices have significant advantages over planar devices, which were already

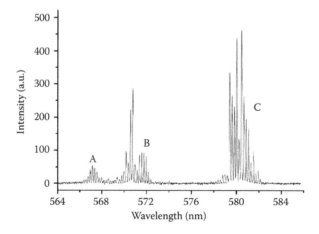

FIGURE 7.15

A Laser emission from a 350-μm diameter knot MLR fabricated from a 3.9-μm MF pumped by a pulsed frequency-double Nd:YAG laser. Typical lasing groups centered around wavelengths 567 nm (Group A), 570 nm (Group B) and 580 nm (Group C) are observed. (From Jiang, X., Song, Q., Xu, L., Fu, J., and Tong, L., *Appl. Phys. Lett.*, 90, 233501, 2007.)

mentioned in the beginning to this Chapter: smaller dimensions, simple low-loss input and output connections, possibility of fabrication with feedback, smaller transmission loss than that of a planar waveguide.

A simple case of an MCR is a higher-order MLR, determined as a coiled MF, which consists of several independent MLRs as, e.g., shown in Figure 7.2b. The design of higher-order MLRs is similar to that of all-pass ring resonators [35]. Alternatively, an MCR may be represented by a uniformly wrapped self-coupling coil shown in Figure 7.2d studied in Ref. [50]. Figure 7.2c illustrates an MCR wrapped on a nonuniform core studied in Refs. [51,52].

Currently, fabrication of MCRs is in its initial stage. The first experimental demonstration of an MCR containing more than one resonant loop was performed in Ref. [23] (see also Ref. [24]). Independently, a MCR, which exhibited single loop resonances, were demonstrated in Refs. [25,26].

7.4.1 Theory of an MCR

Consider propagation of light along a single mode optical MF wrapped on a dielectric rod having refractive index smaller than or equal to the index of the MF. The MF diameter is assumed to be smaller than or comparable to the wavelength of radiation. If the characteristic transversal dimension of the propagating mode is much smaller than the characteristic bend radius, then the adiabatic approximation can be applied. In this approximation and for a relatively small pitch of the MF coil, we take into account weak coupling between co-propagating light in the adjacent turns. Similar to the assumptions of Section 7.2, we ignore coupling between the TE and TM polarization modes. Then, the coupled wave equations, Equation 7.1, can be generalized for the case of an MCR with M turns [11]:

$$\frac{d}{ds}\begin{pmatrix} A_1 \\ A_2 \\ \dots \\ A_m \\ \dots \\ A_{M-1} \\ A_M \end{pmatrix} = i \begin{pmatrix} 0 & \chi_{12}(s) & 0 & \dots & 0 & 0 & 0 \\ \chi_{21}(s) & 0 & \chi_{23}(s) & \dots & 0 & 0 & 0 \\ 0 & \chi_{32}(s) & 0 & \dots & 0 & 0 & 0 \\ \dots & \dots & \dots & \dots & \dots & \dots & \dots \\ 0 & 0 & 0 & \dots & 0 & \chi_{M-1,M-2}(s) & 0 \\ 0 & 0 & 0 & \dots & \chi_{M-2,M-1}(s) & 0 & \chi_{M-1,M}(s) \\ 0 & 0 & 0 & \dots & 0 & \chi_{M,M-1}(s) & 0 \end{pmatrix}\begin{pmatrix} A_1 \\ A_2 \\ \dots \\ A_m \\ \dots \\ A_{M-1} \\ A_M \end{pmatrix} \qquad (7.15)$$

$$\chi_{pq}(s) = \kappa_{pq}(s)\exp\left(i\int_{s_1}^{s} (\beta_p(s)-\beta_q(s))ds \right).$$

Here, S is the length of a single turn and coefficients $A_m(s)$ are defined at the turns $s_1 < s < s_1 + S$ and satisfy the continuity conditions:

$$A_{m+1}(s_1) = A_m(s_1 + S)\exp[i\int_{s_1}^{s_1+S} \beta_m(s)ds], \quad m = 1,2\dots,M-1. \qquad (7.16)$$

Using Equations 7.15 and 7.16 it is straightforward to calculate the resonance spectrum of MCRs with interturn coupling changing along the length of the coiled MF. In Refs. [51,52] these equations were generalized to the case of nonuniform MCR with changing radius of turns. The transmission amplitude of an MCR can be calculated with Equations 7.15 and 7.16 numerically. Here we consider two situations when these equations allow analytical solution and full analysis. In the first case, an MCR is composed of M independent MLR. In the second case the MCR is composed of a uniformly coiled MF with M turns.

7.4.1.1 Higher-Order MLR

Higher-order MLR is a coiled MF, which consists of independent MLRs as illustrated in Figure 7.16a. In this case, coupling between adjacent turns of the MF coil is essential in each section containing an MLR, while it is negligible outside these sections. The transmission amplitude of a high-order MLR, T, is simply a product of transmission amplitudes of individual MLRs, T_m and the total group delay, t_d (see Equation 7.5), is a sum of delay times, t_{dm}, of each MLR:

$$T = \prod_{m=1}^{M} T_m, \quad t_d = \sum_{m=1}^{M} t_{dm}. \tag{7.17}$$

Therefore, the design of the higher-order MLR generating the predetermined group delay dependence is similar to the design of the all-pass ring resonators (see Ref. [35] and references therein).

7.4.1.2 Uniform MCR

For a uniform MCR illustrated in Figure 7.16b, Equation 7.15 is reduced to:

$$\frac{dA_1}{ds} = \kappa A_2$$

$$\frac{dA_m}{ds} = \kappa(A_{m-1} + A_{m+1}), \quad m = 1, 2, ..., M-1 \tag{7.18}$$

$$\frac{dA_M}{ds} = \kappa A_{M-1}$$

where κ is the constant coupling coefficient between adjacent turns. Solution of Equations 7.18 and 7.16 is [50]:

(a)   (b)

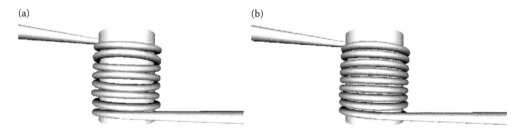

FIGURE 7.16
(a) Higher-order MLR and (b) uniform MCR.

$$A_m(s) = \sum_{n=1}^{M} \bar{a}_{n1} A_{mn}(s), \quad \|\bar{a}_{mn}\| = \|a_{mn}\|^{-1},$$

$$a_{mn} = \exp\left[iS\left(\beta + 2\kappa\cos\frac{\pi n}{M+1}\right)\right]\sin\left(\frac{\pi(m-1)n}{M+1}\right) - \sin\left(\frac{\pi mn}{N+1}\right), \tag{7.19}$$

$$A_{mn}(s) = \exp\left(2i\kappa s\cos\frac{\pi n}{M+1}\right)\sin\left(\frac{\pi mn}{M+1}\right), \quad m,n = 1,2,\ldots,M$$

With this equation, the transmission amplitude is defined as:

$$T = (A_1(0))^{-1}\sum_{n=1}^{M} \bar{a}_{n,1} A_{Mn}(S), \tag{7.20}$$

For the lossless MCR considered in this Section, $|T| = 1$. Equation 7.19 shows that the optical properties of MCR depend on three dimensionless parameters: the number of turns, M, the dimensionless propagation constant, $B = \beta S$, and the dimensionless coupling parameter, $K = \kappa S$. The parameters B and K, which define the eigenmodes of the MCR can be determined from the condition $t_d = \infty$.

The transmission spectrum of a uniform MCR with two turns ($N=2$) is similar to the transmission spectra of the loop and ring resonators [11,34]. In this case, the eigenmodes exist under the conditions of the full interturn light transition, $\kappa S = (\pi/2)(2m-1)$, and the constructive interference of electromagnetic field at the adjacent turns, $\beta S = (\pi/2)(4n-2m+1)$, where m and n are integers. If these conditions are met, the light in the input and output waveguides of the MCR is completely trapped by the MCR. During a roundtrip, the light transfers from one turn to another $2m-1$ times and fully returns to the original turn with the same phase.

For the uniform MCR with $M>2$, more than two turns are coupling simultaneously. This leads to a complex interplay of propagation along the MF and propagation through interturn coupling. Under certain conditions, the MCR with $M>2$ possesses eigenmodes. Figure 7.17 shows the surface relief of the time delay $t_d(B,K)$ for the MCRs with different numbers of turns, M. Because $t_d(B+2\pi,K)=t_d(B,K)$, only one full period of $t_d(B,K)$, $2\pi n < B < 2\pi(n+1)$, where n is an integer, is shown. It is seen that the shape of the MCR spectrum qualitatively depends on the value of K. The top row of plots in Figure 7.17 shows that the spectrum behavior is periodic as a function of coupling parameter, K, only for $M=2$ and $M=3$. For $M \geq 4$, the dependence on coupling parameter, K, is no longer periodic.

Notice the interesting evolution of the group delay spectrum with growth of M, which is shown in the lower rows of plots in Figure 7.17. All features of the spectrum including the eigenvalues, which appear for certain number of turns M in the lower row, do not disappear for larger M. These features move along the straight axial lines towards the point of spectral collapse, $(B_c,K_c)=(\pi(2n-1/2),1/2)$, where n is an integer. The features shrink proportionally to their distance from the collapse point and fully collapse at (B_c,K_c).

7.4.1.3 MCR Transmission Line

An MCR with infinite number of turns, $M \to \infty$, represents an interesting type of the optical transmission line. In this case, the first and last equations in Equation 7.18 can be ignored.

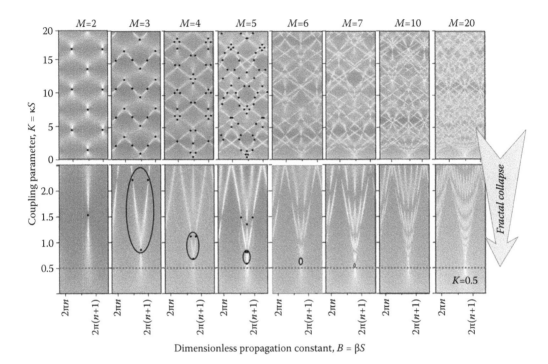

FIGURE 7.17
Surface plots of the time delay in the plane (B,K) for the number of turns M equal to 2 through 7, 10, and 20. Brighter points correspond to larger time delay. For $M=2$, 3, 4, and 5, the points corresponding to MCR eigenmodes are marked by black dots. Upper row of plots show the time delay for $0<K<20$, while the lower plots show the time delay for $0<K<2.5$ with higher resolution. In the lower plots, the circles mark similar features (V-shaped and W-shaped). For $M\to\infty$, all similar features tend to the collapse point. The ordinate of collapse points, $K=0.5$, is marked by a dotted line. (From Sumetsky, M., *Opt. Express*, 13, 4331–4340, 2005.)

Then Equation 7.18 is solved in the form $A_m^{(\pm k)}(s) = \exp[2i\kappa\cos(\xi S)s \pm i\xi Sm]$, where the integer m is a number of a turn and ξ is an effective propagation constant. The continuity condition, Equation 7.16, applied to this solution yields the dispersion relation [50]:

$$\omega(\xi) = \frac{c}{n_f}(\xi - 2\kappa\cos(\xi S)) \tag{7.21}$$

where $\omega=c\beta/n_f$ is the frequency of electromagnetic field, c is the speed of light, and n_f is the effective refractive index of the MF. It can be found from Equation 7.21 that the MCR transmission line does not have stop bands. Three qualitatively different situations occur depending on the value of coupling parameter, K. If $K<1/2$, the dispersion relation is monotonic as shown in Figure 7.18a1. In this case the dispersion relation is qualitatively similar to that of a SCISSOR shown in Figure 7.18c [34]. The crossover situation occurs at $K=1/2$. The corresponding dispersion relation is shown in Figure 7.18a2. In this case, function $\omega(\xi)$ has inflection points at $\xi_n = \pi(2n - \frac{1}{2})/S$, where n is an integer. In the vicinity of these points, $\omega(\xi) \approx \omega(\xi_n) + \alpha(\xi - \xi_n)^3$, and the group velocity is vanishing simultaneously with the inverse group velocity dispersion. Similar phenomenon has been predicted for photonic crystal waveguides [53]. A pulse in the neighborhood of these points experiences dramatic distortion. Finally, for the case of strong coupling illustrated in Figure 7.6a3,

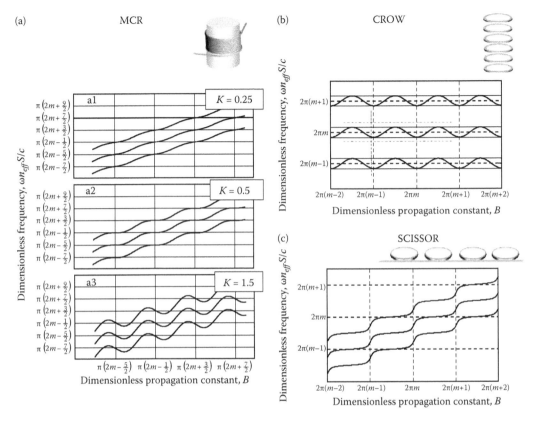

FIGURE 7.18
Comparison of characteristic dispersion relations for an MCR, a CROW, and a SCISSOR. (a) Dispersion relations for an MCR with the coupling parameter $K=0.25$ (a1), $K=0.5$, (a2), and $K=1.5$ (a3); (b) dispersion relation of a CROW; (c) dispersion relation of a SCISSOR. (From Sumetsky, M., *Opt. Express*, 13, 4331–4340, 2005.)

when $K>1/2$, the dispersion relation has the minima and maxima points. At these points, the group velocity is equal to zero. In the neighborhood of these points, propagation of light is similar to propagation near the band edges of photonics crystals and CROWs [34], which is illustrated in Figure 7.18b. However, in contrast to the CROW, the MCR is an all-pass and no bandgap structure.

7.4.2 Experimental Demonstration and Application of MCR

Manipulation of an MF in free space allows creation of a high Q-factor loop MLR reviewed in Section 7.2. However, it is not possible to create a multi-turn MCR by moving the ends of an MF in free space without touching and holding the intermediate part of the MF, which introduces additional transmission losses. Here, we describe methods of MCR fabrication by wrapping an MF on a central rod [23–26].

7.4.2.1 MCR in Air

This Section reviews recent results in fabrication of MCRs in air [24,25]. A setup used in Ref. [24] for wrapping a silica MF on a silica rod is illustrated in Figure 7.19. The MF was a part of 1.7 μm diameter waist of a biconical taper fabricated from a conventional single mode fiber. A 345 µm diameter silica fiber was used as an optical rod. The taper was connected to a broadband light source and to an optical spectrum analyzer, which controlled the MF transmission loss. The transmission spectrum of the free taper is shown in Figure 7.20, curve 1. Before wrapping, the silica rod was oriented perpendicular to the MF and put in contact with it. At this step, the transmission spectrum of the MF exhibited whispering gallery modes resonances generated in the rod, shown in Figure 7.20, curve 2. Then, the MF was wrapped on a silica rod using the rotation and 3D translation stages. In the process of wrapping performed by a rotation stage, the angle between the MF and the rod was

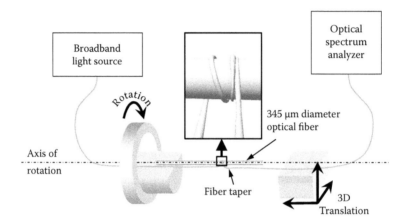

FIGURE 7.19
Illustration of the setup for fabrication of an MCR in air. (From Sumetsky, M., *IEEE J. Lightwave Technol.*, 26, 21–27, 2008.)

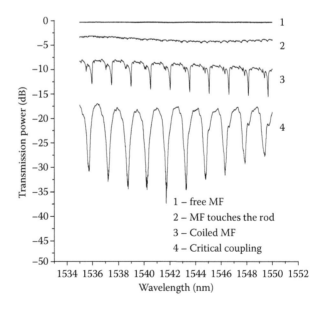

FIGURE 7.20
Transmission spectrum of an MCR fabricated in air. (1) A free MF; (2) a straight MF in contact with the optical rod; (3) an MCR tuned to the highest Q; (4) an MCR tuned to the highest extinction ratio. (From Sumetsky, M., *IEEE J. Lightwave Technol.*, 26, 21–27, 2008.)

controlled by a 3D translation stage. This angle was gradually increased so that it achieved ~90⁰ when the uniform MF part touched the rod (see Figure 7.19). This happened after ~2 turns were made. The angle between the MF and the rod was tuned in order to form a MF closed loop and to arrive at the transmission spectrum resonances with the greatest Q-factor and extinction ratio shown in Figure 7.20, curves 3 and 4. In order to reduce the insertion losses outside the loop section, it was necessary to wrap an MF adiabatically. In particular, the first contact point between the MF and the rod should be at a thicker section of the fiber waist. In addition, the MF should be slowly (adiabatically) reduced in thickness in continuous contact with the rod down to the radius of the MF loop. In the described experiment, these conditions were not fully achieved. Consequently, significant insertion loss was observed. In Figure 7.20, curve 3 is the transmission spectrum of the MF after it was wrapped on the rod and the resonant loop was aligned to maximize the resonance Q-factor of the device. The Q-factor ~20000 has been achieved. The insertion loss of the fabricated MCR was ~9 dB, which was primarily generated outside the resonant loop. The loss was caused by discontinuities in the contact between the rod and the MF, contamination of the silica surface, and nonadiabaticity of wrapping. Curve 4 in Figure 7.20 is the transmission spectrum of the same MF after it was aligned in order to arrive at the condition of critical coupling ($\gamma_a = \gamma_c$ in Equation 7.7), which maximized the extinction ratio of transmission power oscillations. The achieved extinction ratio was greater than 20 dB.

Fabrication of an MCR on a central rod with the refractive index lower than that of an MF may be more preferable because coupling between the MF mode and the rod modes can be better suppressed. In Ref. [25] an MCR was created by wrapping a silica MF on a Teflon AF rod with refractive index 1.29. The maximum achieved Q-factor of MCR was ~10,000, i.e. still less than that achieved for a silica rod. Potentially, the obtained Q-factor can be increased by improving the smoothness and uniformity of the low-index rod.

Thus, to date, fabrication of MCRs in air encountered several problems resulting in considerable propagation loss. The loss is introduced by contamination of the MF and rod surfaces and by discontinuities of the MF-rod contact line. It is believed that further development of MCR fabrication methods in air will allow to improve its Q-factor and reduce the insertion loss.

7.4.2.2 MCR in Low-Index Polymer

The problems in fabrication of MCRs, mentioned at the end of the previous Section, can be solved if the MCR is completely immersed in a liquid or solid matrix, e.g. a polymer [23,24,26]. In particular, the MCR can be immersed in an environment having its refractive index equal to the refractive index of the rod. In this case the MF does not 'see' the rod optically, and the MCR behaves as though in uniform space.

The experimental setup for fabrication of MCRs in liquid used in Ref. [23,24] is shown in Figure 7.21. The dimensions of the biconical taper used in this experiment were the same as described in the previous Section (the MF diameter was 1.7 μm, and the length of its uniform section was 3.5 mm). One of the taper ends was glued to the Γ-shaped leg, which was fixed at the rotating shaft. The other end was fixed at the fiber spring, which maintained the taper in a strained condition. The rod was fabricated of a silica fiber coated with a film of a cured, low-index polymer with low refractive index 1.384. The MF and the rod were immersed in a pool of the same uncured liquid polymer. The convex meniscus at the pool edges, shown in Figure 7.21, allowed free entrance of the MF into the liquid. Due to the small difference between the refractive indices of the cured and uncured polymer, it was assumed that the interface between them was almost invisible for the light, which

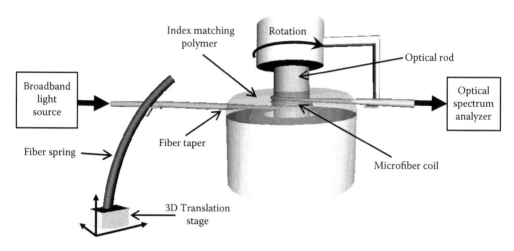

FIGURE 7.21

Illustration of a setup for the fabrication of an MCR in index-matching liquid. (From Sumetsky, M., *IEEE J. Lightwave Technol.*, 26, 21–27, 2008.)

propagates along the MF. The MF was wrapped onto the rod by rotating the shaft and the Γ-shaped leg. Fabrication of the MCR was performed in several steps. After the MF was immersed into the liquid polymer, the transmitted power decreased from ~−0.5 dB in free space to ~−1 dB. (curves 1 and 2 in Figure 7.22). This was caused by scattering of light at the meniscus surface. Next, the MCR was created by wrapping the MF around the rod. In the process of wrapping of the first turn, no noticeable transmission losses were observed. Curves 3 and 4 in Figure 7.22 show the transmission spectrum of the taper at rotation angles 180° and 360°, respectively. At rotation angle of 360° and greater, the relative position of MF ends were manipulated to create a closed loop and to initiate self-coupling. The self-coupling showed itself for the first time at ~450° when oscillations in transmission spectrum appeared (curve 5 in Figure 7.22). At larger angles, a more pronounced resonant spectrum was observed. Curve 6 in Figure 7.22 shows the resonant spectrum exhibiting the Q-factor of 61000, observed at 720°, i.e., for the two-turn MCR. From curve 6, which was fitted using Equation 7.7, the total loss of the resonant loop was estimated to be ~0.06 dB [24]. The resonance structure of this spectrum corresponded to a double loop MCR. This conclusion was based on the fact, that the group of four peaks, which appears periodically in curve 6, corresponds to two resonances of different polarization of the first and second loop of the MCR.

7.4.2.3 MCR Microfluidic Sensor

The advantage of an MF for sensing application compared to other optical waveguide sensors is its larger area of interaction with the ambient medium. On the other hand, the fragileness of free MF devices as, e.g., an MLR sensor considered in Section 7.3.2.1.2, restricts its applications. The design of an optical sensor by assembling MFs with more robust elements and embedding into the curable liquid can increase the robustness of the device as well as broaden its functionality. In Section 7.2 fabrication of an MLR embedded into a polymer matrix was considered [31]. The authors of Refs. [54,55] suggested and

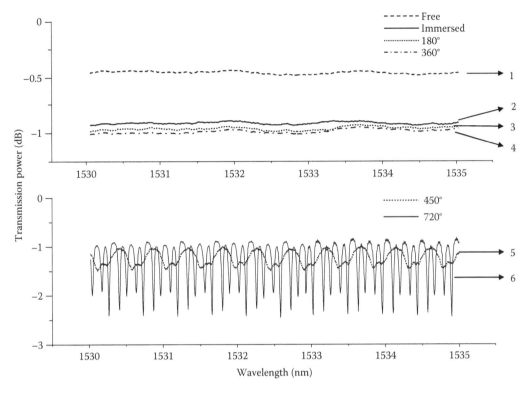

FIGURE 7.22
Transmission spectrum of the MF in the process of MCR fabrication: (1) in air; (2) in liquid polymer; (3) at 180° of rotation; (4) at 360° of rotation; (5) at 450° of rotation; (6) at 720° of rotation. (From Sumetsky, M., *IEEE J. Lightwave Technol.*, 26, 21–27, 2008.)

demonstrated the MCR refractometric sensor that was created from an embedded MCR by removing its supporting rod. This device is an advanced version of an optical liquid ring resonator sensor, composed of an ultra-thin microcapillary coupled to an MF imbedded into a cured polymer [49]. The sensor was fabricated as follows. First, a 50-mm length and 2.5 μm diameter MF was fabricated. Then, the MF was wrapped on a 1 mm diameter polymethylmethacrylate (PMMA) rod. The PMMA refractive index was in the range of 1.49–1.51. The whole structure was repeatedly coated by the low-index Teflon solution 601S1-100-6 (DuPont, USA), which was dried in air. Next, the dried embedded MCL was left soaking into acetone to dissolve the support rod. Finally, the microfluidic MCR sensor with a ~1 mm diameter microchannel was fabricated. The sensor consisted of a MCR with five turns and had a channel inside as illustrated in Figure 7.23. The MCR was very close to the channel. The sensitivity of this device was measured by inserting the sensor in a beaker containing mixtures of isopropyl and methanol with ratios 60%, 61.5%, 63%, 64.3%, 65.5%, 66.7%, and 67.7%. The refractive indexes of pure isopropyl and methanol at 1.5 μm are 1.364 and 1.317, respectively. Figure 7.24 shows the spectra recorded at 1530 nm. The resonator peak shifted towards longer wavelengths with growth of the refractive index. The calculated sensitivity of this MCR sensor was ~40 nm/RIU, which is comparable to the sensitivity ~60 nm/RIU obtained for the embedded ring resonator sensor in Ref. [49].

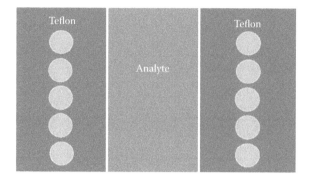

FIGURE 7.23
The cross section of an MCR microfluidic sensor. (From Xu, F., and Brambilla, G., *Appl. Phys. Lett.*, 92, 101126, 2008.)

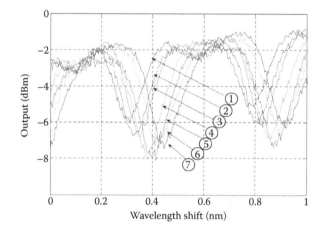

FIGURE 7.24
Output spectrum of the MCR microfluidic sensor in mixtures of isopropyl and methanol. The isopropyl fraction is (1) 60%; (2) 61.5%; (3) 63%; (4) 64.3%; (5) 65.5%; (6) 66.7%, and (7) 67.7%, respectively. (From Xu, F., and Brambilla, G., *Appl. Phys. Lett.*, 92, 101126, 2008.)

7.5 Conclusion

In this Chapter, a novel platform for the fabrication of photonic circuits based on MF assembly in air and in an optical medium was reviewed. The MLR and MCR fabrication technique developed to date is a significant step toward controllable manufacture of MF devices, e.g., delay lines, filters, dispersion compensators, miniature lasers, and sensors. The described approaches of MLR and MCR manufacturing are very simple and far from perfection. Nevertheless, the initial results obtained in theory, fabrication, and first applications of MLRs and MCRs are promising. It is feasible that further development of the theory of MF-based devices and progress in their fabrication will lead to the creation of robust, low-loss, and highly compact MF photonic circuits with numerous potential applications.

References

1. N. S. Kapany. High-resolution fibre optics using sub-micron multiple fibres. *Nature*, 184, 881–883, 1959.
2. N. S. Kapany and J. J. Burke. Fiber optics. IX. Waveguide effects. *J. Opt. Soc. Am.*, 51, 1067–1078, 1961.
3. C. Caspar and E.-J. Bachus. Fibre-optics micro-ring-resonator with 2 mm diameter. *Electron. Lett.*, 25, 1506–1508, 1989.
4. H. S. Mackenzie and F. P. Payne. Evanescent field amplification in a tapered single-mode fibre. *Electron. Lett.*, 26, 130–132, 1990.
5. G. J. Pendock, H. S. MacKenzie, and F. P. Payne. Dye lasers using tapered optical fibers. *Appl. Opt.*, 32, 5236–5242, 1993.
6. W. M. Henry and F. P. Payne. Solid-state tapered optical fibre devices. *J. Opt. Quantum Electron.*, 27, 185–191, 1995.
7. J. C. Knight, G. Cheung, F. Jacques, and T. A. Birks. Phase-matched excitation of whispering-gallery mode resonances by a fiber taper. *Opt. Lett.*, 22, 1129–1131, 1997.
8. J. Bures and R. Ghosh. Power density of the evanescent field in the vicinity of a tapered fiber. *J. Opt. Soc. Am. A*, 16, 1992–1996, 1999.
9. T. A. Birks, W. J. Wadsworth, and P. St. J. Russell. Supercontinuum generation in tapered fibers. *Opt. Lett.*, 25, 1415–1417, 2000.
10. L. Tong, R. R. Gattass, J. B. Ashcom, S. He, J. Lou, M. Shen, I. Maxwell, and E. Mazur. Subwavelength-diameter silica wires for low-loss optical wave guiding. *Nature*, 426, 816–819, 2003.
11. M. Sumetsky. Optical fiber microcoil resonator. *Opt. Express*, 12, 2303–2316, 2004.
12. G. Brambilla, V. Finazzi, and D. J. Richardson. Ultra-low-loss optical fiber nanotapers. *Opt. Express*, 12, 2258–2263, 2004.
13. S. G. Leon-Saval, T. A. Birks, W. J. Wadsworth and P. St. J. Russell. Guidance properties of low-contrast photonic bandgap fibres. *Opt. Express*, 12, 2864–2869, 2004.
14. M. Sumetsky, Y. Dulashko, and A. Hale. Fabrication and study of bent and coiled free silica nanowires: Self-coupling microloop optical interferometer. *Opt. Express*, 12, 3521–3531, 2004.
15. G. Brambilla, F. Koizumi, X. Feng, and D. J. Richardson. Compound-glass optical nanowires. *Electron. Lett.* 41, 400–402, 2005.
16. M. Sumetsky, Y. Dulashko, J. M. Fini, A. Hale, and J. W. Nicholson. Probing optical microfiber nonuniformities at nanoscale. *Opt. Lett.*, 31, 2393–2395, 2006.
17. M. Sumetsky. How thin can a microfiber be and still guide light? *Opt. Lett.*, 31, 870–872, 2006.
18. M. Sumetsky. How thin can a microfiber be and still guide light? Errata, *Opt. Lett.*, 31, 3577–3578, 2006.
19. M. Sumetsky. Optics of tunneling from adiabatic nanotapers. *Opt. Lett.*, 31, 3420–3422, 2006.
20. M. Sumetsky, Y. Dulashko, P. Domachuk, and B. J. Eggleton. Thinnest optical waveguide: experimental test. *Opt. Lett.*, 32, 754–756, 2007.
21. M. Sumetsky, Y. Dulashko, J. M. Fini, and A. Hale. Optical microfiber loop resonator, *Appl. Phys. Lett.*, 86, 161108, 2005.
22. M. Sumetsky, Y. Dulashko, J. M. Fini, A. Hale, and D. J. DiGiovanni. The microfiber loop resonator: theory, experiment, and application. *IEEE J. Lightwave Technol.*, 24, 242–250, 2006.
23. M. Sumetsky, Y. Dulashko, and M. Fishteyn. Demonstration of a multi-turn microfiber coil resonator. In Proceedings of Optical Fiber Communication Conference, Anaheim, CA, 2007. Postdeadline papers, Paper PDP46.
24. M. Sumetsky. Basic elements for microfiber photonics: micro/nanofibers and microfiber coil resonators. *IEEE J. Lightwave Technol.*, 26, 21–27, 2008.
25. X. Fei and G. Brambilla. Manufacture of 3-D microfiber coil resonators. *IEEE Photon. Technol. Lett.*, 19, 1481–1483, 2007.
26. F. Xu and G. Brambilla. Embedding optical microfiber coil resonators in Teflon. *Opt. Lett.*, 32, 2164–2166, 2007.

27. L. M. Tong, J. Y. Lou, R. R. Gattass, S. L He, X. W. Chen, L. Liu, and E. Mazur. Assembly of silica nanowires on silica aerogels for microphotonic devices. *Nano Lett.*, 5, 259–262, 2005.

28. X. Jiang, L. Tong, G. Vienne, X. Guo, A. Tsao, Q. Yang, and D. Yang. Demonstration of optical microfiber knot resonators. *Appl. Phys. Lett.*, 88, 223501, 2006.

29. X. Jiang, Q. Yang, G. Vienne, Y. Li, L. Tong, J. Zhang, L. Hu. Demonstration of microfiber knot laser. *Appl. Phys. Lett.*, 89, 143513, 2006.

30. X. Jiang, Q. Song, L. Xu, J. Fu, and L. Tong. Microfiber knot dye laser based on the evanescent-wave-coupled gain. *Appl. Phys. Lett.*, 90, 233501, 2007.

31. G. Vienne, Y. Li and L. Tong. Microfiber resonator in polymer matrix, *IEICE Trans. Electron.*, E90, 415–421, 2007.

32. L. F. Stokes, M. Chodorow, and H. J. Shaw. All-single-mode fiber resonator. *Opt. Lett.*, 7, 288–230, 1982.

33. W. Snyder and J. D. Love. *Optical Waveguide Theory*. Chapman & Hall, London, 1983.

34. J. Heebner, T. Ibrahim, and R. Grover. *Optical Microresonators: Theory, Fabrication, and Applications.* Springer-Verlag, New York, 2007.

35. G. Lenz, B. J. Eggleton, C. K. Madsen, and R. E. Slusher. Optical delay lines based on optical filters. *IEEE J. Quant. Electron.*, 37, 525–532, 2001.

36. L. D. Landau and E. M. Lifshitz. *Quantum Mechanics*. Pergamon Press, Oxford, 1977.

37. R. B. Smith. Analytic solution for linearly tapered directional couplers. *J. Opt. Soc. Am.*, 66, 882–892, 1976.

38. C. K. Madsen, S. Chandrasekhar, E. J. Laskowski, M. A. Cappuzzo, J. Bailey, E. Chen, L. T. Gomez, A. Griffin, R. Long, M. Rasras, A. Wong-Foy, L. W. Stulz, J. Weld, and Y. Low. An integrated tunable chromatic dispersion compensator for 40 Gb/s NRZ and CSRZ. In Optical Fiber Communication Conference, Anaheim, CA, 2002. Postdeadline papers, Paper FD9.

39. G. Bourdon, G. Alibert, A. Beguin, B. Bellman, and E. Guiot. Ultralow loss ring resonators using 3.5% index-contrast Ge-doped silica waveguides. *IEEE Photon. Technol. Lett.*, 15, 709–711, 2003.

40. B. E. Little, S. T. Chu, P. P. Absil, J. V. Hryniewicz, F. G. Johnson, F. Seiferth, D. Gill, V. Van, O. King, and M. Trakalo. Very high-order microring resonator filters for WDM applications. *IEEE Photon. Technol. Lett.*, 16, 2263–2265, 2004.

41. J. Niehusmann, A. Vörckel, P. H. Bolivar, T. Wahlbrink, W. Henschel, and H. Kurz. Ultrahigh-quality-factor silicon-on-insulator microring resonator. *Opt. Lett.*, 29, 2861–2863 , 2004.

42. K. J. Vahala. Optical microcavities. *Nature*, 424, 839–846, 2003.

43. B. E. Little, T. Chu, H. A. Haus. Second-order filtering and sensing with partially coupled traveling waves in a single resonator. *Opt. Lett.*, 23, 1570–1572, 1998.

44. E. Krioukov, D. J. W. Klunder, A. Driessen, J. Greve, and C. Otto. Integrated optical microcavities for enhanced evanescent-wave spectroscopy. *Opt. Lett.*, 27, 1504–1506, 2002.

45. S. Ashkenazi, C.-Y. Chao, L. J. Guo, and M. O'Donnell. Ultrasound detection using polymer microring optical resonator. *Appl. Phys. Lett.*, 85, 5418–5420, 2004.

46. A. B. Matsko, V. S. Ilchenko. Optical resonators with whispering-gallery modes - Part I: Basics. *IEEE J. Sel. Top Quant. Electron.*, 12, 3–14, 2006.

47. V. S. Ilchenko, A. B. Matsko. Optical resonators with whispering-gallery modes - Part II: Applications. *IEEE J. Sel. Top Quant. Electron.*, 12, 15–32, 2006.

48. M. White, H. Oveys, and X. Fan. Liquid core optical ring resonator sensors. *Opt. Lett.*, 31, 1319–1321, 2006.

49. M. Sumetsky, R. S. Windeler, Y. Dulashko, and X. Fan. Optical liquid ring resonator sensor. *Opt. Express*, 15, 14376–14381, 2007.

50. M. Sumetsky. Uniform coil optical resonator and waveguide: transmission spectrum, eigenmodes, and dispersion relation. *Opt. Express*, 13, 4331–4340, 2005.

51. F. Xu, P. Horak, and G. Brambilla. Conical and biconical ultra-high-Q optical-fiber nanowire microcoil resonator. *Appl. Opt.*, 46, 570–573, 2007.

52. F. Xu, P. Horak, and G. Brambilla. Optimized design of microcoil resonators. *J. Lightwave Technol.*, 25, 1561–1567, 2007.

53. Figotin and I. Vitebskiy. Electromagnetic unidirectionality in magnetic photonic crystals. *Phys. Rev. B*, 67, 165210, 2003.
54. F. Xu, P. Horak, and G. Brambilla. Optical microfiber coil resonator refractometric sensor. *Opt. Express*, 15, 7888–7893, 2007.
55. F. Xu and G. Brambilla. Demonstration of a refractometric sensor based on optical microfiber coil resonator. *Appl. Phys. Lett.*, 92, 101126, 2008.

8

Optofluidic Ring Resonator Biological and Chemical Sensors

Xudong Fan, Ian M. White, Siyka I. Shopova, Hongying Zhu,
Jonathan D. Suter, Yuze Sun, and Gilmo Yang
University of Missouri

CONTENTS

8.1 Introduction

8.1.1 Background

Optical label-free bio/chemical sensors are highly desirable for molecular analysis in many applications such as disease detection, genomics, proteomics, environmental protection, homeland security, and drug discovery. They enable detection of bio/chemical molecules without the laborious fluorophore attachment processes that may interfere with the molecule's bio/chemical functions, and allow for kinetics measurement in real time to provide important insight into molecule–molecule interaction. A well-known example is the surface plasmon resonance based bioanalytical instrument [1,2], as commercialized by Biacore, Inc. [3], which combines advantageous label-free sensing and high detection sensitivity.

Optical waveguides can also be utilized as a label-free bio/chemical sensor [4–7]. A typical label-free waveguide bio/chemical sensor is illustrated in Figure 8.1A. The surface of the waveguide is first immobilized with a layer of bio/chemical recognition molecules such as antibodies. The guided light traveling along the waveguide has an evanescent field extending into the surrounding medium (e.g., water) for approximately 100 nm and is capable of interacting with the bio/chemical molecules near the waveguide surface. A small change in refractive index (RI) near the waveguide surface when target analyte is captured causes modifications in the optical sensing transduction signal such as phase or intensity change at the output.

Waveguide based bio/chemical sensors have been in existence for over 20 years and found applications in many fields [4]. However, this type of biosensor has a fundamental limitation. In the waveguide bio/chemical sensor, the sensing signal accumulates along the waveguide while interacting with the analyte. A longer light–analyte interaction length results in a lower (i.e., better) detection limit. As a consequence, a waveguide as long as a few centimeters is needed to detect trace amount of bio/chemical molecules, which has significantly impaired the sensor multiplexing capability and tremendously increased the sample consumption [4–7].

8.1.2 Optical Ring Resonator Biosensors

Optical microring resonators are an emerging bio/chemical sensing technology that has recently drawn increased attention to circumvent the obstacles faced by the straight waveguide [8–50]. In an optical ring resonator, the light propagates in the form of whispering gallery modes (WGMs) [51–53], which result from the total internal reflection of the light along the curved surface. The WGM circulates along the resonator surface and interacts

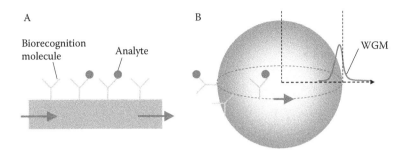

FIGURE 8.1
(A) An optical biosensor based on a straight waveguide. (B) An optical ring resonator in which the light circulates along the resonator surface.

repeatedly with the analytes on its surface through the WGM evanescent field, as illustrated in Figure 8.1B. In contrast to a straight waveguide based bio/chemical sensor, the effective light–analyte interaction length of a ring resonator is no longer determined by the sensor's physical size, but rather by the number of revolutions of the light supported by the resonator, which is characterized by the resonator quality factor, or the Q-factor. The effective length L_{eff} is related to the Q-factor by [51,54]:

$$L_{eff} = \frac{Q\lambda}{2\pi n},$$

(8.1)

where λ is wavelength and n is the RI of the ring resonator. For example, for a ring resonator with a Q-factor of 10^6, $n = 1.45$, and $\lambda = 1550$ nm, L_{eff} can be as long as 17 cm. Consequently, despite the small physical size of a resonator (10–200 μm in diameter), a ring resonator can deliver sensing performance superior to a waveguide while using orders of magnitude less surface area and sample volume. Furthermore, due to the small size of ring resonators, high density sensor integration becomes possible. Optical ring resonators have previously been implemented in the form of dielectric microspheres, microcylinders, and microtoroids [8–17,19–28,46–50,55], as well as planar ring resonator on a substrate [29–45].

Similar to all other optical label-free bio/chemical sensors, the ring resonator utilizes the RI change to perform detection. The WGM spectral position, i.e., resonant wavelength, λ, is related to the RI through the resonant condition [51]:

$$\lambda = \frac{2\pi R n_{eff}}{m},$$

(8.2)

where R is the ring outer radius, n_{eff} is the effective RI experienced by the WGM, and m is an integer that describes the WGM angular momentum. n_{eff} changes when the RI near the ring resonator surface is modified due to the capture of target molecules on the surface, which in turn leads to a shift in the WGM spectral position, as illustrated in Figure 8.2. Thus, by directly or indirectly monitoring the WGM spectral shift, it is possible to obtain both quantitative and kinetic information about the binding of molecules near the surface. This label-free sensing mechanism allows for the detection of bio/chemical molecules in their natural form without the laborious fluorescent labeling process, as well as the detection of nonfluorescent chemical molecules.

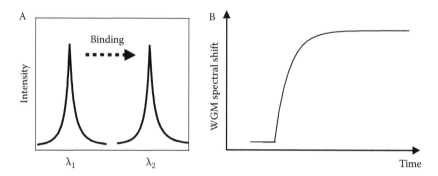

FIGURE 8.2
(A) The WGM resonance shifts in response to the attachment of the target molecules to the sensor surface.
(B) Sensorgram is obtained by recording the WGM spectral position over time.

8.1.3 Opto-Fluidic Ring Resonator (OFRR) Biosensors

Incorporation of microfluidics is crucial in bio/chemical sensor development, as nearly all biosensing and many chemical sensing analyses are performed in an aqueous environment. Efficient microfluidics not only enables sensor miniaturization and multiplexed detection, but also helps reduce sample consumption and detection time. A number of schemes have been implemented to integrate microfluidics with the ring resonator, as exemplified in Figure 8.3A through E [9,19,21,23,25,51,56–68]. However, these designs may not be fully satisfactory. For example, while stand-alone solid microspheres (Figure 8.3A) possess very high Q-factors (~10^8), fluidic integration, mass production, and reproducibility of a microsphere are extremely difficult. Liquid droplets (Figure 8.3B through D) suffer from evaporation and/or are hard to manipulate. Planar ring resonators (Figure 8.3D) on a substrate can be mass-produced relatively easily through lithographic or imprint technologies. However, they usually have relatively low Q-factors (typically 10^4 in water [31,33,67], although $Q = 3 \times 10^6$ was achieved in early 2007 in the ring resonator in air [69]). Moreover, the fluidics need to be fabricated separately and then mounted onto ring resonator sensors through multiple complicated steps [64,67], which may present many design challenges.

Recently, in addition to solid microsphere, liquid droplet, and planar ring resonator, a new class of ring resonator platform, 'opto-fluidic ring resonator (OFRR)', has been developed, which inherently integrates the state-of-the-art photonics with microfluidics for sensing and other applications [70–83]. As illustrated in Figure 8.4, the OFRR employs a piece of fused silica capillary of 30–200 μm diameter. The circular cross section of the capillary forms ring resonators that support the WGMs. The capillary wall is relatively thin (<4 μm) so that the WGMs of high Q-factors (>10^7) are exposed to the core (penetration depth~100 nm) and interact with the sample passing through the capillary. The OFRR architecture thus achieves dual use of the capillary as a sensor and as a fluidic channel. Therefore, it retains the meritorious high Q-factors and small sample volume of ring resonators while exhibiting the excellent fluid handling capability inherent to capillaries [84]. Furthermore, the WGM in the OFRR can be excited or coupled out evanescently through a waveguide [72,76] or a tapered optical fiber [75,85] arranged perpendicularly to and in touch with the OFRR. This configuration allows for independent engineering of fluidics and input/output photonics. The OFRR can be fabricated easily and cost-effectively using a capillary pulling station or a fiber draw tower. In addition, the OFRR can be scalable to a 2-dimensional array, as illustrated in Figure 8.4B with the sample volume for each OFRR on the order of nano-liters.

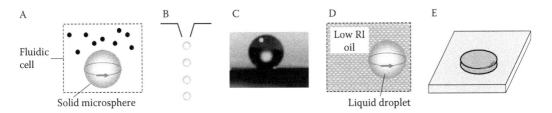

FIGURE 8.3
(A) Solid microsphere in a liquid well [9,19]. (B) Liquid droplets [51,58]. (C) Liquid droplet formed on an ultrahydrophobic surface. (D) Liquid droplet immersed in oil [65]. (E) Planar ring resonators [31,33].

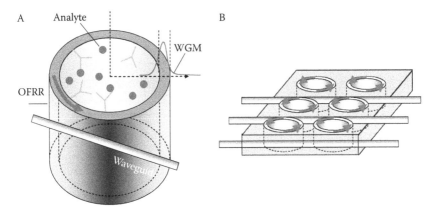

FIGURE 8.4
(A) Conceptual illustration of the OFRR. Analyte flows in the capillary and is captured by bio/chemical recognition molecules immobilized on the OFRR interior surface. The WGM has an evanescent field in the core and interacts with analyte in the core. A waveguide in contact with the OFRR is used to couple the light into (or out of) the OFRR. (B) Conceptual illustration of an OFRR array.

The OFRR has broad applications in refractometry [75,81,83], label-free biosensing [70,76,79], fluorescence-based biosensing [77,80], Raman based sensing [86], capillary electrophoresis [73], liquid and gas chromatography, and microfluidic dye lasers [77,80]. Here we will focus on the label-free biosensor development. In the following sections, we will first present the theoretical analysis of the OFRR as a bio/chemical sensor, followed by the experimental results for various biomolecule detections, and fluidic and waveguide integration.

8.2 Theoretical Analysis

8.2.1 Model

The WGM of the OFRR can fully be described using Mie theory by considering a three-layered radial structure, as shown in Figure 8.5 [52,75]. The radial distribution of the WGM electrical field of an OFRR is governed by:

$$E_{m,l}(r) = \begin{cases} AJ_m(k_{m,l}n_1 r) & (r \leq r_1) \\ BJ_m(k_{m,l}n_2 r) + CH_m^{(1)}(k_{m,l}n_2 r) & (r_1 \leq r \leq r_2), \\ DH_m^{(1)}(k_{m,l}n_3 r) & (r \leq r_2) \end{cases} \tag{8.3}$$

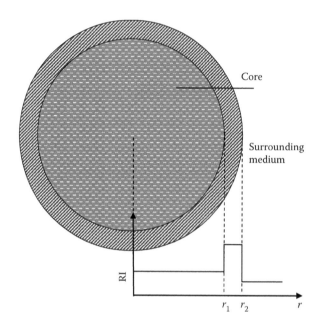

FIGURE 8.5
Three-layer model.

where J_m and $H_m^{(1)}$ are the mth Bessel function and the mth Hankel function of the first kind, respectively. The RI of the core, wall, and the surrounding medium is described by n_1, n_2, and n_3. The terms r_1 and r_2 represent the inner and outer radius of the OFRR, respectively, and $k_{m,l}$ is the amplitude of the wave vector in vacuum for the lth order radial WGM. The WGM has two polarization states with the electric field being along or perpendicular to the OFRR longitudinal direction. Using this three-layer model, the WGM spectral position, the radial distribution of the light, and radiation-limited Q-factor (in case that the OFRR is immersed in a medium other than air) can be obtained as a function of wall thickness, the OFRR size, and angular momentum of light, etc.

8.2.2 Bulk Refractive Index Sensitivity (BRIS)

Label-free optical sensors essentially detect the RI change induced by the changes in bulk solution or by the binding of molecules to the sensor surface. Bulk refractive index sensitivity (BRIS) in units of nm/RIU (RIU: refractive index units) provides a measure to characterize the sensor spectral response to the RI change in bulk solution, which allows us to compare the performance of sensors based on different types of ring resonators, or even different sensing technologies. In addition, as discussed in the next section, the BRIS is directly related to the sensor sensitivity to the binding of molecules. The BRIS S, is related to the fraction of light in solution by [70,81,87]:

$$S = \frac{\lambda_0}{n_{eff}}\eta,$$ (8.4)

where λ_0 and n_{eff} are the wavelength in vacuum and the effective RI of the WGM, respectively. η is the fraction of light energy in the core. When the WGM is mostly confined within the wall

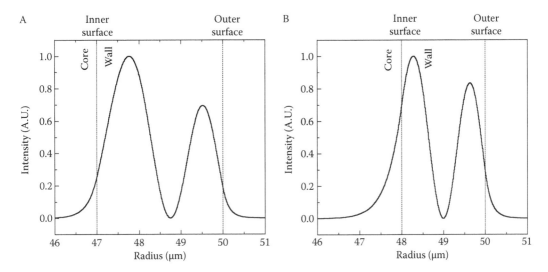

FIGURE 8.6

Radial distribution of WGMs ($m=271$, $l=2$. Wall thickness$=3$ µm) (A) and 2.0 µm (B) OD$=100$ µm. Refractive index: $n_1 = 1.33$, $n_2 = 1.45$, $n_3 = 1.0$. $\lambda \sim 1550$ nm. Fraction of light in the core: $\eta = 3\%$ (A) and 14.6% (B). Polarization is along the OFRR longitudinal direction.

$$S = \frac{\lambda_0}{n_{wall}} \eta,$$

where n_{wall} is the RI of the OFRR wall.

8.2.2.1 Wall Thickness Dependence

The BRIS of the OFRR depends highly on the OFRR wall thickness. A thinner wall results in a larger evanescent field, which enables higher light–matter interaction, and thus higher sensitivity. This phenomenon is illustrated in Figure 8.6, in which the light intensity distribution for an OFRR with 2 and 3 µm wall thickness is plotted using Equation 8.3 and clearly shows that an OFRR with a 2-µm wall has a higher evanescent field in the core.

8.2.2.2 Mode Number Dependence

Sensitivity is further dependent upon the radial mode order. The higher order radial mode has a higher evanescent field in the core of the OFRR, resulting in higher light–matter interaction. Figure 8.7 shows a second and third radial mode for the same OFRR. Note that this trend is consistent with the conventional ring resonators where higher order modes have more evanescent field available for sensing [11,19,27]. Recently, the WGMs with very high radial mode numbers have been employed for sensing [82]. These high order modes are able to penetrate through a relatively thick capillary wall and have significant portion of the light exposed to the core [82,88]. One example of such a mode is plotted in Figure 8.8, showing that the peak of the light penetrates a few micrometers into the core. Although this type of a mode has a large BRIS, it may not be used for typical bio/chemical sensing which usually depend on evanescent field. For those evanescent sensors, the sensing signal is generated only when bio/chemical molecules bind to the sensing surface, thus

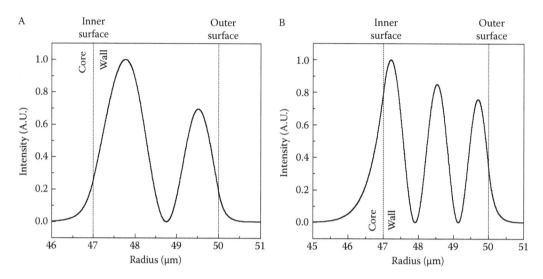

FIGURE 8.7
Radial distribution of WGMs. (A) $m = 271$, $l = 2$, $\lambda = 1554.198$ nm. (B) $m = 256$, $l = 3$, $\lambda = l = 1554.527$ nm. OD = 100 μm. Wall thickness = 3 μm. Refractive index: $n_1 = 1.33$, $n_2 = 1.45$, $n_3 = 1.0$. Fraction of light in the core: $\eta = 3\%$ (A) and 14.9% (B). Polarization is along the OFRR longitudinal direction.

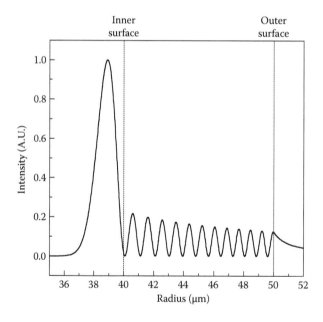

FIGURE 8.8
Radial distribution of a high order WGM. (A) $m = 204$, $l = 13$, $\lambda = 1557.064$ nm. OD = 100 μm. Wall thickness = 10 μm. $n_1 = 1.33$, $n_2 = 1.45$, $n_3 = 1.0$. $\eta = 56.4\%$. Polarization is along the OFRR longitudinal direction.

providing a means to distinguish the target analytes captured by the recognition molecules from those in solution. Nevertheless, these high order modes can be employed for refractometric measurement of solution and biosensing in free-solution that do not rely on the binding to a substrate [89].

8.2.2.3 OFRR Size Dependence

The BRIS is also related to the OFRR size. Figure 8.9 shows the BRIS dependence upon outer diameter while the thickness of the wall is held constant at 4 μm. As the graph shows, the BRIS increases with the increased OFRR diameter, in sharp contrast to the conventional ring resonators such as microspheres where the BRIS is inversely proportional to the ring size [11,19]. This phenomenon can be explained by treating the OFRR as a bent waveguide. Due to bending, the light is redistributed along the radial direction due to centrifugal potential with more light being outside the OFRR and less inside the core, as compared to the straight waveguide case [90,91]. Therefore, a smaller diameter results in a smaller fraction of light in the core, as shown in Figure 8.9. In the limit of infinitely large diameter, the fraction of the light in the core, should approach that of a straight waveguide, as shown by the dashed line in Figure 8.9.

The BRIS, or η, is further determined by operating wavelength (see Equation 8.4) and WGM polarization. The core RI also affects the BRIS, as high RI in the core can pull more light into the core [81,83,92]. However, for most biosensing, the analytes are dissolved in water based buffer with an RI around 1.33. In addition, as discussed in a later section, the BRIS does not provide a complete picture regarding a bio/chemical sensor performance. The detection limit that considers both BRIS and the system resolution needs to be taken into account for sensor performance evaluation.

8.2.3 Noise Analysis

8.2.3.1 Thermally Induced Noise

Temperature has significant impact on the performance of label-free optical sensors. For an OFRR, temperature fluctuation leads to inaccuracies in determination of the WGM spectral shift induced by binding of molecules to the surface. Suter et al. provided a detailed analysis of the WGM shift, $\Delta\lambda$, induced by the temperature change, ΔT, of the OFRR [74], which is given by:

$$\frac{\Delta\lambda}{\lambda} = \alpha\Delta T + \frac{\partial n_{eff}}{\partial n_{wall}}\frac{\kappa_{wall}}{n_{eff}}\Delta T + \frac{\partial n_{eff}}{\partial n_{core}}\frac{\kappa_{core}}{n_{eff}}\Delta T, \tag{8.5}$$

where $\alpha = 1/r\ \partial r/\partial T$ is the OFRR thermal expansion coefficient and $\kappa_{wall(core)} = \partial n/\partial T$ is the thermo-optic coefficient for the OFRR wall (core). The first term on the right side of Equation 8.5 is the noise due to thermal expansion, and the second and the third term represent the respective noise induced by temperature induced refractive index change in the wall and core. For a thick-walled OFRR, the amount of the light in the core is negligible. Therefore, the thermal noise is mainly determined by the first two terms on the right side of Equation 8.5, which leads to a temperature dependent WGM shift of approximately 8 pm/K (or 1 GHz/K) at 1550 nm for silica glass, as shown in Figure 8.10. When the OFRR wall becomes thinner, the evanescent field of the WGM in the core, and hence the third term on the right side of Equation 8.5, can not be ignored. Since the thermo-optic coefficient for the core, which is typically water (or other organic solutions), is negative and large in magnitude in comparison with that for the wall [93,94], the thermally induced noise is suppressed, as demonstrated in Figure 8.10 with the 3.5 μm thick wall. Figure 8.10 further shows that at certain wall thickness, the water effect will completely counterbalance the thermal noise induced by the wall, leading to elimination of the thermal noise to the first order. This phenomenon provides a means for us to optimize the OFRR wall thickness to minimize the thermally

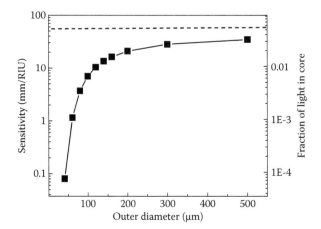

FIGURE 8.9
OFRR sensitivity and fraction of light in core as a function of the OFRR size for the second order WGM. Wall thickness is kept constant at 4 μm. $n_1 = 1.33$, $n_2 = 1.45$, $n_3 = 1.0$. $\lambda \sim 1550$ nm. Dashed line shows the fraction of light in the core in the limit of a straight waveguide of 4 μm wide. Polarization is along the OFRR longitudinal direction.

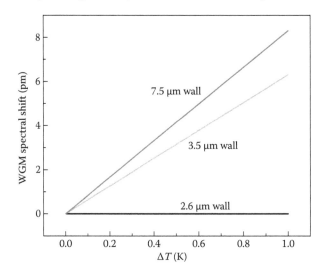

FIGURE 8.10
The WGM spectral response to temperature change. The OFRR is made of fused silica and filled with water. OD (outer diameter) = 100 μm; $\alpha_{wall} = 5 \times 10^{-7}$ K^{-1}; $\kappa_{wall} = 6.4 \times 10^{-6}$ K^{-1}; $\kappa_{core} = -1 \times 10^{-4}$ K^{-1}. Refractive index: $n_1 = 1.33$; $n_2 = 1.45$; $n_3 = 1$. Wavelength: 1550 nm. Wall thickness is labeled on the figure. Zero thermal sensitivity is achieved when the OFRR wall thickness is 2.6 μm. Polarization is along the OFRR longitudinal direction. OFRR wall thickness is 2.6 μm. Polarization is along the OFRR longitudinal direction. (Reused from Fan, X., White, Ian M., Zhu, H.,. Suter, Jonathan D., and Oveys, H. *Proc. SPIE*, 6452, 64520M 1–20, 2007. © SPIE. With permission.)

induced noise. Note that similar phenomenon was also theoretically studied by coating the microspheres with a dielectric layer with a negative thermal-optic coefficient [95].

8.2.3.2 Amplitude Noise

Intensity noise from the laser source and the photodetector will also cause inaccuracies in WGM spectral peak determination, as shown in Figure 8.11A [96]. A Monte Carlo

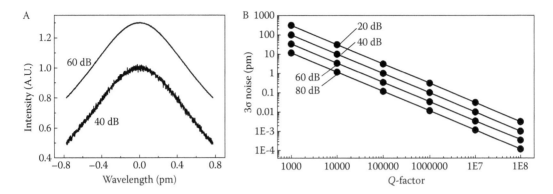

FIGURE 8.11

(A) WGMs with Gaussian noise added for SNR of 40 dB and 60 dB (SNR=peak power of signal divided by variance of noise distribution). Curves are shifted vertically for clarity. (B) Results of Monte Carlo simulations over a range of Q-factors and amplitude noise variances. Vertical axis represents the accuracy in WGM spectral peak determination. WGM wavelength is 1550 nm. (Reused from White, Ian M. and Fan, X., *Optics Express*, 16, 2008. © Optical Society of America. With permission.)

simulation plotted in Figure 8.11B shows the three standard deviations (3σ) value of the determined maximum for signal-to-noise-ratio (SNR) ranging from 20 to 80 dB, and for Q-factors between 10^8 and 10^3. The simulation results reveal that having a high Q-factor is advantageous in reducing the spectral noise of the sensor, as the mode with narrower linewidth filters the spectral noise more effectively. The simulation results show an empirical linear dependence of the standard deviation on the linewidth and an exponential relationship with the SNR [96]:

$$\sigma \approx \frac{\Delta\lambda}{4.5(SNR^{0.25})},$$ (8.6)

where σ is the standard deviation of the resulting spectral variation and $\Delta\lambda$ is the full-width-half-max of the mode amplitude and is related to the Q-factor by $Q=\lambda/\Delta\lambda$. SNR is in linear units in this expression. For $Q=10^6$ and $SNR=10^6$, $3\sigma=0.03$ pm at 1550 nm wavelength, or 1/50 of the WGM linewidth. When SNR is increased to 10^8, 3σ becomes 0.01 pm, or 1/150 of the WGM linewidth.

8.2.3.3 Pressure Induced Noise

The WGM spectral position of an OFRR may potentially be susceptible to the variation in the differential pressure between the liquid in the core and the surrounding medium (e.g., atmosphere), as illustrated in Figure 8.12. The pressure induced WGM change can be expressed as:

$$\frac{\Delta\lambda}{\lambda} = \frac{\Delta r}{r} = \frac{r}{t} \cdot \frac{\Delta P}{E},$$ (8.7)

where E is the OFRR Young's modulus. For fused silica used in OFRRs, $E=7\times10^{10}$ Pa. For an OFRR with $r=100$ μm and $t=2$ μm, Equation 8.7 becomes:

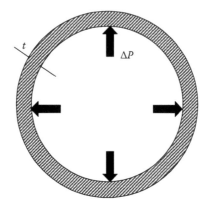

FIGURE 8.12
OFRR radius changes under differential pressure (ΔP) between the core and the surrounding medium.

$$\frac{\Delta\lambda}{\lambda} \approx 10^{-9}\Delta P\,(\text{Pa}). \tag{8.8}$$

In order to keep the pressure induced noise small in comparison with the spectral resolution set by the WGM spectral linewidth (or the Q-factor), restrictions on the external pressure that pushes the liquid sample through the OFRR and on the flow rate may need to be imposed, especially when mechanical pumps such as peristaltic pumps are used. For example, if the Q-factor is 10^6 and if we assume that the spectral resolution is one hundredth of the WGM spectral linewidth, then the variation in differential pressure should be kept below 10 Pa, which can easily be maintained during an experiment. In addition, the pressure noise introduced by a mechanical pump has a well-defined frequency (e.g., a few Hz for a peristaltic pump). Therefore, it can be removed relatively easily during the data processing stage. Furthermore, when electro-kinetic pump such as electro-osmosis method is used, the pressure induced noise can be completely removed.

8.2.3 Relation between BRIS and the Sensitivity to Molecule Binding

For a biosensor that uses evanescent field for detection, it is of importance to characterize the sensing signal change in response to the attachment of bio/chemical molecules to the sensing surface. For the OFRR, it is found that the WGM spectral shift for bio/chemical molecule binding is directly related to the OFRR's BRIS [70]:

$$\frac{\Delta\lambda}{\lambda} = \sigma_p\alpha_{ex}\frac{2\pi\sqrt{n_{wall}^2 - n_{core}^2}}{\varepsilon_0\lambda^2}\frac{n_{wall}}{n_{core}^2}S, \tag{8.9}$$

where σ_p is the surface density of the captured biomolecules and α_{ex} is the excess polarizability of the molecule. ε_0 is vacuum dielectric constant. Equation 8.9 shows that the contribution of each molecule to the WGM spectral shift is weighted by the OFRR BRIS. In addition, since the BRIS measurement is nondestructive and is easy to accomplish (see Section 8.3.4. for experimental results on BRIS measurement), Equation 8.9 provides a method to characterize the OFRR performance in bio/chemical molecule sensing using BRIS. Note that Equation 8.9 is valid regardless of the WGM mode number and polarization; it is also broadly applicable to other types of ring resonators based on microspheres and planar ring resonators where exterior surface may be used for bio/chemical sensing [70].

8.2.4 Detection Limit

A sensor performance is eventually determined by its detection limit (DL). The DL for bulk RI change is given by:

$$DL_{bulk} = \delta/S, \qquad (8.10)$$

where δ is the system spectral resolution in units of pm. Figure 8.13 plots the bulk RI detection limit for various solvent absorption.

The detection limit for molecule density of the OFRR surface can be deduced using Equations 8.9 and 8.10 [70]:

$$DL_{molecule} = \left(\frac{n_{core}^2}{n_{wall}\sqrt{n_{wall}^2 - n_{core}^2}} \frac{\varepsilon_0\lambda}{2\pi\alpha_{ex}} \right) DL_{bulk}. \qquad (8.11)$$

As discussed in Refs. [70,73,96,97], DL_{bulk} is on the order of 10^{-7} RIU, leading to a detection limit for biomolecules (protein and DNA) on the order of 0.1–1 pg/mm^2, which is highly competitive as compared to the state-of-the-art label-free optical sensors [1,3,9,12,30,31,33,34,98–104].

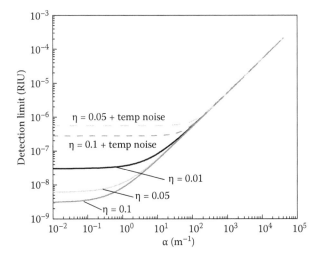

FIGURE 8.13
Calculated DL as a function of solvent absorption coefficient for $\eta=0.01$, 0.05, and 0.1. For solid lines, only the mode amplitude noise is considered. For the dashed lines, temperature-induced spectral fluctuation (0.01 pm) is also considered. $\lambda=1550$, $Q_0=10^8$, SNR$=60$ dB. (Reused from Ian M. White and Xudong Fan, Optics Express 16, 1020–1028, 2008. © Optical Society of America. With permission.)

8.3 Experimental Investigations

8.3.1 OFRR Fabrication

Since the thin-walled capillaries are not commercially available, we have assembled a computer-controlled pulling station (see Figure 8.14) capable of fabricating the OFRR of up to 50 cm (Figure 8.15A) by rapidly stretching a fused silica preform under heat generated by CO_2 lasers. A laser micrometer (0.1 μm accuracy) is used to monitor the capillary OD. The foundation of capillary pulling is mass conservation [105]:

$$\pi[R_0^2 - (R_0 - t_0)^2]V_f = \pi[R_1^2 - (R_1 - t_1)^2]V_p, \tag{8.12}$$

where R_0, t_0, and V_f are the original outer radius, wall thickness, and feed-in speed, respectively. R_1, t_1, and V_p are the final outer radius, wall thickness, and pulling speed, respectively. If we assume that there is no surface tension induced capillary collapse involved, the capillary aspect ratio, i.e., OD/t should remain constant. Therefore, by controlling the ratio between the feed-in and pulling speeds, we are able to control the final size of the OFRR and its wall thickness.

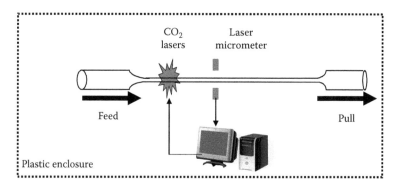

FIGURE 8.14
OFRR pulling station.

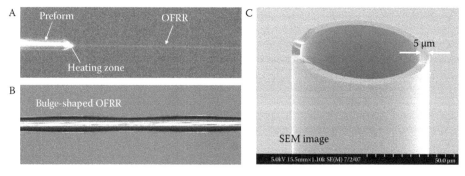

FIGURE 8.15
(A) OFRR made from a preform. (B) SEM image. (C) Bulge-shaped OFRR.

In practice, a certain degree of collapsing may occur due to glass surface tension, as reported in the literature on capillary pulling [105–108], leading to a thicker wall, i.e., a lower aspect ratio than the initial value. The collapsing effect can be mitigated by using a higher pulling speed and/or lower heating temperature. In our experiment, we start with a commercial fused silica capillary (Polymicro TSP530660, OD = 616 µm and wall thickness: t = 40 µm). After systematic investigation in pulling parameters, and quality check of the final size and the wall thickness using an optical microscope, it is found that the original aspect ratio can well be maintained after pulling, as shown in Figure 8.15B. In fact, over 95% of the original aspect ratio was reported in a previous study [105]. In addition, OFRRs of various shapes can also be fabricated by modulating the CO_2 laser or pulling/feed speed during fabrication (Figure 8.15C). The OFRR prepared in this manner have an OD of approximately 50–100 µm and wall thickness of 3–6 µm (see Figure 8.15B).

In order to achieve an even thinner wall, low concentration of hydrofluoric acid (HF) is passed through the OFRR to slightly etch the glass. However, HF etching may result in surface roughness and causes the Q-factor to degrade. Direct fabrication of the thin-walled OFRRs is of great interest to preserve high Q-factors and to reduce fabrication time. Recently, capillaries with sub-micron wall thickness walls have been achieved when the preform is pressurized and tens of meters long thin-walled OFRR have also been fabricated using a commercial draw tower [81,83].

8.3.2 Experimental Setup

The setup for the experimental characterization of the OFRR is shown in Figure 8.16. Light from a tunable diode laser is coupled into the WGM through evanescent coupling by an optical fiber taper or waveguide in contact with the OFRR exterior surface. The tunable laser periodically scans across a wavelength range of about 100 pm (scanning rate 2–10 Hz) while Detector #1 at the output of the waveguide or fiber taper measures the optical power. As shown in Figure 8.16C, when the laser wavelength matches the WGM resonant condition (Equation 8.1), the light couples into the ring resonator and causes the measured transmission power to drop, leaving a spectral dip at Detector #1. In the meantime, the

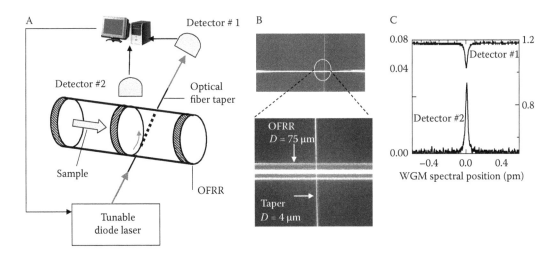

FIGURE 8.16
(A) Experimental setup. (B) A picture of the LCORR in touch with an optical fiber taper. (C) Transmission signal from Detector #1 and scattering signal from Detector #2 indicate the WGM spectral position.

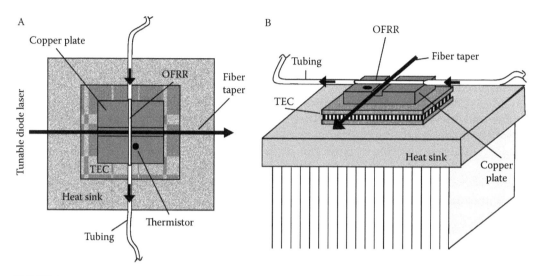

FIGURE 8.17
Top view (A) and side view (B) of the experimental setup with TEC.

light coupled into the OFRR is scattered off the OFRR surface and can be detected as a spectral peak with a detector placed above the OFRR (Detector #2). Either of these measured signals can be used to indicate the WGM spectral position. While the first approach is easier to implement, the second scheme is more suitable for an OFRR array.

As schematically shown in Figure 8.17, the OFRR is then fixed onto a copper plate (1 cm²) that, in turn, sits on top of a thermo-electric cooler (TEC) unit (Marlow Industries, Inc.) connected to a temperature controller (ILX Lightwave LDT-5910B). A thermister is embedded in the copper plate 2 mm away from the coupling region with thermally conductive epoxy (Arctic Silver). A custom fitted acrylic cap is placed on top of the copper plate to protect the system from air current disruption. Fluidic delivery is handled by a syringe pump (Harvard Apparatus). The reagent is sucked through Tygon tubing from a 1-mL polyethylene source well, which can be conveniently filled and emptied via manual pipette during the experiment. All experiments are performed at room temperature.

8.3.3 Q-Factor of the OFRR

The Q-factor of the OFRR is determined by:

$$Q^{-1} = Q_{rad}^{-1} + Q_{wall}^{-1} + Q_{sca}^{-1} + Q_{sol}^{-1}, \tag{8.13}$$

where Q_{rad}, Q_{wall}, and Q_{sca} are the Q-factor determined by the radiation loss, loss in the wall medium, and loss resulting from the surface scattering. $Q_{sol} = 2\pi n_2/\eta\lambda\alpha_{sol}$, where α_{sol} is the absorption coefficient of the solvent. For low order WGMs, $Q_{rad} \gg 10^{11}$. Fused silica absorption and surface roughness induced losses have been investigated in detail with microspheres, which show a Q-factor as high as 8×10^9 [54,109]. Since the OFRR uses the same material and is fabricated under the similar heating condition (before HF etching) as for a microsphere, it should have a Q_{wall} and Q_{sca} well above 10^9. For biosensing, the solvent is basically water, which has low absorption at visible spectrum [110], therefore, the solvent absorption induced Q-factor, Q_{sol} should be in excess of 10^8 at visible spectrum. At a longer wavelength (e.g., 1550 nm), Q_{sol} suffers severely from the high absorption of water. Heavy water, which has a much lower absorption, can be used to replace water for biosensing [28]. Although high Q-factor can be obtained upon fabrication, the Q-factor is degraded due to HF etching, which results in a Q-factor on the order of 10^6. The best Q-factor experimentally achieved is 10^7 at 690 nm [77]. Recent experimental results in OFRR laser (no HF etching involved) studies suggest that the Q-factor of the OFRR can be on the order of 10^9 [111].

8.3.4 BRIS Characterization

The OFRR is initially filled with water and then various concentrations of water–ethanol mixtures with known RI are passed sequentially through the OFRR. The WGM shifts to a longer wavelength in response to the RI increase in the core, as illustrated in the inset of Figure 8.18. The sensitivity is obtained by calculating the slope of the WGM response to the RI change [8,9]. The highest sensitivity reported so far for the water-filled core is approximately 50 nm/RIU [81]. Higher BRIS (~75 nm/RIU) has also been achieved in the lab (unpublished).

8.3.5 Characterization of Thermally-Induced Noise

In Figure 8.19A, we experimentally characterize the OFRR response to the temperature and show that the WGM of the thin-walled OFRR (4 μm wall thickness) drifts at a rate of approximately 5 pm/K. This rate is lower than that for the thick-walled OFRR due to the water effects, as discussed previously. A thinner wall will lead to an even smaller thermal drift rate. Figure 8.19B shows that the standard deviation of the thermally–induced WGM spectral position noise, σ, is 0.0048 pm, which results in a spectral resolution better than 0.02 pm (3σ). In a high-Q-factor system such as the OFRR, this spectral resolution is the limiting factor in the detection limit [96]. One method to further lower the detection limit is to implement a reference channel as shown in Figure 8.20. Common-mode noise such as non–specific binding of molecules and temperature fluctuations can be cancelled out.

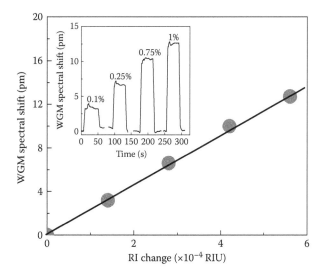

FIGURE 8.18
WGM spectral shift versus RI change in the core. Sensitivity is 22 nm/RIU. The inset is the sensorgram showing various concentrations of ethanol passing through the OFRR.

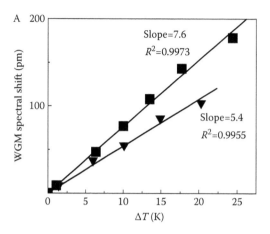
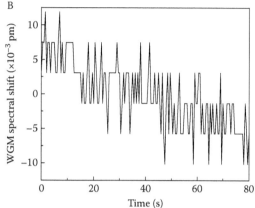

FIGURE 8.19

(A) WGM shift in response to temperature change for thick-walled OFRR (square, wall thickness=8 μm) and thin-walled OFRR (triangle, wall thickness=4 μm). Solid lines are simulation curves based on Equation 8.3. (B) OFRR noise with thermo-electric cooler. σ=0.0048 pm. Wavelength=1550 nm. Wavelength scanning step size=0.004 pm. (Reused from Xudong Fan, Ian M. White, Hongying Zhu, Jonathan D. Suter, and Hesam Oveys, *Proc. SPIE*, 6452, 64520M, 1–20, 2007. © SPIE. With permission.)

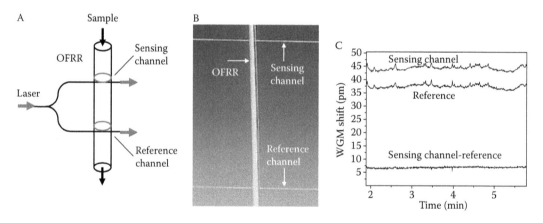

FIGURE 8.20

(A) Schematic of an OFRR with a reference channel; (B) A picture of such a setup; (C) Common-mode noise reduction by subtracting the reference from the actual sensing signal. (Reused from Xudong Fan, Ian M. White, Hongying Zhu, Jonathan D. Suter, and Hesam Oveys, *Proc. SPIE*, 6452, 64520M 1–20, 2007. © SPIE. With permission.)

8.3.6 Surface Activation

In order to capture the target bio/chemical molecules, the interior surface of the OFRR needs to be activated or functionalized so that biorecognition molecules such as antibody, aptamer, or oligonucleotide probes can be immobilized. This can be done by first silanizing the surface to introduce amine (NH_2), carboxyl (COOH), or thiol (SH) group. Then cross-linker is used to covalently link the biorecognition molecules to the surface. Details of the experimental procedures can be found in Refs. [12,21,23,25,70,79]. To immobilize multiple biorecognition molecules onto the inner surface of an OFRR, we use a photo-assisted molecular grafting method [112–114]. As illustrated in Figure 8.21A, the inner surface of the OFRR is first silanized with thiol or amine functional groups. Then a photo-sensitive

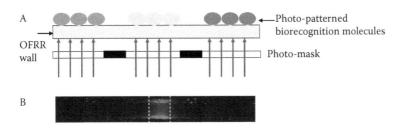

FIGURE 8.21
(A) Molecule patterning through photo-assisted molecular grafting. (B) Results were achieved with a single-window photomask. The molecules are labeled with fluorescent dye molecules to guide the eye. (Reused from Xudong Fan, Ian M. White, Hongying Zhu, Jonathan D. Suter, and Hesam Oveys, *Proc. SPIE*, 6452, 64520M 1–20, 2007. © SPIE. With permission.)

cross-linker such as N-5-Azido-2-nitrobenzoyloxysuccinimide in solution is introduced, followed by exposure to a UV lamp or high intensity Argon laser through a photo-mask. Finally, the biorecognition molecule is introduced and immobilized at the exposed locations. Multiple exposures can be repeated to immobilize different types of biorecognition molecules. Figure 8.21B shows the immobilization result obtained with a single window photo-mask exposed to an Argon laser.

8.3.7 Bio/chemical Molecule Detection

The OFRR can be used for detection of a number of biological and chemical molecules ranging from relatively small molecules such as protein and DNA to large molecules such as virus and bacterium.

8.3.7.1 Protein Detection

We first use bovine serum albumin (BSA) as a model system to test the OFRR's protein detection capability. As shown in Figure 8.22A, upon introduction of BSA into the OFRR, which is initially treated with 5% glutaraldehyde to cross-linker protein molecules, the WGM shifts to a longer wavelength and finally reaches equilibrium. Figure 8.22B plots the equilibrium WGM shift versus BSA concentration for the OFRR under test. With the increased BSA concentration, the equilibrium WGM shift increases and then becomes saturated when BSA concentration is higher than 200 nM. The inset of Figure 8.22B shows the BSA concentration dependent WGM shift in the log-log scale. The lowest concentration used is 150 pM, causing a spectral shift of 2.5 pm, well above the OFRR detection noise level. The theoretical detection limit obtained by extrapolation is estimated to be approximately 3 pM, reflecting the excellent detection capability of the OFRR. Using Equation 8.9, we are able to calculate the BSA density on the OFRR surface. As shown in Figure 8.23, BSA forms a very compact layer on the OFRR surface, in agreement with previous observations [9,10,21].

In addition to detecting the attachment of biomolecules, the OFRR is also capable of detecting molecules removed from the surface, which is important for measurement of enzyme proteolytic activities. Figure 8.24 shows that amino acids are cleaved from BSA on the OFRR surface by trypsin. Upon the injection of trypsin, the WGM shifts to a lower wavelength, corresponding to a mass reduction on the surface, from which the mass removal rate or trypsin proteolytic activity can be deduced [21,70]. Figure 8.24 shows that 80% of

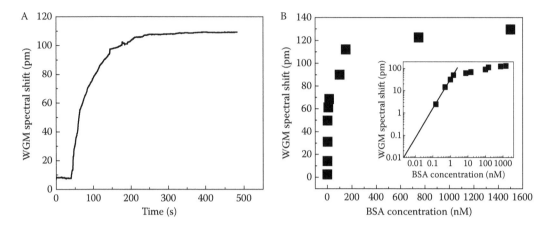

FIGURE 8.22
(A) Sensorgram for BSA binding. (B) WGM spectral shift vs.BSA concentration for the OFRR with BRIS of 31 nm/RIU. The LCORR starts to saturate at 200 nM and the saturation WGM shift is 129 pm. Half saturation concentration: 8 nM. Inset: log–log scale. Solid line is the linear fit in log–log scale. (Reused from Hongying Zhu, Ian M. White, Jonathan D. Suter, Paul S. Dale, and Xudong Fan, *Optics Express*, 15, 9139–9146, 2007. © Optical Society of America. With permission.)

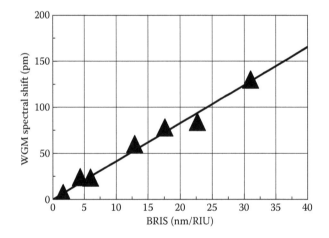

FIGURE 8.23
Saturation WGM spectral shift vs. OFRR BRIS. The solid curve is the theoretical calculation based on Equation 8.9. $n_2 = 1.45$, $n_3 = 1.333$. $\sigma_p = 2.9 \times 10^{12}$ cm^{-2} and $\alpha_{ex} = (4\pi\varepsilon_0) \times (3.85 \times 10^{-21})$ cm^3. (Reused from Hongying Zhu, Ian M. White, Jonathan D. Suter, Paul S. Dale, and Xudong Fan, *Optics Express*, 15, 9139–9146, 2007. © Optical Society of America. With permission.)

BSA molecules are removed from the OFRR. The remaining 20% permanently remains on the OFRR surface due to the lack of available cleavage sites [21,70].

As one of practical applications, we use the OFRR for detection of breast cancer biomarker. Breast cancer is the most frequently diagnosed malignancy in women, ranking second among cancer death in women, especially in Western countries [115]. As it has enormous impact on women's health, a lot of efforts might have been put into breast cancer research than any other malignancy. Early diagnosis and assess treatment prognosis are very important for malignancy treatment. Tissue based assay test of estrogen receptor (ER) and progesterone receptor (PR) are the most routine method for predicting hormone

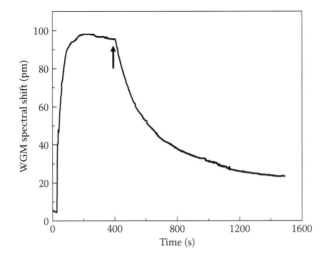

FIGURE 8.24
BSA molecules bind to the OFRR surface, followed by the cleavage by trypsin. BSA concentration: 0.1 mg/mL. Trypsin concentration: 0.1 mg/mL. Arrow indicates the time when trypsin is added. (Reused from Hongying Zhu, Ian M. White, Jonathan D. Suter, Paul S. Dale, and Xudong Fan, *Optics Express*, 15, 9139–9146, 2007. © Optical Society of America. With permission.)

dependence of breast cancer and assess prognosis. However there are very few mature techniques for detecting the sensitive and specific circulating breast cancer biomarker. Herein, we attempt to apply the OFRR for real-time, label-free, rapid detection of CA15-3, one of the most widely used circulating cancer biomarker that is recommended for the monitoring of the course and response to therapy of breast cancer [116,117]. During the experiment, the OFRR surface is first immobilized with Anti-CA15-3 antibodies through the glutaraldehyde. The unoccupied sites on the sensing surface are then blocked by casein blocking buffer. After this, the OFRR is ready for CA15-3 measurement. Figure 8.25 plots the OFRR's response to various concentration of CA15-3, which covers the CA15-3 cut-off point (between 20 and 40 Units/mL according to clinical studies [118]) that indicates the presence of the circulating tumor cells in blood. The detection time for each concentration takes only a few minutes. These results demonstrate the potential of the OFRR as a diagnostic instrument for rapidly detecting the cancer biomarker. For further development of OFRR based device for cancer biomarker detection, real serum samples obtained from patients will be used to test in the future.

8.3.7.2 DNA Detection

The OFRR can also be used for label-free DNA detection [79]. Figure 8.26A shows the OFRR response to target DNA of different strand length when the targets are hybridized to immobilized complimentary probes. Additionally, the OFRR is capable of differentiating target DNA based on mismatched bases, as shown in Figure 8.26B. Since the OFRR is capable of measuring the binding in real time, this provides another tool to differentiate mismatched DNA molecules, as shown in Figure 8.27.

Having the ability to detect a low number of mismatches makes the OFRR useful for highly specific sequencing and for research on diseases of genetic origin. For example, assays may be designed to look for known single-nucleotide polymorphisms responsible for anything from cancer to Alzheimer's disease [119]. As the number of

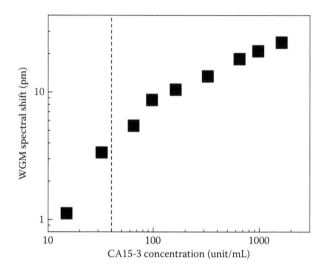

FIGURE 8.25
Detection of CA 15-3 breast cancer marker. Dashed line represents 40 Units/mL.

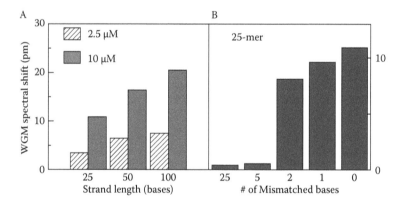

FIGURE 8.26
(A) WGM shift for various concentrations and strand length variables. (B) WGM shift for various strand mismatches. Concentration: 10 μM. (Reused from Jonathan D. Suter, Ian M. White, Hongying Zhu, Huidong Shi, Charles W. Caldwell, and Xudong Fan, Biosensors and Bioelectronics, 23, 1003–1009, 2008. © Elsevier. With permission.)

mismatched bases between probe and target DNA strands increases, the overall probability of binding decreases, resulting in a lower overall spectral shift. Figure 8.26B shows about a 10% drop in signal for a 1-base mismatch, but results from the literature suggest that this result could be improved to 90% depending on the experimental conditions used [12].

Surface coverage is easily analyzed from the spectral response based on Equation 8.9 [70]. We estimate that the OFRR is capable of detecting a DNA surface coverage as small as 6.8×10^9 molecules/cm^2 with 100-base targets [79]. This represents approximately 0.012% of the theoretical maximum surface loading. Fractional coverage, however, is highly dependent upon the size of the adsorbed molecule, as larger molecules interfere sterically with each other, generally leading to reduced densities [79].

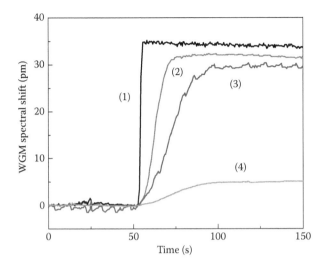

FIGURE 8.27

Sensorgrams for various 25-mer DNA target. (1) 100% complementary; (2) 2-base mismatch; (3) 5-base mismatch; (4) completely mismatch. (Reused from Jonathan D. Suter, Ian M. White, Hongying Zhu, Huidong Shi, Charles W. Caldwell, and Xudong Fan, Biosensors and Bioelectronics, 23, 1003–1009, 2008. © Elsevier. With permission.)

Figure 8.28 shows the net WGM shift in response to a series of target DNA concentrations between 0.5 and 500 nM using a 6.8 nm/RIU OFRR, which is modeled using a Michaelis-Menton type curve. The inset graph shows detection of 10 pM target using a more sensitive OFRR. This demonstrates the ability of the sensor to detect extremely small concentrations and proves that greater performance is possible if higher sensitivities are used.

8.3.7.3 Virus Detection

We demonstrate the sensitive, rapid, on-line virus detection with the OFRR. Here we use M13 filamentous bacteriophage as a safe model system. M13 is about 900 nm long and 10 nm wide. M13 has similar molecular weight to serious human pathogens, yielding valuable information for other types of viral detection. Figure 8.29 shows that the OFRR a detection dynamic range spans over seven orders of magnitude (2×10^3–2×10^{10} pfu/mL, pfu: plaque forming unit, it simply represents the number of viruses). The experimental detection limit is on the order of 1000 pfu/mL, which results in a WGM shift of a few pico-meters, well above the sensor noise level. With further improvement, it is possible to detect the virus concentration down to tens of virus particles per mL. Given the microsized dimension of the OFRR, such a low concentration would correspond to only a few viruses captured to the surface. In addition, the unique microfluidic design of the OFRR enables simple on-line virus detection, allowing for in situ real-time virus monitoring and detection of virus in liquid samples. Furthermore, due to the circular structure of the capillary, virus capture efficiency is increased, enabling rapid virus detection. In our experiment, virus can be detected with 10 minutes after introduction into the OFRR.

8.3.7.4 Bacterium and Whole Cell Detection

Detection of bacterium or the whole cell is challenging for evanescent sensors, as the penetration depth of interrogation light is on the order of only 100 nm whereas the cell is

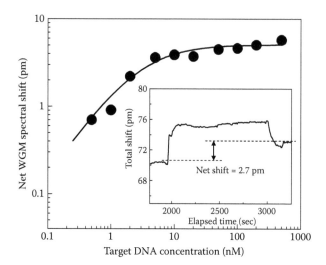

FIGURE 8.28

Concentration dependent WGM shift for 25-mer strands. Inset shows the WGM shift from 10 pM target DNA using a 37 nm/RIU OFRR. (Reused from Jonathan D. Suter, Ian M. White, Hongying Zhu, Huidong Shi, Charles W. Caldwell, and Xudong Fan, Biosensors and Bioelectronics, 23, 1003–1009, 2008. © Elsevier. With permission.)

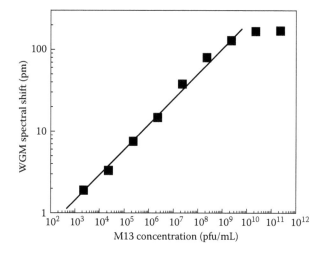

FIGURE 8.29

WGM spectral shift measured for various concentrations of M13 phage in PBS buffer. Concentration varies from 2.3×10^3 to 2.3×10^{11} pfu/mL. (Reused from Hongying Zhu, Ian M. White, Jonathan D. Suter, Mohammed Zourob, and Xudong Fan, Analyst, 132, 356–360, 2008. © The Royal Society of Chemistry. With permission.)

typically a few micrometers in size. Optical label-free detection of bacterium has been achieved using SPR [120–122], waveguide [123,124], and photonic crystal sensors [102]. Recently, microspheres and planar ring resonators were also employed [45,125]. The detection limit ranges from 10^5 to 10^7 cfu/mL (cfu: colony-forming unit, it simply represents the number of bacteria or cells). While the performance of the OFRR is similar to that of other label-free optical sensor (the detection limit is on the order of 10^5 cfu/mL), it benefits from

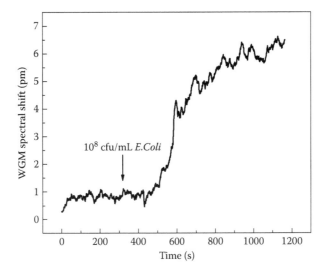

FIGURE 8.30
Sensorgram for 10^8 cfu/mL *E. coli* running through the OFRR.

integrated microfluidics and the circular shape of the capillary for on-line rapid detection of bacteria. Figure 8.30 shows an example of typical OFRR response to bacteria.

8.3.7.5 Pesticide Detection

In addition to biomolecules, the OFRR can also be used for chemical detection, which is important in environment monitoring, food safety, etc. Here we use parathion-methyl as a model system to mimic organophosphorus, a broad-spectrum pesticides used on a wide range of crops including vegetables, fruits, grains and ornamentals and can cause adverse effects to the nervous system in humans. Figure 8.31 shows that the OFRR has a linear working range from 10^{-11} to 10^{-6} M. The detection limit of 3.8×10^{-11} M for parathion-methyl is achieved. Using the OFRR, the analysis time is approximately 0.5 min, ten times faster than the SPR system [126].

8.3.8 Integration with Microfluidics

For further device miniaturization and easy fluid control, we demonstrate the feasibility of using an electro-kinetic pump to deliver the fluidic samples [73]. Figure 8.32 shows the experimental setup where two electrodes are placed in respective reservoirs at each end of the OFRR. Figure 8.33A shows that the analyte repeatedly passes the location where the optical taper is in contact with the OFRR. The electric field needed is less than 100 V/cm, which can readily be obtained with a small on-board voltage control in future miniaturized lab-on-a-chip design. In addition to sample delivery, the OFRR can also perform on-column detection for capillary electrophoresis that separate samples based on their interaction with the capillary surface [84], as shown in Figure 8.33B. The OFRR inherently integrates the state-of-the-art photonic ring resonator with a capillary. Therefore, it will have vast applications in analytical instruments such as capillary electrophoresis and liquid/gas chromatography.

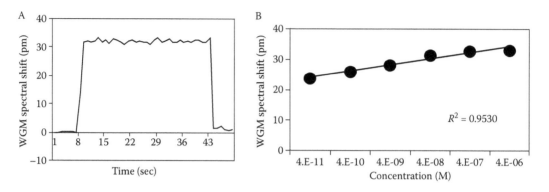

FIGURE 8.31

(A) Sensorgram for pesticide (parathion-methyl) binding and subsequent removal from the OFRR interior surface. (B) WGM shift for various concentrations of pesticide.

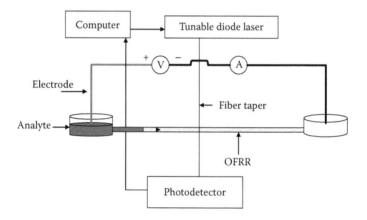

FIGURE 8.32

Schematic of electrophoresis experimental setup.

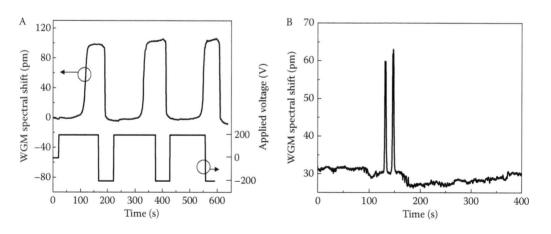

FIGURE 8.33

(A) The WGM spectral shift shows that analyte solution is driven into and out of the OFRR when the voltage supply switches polarity. (B) Electrophoretic separation between sucrose and ribose.

8.3.9 Integration with Waveguides

Previously we have demonstrated the capability of using an optical fiber taper to couple the resonant light into the OFRR, it is important to fully integrate the OFRR with photonic circuits for future device miniaturization and multiplexing. Optical fiber tapers suffer from low reproducibility, low product yield, and lack of fabrication control. Furthermore, fiber tapers are mechanically fragile. Also it is difficult to position an array of tapers to the desired locations on the OFRR. Here we present three approaches to address the above problems.

8.3.9.1 Integration with Antiresonant Reflecting Optical Waveguide (ARROW)

As compared to a fiber taper, lithographically fabricated waveguide arrays on a wafer are advantageous owing to mass production capability, well-controlled fabrication processes, and high channel density. Moreover, they are also highly compatibility with optoelectronics, which enables integration with electronics and on-chip lasers in the future. In our experiment, we use a special type of waveguide, called antiresonant reflecting optical waveguide (ARROW) [127–131], to excite the WGM in an OFRR. An ARROW, as schematically shown in Figure 8.34, consists of a low refractive index core and a high–low refractive index pair of cladding layers. The thickness of the high index layer has to meet the following condition for the core to guide the light [127–131]:

$$t = \frac{\lambda}{4n_2}(2N+1)\left(1 - \frac{n_1^2}{n_2^2} + \frac{\lambda^2}{4n_2^2h^2}\right)^{1/2} \quad (N = 0,1,2,\ldots), \tag{8.13}$$

where λ is the laser wavelength in vacuum, n_1 and n_2 are the refractive index of the core and high index layer, respectively, and h is the thickness of the core. Under this antiresonance condition, the ARROW prevents the light in the core from leaking into the substrate, and in the meantime presents a sufficient evanescent field above the core for the coupling between the ARROW and the OFRR. In an ARROW, only TE-polarized light will propagate whereas the TM-mode is very lossy and can not be transmitted through the ARROW.

Similar to the coupling between the ARROW and a microsphere ring resonator that has previously been achieved [129–131], the coupling between the ARROW and the OFRR can be accomplished by placing the OFRR in touch with the ARROW channels, as shown in

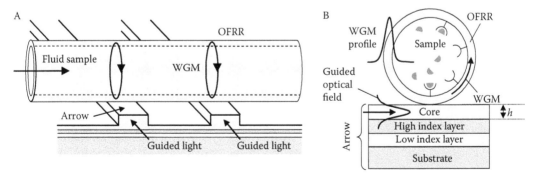

FIGURE 8.34
(A) OFRR-ARROW system. (B) OFRR cross section on top of an ARROW structure. Light polarization is along the OFRR longitudinal direction. (Reused from Ian M. White, Hesam Oveys, Xudong Fan, Terry L. Smith, and Junying Zhang, *Applied Physics Letters*, 89, 191106, 2006. © American Institute of Physics. With permission.)

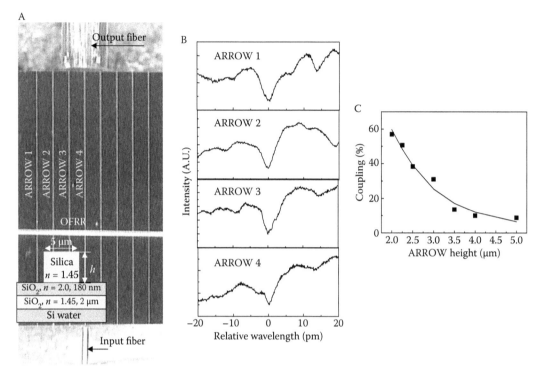

FIGURE 8.35
(A) Experimental setup for simultaneous excitation of the WGMs along an OFRR with an ARROW. The inset in the picture shows the detail of the ARROW structure designed to for 980 nm. (B) WGM spectra at the ARROW output. (C) Coupling depth for various ARROW heights (h). (Reused from Ian M. White, Hesam Oveys, Xudong Fan, Terry L. Smith, and Junying Zhang, *Applied Physics Letters*, 89, 191106, 2006. © American Institute of Physics. With permission.)

Figure 8.35A and B. Furthermore, Figure 8.35C shows that the coupling can be adjusted by changing the ARROW core height [132,133]. Our results will enable the development of integrated high throughput sensor arrays.

8.3.9.2 Integration with Metal-Clad Waveguide

The ARROW structure is highly wavelength dependent because it relies on the resonant nature of reflections from multiple dielectric layers in the cladding. As a result, the ARROW planar waveguide is only effective for a specific wavelength region. To circumvent this problem, we use a simple structure based on a gold-clad planar waveguide to excite the WGMs in a broad range of wavelength (600–1550 nm) [72]. The metal-clad waveguide structure is illustrated in Figure 8.36. A thin gold cladding layer acts as a mirror that prevents light leakage into the substrate over a large range of wavelengths. Meanwhile, a high evanescent field is presented at the top of the waveguide for sufficient excitation of the ring resonator WGMs. The TE mode has a low loss from visible to near infrared. Additionally, effective coupling between the gold-clad waveguide and the OFRR has been experimentally achieved (see Figure 8.37A). Similar to the ARROW structure, the coupling strength can be adjusted by changing the core height (see Figure 8.37B).

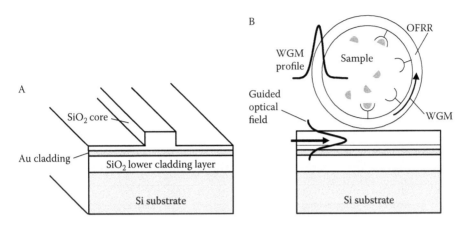

FIGURE 8.36
(A) Cross-section of waveguide chip structure. (B) Ring resonator in contact with waveguide for WGM excitation. (Reused from Ian M. White, Jonathan D. Suter, Hesam Oveys, Xudong Fan, Terry L. Smith, Junying Zhang, Barry J. Koch, and Michael A. Haase, *Optics Express*, 15, 646–651, 2007. © Optical Society of America. With permission.)

8.3.9.3 *Integration with Microfiber and Low Index Polymer*

A third approach is to embed the OFRR-taper system in a curable low RI polymer, as illustrated in Figure 8.38 [81]. This approach significantly improves the device's mechanical strength and protects the OFRR and the taper from contamination. It should be noted that although the taper is immersed in the polymer, the transmission is not compromised [81,134]. Furthermore, relatively high Q-factors can still be maintained (see Figure 8.38B).

8.4 Summary and Future Work

The OFRR is an emerging technology that marries photonics and microfluidics. In addition to label-free sensing, fluorescence-based and Raman-based sensing [86] can be performed using the same platform. Furthermore, the OFRR can be used for development of novel analytical instruments such as capillary electrophoresis and liquid/gas chromatography [73,135]. Microfluidic lasers [77,80,111] and opto-fluidic components such as filters and delay lines will also be very attractive. Since the OFRR is still in its infancy, significant amount of experimental and theoretical work is needed to fully exploit the unique advantages that the OFRR can offer.

Acknowledgments

The authors acknowledge the financial support from the 3M Non-Tenured Faculty Award, the University of Missouri Research Board Award (05-013), the University of Missouri

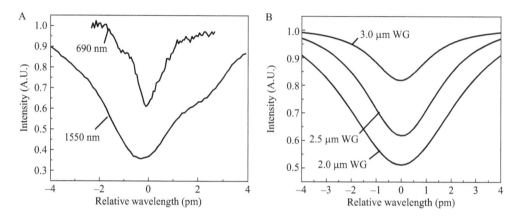

FIGURE 8.37

(A) WGMs in an OFRR of 140 μm in diameter for 690 nm and 1550 nm that are excited by a gold-clad waveguide of 2.5 μm in height. (B) Lorentzian fits of recorded WGMs in OFRR excited by 1550 nm light coupled from gold-clad waveguides of heights of 2.0, 2.5, and 3.0 μm. (Reused from Ian M. White, Jonathan D. Suter, Hesam Oveys, Xudong Fan, Terry L. Smith, Junying Zhang, Barry J. Koch, and Michael A. Haase, *Optics Express*, 15, 646–651, 2007. © Optical Society of America. With permission.)

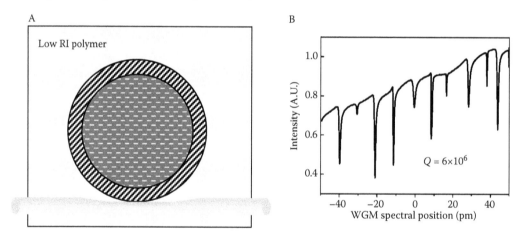

FIGURE 8.38

(A) OFRR-microfiber system is embedded in a low index polymer. (B) High *Q*-factors can still be obtained for the system in (A).

Bioprocessing and Biosensing Center, University of Missouri Research Council (URC-06-035), American Chemistry Society Petroleum Research Fund (43879-G10), the Wallace H. Coulter Early Career Award, MU Life Sciences Fellowship Program, the NIH Postdoctoral Fellowship (1K25 EB006011), and the NSF (ECCS-0729903 and CBET-0747398).

References

1. J. Homola, S. S. Yee, and G. Gauglitz. Surface plasmon resonance sensors: review. *Sensors and Actuators B* 54, 3–15, 1999.
2. C. L. Baird, and D. G. Myszka. Current and emerging commercial optical biosensors. *Journal of Molecular Recognition*, 14, 261–268, 2001.

3. http://www.biacore.com/lifesciences/index.html. November 8, 2008.

4. R. Narayanaswamy, and O. S. Wolfbeis. *Optical Sensors*. Springer, New York, 2004.

5. L. M. Lechuga, A. T. M. Lenferink, R. P. H. Kooyman, and J. Greve. Feasibility of evanescent wave interferometer immunosensors for pesticide detection: chemical aspects. *Sensors and Actuators B*, 25, 762–765, 1995.

6. C. R. Lavers, K. Itoh, S. C. Wu, M. Murabayashi, I. Mauchline, G. Stewart, and T. Stout. Planar optical waveguides for sensing applications. *Sensors and Actuators B*, 69, 85–95, 2000.

7. A. Ymeti, J. Greve, P. V. Lambeck, T. Wink, S. W. F. M. v. Hovell, T. A. M. Beumer, R. R. Wijn, R. G. Heideman, V. Subramaniam, and J. S. Kanger. Fast, ultrasensitive virus detection using a Young interferometer sensor. *Nano Letters*, 7, 394–397, 2007.

8. A. Serpenguzel, S. Arnold, and G. Griffel. Excitation of resonances of microspheres on an optical fiber. *Optics Letters*, 20, 654–656, 1995.

9. F. Vollmer, D. Braun, A. Libchaber, M. Khoshsima, I. Teraoka, and S. Arnold. Protein detection by optical shift of a resonant microcavity. *Applied Physics Letters*, 80, 4057–4059, 2002.

10. S. Arnold, M. Khoshsima, I. Teraoka, S. Holler, and F. Vollmer. Shift of whispering-gallery modes in microspheres by protein adsorption. *Optics Letters*, 28, 272–274, 2003.

11. I. Teraoka, S. Arnold, and F. Vollmer. Perturbation approach to resonance shifts of whispering-gallery modes in a dielectric microsphere as a probe of a surrounding medium. *Journal of the Optical Society of America B*, 20, 1937–1947, 2003.

12. F. Vollmer, S. Arnold, D. Braun, I. Teraoka, and A. Libchaber. Multiplexed DNA quantification by spectroscopic shift of two microsphere cavities. *Biophysical Journal*, 85, 1974–1979, 2003.

13. M. Noto, M. Khoshsima, D. Keng, I. Teraoka, V. Kolchenko, and S. Arnold. Molecular weight dependence of a whispering gallery mode biosensor. *Applied Physics Letters*, 87, 223901, 2005.

14. M. Noto, F. Vollmer, D. Keng, I. Teraoka, and S. Arnold. Nanolayer characterization through wavelength multiplexing of a microsphere resonator. *Optics Letters*, 30, 510–512, 2005.

15. I. Teraoka, and S. Arnold. Theory of resonance shifts in TE and TM whispering gallery modes by nonradial perturbations for sensing applications. *Journal of the Optical Society of America B*, 23, 1381–1389, 2006.

16. I. Teraoka, and S. Arnold. Enhancing the sensitivity of a whispering-gallery mode microsphere sensor by a high-refractive-index surface layer. *Journal of the Optical Society of America B*, 23, 1434–1441, 2006.

17. X. Fan, P. Palinginis, S. Lacey, H. Wang, and M. C. Lonergan. Coupling semiconductor nanocrystals to a fused-silica microsphere: a quantum-dot microcavity with extremely high Q factors. *Optics Letters*, 25, 1600–1602, 2000.

18. I. M. White, and X. Fan. Demonstration of composite microsphere cavity and surface enhanced Raman spectroscopy for improved sensitivity. In *Chemical and Biological Sensors for Industrial and Environmental Security*, A. J. Sedlacek, S. D. Christesen, R. J. Combs, and T. Vo-Dinh SPIE, Bellingham, WA, 2005, 59940G.

19. N. M. Hanumegowda, C. J. Stica, B. C. Patel, I. M. White, and X. Fan. Refractometric sensors based on microsphere resonators. *Applied Physics Letters*, 87, 201107, 2005.

20. I. M. White, N. M. Hanumegowda, and X. Fan. Subfemtomole detection of small molecules with microsphere sensors. *Optics Letters*, 30, 3189–3191, 2005.

21. N. M. Hanumegowda, I. M. White, H. Oveys, and X. Fan. Label-free protease sensors based on optical microsphere resonators. *Sensor Letters*, 3, 315–319, 2005.

22. I. M. White, N. M. Hanumegowda, H. Oveys, and X. Fan. Tuning whispering gallery modes in optical microspheres with chemical etching. *Optics Express*, 13, 10754–10759, 2005.

23. N. M. Hanumegowda, I. M. White, and X. Fan. Aqueous mercuric ion detection with microsphere optical ring resonator sensors. *Sensors and Actuators B: Chemical*, 120, 207–212, 2006.

24. I. M. White, H. Oveys, and X. Fan. Increasing the enhancement of SERS with dielectric microsphere resonators. *Spectroscopy*, 21, 36–42, 2006.

25. H. Zhu, J. D. Suter, I. M. White, and X. Fan. Aptamer based microsphere biosensor for thrombin detection. *Sensors*, 6, 785–795, 2006.

26. J. P. Rezac, and A. T. Rosenberger. Locking a microsphere whispering-gallery mode to a laser. *Optics Express*, 8, 605–610, 2001.

27. J. L. Nadeau, V. S. Ilchenko, D. Kossakovski, G. H. Bearman, and L. Maleki. High-Q whispering-gallery mode sensor in liquids. In *Laser Resonators and Beam Control V*, A. V. Kudryashov, ed. Bellingham, WA, 2002, 4629, 172–180.

28. A. M. Armani, and K. J. Vahala. Heavy water detection using ultra-high-Q microcavities. *Optics Letters*, 31, 1896–1898, 2006.

29. H. Quan, and Z. Guo. Simulation of whispering-gallery-mode resonance shifts for optical miniature biosensors. *Journal of Quantitative Spectroscopy & Radiative Transfer*, 93, 231–243, 2005.

30. C.-Y. Chao, and L. J. Guo. Biochemical sensors based on polymer microrings with sharp asymmetrical resonance. *Applied Physics Letters*, 83, 1527–1529, 2003.

31. C.-Y. Chao, W. Fung, and L. J. Guo. Polymer microring resonators for biochemical sensing applications. *IEEE Journal of Selected Topics in Quantum Electronics*, 12, 134–142, 2006.

32. J. Yang, and L. J. Guo. Optical sensors based on active microcavities. *IEEE Journal of Selected Topics in Quantum Electronics*, 12, 143–147, 2006.

33. A. Yalcin, K. C. Popat, O. C. Aldridge, T. A. Desai, J. Hryniewicz, N. Chbouki, B. E. Little, O. King, V. Van, S. Chu, D. Gill, M. Anthes-Washburn, M. S. Unlu, and B. B. Goldberg. Optical sensing of biomolecules using microring resonators. *IEEE Journal of Selected Topics in Quantum Electronics*, 12, 148–155, 2006.

34. A. Ksendzov, and Y. Lin. Integrated optics ring-resonator sensors for protein detection. *Optics Letters*, 30, 3344–3346, 2005.

35. A. Ksendzov, M. L. Homer, and A. M. Manfreda. Integrated optics ring-resonator chemical sensor with polymer transduction layer. *Electronics Letters*, 40, 63–65 2004.

36. E. Krioukov, D. Klunder, A. Driessen, J. Grevea, and C. Otto. Two-photon fluorescence excitation using an integrated optical microcavity: a promising tool for biosensing of natural chromophores. *Talanta*, 65, 1086–1090, 2005.

37. E. Krioukov, J. Greve, and C. Otto. Performance of integrated optical microcavities for refractive index and fluorescence sensing. *Sensors and Actuators B*, 90, 58–67, 2003.

38. E. Krioukov, D. J. W. Klunder, A. Driessen, J. Greve, and C. Otto. Integrated optical microcavities for enhanced evanescent-wave spectroscopy. *Optics Letters*, 27, 1504–1506, 2002.

39. R. W. Boyd, and J. E. Heebner. Sensitive disk resonator photonic biosensor. *Applied Optics*, 40, 5742–5747, 2001.

40. S. Blair, and Y. Chen. Resonant-enhanced evanescent-wave fluorescence biosensing with cylindrical optical cavities. *Applied Optics*, 40, 570–582, 2001.

41. J. Scheuer, W. M. J. Green, G. A. DeRose, and A. Yariv. InGaAsP annular Bragg lasers: theory, applications, and modal properties. *IEEE Journal of Selected Topics in Quantum Electronics*, 11, 476–484, 2005.

42. T. Baehr-Jones, M. Hochberg, C. Walker, and A. Scherer. High-Q ring resonators in thin silicon-on-insulator. *Applied Physics Letters*, 85, 3346–3347, 2004.

43. T. Baehr-Jones, M. Hochberg, C. Walker, and A. Scherer. High-Q optical resonators in silicon-on-insulator-based slot waveguides. *Applied Physics Letters*, 86, 081101, 2005.

44. V. R. Almeida, Q. Xu, C. A. Barrios, and M. Lipson. Guiding and confining light in void nanostructure. *Optics Letters*, 29, 1209–1211, 2004.

45. A. Ramachandran, S. Wang, J. Clarke, S. J. Ja, D. Goad, L. Wald, E. M. Flood, E. Knobbe, J. V. Hryniewicz, S. T. Chu, D. Gill, W. Chen, O. King, and B. E. Little. A universal biosensing platform based on optical micro-ring resonators. *Biosensors and Bioelectronics*, 23, 939–944, 2008.

46. J. Topolancik, and F. Vollmer. Photoinduced transformations in bacteriorhodopsin membrane monitored with optical microcavities. *Biophysical Journal*, 92, 2223–2229, 2007.

47. F. Vollmer, and P. Fischer. Ring-resonator-based frequency-domain optical activity measurements of a chiral liquid. *Optics Letters*, 31, 453–455, 2006.

48. S. Arnold, R. Ramjit, D. Keng, V. Kolchenko, and I. Teraoka. MicroParticle photophysics illuminates viral bio-sensing. *Faraday Discussions, 137, 65–83, 2008*.

49. D. Keng, S. R. McAnanama, I. Teraoka, and S. Arnold. Resonance fluctuations of a whispering gallery mode biosensor by particles undergoing Brownian motion. *Applied Physics Letters*, 91, 103902, 2007.

50. M. Noto, D. Keng, I. Teraoka, and S. Arnold. Detection of protein orientation on the silica microsphere surface using transverse electric/transverse magnetic whispering gallery modes. *Biophysical Journal*, 92, 4466–4472, 2007.

51. R. K. Chang, and A. J. Campillo. *Optical Processes in Microcavities*. World Scientific, Singapore, 1996.

52. C. F. Bohren, and D. R. Huffman. *Absorption and Scattering of Light by Small Particles*. John Wiley & Sons, New York, 1998.

53. K. Vahala. *Optical Microcavities*. World Scientific, Singapore, 2005.

54. M. L. Gorodetsky, A. A. Savchenkov, and V. S. Ilchenko. Ultimate Q of optical microsphere resonators. *Optics Letters*, 21, 453–455, 1996.

55. H.-J. Moon, Y.-T. Chough, and K. An. Cylindrical Microcavity Laser Based on the Evanescent-Wave-Coupled Gain. *Physical Review Letters*, 85, 3161–3164, 2000.

56. H.-M. Tzeng, K. F. Wall, M. B. Long, and R. K. Chang. Laser emission from individual droplets at wavelengths corresponding to morphology-dependent resonances. *Optics Letters*, 9, 499–501, 1984.

57. J. B. Snow, S.-X. Qian, and R. K. Chang. Stimulated Raman scattering from individual water and ethanol droplets at morphology-dependent resonances. *Optics Letters*, 10, 37–39, 1985.

58. H.-B. Lin, A. L. Huston, B. L. Justus, and A. J. Campillo. Some characteristics of a droplet whispering-gallery-mode laser. *Optics Letters*, 11, 614–616, 1986.

59. H.-B. Lin, J. D. Eversole, and A. J. Campillo. Frequency pulling of stimulated Raman scattering in microdroplets. *Optics Letters*, 15, 387–389, 1990.

60. A. S. Kwok, and R. K. Chang. Stimulated resonance Raman scattering of Rhodamine 6G. *Optics Letters*, 18, 1703–1705, 1993.

61. J. Feng, G. Shan, B. D. Hammock, and I. M. Kennedy. Fluorescence quenching competitive immunoassay in micro droplets. *Biosensors and Bioelectronics*, 18, 1055–1063, 2003.

62. M. Nichkova, J. Feng, F. Sanchez-Baeza, M.-P. Marco, B. D. Hammock, and I. M. Kennedy. Competitive quenching fluorescence immunoassay for chlorophenols based on laser-induced fluorescence detection in microdroplets. *Analytical Chemistry*, 75, 83–90, 2003.

63. M. Tanyeri, D. Dosev, and I. M. Kennedy. Chemical and biological sensing through optical resonances in pendant droplets. In *Nanosensing: Materials and Devices II*, M. S. Islam, and A. K. Dutta, eds. SPIE, Boston, MA, 2005, 60080Q.

64. U. Levy, K. Campbell, A. Groismanb, S. Mookherjea, and Y. Fainman. On-chip microfluidic tuning of an optical microring resonator. *Applied Physics Letters*, 88, 111107, 2006.

65. M. Hossein-Zadeh, and K. J. Vahala. Fiber-taper coupling to whispering-gallery modes of fluidic resonators embedded in a liquid medium. *Optics Express*, 14, 10800–10810, 2006.

66. M. Tanyeri, M. Nichkova, B. D. Hammock, and I. M. Kennedy. *Chemical and Biological Sensing through Optical Resonances in Microcavities*, D. V. Nicolau, J. Enderlein, R. C. Leif, D. L. Farkas, and R. Raghavachari, eds. SPIE, Boston, MA, 2005, 227–236.

67. K. D. Vos, I. Bartolozzi, E. Schacht, P. Bienstman, and R. Baets. Silicon-on-insulator microring resonator for sensitive and label-free biosensing. *Optics Express*, 15, 7610–7615, 2007.

68. M. D. Barnes, K. C. Ng, W. B. Whitten, and J. M. Ramsey. Detection of single rhodamine 6G molecules in levitated microdroplets. *Analytical Chemistry*, 65, 2360–2365, 1993.

69. M. Soltani, S. Yegnanarayanan, and A. Adibi. Ultra-high Q planar silicon microdisk resonators for chip-scale silicon photonics. *Optics Express*, 15, 4694–4704, 2007.

70. H. Zhu, I. M. White, J. D. Suter, P. S. Dale, and X. Fan. Analysis of biomolecule detection with optofluidic ring resonator sensors. *Optics Express*, 15, 9139–9146, 2007.

71. X. Fan, I. M. White, H. Zhu, J. D. Suter, and H. Oveys. Overview of novel integrated optical ring resonator bio/chemical sensors (invited paper). *SPIE Laser Resonators and Beam Control*, 6452, 64520M, 2007.

72. I. M. White, J. D. Suter, H. Oveys, X. Fan, T. L. Smith, J. Zhang, B. J. Koch, and M. A. Haase. Universal coupling between metal-clad waveguides and optical ring resonators. *Optics Express*, 15, 646–651, 2007.

73. H. Zhu, I. M. White, J. D. Suter, M. Zourob, and X. Fan. An integrated refractive index optical ring resonator detector for capillary electrophoresis. *Analytical Chemistry*, 79, 930–937, 2007.

74. J. D. Suter, I. M. White, H. Zhu, and X. Fan. Thermal characterization of liquid core optical ring resonator sensors. *Applied Optics*, 46, 389–396, 2007.

75. I. M. White, H. Oveys, and X. Fan. Liquid core optical ring resonator sensors. *Optics Letters*, 31, 1319–1321, 2006.

76. I. M. White, H. Oveys, X. Fan, T. L. Smith, and J. Zhang. Integrated multiplexed biosensors based on liquid core optical ring resonators and antiresonant reflecting optical waveguides. *Applied Physics Letters*, 89, 191106, 2006.

77. S. I. Shopova, H. Zhu, X. Fan, and P. Zhang. Optofluidic ring resonator based dye laser. *Applied Physics Letters*, 90, 221101, 2007.

78. H. Zhu, I. M. White, J. D. Suter, M. Zourob, and X. Fan. Opto-fluidic micro-ring resonator for sensitive label-free detection. Analyst, 132, 356–360, 2008.

79. J. D. Suter, I. M. White, H. Zhu, H. Shi, C. W. Caldwell, and X. Fan. Label-free quantitative DNA detection using the liquid core optical ring resonator. *Biosensors and Bioelectronics*, 23, 1003–1009, 2008.

80. S. I. Shopova, J. M. Cupps, P. Zhang, E. P. Henderson, S. Lacey, and X. Fan. Opto-fluidic ring resonator lasers based on highly efficient resonant energy transfer. *Optics Express*, 15, 12735–12742, 2007.

81. M. Sumetsky, R. S. Windeler, Y. Dulashko, and X. Fan. Optical liquid ring resonator sensor. *Optics Express*, 15, 14376–14381, 2007.

82. T. Ling and L. J. Guo. A unique resonance mode observed in a prism-coupled micro-tube resonator sensor with superior index sensitivity. Optics Express, 15, 17424–17432, 2007.

83. V. Zamora, A. Díez, M. V. Andrés, and B. Gimeno. Refractometric sensor based on whispering gallery modes of thin capillaries. *Optics Express*, 15, 12011–12016, 2007.

84. J. P. Landers. *Handbook of Capillary Electrophoresis*. Boca Raton, FL.

85. M. Cai, and K. Vahala. Highly efficient optical power transfer to whispering-gallery modes by use of a symmetrical dual-coupling configuration. *Optics Letters*, 25, 260–262, 2000.

86. I. M. White, J. Gohring, and X. Fan. SERS-based detection in an optofluidic ring resonator platform. *Optics Express*, 15, 17433–17442, 2007.

87. N. A. Mortensen, S. Xiao, and J. Pedersen. Liquid-infiltrated photonic crystals: enhanced light–matter interactions for lab-on-a-chip applications. *Microfluidics and Nanofluidics*, 3, 1613–4982, 2007.

88. H.-J. Moon, and K. An. Interferential coupling effect on the whispering-gallery mode lasing in a double-layered microcylinder. *Applied Physics Letters*, 80, 3250–3252, 2002.

89. D. J. Bornhop, J. C. Latham, A. Kussrow, D. A. Markov, R. D. Jones, and H. S. Sørensen. Free-solution, label-free molecular interactions studied by back-scattering interferometry. *Science*, 317, 1732–1736, 2007.

90. M. K. Chin, and S. T. Ho. Design and modeling of waveguide-coupled single-mode microring resonators. *Journal of Lightwave Technology*, 16, 1433–1446, 1998.

91. B. R. Johnson. Theory of morphology-dependent resonances: shape resonances and width formulas. *Journal of the Optical Society of America A.*, 10, 343–352, 1993.

92. I. M. White, J. Gohring, Y. Sun, G. Yang, S. Lacey, and X. Fan. Versatile waveguide-coupled optofluidic devices based on liquid core optical ring resonators. *Applied Physics Letters*, 91, 241104, 2007.

93. D. R. Linde. *The CRC Handbook of Chemistry and Physics*. Boca Raton, FL, 2005.

94. J. P. Longtin, and C.-H. Fan. Precision laser-based concentration and refractive index measurement of liquid. *Microscale Thermophysical Engineering*, 2, 261–272, 1998.

95. M. Han, and A. Wang. Temperature compensation of optical microresonators using a surface layer with negative thermo-optic coefficient. *Optics Letters*, 32, 1800–1802, 2007.

96. I. M. White, and X. Fan. On the performance quantification of resonant refractive index sensors. *Optics Express*, 16(2), 1020–1028, 2008.

97. I. M. White, H. Zhu, J. D. Suter, X. Fan, and M. Zourob. Label-free detection with the liquid core optical ring resonator sensing platform. In A. Rasooly and K. E. Herold, *Biosensors and Biodetection: Methods and Protocols,* Volume 1: Optical-Based Detectors (Methods in Molecular Biology) Humana Press, New York.

98. B. Cunningham, P. Li, B. Lin, and J. Pepper. Colorimetric resonant reflection as a direct biochemical assay technique. *Sensors and Actuators B*, 81, 316–328, 2002.

99. B. Cunningham, J. Qiu, P. Li, and B. Lin. Enhancing the surface sensitivity of colorimetric resonant optical biosensors. *Sensors and Actuators B*, 87, 365–370, 2002.

100. B. T. Cunningham, P. Li, S. Schulz, B. Lin, C. Baird, J. Gerstenmaier, C. Genick, F. Wang, E. Fine, and L. Laing. Label-free assays on the BIND system. *Journal of Biomolecular Screening*, 9, 481–490, 2004.

101. L. L. Chan, B. T. Cunningham, P. Y. Li, and D. Puff. A self-referencing method for microplate label-free photonic-crystal biosensors. *Sensors Journal, IEEE*, 6, 1551, 2006.

102. B. Lin, P. Li, and B. T. Cunningham. A label-free biosensor-based cell attachment assay for characterization of cell surface molecules. *Sensors and Actuators B: Chemical*, 114, 559–564, 2006.

103. M. M. Varma, D. D. Nolte, H. D. Inerowicz, and F. E. Regnier. Spinning-disk self-referencing interferometry of antigen–antibody recognition. *Optics Letters*, 29, 950–952, 2004.

104. M. M. Varma, H. D. Inerowicz, F. E. Regnier, and D. D. Nolte. High-speed label-free detection by spinning-disk micro-interferometry. *Biosensors and Bioelectronics*, 19, 1371–1376, 2004.

105. A. D. Fitt, K. Furusawa, T. M. Monro, and C. P. Please. Modeling the fabrication of hollow fibers: Capillary drawing. *Journal of Lightwave Technology*, 19, 1924–1931, 2001.

106. S. C. Xue, R. I. Tanner, G. W. Barton, R. Lwin, M. C. J. Large, and L. Poladian. Fabrication of microstructured optical fibers-part I: problem formulation and numerical modeling of transient draw process. *Journal of Lightwave Technology*, 23, 2245–2254, 2005.

107. S. C. Xue, R. I. Tanner, G. W. Barton, R. Lwin, M. C. J. Large, and L. Poladian. Fabrication of microstructured optical fibers—Part II: numerical modeling of steady-state draw process. *Journal of Lightwave Technology*, 23, 2255–2266, 2005.

108. S. C. Xue, M. C. J. Large, G. W. Barton, R. I. Tanner, L. Poladian, and R. Lwin. Role of material properties and drawing conditions in the fabrication of microstructured optical fibers. *Journal of Lightwave Technology*, 24, 853–860, 2006.

109. D. W. Vernooy, V. S. Ilchenko, H. Mabuchi, E. W. Streed, and H. J. Kimble. High-Q measurements of fused-silica microspheres in the near infrared. *Optics Letters*, 23, 247–249, 1998.

110. L. Kou, D. Labrie, and P. Chylek. Refractive indices of water and ice in the 0.65- to 2.5-Mum spectral range. *Applied Optics*, 32, 3531–3540, 1993.

111. S. Lacey, I. M. White, Y. Sun, S. I. Shopova, J. M. Cupps, P. Zhang, and X. Fan. Versatile optofluidic ring resonator lasers with ultra-low threshold. *Optics Express*, 15, 15523–15530, 2007.

112. F. S. Ligler, M. Breimer, J. P. Golden, D. A. Nivens, J. P. Dodson, T. M. Green, D. P. Haders, and O. A. Sadik. Integrating waveguide biosensor. *Analytical Chemistry*, 74, 713–719, 2002.

113. M. Y. Balakirev, S. Porte, M. Vernaz-Gris, M. Berger, J.-P. Arie, B. Fouque, and F. Chatelain. Photochemical patterning of biological molecules inside a glass capillary. *Analytical Chemistry*, 77, 5474–5479, 2005.

114. M. A. Holden, S.-Y. Jung, and P. S. Cremer. Patterning enzymes inside microfluidic channels via photoattachment chemistry. *Analytical Chemistry*, 76, 1838–1843, 2004.

115. *Cancer Facts and Figures*. American Cancer Society, Atlanta, Georgia Publisher, American Cancer Society 2007.

116. F. Safi, I. Kohler, E. Röttinger, and H. Beger. The value of the tumor marker CA 15-3 in diagnosing and monitoring breast cancer. A comparative study with carcinoembryonic antigen. *Cancer*, 68, 574–582, 1991.

117. M. J. Duffy. CA 15-3 and related mucins as circulating markers in breast cancer. *Annals of Clinical Biochemistry*, 36, 579–586, 1999.
118. D. Aguiar-Bujanda. False elevation of serum CA 15-3 levels in patients under follow-up for breast cancer. *The Breast Journal*, 10, 375–376, 2004.
119. M. Cargill, D. Altshuler, J. Ireland, P. Sklar, K. Ardlie, N. Patil, C. R. Lane, E. P. Lim, N. Kalyanaraman, J. Nemesh, L. Ziaugra, L. Friedland, A. Rolfe, J. Warrington, R. Lipshutz, G. Q. Daley, and E. S. Lander. Characterization of single-nucleotide polymorphisms in coding regions of human genes. *Nature Genetics*, 22, 231–238, 1999.
120. B. Barlen, S. D. Mazumdar, O. Lezrich, P. Kämpfer, and M. Keusgen. Detection of *Salmonella* by surface plasmon resonance. *Sensors*, 7, 1427–1446, 2007.
121. V. Nanduri, A. K. Bhunia, S.-I. Tu, G. C. Paoli, and J. D. Brewster. SPR biosensor for the detection of *L. monocytogenes* using phage-displayed antibody. *Biosensors and Bioelectronics*, 23, 248–252, 2007.
122. A. D. Taylor, J. Ladda, Q. Yu, S. Chena, J. Homola, and S. Jiang. Quantitative and simultaneous detection of four foodborne bacterial pathogens with a multi-channel SPR sensor. *Biosensors and Bioelectronics*, 22, 752–758, 2006.
123. M. Zourob, S. Mohr, B. J. T. Brown, P. R. Fielden, M. B. McDonnell, and N. J. Goddard. An integrated metal clad leaky waveguide sensor for detection of bacteria. *Analytical Chemistry*, 77, 232–242, 2005.
124. R. Horváth, H. C. Pedersen, N. Skivesen, D. Selmeczi, and N. B. Larsen. Optical waveguide sensor for on-line monitoring of bacteria. *Optics Letters*, 28, 1233–1235, 2003.
125. H.-C. Ren, F. Vollmer, S. Arnold, and A. Libchaber. High-Q microsphere biosensor - analysis for adsorption of rodlike bacteria. *Optics Express*, 15, 17410–17423, 2007.
126. G. Yang, I. M. White, and X. Fan. An opto-fluidic ring resonator biosensor for the detection of organophosphorus pesticides, under review. *Sensors and Actuators B*, 133, 105–112, 2008.
127. M. A. Duguay, Y. Kokubun, T. L. Koch, and L. Pfeiffer. Antiresonant reflecting optical waveguides in SiO_2-Si multilayer structures. *Applied Physics Letters*, 49, 13–15, 1986.
128. T. Baba, Y. Kokubun, T. Sakaki, and K. Iga. Loss reduction of an ARROW waveguide in shorter wavelength and its stack configuration. *Journal of Lightwave Technology*, 6, 1440–1445, 1988.
129. B. E. Little, J.-P. Laine, D. R. Lim, H. A. Haus, L. C. Kimerling, and S. T. Chu. Pedestal antiresonant reflecting waveguides for robust coupling to microsphere resonators and for microphotonic circuits. *Optics Letters*, 25, 73–75, 2000.
130. J.-P. Laine, B. E. Little, D. R. Lim, H. C. Tapalian, L. C. Kimerling, and H. A. Haus. Microsphere resonator mode characterization by pedestal anti-resonant reflecting waveguide. *IEEE Photonics Technology Letters*, 12, 1004–1006, 2000.
131. J. P. Laine, B. E. Little, D. R. Lim, H. C. Tapalian, L. C. Kimerling, and H. A. Haus. Planar integrated wavelength-drop device based on pedestal antiresonant reflecting waveguides and high-Q silica microspheres. *Optics Letters*, 25, 1636–1638, 2000.
132. I. M. White, H. Oveys, X. Fan, T. L. Smith, and J. Zhang. Integrated multiplexed biosensors based on liquid core optical ring resonators and antiresonant reflecting optical waveguides. *Applied Physics Letters*, 89, 191106, 2006.
133. I. M. White, H. Oveys, X. Fan, T. L. Smith, and J. Zhang. Demonstration of a liquid core optical ring resonator sensor coupled with an ARROW waveguide array. In *SPIE Integrated Optics: Devices, Materials, and Technologies XI*, Y. Sidorin, and C. A. Waechter, eds., 2006, 6475–6403.
134. M. Sumetsky. Basic Elements for Microfiber Photonics: Micro/Nanofibers and Microfiber Coil Resonators. *Journal of Lightwave Technology*, 26, 21–27, 2008.
135. On-Column Micro Gas Chromatography Detection with Capillary-Based Optical Ring Resonators Analytical Chemistry 80, 2232–2238, 2008.

9

A Non-Electronic Wireless Receiver with Immunity to Damage by Electromagnetic Pulses

Bahram Jalali and Ali Ayazi
University of California, Los Angeles

Rick C. J. Hsu
Broadcom Corporation

Andrew Yick and William H. Steier
University of Southern California, Los Angeles

Gary Betts
Photonic Systems Inc.

CONTENTS

9.1 Introduction

The trend toward circuits with smaller sizes and voltages renders modern electronics highly susceptible to damage from high power electromagnetic sources. Radio frequency (RF) communication systems are particularly vulnerable because the antenna provides a direct port of entry for electromagnetic radiation. The threats to civil society posed by such weapons are viewed as a grim but real possibility in the post September 11 world [1–3].

This chapter describes a new type of RF receiver front-end that offers protection against electromagnetic weapons. The receiver features a complete absence of electronic circuitry and metal interconnects, the traditional "soft spots" of a conventional RF receiver. It exploits a dielectric resonator antenna (DRA) to capture and deliver the RF signal onto an electro-optic field sensor. The dielectric approach has an added benefit: it reduces physical size of the front-end—an important benefit in mobile applications. Also, the resonant enhancement of the RF field in the dielectric antenna enhances the receiver sensitivity.

It has been known that electronic equipment will stop working when brought into close proximity of TV or radio towers. Separately, it was found in the early 1960s that a high altitude nuclear explosion can destroy electronics in satellites, ships and homes hundreds of miles away. These observations served as motivation for development of microwave based directed energy weapons, a recent example of which is U.S. Navy's NIRF (neutralizing improvised explosive devices using RF). Bordering the worlds of science and science fiction, electromagnetic weapons have fascinated the public and have been the subject of many movies, including *A View to a Kill* (1985) and *Ocean's Eleven* (2001)—to name a few.

In the real world, there is great concern that these sources may become widely available and accessible to unconventional adversaries through the use of flux compression generators (FCG)—a low cost method for creating intense bursts of electromagnetic energy [4]. These generators produce a power surge capable of destroying or damaging sensitive circuitry in electronic systems. They can shut down telecommunications networks without leaving behind a trace of the attacker [2].

High power electromagnetic weapons include narrow band sources known as high power microwaves (HPM) and broadband (pulse) sources known as electromagnetic pulse (EMP) devices. Ironically, the danger caused by HPM and EMP sources is exacerbated by the impact of the celebrated Moore's law, which describes the electronic industry's relentless pursuit of miniaturization. The scaling has resulted in a continuing decrease in the maximum voltage that such circuits can tolerate, rendering modern electronics highly susceptible to damage from high power electromagnetic sources. For example the breakdown voltage of the 1.5-nm gate oxide (used in state-of-the-art integrated circuits) during a 100 ns pulse has been measured to be less than 5 V [5]. This value should be compared to the capabilities of HPM weapons, which can deliver transient potentials of several kilovolts inside a circuit.

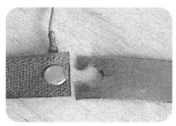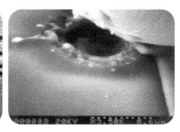

FIGURE 9.1
Example of damage suffered by metallic wiring and interconnects caused by high electric fields.

The breakdown of dielectrics is only one of the failure modes of electronic circuits. Catastrophic damage also results from melting and arcing of metallic interconnects by the large current surge produced by an HPM source [2]. The close spacing of interconnects in modern integrated circuit exacerbates this problem. Integrated circuit components (transistors, diodes, inductors, capacitors) and metallic interconnects represent the "soft spots" of integrated circuits. An example of damage to metallic components is shown in Figure 9.1.

In an RF front-end, the receiver is more vulnerable and more difficult to protect than the transmitter because of the sensitive electronics (e.g., the low noise amplifier) it contains. Presently, there are few practical options for protecting high frequency receivers from high power electromagnetic radiation (a review of conventional protection technologies appears later in this chapter). The All-Dielectric Non-Electronic Radio Front-End Technology (ADNERF) is aimed at addressing this need.

9.2 The ADNERF Concept

The vulnerability to HPM can be mitigated by eliminating all metallic components and transistors in the RF front-end, and by providing charge isolation between the front-end and down stream electronic data-processing circuitry. We create such a device by combining a dielectric antenna with an electro-optic crystal field sensor. The dielectric antenna is a leaky resonator that is fabricated from a material with low loss and high relative permittivity. The incoming signal, a free space electromagnetic field, excites a resonant mode of the antenna resulting in a desired build-up of field inside the structure. The device exploits this enhanced electric field to modulate an optical carrier in a linear electro-optic crystal. To further enhance the receiver sensitivity, the crystal can be formed into a resonator. An optical waveguide, such as a single mode fiber, transfers the signal to the system back-end where a photodetector converts it to an electrical signal that is subsequently processed using conventional electronics. This allows the sensitive electronics to be inside a protected enclosure with only dielectrics connecting it to the antenna—a valuable feature in an electromagnetically hostile environment (Figure 9.2).

The novelty behind this new technology is multifold. First, the antenna itself is made of only dielectric materials. With no metallic electrodes or interconnects present in the front-end, the high field damage caused by arcing or melting of metallic components is eliminated. Second, the electrical-optical-electrical conversion provides *complete charge isolation*.

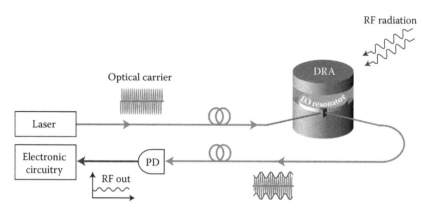

FIGURE 9.2
Concept of the photonic-assisted all-dielectric RF front-end technology. An EO powered dielectric antenna captures the free-space RF signal. The optical link provides complete electrical isolation between the air interface and the electronic circuitry, which is located only after the photodetector (PD) [8,9].

The ADNERF device isolates the electronics in a manner similar to an optocoupler—a common device that creates electrical isolation between two nodes of a low-speed digital circuit (by using an LED and a photodetector to create a short optical link between them). Therefore the ability of optics to protect sensitive electronics is already proven. However, there are critical differences between the optocoupler and the new technology. The traditional optocoupler uses semiconductor devices, metallic interconnects and electrodes, so it does not meet our objectives. Furthermore, its utility is limited to the situations where standard digital logic signals are used. In other words, the optoisolator lacks the high frequency response and the sensitivity and linearity that is required in an RF receiver. The third unique feature of the ADNERF device originates from the fact that the dielectric antenna is made of high permittivity materials and is smaller than the conventional metallic antennas, resulting in a highly compact front-end. In fact, antenna size reduction is the main motivation for use of dielectric antennas in conventional applications [6].

Yet, another advantage is the higher qualify factor of the dielectric antenna compared to a metallic antenna, in particular at high frequencies. This leads to superior noise figure and sensitivity at high frequencies.

Under a DARPA funded project that began in 2005, a prototype device was demonstrated at UCLA [7–9] (Figure 9.3). The device uses a cylindrical geometry for the DRA which is designed using finite element simulations. The optimum location for the electro-optic (EO) crystal is the position inside the DRA where the maximum RF field enhancement occurs. One of the objectives of finite element simulations is to identify this location, which in turn depends on the resonant mode of the DRA. A convenient mode is the $TM_{011+\delta}$ mode of the cylinder because it provides a high electric field inside the DRA with field lines that are perpendicular to the top and bottom surfaces (Figure 9.4). The EO crystal can then be placed inside the cylinder as shown in Figure 9.5.

Figure 9.6 shows the received signal power on the spectrum analyzer with and without the dielectric antenna. As the DRA is removed away from the $LiNbO_3$ crystal, signal level decreases by 55 dB. The RF spectrum analyzer's resolution bandwidth was set to 10 kHz. The observed noise floor of −134 dBm/Hz (−94 dBm over 10 kHz instrument bandwidth) is generated by the RIN of the source laser used in the experiments which had a RIN of −160 dB/Hz. It will be shown later in this chapter that much higher sensitivities can be achieved with lasers that have better RIN performance.

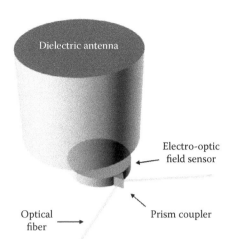

FIGURE 9.3
Proof of concept demonstration ADNERF receiver.

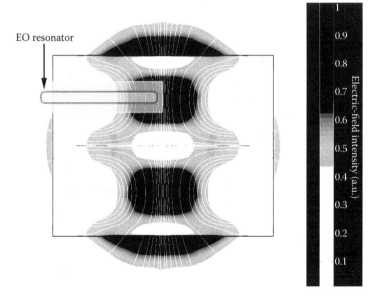

FIGURE 9.4
Simulated received RF electric field inside the dielectric antenna and EO resonator for the structure shown in Figure 9.5. Note the high intensity locations occur in pairs that are 180 degrees out of phase. We will exploit this property of dielectric antenna to achieve differential modulation and detection [8].

9.3 Immunity to High EM Fields

The dielectric strengths of antenna ceramic and LiNbO$_3$ materials are in the 10^6 V/cm range, similar to those in semiconductors and dielectrics used in integrated circuit manufacturing. However, in a typical integrated circuit, voltage surge appears across the gate insulator of the front-end transistor, which is on the order of 100 nm or less in a high speed transistor. In the dielectric antenna, this dimension is on the order of the free space RF

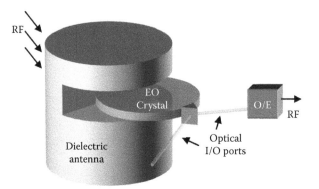

FIGURE 9.5
The all-dielectric RF front-end. A resonant ceramic antenna captures the RF signal and concentrates it onto an EO microresonator, resulting in RF modulation of an optical carrier. A photodetector suitably located at a remote or shielded enclosure delivers the RF output to an LNA. The use of optics creates complete charge isolation between the LNA and air. The doubly resonant structure (RF and optical) offers extremely high sensitivity, and can amplify the RF before the LNA [7,8].

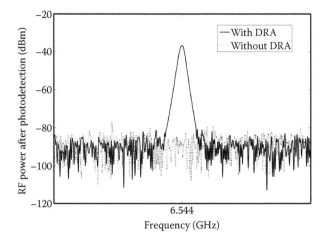

FIGURE 9.6
Demodulated microwave spectrum after photodetection with (solid line) and without (dashed line) the dielectric resonator antenna. Horn antenna feed power is 10 dBm and resolution BW=10 KHz [7–9].

wavelength divided by the square root of the antenna dielectric constant, which leads to a value of about 5 mm (10 GHz signal and a dielectric constant of 36). Even if the DRA amplifies the field from a 100 MW/m² pulse by a factor of 100, the electric field inside the DRA (and the LiNbO$_3$) will be around 10^5 V/cm, well below the breakdown field of the materials. The field occurring in the front-end transistor would be 10^3–10^4 times larger, which would certainly destroy the transistor.

9.4 Thermal Consideration

The temperature rise due to the HPM pulse should be considered so as not to damage the material nor degrade the performance of the system. The dielectric power dissipation in a material is proportional to the dielectric loss factor of the material. A dielectric heating equation relating power dissipation, applied electric field properties and dielectric properties of the material can be derived from Maxwell's equation [10],

$$P=2\pi f\varepsilon_0\varepsilon''E^2 \tag{9.1}$$

Where P is the power dissipation (W m^{-3}), f is the electric field frequency, ε_0 is the permittivity of free space ($8.85\cdot10^{-12}$ F m^{-1}), ε'' is the dielectric loss factor of the material ($\varepsilon''=\varepsilon'$ tanδ) and E is the average electric field strength (V m^{-1}). Equation 9.1 assumes that the electric field is uniform throughout the material. The sample dimensions, therefore, should be sufficiently small so that the effect of penetration depth, which is inversely proportional to the dielectric loss factor, is negligible. The dissipation of microwave power within the material results in an increase in the temperature of the material. The heating rate is given by the equation,

$$\frac{dT}{dt}=\frac{P}{\rho C_p} \tag{9.2}$$

Where dT/dt is the heating rate (°C s^{-1}), P is the power dissipation (W m^{-3}), ρ is the density of the material (Kg m^{-3}) and C_p is the specific heat capacity of the material (J g^{-1} °C^{-1}).

For the DRA of our experiment, power dissipation of Equation 9.1 should be calculated based on the peak electric field inside the resonator to give an idea of the maximum temperature rise occurring inside it. Moreover, it should be noted that the ratio of the peak electric field inside the resonator to that of the incident wave is defined as *field enhancement factor*. Therefore Equation 9.1 is modified in this fashion:

$$P=2\pi f\varepsilon_0\varepsilon''\cdot|E_{DRA}|^2=2\pi f\varepsilon_0\varepsilon''\cdot|E_0|^2\cdot\beta^2 \tag{9.3}$$

Where β is the field enhancement factor—the ratio of the field inside the DRA to that outside of it. On the other hand, incident wave electric field magnitude is related to its irradiance through $I=1/2\cdot c\varepsilon_0|E_0|^2$ giving rise to the new equation for power dissipation:

$$P=\frac{4\pi f\varepsilon''\cdot I\cdot\beta^2}{c} \tag{9.4}$$

Using Equations 9.2 and 9.4, heating rate becomes:

$$\frac{dT}{dt}=\frac{4\pi f\varepsilon''\cdot I\cdot\beta^2}{\rho\cdot c\cdot c_p} \tag{9.5}$$

Now, for a short pulse (Δt) of HPM, maximum temperature rise inside the resonator (ΔT) can be written as:

$$\Delta T=\frac{4\pi f\varepsilon''\cdot\beta^2\cdot(I\cdot\Delta t)}{\rho\cdot c\cdot c_p} \tag{9.6}$$

But $I\cdot\Delta t$ is called *single pulse fluence* (with the dimension of J/m^2) and is a measure of the total energy (per unit area) that is incident on the antenna.

TABLE 9.1

Material Parameters for LiNbO$_3$ and DRA

	ε'	tanδ	ρ (kg·m^{-3})	c_p (J·kg^{-1}·K^{-1})	β
DRA	100	4.5×10^{-4}	4380	600	20
LiNbO$_3$	30	1.5×10^{-3}	4650	630	67

Given that the single pulse fluence in air is limited by the air's dielectric breakdown threshold, it is possible to estimate the highest power density and energy fluence that can be beamed on a target a given distance away by an HPM source of given size, frequency and pulse width.

Taking the dielectric breakdown threshold into account, the theoretical maximum of single pulse fluence at 300 m from the source is around 10^3 mJ/cm^2 for a 1 µs pulse [11].

The fluence of existing EMP sources are believed to be still several orders of magnitude below this breakdown threshold. Assuming a source of one order of magnitude below this threshold and using Table 9.1 for the material parameters, Equation 9.6 leads us to a peak temperature rise of about 2°C inside the DRA and about 13°C inside the LiNbO$_3$. Both of which are tolerable by the materials.

9.5 Temperature Dependence of Dielectric Antenna

The temperature sensitivity of microwave dielectrics is well characterized because they are widely employed as microwave resonators in military systems. In our application, the change in dielectric constant with temperature shifts the antenna center frequency (dielectric antenna is a leaky resonator). This phenomenon plays a dual role in our receiver. The material must have low enough temperature sensitivity for normal operation. At the other hand, the temperature increase that results under an EMP/HPM attack will be beneficial because it causes the antenna's center frequency to "walk away" from the source. This acts as a built-in limiter and provides another layer of protection for the receiver.

9.6 Receiver Sensitivity

In this section we present a model to compute the sensitivity of the ADNERF receiver, otherwise known as the minimum detectable RF power. We consider a DRA with an electro-optic microdisk resonator placed at the location of the peak RF field (Figure 9.7). This model therefore combines both microwave and optical resonators. The incident RF power at which the carrier to noise ratio (CNR) goes to one defines the sensitivity.

The incident RF power, P_{rf}^{inc} is related to the incident *peak to peak* RF electric field, E_{rf}^{inc}, by $P_{rf}^{inc} = \left(E_{rf}^{inc}\right)^2 (A_e/8\eta)$ where A_e is the effective receiving area of the DRA and η is the impedance of free space. We define $\beta = E_{rf}^{DRA}/E_{rf}^{inc}$ as the ratio of the field inside the DRA to that outside of it. This parameter is the antenna's field enhancement factor and depends on its dielectric constant and geometry.

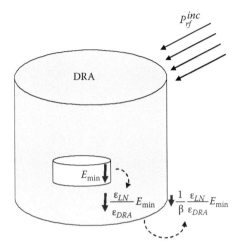

FIGURE 9.7
ADNERF receiver considered in the sensitivity calculations.

The field in the DRA is related to the field experienced by the EO modulator (resonator) by

$$\frac{E_{rf}^{DRA}}{E_{rf}^{m}} = \frac{\varepsilon_m}{\varepsilon_{DRA}}$$

where ε_m and ε_{DRA} are the dielectric constants of the EO material and DRA. The depth of modulation, m, of the light in the resonant modulator can be written as

$$m = \frac{\pi E_{rf}^{m}}{E_{\pi}^{eff}} \frac{P_i}{P_o}$$

where

$$E_{\pi}^{eff} = \frac{4\sqrt{3}\pi^2\lambda}{9Fn^3 r_{33}L}$$

and F=EO resonator finesse, and $L=2\pi r$. We assume the resonant modulator is biased at 25% transmission point (see Appendix A).

If we let P_o be the optical power at the detector, then the RF power from the detector, P_D^{rf}, is

$$P_D^{rf} = \frac{\left(R P_o^{rf}\right)^2 R_d}{8}$$

where R=receiver responsivity, and R_d=detector impedance. Here, $P_o^{rf}=mP_o$ is the peak-to-peak RF modulation of the optical carrier.

The CNR, over the RF bandwidth, B, is:

$$CNR = \frac{(mRP_0)^2/8}{\sigma_{RIN}^2 + \sigma_{shot}^2 + \sigma_{thermal}^2} \quad \sigma_{RIN}^2 = (RIN)\frac{(RP_0)^2}{2}B \quad \sigma_{shot}^2 = 2q(RP_0)B$$

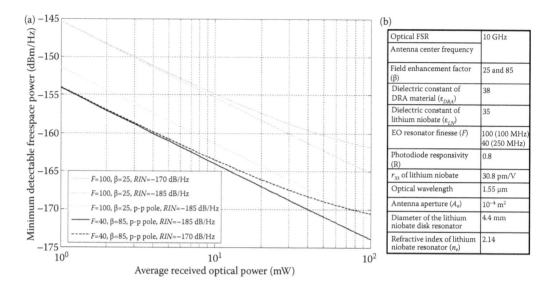

FIGURE 9.8
Minimum detectable power into antenna aperture for a resonant LiNbO$_3$ modulator receiver. The notation p-p indicates the modulator has been push–pull poled. The table shows the parameters used in the calculations.

Where the terms in the denominator describe the RIN, shot, and thermal noise powers. For an optimum receiver, the received power is high enough to overcome the thermal noise, so we can ignore this term.

Setting CNR to one, and relating m to the incident RF power, the minimum detectable power per unit bandwidth, or equivalently the receiver sensitivity, is calculated to be:

$$P_{rf}^{min} = \left(\frac{1}{\beta \cdot F}\right)^2 \left[\frac{2qR}{P_o} + (RIN)\frac{R^2}{2}\right] \left(\frac{\varepsilon_m}{\varepsilon_{DRA}}\right)^2 \left(\frac{2\sqrt{3}\pi\lambda}{9n^3 r_{33} L}\right)^2 \left(\frac{A_e}{R^2 \cdot \eta}\right)$$

It is readily seen that high finesse, F, high β, and high optical power, P_o, all result in better sensitivity. In particular, the sensitivity improves with the square of β and F. The sensitivity improves linearly with P_o in the shot noise limited case (first term in square brackets) and becomes independent of P_o when RIN noise dominates (second term).

Figure 9.8 is a plot of minimum detectable power as a function of average optical power. A LiNbO$_3$ resonator is considered and the parameters used in the calculations are shown in the accompanying table.

9.7 Choice of EO Material

The leading candidates for the EO material are LiNbO$_3$ and the EO polymers. LiNbO$_3$ is a well developed and robust material with $r_{33} = 30$ pm/V. The EO polymers, while offering much higher EO coefficients (r_{33} up to 300 pm/V), are still in the R&D stage. The latest EO polymers, developed at the University of Washington under the DARPA MORPH program, have $r_{33} > 300$ pm/V and therefore their $n^3 r_{33}$ is ~5.5 times that of LiNbO$_3$. Very high finesse (10^3) LiNbO$_3$ resonators have been demonstrated and modest finesse (10) polymer ring resonators have been reported. For a review of $LiNbO_3$ microdisk resonators and their use as EO modulators see Ref. [12]. Polymer EO materials have been demonstrated as ring waveguide resonant modulators operating at up to six times the FSR (162 GHz).

A high $n^3 r_{33}$, and a low ε_r are advantageous. In addition high optical power is advantageous to drive the sensitivity to the RIN limit. Table 9.2 shows the trade-off between power handling capacity and sensitivity.

9.8 Receiver Dynamic Range

One of the unique properties of resonant modulators, ring, disk, or Fabry–Perot, is that their linearity is a function of the optical bias point and this can be used to produce a linear modulator. In the terminology of resonant EO modulators, the term "biasing" describes tuning the optical wavelength to a given point along the resonator's transmission spectrum (Figure 9.9). The maximum slope of the transmission function, for a critically coupled EO resonator, occurs at 25% transmission point. On the other hand, if the device is biased at the 50% transmission point, third order distortion vanishes at the expense of a 1.1 dB increase in the half wave field. In our experiment, a feedback control

TABLE 9.2

Material Composition

EO Material	$n^3 r_{33}/\varepsilon_r$ (pm/V)	I_{max} (W/μm²)
$LiNbO_3$	11	0.3
Polymer	5.0×10^3	5×10^{-3}

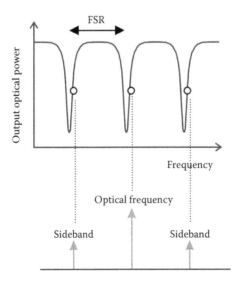

FIGURE 9.9

Transmission spectrum of an EO resonator (top) and the spectrum of double-side-band modulated optical carrier (bottom). The RF modulation frequency must be equal to the free spectral range (FSR) of the resonator. The modulator is "biased" by tuning the optical frequency to an optimum point on the transmission spectrum.

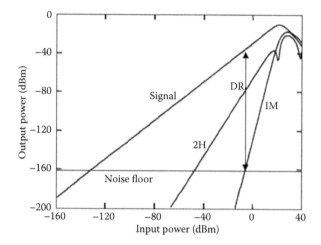

FIGURE 9.10
Dynamic range (DR) calculation of the resonant modulator biased for maximum linearity. Lines show the output signal power along with 2nd harmonic (2H) and odd-order intermodulation (IM) distortion. The receiver achieves a third-order spur free dynamic range (SFDR) of 127.7 dB/Hz$^{4/5}$.

loop is used to lock the tunable laser at a desired laser frequency to the EO resonator for maximum slope of the transmission function. The mathematical model describing these features of resonant EO modulators is described in Appendix A. The results, as they relate to the dynamic range goal of this program, are shown in Figure 9.10. The plot shows the dynamic range (DR) for a LiNbO$_3$ resonant modulator biased for maximum linearity. The lines show the output power along with 2nd harmonic (2H) and odd-order intermodulation (IM) distortion. The receiver achieves a third-order spur free dynamic range (SFDR) of −127.7 dB/Hz$^{4/5}$.

9.9 RF Gain in the Optical Front-End

An external modulation optical link, such as that embedded in the ADNERF approach, can have RF gain. In other words, the front-end can amplify the signal without an actual optical or electrical amplifier! The operation of the link is analogous to a field effect transistor (FET). This analogy exists because the optical signal is modulated by the voltage applied to the EO modulator. The modulated optical signal is then converted back to RF at the photodetector. The photodetector produces RF power by using its d.c. power supply as the power source and the modulated optical signal as the control input. The overall process of modulation and detection is that an input RF voltage controls an output current, and the output current draws its power from an external power supply, not from the input RF power; this is the same general type of operation as the FET although the internal physical mechanisms are different. The RF gain is proportional to the sensitivity of the EO modulator (the amount of modulation per volt) measured by V_π or E_π and to the optical power. Although detector saturation and photorefractive damage to LiNbO$_3$ may limit the maximum optical power and hence will limit the amount of gain that can be achieved, the doubly resonant (RF and optical) design of our front-end provides an excellent chance of achieving gain. Appendix B of this chapter describes in detail how the receiver can achieve RF gain.

9.10 Heterogeneous Dielectric Antenna for Wideband Operation

The receiver performance is dependent on how efficiently the incoming RF signal can modulate the optical carrier. This depends on the modulation efficiency of the EO modulator described by its E_π (lower E_π is preferred), but also on the dielectric antenna's field enhancement factor (the term β in the equation for minimum detectable power). This is the factor by which the electric field at the location of the EO modulator is enhanced compared to the field in free space. Therefore, it is necessary to have an antenna which provides the highest possible enhancement factor at the location of the EO crystal. The antenna must provide this enhancement factor over the entire RF bandwidth, namely up to 1 GHz.

Bandwidth and field enhancement factor are inversely related parameters. In a simple cylindrical dielectric antenna, loss tangent, dielectric constant and aspect ratio (height over radius) of the cylinder are the key factors that determine the enhancement factor. It is possible to modify these parameters locally, in order to optimize the characteristics. As an example, we have already considered heterogeneous cylindrical configurations consisting of a core cylinder that has the highest permittivity and outer shells with successively lower permittivities (Figure 9.11). Preliminary simulations have shown that a gain-bandwidth product improvement of 2x is possible by enclosing a $\varepsilon=38$ core with a single outer shell of $\varepsilon=20$. This suggests that more complex structures will offer even larger improvements.

Another way to enhance the antenna characteristics, particularly its bandwidth, is the stacked disk resonator. For instance, in place of a single-piece solid dielectric resonator, three thinner disks can be stacked. The center disk has either a higher or lower permittivity than the upper and lower disks. The disks may have identical or different radial dimensions. This geometry provides better impedance matching to free space. To demonstrate usefulness of such a design, we have carried out simulation (without optimization) to compare its bandwidth-field enhancement factor product compared to a standard structure. The results shown in Figure 9.12 indicate that even a simple and nonoptimized stacked structure provides improvement of about 40% in terms of bandwidth-enhancement factor product. Again, more complex and optimized structures are expected to offer even larger improvements.

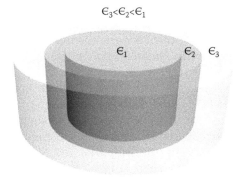

FIGURE 9.11
Concentric heterogenous dielectric antenna structure. Such structures offer a larger sensitivity-bandwidth product than a uniform structure.

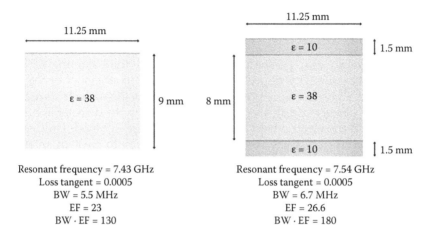

FIGURE 9.12
Stacked dielectric antenna structure and results of preliminary (nonoptimized) simulations. The structure on the right offers larger sensitivity-bandwidth combination than a uniform structure and offers a path toward receiver optimization. EF: field enhancement factor.

9.11 Microwave Ceramics for Dielectric Antenna

The heterogeneous dielectric antenna designs require both low and high dielectric constants. While low dielectric constant materials ($\varepsilon < 40$) are readily available, high dielectric constant materials that also exhibit low loss are challenging. This is best achieved using ceramic-based materials. There are a large number of candidate material systems for microwave dielectric materials with permittivity values of 100 or greater. Commercial companies offer materials which are suitable for initial trials; these materials have high dielectric constants and acceptable loss. Some of the high dielectric materials are listed in Table 9.3 along with their relevant parameters.

9.12 EO Resonator Design: Whispering Gallery Versus Fabry–Perot

The field sensor is an EO modulator that converts the microwave signal, captured by the dielectric antenna, to an intensity modulated optical beam. For the optical front-end to achieve gain and for the receiver to have high sensitivity, the device must exhibit efficient EO modulation. This requires a material with a high EO coefficient (r_{33}) as well as long interaction length between the EM field of the dielectric antenna and the optical beam. Optical resonators are desirable because the multiple pass ensures long interaction between the RF and optical fields. This is characterized by a receiver sensitivity that improves quadratically with the finesse of the cavity. To overcome shot and thermal noises, a large optical power is also desired. However, the resonant enhancement of the optical intensity (optical power divided by optical mode area) can damage the EO material. To maximize the optical power while minimizing the intensity, a large optical mode area is desired. Combined, these considerations imply that the ideal EO resonator must have (1) a large modal area and hence high optical power handling capability; (2) achieve high finesse over the operating bandwidth (1 GHz); and (3) be fabricated in a material with high r_{33}.

TABLE 9.3

Materials with Large Dielectric Constant

Material	Dielectric Constant	Loss Tangent (1/Q)	Temperature Coefficient coefficient of Dielectric Constant	Comments
D-100 (TiO$_2$)	100	<0.0010	–575 E-6	Limited or no compositional variability.
MCT-100 through MCT-140	100–140	<0.0015	–1070 E-6	Composition variable. Higher permittivities lead to more negative temperature coefficients.
Ba-Ln-Ti-O Tetragonal Tungsten Bronze (Ln=La,Nd,Sm and Gd)	90–140	<0.001	Can be tuned by composition to near 0 for permittivity of 90.	Extremely anisotropic material for both permittivity and temperature coefficient.
Ba-Ln-Ti-Nb(Ta)-O Tetragonal Tungsten Bronze	120–200	<0.001	Temperature coefficient depends on composition.	Extremely anisotropic material for both permittivity and temperature coefficient.

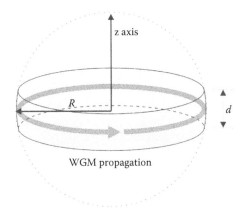

FIGURE 9.13
The geometry of a microdisk resonator which supports whispering gallery mode (WGM) propagation.

For the EO sensors based on optical resonance, one immediate path to increasing the EO sensitivity is to push–pull pole the EO material. Because the RF frequency matches the FSR of the resonator, the resonator can only be exposed to the RF field over one half of the optical path. Poling the LiNbO$_3$ or the polymer in opposite directions over each half of the optical path allows the full resonator to be exposed to the RF field which decreases the E_π by a factor of 2 and increases the sensitivity by 6 dB.

The proof of concept ADNERF demonstration made use of LiNbO$_3$ microdisk resonators, such as the one shown in Figure 9.13. This approach combines LiNbO$_3$, a proven material for EO modulation, with high finesse achievable in microdisk structures. While this approach can achieve high sensitivity and dynamic range, it suffers from the limited bandwidth and limited optical power that can be used with high finesse disk resonators. To overcome these limitations, alternate approaches to EO field sensors can be considered.

One such approach to high finesse combined with high optical power is the Fabry–Perot resonator. The mode size in the Fabry–Perot can be more than an order of magnitude larger than the ~3 μm^2 mode size in the microdisk resonators. The small mode volume in the microdisk is a manifestation of the side wall's vertical curvature (Figure 9.13). For the same finesse and LiNbO$_3$ thickness (~1 mm), this device will have a much larger mode area and hence can handle orders of magnitude more optical power. It is unlikely that these devices will have the ultra-high finesse possible (>1000) in the microdisk, however a lower finesse implies a larger bandwidth. Therefore, the choice between Fabry–Perot and microdisk depends on the bandwidth and the required optical power at the detector that are needed to achieve the system performance.

9.13 Optical Power Limit in a Resonant Field Sensor

The maximum optical power will be limited by detector saturation or nonlinear effects and/or optical damage in the resonant modulators. EO polymers show optical bleaching at high optical power and high power in LiNbO$_3$ can result in Raman scattering and photorefraction. In the resonant modulators with high finesse the circulating optical intensity can be significantly higher than the incident optical intensity. At critical coupling the ratio of the inter-resonator intensity, I_4, to the incident intensity, I_1, can be written as:

$$\frac{I_4}{I_1} = \frac{\lambda_o Q_L}{2\pi^2 a R n_{eff}} = \frac{F}{a\pi}$$

Where Q_L=loaded Q; $a=\exp(-\alpha\pi R)=E$ field transmission for one round trip in the resonator; α=power absorption coefficient; R=ring radius; n_{eff}=effective index of refraction; and F=finesse. Critical coupling and λ tuned to resonance is assumed. When λ is tuned to $T=0.5$ (for maximum linearity), I_4 is reduced by 0.5.

The critical issue is the optical power, P_o, which is incident on the detector. If this power is limited by the maximum circulating power within the resonator, P_{max}, then the maximum power on the detector, $P_o(max)$ is

$$P_o(\text{max}) = \frac{0.25 P_{max} \dfrac{I_1}{I_4}}{0.75} = \frac{P_{max} a\pi}{3F}$$

when the transmission is set to $T=0.25$.

9.14 Maximum Power in Disk and Ring Resonators

Ignoring the losses ($a=1$) the above equation yields an intuitively satisfying result: the power inside the resonator is the power transmitted multiplied by the finesse. The question is whether the resonator can handle such high powers before optical damage occurs.

In high performance analog optical links, Mach–Zehnder modulators operates at power levels as high as 3 W. However the mode size in the LiNbO$_3$ disk resonators (~3 μm^2) is approximately a factor of ten smaller than in the modulators (~30 μm^2). The resulting optical intensity will therefore be 10 times higher in the microdisk. So damage may occur. At this point, no conclusive data exists for the maximum intensity in $LiNbO_3$. Any observable damage may be related to the epoxy used in the fiber connections and not the crystal itself. Also, excessive stimulated Raman scattering, or photorefractive effect may limit the power before crystal damage occurs.

9.15 Fabry–Perot LiNbO$_3$ Resonant Modulator

The power limitations on the $LiNbO_3$ disk resonator are largely due to the small cross-section size of the circulating mode and the consequent high optical intensity. Fabry–Perot resonators can be fabricated with a finesse of 100 or greater with the mode cross-section significantly larger than in the disk and consequently these resonant modulators should be able to handle significantly higher optical power. For example, a Fabry–Perot resonator with a mode area of 100 μm^2 would be able to handle ~10 W of circulating optical power. For $F = 100$ the output power will be 240 mW; more than enough for a single detector. Since the modulating RF field generated by the dielectric antenna is larger that 100 μm^2 there is no penalty for the larger mode size. Since the propagation loss in $LiNbO_3$ is low, the finesse, F, is determined by the mirror reflection, R. For $F = 100$, a reflectivity $R = 0.97$ will be needed ($F = \pi\sqrt{R}/(1-R)$). This is a reasonable reflectivity for dielectric coatings.

9.16 Other Applications of the ADNERF Technology

Field probes are devices that characterize the electric field inside electronic systems. They are used to measure the amount of HPM/EMP power that leaks inside shielded equipment, missile, and aircraft, etc. They are also used to characterize HPM/EMP sources. These devices require survivability under direct exposure and they are supposed to be minimally invasive. In other words, the probe should not disturb the field that is to be measured. Metallic antennas and other components pose a significant problem. The ADNERF technology is uniquely applicable to field probes. In addition to high field tolerance, the all-dielectric front-end offers a minimally intrusive solution.

9.17 Competing Technologies

To protect against the HPM/EMP threat, one may consider using an approach similar to conventional electro-static discharge (ESD) protection techniques. These techniques aim to protect the circuits against electrostatic discharge, as defined by the human body model (HBM) standard, and exemplified by recent demonstration of a 5 GHz CMOS LNA with up to 5.5 KV ESD protection [5]. While the literature on this topic is vast, the basic technique is rather simple and is described in Figure 9.14. ESD is primarily a low frequency phenomenon and most of its power is out of band of the LNA. Providing a short circuit to ground at d.c. and very low frequencies, a shunt inductor protects the input node of the LNA. The desired RF signal, being at higher frequency, avoids the inductor and is a.c. coupled by the coupling capacitor to into the LNA. To protect against residual ESD that gets passed the inductor, diodes are used to clamp the LNA input, preventing its voltage from exceeding the gate breakdown voltage. In practice, diodes with both polarities must be used although for simplicity we only show one polarity in Figure 9.14.

These schemes have several limitations. First, the capacitive loading caused by the presence of extra components at the input will compromise the noise figure. Because of capacitive loading, this technique does not scale well into very high frequencies, namely beyond 10 GHz. To get a high-level ESD protection a large size diode is needed to conduct the ESD current, which results in a large capacitive load at the input/output (I/O) pins. Using several diodes in series can reduce the capacitance; however, the cascaded voltage drops across the diode string will exceed the breakdown voltage of the device in the core circuit [13]. At or near resonance frequency, the shunt circuit's equivalent capacitance increases and its impedance decreases, significantly degrading circuit performance. To avoid this, the resonant frequency must be much higher than the operating frequency. For example, a resonant frequency greater than 20 GHz had to be used for a 10 GHz circuit [13]. This requires small diode capacitances resulting in low levels of protection.

Although these techniques can and are being further optimized and may be applied to protection against HPM/EMP, their fundamental limitations are inescapable. They work best when the threat is out of band. Performance is compromised by the extra circuitry added to the signal path with the penalty becoming more severe for large and in-band threats. Also they don't scale well to very high frequency (due to capacitive loading). Ironically, they rely on integrated circuit components with similar vulnerabilities to those they are supposed to protect! Electronic protection techniques will be unable to protect against future threats which, undoubtedly, will be designed to exploit the fundamental limitations of the technologies.

Plasma limiters are another class of solutions (Figure 9.15). Here, a hollow metallic waveguide filled with an inert gas undergoes breakdown when an EMP pulse arrives. The resulting conductive plasma reflects back the incoming EMP pulse. These devices have

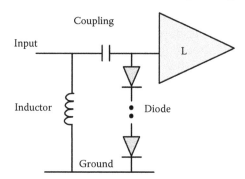

FIGURE 9.14
ESD protection circuitry. The shunt inductor directs low frequency out-of-band portion of ESD to ground. The coupling capacitor a.c. couples the incoming RF signal to the LNA. Diodes clamp the LNA input preventing it exceed gate breakdown voltage. Such solutions work best for out of band threats. The additional components tied to the LNA input compromise its noise Figure and frequency response.

FIGURE 9.15

A plasma limiter consists of a gas filled hollow metallic waveguide. When an intense EM field arrives, the gas undergoes discharge. The conductive gas reflects the pulse. Protection is only triggered when pulse power reaches the kW level. Also, there is a delay associated with discharge build up. At lower levels and shorter times, no protection exists, rendering this technology unsuitable for protecting very sensitive electronics (www.accurate-automation.com).

high power handling capability and hence are superior to ESD technologies. However, they do not provide continuous protection, in fact, there is no protection before the plasma is activated. State-of-the-art devices known to us (www.accurate-automation.com) have a rise time of about ~1 ns and a trigger level ~5 kW. Until these levels are reached, electromagnetic energy leaks in and can destroy the LNA. Therefore, they are not adequate for protecting very sensitive electronics, and do not protect against very short pulse threats. Their bulky size notwithstanding, gas discharges are inherently unstable. For that reason, some means of "priming" the gas is necessary in order to ensure stable activation on each and every pulse. Older designs employ a small electrode, called a "keep-alive", for this purpose, which requires application of a very large d.c. voltage. In addition to the inconvenience of requiring a power supply, the keep-alive also limits useful tube life and is a source of a small amount of excess noise [14]. In order to overcome these problems, radioactive priming was developed. The radioactive source provides a source of electrons for a more stable activation of plasma, and has been the standard approach in modern devices. These issues notwithstanding, the leakage of gas during long periods in the field, and the inability to monitor the gas pressure, pose reliability concerns. In other words, the gas discharge may not activate when needed.

The ferrite limiter is another type of limiter that requires activation [14]. It employs ferrite material which is mounted along waveguide walls of a metallic waveguide. The ferrite is magnetically biased with permanent magnets. The electrons in the ferrite will precess around the magnetic field lines. When the input power reaches a critical threshold, the RF energy will couple into the precession motion causing the ferrite to absorb power as the

RF passes down the waveguide. The absorbed energy is converted to heat. Its advantage is that it has fast (but not zero) activation and recovery time. Its main disadvantages are that it exhibits relatively high insertion loss and its performance is very sensitive to changes in ambient temperature. Also, its leakage is too high for a sensitive receiver to sustain. Thus, it is normally followed by a diode limiter [14].

Another limiter technology is the multipactor; a vacuum device which employs secondary emission materials to absorb RF power. When the input RF power reaches a critical threshold, multipaction, enhanced by the presence of secondary emission material, takes place. This causes power to be absorbed as the RF energy passes down the waveguide. The absorbed energy is converted to heat. It requires a number of d.c. supplies to operate certain of its key components. In addition, it must be liquid cooled, and it has a limited lifetime. It has large leakage and thus must be followed by a diode limiter to clean up the leakage.

To protect sensitive RF front-ends against HPM/EMP treats with arbitrary spectrum, power levels, and temporal profiles, an entirely new technology is clearly needed. The technology must eliminate metallic interconnects, metallic based devices such as inductors and capacitors, and transistors. It must provide complete charge isolation between the air interface and any electronic circuitry that eventually appear in the signal path. In addition, a reliable protection is one that is always on, i.e. it does not need to be activated when certain conditions are reached. Neither plasma/ferrite limiters nor diode/inductor circuits represent a true and complete solution. They do not provide leakage-free, and activation-less protection. On the other had, the ADNERF technology does. The success of the venerable optoisolator in protecting digital circuitry is a testament to the potential of this solution.

9.18 Summary

The trend towards reduced geometry and voltage has rendered modern electronics highly susceptible to damage from EMP attack. The failure mechanisms in integrated circuits are two-fold: breakdown of dielectrics used in transistor gates and capacitors, and melting and arcing in metallic electrodes and interconnects.

ADNERF is a fundamentally new RF receiver technology that is inherently immune to electromagnetic induced damage. It intends to eliminate electronic devices, metallic wiring and interconnects from the RF front-end—the elements that constitute the "soft spots" in a conventional receiver. The new receiver can match conventional receivers in terms of bandwidth, sensitivity and dynamic range. It is compact and offers the potential for chip scale integration.

The approach is to use a dielectric antenna coupled to an EO crystal to create an all-dielectric RF front-end. The dielectric antenna captures the incoming RF and concentrates it onto the EO crystal. The RF is modulated onto an optical carrier which delivers it to a photodetector followed by an LNA. The optical link so created provides complete charge isolation between the air interface and the LNA. It also makes it possible to remote the LNA and subsequent circuitry away from the front-end. The doubly resonant design (RF resonant in the dielectric antenna and optical resonant in the EO microdisk) offers extremely high receiver sensitivity. The front-end features a complete absence of any electronic and metallic components, and can amplify the RF even before it reaches the LNA.

There are other approaches to protect against the electromagnetic pulse threat such as the diode limiter, the plasma limiter, or the ferrite limiter. None of these competing technologies represents a complete and reliable solution. Protection does not exist at all times; it is only activated when certain conditions are reached. None provides leakage-free and activation-less (continuous) protection. They do not eliminate metallic components from the front-end. The ADNERF technology offers all these features.

The ability of optics to protect electronics against electrical transients is already proven, as evidenced by the popularity of the venerable Optoisolator, the de facto standard in protecting *digital* logic circuitry against high voltage transients. ADNERF is a modern optoisolator; one that has no electronic and metallic components, is *analog*, and has the sensitivity and dynamic range required by communication and radar systems.

Acknowledgment

This work was supported by DARPA, and by U.S. Army through the DETEC program. The LiNbO$_3$ microdisk resonator used in the prototype device was fabricated by Dr. Lute Maleki's group at JPL.

Appendix A: Theory of Resonant EO Field Sensors

A.1 Equivalent E_π of Resonant Modulators

The advantage of the resonant modulator is the high slope of the transmission, T, versus the applied voltage V. For a critically coupled resonator the slope is maximum when $T=0.25$. The effect of the high slope can be seen by calculating an equivalent V_π. The V_π of a Mach–Zehnder (MZ) biased at quadrature relates to the slope of the transmission, T, versus the applied voltage, V.

$$\frac{dT}{dV} = \frac{\pi}{2V_\pi}$$

In a similar way, the effective V_π^{eff} of a resonant modulator is based on the slope at the optical bias point selected. For the MZ modulator, V_π can be expressed as

$$V_\pi = \frac{\pi}{2}\left(\frac{dT}{dV}\right)^{-1}$$

The ring equivalent equation is

$$V_\pi^{eff} = \frac{\pi}{2}\left(\frac{dT}{d\theta}\frac{d\theta}{dV}\right)^{-1}$$

where

$$\theta = \theta_o + \frac{\pi n^3 r_{33} V L}{\lambda g},$$

g is the electrode gap, L is the circumference of the ring, and θ_0 is the optical bias point.

The transmission of a ring when critically coupled can be written as

$$T(\theta) = 1 - \frac{1}{1 + \left(\frac{2F}{\pi}\right)^2 \sin^2 \theta/2}$$

where F = finesse and $\theta = 0$ is the resonance point.

Taking the derivative of T with θ and assuming F is large and θ is small, the maximum of $dT/d\theta$ occurs at

$$\sin \frac{\theta_o}{2} = \frac{\pi}{2\sqrt{3}F}$$

where $T = 0.25$. Using this value for $\sin(\theta_0/2)$ we arrive at

$$V_\pi^{eff} = \frac{4\sqrt{3}\pi\lambda g}{9Fn^3 r_{33}L}.$$

In the ADNERF application, we are concerned about detecting a microwave E field and it is more appropriate to consider E_π^{eff} where

$$E_\pi^{eff} = \frac{4\sqrt{3}\pi\lambda}{9Fn^3 r_{33}L}.$$

In ADNERF, it is assumed that the microwave frequency matches the FSR of the resonator and that only half the ring is exposed to the microwave E field. Therefore L becomes L/π. Alternately the EO effect in the ring in reversed by poling in each half of the ring and the entire ring is exposed to the E field. In this case L becomes $2L/\pi$.

The advantage provided by the optical resonance can be seen by comparing it to an equivalent MZ modulator with the same E_π^{eff}. For the same E_π^{eff}, the interaction length of the MZ must be increased by the factor of $9F/(4\sqrt{3})$. The resonant modulator with a FSR of 10 GHz will have a circumference of ~1.5 cm. For $F = 100$, an MZ modulator with same E_π^{eff} would require an interaction length of 62 cm! The advantage of the resonant optical modulator can be clearly seen, particularly for the case of high finesse resonators.

A.2 Dynamic Range of Resonant EO Field Sensors

One of the unique properties of resonant modulators, ring, disk, or Fabry–Perot, is that their linearity is a function of the optical bias point and this can be used to produce a linear modulator to increase the SFDR of the ADNERF link.

A.3 Biasing for Maximum Signal

In the prior section, it was shown that the maximum sensitivity is when the resonator is biased at $T=0.25$. At his point the optical transmission can be expanded as a Taylor's series expansion

$$T = \frac{1}{4} + \frac{1}{2}\delta - c_3\delta^3 + c_4\delta^4 - c_6\delta^6 \ldots$$

$$\delta = \frac{\pi E_m}{E_\pi^{eff}}$$

At this bias point there is no second order distortion and E_π^{eff} is minimized. For the Fabry–Perot resonator all of the above results apply except the output is biased to 75% transmission.

A.4 Biasing for Minimum Distortion

If the bias point is set to $\Delta f_{FWHM}/2$ where $T=0.5$, the optical transmission can be written as

$$T = \frac{1}{2} + \frac{1}{2}\delta - \frac{5}{22}\delta^2 + \frac{5}{11}\delta^4 - \frac{10}{11}\delta^5 \ldots$$

where

$$\delta = \frac{\pi E_m}{E_\pi^{eff}}.$$

At this bias there are no third or seventh-degree terms and

$$E_\pi^{eff} = \frac{\pi\lambda}{Fn^3 r_{33}L}.$$

Hence at this bias point, the third order intermode is reduced but E_π^{eff} increases by 1.1 dB and the d.c. optical power level increases by 3 dB. The 3 dB increase in optical power at the detector increases CNR when shot noise is limiting. In this case also, L becomes L/π or L becomes $2L/\pi$ as explained above.

For the Fabry–Perot resonator all of the above results apply and the output is biased to 50% transmission.

In sub-octave bandwidth systems, the even order terms are not of importance and if the optical bias point is set at $\Delta f_{FWHM}/2$ to eliminate the third-degree term, the remaining distortion is fifth-degree. By assuming reasonable link parameters ($P_L=100$ mW, laser (RIN)$=-165$ dB/Hz, and detector responsivity$=0.7$ A/W) a third-order SFDR of 127.7 dB/Hz$^{4/5}$ can be expected with micro-resonant modulators as shown in Figure 9.A.1. In Figure 9.A.1 the horizontal axis is the RF power in the resonant modulator and does not include the E_{rf} enhancement of the DRA and the F of the resonator.

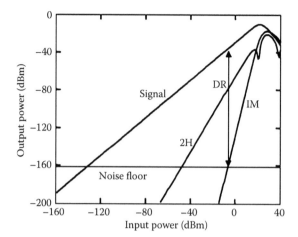

FIGURE 9.A.1
SFDR calculation of the resonant modulator biased for maximum linearity. SFDR=127.7 dBHz$^{4/5}$.

The bandwidth of the linearity is an issue. The linearity bandwidth will be limited because the sidebands of the modulation will fall at different points on the resonator transmission curve and therefore will be frequency filtered by the resonator response curve.

Appendix B: RF Gain in the Optical Front-End

To see how an RF optical link can achieve gain consider the following. The gain, the ratio of the intercepted RF power, P_{rf}, to the detector RF power, P_{det}, can be calculated as follows. Let E_{rf} be the incident RF field, E_{DRA} the field inside the DRA, and E_{mod} the field at the EO modulator. Then,

$$E_{rf} = \frac{1}{\beta}\frac{\varepsilon_M}{\varepsilon_{DRA}}E_{mod} \quad \beta = \frac{E_{DRA}}{E_{rf}}$$

$$P_{rf} = \frac{E_{rf}^2}{2\eta_o}A_{DRA} \quad A_{DRA} = \text{Effective area of the DRA}$$

Here, β is the field enhancement factor of the DRA, ε is the dielectric constant and η_o the free space impedance. Combining these equations,

$$E_{mod}^2 = \frac{2\eta_o P_{rf}}{A_{DRA}}\left(\beta\frac{\varepsilon_{DRA}}{\varepsilon_M}\right)^2$$

Given a modulation depth m, the transmission of the EO resonant modulator biased for maximum sensitivity is,

$$T = \frac{1}{4} + \frac{1}{2}m \quad m = \frac{\pi E_{rf}^m}{E_\pi^{eff}}\frac{P_i}{P_o}$$

where E_π^{eff} is the effective half wave voltage of the EO modulator. Now, let i=detector current$= (m/2)P_oR$, where R is the detector responsivity, and P_o is the optical power at the detector. This leads to an RF power at the detector,

$$P_{det} = \frac{1}{2}i^2R_D \quad R_D = \text{detector resistance}$$

$$= \frac{R_D}{4}(P_iR)^2\left(\frac{\pi}{E_\pi^{eff}}\right)^2 \frac{\eta_o P_{rf}}{A_{DRA}}\left(\beta\frac{\varepsilon_{DRA}}{\varepsilon_M}\right)^2$$

B.1 Example: LiNbO₃ Resonant Modulator

Let $F = 100$, $n^3r = 280\times10^{-12}m/v$, $L = 1.5\times10^{-2}m$, $\lambda = 1.55\times10^{-6}\ m$

When the EO modulator is biased for maximum sensitivity: $E_\pi^{eff} = 2.8\times10^4\ V/m$

$$\eta_o = 377, \quad A_{DRA} = 10^{-4}m^2, \quad \beta = 25, \quad \frac{\varepsilon_{DRA}}{\varepsilon_M} = \frac{38}{35},$$

$$R_D = 50\Omega, \quad R = 0.8\frac{A}{W}, \quad P_{det} = 2.8\times10^2\,P_o^2P_{rf}$$

If P_o=0.1 W, then P_{det}=11.2P_{rf}. In other words, the system has an RF power gain of 11.2. If the resonator is push–pull poled, this decreases E_π^{eff} by ½ and increases the gain by an additional factor of 4.

References

1. Special Issue on High-Power Electromagnetics (HPEM) and Intentional Electromagnetic Interference (IEMI), *IEEE Trans. EMC*, 46, 2004.
2. M. Abrams, Dawn of the e-bomb. *IEEE Spectrum*, 40, 24–30, 2003.
3. Report of the Commission to Assess the Threat to the United States from Electromagnetic Pulse (EMP) Attack, Volume 1: Executive Report 2004, http://www.globalsecurity.org/wmd/library/congress/2004_r/04-07-22emp.pdf
4. C. M. Fowler, R. S. Caird, and W. B. Garn. *An introduction to explosive magnetic flux compression generators.* Los Alamos National Laboratory, Report LA-5890-MS, 1975.
5. D. Linten, et al. An integrated 5 GHz low-noise amplifier with 5.5 kV HBM ESD protection in 90 nm RF CMOS. *2005 Symposium on VLSI Circuits*, 2005. Digest of Technical Papers. pp. 86–89, Kyoto, Japan, June 2005.
6. R. Dettmer, Dielectric antenna make waves. *IEE Review*, 49, 28–31, 2003.
7. R. C. J. Hsu, A. Ayazi, B. Houshmand, and B. Jalali. All-dielectric wireless receiver. *International Microwave Symposium*, Honolulu, HI, 221–224, June 2007.
8. R. C. J. Hsu, A. Ayazi, B. Houshmand, and B. Jalali. All-dielectric photonic-assisted radio front-end technology. *Nature Photonics*, Vol. 1, 535–538, 2007.

9. A. Ayazi, R. C. J. Hsu, B. Houshmand, and B. Jalali. All-dielectric photonic-assisted wireless receiver. *LEOS Annual Meeting,* Paper #ME3, Lake Buena Vista, FL, October 2007.
10. R. Meredith. *Engineers' Handbook of Industrial Microwave Heating.* The Institution of Electrical Engineers, London, UK, 1998.
11. H. K. Florig. The future battlefield: A blast of gigawatts? *IEEE Spectrum,* 25, 50–54, 1988.
12. V. S. Ilchenko, A. A. Savchenkov, A. B.Matsko and L. Maleki. Whisper-gallery-mode electro-optic modulator and photonic microwave receiver. *J. Opt. Soc. Am. B,* 20, 333–341, 2003.
13. Maoyou Sun and Yicheng Lu. A new ESD protection structure for high-speed GaAs RF ICs. *Electron Device Lett., IEEE,* 26(3), 133–135, 2005.
14. R. F. Bilotta. Receiver Protectors: A Technology Update. *Microwave Journal,* 40, August 1997.

10

*Cavity Opto-Mechanics**

Tobias Jan Kippenberg
Max Planck Institut fur Quantenoptik

Kerry J. Vahala
California Institute of Technology

CONTENTS

10.1 Introduction

Recent years have witnessed a series of developments at the intersection of two, previously distinct subjects. Optical (micro) cavities [2] and micro (nano) mechanical resonators [3], each a subject in their own right with a rich scientific and technological history, have, in a

---

* This chapter was originally published in *Optics Express* [1]. [Kippenberg, T. J., and Vahala, K., *Optics Express*, 15, 17172, 2007.]

sense, become entangled experimentally by the underlying mechanism of optical, radiation pressure forces. These forces and their related physics have been of major interest in the field of atomic physics [4–7] for over five decades and the emerging opto-mechanical context has many parallels with this field.

The opto-mechanical coupling between a moving mirror and the radiation pressure of light first appeared in the context of interferometric gravitational wave experiments. Owing to the discrete nature of photons, the quantum fluctuations of the radiation pressure forces give rise to the so called *standard quantum limit* [8–10]. In addition to this *quantum back-action* effect, the pioneering work of Braginsky [11] predicted that the radiation pressure of light, confined within an interferometer (or resonator), gives rise to the effect of dynamic back-action (which is a classical effect, in contrast to the aforementioned quantum back-action) owing to the finite cavity decay time. The resulting phenomena, which are the parametric instability (and associated mechanical oscillation) and opto-mechanical back-action cooling represent *two sides of the same underlying "dynamic back-action" mechanism*. Later, theoretical work proposed using the radiation-pressure coupling to realize quantum nondemolition measurements of the light field [12] and as a means to create nonclassical states of the light field [13] and the mechanical system [14]. It is noted that the effect of dynamic back-action is of rather general relevance and can occur outside of the opto-mechanical context. Indeed, the same dynamic back-action phenomena have been predicted to occur in systems where a mechanical oscillator is parametrically coupled to an electromagnetic resonant system, and have indeed been observed in an electronic resonance of a superconducting single electron transistor coupled to a mechanical oscillator [15,16] (or LC circuit [11,17]).

The experimental manifestations of opto-mechanical coupling by radiation pressure have been observable for some time. For instance, radiation pressure forces were observed in the pioneering work of Walther at the Max Planck Institut of Quantum Optics (MPQ), manifesting themselves in radiation pressure (pondermotive) bistability [18], while even earlier work in the microwave domain had been carried out by Braginsky [11]. Moreover the modification of mechanical oscillator stiffness caused by radiation pressure, the *"optical spring"*, have also been observed [19]. However, in contrast to these *static* manifestations of radiation pressure, the *dynamic* manifestations of radiation pressure forces on micro- and nano-mechanical objects have only recently become an experimental reality. Curiously, while the theory of dynamic back action was motivated by consideration of precision measurement in the context of gravitational wave detection using large interferometers [20], the first observation of this mechanism was reported in 2005 at a vastly different size scale in microtoroid cavities [21–23]. These observations were focused on the radiation pressure induced parametric instability [20]. Subsequently, the reverse mechanism [24] (back-action cooling) has been exploited to cool cantilevers [25–27], microtoroids [28] and macroscopic mirror modes [29,30] as well as mechanical nano-membranes [31]. We note that this technique is different than the earlier demonstrated radiation-pressure feedback cooling [29,32], which uses electronic feedback analogous to "stochastic cooling" [33] of ions in storage rings and which can also provide very efficient cooling as demonstrated in recent experiments [27,34,35]. Indeed, research in this subject has experienced a remarkable acceleration over the past 3 years as researchers in diverse fields such as optical microcavities [2], micro and nano-mechanical resonators [3] and quantum optics pursue a common set of scientific goals set forward by a decade-old theoretical framework. Indeed, there exists a rich theoretical history that considers the implications of optical forces in this new context. Subjects ranging from entanglement [36,37]; generation of squeezed states of light [13]; to measurements at or beyond the standard quantum limit [12,38,39]; and even to tests of quantum theory itself are in play here [40]. On the practical side, there are opportunities to harness these forces for new metrology tools [35] and even for new functions on a semiconductor chip (e.g., oscillators [21,22], optical mixers [41], and tuneable optical filters and switches [42,43]. It seems clear that a new field of *cavity opto-mechanics* has emerged, and will soon evolve into *cavity quantum opto-mechanics* (cavity QOM) whose goal is the observation and exploration of quantum phenomena of mechanical systems [44] as well as quantum phenomena involving both photons and mechanical systems.

The realization of dynamical, opto-mechanical coupling in which radiation forces mediate the interaction, is a natural outcome of underlying improvements in the technologies of optical (micro) cavities and mechanical micro (nano-) resonators. Reduction of loss (increasing optical and mechanical Q) and reductions in form factor (modal volume) have enabled a regime of operation in which optical forces are dominant [21,22,26,28,30,31,34,35,43]. This coupling also requires coexistence of both high-Q optical and high-Q mechanical modes. Such coexistence has been achieved in the geometries illustrated in Figure 10.1. It also seems likely that other optical microcavity geometries such as photonic crystals [45] can exhibit the dynamic back-action effect provided that structures are modified to support high-Q mechanical modes.

To understand how the coupling of optical and mechanical degrees of freedom occurs in any of the depicted geometries, one need only consider the schematic in the upper panel of Figure 10.1. Here, a Fabry–Perot optical cavity is assumed to feature a mirror that also functions as a mass-on-a-spring (i.e. is harmonically suspended). Such a configuration can indeed be encountered in gravitational wave laser interferometers (such as LIGO) and is also, in fact, a direct representation of the "cantilever mirror" embodiment in the lower panel within Figure 10.1. In addition it is functionally equivalent to the case of a

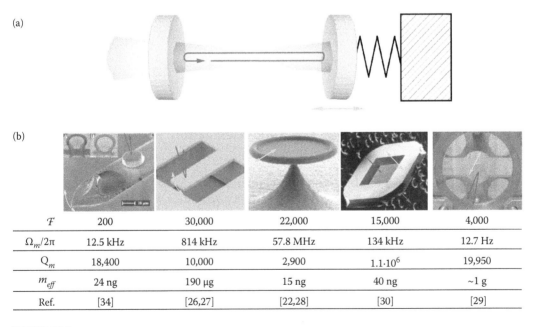

(a)

(b)

$\mathcal{F}$	200	30,000	22,000	15,000	4,000
$\Omega_m/2\pi$	12.5 kHz	814 kHz	57.8 MHz	134 kHz	12.7 Hz
Q_m	18,400	10,000	2,900	$1.1 \cdot 10^6$	19,950
m_{eff}	24 ng	190 μg	15 ng	40 ng	~1 g
Ref.	[34]	[26,27]	[22,28]	[30]	[29]

FIGURE 10.1
(a) A cavity opto-mechanical system consisting of a Fabry–Perot cavity with a harmonically bound end mirror. (b) Different physical realizations of cavity opto-mechanical experiments employing cantilevers, micromirrors, microcavities, nano-membranes and macroscopic mirror modes. Red and green arrows represent the optical trajectory and mechanical motion (From Optics Express. With permission.).

microtoroid embodiment (also shown in the lower panel), where the toroid itself provides both the optical modes as well as mechanical breathing modes (see Figure 10.1 and discussion below). Returning to the upper panel, incident optical power that is resonant with a cavity mode creates a large circulating power within the cavity. This circulating power exerts a force upon the "mass-spring" mirror, thereby causing it to move. Reciprocally, the mirror motion results in a new optical round trip condition, which modifies the detuning of the cavity resonance with respect to the incident field. This will cause the system to simply establish a new, static equilibrium condition. The nonlinear nature of the coupling in such a case can manifest itself as a hysteretic behavior and was observed over two decades ago in the work by Walther et al. [18]. However, if the mechanical and optical Qs are sufficiently high (as is further detailed in what follows, such that the mechanical oscillation period is comparable or exceeds the cavity photon lifetime) a new set of dynamical phenomena can emerge, related to mechanical amplification and cooling.

In this Chapter, we will give the first unified treatment of this subject. Although microtoroid optical microcavities will be used as an illustrative platform, the treatment and phenomena are universal and pertain to any cavity opto-mechanical systems supporting high Q optical and mechanical modes. In what follows we begin with a theoretical framework through which dynamic back-action can be understood. The observation of micromechanical oscillation will then be considered by reviewing both old and new experimental results that illustrate the basic phenomena [21–23]. Although this mechanism has been referred to as the parametric instability, we show that it is more properly defined in the context of mechanical amplification and regenerative oscillation. For this reason, we introduce and define a mechanical gain, its spectrum, and, correspondingly, a threshold for oscillation. Mechanical cooling is introduced as the reverse mechanism to amplification. We will then review the experimental observation of cooling by dynamic back-action [28,46] and also the quantum limits of cooling using back action [47,48].

Finally, we will attempt to discuss some of the possible future directions for this new field of research.

10.2 Theoretical Framework of Dynamic Back-action

10.2.1 Coupled Mode Equations

Dynamic back-action is a modification of mechanical dynamics caused by nonadiabatic response of the optical field to changes in the cavity size. It can be understood through the coupled equations of motion for optical and mechanical modes, which can be derived from a single Hamiltonian [49].

$$\frac{da}{dt} = -i\Delta(x)a - \left(\frac{1}{2\tau_0} + \frac{1}{2\tau_{ex}}\right)a + i\sqrt{\frac{1}{\tau_{ex}}}s \tag{10.1}$$

$$\frac{d^2x}{dt^2} + \frac{\Omega_m}{Q_m}\frac{dx}{dt} + \Omega_m^2 x = \frac{F_{RP}(t)}{m_{eff}} + \frac{F_L(t)}{m_{eff}} = \frac{\zeta}{cm_{eff}}\frac{|a|^2}{T_{rt}} + \frac{F_L(t)}{m_{eff}} \tag{10.2}$$

Aside from the *x*-dependence, the first equation governs the dynamics of the optical field according to the formalism of Haus [50], i.e., $|a|^2$ is the stored cavity energy, whereas

$|s|^2$ denotes the launched input power upon the cavity system. Moreover the optical field decays with a rate $1/2\tau = (1/2\tau_0) + (1/2\tau_{ex})$ and $\Delta(x) = \omega - \omega_0(x)$ accounts for the detuning of the pump laser frequency ω with respect to the cavity resonance $\omega_0(x)$ (which, as shown below, depends on the mechanical coordinate, x). The power coupling rate into outgoing modes is described by the rate $1/\tau_{ex}$, whereas the intrinsic cavity loss rate is given by $1/\tau_0$. In the subsequent discussion, the photon decay rate is also used $\kappa \equiv 1/\tau$.

The second equation describes the mechanical coordinate (x) accounting for the movable cavity boundary (i.e. mirror), which is assumed to be harmonically bound and undergoing harmonic oscillation at frequency Ω_m with a power dissipation rate $\Gamma_m = (\Omega_m/Q_m)$. Moreover, m_{eff} is the effective mass of the mirror mode and will be discussed in a later section. This mass describes in large part the strength of the coupling between optical and mechanical mode. For an excellent treatment of its determination and derivation the reader is referred to Ref. [51]. The radiation pressure forcing function is given by $F_{RP}(t) = (\zeta/c)(|a|^2/T_{rt})$, where the dimensionless parameter ζ takes on the value $2\pi n$ for a whispering gallery mode microcavity (consisting of a dielectric material with refractive index n), and $\zeta = 2$ for a Fabry–Perot cavity; and where T_{rt} is the cavity round trip time (note that the intracavity circulating power is given by $|a|^2/T_{rt}$). Moreover the term $F_L(t)$ denotes the random Langevin force, and obeys $\langle F_L(t)F_L(t')\rangle = \Gamma_m k_B T_R m_{eff}\delta(t - t')$, where k_B is the Boltzmann constant and T_R is the temperature of the reservoir. The Langevin force ensures that the fluctuation dissipation theorem is satisfied, such that the total steady state energy in the (classical) mechanical mode E_m (in the absence of laser radiation) is given by $E_m = \int \Omega^2 m_{eff}|\chi(\Omega)F_L(\Omega)|^2 d\Omega = k_B T_R$, where the mechanical susceptibility $\chi(\Omega) = m_{eff}^{-1}/(i\Omega\Gamma_m + \Omega_m^2 - \Omega^2)$ has been introduced. Of special interest in the first equation is the optical detuning $\Delta(x)$ which provides coupling to the second equation through the relation:

$$\Delta(x) = \Delta + \frac{\omega_0}{R}x \tag{10.3}$$

This relation assumes that, under circumstances in which the mass-spring is at rest ($x=0$), the optical pump is detuned by Δ relative to the optical mode resonance. Two cases of interest, both illustrated in Figure 10.1, will emerge: blue detuned ($\Delta > 0$) and red-detuned ($\Delta < 0$) operation of the pump-wave relative to the cavity resonance. It is important to note that quadratic coupling can also be realized, e.g., Ref. [31], however, this case will not be considered here.

Before discussing the physics associated with the optical delay (which gives rise to dynamical back-action), we briefly divert to a *static* phenomena that is associated with the steady state solutions of the above coupled equations: the mirror bistability [18] and multi-stability [52]. Indeed, as noted earlier, the radiation-pressure-induced bi-stability was observed over two decades ago [18] using a harmonically-suspended mirror. In brief, considering purely the static solutions for the displacement (i.e., $\bar{x}$) of the above set of equations, it becomes directly evident that the equilibrium position of the mechanical oscillator will depend upon the intra-cavity power. Since the latter is again coupled to the mechanical displacement, this leads to a cubic equation for the mirror position $\bar{x}$ as a function of applied power:

$$\frac{\tau^2}{\tau_{ex}}|s|^2 = \frac{T_{rt}cm_{eff}}{4\zeta}\Omega_m^2\bar{x}\left(4\tau^2\left(\Delta + \frac{\omega_0}{R}\bar{x}\right)^2 + 1\right) \tag{10.4}$$

For sufficiently high power, this leads to bi-stable behavior (namely for a given detuning and input power the mechanical position can take on several possible values).

10.2.2 Modifications due to Dynamic Back-action: Method of Retardation Expansion

The circulating optical power will vary in response to changes in the coordinate x. The delineation of this response into adiabatic and nonadiabatic contributions provides a starting point to understand the origin of dynamic back-action and its two manifestations. This delineation is possible by formally integrating the above equation for the circulating field, and treating the term x within it as a perturbation. Furthermore, if the time variation of x is assumed to be slow on the time scale of the optical cavity decay time (or equivalently if $\kappa \gg \Omega_m$) then an expansion into orders of retardation is possible. Keeping only terms up to dx/dt in the series expansion, the circulating optical power (and hence the forcing function in mechanical oscillator equation) can be expressed as follows,

$$P_{cav} = P_{cav}^0 - P_{cav}^0 \left(\frac{8\Delta\tau^2}{\left(4\tau^2\Delta^2 + 1\right)^2} \right) \frac{\omega_0}{R} x \tag{10.5}$$

$$+ \tau P_{cav}^0 \left(\frac{64\Delta\tau^3}{\left(4\tau^2\Delta^2 + 1\right)^2} \right) \frac{\omega_0}{R} \frac{dx}{dt} \tag{10.6}$$

Here the power $P_{cav}^0 = (|a|^2/T_{rt}) = (\tau/T_{rt})((4\tau/\tau_{ex})/(4\Delta^2\tau^2 + 1))|s|^2$ denotes the power in the cavity for zero mechanical displacement. The circulating power can also be written as $P_{cav}^0 = 2\mathcal{F}/\pi \cdot C \cdot |s|^2$, where $C = (\tau/\tau_{ex})/(4\Delta^2\tau^2 + 1)$ and where $\mathcal{F} = 2\pi(\tau/T_{rt})$ denotes the cavity finesse. The first two terms in this series provide the adiabatic response of the optical power to changes in position of the mirror. Intuitively, they give the instantaneous variations in coupled power that result as the cavity is "tuned" by changes in "x". It is apparent that the x-dependent contribution to this adiabatic response (when input to the mechanical-oscillator equation-of-motion through the forcing function term) provides an optical-contribution to the stiffness of the mass-spring system. This so-called *"optical spring"* effect (or *"light induced rigidity"*) has been observed in the context of LIGO [19] and also in microcavities [53]. The corresponding change in spring constant leads to a frequency shift, relative to the unpumped mechanical oscillator eigenfrequency, as given by (where $P = |s|^2$ denotes the input power):

$$\Delta\Omega_m \underset{\kappa \gg \Omega_m}{=} \mathcal{F}^2 \frac{8n^2\omega_0}{\Omega_m m_{eff} c^2} C \cdot \left[\frac{2\Delta\tau}{\left(4\tau^2\Delta^2 + 1\right)} \right] P \tag{10.7}$$

The nonadiabatic contribution in Equation 10.5 is proportional to the velocity of the mass-spring system. When input to the mechanical-oscillator equation, the coefficient of this term is paired with the intrinsic mechanical damping term and leads to the following damping rate given by (for a WGM cavity of radius R):

$$\Gamma \underset{\kappa \gg \Omega_m}{=} -\mathcal{F}^3 \frac{16n^3\omega_0 R}{\Omega_m m_{eff} c^3} C \cdot \left[\frac{8\Delta\tau}{\left(4\tau^2\Delta^2 + 1\right)^2} \right] P \tag{10.8}$$

Consequently, the modified (effective) damping rate of the mechanical oscillator is given by:

$$\Gamma_{eff} = \Gamma + \Gamma_m \tag{10.9}$$

In both Equations 10.7 and 10.8, we have stressed the fact that these expressions are valid only in the weak retardation regime in which $\kappa \gg \Omega_m$. The sign of Γ (and the corresponding direction of power flow) depends upon the relative detuning of the optical pump with respect to the cavity resonant frequency. In particular, a red-detuned pump ($\Delta < 0$) results in a sign such that optical forces augment intrinsic mechanical damping, while a blue-detuned pump ($\Delta > 0$) reverses the sign so that damping is offset (negative damping or amplification). It is important to note that the cooling rate in the weak-retardation regime depends strongly ($\propto \mathcal{F}^3$) on the optical finesse, which has been experimentally verified as discussed in Section 10.4.2. Note also that maximum cooling or amplification rate for given power occurs when the laser is detuned to the maximum slope of the Lorenzian cavity; these two cases are illustrated in Figure 10.1. These modifications have been first derived by Braginsky [11] more than 3 decades ago and are termed *dynamic back-action*. Specifically, an optical probe used to ascertain the position of a mirror within an optical resonator, will have the side-effect of altering the dynamical properties of the mirror (viewed as a mass-spring system).

The direction of power-flow is also determined by the sign of the pump frequency detuning relative to cavity resonant frequency. Damping (red tuning of the pump) is accompanied by power flow from the mechanical mode to the optical mode. This flow results in cooling of the mechanical mode. Amplification (blue tuning of pump) is, not surprisingly, accompanied by net power flow from the optical mode to the mechanical mode. This case has also been referred to as *heating*, however, it is more appropriately referred to as *amplification* since the power flow in this direction performs work on the mechanical mode. The nature of power flow between the mechanical and optical components of the system will be explored here in several ways, however, one form of analysis makes contact with the thermodynamic analogy of cycles in a Clapeyron or Watt diagram (i.e., a pressure–volume diagram). In the present case—assuming the mechanical oscillation period to be comparable to or longer than the cavity lifetime—such a diagram can be constructed to analyze power flow resulting from cycles of the coherent radiation gas interacting with a movable piston-mirror [54] (see Figure 10.2). In particular, a plot of radiation pressure exerted on the piston-mirror versus changes in optical mode volume provides a coordinate space in which it is possible to understand the origin and sign of work done during one oscillation cycle of the piston mirror. Considering the oscillatory motion of the piston-mirror at some eigenfrequency Ω_m, then because pressure (proportional to circulating optical power) and displaced volume (proportional to x) have a quadrature relationship (through the dynamic back-action term involving the velocity dx/dt), the contour for a PV cycle will encompass a nonzero area, giving the net-work performed during one cycle of mechanical oscillation. Note that the area reduces as the photon lifetime shortens (i.e., as retardation is weakened). Also, the sense in which the PV cycle is traversed is opposite for the two cases of red and blue detuning of the pump (i.e., the area and hence work-done changes sign). For blue detuning, the radiation gas does net work on the piston while the reverse is true for red detuning. The fact that cooling is possible for the case of a red detuned pump is the result of both: the sign of work in this case being such that the piston does positive work on the photon gas, and, equally important, that the photon gas (if in a coherent state) provides only quantum, back-action on the piston. This makes the photon gas effectively very cold.

The cooling and amplification processes can also be understood in terms of the creation of stokes and anti-Stokes sidebands [21]. Oscillatory motion of the cavity mirror will create sidebands on the probe wave as the circulating optical power is Doppler-shifted by the mirror's motion (or equivalently the expansion and contraction of the

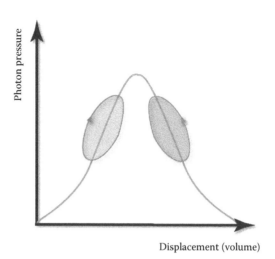

FIGURE 10.2
Work done during one cycle of mechanical oscillation can be understood using a PV diagram for the radiation pressure applied to a piston-mirror versus the mode volume displaced during the cycle. In this diagram the cycle follows a contour that circumscribes an area in PV space and hence work is performed during the cycle. The sense in which the contour is traversed (clockwise or counterclockwise) depends upon whether the pump is blue or red detuned with respect to the optical mode. Positive work (amplification) or negative work (cooling) are performed by the photon gas on the piston mirror in the corresponding cases (From Optics Express. With permission.).

whispering gallery in the case of a toroid or disk resonator). As shown in Figure 10.3, these motional sidebands will have asymmetric amplitudes owing to the fact that the pump wave frequency is detuned from the cavity resonance. The two cases of interest (red and blue detuning) are illustrated in the Figure 10.2 and produce opposite asymmetry. Intuitively, the sideband that is closer to the cavity resonance has its amplitude resonantly enhanced. This will be rigorously derived in the next Section. This asymmetry indicates a net deficit (blue detuned) or surplus (red detuned) of power in the transmitted pump wave. The sideband asymmetry also produces strong amplitude modulation that can be measured as a photocurrent upon detection of the transmitted power from the resonator.

Amplification and damping (i.e., cooling) have been verified experimentally. In particular, a mechanical mode subjected to dynamic backaction through a properly detuned optical pump wave will exhibit a thermal noise spectrum whose line-shape is modified by the presence of added optical damping or gain (negative damping). In either case, the damping rate will depend in a linear fashion on the coupled optical pump power for fixed pump detuning. Before proceeding with the further development of the theory, it is useful to consider the measurement of this damping dependence in an actual system (further details regarding the realization of amplification and cooling will be the subject of Sections 10.4 and 10.5). Measurement of this behavior is possible by probing the motion of the cavity mirror using the transmitted optical pump wave. If the pump wave is detuned relative to the cavity resonance, it is power-modulated by the resonator since the mechanical motions vary in time the cavity resonance. (Equivalently, the asymmetric motional sidebands on the transmitted pump are detected.) The detected amplitude modulation contains information about the underlying mechanical motion. If, for example, the mirror is undergoing regenerative oscillations (see Section 10.4), then spectral analysis of the photocurrent will reveal the oscillation frequency (and even the effective temperature) and measurement

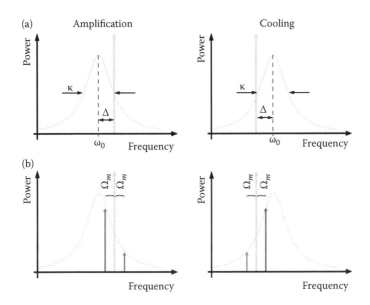

FIGURE 10.3
The two manifestations of dynamic back-action: blue-detuned and red-detuned pump wave (green) with respect to optical mode line-shape (blue) provide mechanical amplification and cooling, respectively. Also shown in the lower panels are motional sidebands (Stokes and anti-Stokes fields) generated by mirror vibration and subsequent Doppler-shifts of the circulating pump field. The amplitudes of these motional sidebands are asymmetric owing to cavity enhancement of the Doppler scattering process (From Optics Express. With permission.).

of the modulation power can be used to ascertain the mechanical oscillation amplitude. In cases where the mirror is excited only by thermal energy, the spectrally-broad thermal excitation (upon measurement as a photocurrent spectral density) provides a way to observe the oscillator line-shape and thereby determine its linewidth and effective damping rate. If the optical pump wave is weak, this damping rate will reflect the intrinsic loss (intrinsic mechanical $Q_m = \Omega_m/\Gamma_m$) of the mechanical mode. On the other hand, as the probe power is increased, it will modify the damping rate causing the line to narrow (blue detuned) or broaden (red detuned) in accordance with the above model.

Figure 10.4 presents such line-shape data taken using a microtoroid resonator. The power spectral density of the photocurrent is measured for a mechanical mode at an eigenfrequency of 40.6 MHz; three spectra are shown, corresponding to room temperature intrinsic motion (i.e., negligible pumping), mechanical amplification and cooling. In addition to measurements of amplification and damping as a function of pump power (for fixed detuning), the dependence of these quantities on pump detuning (with pump power fixed) can also be measured [28]. Furthermore, pulling of the mechanical eigenfrequency (caused by the radiation pressure modification to mechanical stiffness) can also be studied [53]. A summary of such data measured using a microtoroid in the regime where $\kappa \gtrsim \Omega$ is provided in Figure 10.5. Both the case of red- (cooling) and blue- (amplification) pump detuning are shown. Furthermore, it can be seen that pump power was sufficient to drive the mechanical system into regenerative oscillation over a portion of the blue detuning region (section of plot where linewidth is nearly zero). For comparison, the theoretical prediction is shown as the solid curve in the plots. Concerning radiation-pressure-induced stiffness, it should be noted that for red-detuning, the frequency is pulled to lower frequencies (stiffness is reduced) while for blue-detuning the stiffness increases and the mechanical eigenfrequency shifts to higher values. This is in dramatic contrast to similar changes that will

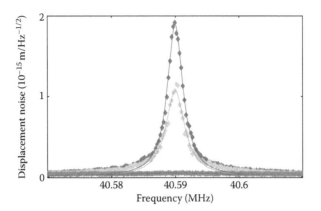

FIGURE 10.4

Dynamics in the weak retardation regime. Experimental displacement spectral density functions for a mechanical mode with eigenfrequency 40.6 MHz measured using three, distinct pump powers for both blue and red pump detuning. The mode is thermally excited (green data) and its linewidth can be seen to narrow under blue pump detuning on account of the presence of mechanical gain (not sufficient in the present measurement to excite full, regenerative oscillations); and to broaden under red pump detuning on account of radiation pressure damping (From Optics Express. With permission.).

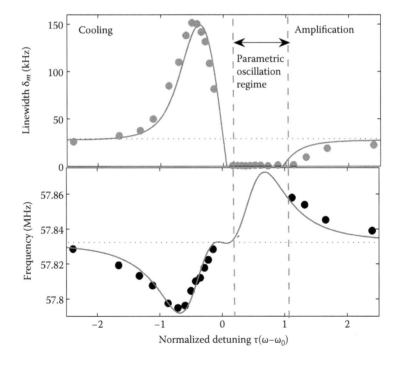

FIGURE 10.5

Upper panel shows mechanical linewidth ($\delta_m = \Gamma_{eff}/2\pi$) and shift in mechanical frequency (lower panel) measured versus pump wave detuning in the regime where $\Omega_m \ll \kappa$. For negative (positive) detuning cooling (amplification) occurs. The region between the dashed lines denotes the onset of the parametric oscillation, where gain compensates mechanical loss. Solid curves are theoretical predictions based on the sideband model (see Section 10.2.2). (From Schliesser, A., Del'Haye, P., Nooshi, N., Vahala, K. J., and Kippenberg, T. J., *Physical Review Letters*, 97, 243905, 2006) (From Optics Express. With permission.).

be discussed in the next Section. While in Figure 10.5 the absolute shift in the mechanical eigenfrequency is small compared to Ω_m, it is interesting to note that it is possible for this shift to be large. Specifically, statically unstable behavior is possible if the total spring constant reaches a negative value. This has indeed been observed experimentally in gram scale mirrors coupled to strong intra-cavity fields by the LIGO group at MIT [30].

While the above approach provides a convenient way to understand the origin of gain and damping and their relationship to nonadiabatic response, a more general understanding of the dynamical behavior requires an extension of the formalism. Cases where the mechanical frequency, itself, varies rapidly on the time scale of the cavity lifetime cannot be described correctly using the above model. From the viewpoint of the sideband picture mentioned above, the modified formalism must include the regime in which the sideband spectral separation from the pump can be comparable or larger than the cavity linewidth.

10.2.3 Sideband Formalism

It is important to realize that the derivation of the last section only applies to the case where the condition $\kappa \gg \Omega_m$ is satisfied. In contrast, a perturbative expansion of the coupled mode equations (Equations 10.1 and 10.2) gives an improved description that is also valid in the regime where the mechanical frequency is comparable to or even exceeds the cavity decay rate (where $\Omega_m \gg \kappa$ is the so called *resolved sideband regime*). The derivation that leads to these results [28] is outlined here. For most experimental considerations (in particular for the case of cooling) the quantity $\varepsilon = (x/2R)$ is very small, and a perturbative expansion of the intracavity field in powers of this parameter is possible, i.e.

$$a \equiv \sum_{n=0}^{\infty} \varepsilon^n a_n$$

Here, the zeroth order perturbation amplitude (a_0) describes the cavity build-up factor in the absence of the opto-mechanical coupling

$$a_0(t) = \frac{i\sqrt{\frac{1}{\tau_{ex}}}}{i\Delta + \frac{1}{2\tau}} s \tag{10.10}$$

where the detuning is given by Δ. Making the assumption that the mechanical system is undergoing harmonic motion $x(t) = x \cdot \cos(\Omega_m t)$, it is possible to solve for the first-order perturbation term:

$$a_1(t) = i a_0(t) 2\tau \omega_0 \left(\frac{e^{i\Omega_m t}}{2i(\Delta + \Omega_m)\tau + 1} + \frac{e^{-i\Omega_m t}}{2i(\Delta - \Omega_m)\tau + 1} \right)$$

Inspection shows that this 1st order term consists of two, independent fields, a frequency upshifted anti-Stokes sideband ($\omega_{AS} = \omega + \Omega_m$), and a frequency down shifted ($\omega_S = \omega - \Omega_m$) Stokes sideband produced by the harmonic mechanical motion. These fields account for Doppler shifting of the circulating pump field caused by the motion of the mirror or dielectric cavity. The parameter ε is sufficiently small to neglect the higher-order terms in the perturbation. The radiation pressure force acting on the mechanical oscillator in the toroidal (or other whispering gallery mode) cavity is $F_{RP}(t) = (2\pi n/c)(|a|^2/T_{rt})$ (and for a Fabry–Perot cavity $F_{RP}(t) = (2/c)(|a|^2/T_{rt})$). Here, T_{rt} is the cavity round trip time and

$|a|^2$ and is to 1st order by $|a|^2 \cong |a_0|^2 + 2\varepsilon\Re(a_0^* a_1)$. The radiation pressure force can now be expressed in terms of *in-phase* and *quadrature* components (with respect to the harmonic displacement, assumed above):

$$F_{RP}(t) = \cos(\Omega_m t)F_I + \sin(\Omega_m t)F_Q$$

Where the in-phase component is given by:

$$F_I = +\varepsilon \frac{16\pi n}{cT_{rt}} \frac{\omega_0 \tau^3/\tau_{ex}}{4\Delta^2\tau^2 + 1} \left(\frac{2(\Delta+\Omega_m)\tau}{4(\Delta+\Omega_m)^2\tau^2 + 1} + \frac{2(\Delta-\Omega_m)\tau}{4(\Delta-\Omega_m)^2\tau^2 + 1} \right) P \qquad (10.11)$$

and the quadrature component takes on the form:

$$F_Q = -\varepsilon \frac{8\pi n}{cT_{rt}} \frac{\omega_0 \tau^2/\tau_{ex}}{4\Delta^2\tau^2 + 1} \left(\frac{2\tau}{4(\Delta+\Omega_m)^2\tau^2 + 1} - \frac{2\tau}{4(\Delta-\Omega_m)^2\tau^2 + 1} \right) P \qquad (10.12)$$

The in-phase part of this force is responsible for changes in rigidity whereas the quadrature part is responsible for changes in the damping factor, which leads to cooling (or amplification). From the force, the net power (P_m) transferred from the mechanical mode to (or from) the optical mode can be calculated via the relation $\langle P_m \rangle = \langle F_Q \cdot \sin(\Omega_m t) \cdot \dot{x} \rangle$ yielding:

$$\langle P_m \rangle = \langle x^2 \rangle \frac{8\pi n}{cT_{rt}} \frac{\omega_0 \tau^3/\tau_{ex}}{4\tau^2\Delta^2 + 1} \frac{\Omega_m}{2R} \left(\frac{2\tau}{4(\Delta+\Omega_m)^2\tau^2 + 1} - \frac{2\tau}{4(\Delta-\Omega_m)^2\tau^2 + 1} \right) P \qquad (10.13)$$

This quantity can be recognized as the difference in intra-cavity energy of anti-Stokes and Stokes modes, i.e., $\langle P_m \rangle = (|a_{AS}|^2 - |a_S|^2)/T_{rt}$, as expected from energy conservation considerations. Consequently, this analysis reveals how the mechanical mode extracts (amplification) or loses energy (cooling) to the optical field, despite the vastly different frequencies. Cooling or amplification of the mechanical mode thus arises from the fact that the two sidebands created by the harmonic mirror motion are *not equal* in magnitude, due to the detuned nature of the excitation (see Figure 10.2). In the case of cooling, a red detuned laser beam will cause the system to create more anti-Stokes than Stokes photons entailing a net power flow from the mechanical mode to the optical field. Conversely, a blue-detuned pump will reverse the sideband asymmetry and cause a net power flow from the optical field to the mechanical mode, leading to amplification. We note that in the case of cooling, the mechanism is similar to the cooling of atoms or molecules inside an optical resonator via "coherent scattering" as theoretically proposed [55] and experimentally observed [56]. From the power calculation, the cooling or mechanical amplification rate $\Gamma = (\langle P_m \rangle / 1/2 m_{eff}\Omega_m^2 \langle x^2 \rangle)$ can be derived and expressed as [28]:

$$\Gamma = -\mathcal{F}^2 \frac{8n^2\omega_0}{\Omega_m m_{eff}c^2} C \cdot \left(\frac{1}{4(\Delta-\Omega_m)^2\tau^2 + 1} - \frac{1}{4(\Delta+\Omega_m)^2\tau^2 + 1} \right) P \qquad (10.14)$$

Here the finesse $\mathcal{F}$ has been introduced (as before) as well as the previously introduced dimensionless coupling factor $C = (\tau/\tau_{ex})/(4\Delta^2\tau^2 + 1)$ which takes on a value between 0 and 1. Analysis of the above formula allows derivation of the optimum detuning, for which the cooling or amplification rate is maximum. Note that in the limit of $\kappa \gg \Omega_m$ one recovers the result of the previous section (Equation 10.8), as

$$\lim_{\Omega_m \to 0} \frac{1}{2\Omega_m}\left(\frac{1}{4(\Delta-\Omega_m)^2\tau^2+1}-\frac{1}{4(\Delta+\Omega_m)^2\tau^2+1}\right)=\frac{8\Delta\tau^2}{\left(4\Delta^2\tau^2+1\right)^2} \qquad (10.15)$$

Hence as noted before, in the weak retardation regime, the maximum cooling (or amplification) rate for a given power occurs when the laser is detuned to the maximum slope of the cavity Lorenzian, i.e., $|\Delta|=\kappa/2$, in close analogy to Doppler cooling [7] in atomic physics. On the other hand, in the resolved sideband regime optimum detuning occurs when the laser is tuned either to the lower or upper "motional" sideband of the cavity, i.e., $\Delta=\pm\Omega_m$. This behavior is also shown in Figure 10.6. The above rate modifies the dynamics of the mechanical oscillator. In the case of amplification (blue detuned pump) it offsets the intrinsic loss of the oscillator and overall mechanical damping is reduced. Ultimately, a "threshold condition" in which mechanical loss is completely offset by gain occurs at a particular pump power. Beyond this power level, regenerative mechanical oscillation occurs. This will be studied in Section 10.4.

For red detuning, the oscillator experiences enhanced damping. Beyond the power flow analysis provided above, a simple classical analysis can also be used to understand how such damping can result in cooling. Specifically, in the absence of the laser, the mean energy (following from the equation for x) obeys $\langle E_m\rangle/dt = -\Gamma_m\langle E_m\rangle + k_B T_R \Gamma_m$, implying that the mean energy is given by the reservoir temperature $\langle E_m\rangle = k_B T_R$. When considering the modifications to this equation resulting from back-action damping, an additional loss term appears in the equation for the average energy:

$$\frac{d}{dt}\langle E_m\rangle = -(\Gamma_m+\Gamma)\langle E_m\rangle + k_B T_R \Gamma_m \qquad (10.16)$$

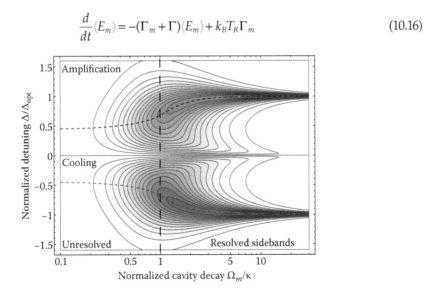

FIGURE 10.6
The mechanical gain and cooling rate as a function of detuning and inverse cavity decay rate (normalized with respect to the fixed mechanical frequency). Also shown is the optimum amplification and cooling rate for fixed frequency (dotted lines). In the simulation, pump power and cavity dimension are fixed parameters (From Optics Express. With permission.).

Note that within this classical model, the cooling introduced by the laser is what has been described by some authors [29] as *"cold damping"*: the laser introduces a damping without introducing a modified Langevin force (in contrast to the case of intrinsic damping). This is a key feature and allows the enhanced damping to reduce the mechanical oscillator temperature, yielding as a final effective temperature for the mechanical mode under consideration:

$$T_{eff} \cong \frac{\Gamma_m}{\Gamma_m + \Gamma} T_R \tag{10.17}$$

It deserves notice that the above formula predicts that arbitrarily low temperatures can be attained. As discussed later, the laser damping also introduces a small noise term, due to the quantum nature of light, which adds a further Langevin force to previous equations. This will be considered in the last section and is shown to provide the ultimate cooling limit of this technique. Moreover, it is noted that the above formula is only valid [47,48] as long as $\Gamma \ll \kappa$ and as $(\Gamma_m/\Gamma_m + \Gamma) > (1/Q_m)$.

For completeness, the in-phase component of the radiation pressure force is also investigated. This component of the force causes a change in the mechanical oscillator's rigidity, and its adiabatic contribution is the well-known optical spring effect described earlier. Specifically, the change in mechanical resonance frequency (from its intrinsic value) is given by [28]:

$$\Delta\Omega_m = \mathcal{F}^2 \frac{8n^2\omega_0}{\Omega_m m_{eff} c^2} C\tau \cdot \left(\frac{\Delta - \Omega_m}{4(\Delta - \Omega_m)^2\tau^2 + 1} + \frac{\Delta + \Omega_m}{4(\Delta + \Omega_m)^2\tau^2 + 1} \right) P \tag{10.18}$$

Note that in the regime where the mechanical frequency is comparable to, or exceeds the cavity decay rate it's behavior is quite different from that described earlier for the conventional adiabatic case; and was only recently observed experimentally [28]. As noted earlier, in the adiabatic regime, the mechanical frequency is always downshifted by a red-detuned laser (i.e. a reduced rigidity). However, when $\Omega_m > 1/2\kappa$, an interesting phenomena can occur. Specifically, when the pump laser detuning is relatively small, the mechanical frequency shift is opposite in sign to the conventional, adiabatic spring effect. The same behavior can occur in the case of amplification ($\Delta > 0$). Furthermore, a pump detuning exists where the radiation-pressure induced mechanical frequency shift is zero, even while the damping/amplification rate is nonzero. The latter has an important meaning as it implies that the entire radiation pressure force is *viscous* for red detuning, contributing only to cooling (or to amplification for blue detuning). Figure 10.6 shows the attained cooling rate (as a contour plot) for fixed power and mechanical oscillator frequency as a function of normalized optical cavity decay rate and the normalized laser detuning (normalized with respect to $\Delta_{opt} = \sqrt{\Omega_m^2 + \kappa^2/4}$. Evidently, the highest cooling/amplification rates are achieved in this resolved sideband regime, provided the pump laser is detuned to $+\Omega_m$ or $-\Omega_m$ (i.e. corresponding to the cavities lower and upper motional sideband).

An important feature of cooling and amplification provided by dynamic back-action is the high level of mechanical spectral selectivity that is possible. Since, the cooling/amplification rates depend upon asymmetry in the motional sidebands, the optical linewidth and pump laser detuning can be used to select a particular mechanical mode to receive the maximum cooling or amplification. In effect, the damping/amplification rates shown above have a spectral shape (and spectral maximum) that can be controlled in an experimental setting as given by horizontal cuts of Figure 10.6. This feature is important since it can provide a method to control oscillation frequency in cases of regenerative oscillation on the blue detuning of the pump. Moreover, it restricts the cooling power to only one or a relatively small number of mechanical modes. In cases of cooling, this means that the overall mechanical structure can remain at room temperature while a target mechanical mode is refrigerated.

The full-effect of the radiation-pressure force (both the in-phase component, giving rise to a mechanical frequency shift, and the quadrature-phase component which can give rise to mechanical amplification/cooling) were studied in a "detuning series", as introduced in the last Section. The predictions made in the $\Omega_m > 0.5\kappa$ regime have been verified experimentally as shown in Figure 10.7. Specifically, the predicted change in the rigidity of the oscillator was experimentally observed as shown in Figure 10.7 and is in excellent agreement with the theoretical model (solid red line). Keeping the same sample but using a different optical resonance with a linewidth of 113 MHz (57.8 MHz mechanical resonance), the transition to a pure increasing and decreasing mechanical frequency shift in the cooling and amplification regimes was observed as shown in the preceding Section, again confirming the validity of our theoretical model based on the motional sidebands.

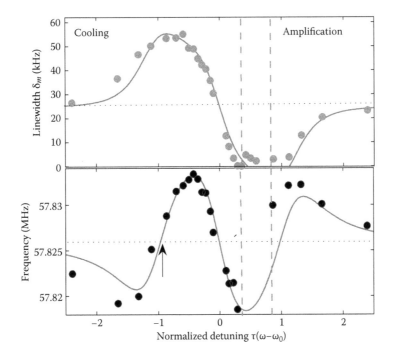

FIGURE 10.7

Dynamics in the regime where $\Omega_m \sim \kappa$ as reported in Ref. [28]. Upper panel shows the induced damping/amplification rate ($\delta_m = \Gamma_{eff}/2\pi$) as a function of normalized detuning of the laser at constant power. The points represent actual experiments on toroidal microcavities, and the solid line denotes a fit using the sideband theoretical model (Equations 10.14 and 10.18). Lower panel shows the mechanical frequency shift as a function of normalized detuning. Arrow denotes the point where the radiation pressure force is entirely viscous causing negligible in phase, but a maximum quadrature component. The region between the dotted lines denotes the onset of the parametric instability (as discussed in Section 10.3). (From Schliesser, A., Del'Haye, P., Nooshi, N., Vahala, K. J., and Kippenberg, T. J., *Physical Review Letters*, 97, 243905, 2006). (From Optics Express. With permission.).

10.3 Opto-mechanical Coupling and Displacement Measurements

10.3.1 Mechanical Modes of Optical Microcavities

The coupling of mechanical and optical modes in a Fabry–Perot cavity with a suspended mirror (as applying to the case of gravitational wave detectors) or mirror on a spring (as in the case of a cantilever) is easily seen to result from momentum transfer upon mirror reflection and can be described by a Hamiltonian formalism [49]. It is, however, important to realize that radiation pressure coupling can also occur in other resonant geometries: notably in the class of optical whispering gallery mode (WGM) microcavities such as microspheres [57], microdisks [58] and microtoroids [59] (see Figure 10.8). Whereas in a Fabry–Perot cavity the momentum transfer occurs along the propagation direction of the confined photons [11], the mechanism of radiation pressure coupling in a WGM cavity occurs normal to the optical trajector [21–23]. This can be understood to result from momentum conservation of the combined resonator-photon system as photons, trapped within the whispering gallery by continuous total-internal reflection, execute circular orbits. In particular, their orbital motion necessitates a radial radiation pressure exerted onto the cavity boundary. The latter can provide coupling to the resonator's mechanical modes.

While the mechanical modes of dielectric microspheres are well known and described by analytic solutions (the Lamb theory), numerical finite-element simulations are required to calculate the frequency, as well as strain and stress fields, of more complicated structures (exhibiting lower symmetry) such as micro-disks and microtoroids. In the case of a sphere [60], two classes of modes exist. Torsional vibrations exhibit only shear stress without volume change, and therefore no radial displacement takes place in these modes. Consequently these modes cannot be excited using radiation pressure, which relies upon a change in the optical path length (more formally, these modes do not satisfy the selection rule of opto-mechanical coupling, which requires that the integral of radiation pressure force and strain does not vanish along the optical trajectory). In contrast, the class of modes for which volume change is present is referred to as spheroidal modes. An example of this type of mode is the radial breathing mode of a microsphere. In the case of a toroid, this mode's equivalent is shown in Figure 10.8 and exhibits a radially-symmetric, mechanical displacement of the torus. Note that in a rotationally symmetric WGM microcavity, efficient opto-mechanical coupling requires that the mechanical modes exhibit the same rotational symmetry (i.e. satisfy the opto-mechanical selection rule). Deviations may occur in cases where the symmetry is lowered further due to eccentricities in the shape of the resonator, and in such cases excitation of modes of lowered symmetry is possible. Excitation of eccentricity split modes has been observed in microtoroids [61], but these cases are not considered here for the sake of simplicity.

A unique aspect of the microcavities, in contrast to other opto-mechanical platforms, is that their fundamental mechanical breathing modes can exhibit high mechanical frequency. The radial breathing mode of a 60-μm diameter microsphere is equal to ~55 MHz, and, indeed, microwave-rate regenerative opto-mechanical oscillation has been achieved in spheroidal microresonators [62]. This feature is important in the context of ground-state cooling as described later. It is also potentially important should these devices find applications as high-frequency oscillators. Another important aspect of the mechanical modes in these structures is the level of dissipation, which is governed by several loss mechanisms (clamping losses, thermoelastic losses, etc.). While detailed understanding of dissipation in dielectric microcavities is presently being established, quality factors as high as 50,000

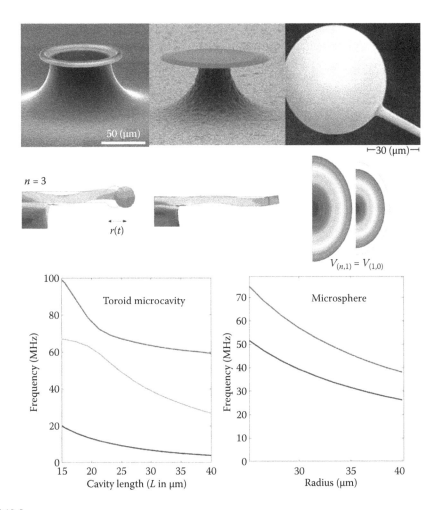

FIGURE 10.8
Upper panel: SEM images and mechanical modes of several types of whispering gallery mode microcavities: toroid microcavities [59] microdisks [58] and microspheres [57]. Also shown are the stress and strain field in cross section of the fundamental radial breathing modes, which include radial dilatation of the cavity boundary. Lower panel: the dispersion diagram for the lowest lying, rotationally symmetric mechanical modes for a toroid (as a function of its undercut) and for a microsphere (as a function of radius) (From Optics Express. With permission.).

have been observed at 50 MHz and room temperature, comparing favorably with the best nano-mechanical resonators [16] to date at low temperatures.

While the micromechanical modes coexist within the same physical structure as the optical WGM, it is important to note that there is high level of spatial separation between the modes of these physical systems. In fact, the optical whispering gallery confines optical energy to the extreme periphery of the device, while the mechanical mode impacts the entire structure. It is therefore possible to affect changes in the mechanical Q and eigenfrequency spectrum by introduction of micromechanical probes without affecting in any way the optical properties of the resonator. This feature can provide additional ways to investigate the physics of these devices. For example, a micromechanical probe, when scanned across the surface of a microtoroid, is found to modify the eigenfrequency of a mechanical mode in proportion to the amplitude-squared of the mode function.

By measuring the mechanical eigenfrequency during such a scan (using the optical probe technique described above), the underlying mechanical mode can be "imaged". Figure 10.9 provides images taken of both a fundamental and first-excited mode in a microtoroid.

10.3.2 Measuring the Opto-mechanical Response

The mechanical properties of WGM microcavities can be probed by coupling resonant laser radiation into the microcavities using tapered optical fibers [63]. Such probing will detect mirror or cavity motion as a modulation in the power transmitted past the resonator. This modulation can, in turn, be measured as a photocurrent upon detection with a photodiode. The continuous, optical probe wave, itself, can also be used to affect changes in the mirror dynamics via the back-action effect. A schematic of the experiment, which can be used to study opto-mechanical phenomena, is shown in the Figure 10.10. It consists of a continuous-wave pump laser (here, a 965-nm diode laser) which is coupled into a standard, single-mode optical fiber. This fiber enters the experimental chamber, where it also contains a tapered region used to enable evanescent coupling between the tapered fiber and various types of microcavities. The output fiber is

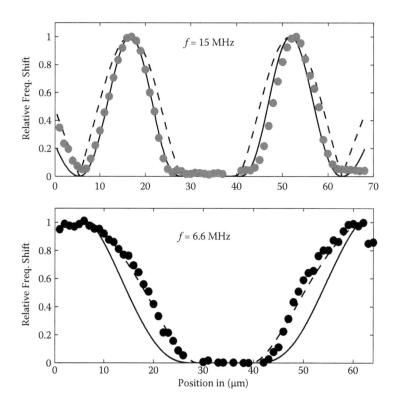

FIGURE 10.9

Scanning probe microscopy of the two lowest lying micromechanical resonances of a toroid microcavity. Lower graph: The normalized mechanical frequency shift for the first mode as a function of position. Upper graph: The normalized frequency shift for the second mechanical mode as a function of scanned distance across the toroid. Superimposed is the scaled amplitude (solid line) and the amplitude squared (dotted line) of the mechanical oscillator modes obtained by finite element simulation of the exact geometry parameters (as inferred by SEM) (From Optics Express. With permission.).

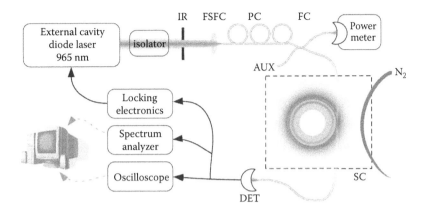

FIGURE 10.10
Experimental setup for the observation of cavity cooling or oscillation of a mechanical oscillator. All relevant data from the electronic spectrum analyzer and the oscilloscope are transferred to a computer controlling the experiment. More details in the text. IR: iris, FSFC: free-space to fibre coupler, PC: fibre polarization controller, FC: fibre coupler, AUX: auxiliary input, SC: sealed chamber, DET: fast photoreceiver (From Optics Express. With permission.).

connected to different analysis instruments, including an electrical spectrum analyzer and oscilloscope. Note that moreover locking electronics is used to ensure operation at a fixed detuning.

10.3.3 Displacement Sensitivity

Both the damping rate as well the effective temperature of a mechanical mode are observed by measuring the calibrated displacement noise spectra as a function of power. Despite their small amplitude, these thermally excited oscillations are readily observable in the transmitted light. Indeed, optical interferometers are among the most sensitive monitors for displacement. For quantum-noise-limited homodyne detection the shot noise limited displacement sensitivity of a cavity opto-mechanical system is given by:

$$\delta x_{\min} \cong \frac{\lambda}{8\pi\mathcal{F}\sqrt{\eta P/\hbar\omega}} \tag{10.19}$$

For numbers typical of the toroidal microcavity work ($\mathcal{F} \approx 4000$, $\eta \approx 1\ \mu W$ $\lambda = 1064$ nm) this implies a displacement sensitivity of $\delta x_{\min} \cong 5 \cdot 10^{-19}$ m/$\sqrt{\text{Hz}}$. Note that this is a remarkably low level, which has been experimentally achieved at the LKB [35] and also for monitoring of the toroid cavity modes [46,64]. Interestingly, it is, and in principle, even sufficient to detect the zero point motion $\delta x_{\min} \cong 5 \cdot 10^{-16}$ m within a 1-kHz resolution bandwidth. In practice, however, such a value can only be achieved in cases where true quantum-limited-readout is present, necessitating lasers which operate with quantum limited amplitude and phase noise, i.e. do not have excess classical noise in either of the two quadratures (the latter is the case of Nd:YAG lasers for frequencies above ca. 1 MHz). For the experimental measurements described [28] herein the actual displacement sensitivity did not achieve this level owing to the fact that the diode lasers exhibited excess phase noise, and limited the sensitivity to a value of about $\delta x_{\min} \cong 5 \cdot 10^{-18}$ m/$\sqrt{\text{Hz}}$. Recent work, however, has also obtained higher sensitivity by employing low noise lasers [46]. A typical calibrated

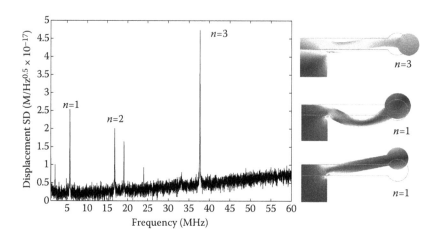

FIGURE 10.11
Calibrated [46] displacement spectral density as measured by the setup shown in Figure 10.10. The peaks denote different mechanical eigenmodes of the toroidal microcavity. The probe power is sufficiently weak such that the mechanical modes amplitude is dominated by Brownian motion at room temperature and backaction effects are negligible. Cross-sectional representations of the $n=1,2,3$ modes and their corresponding spectral peaks are also given as inferred by finite element simulations. (From Optics Express. With permission.).

and broadband displacement spectrum that can be attained with diode lasers is shown in Figure 10.11. It reveals several mechanical modes, which can be accurately assigned via finite element modeling.

10.4 Blue Detuning: Mechanical Gain and Parametric Oscillation Instability

10.4.1 Threshold and Mode Selection Mechanisms

Regenerative oscillation occurs in the case of a blue-detuned optical pump wave and at pump power levels above a threshold pump power (P_{thresh}). The oscillation threshold occurs when mechanical gain balances mechanical loss. By equating the expression for the mechanical amplification rate Γ of a specific mechanical mode (Equation 10.14) to its intrinsic loss ($\Gamma_m = \Omega_m/Q_m$), the following expression for threshold optical pump power results.

$$P_{thresh} = \frac{\Omega_m^2}{Q_m} \frac{m_{eff}c^2}{\omega_0 \mathcal{F}^2 8n^2 C} \cdot \left(\frac{1}{4(\Delta - \Omega_m)^2 \tau^2 + 1} - \frac{1}{4(\Delta + \Omega_m)^2 \tau^2 + 1} \right)^{-1} \tag{10.20}$$

In the case of weak retardation ($\kappa \gg \Omega_m$) this result simplifies to:

$$P_{thresh} \underset{\kappa \gg \Omega_m}{=} \frac{\Omega_m}{Q_m} \frac{m_{eff}c^3}{\omega_0 \mathcal{F}^3 16n^3 RC} \left(\frac{8\Delta\tau}{(4\Delta^2\tau^2 + 1)^2} \right) \tag{10.21}$$

It is worth noting that even though cooling does not exhibit a similar threshold condition, the above condition, when expressed for the case of red-detuning of the pump wave, gives the condition in which the radiation pressure cooling rate equals the heating rate of the mechanical mode. As such, a factor of two in temperature reduction (cooling) already requires that parametric oscillation is observable for the corresponding blue detuning.

An important feature of the mechanical gain is its dependence with respect to the mechanical eigenfrequency. Mechanical modes whose eigenfrequencies fall near the peak of this curve have the lowest threshold pump power for oscillation. The general shape of this curve can be inferred from Figure 10.6, which shows a contour plot of both the gain and the cooling rate versus the normalized mechanical eigenfrequency and the normalized detuning. As noted earlier, horizontal slices of this plot give the gain (or cooling rate) spectral shapes. Experimental control of the spectral peak of the gain (cooling rate) is obviously important since it determines which mechanical modes oscillate or receive maximum cooling. The contour of maximum gain appears as a dashed contour in the plot (likewise there is a corresponding contour for maximum cooling rate). The unresolved sideband case (case of weak retardation) in the figure provides a convenient physical limit in which to illustrate one form of spectral control. In this case, as noted before, the contour of maximum gain (cooling) occurs when the pump wave is detuned to the half-max position of the optical lineshape function (cf. the vertical axis value of the contour in unresolved sideband regime). The maximum mechanical gain increases along this contour as the parameter Ω_m/κ increases towards 0.5 and then diminishes beyond this value (for pump-wave detuning fixed). (Beyond this value, even somewhat before it is reached, the pump detuning must be adjusted continuously in concert with increases in Ω_m/κ to remain on the contour of maximum gain. This is a result of transition into the sideband resolved regime as described earlier). This behavior can be understood in the context of the motional sideband (Stokes/anti-Stokes waves) description provided earlier (see Figure 10.2). Specifically, the case of a pump wave detuned to the half-max point is diagrammed in Figure 10.12 for three values of the parameter Ω_m/κ ($<,=,>0.5$). The corresponding sideband configuration in each of these cases is also illustrated for comparison. Since mechanical gain is largest when the sideband asymmetry is maximum, the intermediate case of $\Omega_m/\kappa=0.5$ will exhibit maximum gain in the scenario depicted in Figure 10.12. Through adjustments of the optical mode linewidth (as can be done by controlling waveguide loading of the microresonator) an optimum Ω_m/κ can be set experimentally. This method has, in fact, been used to provide targeted oscillation of mechanical modes (even into the microwave regime) through control of optical waveguide loading. It is important to note however that the above considerations assume a constant normalized detuning. If the detuning is allowed to vary as well, maximum cooling or amplification rates always occurs in the resolved sideband regime, when the detuning equates to $\Delta=\pm\Omega_m$ (cf. Figures 10.6 and 10.12). The inherent advantages of this regime are that it enables a higher level of asymmetry in the sidebands. As described in the next section, sideband asymmetry takes on even greater significance in the context of ground state cooling.

10.4.2 Threshold Dependence on Optical Q and Mechanical Q

There are several additional features of the threshold equation that are worth noting. First, it exhibits a classic inverse dependence on mechanical Q. This is a signature for any regenerative system. Second, in the unresolved sideband regime, the threshold exhibits an inverse-cubic dependence on the optical Q factor (and correspondingly also finesse).

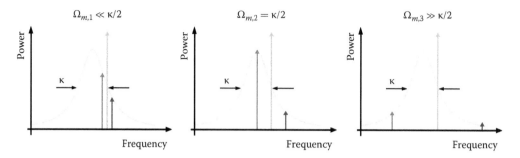

FIGURE 10.12
Back-action tuning for mode selection for a fixed laser detuning corresponding to $\Delta = \kappa/2$. The target mode that receives maximum gain (or optimal cooling for $\Delta = -\kappa/2$) can be controlled by setting the cavity linewidth to produce maximum sideband asymmetry for that particular mechanical mode. In this schematic, three mechanical modes (having frequencies $\Omega_{m,i}$ $i=1,2,3$ interact with an optical pump interact with the optical cavity mode, however, in the present scenario only the intermediate mode experiences maximum gain (or cooling) since its sideband asymmetry is maximal (since $\Omega_{m,2}/\kappa = 0.5$). It is important to note however, that if the laser detuning is allowed to vary as well, the highest frequency mode would experience the largest gain if $\Delta = \Omega_{m,2}$ was chosen (From Optics Express. With permission.).

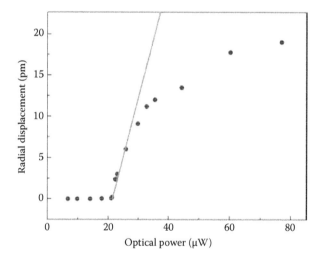

FIGURE 10.13
Regenerative oscillation amplitude plotted versus pump power. The threshold knee is clearly visible. In this case a threshold of 20 µW is observed. (From Kippenberg, T. J., Rokhsari, H., Carmon, T., Scherer, A., and Vahala, K. J., *Physical Review Letters*, 95, 033901, 2005) (From Optics Express. With permission.).

The measurement of these inherent dependences provides further evidence of the underlying nature of the interaction. Measurement of the threshold is straightforward and involves monitoring the photocurrent of the detected transmission either in the time domain or on a spectrum analyzer (as described earlier). The amplitude of oscillations at a particular mechanical eigenfrequency will exhibit a "threshold knee" when plotted versus the coupled optical pump power. This knee is an easily measurable feature and one example is provided in Figure 10.13.

The first observation of radiation pressure parametric oscillation instability, as predicted by Braginsky, was made in toroidal microcavities [21–23]. The setup employed a tapered optical fiber coupling such that optical loading could be controlled during the experiment. This enabled control of the optical linewidth as described above to effect control of the specific mechanical mode designated for oscillation. When exciting the cavity using a blue-detuned laser pump, an oscillatory output of the cavity could be observed, indicative of the excitation of mechanical modes. The oscillation is readily observable in microtoroids with threshold values in the microwatt range. Indeed, by using typical parameters in the above threshold equation ($F \approx 10^5$, $m_{eff} \approx 10^{-11}$ kg, $Q_m \approx 10^4$) a threshold in the range of a few microwatts is predicted (which is even below the threshold for Raman [65] and parametric oscillations [66]). The fact that this process can be observed in a reproducible manner allows probing of fundamental metrics of these phenomena including the above-noted mechanical and optical Q-dependences.

Measurement of the dependence of the parametric oscillation threshold on the mechanical Q is shown in Figure 10.14. In this experiment, first reported in Ref. [21], the mechanical Q-factor was varied while the optical Q factor was left unchanged. In a variation on the mechanical probing technique described above to provide image scans of the mechanical mode, a microprobe, in the form of a sharp silica fiber tip, was brought into contact, at a fixed position, with the interior of the microcavity. This caused dissipative coupling of the mechanical mode, decreasing its value from an initial, room temperature and ambient pressure Q-value of 5000 to below 50. As noted above, while the mechanical probe modifies the Q (and weakly, the eigenfrequency) of the mechanical mode, it has no effect on the optical performance of the whispering gallery. Therefore, by using this method, the dependence of threshold on mechanical Q can be probed in a nearly ideal way. Gradual change could be induced by variation of the tip pressure. Figure 10.14 shows the result of this measurement for the $n=1$ flexural mode. As is evident, there is excellent agreement with the theoretical prediction.

Next, we examine the measured oscillation threshold dependence on the optical Q factor. To illustrate behavior occurring in the sideband resolved and unresolved regimes

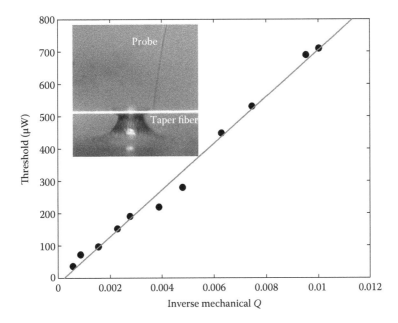

FIGURE 10.14

Main figure: The observed threshold for the parametric oscillation (of an $n=1$ mode) as a function of inverse mechanical quality factor. In the experiment, variation of Q factor was achieved by placing a fiber tip in mechanical contact with the silica membrane, which thereby allowed reduction of the mechanical Q (cf. inset). The mechanical mode was a 6 MHz flexural mode. (From Optics Express. With permission.).

two mechanical modes were measured: a fundamental ($n=1$) at 4.4 MHz and a third-order mode ($n=3$) at 49 MHz. The optical Q factor was adjusted by exciting different radial and transverse optical modes. For lower optical Q, wherein the mechanical oscillation frequency falls within the cavity bandwidth (i.e. the adiabatic regime), a rapid dependence $1/Q^3$ is observed, which is in agreement with the threshold equation (i.e. $P_{thresh} \propto 1/\mathcal{F}^3$ cf. Equation 10.21). However, the scaling of threshold changes once a transition from the unresolved (weak retardation) to the resolved sideband limit occurs. Indeed, for $\Omega_m \gtrsim \kappa$, the threshold dependence on optical Q (and finesse) weakens and eventually approaches an asymptotic value for $\Omega_m \gg \kappa$. The deviation from the cubic dependence is indeed observed experimentally as shown in Figure 10.15. The solid line in the figure is a prediction based on the threshold Equation 10.20 with effective mass as an adjustable parameter (where optimum detuning is assumed, and optimum coupling for each optical Q value). The inset of Figure 10.15 shows the threshold behavior for the $n=3$ mechanical mode which for which $\Omega_m \gtrsim \kappa$ is satisfied for the entire range of Q values. As expected, a much weaker dependence on optical Q-factor is found as predicted theoretically. Direct comparison with the $n=1$ mode data shows that oscillation on the $n=1$ mode is preferred for lower optical Qs. Indeed, preference to the $n=3$ mode was possible by increased waveguide loading of the microcavity in agreement with theory. The solid curve in the inset gives the single-parameter fit to the $n=3$ data yielding, $m_{eff} = 5 \times 10^{-11}$ kg which is a factor of 660 lower than the mass of the $n=1$ mode.

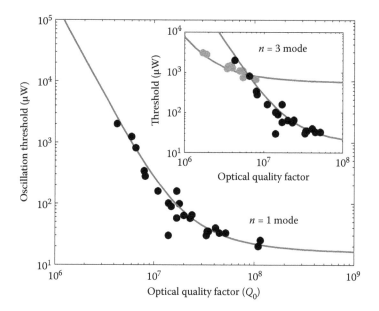

FIGURE 10.15

Main panel shows the measured mechanical oscillation threshold (in microwatts) plotted versus the optical Q factor for the fundamental flexural mode ($n=1, \Omega_m/2\pi=4.4$ MHz, $m_{eff} \approx 3.3 \times 10^{-8}$ kg, $Q_m \approx 3500$). The solid line is a one-parameter theoretical fit obtained from the minimum threshold equation by first performing a minimization with respect to coupling (C) and pump wavelength detuning (Δ), and then fitting by adjustment of the effective mass. Inset: The measured threshold for the 3rd order mode ($n=3, \Omega_m/2\pi=49$ MHz, $m_{eff} \approx 5 \times 10^{-11}$ kg, $Q_m \approx 2800$) plotted versus optical Q. The solid line gives again the theoretical prediction. The $n=1$ data from the main panel is superimposed for comparison. (From Kippenberg, T. J., Rokhsari, H., Carmon, T., Scherer, A., and Vahala, K. J., *Physical Review Letters*, 95, 033901, 2005) (From Optics Express. With permission.).

10.4.3 Oscillation Linewidth

A further important characteristic of the regenerative mechanical oscillation is the linewidth of the mechanical oscillation frequency. Theoretically, the limit for the linewidth in the case of temperatures $k_B T_R > \hbar\Omega_m$ is set by classical, thermal noise and obeys the relationship [53,67]:

$$\Delta\Omega_m \cong \frac{k_B T_R}{P}\left(\frac{\Omega_m}{Q_m}\right)^2$$

Here, P is the power dissipated in the mechanical oscillator, i.e. $P = \Gamma_m m_{eff}\Omega_m^2\langle x^2\rangle$. Consequently the equation predicts that the linewidth and mechanical amplitude satisfy $\Delta\Omega_m \propto 1/\langle x^2\rangle$. Indeed, experimental work has confirmed this scaling in toroidal microcavities [67,68]. The measured, inverse-quadratic dependence (as first reported in Refs. [53,67] is presented in Figure 10.16. Fundamental linewidths that are sub Hertz have been measured, however, with improvements in mechanical Q-factor, these values are expected to improve, owing to the inverse-quadratic dependence appearing in the above formula.

10.5 Red Detuning: Radiation Pressure Cooling

10.5.1 Experimental Setup

From the theoretical section it is clear that both cooling and amplification are indeed the same effect, albeit with opposite signs in the work performed on the mechanical system. Cooling was first theoretically noted by Braginksy [24], who proposed that red detuning

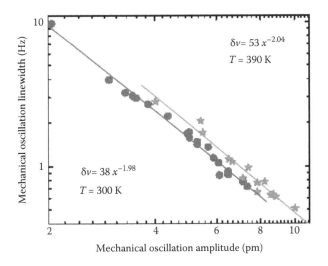

FIGURE 10.16
Linewidth measurements from Refs. [67,53] of the opto-mechanical oscillator for different amplitudes of oscillation plotted in picometers. The measurement is done at room temperature (dots) and at temperature 90°C above room temperature (stars). The solid lines and the corresponding equations are the best fits to the log–log data. Solid line denotes theoretically expected behavior (From Optics Express. With permission.).

could be used to "tranquilize" the mechanical modes of an interferometer. It has subsequently been considered in other work [21,69]. While the manifestations of parametric amplification are readily observed (leading to a strong periodic modulation of the cavity transmission), the effect of cooling is more subtle as it decreases the Brownian motion of the mechanical motion. Consequently, the observation of cooling requires both a careful calibration of the mechanical displacement spectra and sufficient signal-to-noise to detect the *cooled* mechanical mode. Indeed, cooling was observed only after the parametric oscillation in a series of three experiments reported at the end of 2006, which employed coated micromirrors (in the Paris [25] and the Vienna [26] experiments) and toroidal microcavities [28] (at the MPQ in Garching). As mentioned earlier, there is no threshold condition for cooling (as in the case of the parametric oscillation), however, to achieve efficient temperature reduction the power injected when red detuning, must greatly exceed the threshold of the parametric oscillation instability for blue detuning.

Cooling in toroidal microcavities was observed by pumping with an external diode laser that is red detuned with respect to the optical resonance. It deserves notice that this detuning is intrinsically unstable owing to the thermal bi-stability effect [70]. The cavity absorption induced heating causes a change of the cavity path length, which subsequently causes a red-shift of the cavity resonance. This feature leads to thermal, self-locking on the blue-sideband used to observe parametric amplification and oscillation, but simultaneously destabilizes locking on the red sideband. This has necessitated the implementation of fast electronic feedback to the laser, in order to be able to observe cooling. As an error signal, either the detected transmission signal was used directly or a signal derived from this by a frequency modulation technique [71]. The error signal is pre-amplified with a low-noise amplifier (DC –1 MHz), the two outputs of which are fed to two, custom-built proportional-integral controllers with bandwidths on the order of 1 kHz and 1 MHz. Both controllers allow us to apply an offset to the error signal input, enabling continuous variation of the control set-point and thus constant detuning from line center. Without further amplification, the output of the slower controller is applied to a piezoelectric element actuating the grating in the laser to tune the laser emission frequency. For the compensation of fast fluctuations, the output from the faster controller is applied to a field effect transistor parallel to the laser diode. The consequent temporary change of diode current leads to the desired laser frequency adjustment. Laser emission power is affected only on the order of 5% and, since the output of the fast controller is high-pass filtered (cut-off $\gg$10 Hz), it remains unmodified on average.

The described locking technique was used in Ref. [28] to observe back-action cooling by radiation pressure. In the experiment the laser was detuned to $\Delta \cdot \tau \approx -0.5$. The mechanical mode under consideration in this study is a radial breathing mode with resonance frequency of 57.8 MHz. Since the mechanical mode is in thermal equilibrium with the environment at room temperature (T=300 K), it follows from the classical equipartition theorem that the mechanical oscillator undergoes Brownian motion.

10.5.2 Experimental Observation of Cooling

Given the high displacement sensitivity, it is possible to record important characteristics of the mechanical mirror motion, such as the resonance frequency $\Omega_m/2\pi$, damping rate and displacement spectral density $S_x(\omega)$. The determination of the intrinsic mechanical properties was carried out by injecting very-low power into the cavity, and recording the power spectral density with an electrical spectrum analyzer. Upon sufficiently low injected power, this leads to a nearly identical noise spectra for both blue and red detuning for $\Delta=\pm\kappa/2$ (since the mechanical amplification or cooling rate is much smaller than the intrinsic mechanical damping rate). Conversion to displacement spectral density then requires knowledge of the effective mass [51] of the mechanical mode which could be independently determined in two ways. First, from finite-element simulation of the actual cavity geometry parameters, as inferred from scanning electron microscopy images. As detailed in Ref. [21] this is accomplished via the relation $m_{eff} = (E_m/\Omega_m^2(\delta R)^2)$, where δR is the mechanical energy causing a displacement in the radial direction. Second, the effective mass was determined by experimental measurements, by recording the threshold for parametric oscillation for blue detuning and subsequently inverting the threshold equation for the effective mass. Note that in the described experiments, both techniques agree very well, from which the effective mass of the radial breathing mode of Figure 10.17 is inferred to be $m_{eff}=1.5\times10^{-11}kg$. Correspondingly, the rms motion caused by Brownian motion of the radial breathing mode at room temperature is on the order of $\langle x^2\rangle^{1/2}\approx5\times10^{-14}m$.

Figure 10.18 shows the displacement spectral density for the radial breathing mode under conditions of constant red detuning ($\Delta=-\kappa/2$) for varying input power levels P. By extrapolating the resonant frequency and linewidth to zero power (cf. inset) the intrinsic resonance frequency and an intrinsic mechanical Q factor of 2890 were extracted. Note that much higher mechanical Q (>50,000) are possible in an evacuated chamber [60] and by optimizing the microcavity shape to reduce dissipative clamping losses. In the reported experiment the optical linewidth was 50 MHz, equivalent to an optical Q factor of 4.4×10^6. When varying the pump power, a clear reduction of the noise spectra is observed as shown

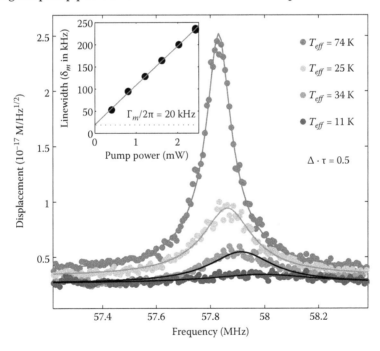

FIGURE 10.17

Main figure shows the normalized, measured noise spectra around the mechanical breathing mode frequency for $\Delta\cdot\tau\approx-0.5$ and varying power (0.25, 0.75, 1.25, and 1.75 mW). The effective temperatures were inferred using mechanical damping, with the lowest attained temperature being 11 K. (b) Inset shows increase in the linewidth (damping $\delta_m=\Gamma_{eff}/2\pi$) of the 57.1–MHz mode as a function of launched power, exhibiting the expected linear behavior as theoretically predicted. (From Schliesser, A., Del'Haye, P., Nooshi, N., Vahala, K. J., and Kippenberg, T. J., *Physical Review Letters*, 97, 243905, 2006) (From Optics Express. With permission.).

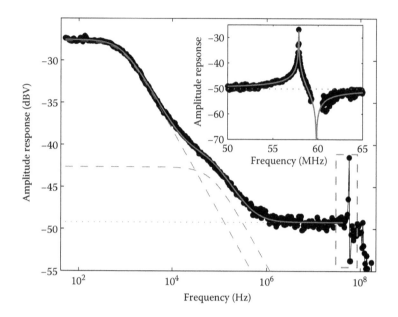

FIGURE 10.18

The frequency response from 0 to 200 MHz of a toroidal opto-mechanical system. The plateau occurring above 1MHz is ascribed to the (instantaneous) Kerr nonlinearity of silica (dotted line). The high-frequency cutoff is due to both detector and cavity bandwidth. The response poles at low frequency are thermal in nature. Inset: Data in the vicinity of mechanical oscillator response which shows the interference of the Kerr nonlinearity and the radiation pressure-driven micromechanical resonator (which, on resonance, is $\pi/2$ out-of-phase with the modulating pump and the instantaneous Kerr nonlinearity). From the fits (solid lines) it can be inferred that the radiation pressure response a factor of 260 larger than the Kerr response and a factor of ×100 larger than the thermomechanical contribution (Adapted from Schliesser, A., Del'Haye, P., Nooshi, N., Vahala, K. J., and Kippenberg, T. J., *Physical Review Letters*, 97, 243905, 2006) (From Optics Express. With permission.).

in the main panel. The shape of the spectra changes in two dramatic ways. First the linewidth of the mechanical spectra increase, owing to the fact that the light field provides a viscous force, thereby increasing the damping rate. Note that this rate varies linearly with applied power (cf. inset). Moreover, and importantly, the areas of the mechanical curves also reduce, which are a direct measure of the mechanical breathing modes (RBM) temperature. Indeed the effective temperature of the RBM is given by $k_B T_{eff} = \int \Omega^2 m_{eff} S_x(\Omega) \, d\Omega$ and correspondingly the peak of the displacement spectral density is reduced in a quadratic fashion with applied laser power. While measurement of the calibrated noise spectra is the most accurate way to determine the effective temperature, the simplified analysis presented earlier (and neglecting any other heating mechanism) yields a temperature (cf. Equation 10.17) given by the ratio of damping rates (with and without the pump laser):

$$T_{eff} \cong \frac{\Gamma_m}{\Gamma_m + \Gamma} T_R \qquad (10.22)$$

As noted before this formula only proves valid when $(\Gamma_m/\Gamma_m + \Gamma) < 1/Q_m$ and for cooling rates satisfying $\Gamma \ll \kappa$. Note that the maximum temperature reduction factor is bound by the cavity decay rate $\sim \Gamma/\kappa$. Thus, the highest temperature reduction factor which can be attained is given by $\sim \kappa/\Gamma_m$. Furthermore, additional modifications of this expression are necessary when entering the limit of temperatures, which correspond to only a few quanta, and will be considered in the next section.

For the highest pump power (2 mW and 970 nm), the effective temperature was reduced from 300 K to 11 K. This experiment, reported in Ref. [28], along with two earlier publications [25,26] represent the first demonstration of radiation-pressure back-action cooling. It is noted for completeness that the experiment described in Ref. [26] also attributed an appreciable cooling effect due to thermal effects in the mirror coatings. Indeed, of the physical mechanisms that can create opto-mechanical coupling temperature is another possibility. Temperature variations introduced by absorption, for example, create a well-known, trivial coupling, by way of thermal expansion (and absorption of photons). Indeed both mechanical amplification [72] and cooling [69] have been demonstrated using this mechanism. In the present, microtoroid studies, thermal effects are negligible. This is known, first, because of the parametric-instability studies of the previous section, where both the observed threshold dependence on optical Q as well as the magnitude of the threshold power, itself, are in excellent agreement with the theory of radiation-pressure-induced coupling. Second, the recent observation of microwave-rate parametric oscillations [62] confirms the broadband nature of the underlying mechanism, greatly exceeding bandwidths possible by thermal coupling mechanisms.

To *quantitatively* assess the contributions of thermal and radiation-pressure induced effects in the present context, pump-probe-type response measurements were performed using a second laser coupled to the cavity and operating in the wavelength region around 1550 nm. This laser serves as *pump*, providing a sinusoidally modulated input power $P(t) = P_0(1 + \varepsilon \sin(\Omega t))$, which in turn causes the optical resonances to periodically shift via both thermal effects and radiation-pressure-induced mechanical displacement, but also via the Kerr-nonlinearity of fused silica. These shifts are then "read out" with the 965 nm probe laser tuned to the wing of an optical resonance. Modulating the power of the pump laser and demodulating the detected probe power with the same frequency, it is possible to measure the microcavity response caused by all three nonlinearities (thermal, Kerr, and radiation pressure). Note that due to their different spectral response, the three nonlinearities can be readily differentiated when performing a frequency sweep. While electronic modulation and demodulation are conveniently accomplished with an electronic network analyzer, we use a fibre-coupled interferometric $LiNbO_3$ amplitude modulator to generate the modulated pump light. Data were taken on a logarithmic frequency scale between 50 Hz and 200 MHz, and, subsequently, on a linear frequency scale in the interval 50–65 MHz. From finite-element simulation, this resonance can be identified as the radial breathing mode. The result of this measurement, as first reported in Ref. [28], is shown in Figure 10.18. Several features are evident from the graph and are now discussed in detail.

First, the plateau in the response, for frequencies beyond 1 MHz and up to the cavity (and detector) cut-off at 200 MHz, is attributed to the intensity-dependent refractive index (Kerr effect) of fused silica [73,74]. Importantly, since both thermal and mechanical responses exceed the Kerr nonlinearity in some frequency domains, the Kerr effect (plateau) provides a precisely known reference for all observed nonlinearities. Second, the response up to a frequency of about 1 MHz can be well fitted assuming the sum of two single-poled functions with cut-off frequencies of 1.6 kHz and 119 kHz. It has been attributed to the response related to convective and conductive heat exchange of the cavity mode with its environment [28,57]. Concerning this thermal-related response function, it is important to note that a temperature change to the silica microcavity causes resonant frequency shifts via both a change in the refractive index and a displacement due to thermo-mechanical expansion. However, temperature-induced index changes dominate over thermal expansion by a factor of at least 15 for glass. Therefore, the thermo-mechanical contribution to the low-frequency-response shown in Figure 10.18 is, at least, one order-of-magnitude smaller than the total thermal response.

Finally, at 50 MHz in the response spectrum, the radiation-pressure-driven mechanical response is observable, which (inset Figure 10.18) is a factor of ×260 stronger than the Kerr nonlinearity in the present device. It is emphasized that this ratio is in quantitative agreement with the theoretically predicted Kerr-to-radiation pressure ratio. From the well-identified frequency response of the aforementioned thermal effects it is possible to conclude that the thermal effect contribution to the interaction of the cavity field with the 58 MHz radial breathing mode is at least two orders of magnitude too weak to explain the observed effects. Consequently, the response measurements give unambiguous proof that such thermal effects contribute less than 1 part in 100 to the observed cooling (or amplification) rate.

At a basic level, it is of significant importance that *optical forces* are dominant. These forces are both conservative and broadband, allowing for a Hamiltonian formulation [49], features that are fundamental to cavity-QOM physics as well as to potential applications of cavity-OM as a technology.

10.5.3 Quantum Limits of Radiation Pressure Back-action Cooling

An important question is what limiting temperature is achievable with the radiation pressure back-action cooling technique as described above. Indeed, ground state cooling of harmonically bound ions and atoms has first been demonstrated by Wineland [75] almost two decades ago and has led to a remarkable set of advances in atomic physics, such as the generation of nonclassical states of ion motion [76] or Schrödinger cat states [77]. Note that ground state cooling of mechanical oscillators is particularly challenging as the temperature corresponding to a single phonon corresponds to 50 μK for a 1 MHz oscillator, or 50 mK for a 1 GHz oscillator, which are only accessible by dilution refridgerators. Radiation-pressure back-action cooling might provide a route to achieve yet lower temperatures.

However, even the early work of Braginsky did not address the fundamental question of whether ground-state cooling is possible using the described mechanisms. Recently, two theoretical papers [47,48] have extended the classical theory of radiation-pressure back-action cooling to the quantum regime and shown the close relationship that cavity back-action cooling has with the laser cooling of harmonically bound atoms and ions. While a detailed description of the theoretical framework can be found in Refs. [47,48] we here briefly recapitulate the main results in terms of the fundamental temperature limit.

In the quantum mechanical treatment, the mechanical oscillator is described in terms of its average occupancy n, where $n=0$ designates the quantum ground state (in which the harmonic oscillator energy only contains the zero-point energy contribution). In brief and as shown in Refs. [47,48], the quantum mechanical cooling limit is due to the fact that cooling proceeds both by motional increasing and decreasing processes. Motional decreasing (increasing) processes occur with a rate $R^{BSB} \propto \eta^2 A^- n$ ($R^{RSB} \propto \eta^2 A^+ (n+1)$), where n is the phonon occupancy, $A^\pm \propto (1/4(\Delta \mp \Omega_m)^2 + \kappa^2)$ are the Lorentzian weights of the Stokes and anti-Stokes sidebands and η ($\omega_0 x_0 / \Omega_m R$) an effective parameter [48] with $x_0 = \sqrt{\hbar/m_{eff}\Omega_m}$ being the zero point motion of the mechanical oscillator mode. Detailed balance [78] of the motional increasing and decreasing processes then yields the minimum final occupancy (neglecting reservoir heating):

$$n_f = \frac{A^+}{A^- - A^+} \tag{10.23}$$

In the unresolved $\kappa \gg \Omega_m$ sideband case, (which has been referred to as "weak binding" in atomic laser cooling [79]) this prevents ground state cooling as:

$$n_f \approx \frac{\kappa}{4\Omega_m} \gg 1 \qquad (10.24)$$

Converting this expression into an equivalent temperature $T_D \approx (\hbar\kappa/4k_B)$, yields the *Doppler temperature* (as known in atomic laser cooling). On the other hand, occupancies well below unity can be attained in the resolved sideband case $\Omega_m \gg \kappa$ yielding:

$$n_f \approx \frac{\kappa^2}{16\Omega_m^2} \ll 1 \qquad (10.25)$$

If also the contribution from reservoir heating is included, the final occupancy that can be achieved then takes the form

$$n_f = \frac{A^+}{A^- - A^+} + \frac{\Gamma_m}{\Gamma_m + \Gamma} n_R$$

Here the n_R average denotes the average occupancy of the harmonic oscillator prior to applying radiation pressure cooling. As before, this limit is valid over the range where the cooling satisfies $(\Gamma_m + \Gamma/\Gamma_m) < n_R$ and $\Gamma < \kappa$. Note that from the above expression it becomes clear that pre-cooling of the mechanical oscillators is advantageous and is indeed presently undertaken by several groups [31]. Resolved sideband cooling has also recently been demonstrated experimentally [46].

10.5.4 Physical Interpretation of the Quantum Limits of Back-action Cooling

We next give some physical intuition into the above limits. It is worth noting that the above introduced temperature limits are in fact a direct manifestation of the *Heisenberg uncertainty principle* and are of rather general validity. This can be understood as follows. As shown

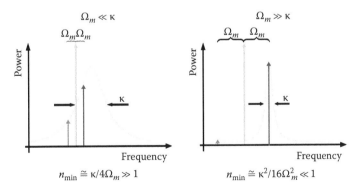

FIGURE 10.19
Quantum limits of radiation pressure cooling. Left panel shows the regime of unresolved sidebands whereas the right panel shows the resolved sideband case. Note that only in the resolved sideband regime average final occupancies below unity can be attained (i.e. ground state cooling). (Adapted from Schliesser, A., Riviere, R., Anetsberger, G., Arcizet, O., and Kippenberg, T. J., *Nature Physics*, doi:10.1038/nphys939, 2008) (From Optics Express. With permission.).

in Section 10.2 the mechanical motion produces blue-shifted sidebands which remove mechanical quanta from the mechanical oscillator. These generated photons decay during the cavity lifetime $1/\kappa$ into the outside world (i.e. into the tapered optical fiber waveguide, or through the finite reflection of the mirror in the case of a Fabry–Perot). The finite decay time of the cavity however entails that the energy that is carried away by a blue-shifted photon has a Heisenberg limited uncertainty of $\Delta E = \hbar / \Delta t \approx \hbar \kappa$. Consequently, the mean energy of the mechanical oscillator $E = \Omega_m(n + 1/2)$ cannot be lower than this limit, implying a final Doppler temperature of $T_D \approx \hbar \kappa / k_B$ in entire analogy to laser cooling in atomic physics [7]. This Doppler temperature entails that the final occupancy is much larger than unity in the case where the mechanical oscillator frequency is smaller than the cavity linewidth (since $\kappa \gg \Omega_m$, see Figure 10.19). In contrast, ground state cooling is possible when then energy quantum of the mechanical oscillator ($\hbar\Omega_m$) is made large in comparison with the energy scale set by the Doppler temperature. This required the "strong binding condition" to be satisfied, i.e. $\Omega_m \gg \kappa$ (Figure 10.19).

A second consideration that can be used to successfully estimate the quantum limits of back-action cooling is to consider the mechanical mirror as performing a measurement on the intra-cavity photons. As in the case of photo-detector shot noise, the random arrival time of photons onto the mirror will entail a fluctuation of the radiation pressure force that will heat the mechanical oscillator. If N is the average number of photons in the cavity, the fluctuation of photon numbers (governed by Poissonian statistics) is given by $\sqrt{N}$, entailing a radiation pressure force fluctuation of $\langle \Delta F_{RP}^2 \rangle^{1/2} = \sqrt{N}(2\hbar\kappa/T_{rt})$. Approximating this fluctuation as a white noise spectrum over the cavity bandwidth, the corresponding force spectral density of the radiation pressure quantum noise is given by $\delta F_{RP}^2(\Omega) = \langle \Delta F_{RP}^2 \rangle / \kappa$. This effectively white spectrum provides a driving force for the mechanical oscillator, and can be used to determine the final temperature via the relation $k_B T_F = \int \Omega^2 m_{eff} |\chi_{eff}(\Omega) \delta F_{RP}(\Omega)|^2 d\Omega = \Gamma^{-1} m_{eff} |\delta F_{RP}(\Omega_m)|^2$ (where $\chi_{eff}(\Omega)$ is the modified mechanical oscillators susceptibility). Two limits can be derived. First, assuming that the cavity linewidth exceeds the mechanical oscillator frequency, the noise spectrum is approximately given by $\delta F_{RP}^2(\Omega) \approx (P/\hbar\omega)(1/2)\tau(\hbar k/T_{rt})^2 = (P\hbar\omega/2L^2)\tau^2$ which yields, i.e. $k_B T_f \approx \hbar\kappa/4$ (i.e. $n_f \approx \kappa/4\Omega_m$ recovering the result derived in Refs. [47,48]). Here we have assumed a Fabry–Perot cavity of length L. On the other hand considering the resolved sideband regime (i.e. assuming laser and $\Delta = -\Omega_m$) the spectral density of the radiation pressure quantum fluctuations at the frequency of the mechanical oscillator are given $\delta F_{RP}^2(\Omega_m) \approx (P\hbar\omega/2L^2\Omega_m)$ yielding the result $k_B T_f = (\hbar\Omega_m/2)(1 + \kappa^2/8\Omega_m^2)$ or $n_f \approx \kappa^2/16\Omega_m^2$. Thus, the final occupancy in the unresolved and resolved sideband case can be understood as arising from the quantum fluctuation of the intra-cavity field and are in agreement with the results of a rigorous calculation.

10.6 Summary and Outlook

In summary, we have described dynamical effects of radiation pressure in cavity opto-mechanics. Specifically, both amplification and cooling of mechanical eigenmodes have been described as manifestations of finite-cavity-delay on mechanical oscillator's dynamics. Beyond the described phenomena of dynamical back-action, the Physics that can be studied in the field of cavity opto-mechanics encompasses several other areas of investigation. For instance, the classical dynamics of cavity Opto-mechanical systems exhibits a wide range of phenomena, ranging from dynamical multistability [52], static bistability

[18] to chaotic regimes [61], some of which have already been observed in an experimental setting and allow one to create switchable tunable optical filters [43]. Furthermore, parametric amplification of mechanical modes constitutes an entirely new way of creating the analogue of a "photonic quartz oscillator" which is driven purely by the radiation pressure of light and whose linewidth is limited by thermal noise. With continued improvements in mechanical Q that are already underway to address the requirements of cooling-related-research, there is the potential for realization of a new class of ultra-stable, narrow linewidth RF oscillators.

On the cooling side, there is a rich history of theoretical proposals pertaining to entangling mechanical oscillators with a light field using radiation pressure [13,36,37]. Most of these will require achieving temperatures at which the mechanical system is close-to, or at, the quantum ground state. With the rapid progress towards realization of ground-state cooling, it now seems likely that many of these ideas and proposals can be tested over the next decade. Similarly, macroscopic mechanical modes can provide a new medium in which to explore quantum information phenomena, as has been true in the rich scientific arena of cavity QED [80]. Indeed, one can view cavity QOM as the logical extension of cavity QED into the macroscopic realm.

Another interesting goal is to reach a regime where cavity QOM phenomena become observable. For example, a regime where the back-action from the quantum noise of the radiation pressure dominates the thermal noise [9,10] would enable experimental realizations of proposals such as the quantum nondemolition measurement of photon number or pondermotive squeezing [13]. Moreover, recent work from Yale [31] has demonstrated cavity OM systems which realize a quadratic coupling to the mechanical coordinate, thereby lending themselves to perform QND measurements of the mechanical eigenstate.

As originally noted, both cooling [24] ("tranquilizer") and parametric instability [81] were first conceived theoretically in the context of gravitational-wave detection by Braginsky. No doubt, the better understanding of these phenomena gained by their demonstration in the microscale will benefit this important field. Along these lines, since the initial observation of radiation pressure parametric instability in microtoroids [21–23] in 2005, the MIT gravitational-wave group has reported observation of parametric instability [82]. Even more recently they have also reported the cooling of gram scale mirror modes [30]. These results bode well for further progress in this field.

From a practical point-of-view, the ability to achieve cooling and oscillation of micromechanical modes on a semiconductor chip bodes well for realization of new technologies that could leverage these new tools. Specifically, miniaturization and integration of these functions with electronics and other optical functions is already possible because of this microfabricated, chip-based platform. Also significant is that radiation-pressure cooling through dynamic back action, as already noted, is a highly targeted form of cooling in which a selected mechanical mode(s) can be precisely defined to receive the benefit of cooling, while other modes remain at elevated or even at room temperature. Indeed, the first demonstrations of back-action cooling were to temperatures in the range of 10 K, but featured mechanical structures that were otherwise uncooled and at room temperature. As a result, this novel form of cooling offers ultra-low temperature performance with relatively low power requirements (milli-Watts) and without the need for cryogenics, vacuum handling, or any of the other necessities of conventional refrigeration. This feature, above all others, would seem to offer the greatest advantages in terms of new technologies.

Finally, the range of phenomena that have been described here extend over the entire electromagnetic spectrum. Indeed, the concept of dynamic back action was conceived-of first in the microwave realm [11]. It is also important to note that there are electro-mechanical systems that provide an analogous form of back-action cooling [16,17]. These systems can potentially provide a means to achieve the quantum ground-state.

In summary, cavity (quantum) opto-mechanics represents many of the concepts of atomic and molecular physics, however embodied in an entirely different macroscale system. It is currently experiencing rapid experimental and theoretical success in various laboratories worldwide and offers entire new inroads for basic science and potentially new technologies. It seems clear that this field is entering an exciting period of experimental science.

Acknowledgments

KJV acknowledges the Caltech Lee Center and DARPA for supporting this work. TJK acknowledges support via a Max Planck Independent Junior Research Group, a Marie Curie Excellence Grant (MEXT-CT-2006-042842) and the Nanosystems Initiative Munich (NIM). The authors kindly thank Albert Schliesser, Olivier Arcizet, Jens Dobrindt , Stefan Weis, Emanuel Gavartin and Mani Hossein-Zadeh and Jiang Li for contributions to this review.

References

1. T. J. Kippenberg and K. Vahala, *Optics Express,* 15, 17172, 2007.
2. K. J. Vahala. *Nature,* 424, 839, 2003.
3. H. G. Craighead. *Science,* 290, 1532, 2000.
4. T. W. Hänsch and A. L. Schawlow. *Optics Communications,* 13, 68, 1975.
5. D. J. Wineland, R. E. Drullinger, and F. L. Walls. *Physical Review Letters,* 40, 1639, 1978.
6. S. Chu, L. Hollberg, J. E. Bjorkholm, A. Cable, and A. Ashkin. *Physical Review Letters,* 55, 48, 1985.
7. S. Stenholm. *Reviews of Modern Physics,* 58, 699, 1986.
8. C. M. Caves. *Physical Review D,* 23, 1693, 1981.
9. K. Jacobs, I. Tittonen, H. M. Wiseman, and S. Schiller. *Physical Review A,* 60, 538, 1999.
10. I. Tittonen, G. Breitenbach, T. Kalkbrenner, T. Muller, R. Conradt, S. Schiller, E. Steinsland, N. Blanc, and N. F. de Rooij. *Physical Review A,* 59, 1038, 1999.
11. V. B. Braginsky. *Measurement of Weak Forces in Physics Experiments.* University of Chicago Press, Chicago, IL, 1977.
12. V. B. Braginsky and F. Khalili. *Quantum Measurement,* Cambridge University Press, Cambridge, England, 1992.
13. S. Mancini and P. Tombesi. *Physical Review A,* 49, 4055, 1994.
14. S. Bose, K. Jacobs, and P. L. Knight. *Physical Review A,* 56, 4175, 1997.
15. M. D. LaHaye, O. Buu, B. Camarota, and K. C. Schwab. *Science,* 304, 74, 2004.
16. Naik, O. Buu, M. D. LaHaye, A. D. Armour, A. A. Clerk, M. P. Blencowe, and K. C. Schwab. *Nature,* 443, 193, 2006.
17. K. Brown, J. Britton, R. Epstein, J. Chiaverini, D. Leibfried, and D. Wineland. *Physical Review Letters,* 99, 137205, 2007.
18. Dorsel, J. D. McCullen, P. Meystre, E. Vignes, and H. Walther. *Physical Review Letters,* 51, 1550, 1983.
19. S. Sheard, M. B. Gray, C. M. Mow-Lowry, D. E. McClelland, and S. E. Whitcomb. *Physical Review A,* 69, 2004.

20. V. B. Braginsky, S. E. Strigin, and S. P. Vyatchanin. *Physics Letters A*, 287, 331, 2001.
21. T. J. Kippenberg, H. Rokhsari, T. Carmon, A. Scherer, and K. J. Vahala. *Physical Review Letters*, 95, 033901, 2005.
22. H. Rokhsari, T. J. Kippenberg, T. Carmon, and K. J. Vahala. *Optics Express*, 13, 5293, 2005.
23. T. Carmon, H. Rokhsari, L. Yang, T. J. Kippenberg, and K. J. Vahala. *Physical Review Letters*, 94, 2005.
24. V. B. Braginsky and S. P. Vyatchanin. *Physics Letters A*, 293, 228, 2002.
25. O. Arcizet, P. F. Cohadon, T. Briant, M. Pinard, and A. Heidmann. *Nature*, 444, 71, 2006.
26. S. Gigan, H. R. Bohm, M. Paternostro, F. Blaser, G. Langer, J. B. Hertzberg, K. C. Schwab, D. Bauerle, M. Aspelmeyer, and A. Zeilinger. *Nature*, 444, 67, 2006.
27. M. Poggio, C. L. Degen, H. J. Mamin, and D. Rugar. *Physical Review Letters*, 99, 2007.
28. Schliesser, P. Del'Haye, N. Nooshi, K. J. Vahala, and T. J. Kippenberg. *Physical Review Letters*, 97, 243905, 2006.
29. P. F. Cohadon, A. Heidmann, and M. Pinard. *Physical Review Letters*, 83, 3174, 1999.
30. T. Corbitt, Y. B. Chen, E. Innerhofer, H. Muller-Ebhardt, D. Ottaway, H. Rehbein, D. Sigg, S. Whitcomb, C. Wipf, and N. Mavalvala. *Physical Review Letters*, 98, 150802, 2007.
31. J. D. Thompson, B. M. Zwickl, A. M. Yarich, F. Marquardt, S. M. Girvin, and J. Harris. arXiv:0707.1724, 2007.
32. S. Mancini, D. Vitali, and P. Tombesi. *Physical Review Letters*, 80, 688, 1998.
33. S. Vandermeer. *Reviews of Modern Physics*, 57, 689, 1985.
34. D. Kleckner and D. Bouwmeester. *Nature*, 444, 75, 2006.
35. O. Arcizet, P. F. Cohadon, T. Briant, M. Pinard, A. Heidmann, J. M. Mackowski, C. Michel, L. Pinard, O. Francais, and L. Rousseau. *Physical Review Letters*, 97, 133601, 2006b.
36. V. Giovannetti, S. Mancini, and P. Tombesi. *Europhysics Letters*, 54, 559, 2001.
37. S. Mancini, V. Giovannetti, D. Vitali, and P. Tombesi. *Physical Review Letters*, 88, 120401, 2002.
38. J. M. Courty, A. Heidmann, and M. Pinard. *Physical Review Letters*, 90, 2003.
39. O. Arcizet, T. Briant, A. Heidmann, and M. Pinard. *Physical Review A*, 73, 2006.
40. W. Marshall, C. Simon, R. Penrose, and D. Bouwmeester. *Physical Review Letters*, 91, 2003.
41. M. Hossein-Zadeh and K. J. Vahala. Volume 20, No. 4, February 2008, Page(s): 234–236.
42. M. L. Povinelli, J. M. Johnson, M. Loncar, M. Ibanescu, E. J. Smythe, F. Capasso, and J. D. Joannopoulos. *Optics Express*, 13, 8286, 2005.
43. M. Eichenfeld, C. Michael, R. Perahia, and O. Painter. *Nature Photonics*, 1, 416, 2007.
44. K. C. Schwab and M. L. Roukes. *Physics Today*, 58, 36, 2005.
45. Y. Akahane, T. Asano, B. S. Song, and S. Noda. *Nature*, 425, 944, 2003.
46. A. Schliesser, R. Riviere, G. Anetsberger, O. Arcizet, and T. J. Kippenberg. *Nature Physics*, doi:10.1038/nphys939, 2008.
47. F. Marquardt, J. P. Chen, A. A. Clerk, and S. M. Girvin. *Physical Review Letters*, 99, 093902, 2007.
48. I. Wilson-Rae, N. Nooshi, W. Zwerger, and T. J. Kippenberg. *Physical Review Letters*, 99, 093902, 2007.
49. K. Law. *Physical Review A*, 51, 2537, 1995.
50. H. A. Haus. *Electromagnetic Fields and Energy*. Prentice Hall, Englewood Cliff, NJ, 1989.
51. M. Pinard, Y. Hadjar, and A. Heidmann. *European Physical Journal D*, 7, 107, 1999.
52. F. Marquardt, J. Harris, and S. Girvin. *Physical Review Letters*, 96, 2006.
53. M. Hossein-Zadeh and K. J. Vahala. *Optics Letters*, 32, 1611, 2007.
54. K. Karrai. *Nature*, 444, 41, 2006.
55. V. Vuletic and S. Chu. *Physical Review Letters*, 84, 3787, 2000.
56. P. Maunz, T. Puppe, I. Schuster, N. Syassen, P. W. H. Pinkse, and G. Rempe. *Nature*, 428, 50, 2004.
57. V. B. Braginsky, M. L. Gorodetsky, and V. S. Ilchenko. *Physics Letters A*, 137, 393, 1989.
58. T. J. Kippenberg, S. M. Spillane, D. K. Armani, and K. J. Vahala. *Applied Physics Letters*, 83, 797, 2003.
59. K. Armani, T. J. Kippenberg, S. M. Spillane, and K. J. Vahala. *Nature*, 421, 925, 2003.

60. R. Ma, A. Schliesser, P. Del'Haye, A. Dabirian, G. Anetsberger, and T. J. Kippenberg. *Optics Letters*, 32, 2200, 2007.
61. T. Carmon, M. C. Cross, and K. J. Vahala. *Physical Review Letters*, 98, 0031-9007, 2007.
62. T. Carmon and K. J. Vahala. *Physical Review Letters*, Volume 98, 123901, March 2007.
63. S. M. Spillane, T. J. Kippenberg, O. J. Painter, and K. J. Vahala. *Physical Review Letters*, vol 91, 043902, July 2003.
64. A. Schliesser, R. Riviere, G. Anetsberger, O. Arcizet, and T. J. Kippenberg. Nature Physics, doi:10.1038/nphys939, 2008.
65. S. M. Spillane, T. J. Kippenberg, and K. J. Vahala. *Nature*, 415, 621, 2002.
66. T. J. Kippenberg, S. M. Spillane, and K. J. Vahala. *Physical Review Letters*, 93, No. 8, 083904 2004.
67. M. Hossein-Zadeh, H. Rokhsari, A. Hajimiri, and K. J. Vahala. *Physical Review A*, Volume 74, Art. No. 023813, July 2006.
68. H. Rokhsari, M. Hossein-Zadeh, A. Hajimiri, and K. J. Vahala. *Applied Physics Letters*, 89, Art. No 261109, December 2006.
69. H. Metzger and K. Karrai. *Nature*, 432, 1002, 2004.
70. T. Carmon, L. Yang, and K. J. Vahala. *Optics Express*, 12, 4742, 2004.
71. G. Bjorklund, M. Levenson, W. Lenth, and C. Ortiz. *Applied Physics B Lasers and Optics*, 32, 145, 1983.
72. M. Zalalutdinov, A. Zehnder, A. Olkhovets, S. Turner, L. Sekaric, B. Ilic, D. Czaplewski, J. M. Parpia, and H. G. Craighead. *Applied Physics Letters*, 79, 695, 2001.
73. F. Treussart, V. S. Ilchenko, J. F. Roch, J. Hare, V. Lefevre-Seguin, J. M. Raimond, and S. Haroche. *European Physical Journal D*, 1, 235, 1998.
74. H. Rokhsari and K. J. Vahala. *Optics Letters*, 30, 427, 2005.
75. F. Diedrich, J. C. Bergquist, W. M. Itano, and D. J. Wineland. *Physical Review Letters*, 62, 403, 1989.
76. M. Meekhof, C. Monroe, B. E. King, W. M. Itano, and D. J. Wineland. *Physical Review Letters*, 76, 1796, 1996.
77. C. Monroe, D. M. Meekhof, B. E. King, and D. J. Wineland. *Science*, 272, 1131, 1996.
78. D. Leibfried, R. Blatt, C. Monroe, and D. Wineland. *Reviews of Modern Physics*, 75, 281, 2003.
79. J. Wineland and W. M. Itano. *Physical Review A*, 20, 1521, 1979.
80. J. Kimble. *Physica Scripta*, T76, 127, 1998.
81. V. B. Braginsky, S. E. Strigin, and S. P. Vyatchanin. *Physics Letters A*, 305, 111, 2002.
82. T. Corbitt, D. Ottaway, E. Innerhofer, J. Pelc, and N. Mavalvala. *Physical Review A*, 74, 2006.

11

Optical Frequency Comb Generation in Monolithic Microresonators

**Olivier Arcizet, Albert Schliesser, Pascal Del'Haye,
Ronald Holzwarth, and Tobias Jan Kippenberg**
Max Planck Institüt für Quantenoptik, Garching, Germany

CONTENTS

11.1 Introduction to Optical Frequency Combs

Optical frequency combs [1–3] provide equidistant frequency markers in the infrared, visible and extreme ultra-violet [4,5]. Early work by T.W. Hänsch recognized that the periodic pulse train of a mode-locked laser [6,7], which intrinsically provides in Fourier domain an output spectrum that constitutes an optical frequency comb, can be used to measure unknown optical frequencies. An optical frequency comb is generally characterized by its two degrees of freedom, the mode spacing f_{rep}, which in a pulsed laser is given by the pulse repetition rate [8], as well as the carrier envelope offset frequency f_{ceo}, which determines the frequency offset of the comb teeth from integer multiples of the repetition rates.

In time domain, f_{ceo} describes the phase slippage of the pulse envelope with respect to the carrier. Thus, any comb line of the resulting output spectrum can be expressed by $f_m = f_{ceo} + m f_{rep}$, where m is an integer. Measurement and control of f_{rep} and f_{ceo} allows to phase-coherently link optical frequencies across the entire spectrum spanned by the comb [8,9]. In this case, every comb tooth is uniquely determined by the experimentally controlled quantities f_{ceo} and f_{rep}. Therefore, any arbitrary optical frequency within the spectrum of the comb may be synthesized, or, vice versa, any given optical frequency within this spectrum may be phase-coherently compared to the frequencies f_{ceo} and f_{rep} in the RF domain, which can be referenced to microwave time standard [10,11]. This unique ability has made the frequency comb a revolutionary tool in metrology and laser spectroscopy, and, serving as a clockwork mechanism, has enabled the development and practical use of time standards in the optical domain [2,3,12,13]. Beyond these advances in metrology and precision measurements, frequency combs have given new impetus to applications such as broadband laser-based gas sensing [14–16] or cavity ring down spectroscopy [17].

In an earlier work [18], Kourogi generated optical frequency combs by inserting a phase modulator inside a cavity. Injection of a continuous wave laser resonant with one of the cavity mode lead to the generation of multiple sidebands when the modulator driving frequency is matched to a multiple of the cavity's modespacing, in this case, the bandwidth of the comb was limited by the dispersion of the cavity and phase modulator. Combs generated in this way could span several terahertz. This work was refined in later research at the Joint Institute for Laboratory Astrophysics (JILA) [19,20].

Prior to this work, the ability to link an optical frequency down to the RF domain in a phase-coherent way was pursued in experiments at the Physikalisch-Technische Bundesanstalt (PTB) in Braunschweig, Germany, following an approach taken at the National Bureau of Standards (NBS, now National Institute of Standard and Technology, NIST) in Boulder, CO. Nonlinear optical processes, in conjunction with a plethora of auxiliary oscillators, were used to phase-coherently link a Cs atomic clock reference to the optical domain [21]. For the task of bridging a large frequency interval, researchers at the Max Planck Institute of Quantum Optics (MPQ) in Garching, Germany, had used another approach: in their so-called optical interval divider chain, the interval to be measured was successively divided using several identical compact semiconductor lasers, until a direct microwave signal could be measured [22].

A revolutionary and extremely powerful tool—that made frequency chains obsolete—was realized with the demonstration that the optical frequency combs from a femtosecond laser can be fully stabilized and phase-coherently compared to the RF standards. While mode-locked lasers naturally generate an optical frequency comb owing to the periodic nature of the optical emission, a key ability of a frequency comb is to stabilize the optical spectrum in terms of the absolute frequency of each comb tooth. While locking of the mode spacing of a mode-locked laser can be readily implemented, full stabilization requires measuring and locking the carrier envelope frequency (f_{ceo}). A powerful technique to measure the carrier envelope frequency and thereby creating an optical frequency synthesizer is the use of an $f - 2f$ (or $2f - 3f$) interferometer, which requires in the former case however an optical spectrum that spans a full octave. With the advent of photonic crystal fibers, that exhibited a zero dispersion wavelength in the 800 nm range, it became possible to broaden the output spectrum of a Ti:Sapphire laser system to a full octave and thereby measure (and stabilize) for the first time the carrier envelope offset frequency [10,11,13]. This marked the beginning of a new era of frequency metrology, and led to new advances in diverse fields such as optical frequency metrology and precision measurements and in applications such as broadband laser-based gas sensing [17] and molecular fingerprinting [14–16]. Moreover frequency combs are an integral part of a new era of atomic clocks, based on optical transitions (exhibiting very narrow transition linewidth and high frequency) such as those afforded by a single trapped ^{133}Hg$^+$ ion [23], or Sr atoms confined in an optical lattice [24].

11.2 Frequency Comb Generation from a Monolithic Silica Microresonator

11.2.1 Physics of the Comb Generation Process

In addition to the above mentioned ways of creating optical frequency combs, an entirely novel comb generator was reported at the Max Planck Institute of Quantum Optics in 2007, which allowed a dramatic reduction in size and enabled access to ultra-high repetition rates, exceeding 40 GHz. The approach is based on χ^3 nonlinearity parametric frequency conversion* in ultra-high-Q optical microresonators [26]. Their high optical finesse and small mode volume make them uniquely suited for optical frequency conversion based on third-order nonlinearities [27] and has led to a significant reduction in the threshold of nonlinear optical processes.

A microscope image of the silica toroidal microcavity [26] used in this work is shown in Figure 11.1. The 75-μm diameter structure supports optical whispering gallery type modes (WGM) in which the electromagnetic field is confined by continuous total internal reflection at the cavity boundary. The micro-fabrication process proceeds as described in Ref. [26], and uses a CO_2-laser assisted reflow, to ensure a very low surface roughness and reduced surface scattering losses. Typical resonance quality (Q) factors exceed 10^8, mainly limited by water absorption. Due to the total internal reflection, the optical mode exhibits an evanescent tail outside the silica, which is exploited to couple light in and out of the toroid by means of an evanescent coupling to a tapered optical fiber [27,28] as shown in Figure 11.1. If the position, parallelism and diameter (of the order of the light wavelength) of the fiber are carefully adjusted, one can couple light very efficiently to the optical microresonator, and in particular one can tune the coupling strength by varying the distance between the fiber taper and the rim of the toroid.

When pumping an ultra high-Q ($>10^8$) toroid optical resonance with a continuous, intense (>10 mW) and monochromatic laser in the 1550-nm range, it is possible to observe in transmission of the fiber after the microcavity, a strongly nonmonochromatic comb-like spectrum (cf. Figure 11.1). The spectrum was measured using an optical spectrum analyzer, while pumping the toroid with 60 mW of power at 1550-nm, which corresponds to an intracavity intensity exceeding ~1 GW/cm^2. It consists of several bright emission lines, spaced by ca. 7 nm, corresponding to the cavity free spectral range (FSR) $\nu_{FSR} = c/(2\pi R n_{eff})$, where R is the cavity radius ($R = 38$ μm in the shown case) and n_{eff} the effective refractive index. The observed process is very efficient, with a typical conversion efficiency above 50% from the pump to the sidebands, which exhibit a high power per comb mode (ca. 1 mW) and can span over several hundreds of nanometers.

The mechanism responsible for this broadband comb generation is a cavity-enhanced cascaded four-wave mixing (FWM) process. The parametric frequency conversion is based

---

* This process has also been referred to as *hyper*parametric frequency conversion [25].

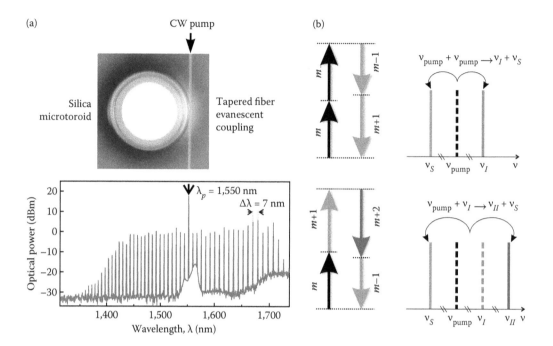

FIGURE 11.1

Optical frequency comb generation in a microresonator. Panel (a) shows the measured comb spectrum via the output of the tapered optical fiber. The individual comb modes are spaced by the free spectral range of the cavity. The change in the baseline observed close to the pump wavelength is due to the erbium doped fiber amplifier used to enhance the launched pump power. Panel (b): Physical principle of the comb generation process. Degenerate and nondegenerate four-wave mixing allow conversion of a CW laser into an optical frequency comb. (Adapted from Del'Haye, P., Schliesser, A., Arcizet, O., Wilken, T., Holzwarth, R., and Kippenberg, T. J., *Nature*, 450, 1214, 2007.)

on an FWM process mediated by the intensity dependent refractive index of silica (the Kerr nonlinearity, $n_2 = 2.3 \times 10^{-16}$ cm²/W) that initiates the process and results in the creation of coherent signal and idler photons from two pump photons: $\nu_{pump} + \nu_{pump} \rightarrow \nu_I + \nu_S$ (see Figure 11.1b). The conservation of energy ensures that the generated photon pairs are symmetric in frequency with respect to the pump. This mechanism is resonantly enhanced by the cavity if idler, signal and pump frequencies all coincide with optical modes of the microresonator. Parametric oscillations from the pump into signal and idler sidebands had first been observed in optical microcavities in 2004, in both silica microtoroidal resonators [30] and crystalline CaF₂ resonators [31] with very low thresholds (tens of μW), as a consequence of the cavity enhancement. CaF₂ resonators are typically larger and consequently have a smaller free spectral range in the order of 10 GHz. The larger mode volume in this kind of resonators is compensated by a higher optical quality factors of more than 10¹⁰ [32].

It is important to note that this process can cascade via nondegenerate FWM among pump and first-order sideband, to produce higher order sidebands (for example $\nu_{pump} + \nu_I \rightarrow \nu_{II} + \nu_S$) as can be seen in Figure 11.1. This cascaded mechanism ensures that the frequency difference of pump and first-order sidebands $|\nu_{pump} - \nu_I| = |\nu_S - \nu_{pump}|$ is exactly transferred to all higher-order inter-sideband spacings. Thus, provided that the cavity exhibits a sufficiently equidistant mode spacing, successive FWM to higher orders intrinsically leads to the generation of phase-coherent sidebands with equal spacing, that is, an *optical frequency comb* (termed *Kerr comb* in the remaining discussion). Indeed, the process

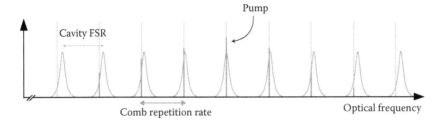

FIGURE 11.2
The role of dispersion in broadband comb generation. Due to the dispersion in the cavity, its free spectral range (FSR) varies with the optical frequency, so that the cold cavity resonances (blue Lorentzians) are not spaced equidistantly in frequency space. The generated optical frequency comb (red lines), in contrast, is perfectly equidistant. Therefore the generated sidebands walk off from the cavity resonances with increasing sideband order, reducing the cavity enhancement of the four-wave mixing process. As a consequence, uncompensated cavity dispersion can eventually limit the comb bandwidth.

can cascade over a wide frequency span, determined by the condition that the frequency of higher order converted photons must be sufficiently close to a microcavity optical resonance so that the process can still be resonantly enhanced by the optical cavity [31] (cf. Figure 11.2). Therefore, the comb generation process is naturally limited by the fact that dispersion renders the cavity modes intrinsically nonequidistant.

Dispersion in microcavities arises from two contributions resulting from material and resonator waveguide dispersion. In the case of silica microcavities these two contributions have opposite signs allowing partial compensation of dispersion in the 1550-nm window. Indeed, measurements have verified that the optical modes exhibit only around 20 MHz deviation from equidistance for the cold cavity over a 100-nm wide span. This value is comparable to the frequency bandwidth of the individual optical resonances, of the order of 10 MHz for the considered sample when overcoupled [33], which explains the broadband frequency comb generation. Note that self- and cross-phase modulation mechanisms also help to compensate the residual dispersion [30].

It is important to note that the parametric process itself could in principle produce couples of signal/idler sidebands that are only pair-wise equidistant but not mutually equidistant as required for a comb. It is thus required to verify the equidistance of the spectrum. In the next section, a measurement is described which has allowed to prove the equidistance between the different lines at the mHz level, demonstrating the generation of a frequency comb from a continuously pumped monolithic high-Q microcavity [29].

Finally, in addition to energy conservation, parametric oscillations also require angular momentum conservation to be satisfied for a rotational invariant geometry. In a microcavity of the whispering gallery type, the optical eigenmodes are also eigenstates of the angular momentum and have discrete propagation constants given by $\beta_m = m/R$ as a consequence of the periodic boundary condition ($e^{i2\pi\beta_m R} = 1$), where m is the angular mode number. Consequently the angular phase matching condition is intrinsically satisfied when the signal and idler photons correspond to adjacent cavity modes, since in this case $2\beta_{pump} = \beta_I + \beta_S$. Analogous reasoning also applies to the case of nondegenerate FWM (e.g. $\beta_{pump} + \beta_I = \beta_{II} + \beta_S$). It is important to note that in contrast to the second order nonlinearity, where the phase matching condition has to be ensured for frequency differences in the optical range, the FWM only requires phase matching condition over the frequency range occupied by the optical modes involved in the parametric frequency conversion process.

11.2.2 Verification of the Comb Spectrum Equidistance

To prove that the emitted spectrum of the microcavity is indeed an optical frequency comb, it is necessary to verify that the generated modes are mutually equidistant. At present, the most accurate references for optical frequency differences are provided by optical frequency combs derived from mode-locked femtosecond lasers [2,3,34]. In the following, we describe two experiments in which the mode spacing of a microcavity frequency comb is compared against such a femtosecond laser frequency comb to verify the equidistance of the mode spacing [29].

11.2.2.1 Multiheterodyne Spectroscopy

One possible measurement approach is a multi-heterodyne scheme [15]. In such a measurement, two light beams both containing a certain number of optical fields oscillating at different frequencies (such as two frequency combs) are brought to interference on a beam splitter, giving rise to power modulation (a *beat*) of the output beams at the various difference frequencies of the incident modes. All beats can be simultaneously recorded with a single detector of sufficient bandwidth and individually analyzed in the Fourier domain. In particular, if both input fields are equidistant frequency combs, the frequencies of the resulting beats are also expected to be equidistant. The detected signal thus constitutes a frequency comb in the RF domain. Deviations from equidistance of either optical input are directly apparent in the RF domain. Using an erbium-fiber-based mode-locked femtoscond laser [35] (Menlo-Systems GmbH) as a reference frequency comb, for which equidistant mode spacing is established, it has been possible to verify the equidistance of the microcavity frequency comb (Figure 11.3).

The reference fiber frequency comb produces a spectrum with frequencies $f_{ceo} + nf_r$, where f_r designates the repetition rate of ~100 MHz, f_{ceo} is the carrier envelope offset frequency and n an integer number of order 2×10^6. The microcavity comb produces frequencies $\nu_0 + m\Delta\nu$ (m integer). Adjusting the reference comb's repetition rate such that a multiple of it is close to the microcavity comb mode spacing, i.e. $m_0 f_r \approx \Delta\nu$ (with an integer m_0) entails that the N different microcavity comb lines will generate N different RF beat notes which will again be evenly spaced (if the Kerr comb is indeed equidistant), i.e. their frequencies will be $f_0 + k\Delta f$ with $\Delta f = (\Delta\nu \bmod f_r)$ and k integer.

Figure 11.3 shows the experimental setup for such a measurement. In the experiment, a 1550-nm external cavity diode laser (ECDL) is coupled to a high-quality WGM in a silica toroidal microresonator. A cascade of sidebands is generated in the resonator through FWM, all of which can be coupled back into a standard single mode fiber using the same tapered fiber. Using fused couplers, the output is split into several branches to monitor the transmitted power, as well as the spectrum using an optical spectrum analyzer. One of the outputs is sent to a *beat detection unit* (BDU) for the purpose of comparison with the reference frequency comb [36]. In the BDU, the inputs from microcavity and reference combs are first converted to free-space beams using collimators. The two beams are then adjusted to orthogonal polarizations and spatially combined in a polarizing beam splitter (PBS), such that they exit the PBS at the same port. A $\lambda/2$-retarder plate and a second PBS enforce interference between the two inputs with an adjustable power ratio of the input beams. A subsequent grating allows to extract spectral regions of interest, before sending the beam to a fast (~200 MHz-bandwidth) InGaAs-photoreceiver (Menlo-Systems FPD 510). The signal from this receiver is analyzed using a fast-Fourier transform (FFT) or an electronic spectrum analyzer. By adjusting the repetition rate f_r of the reference comb it is

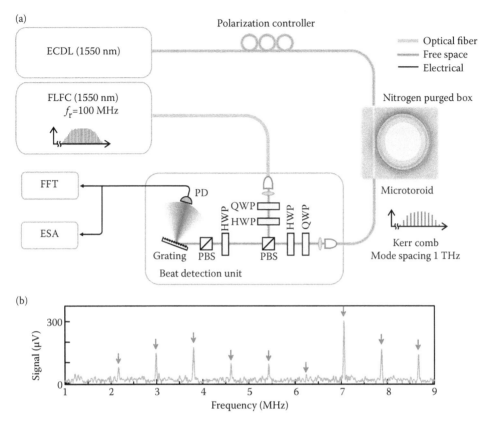

FIGURE 11.3

Multi-heterodyne beat note detection. (a) Experimental setup employed. ECDL, external cavity diode laser; FLFC, femtosecond laser frequency comb; PBS, polarizing beam splitter; HWP, half-wave retarder plate; QWP, quarter-wave retarder plate; PD, photodiode; FFT, fast-Fourier transform analyzer; ESA, electronic spectrum analyzer. (b) Radio frequency spectrum of nine simultaneously oscillating microcavity modes beating against reference comb modes. (Data from Del'Haye, P., Schliesser, A., Arcizet, O., Wilken, T., Holzwarth, R., and Kippenberg, T. J., *Nature*, 450, 1214, 2007.)

possible to make the spacing Δf of the resulting beats as small as ~1 MHz, so that all beat frequencies $f_0 + k\Delta f$ of interest fall in the window between DC and $f_r/2$.

The results for such a measurement on nine microcavity comb lines incident on a single BDU, covering a span of 50 nm or 6.5 THz, are shown in Figure 11.3c. No deviation from equidistance can be found at the level of 5 kHz, limited by the acquisition time. This proves the equidistance for all nine lines at a level of 10^{-9} (or 10^{-11} when referenced to the optical carrier). While higher resolutions and signal-to-noise ratios (SNR) would in principle be available by extending the measurement time, the mutual fluctuations of the two free-running comb generators however prevent such improvements in practice.

11.2.2.2 Proving the Equidistance of the Mode Spacing at the mHz Level

Verification of the equidistance with higher resolution necessitates reducing the mutual fluctuations of the microcavity and reference comb. This can be accomplished by locking the EDCL generating the microcavity comb to the reference comb using an offset lock.

To this end, a first BDU is used to detect the beat between the ECDL and the closest reference comb mode. The phase of the resulting RF-beat is compared to a stable RF-reference at the "offset-frequency" f_0, obtained from a synthesizer referenced to the MPQ in-house hydrogen maser. By feeding back on the frequency of the pumping laser, the two RF signals are held in phase. On time scales longer than the feedback's response time the frequency of the pump laser is thus rigidly connected to the closest reference frequency comb mode, with a frequency offset of f_0. The stability of the measurement is further improved by locking the repetition rate of the reference comb to around 100 MHz. These improvements enable measurements of much longer durations, up to several hundreds of seconds, without degradation of the beat signals due to mutual fluctuations. Measurement of radio frequencies on such a long timescale can be accomplished using radio-frequency counters. In order to achieve the SNR required for reliable counting, it is advantageous to detect each beat in a dedicated BDU. This allows an individual optimization of the polarization and power ratio of the input beams as well as a selection of narrow spectral regions reducing the amount of background light incident on the photoreceiver. The achieved SNR exceeding 30 dB in a 500-kHz bandwidth proved sufficient, in spite of still lacking about 26 dB to the shot-noise limit [36] of $\mathrm{SNR}_{opt} \simeq \eta P_{rcm}/h\nu \cdot \mathrm{RBW}$, detection efficiency $\eta \sim 0.5$, power of the reference comb mode beating with the microcavity Kerr comb mode $P_{rcm} \simeq 50$ nW, optical frequency $\nu \simeq 200$ THz and detection bandwidth RBW = 500 kHz.

To verify the equidistance of the microcavity Kerr comb, at least three modes have to be measured simultaneously. One mode of the microcavity comb is the pump laser itself; its offset to the closest reference comb mode is locked to f_0. It is therefore sufficient to measure two more beats of the microcavity comb against the reference comb. Figure 11.4 shows the setup employed in the most accurate measurement of microcavity combs performed so far [29]. The two other counted beats f_1 and f_2 correspond to the microcavity comb lines which are located at five and seven FSRs from the pump laser. A measure for the deviation from a perfectly equidistant mode spacing is the difference of the average mode spacing between the three measured lines $\varepsilon = (f_2 - f_1)/(7-5) - (f_1 - f_0)/5$. For reproducible results, all counters were referenced to the same hydrogen maser and triggered by a common trigger signal to ensure simultaneous measurements.

A slightly modified RF setup could even make one counter obsolete, by directly counting a frequency ratio instead of absolute counting of two radio frequencies. This was achieved by mixing the outputs of both BDU2 and BDU3 with the reference frequency f_0 in order to directly obtain $f_1 - f_0$ and $f_2 - f_0$ at the outputs of the mixers. Counting the ratio $R = (f_2 - f_0)/(f_1 - f_0)$ allows direct calculation of the deviation ε according to $\varepsilon = (R - 7/5) \cdot (f_1 - f_0)/2$, where $f_1 - f_0$ provides only a scale factor and needs not to be known very accurately. This method yielded slightly better results, which is attributed to latencies in the counters degrading temporal overlap of the counting intervals in spite of a common trigger signal.

Figure 11.5 illustrates the results and statistics of a series of measurements in the two-counter-configuration. The results obtained for ε were found to be normally distributed around a mean of (-0.91 ± 5.5) mHz with a standard deviation of 322 mHz. Together with many more measurements in the ratio-counting configuration, accumulating to a total measurement time of several hours, the mean deviation from equidistance could be bracketed to (-0.8 ± 1.4) mHz, confirming the uniformity of the comb spacing to a level of 7.3×10^{-18} relative to the optical carrier, and 5.2×10^{-16} relative to the 2.1 THz span of the measured microcavity comb line [29].

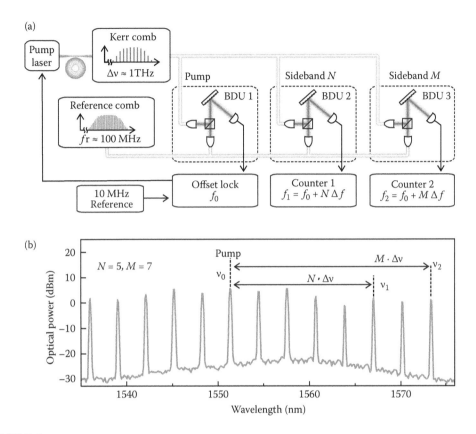

FIGURE 11.4
High-resolution measurement of the microcavity comb structure. (a) Simplified schematic of the setup. Three independent BDUs are used. With the signal from BDU 1, the pump laser is locked to one comb-line of the reference comb with an offset frequency f_0. BDU 2 and BDU 3 count the beats of the 5th and 7th sideband of the microcavity comb spectrum shown in (b) against the reference comb. (Adapted from Del'Haye, P., Schliesser, A., Arcizet, O., Wilken, T., Holzwarth, R., and Kippenberg, T. J., *Nature*, 450, 1214, 2007.)

11.2.3 Dispersion in Toroidal Microresonators

Since the resonant enhancement of FWM requires the microcavity frequency comb modes to coincide in frequency with the high-quality WGM of the microresonator, the resonator's dispersion is of crucial importance to the generation of combs [30,31,37]. In the following, the two contributions to resonator's dispersion are assessed both theoretically and experimentally.

The first contribution to dispersion arises from the inherent variation of the free spectral range due to the resonator geometry. In the weak compression regime, it is possible to approximate the resonances of a toroid by those of a microsphere [38]. In this case the resonance frequency of the fundamental mode of a microsphere is approximately given by [39]:

$$\nu_m = \frac{c}{2\pi nR}\left(m+1/2+\eta_1\left(\frac{m+1/2}{2}\right)^{1/3}+\dots\right), \tag{11.1}$$

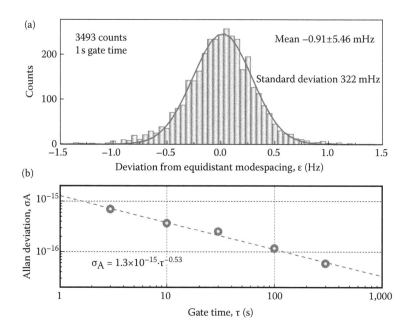

FIGURE 11.5
Verification of equidistant mode spacing. (a) The statistics of the measurement results for the deviation from equidistance ε in a 1-s gate time. The results are normally distributed. (b) Allan deviation as a function of gate time, displaying an approximately inverse square-root dependence. (Adapted from Del Haye, P., Schliesser, A., Arcizet, O., Wilken, T., Holzwarth, R., and Kippenberg, T. J., *Nature*, 450, 1214, 2007.)

where c is vacuum speed of light, n the refractive index, R the cavity radius and $-\eta_1$ the first zero of the Airy function ($\eta_1 \approx 2.34$). As a consequence, the variation of the free spectral range $\Delta\nu_{FSR} = (\nu_{m+1} - \nu_m) - (\nu_m - \nu_{m-1}) \approx \partial^2\nu_m / \partial m^2$ is given by

$$\Delta\nu_{FSR} = -\frac{c}{2\pi nR} \cdot \frac{\eta_1}{18}\left(\frac{m+1/2}{2}\right)^{-5/3} \approx -0.41\frac{c}{2\pi nR}m^{-5/3} < 0 \tag{11.2}$$

Evidently, the free spectral range *reduces* with increasing frequency corresponding to a negative group velocity dispersion (GVD). Low frequency modes exhibit a shorter round trip time than high frequency modes (i.e. the dispersion is normal). This behavior can be understood by noting that higher frequency modes spatially move progressively closer to the cavity boundary, making the classical optical trajectory larger.

A second contribution is due to the material dispersion of the fused silica material constituting the resonator. Considering the fact that the refractive index n is actually a function of frequency (and therefore mode number m), $n \equiv n(m)$, the GVD of fused silica alone would lead to a FSR variation of

$$\Delta\nu_{FSR} \approx \frac{\partial^2}{\partial m^2}\left(\frac{c}{2\pi n(m)R} \cdot m\right) \approx \frac{c^2\lambda^2}{4\pi^2 n^3 R^2} \cdot GVD, \tag{11.3}$$

where $GVD = -(\lambda/c)(\partial^2 n / \partial\lambda^2)$ is the group velocity dispersion of fused silica. This material parameter is well-known to change its sign in the 1300-nm wavelength region from

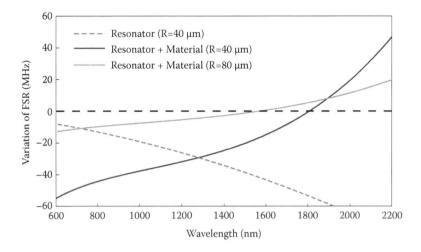

FIGURE 11.6
Variation of the free spectral range of two whispering-gallery microsphere resonators of radii 40 μm and 80 μm including the effect of silica dispersion via the Sellmeier equation. The red curve shows only the geometric dispersion.

about −100 ps/(nm.km) at 800 nm (normal dispersion) to +20 ps/(nm.km) at 1550 nm (anomalous dispersion). The positive sign of the GVD cancels the geometric dispersion to some extent, rendering the FSR nearly constant over a wide frequency span. Figure 11.6 displays the FSR variation for an 80- and 160-μm diameter microsphere, considering both material and geometric dispersion. Importantly, a zero dispersion point close to the operating wavelength of 1550 nm occurs [29]. As finite element simulations performed for microtoroids [40] suggest shorter resonance wavelengths for the same mode number m, the zero dispersion point may therefore be expected to shift to shorter wavelengths in toroids as compared to spheres [41].

To measure the dispersion of toroidal microresonators, the fiber femtosecond laser frequency comb can again be employed as a frequency ruler, measuring the FSR as a function of wavelength. Such a measurement [29], performed on an 80-μm diameter toroidal resonator with a FSR of ca. 0.96 THz yielded a change of FSR from mode pair to mode pair of $\Delta\nu_{FSR} \approx 2.6$ MHz between 1577 nm and 1584 nm. In other words only a small amount of uncompensated positive dispersion was found. We anticipate that accurate, fast and reliable assessment of resonator dispersion in both theory and experiment will be a key ingredient to future research pertaining to the generation of microresonator frequency combs spanning an optical octave.

11.3 Stabilization of the Comb

The cascaded FWM process occurring in a toroidal silica microcavity has the potential for generating a broad and equally spaced optical frequency comb. However, modes of a microcavity are not a priori stable (cf. Figure 11.7), owing to size and refractive index fluctuations, which depend on temperature and intracavity power [42,43]. In order to allow its use in frequency metrology, an important prerequisite is that the generated

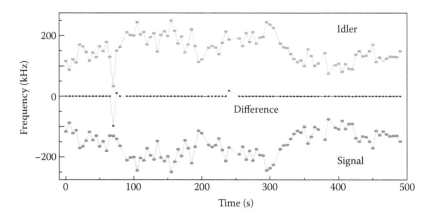

FIGURE 11.7
Temporal variation of the signal and idler sidebands and their frequency difference measured on two different Beat Detection Units (5 s gate time), while the pump frequency is locked on the fiber reference comb. The two comb lines remain equidistant, but drift in time.

comb spectrum can be controlled in terms of its two degrees of freedom (mode spacing and frequency offset). In the following sections it is shown that monolithic microresonator based frequency combs fulfill both of these requirements, allowing full control of the optical mode spectrum and generation of mode spacings in the microwave frequency range. We explain how to gain control on the frequency comb spectrum generated and describe the principle of the stabilization, its implementation and discuss the locking properties.

11.3.1 Principle

An optical comb spectrum can be entirely described by two parameters, namely the absolute frequency of one arbitrary mode and its mode spacing. In the case of a micro-cavity comb the pump frequency ν_{pump} is directly part of the generated spectrum, since the equidistance was checked when using both the pump beam or the corresponding comb line (i.e. before and after the microcavity). Therefore the generated spectrum can be described as: $\nu_m = \nu_{pump} + m f_{rep}$ $m \in \mathbb{Z}$, consequently, full stabilization of the optical frequency comb requires both control of the pump frequency and the comb's mode spacing. The latter can be achieved by identifying the pump power and frequency as two actuators.

In standard femtosecond comb generators, the control of the comb's repetition rate can be accomplished by tuning the cavity length, in general via a piezoelectric element supporting one cavity mirror. This control of the cavity length can obviously not be implemented in the present case since there is no movable part in the monolithic microcavity. It is however possible to control the cavity optical path length by exploiting the temperature dependent refractive index of glass: changing the intracavity intensity modifies the cavity temperature through photon absorption, which in turn affects the refractive index. In the following section it is described how this locking technique is implemented relying on the thermorefractive effect. Note that the different additional nonlinear mechanisms that define the actual mode spacing of the Kerr comb (e.g. self/cross phase modulation [30]) also contribute to its intensity dependence.

11.3.2 Implementation

The setup used in this stabilization experiment (cf. Figure 11.8) is similar to the one described in the previous section, except that the second beat note is now used to derive an error signal, fed back in order to control the pump intensity. The latter is achieved with a tunable gain erbium doped fiber amplifier (EDFA).

In such a configuration, the signal used to control the repetition rate is not directly proportional to the comb mode spacing since it is also affected by the pump frequency drifts. However since the latter is independently locked on an optical reference (one mode of the fully-stabilized fiber comb), one can perfectly use the second beat for testing the viability of the locking principle. In the following, it is shown that it is possible to directly measure the comb repetition rate and use it for locking.

To create an error signal suitable for locking the mode spacing, the beat generated on BDU2 (see Figure 11.8) is compared to a reference signal with a phase detector whose output serves as an error signal. It is then amplified, filtered and injected on the current input of the 980-nm diode laser pumping the EDFA, without affecting the 1550-nm pump frequency, as a direct feedback on the pump laser current would have induced. Once locked, one can tune the mode spacing by simply modifying the frequency of the reference signal.

Figure 11.8b represents the time dependence of the beat between the Kerr comb's 7th sideband with the respective reference comb line, counted with a 1-s gate time and expressed in equivalent mode spacing, by dividing the measured fluctuations by a factor of 7. During the first 65 min, the lock is disabled and one thus simply measures the free drift of the mode spacing. It exhibits fluctuations on a 5-s timescale (see inset) which is attributed to

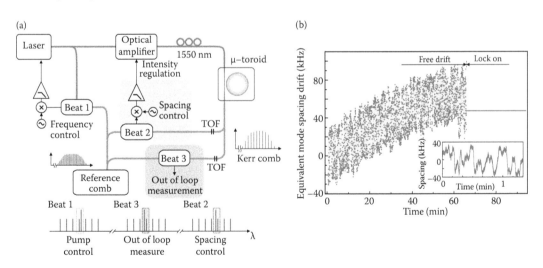

FIGURE 11.8

(a) Experimental setup used to stabilize the Kerr comb. The mode spacing is inferred from the beat of the 7th sideband with the reference comb and is controlled via an intensity regulation implemented with the optical amplifier TOF: tunable optical filter. (b) The fluctuations of the second beat note frequency, counting the beat between the 7th Kerr sideband and an adjacent reference comb mode, can be efficiently reduced by the lock, demonstrating the capability to gain control on the comb spectrum. They are expressed in equivalent mode spacing fluctuations by dividing by a factor 7, the pump frequency remaining locked to one line of the reference comb. Out-of-loop measurements on the 5th sideband confirmed the stabilization effect. (From Del'Haye, P., Arcizet, O., Schliesser, A., Holzwarth, R., and Kippenberg, T.J., Full Stabilization of a microresonator-based optical frequency comb. *Phys. Rev. Lett.*, 101, 053903, 2008.)

a change in the pump intensity and coupling strength and longer time drifts due to environment temperature evolution. After 65 min, the lock is activated, and one can observe a net suppression of the mode spacing fluctuations, without affecting the pump frequency lock. This behavior can also be observed on an independent out-of-loop signal originating from the beat of the 5th sideband of the Kerr comb with the reference comb. This shows the potential of the technique described to gain control over the entire the Kerr comb. Note that during the experiments, the microcavity remains locked to the pump frequency, in particular because of the so called thermal "self-locking mechanism" [45] which will be explained later in the chapter.

11.3.3 Characterization of the Locking Mechanism

We characterized the stability of the comb by counting simultaneously the three beats described in Figure 11.8, namely the two in-loop signals of the pump frequency and the mode spacing locks and the out-of-loop beat 3. When recording simultaneously the beats with a 1-s gate time the two in-loop signals displayed fluctuations with a root-mean-square deviation of the order of 1 mHz, which demonstrates the efficiency of the actuators. The out-of-loop measurement of the 5th sideband presents higher noise, at the 1-Hz level, nevertheless clearly demonstrating that the two locks used in this experiment allow to gain control over the full comb spectrum. This excess of noise has different origins, one of which being only coarsely adjusted feedback parameters.

   Subsequently, the tunability properties of the stabilized Kerr comb were studied. In particular the dependence of the mode spacing on the pump intensity was tested. To this end, the pump laser is brought close to one of the microcavity resonances. The high optical power used in the experiment induces significant thermal bistability. When the pump frequency approaches the microcavity resonance from the blue side, the intensity circulating inside the toroid increases and the induced heating expands the cavity effective length which makes the cavity resonance being pushed away from the laser frequency. This mechanism is responsible for a so-called "thermal self-locking" [45] of the microcavity onto the laser frequency and can be exploited for keeping the microcavity quasi resonant with the pump frequency over several tens of GHz, the effective detuning remaining of the order of the cold cavity linewidth when no other nonlinear mechanism is involved. This broad tunability of the pump frequency enables the locking of the pump laser on one of the reference comb modes (spaced by 100 MHz), without losing the microcavity optical resonance.

   Once the comb is generated, the aforementioned thermal self locking also allows variation of the the pump power without losing the microcavity resonance nor terminating comb generation. We subsequently measured the influence of the pump power on the comb repetition rate. The result is reproduced in Figure 11.9. It was possible to scan the repetition rate over a 5 MHz range by changing the pump power P_{pump} by ca. 3 mW around a mean value of approximately 80 mW. Due to the approximately linear response, the comb mode spacing can be described by the equation: $f_{rep} = f_{rep}^0 + \gamma P_{pump}$ with an experimentally determined $\gamma = -196$ kHz/mW, f_{rep}^0 being the extrapolated cold cavity repetition rate. In general, the linear parameter γ depends on the pump frequency since the pump laser is slightly blue detuned from the cavity resonance, and the way the pump power changes affect the intracavity intensity obviously depends on the laser to cavity detuning. It would thus be more appropriate to write $\gamma(\nu_{pump})$. However in practice the effective detuning of the laser with respect to the cavity resonance is small, such that a change in the pump laser frequency does not affect the intracavity intensity significantly. Indeed it has been

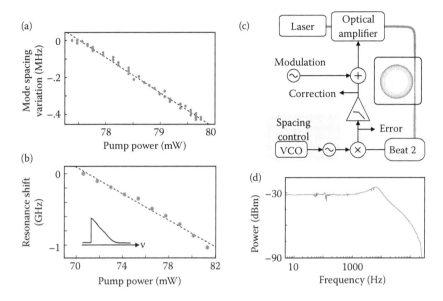

FIGURE 11.9
(a) Static mode spacing dependence on the pump power. (b) Frequency of the thermally induced bistable turning point for different injected intensities, when scanning the laser frequency sufficiently slowly for letting it thermalize. (c) Setup used for characterizing the lock properties: an external perturbation can be applied in order to measure how the system reacts. (d) Frequency response of the correction signal to an external perturbation: the actuation bandwidth is of the order of 10 kHz. Similar characterizations have been performed on microwave repetition rate comb generators (From Del'Haye, P., Arcizet, O., Schliesser, A., Holzwarth, R., and Kippenberg, T.J., Full Stabilization of a microresonator-based optical frequency comb. *Phys. Rev. Lett.*, 101, 053903, 2008.)

experimentally measured that a 1 GHz change in the pump frequency linearly affects the mode spacing by only 1 MHz.

Next, the origin of the intensity dependence of the mode spacing was investigated. From the microtoroid's high quality factors and small mode volumes, several nonlinear mechanisms can contribute to this effect by changing its geometric and refractive properties. It has been shown that the cavity static and low frequency response (below 10 kHz) to an intensity change is governed by the absorption-induced heating and concomitant refractive index change and thermal expansion, which are dominant compared to the Kerr or radiation pressure contributions [46]. The corresponding variation of the microcavity's repetition rate due to a temperature change δT can be written: $\delta f_{rep}/f_{rep} = (\alpha_l + \alpha_n)\delta T$. The comparison of $\alpha_n = 1.0 \times 10^{-5}\,\mathrm{K}^{-1}$, the thermorefractive coefficient of silica, to $\alpha_l = 5.5 \times 10^{-7}\,\mathrm{K}^{-1}$ its thermal expansion coefficient, allows to conclude that the main contribution arises from the temperature dependence of the refractive index. As a consequence, a 0.1 K change in the toroid temperature induces a 400 kHz change of the repetition rate.

The same nonlinear mechanism is also responsible for the optical bistability mentioned above, observed when scanning the microcavity resonance with the laser. The frequency of the bistability turning point depends on the injected power, its dependence on the pump power is represented in Figure 11.9b. It exhibits a characteristic linear behavior with a slope of $\Gamma = -93\,\mathrm{MHz/mW}$. This measurement was taken by scanning the laser frequency sufficiently slowly such that thermal equilibrium of the toroidal cavity was reached for each detuning value. It is possible to estimate the toroid thermalization time with the characteristic diffusion time of a thermal wave from the center of the optical mode to the toroid rims[†] (approximately $\delta = 3$ μm away), as given by: $\tau = \delta^2/D \approx 10\,\mu s$ where $D = 9.5 \times 10^{-7}$ m²/s is the heat diffusion coefficient of silica.

The change in the toroid temperature modifies the effective refractive index n_{eff} but differently affects the individual cavity resonances: $\delta v_m \propto -m\delta n_{\text{eff}}$ for large m. Therefore the thermally induced nonlinearity will affect the mode spacing by a quantity $\delta v_{\text{FSR}} \propto \delta n_{\text{eff}}$. The bistable turning point induced by the thermal displacement δv_M of the pumped cavity mode ($m = M$) is then expected to show a pump power dependence a factor M higher than the mode spacing static dependence δv_{FSR} if the same mechanism is responsible for the two effects. In practice, we pump the $M \approx 500$th cavity mode ($v_{\text{pump}} \approx 200$ THz and FSR ≈ 400 GHz), and measured a ratio $\Gamma/\gamma \approx 474$. The good agreement between these numbers indicates that one can mainly explain the control gained over the comb mode spacing by the contribution of the thermal dependence of the refractive index. However several of the previously mentioned nonlinear mechanisms, such as self/cross phase modulation among the different comb lines (which only affect the oscillating modes) also contribute to the observed intensity dependence and are under current investigation. As mentioned before, the input power dependence of the mode spacing γ also depends on the pump frequency detuning, and the previous treatment is valid only when the pump frequency is sufficiently close to the cavity resonance. As explained before, the comb mode spacing presents a strong dependence on the pump power and one can even generate with the same microcavity comb spectra with higher-order mode spacings, corresponding to multiples of the cavity FSR (cf. Figure 11.10). As a conclusion, even if the exploited mechanism still has to be completely clarified, it is in practice sufficiently nonparallel to the pump frequency actuator to allow a full stabilization of the Kerr comb.

11.3.4 Actuation Properties

An important figure of merit for a feedback loop is its actuation bandwidth. The feedback mechanism we used relies on thermal effects which are very fast, of the order of 10 μs, due to the toroid micron-scale dimensions. To characterize the dynamical properties of the locking mechanism, the reference signal used to phase lock the second beat signal to, was frequency modulated (it is also possible to add an in-loop perturbation, see Figure 11.9c). When changing the reference signal frequency, the servo regulates the power sent into the microcavity to adjust the repetition rate of the Kerr comb. It was then possible to dynamically tune the repetition rate over a MHz range while remaining locked. For a higher tuning range, it is also possible to change the toroid temperature, the measured -2 GHz/K shift of the optical resonance at room temperature corresponds to a -4 MHz/K change for the mode spacing. Such a performance is on the same order of magnitude as the actuation range of the conventional fiber frequency comb.

In order to test the actuation bandwidth, the reference signal was dithered with a signal of constant amplitude, swept in frequency. By simultaneously measuring the correction signal, the frequency range over which the implemented lock is efficient was determined. The results measured with a network analyzer are reported in Figure 11.9d. The correction signal displays a flat response up to ca. 10 kHz, implying that the lock is sufficiently robust for compensating the injected perturbation, while at higher frequencies the lock's ability to compensate the external perturbation diminishes. The measured cutoff is of the same order of magnitude as the characteristic thermal geometric cutoff frequency given by

---

[†] Note however that the specific geometry of the toroid induces a more complex thermal response than this simple estimation.

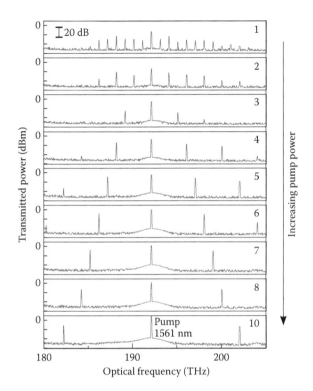

FIGURE 11.10
Spectra obtained with the same microcavity, for different input powers ranging from 400 μW to 70 mW. For increasing pump power it is possible to observe higher order comb repetition rate, corresponding to multiples (up to a factor of 10) of the cavity FSR.

$\Omega_c = D/\delta^2 \approx 2\pi \times 16\,\text{kHz}$, and is also comparable to the efficiency range of the locking electronics that we used. Thus the above measurements have revealed that the locking technique has the potential to reach high frequencies of actuation (up to 10 kHz), which is on par with the actuation range of standard comb generators based on piezo-mounted mirrors.

11.4 Generation of a Stabilized Microwave Repetition Rate Frequency Comb

An important prerequisite for an optical frequency comb to be suitable for metrology is their ability to establish a link from the optical to the RF domain. The latter necessitates repetition rates in the microwave or RF range, amenable to direct electronic detection. In this section we present results on microwave repetition rates obtained with monolithic comb generators which employ larger, mm-scale microcavities. Note that the achieved microwave repetition rates have so far not been reached by femtosecond laser based combs [47], and can prove useful in a variety of applications, such as astronomical spectrometer calibration, and systems where the individual addressing of the comb lines plays an important role, like pulse shaping.

11.4.1 Monolithic Frequency Comb Generators with Microwave Repetition Rate

Aiming at producing microcavities having a repetition rate below 100 GHz, at frequencies which can be directly measured using fast telecom photodiodes, requires a cavity diameter of more than 660 μm. Taking into account the diameter reduction occurring during the reflow process, where the CO_2 laser evaporates part of the silica [26], the usual micro-fabrication process was followed for creating 1-mm diameter disks. The reflow procedure had to be adapted, since the laser intensity was not sufficient when trying a direct and single shot melting of the disk because of the necessary defocusing needed to cover the full disk surface with the laser. We opted for a different procedure where the CO_2 laser was strongly focused and manually scanned around the periphery of the disk. This procedure allowed fabrication of millimetric sized toroids, shown in Figure 11.11, at the expense of a stronger diameter reduction since the laser illumination was increased.

The lower part of Figure 11.11 represents the transmission of the optical fiber when the laser frequency is scanned over several free spectral ranges of the microcavity. Several classes of modes are evident, corresponding to the higher order transverse modes. The mode spacing of the 750-μm diameter cavity corresponds to 86 GHz. The Q-factor for these samples (which was not optimized yet) reached values of 2×10^7. While this value is more

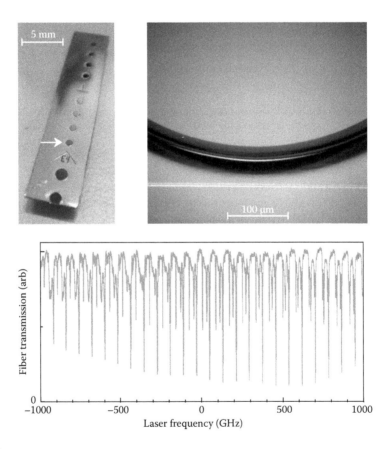

FIGURE 11.11
Upper left: Photograph of a chip supporting several millimeter sized optical resonators; Upper right: Optical microscope view of a 750-μm diameter toroid and the tapered fiber. Lower panel: Transmission spectrum of 660-μm diameter toroidal microcavity coupled to a tapered optical fiber, exhibiting a mode spacing of 86 GHz.

than 10 times lower than what has been observed in smaller toroidal microcavities (and >100 times lower than in crystalline microresonators), it was sufficient to observe parametric oscillations at a threshold value of ca. 10 mW.

Using a 750-μm diameter sample it was possible to generate a frequency comb with a 86 GHz repetition rate, by pumping the toroid at 1550 nm with a 150-mW continuous-wave laser. The typical optical spectrum measured after the cavity is represented in Figure 11.12a. The comb spans over 60 nm which is lower than what was demonstrated with smaller cavities [29]. This is a consequence of the increased dimensions of the toroid, which imply a larger minimum threshold for parametric oscillations [30]:

$$P_{\text{thres}}^{\min} = Q_0^{-2} \frac{(1+K)^3}{K} \frac{\pi^2 n_{\text{eff}}^2}{2} \frac{RA_{\text{eff}}}{\lambda_0 n_2}, \tag{11.4}$$

The expression presents a linear dependence in the effective mode volume $2\pi RA_{\text{eff}}$ ($K = Q_0/Q_{\text{ex}}$ is the coupling parameter, Q_0 and Q_{ex} the intrinsic and external (related to the coupling) optical quality factor, and n_2 the intensity dependent refractive index). For the 750-μm cavities, the diameter is responsible for a ten-fold increase in threshold, and the effective mode area is also expected to increase from the higher inner toroid diameter. Those effects directly increase the threshold and similarly affect the strength of the induced nonlinearities. Note that there is no fundamental limit which prevents higher Q factors. Indeed, $Q > 10^9$ have been measured in millimeter sized microspheres [48].

11.4.2 Stabilization and Characterization of a Microwave Frequency Comb

For the 750-μm diameter samples, we verified that the comb emission allowed direct measurement of the repetition rate. To this end, telecom detectors from u²t Photonics were employed in conjunction with a harmonic mixer fed with a ca. 14.3 GHz signal

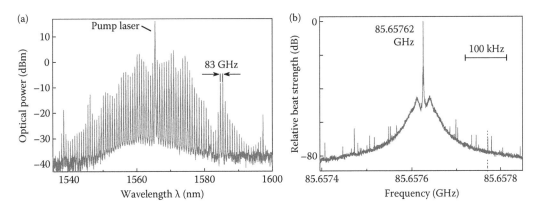

FIGURE 11.12

(a) Typical optical spectrum measured after the millitoroid. The comb exhibits a 83 GHz comb spacing whose intensity beat can be directly measured using a fast photodiode. (b) Typical locked microwave beat (measured by downconverting the signal to 30 MHz using a harmonic mixer). The width of the coherent spike is determined by the resolution bandwidth of the measurement (10 Hz). (From Del'Haye, P., Arcizet, O., Schliesser, A., Holzwarth, R., and Kippenberg, T. J., Full Stabilization of a microresonator-based optical frequency comb. *Phys. Rev. Lett.*, 101, 053903, 2008.

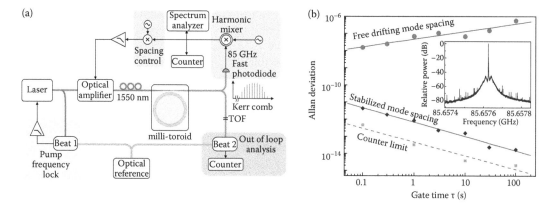

FIGURE 11.13
(a) Setup used for characterizing and stabilizing the microwave frequency comb. The use of a fast photodiode combined with a harmonic mixer allows to directly measured the 86 GHz beat corresponding to the mode spacing of the comb generated from our millimetric sized cavity. TOF: tunable optical filter. (b) Measured Allan deviation of the comb repetition rate as a function of gate time for the free running comb and once locked. (From Del'Haye, P., Arcizet, O., Schliesser, A., Holzwarth, R., and Kippenberg, T. J., Full Stabilization of a microresonator-based optical frequency comb. *Phys. Rev. Lett.*, 101, 053903, 2008.)

whose 6th harmonics falls in the vicinity of the 86 GHz signal to be measured in order to down-convert the signal to countable frequencies, of the order of 30 MHz.

In order to use the beat signal for locking the comb, it was phase compared with a reference signal and the error signal provided was similarly used as feedback to the optical amplifier pump current. The setup is described in Figure 11.13. The technique allowed direct locking of the repetition rate of the comb without relying on the fiber reference comb. The typical locked 86 GHz signal is represented in Figure 11.12b, revealing the coherent spike and the ca. 10 kHz wide noise reduction accomplished by the feedback loop. This noise pedestal reflects the cavity optical path fluctuations above the frequency range where the lock is compensating. Many mechanisms contribute to this noise [42,49], in particular the laser's classical frequency noise, which is not distinguishable in this measurement, the laser's shot noise, the thermally and optically induced noise processes that affect the microcavity's optical path length through its refractive index, expansion and higher frequency mechanical vibrations. Note that even if the opto-mechanical properties of the millitoroid have been intentionally deteriorated, by maintaining a very small undercut and leaving it in air for providing viscous damping that decreased its mechanical Q factor, the optomechanical instability [50,51] i.e. radiation pressure driven regenerative mechanical oscillations, is however very easily observed when the laser detuning is significant, and in general prevents from generating a stable optical comb (for a laser strongly blue detuned). In addition, to limit this optomechanical instability, the fiber was in general intentionally placed in contact to the microcavity for suppressing these radiation-pressure-induced mechanical oscillations.

The long term stability of the microwave beat was investigated using an RF counter, the Allan deviation measurements for the free running beat and the locked signal are shown in Figure 11.13b. The influence of the lock is observable, dramatically increasing the long term stability of the beat and yielding a repetition rate stability at the level of 10 mHz in 1 s, corresponding to an Allan deviation of 10^{-12} for the in-loop measurement. This number reflects the large stability achieved on the optical path length of the toroid. The out-of-loop measurements implemented on the first idler sideband yielded rms fluctuations $\Delta\nu_1$ below 1 Hz for a 1 hour measurement with a 1-s gate time [44]. The measured fluctuation

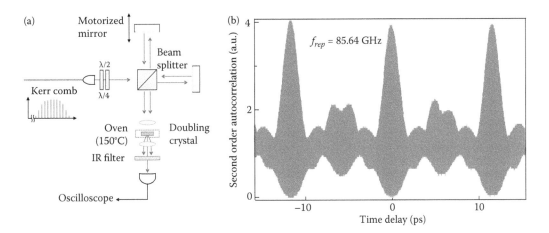

FIGURE 11.14
(a) Autocorrelator based on a periodically poled LiNbO$_3$ non linear crystal. (b) Second order crosscorrelation trace obtained with a 750 µm diameter microcavity.

expressed in an equivalent fluctuation of cavity optical path length, gives a relative stability of $\Delta L_{opt} = 2\pi R\Delta v_1/v_1 \approx 10$ am in 1 s. The demonstrated capacity to actively control the resonator effective optical length, combined with their intrinsic stability, could allow microcavities to become a future promising tool for time transfer. [43].

11.5 Conclusion

In summary, this chapter has introduced a radically different frequency comb generator approach, unique in several ways. First, the approach offers unprecedented form factor, allowing a monolithic approach to optical frequency comb generation. Due to the high Q-factor in conjunction with tapered optical fiber coupling, the approach is moreover highly efficient, with conversion efficiencies frequently observed above 80%. Second, the approach does not rely on the use of atomic or molecular resonances, and therefore can achieve broadband gain in a frequency window only limited by material absorption or dispersion. Owing to the large transparency window of glass, the frequency window extends from the UV to 2 µm. Third, the use of monolithic microresonators allows to significantly augment the repetition rate. While the highest repetition rates of conventional femtosecond lasers so far are approaching 10 GHz, the equivalent cavity length of 3 cm is difficult to reduce further. Moreover, high repetition rate combs have so far only been achieved in Ti:Sa laser systems, whereas for 1.5 µm fiber based frequency comb repetition rates have been generally much lower. The present approach can operate at previously unattainable repetition rates exceeding 10 GHz, and with a high power per individual comb mode, which is useful in applications where the access to individual comb modes is required, such as optical waveform synthesis, high capacity telecommunications, spectrometer calibration [52] or direct comb spectroscopy. Finally, several new prospects exist that bode well for future work and for the future of this technology.

Since this work, the generation of frequency combs in this manner has also been reported in CaF$_2$ resonators [49]. This work allowed reducing the repetition rate to

10 GHz and moreover showed the promise of this technology to serve as low phase noise microwave oscillator. These developments show that microresonator-based frequency comb generation bear promise for the future, possibly providing a route to all chip scale atomic clocks.

Acknowledgments

We thank T. W. Hänsch for discussions and suggestions. T. J. K. acknowledges support via an Independent Max Planck Junior Research Group. This work was funded as part of a Marie Curie Excellence Grant (RG-UHQ) and the DFG-funded Nanosystems Initiative Munich (NIM). We thank J. Kotthaus for access to clean room facilities for sample fabrication.

References

1. T. W. Hänsch. Passion for precision (Nobel lecture). *ChemPhysChem*, 7: 1170–1187, 2006.
2. S. T. Cundiff and J. Ye. Colloquium: Femtosecond optical frequency combs. *Rev. Mod. Phys.*, 75: 325–342, 2003.
3. T. Udem, R. Holzwarth, and T. W. Hänsch. Optical frequency metrology. *Nature*, 416: 233–237, 2002.
4. R. J. Jones, K. D. Moll, M. J. Thorpe, and J. Ye. Phase-coherent frequency combs in the vacuum ultraviolet via high-harmonic generation inside a femtosecond enhancement cavity. *Phys. Rev. Lett.*, 94: 193201, 2005.
5. C. Gohle, T. Udem, M. Herrmann, J. Rauschenberger, R. Holzwarth, H. A. Schuessler, F. Krausz, and T. W. Hänsch. A frequency comb in the extreme ultraviolet. *Nature*, 436: 234–237, 2005.
6. U. Keller. Recent developments in compact ultrafast lasers. *Nature*, 424:831–838, 2003.
7. J. N. Eckstein, A. I. Ferguson, and T. W. Hänsch. High-resolution two-photon spectroscopy with picosecond light pulses. *Phys. Rev. Lett.*, 40: 847–850, 1978.
8. Th. Udem, J. Reichert, R. Holzwarth, and T. W. Hänsch. Accurate measurement of large optical frequency differences with a mode-locked laser. *Optics Lett.*, 24: 881–883, 1999.
9. Th. Udem, J. Reichert, R. Holzwarth, and T. W. Hänsch. Absolute optical frequency measurement of the cesium D1 line with a mode-locked laser. *Phys. Rev. Lett.*, 82: 3568–3571, 1999.
10. D. J. Jones, S. A. Diddams, J. K. Ranka, A. Stentz, R. S. Windeler, J. L. Hall, and S. T. Cundiff. Carrier-envelope phase control of femtosecond mode-locked lasers and direct optical frequency synthesis. *Science*, 288: 635–639, 2000.
11. R. Holzwarth, T. Udem, T. W. Hänsch, J. C. Knight, W. J. Wadsworth, and P. S. J. Russell. Optical frequency synthesizer for precision spectroscopy. *Phys. Rev. Lett.*, 85: 2264–2267, 2000.
12. J. Reichert, M. Niering, R. Holzwarth, M. Weitz, Th. Udem, and T. W. Hänsch. Phase coherent vacuum-ultraviolet to radio frequency comparison with a mode-locked laser. *Phys. Rev. Lett.*, 84: 3232–3235, 2000.
13. S. A. Diddams, D. J. Jones, J. Ye, S. T. Cundiff, J. L. Hall, J. K. Ranka, R. S. Windeler, R. Holzwarth, T. Udem, and T. W. Hänsch. Direct link between microwave and optical frequencies with a 300 THz femtosecond laser comb. *Phys. Rev. Lett.*, 84: 5102–5105, 2000.

14. F. Keilmann, C. Gohle, and R. Holzwarth. Time-domain mid-infrared frequency-comb spectrometer. *Optics Letters*, 29: 1542–1544, 2004.
15. A. Schliesser, M. Brehm, F. Keilmann, and D. W. van der Weide. Frequency-comb infrared spectrometer for rapid, remote chemical sensing. *Optics Express*, 13: 9029–9038, 2005.
16. S. A. Diddams, L. Hollberg, and V. Mbele. Molecular fingerprinting with the resolved modes of a femtosecond laser frequency comb. *Nature*, 445: 627–630, 2007.
17. M. J. Thorpe, K. D. Moll, J. J. Jones, B. Safdi, and J. Ye. Broadband cavity ringdown spectroscopy for sensitive and rapid molecular detection. *Science*, 311: 1595–1599, 2006.
18. M. Kourogi, K. Nakagawa, and M. Ohtsu. Wide-span optical frequency comb generator for accurate optical frequency difference measurement. *IEEE J. Quantum Electron.*, 29: 2693–2701, 1993.
19. S. A. Diddams, L. S. Ma, J. Ye, and J. L. Hall. Broadband optical frequency comb generation with a phase-modulated parametric oscillator. *Optics Lett.*, 24: 1747–1749, 1999.
20. J. Ye, L. S. Ma, T. Day, and J. L. Hall. Highly selective terahertz optical frequency comb generator. *Optics Letters*, 22: 746–746, 1997.
21. H. Schnatz, B. Lipphardt, J. Helmcke, F. Riehle, and G. Zinner. First phase-coherent frequency measurement of visible radiation. *Phys. Rev. Lett.*, 76: 18–21, 1996.
22. T. Udem, A. Huber, B. Gross, J. Reichert, M. Prevedelli, M. Weitz, and T. W. Hänsch. Phase-coherent measurement of the hydrogen 1S-2S transition frequency with an optical frequency interval divider chain. *Phys. Rev. Lett.*, 79:2646–2649, 1997.
23. S. A. Diddams, T. Udem, J. C. Bergquist, E. A. Curtis, R. E. Drullinger, L. Hollberg, W. M. Itano, W. D. Lee, C. W. Oates, K. R. Vogel, and D. J. Wineland. An optical clock based on a single trapped ^{199}Hg$^+$ ion. *Science*, 293: 825–828, 2001.
24. M. Takamoto, F. Hong, R. Higashi, and H. Katori. An optical lattice clock. *Nature*, 435: 321–324, 2005.
25. D. K. Armani, T. J. Kippenberg, S. M. Spillane, and K. J. Vahala. Ultra-high-Q toroid microcavity on a chip. *Nature*, 421: 925–928, 2003.
26. K. J. Vahala. Optical microcavities. *Nature*, 424: 839–846, 2003.
27. M. Cai, O. Painter, and K. J. Vahala. Observation of critical coupling in a fiber taper to a silica-microsphere whispering-gallery mode system. *Phys. Rev. Lett.*, 85: 74–77, 2000.
28. P. Del Haye, A. Schliesser, O. Arcizet, T. Wilken, R. Holzwarth, and T. J. Kippenberg. Optical frequency comb generation from a monolithic microresonator. *Nature*, 450: 1214, 2007.
29. T. J. Kippenberg, S. M. Spillane, and K. J. Vahala. Kerr-nonlinearity optical parametric oscillation in an ultrahigh-Q toroid microcavity. *Phys. Rev. Lett.*, 93: 083904, 2004.
30. A. A. Savchenkov, A. B. Matsko, D. Strekalov, M. Mohageg, V. S. Ilchenko, and L. Maleki. Low threshold optical oscillations in a whispering gallery mode CaF$_2$ resonator. *Phys. Rev. Lett.*, 93: 243905, 2004.
31. A. A. Savchenkov, V. S. Ilchenko, A. B. Matsko, and L. Maleki. Kilohertz optical resonances in dielectric crystal cavities. *Physical Review A*, 70:051804, 2004.
32. S. M. Spillane, T. J. Kippenberg, O. J. Painter, and K. J. Vahala. Ideality in a fiber-taper-coupled microresonator system for application to cavity quantum electrodynamics. *Phys. Rev. Lett.*, 91: 043902, 2003.
33. J. Ye and S. T. Cundiff. *Femtosecond Optical Frequency Comb: Principle, Operation and Applications.* Springer, New York, 2005.
34. P. Kubina, P. Adel, F. Adler, G. Grosche, T. W. Hänsch, R. Holzwarth, A. Leitenstorfer, B. Lipphardt, and H. Schnatz. Long term comparison of two fiber based frequency comb systems. *Optics Express*, 13: 904–909, 2005.
35. J. Reichert, R. Holzwarth, Th. Udem, and T.W. Hänsch. Measuring the frequency of light with mode-locked lasers. *Optics Commun.*, 172: 59–68, 1999.
36. A. B. Matsko, A. A. Savchenkov, D. Strekalov, V. S. Ilchenko, and L. Maleki. Optical hyperparametric oscillations in a whispering-gallery-mode resonator: Threshold and phase diffusion. *Physical Rev., A*, 71: 033804, 2005.

37. B. Min, L. Yang, and K. J. Vahala. Perturbative analytic theory of an ultrahigh-Q toroidal microcavity. *Physical Rev., A*, 76: 013823, 2007.

38. S. Schiller. Asymptotic-expansion of morphological resonance frequencies in Mie scattering. *Appl. Optics*, 32(12): 2181–2185, 1993.

39. Tobias Kippenberg. *Nonlinear Optics in Ultra-high-Q Whispering-Gallery Optical Microcavities.* PhD thesis, California Institute of Technology, Pasadena, 2004.

40. I. H. Agha, Y. Okawachi, M. A. Foster, J. E. Sharping, and Gaeta A. L. Four-wave-mixing parametric oscillations in dispersion-compensated high-Q silica microspheres. *Physical Rev., A*, 76: 043837, 2007.

41. A. B. Matsko, A. A Savchenkov, N. Yu, and L. Maleki. Whispering-gallery-mode resonators as frequency references. I. Fundamental limitations. *J. Opt. Soc. Am. B*, 24: 1324–1335, 2007.

42. A. A. Savchenkov, A. B. Matsko, V. S. Ilchenko, N. Yu, and L. Maleki. Whispering-gallery-mode resonators as frequency references. II. Stabilization. *J. Opt. Soc. Am. B*, 12: 2988–2997, 2007.

43. P. Del'Haye, O. Arcizet, A. Schliesser, R. Holzwarth, and T. J. Kippenberg. Full stabilization of a microresonator-based optical frequency comb. *Phys. Rev. Lett.*, 101, 053903, 2008.

44. T. Carmon, L. Yang, and K. J. Vahala. Dynamical thermal behavior and thermal self-stability of microcavities. *Optics Express*, 12: 4742–4750, 2004.

45. A. Schliesser, P. Del'Haye, N. Nooshi, K. J. Vahala, and T. J. Kippenberg. Radiation pressure cooling of a micromechanical oscillator using dynamical backaction. *Phys. Rev. Lett.*, 97: 243905, 2006.

46. A. Bartels, C. W. Oates, L. Hollberg, and S. A. Diddams. Stabilization of femtosecond laser frequency combs with subhertz residual linewidths. *Optics Lett.*, 29: 1081, 2004.

47. D.W. Vernooy, V.S. Ilchenko, H. Mabuchi, E.W. Streed, and H.J. Kimble. High-Q measurements of fused-silica microspheres in the near infrared. *Optics Lett.*, 23: 247–249, 1998.

48. A. A. Savchenkov, A. B. Matsko, V. S. Ilchenko, I. Solomatine, D. Seidel, and Lute Maleki. Tunable optical frequency comb with a crystalline whispering gallery mode resonator. *Phys. Rev. Lett.* 101: 093902, 2008.

49. T. J. Kippenberg, H. Rokhsari, T. Carmon, A. Scherer, and K. J. Vahala. Analysis of radiation-pressure induced mechanical oscillation of an optical microcavity. *Phys. Rev. Lett.*, 95: 033901, 2005.

50. O. Arcizet, P. F. Cohadon, T. Briant, M. Pinard, and A. Heidmann. Radiation-pressure cooling and optomechanical instability of a micromirror. *Nature*, 444: 71–74, 2006.

51. M.T. Murphy, T. Udem, R. Holzwarth, A. Sizmann, L. Pasquini, C. Araujo-Hauck, H. Dekker, S. Odorico, M. Fischer, T. W. Hänsch, and A. Manescau. High-precision wavelength calibration with laser frequency combs. *Mon. Not. Roy. Astron. Soc.*, 380: 839–847, 2007.

12

Bit Rate Limitations in Single and Coupled Microresonators

Jacob Khurgin

Johns Hopkins University

CONTENTS

12.1 Introduction

Recent years have seen spectacular progress in development of linear and nonlinear optical devices based on microresonators. As one can observe from reading other chapters in this volume, high Q microresonators made from various materials have been successfully fabricated and used for such diverse applications as delay lines [1], microwave photonics [2,3], sensors [4,5] and others. Especially impressive have been recent advances in active microresonators made from electro-optic or nonlinear materials. Active microresonators are expected to be employed as miniature optical modulators [6], switches [7], tunable filters [8] and delay lines [9], that are amenable to large scale integration culminating one day in development of photonic integrated circuits.

The main advantage of optical methods of information transmission is, of course, their large capacities that can reach tens of terabits per second (TBps) in a single optical fiber. Processing of information at such speeds electronically is difficult to fathom not only with today's semiconductor-based electronics, but even in the foreseeable future. This discrepancy between optical transmission and electronic processing results in traffic jams on the information superhighway occurring wherever the information carried by photons needs to be switched and routed electronically. Therefore, optical methods of processing information, at least at some rudimentary level, are desperately needed to relieve this congestion. So far, unfortunately, optical devices performance has lagged behind that of their electronic counterparts. This disadvantage is rooted in the most basic features distinguishing photons from electrons: photons move with high speed, carry no charge, and are not subject to the Pauli principle, which makes all interactions involving photons relatively weak compared to electronic interactions. Thus photons must propagate over long distances, measured in millimeters, to carry out the same tasks that can be accomplished electronically over a few micrometers. Clearly, until effective means of confining photons into tighter spaces are found and perfected, photonic integration is simply not going to happen. Microresonator (MR) technology is ideally suited for effective photon confinement is small volumes and that is why development of microresonators is pursued with such dynamism.

As with any promising technology, MR technology is not without limitations. As the term "microresonator" itself suggests, the ability of photon confinement rests upon the existence of a strong resonance in its characteristics, with all the desirable properties confined to a relatively narrow band near this resonance. This "gain-bandwidth product" limitation affects the performance of the MR-based devices. Understanding of bandwidth limitations of MRs is important because it allows one to decide on applicability of MR technology for a given application with subsequent optimization of the MR design for a particular task. In this chapter we consider the bandwidth limitations of MRs in potential applications as optical delays and optical switches. We compare single MRs with multiple coupled ones and show that, while multiple MRs are the best choice for delay lines, in the optical switches the choice should depend on both bit rate and the strength of index modulation. Then we study delay lines based on multiple coupled MRs and show that their performance at large bandwidths is equally affected by two factors—dispersion of loss and dispersion of group delay. Finally we describe some ideas for the dispersion mitigation.

12.2 Single and Coupled Microresonators as Optical Delay Elements

To begin, we consider a simple MR and estimate its performance as an optical delay element and a modulation or switching device. We shall use a ring resonator as an example but one can consider many alternative schemes, such as Fabry–Perot cavities, photonic crystal resonators, whispering gallery mode micros-disk resonators etc. A ring resonator can be used in two configurations—as an all-pass filter (APF) [8] when coupled to a single optical input–output bus (Figure 12.1a) or as an add-drop filter (ADF) [10,11] (Figure 12.1b), with two coupled waveguides. The resonator is characterized by a circumference L, effective index $\bar{n}$, and a coupling coefficient κ. The transmission function of the APF is

$$\frac{E_{\text{out}}(\omega)}{E_{\text{in}}(\omega)} = \frac{\rho - e^{j\omega\tau}}{1 - \rho e^{j\omega\tau}} \equiv e^{j\Phi(\omega)} \tag{12.1}$$

where $\rho^2 = 1 - \kappa^2$ and $\tau = L\bar{n}/c$ is a ring round trip time. As the name "all-pass" implies, the amplitude transmission is unity for all frequencies, and only the phase is affected as

$$\tan\Phi(\omega) = \frac{\kappa^2 \sin(\omega - \omega_0)\tau}{(1+\rho^2)\cos(\omega - \omega_0)\tau - 2\rho} \tag{12.2}$$

where $\omega_0 = 2\pi m\tau^{-1}$ is a resonance frequency. The group delay of the APF is

$$T_d(\omega) = \frac{\partial\Phi(\omega)}{\partial\omega} = T_d^{(0)} + T_d^{(2)}(\omega - \omega_0)^2 + \dots \tag{12.3}$$

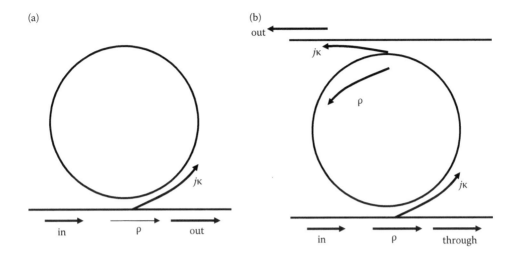

FIGURE 12.1
(a) An all-pass filter (APF) based on a single ring microresonator. (b) An add-drop filter (ADF) based on a single ring microresonator.

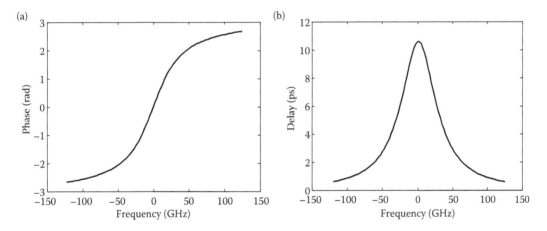

FIGURE 12.2
(a) Phase characteristics of the APF of Figure 12.1a. (b) Group delay in the APF of Figure 12.1a

where the group delay at the resonant frequency is

$$T_d^{(0)} = \tau \frac{1+\rho}{1-\rho} \tag{12.4}$$

and the group delay dispersion (GDD) is

$$T_d^{(2)} = -\tau^3 \frac{(1+\rho)\rho}{(1-\rho)^3} = -\left[T_d^{(0)}\right]^3 \frac{\rho}{(1+\rho)^2} \tag{12.5}$$

In Figure 12.2a and b the phase and group delay characteristics of an APF with $L = 36\mu m$, $\bar{n} = 2.1$ ($\tau = 250$ fs) and $\kappa = 0.3$ [1] are shown. As one can see, large group delays (exceeding τ by a factor of 40) can be achieved in a relatively narrow bandwidth near ν_0.

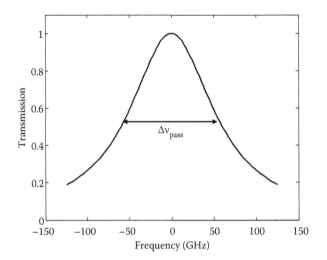

FIGURE 12.3
Transmission characteristics of the ADF of Figure 12.1b.

The transmission coefficient of the ADF of Figure 12.1b

$$\frac{E_{\text{out}}(\omega)}{E_{\text{in}}(\omega)} = \frac{\kappa^2}{1-\rho^2 e^{j\omega\tau}} \tag{12.6}$$

has a Fabry–Perot like spectrum of intensity transmission shown in Figure 12.3 and can be characterized by a full width at half maximum (FWHM)

$$\Delta\omega_{\text{pass}} = 2\kappa^2\tau^{-1} \tag{12.7}$$

The delay time of the signal, i.e. the time spent inside the resonator is

$$T_d = \frac{\partial\Phi}{\partial\omega} = \tau\left(\kappa^{-2} - \frac{1}{2}\right) \tag{12.8}$$

and we obtain a simple relationship between delay and width of the pass-band

$$T_d\Delta\omega_{\text{pass}} = 2 - \kappa^2 \tag{12.9}$$

Irrespective of the application or MR, the longer the light spends inside the MR, the stronger is the effect one can impose on the light. Therefore, Equation 12.9 would lead to a gain-bandwidth product for any application. In this chapter we shall limit ourselves to digital signals; thus we shall consider bit rate limitations, but the implications for bandwidth limitations of analog signals are obvious.

We consider an on-off-keyed signal with a Gaussian profile whose FWHM is equal to one half of the bit interval B. The temporal profile of such a signal is $P(t) = \exp(-16\ln(2)B^2t^2)$, while the power spectrum of signal is also Gaussian with FWHM

$$\Delta\omega_{\text{sig}} = 8\ln(2)B \tag{12.10}$$

Since one needs to maintain $\Delta\omega_{sig} \leq \Delta\omega_{pass}$ then, according to Equation 12.9, the delay-bit rate product, or the number of stored bits in a single resonator is

$$N = T_d B \leq \frac{2-\kappa^2}{8\ln(2)} \approx 0.36 \qquad (12.11)$$

Although the APF is transparent at any frequency, one can still talk about the useful bandwidth of the APF defining is as the bandwidth at which the group delay changes by a factor of two

$$T_d\Delta\omega_{pass} = \sqrt{\frac{2}{\rho}}(1+\rho) \qquad (12.12)$$

For any resonator with a reasonably high Q, ρ approaches unity and a relation almost identical to Equation 12.11 can be obtained, indicating that one cannot possibly store even one bit of information in a single resonator without too much distortion. For this reason one should consider combining the resonators. Just as with single resonators two different arrangements can be made—the all pass arrangement, or the side-coupled integrated spaced sequence of resonators (SCISSOR) [12,13] shown in Figure 12.4a. The dispersion relation of SCISSORS is the same as Equation 12.3, but the relation between the group delay and GDD is changed from Equations 12.4 and 12.5 and can be written as

$$T_{d,S}^{(0)} = N_r\tau\frac{1+\rho}{1-\rho} = N_r\frac{\tau}{\kappa_1} \qquad (12.13)$$

and

$$T_{d,S}^{(2)} = -N_r\tau^3\frac{(1+\rho)\rho}{(1-\rho)^3} = -N_r^{-2}\left[T_{d,S}^{(0)}\right]^3\frac{1-\kappa_1^2}{4} \approx -\frac{1}{4}N_r^{-2}\left[T_{d,S}^{(0)}\right]^3 \qquad (12.14)$$

where we have introduced

$$\kappa_1 = \frac{1-\rho}{1+\rho} \qquad (12.15)$$

As an alternative to SCISSORS one can consider a different arrangement, shown in Figure 12.4b, a so-called coupled resonator optical waveguide (CROW) [14,15] which, just like the single resonator of Figure 12.1b has a distinct transmission band.

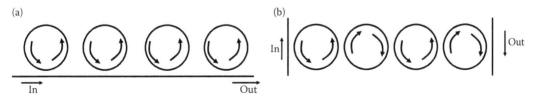

(a)　　　　　　　　　　　　　　　　　　(b)

In　　　　　　　　　　Out　　　　　In　　　　　Out

FIGURE 12.4
Coupled resonator structures (CRS). (a) SCISSOR. (b) CROW.

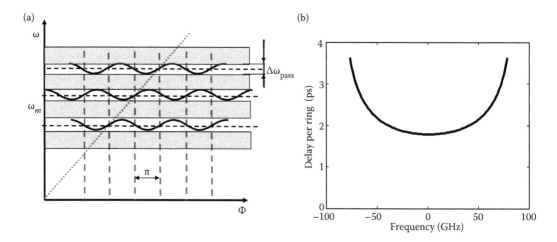

FIGURE 12.5
(a) Dispersion and (b) group delay (per resonator) of a CROW.

The dispersion relation of CROW can be written as

$$\sin(\omega\tau/2) = \kappa \cos(\Phi/N_r) \tag{12.16}$$

The dispersion curve is shown in Figure 12.5a and consists of the series of pass bands around resonant frequencies $\omega_m = 2\pi m/\tau$ separated by the wide gaps. The width of the pass band is

$$\Delta\omega_{pass} = 4\tau^{-1}\sin^{-1}(\kappa) \approx 4\tau^{-1}\kappa$$

and the dispersion can be approximated as

$$\Phi = \Phi_0 + \frac{\partial\Phi}{\partial\omega}(\omega - \omega_0) + \frac{1}{6}\frac{\partial^3\Phi}{\partial\omega^3}(\omega - \omega_0)^3 + \ldots = $$
$$= \Phi_0 + T_{d,C}^{(0)}(\omega - \omega_0) + \frac{1}{3}T_{d,C}^{(2)}(\omega - \omega_0)^3 \tag{12.17}$$

where the delay time at ω_0 is

$$T_{d,C}^{(0)} = \frac{\partial\Phi}{\partial\omega} = N_r\frac{\tau}{2\kappa} \tag{12.18}$$

and the second order dispersion of the delay time is

$$T_{d,C}^{(2)} = \frac{1}{2}\frac{\partial^3\Phi}{\partial\omega^3} = \frac{1}{2}N_r\left(\frac{\tau}{2\kappa}\right)^3(1 - \kappa^2) \approx \frac{1}{2}N_r^{-2}\left[T_{d,C}^{(0)}\right]^3 \tag{12.19}$$

The group delay per one ring in a CROW with the same $\tau = 250$ fs [17] is plotted in Figure 12.5b. The relation (Equation 12.19) is almost identical (except the factor of 2) in magnitude but opposite in sign to the SCISSOR relation (Equation 12.14). We shall return [16] to the difference in sign later on, but for the time being we shall concentrate on the salient feature of the coupled resonators, the fact that the pass band in either CROW or SCISSOR is independent of the number of resonators. The delay time $T_d^{(0)}$ can be increased without affecting the width of pass band, and the delay-bit rate product, and the delay-bandwidth product can be increased compared to a single resonator. One should remember, however, that the GDD (Equation 12.14 or 12.19) will still limit this product.

There can be different ways to gauge the GDD limitation—the one we adapt here is rather simple. We want the differential group delay within the FWHM (Equation 12.10) of the signal spectrum to be less than one half of the bit interval B^{-1} to avoid excessive inter symbol interference (ISI). Thus the GDD limitation is

$$2[4\ln(2)]^2 T_{d,C}^{(2)} B^3 \approx B^3 [4\ln(2)]^2 N_r^{-2} \left[T_{d,C}^{(0)}\right]^3 \le 1 \tag{12.20}$$

If we combine this relation with the definition of delay-bandwidth product $N = T_{d,C}^{(0)} B$ we immediately obtain

$$[4\ln(2)]^2 N_r^{-2} N^3 \le 1 \tag{12.21}$$

independent of the bit rate and the delay per ring [17]. Thus, to store N bits of information in CROW without incurring too much distortion that can cause intersymbol interference, one should use at least $N_{r,C} \approx 2.8 N^{3/2}$, while for SCISSOR it is at least $N_{r,S} \approx 2.2 N^{3/2}$. The difference is insignificant so it makes sense to use a "rule of thumb" relation

$$N_r \approx 3N^{3/2} \tag{12.22}$$

Using this relation one can immediately get the required delay per resonator for a given bit rate as

$$T_d^{(0)} = N/BN_r \approx \frac{1}{3N^{1/2}B} \tag{12.23}$$

From this relation, with a given resonator dimensions, one can determine the coupling coefficients using Equation 12.4 or 12.18 for SCISSOR and CROW, respectively. Thus for the linear task of storage it is always preferable to use multiple MR structures rather than single MR's. The analysis performed so far has not included the influence of loss incurred in the MR's. We shall return to the issue of loss in the delay line later on, but next we turn our attention to optical switching and how it can be improved with MR's.

12.3 Single and Coupled Microresonators as Optical Switches

In order to gauge the advantages offered by MR's in optical switching, we shall define a point of reference with which the prospective MR-based switches and modulators can be compared. In this work we shall choose the most widely used switch based on a Mach-Zehnder interferometer (MZI) as shown in Figure 12.6. One or both arms in the MZI

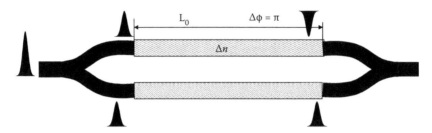

FIGURE 12.6
Optical switch based on MZI.

contain material whose refractive index can be varied by current injection, as in Si devices, by a low frequency electric field (electro-optic effect), or by an optical field (nonlinear optical effect). With regard to the all-optical switches, we shall consider here only the case when the switching (pump) optical field is not resonant with the MR, while the signal is resonant. The double-resonant case, which is more difficult to implement in practice, is treated in Refs. [18,19]. No matter what mechanism causes the refractive index change Δn, the effect can be characterized by the maximum attainable index change Δn_{max} and by the speed characterized by the cut-off frequency F_{max}. From a more fundamental point of view the index change is limited by the material damage resulting from excessive electric or optical fields or from an extremely high current injection level. From more practical considerations the index change can be limited by voltage, current or optical power levels at the disposal of the designer. Speed limitations can be intrinsic (recombination time in current injection schemes) or extrinsic (parasitic capacitances and losses at high frequencies in electro-optic devices). The cut-off frequencies range from the GHz range in Si current injection devices [7,20] to tens of THz in nonlinear optical devices based on the Kerr effect [21]. But as a rule, the faster the speed of the effect, the weaker it is, because the slow index change mechanisms usually rely upon the accumulation of carriers (or temperature rise) while the faster mechanisms have to depend on the instantaneous values of the input. As a result, the product of $\Delta n_{max} F_{max}$ (by using analogy from electronics, one can call it "gain bandwidth product") does not vary much. For instance in a Si-based scheme with carrier injection, the refractive index change [7] $\Delta n_{max} \sim 10^{-3}$ and $F_{max} \sim 1\text{GHz}$, while in the LiNbO$_3$ the bandwidth can be as high as $F_{max} \sim 100\text{GHz}$, but the index change is much smaller, $\Delta n_{max} \sim 10^{-5}$ [22]. In a typical ultra-fast nonlinear optical switch the index changes also do not exceed 10^{-5} [21] but the bandwidth is measured in THz.

To attain switching in the MZI of Figure 12.6a, an optical path length change equal to one half wavelength must be induced in one of the arms and this gives us a switching condition that the required product of index change and length be

$$(\Delta n L)_{MZI} = \lambda_0/2 \tag{12.24}$$

If one considers other nonresonant switching schemes, for instance the ones relying upon directional couplers or Bragg gratings, they all have constraints on the length-index change product that are very similar to Equation 12.24.

Since the index change is limited, one can introduce the "minimum switching distance" in the MZI interferometer as

$$L_0 = \lambda_0 \Delta n_{max}/2 \tag{12.25}$$

Now the "gain" of the resonant scheme is its ability to reduce ΔnL, thus reducing the size and power consumption of the device. Let us see how it can be done and what the bit rate limitations are.

When the effective index of refraction is changed via either an electro-optic or a nonlinear optical process

$$n = \bar{n} + \Delta n \tag{12.26}$$

the round trip time changes as

$$\Delta\tau = \tau\frac{\Delta n}{\bar{n}} \tag{12.27}$$

and the resonant frequencies change accordingly

$$\Delta\omega_0 = -\omega_0\frac{\Delta n}{\bar{n}} \tag{12.28}$$

as shown in Figure 12.7a. Consider now a single resonator of Figure 12.1b. If the pass band width $\Delta\omega_{pass}$ of Equation 12.7 is equal to the signal bandwidth $\Delta\omega_{sig}$, then shifting the resonance frequency by $\Delta\omega_0 = \Delta\omega_{pass} = \Delta\omega_{sig}$ will cause switching in this resonance-shifting scheme, and combining Equation 12.28 with Equation 12.10 one obtains the switching condition.

$$(\Delta nL)_r = \frac{c}{\bar{n}}\Delta n\tau = \frac{c}{\omega_0}8\ln(2)B\tau = \frac{8\ln(2)}{\pi}B\tau\frac{\lambda_0}{2} \approx 1.7B\tau\frac{\lambda_0}{2} \tag{12.29}$$

The gain in the resonance-shifting scheme of Figure 12.7a is then

$$G_r = \frac{(\Delta nL)_{MZI}}{(\Delta nL)_r} \approx \frac{0.7}{B\tau} = \frac{B_{cut}}{B} \tag{12.30}$$

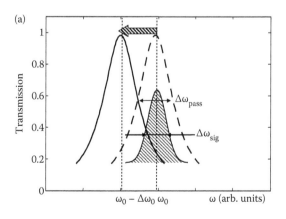

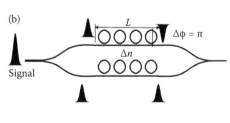

FIGURE 12.7
(a) Optical switch based on resonance shift in a single MR. (b) Optical switch based on coupled multiple MR's inside an MZI slow light switch.

and one can define the gain cut-off bandwidth as

$$B_{cut,r} \approx 0.7\tau^{-1} \tag{12.31}$$

It would appear that one can increase the gain using smaller and smaller radii, but, since maximum index change is limited, one can also define the maximum switching bit rate as

$$B_{max,r} = \frac{\omega_0}{8\ln(2)}\left(\frac{\Delta n}{n}\right)_{max} = 1.1\nu_0\left(\frac{\Delta n}{n}\right)_{max} \tag{12.32}$$

The resonant gain at maximum bit rate in the resonance shifting single resonator scheme is then

$$G_r(B_{max,r}) = \frac{B_{cut,r}}{B_{max,r}} \approx 0.65\left[\nu_0\tau\frac{\Delta n_{max}}{\bar{n}}\right]^{-1} \approx \frac{L_0}{L} \tag{12.33}$$

Since L_0 even for the largest value of $\Delta n_{max} \sim 10^{-3}$ is in the mm range, and the rings can have circumference of less than 100 μm, the maximum bit rate is always at least an order of magnitude smaller than the cut-off bit rate. In Figure 12.8 we have plotted as dashed lines the resonant gain as a function of bandwidth for two different schemes—one is a narrow-band Si current injection scheme with $\Delta n_{max} \sim 10^{-3}$ ($L_0 \sim 0.7$ mm @1550 nm) [7] and $F_{max} \sim 10$ GHz and the other is a GaAs photonic switch based on the optical Kerr effect [21] with $\Delta n_{max} \sim 3 \times 10^{-5}$ ($L_0 \sim 2.5$ cm @1550 nm) and $F_{max} > 10$ THz. We assumed that the ring resonators have about 36 μm circumference or $\tau \sim 250$ fs. For the Si scheme of Figure 12.8a we obtain $B_{max,r} = 110$ GBps which is in fact much faster than F_{max}; hence the speed of Si is intrinsically limited to about 10 GBps (this is shown by thick line). But at this fairly low bit rate one can achieve a tremendous gain of about 300 compared to the MZI. In the GaAs photonic switch (Figure 12.8b), on the other hand, $B_{max,r} \sim 3$ GHz which is orders of magnitude less than its intrinsic potential of almost unlimited bandwidth. Therefore one should consider the

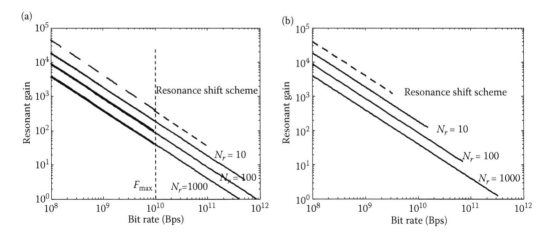

FIGURE 12.8
Resonant gain (decrease in ΔnL relative to MZI) for two different optical switches (a) Si switch with current injection. (b) GaAs all optical switch.

means to improve the speed of these devices using coupled MR's. It is obvious that combining MR's in CROW or SCISSORS and still using the resonance shifting scheme will simply use more rings to achieve the same shift of ω_0 and thus will only make things worse. For this reason we shall consider a different scheme that takes advantage of longer propagation times in the coupled MR's while still relying on MZI geometry to achieve switching. We refer to this scheme shown in Figure 12.7b as the "slow light scheme" [18,19].

With the phase characteristic of CROW or SCISSOR described by Equation 12.17 the phase change caused by the change of index is

$$\Delta\Phi = -\frac{\partial\Phi}{\partial\omega}\Delta\omega_0 = T_{dC}^{(0)}\omega_0\frac{\Delta n}{\bar{n}} = \pi \tag{12.34}$$

while the GDD limitation (Equation 12.20)

$$B^3[4\ln(2)]^2 N_r^{-2}\left[T_{d,C}^{(0)}\right]^3 \le 1 \tag{12.35}$$

must be still enforced to avoid excessive ISI. Then we quickly obtain the required index change

$$\frac{\Delta n}{\bar{n}} = N_r^{-2/3}\frac{B}{2v_0}[4\ln(2)]^{2/3} \tag{12.36}$$

The required index change thus is reduced compared to a single MR by a factor of roughly $N_r^{-2/3}$, but the total length is increased by a factor of N_r. Therefore the length-index change product increases as roughly $N_r^{1/3}$.

$$(\Delta nL)_{N_r} = N_r\frac{c}{\bar{n}}\Delta n\tau_r = N_r^{1/3}\frac{cB}{2v_0}[4\ln(2)]^{2/3}\tau_r \approx 2N_r^{1/3}B\tau_r\frac{\lambda_0}{2} \tag{12.37}$$

and the gain is actually reduced by a factor of $N_r^{1/3}$ compared to a single resonator resonance shifting scheme (Equation 12.30).

$$G_r^{(N_r)} = \frac{(\Delta nL)_{MZI}}{(\Delta nL)_{N_r}} \approx \frac{0.5}{B\tau_r}N_r^{-1/3} \approx G_rN_r^{-1/3} \tag{12.38}$$

The gain cut-off bit rate is then also reduced by $N_r^{1/3}$.

$$B_{cut}^{(N_r)} \approx 0.5\tau_r^{-1}N_r^{-1/3} \approx B_{cut,r}N_r^{-1/3} \tag{12.39}$$

But the main consequence is the fact that the maximum bit rate increases according to Equation 12.36.

$$B_{max}^{(N_r)} \approx N_r^{2/3}v_0\left(\frac{\Delta n}{\bar{n}}\right)_{max} \approx N_r^{2/3}B_{max,r} \tag{12.40}$$

which allows us to fully use the potential of fast-switching materials. We have plotted the gain versus bit rate in the slow light scheme as solid lines in Figure 12.8 for different numbers of resonators. As one can see from Figure 12.8a for the Si-based scheme where the index change is large and intrinsically slow, using coupled resonators offers no advantage relative to a simple resonance-shifting scheme with a single resonator. The increase in the number of resonators causes a drop in the resonant gain and thus raises required switching power. This increased power demand comes with no switching speed benefit because the switching speed is determined by the intrinsic speed of the device. But for the relatively weak and fast scheme, such as the GaAs nonlinear all-optical switch of Figure 12.8b, the coupled MR scheme (in either SCISSOR or CROW configuration) greatly enhances the speed of the device from GBps for a single MR to hundreds of GBps, and still reduces the power requirements compared to a simple MZI by a factor of up to 100 or more. Therefore, the decision of whether to use a single (or a few) MR scheme with resonance shift or a multiple coupled MR slow light scheme should be based on the relation between the intrinsic limitations of the material and required operational speed.

12.4 Loss and GDD Limitations in CRS Delay Lines

We now return to the CRS delay lines for which we earlier obtained a very simple relation (Equation 12.22) between the delay capacity N and number of resonators N_r. That relation, however, was obtained without taking into consideration the loss in the resonators. Presence of loss in CRS affects performance via three mechanisms. First if the loss per individual MR approaches the coupling between the resonators, the dispersion characteristics of the CRS become affected. Further on we shall show that this mechanism is not important for delay lines with reasonably large delay capacity. Second, the absolute value of loss causes signal deterioration. This loss can be effectively compensated by the embedded gain, and, at least for as long as the signal to noise ratio is not depleted by amplified spontaneous emission [23], the delay capacity is limited only by the higher order GDD. The third and more intractable effect is the frequency dependence (dispersion) of loss that results in suppression of some frequency components of the signal, corresponding to the changes in time domain that can cause ISI. Depending on whether the GDD or loss dispersion or their combination is the limiting factor, the bandwidth dependence of the SL delay line performance has a different character. Therefore, in order to develop efficient CRS delay lines, it is important to establish where (as a function of the bandwidth) lays the boundary between the loss-limited and GVD-limited regions.

The loss in the CRS is associated with coupling loss bending losses and is proportional to the distance traveled in the CRS, i.e. to the group delay. Therefore, the spectrum of the original Gaussian OOK signal with original spectral width (Equatoin 12.10), changes and becomes [24,25]:

$$P_{\text{out}}(\omega) = P(\omega)e^{-\alpha\frac{c}{n}T_d(\omega)} = e^{-\alpha\frac{c}{n}T_d(\omega_0)}P(\omega)e^{-\alpha\frac{c}{n}\Delta T_d} \tag{12.41}$$

where we have introduced the differential group delay

$$\Delta T_{\text{d}} = T_d(\omega) - T_d(\omega_0) \tag{12.42}$$

Now the signs of ΔT_d are different for CROW and SCISSOR designs and the loss implications are different. For CROW

$$\Delta T_{dC} \approx T_{dC}^{(2)}(\omega - \omega_0)^2 > 0 \tag{12.43}$$

indicating that the high frequencies in the signal spectrum are attenuated and the signal broadens in time causing ISI. Substituting Equation 12.43 into Equation 12.41 one obtains the expression for the output signal power spectrum

$$P_{\text{out}}(\omega) \sim \exp\left[-\frac{4\ln 2(\omega - \omega_0)^2}{\left(\Delta\omega_{\text{sig}}\right)^2}\left(1 + \frac{16\ln(2)\alpha c B^2}{n}T_{dC}^{(2)}\right) \right] \tag{12.44}$$

The pulse is Gaussian with a new, narrow spectral FWHM

$$\Delta\omega_{\text{sig,out}} = \Delta\omega_{\text{sig}}\left(1 + \frac{16\ln(2)\alpha c B^2}{\bar{n}}T_{dC}^{(2)}\right)^{-1/2} \tag{12.45}$$

and the temporal FWHM is wider by the same factor. Limiting the temporal broadening to a factor of $2^{1/2}$ we obtain a loss-related bit-rate limitation

$$16\ln(2)\alpha c\bar{n}^{-1}B^2 T_{dC}^{(2)} \le 1 \tag{12.46}$$

Now, this condition must be added to the GDD condition (Equation 12.20)

$$2[4\ln(2)]^2 T_{d,C}^{(2)}B^3 \le 1 \tag{12.47}$$

Consider the Gaussian pulse with FWHM of 50 ps (Figure 12.9a) propagating through the CROW delay line with $T_{d,C}^{(2)} = 3\times 10^4$ ps^3 and loss $\alpha c\bar{n}^{-1} = 0.01$ ps^{-1}. In Figure 12.9b we show the impact of GDD only on the shape of output pulse. The pulse is somewhat broadened and there are few decaying oscillations following the pulse which come from the higher frequency components that had longer delay times. In Figure 12.9c we also include the loss into the consideration, and the impact of the loss is twofold. The signal broadens as a result of spectral narrowing, and at the same time the oscillations following the pulse get dampened, because high frequency components get attenuated more. Overall one combines two conditions (Equations 12.46 and 12.47) into a single one by insisting that

$$2[4\ln(2)]^2 T_{d,C}^{(2)}B^3 \sqrt{1 + \frac{B_{\text{loss}}^2}{B^2}} \le 1 \tag{12.48}$$

where we have introduced the "loss bit rate" as $B_{\text{loss}} = \alpha c/2\bar{n}\ln(2)$ the bit rate at which the signal will experience 3 dB loss per 1/4 of the bit interval.

We can re-normalize it to the loss per ring $\alpha_R = \alpha c\tau/\bar{n}$ and obtain

$$B_{\text{loss}} = \alpha_R (2\ln(2)\tau)^{-1} \tag{12.49}$$

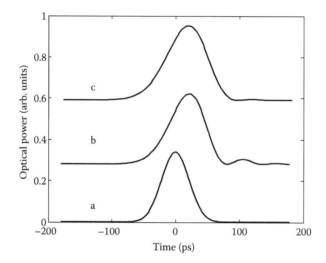

FIGURE 12.9
Evolution of the Gaussian signal through the CROW delay line. (a) Input pulse. (b) Output pulse disregarding loss. (c) Output pulse with loss.

Clearly at bit rates less than B_{loss} the loss distortion dominates while at higher rates GVD becomes the overriding factor. Using Equations 12.18 and 12.19 we express the GDD as

$$T_{d,C}^{(2)} = \frac{1}{2} N_r^{-2} \left[T_{d,C}^{(0)} \right]^3 \left(1 - \kappa^2 \right) = \frac{1}{2} N^3 B^{-3} N_r^{-2} \left[1 - \frac{N_r^2}{N^2} \left(B \frac{\tau}{2} \right)^2 \right] \tag{12.50}$$

and substitute it into Equation 12.48 to obtain

$$[4\ln(2)]^2 \, N^3 N_r^{-2} \, \frac{\sqrt{1 + \dfrac{B_{\text{loss}}^2}{B^2}}}{\left(1 + \left(\dfrac{B}{B_{\text{cut}}^{(N)}} \right)^2 \right) \sqrt{1 + \dfrac{B_{\text{loss}}^2}{B^2}}} \leq 1 \tag{12.51}$$

where we have introduced cut-off bit rate

$$B_{\text{cut}}^{(N)} = (2\ln 2\tau)^{-1} N^{-1/2} = \alpha_R^{-1} B_{\text{loss}} \tag{12.52}$$

which can be understood as the bit rate at which the CRS offers only a $\sqrt{2}$ length advantage over a simple straight waveguide. Thus in lieu of Equation 12.22 we obtain a more refined important relation for the required number of resonators which includes bit rate dependence.

$$N_r(B,N) \approx 2.8 N^{3/2} \, \frac{\sqrt[4]{1 + B_{\text{loss}}^2 / B^2}}{\sqrt{1 + \left[B / B_{\text{cut}}^{(N)} \right]^2} \sqrt{1 + B_{\text{loss}}^2 / B^2}}, \tag{12.53}$$

The dependence (Equation 12.53) is plotted in Figure 12.10 for four different values of N using an example of coupled ring resonators from resonators having about 36 μm

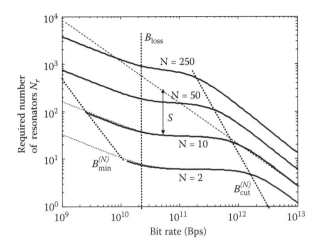

FIGURE 12.10
Number of resonators required to store N bits in a CROW delay line as a function of bit rate.

circumference or $\tau \sim 250$ fs and a 0.04 dB ($\alpha_R = 0.01$ that yields $B_{loss} \sim 25$ GBps) [1,26]. This value indicates that with the present day technology both GDD and loss distortion play roughly equal roles for 10–40 GBps signals, while at higher rates GDD is expected to dominate. Clearly the plot can be split into three distinct regions

$$N_r(N,B) \approx \begin{cases} 2.8(B_{loss}/B)^{1/2} N^{3/2} & B \ll B_{loss} \\ N_r \approx 2.8 N^{3/2} & B_{loss} \ll B \ll B_{cut}^{(N)} \\ 2N/B\tau & B \gg B_{cut}^{(N)} \end{cases} \tag{12.54}$$

In the first region, the loss-induced distortion dominates and the required number of resonators is inversely proportional to the square root of the bit rate. In the second region the required number of resonators is determined by dispersion and remarkably does not depend on the bit rate, Finally, at high bit rates the coupling coefficient κ becomes so close to unity that the CROW delay line becomes essentially equivalent to a simple meandering waveguide consisting of N_r half-circles and having a total length of $c\bar{n}^{-1}NB^{-1}$.

It should be noted that the condition for the existence of the second, GVD-limited, region is, $B_{loss} \ll B \ll B_{cut}^{(N)}$ or $\alpha_R \ll N^{1/2}$; hence for the present structure the region should exist for delay lines with a few hundred bits of storage capacity, albeit at very high bit rates. As one can see, the "flat" region is clearly present in Figure 12.10 for $N=2$ bits and $N=10$ bits, but it is far less prominent for $N=50$ bits and barely visible at $N=250$ bits.

We can now determine he slow down factor $S=\kappa^{-1}$ in different regions. Note that the slow down factor is the distance between the actual curve $N_r(N,B)$ and the dashed line $2N/B\tau$ for the simple meandering waveguide. Thus we obtain

$$S(N,B) \approx \begin{cases} \alpha_R^{-1} N^{-1/2} (B_{loss}/B)^{1/2} & B \ll B_{loss} \\ N_r \approx (B_{cut}^{(N)}/B) N^{-1/2} & B_{loss} \ll B \ll B_{cut}^{(N)} \\ 1 & B \gg B_{cut}^{(N)} \end{cases} \tag{12.55}$$

As one can see from Figure 12.10 it is feasible to obtain the slow down factor on the order of few tens for 10 GBps bit rate and $N=10$ bits.

It is now important to return to the issue of the influence of ring loss on the dispersion characteristics by imposing the condition that round trip loss α_R should be much smaller that coupling coefficient $\kappa = S^{-1}$. Since κ decreases with bit rate we need to consider only the low bit range of $B \ll B_{loss}$ where one immediately obtains from Equation 12.55.

$$\kappa = \alpha_R \, (BN/B_{loss})^{1/2} \tag{12.56}$$

Thus for as long as $B > B_{min}^{(N)} = B_{loss}/N$, the optimized CROW delay line remains operational. As one can see from Figure 12.10, this range encompasses bit rates in excess of a few GBps for $N > 10$; i.e. the ring loss does not change the dispersion itself appreciably for the cases of practical interest.

We shall now turn our attention to the SCISSOR delay line with the group delay dispersion having a negative sign. The high frequency components spend less time in the MR's and therefore suffer from relatively low loss. This means that the spectral width of the signal actually broadens as

$$\Delta\omega_{sig,out} = \Delta\omega_{sig}\left(1 - \frac{16\ln(2)\alpha c B^2}{\bar{n}} T_{dS}^{(2)}\right)^{-1/2} \tag{12.57}$$

By itself, spectral broadening should not be a problem because it may mean that the pulse gets compressed in time domain; however, once the dispersion is taken into account, the pulse shape deteriorates dramatically because the high frequency components propagate much faster than the low frequency ones. Consider what happens with the same 50 ps long Gaussian pulse of Figure 12.11a as it propagates through the SCISSOR with $T_{d,C}^{(2)} = -3 \times 10^4$ ps^3 and the same loss $\alpha c \bar{n}^{-1} = 0.01$ ps^{-1}. In Figure 12.11b, we show the impact of GDD only on the shape of the output pulse. The pulse is somewhat broadened and there are few decaying "precursor" oscillations ahead of the pulse comprised by the higher frequency components of the pulse that have shorter delay times. In Figure 12.11c, we also include the loss and see that, while the main pulse does get narrower, the precursor pulses get amplified

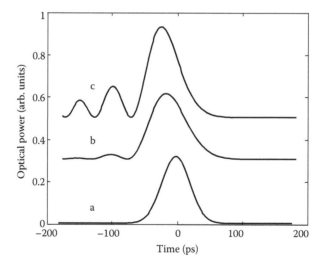

FIGURE 12.11
Evolution of the Gaussian signal through the SCISSOR delay line. (a) Input pulse. (b) Output pulse disregarding loss. (c) Output pulse with loss.

thus greatly distorting the signal. Therefore, to avoid excessive ISI the group delay over the expanded bandwidth $\Delta\omega_{\text{sig,out}}$ should be less than one half of the bit interval, i.e

$$\frac{2[4\ln(2)]^2 T_{d,S}^{(2)} B^3}{1-16\ln(2)\alpha c\bar{n}^{-1}B^2 T_{d,S}^{(2)}} \leq 1 \tag{12.58}$$

From this equation and using Equations 12.13 and 12.14 we obtain

$$N_r(B,N) \approx 2.2 N^{3/2} \frac{\sqrt{1+B_{\text{loss}}^2/B^2}}{\sqrt{1+\left[B/B_{\text{cut}}^{(N)}\right]^2\left(1+B_{\text{loss}}^2/B^2\right)}} \tag{12.59}$$

where B_{loss} is still defined as Equation 12.49 while the cut-off bit rate for SCISSOR is $2^{1/2}$ larger than the CROW cut-off rate

$$B_{\text{cut}}^{(N)} = \left(\sqrt{2}\ln 2\tau\right)^{-1} N^{-1/2} \tag{12.60}$$

The dependence (Equation 12.59) is plotted in Figure 12.12 for four different values of N using the same rings as in CROW example. Once again the plot can be split into three distinct regions

$$N_r(N,B) \approx \begin{cases} 2.2\left(B_{\text{loss}}/B\right)N^{3/2} & B \ll B_{\text{loss}} \\ N_r \approx 2.2N^{3/2} & B_{\text{loss}} \ll B \ll B_{\text{cut}}^{(N)} \\ 2N/B\tau & B \gg B_{\text{cut}}^{(N)} \end{cases} \tag{12.61}$$

The main difference from CROW is in the B^{-1} dependence of the required number of resonators on the bit rate for bit rates below B_{loss}. Therefore, the slow down factor S for SCISSOR is constant for $B < B_{\text{loss}}$ and equal to roughly

$$S_{\text{max}} \sim \alpha_R^{-1} N^{-1/2} \tag{12.62}$$

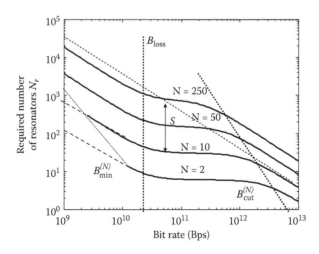

FIGURE 12.12
Number of resonators required to store N bits in a SCISSOR delay line as a function of bit rate.

This is a very practical limitation on the slow down effect in SCISSOR, indicating that with about 1% loss per ring one can expect that, independent of bandwidth, 100 bits can be stored with a slow down factor of 10. We shall conclude our comparison of loss and group delay dispersion in SCISSOR and CROW with the statement that for low loss case SCISSOR offers an advantage while for the higher loss the benefits of CROW become more pronounced. In the end, though the choice may have to be hinged on the technology because with ring resonators both CROW and SCISSOR can be implemented with equal ease, while photonic crystals CROW can be fabricated much easier than SCISSOR.

12.5 Mitigation of GDD

When it comes to GDD mitigation, there can be different pathways. One of them uses the fact that the lowest order GDD terms $T_d^{(2)}$ in SCISSOR (Equation 12.14) and CROW (Equation 12.19) have different signs and thus can be canceled. This is the method first suggested in Ref. [16] where a structure combining CROW and SCISSOR shown in Figure 12.13a was proposed. The structure [16] consists of N_r identical MR's with N_C of them arranged in a CROW configuration resonators and remaining N_S MR's in a SCISSOR arrangement. The compensation condition is easily obtained as

$$T_d^{(2)} = T_{d,C}^{(2)} + T_{d,S}^{(2)} = \frac{1}{2} N_C \left(\frac{\tau}{2\kappa}\right)^3 - \frac{1}{4} N_S \left(\frac{\tau}{\kappa_1}\right)^3 = 0; \qquad (12.63)$$

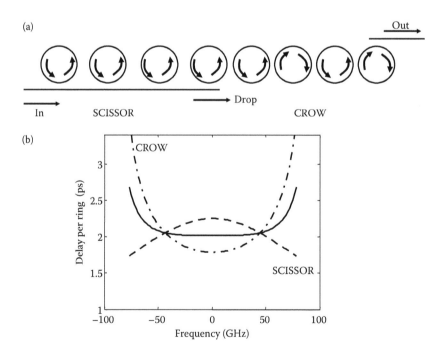

FIGURE 12.13
(a) Mixed SCISSOR–CROW delay line. (b) Group delay (per resonator) in the mixed delay line (solid line) and its components (dashed lines).

where κ_1 was defined in Equation 12.15. For $N_c=N_s$ one readily obtains $\kappa_1=2^{2/3}\kappa$. In Figure 12.13b the results of dispersion compensation are shown. To keep things in prospective we have normalized the group delay to the total number of rings. One can see that the bandwidth gets enlarged, but, unfortunately, the dispersion compensation is not complete, since the next order GDD terms $T_d^{(4)}$ have the same sign in CROW and SCISSOR and one is left with the GDD of the form

$$T_d(\omega) = T_d^{(0)} + T_d^{(4)}(\omega - \omega_0)^4 \tag{12.64}$$

where

$$T_d^{(4)} = \frac{N_C}{24}\left(\frac{\tau}{2\kappa}\right)^5 + \frac{N_S}{8}\left(\frac{\tau}{\kappa_1}\right)^5 \approx \frac{N_r}{12}\left(\frac{\tau}{2\kappa}\right)^5 = \frac{1}{12}N_r^{-4}\left(T_d^{(0)}\right)^5 \tag{12.65}$$

We then evoke once again the GDD limitation in the absence of losses and obtain in lieu of Equation 12.20:

$$2[4\ln(2)]^4 T_d^{(4)}B^5 \approx B^5[4\ln(2)]^2\frac{N_r^{-4}}{6}\left[NB^{-1}\right]^5 \le 1 \tag{12.66}$$

From which we obtain the expression for the number of resonators required to store N bits

$$N_r \approx 2N^{5/4} \tag{12.67}$$

This is indeed a substantial improvement over the $N_r \sim 3N^{3/2}$ dependence in the absence of compensation. For instance, without compensation, storing 50 bits would require 1000 MR's, but with compensation 260 MR's would suffice. The cut-off bit rate expression with compensation becomes

$$B_{\text{cut}}^{(N)} \sim \tau^{-1}N^{-1/4} \tag{12.68}$$

i.e. the operational bandwidth with compensation expands by about $N^{1/4}$.

Another method of dispersion compensation relies exclusively on SCISSORS with two resonances detuned from the central frequency ω_0—with odd resonators having a resonance frequency of $\omega_0+\delta\omega_0$ and even resonators having a resonance frequency $\omega_0-\delta\omega_0$. This arrangement is shown in Figure 12.14a. The lowest order GDD $T_d^{(2)}$ gets cancelled when $\delta\omega_0 \sim \kappa_1/\sqrt{3}\tau$, as shown in Figure 12.14b for the example of Si SCISSOR with a 36 μm circumference. Note that the required relative change in the resonant frequency

$$\frac{\delta\omega_0}{\omega_0} = \frac{\delta\tau}{\tau} = \frac{\kappa_1}{\sqrt{3}\omega_0\tau} \approx 10^{-4} \tag{12.69}$$

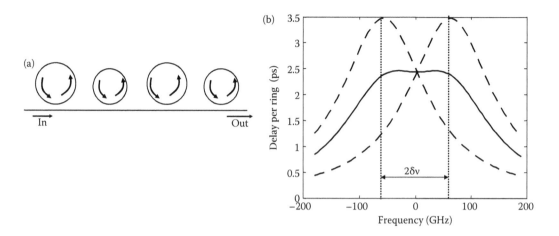

FIGURE 12.14
(a) Balanced SCISSOR delay line. (b) Group delay (per resonator) in the balanced SCISSOR line (solid line) and its components (dashed lines).

can be accomplished by modulating the refractive index either thermally or by carrier injection. Therefore, the "balanced SCISSOR" arrangement of Figure 12.14b has the additional benefit of being tunable.

12.6 Conclusions

In this chapter, we have analyzed limitations imposed on single and multiple coupled MR's (CRS) by group delay dispersion combined with loss. We have considered applications of MR's as delay lines and as optical switches. In case of delay lines we have shown that in order to avoid distortion and ISI one has to use CRS with a large number of MR's in CROW or SCISSOR arrangements and that the required number of resonators N_r increases super linearly with the number of stored (delayed) bits N as $N_r \sim N^{3/2}$. We have also demonstrated that there exists a cut-off bit rate $B_{cut}^{(N)} \sim N^{-1/2}$ beyond which CRS offers no advantage over a simple waveguide as an optical delay element. In addition we have explained that the group delay dispersion distorts the signal shape both directly and also via loss-induced changes in the spectrum. With the current state of MR technology both direct and loss-related bit rate limitations are of equal importance. We have also reviewed the means of mitigating the GDD.

For optical switching applications we have considered two alternative schemes. One is a single-resonator optical switch in which the resonance frequency gets shifted and causes the change in transmission. The other is the slow light scheme where the phase shift gets accumulated as the light passes through CRS. We have demonstrated that for each scheme there exists a maximum bit rate beyond which the MR offers no advantage over the standard MZI switch. We have shown that for relatively low speed switches (up to 10 GBs) the single resonator scheme works as good if not better that the one with CRS, but for the high bit rates the slow light scheme has a distinct advantage.

With the results obtained in this chapter one should be able to decide which MR scheme can offer more benefits for each specific application and material.

References

1. F. Xia, L. Sekaric, and Y. Vlasov. Ultracompact optical buffers on a silicon chip. *Nature Photonics*, 1, 65–71, 2006.
2. H. Tazawa, Y-H. Kuo, I. Dunayevskiy, J. Luo, Al K.-Y. Jen, H. R. Fetterman, and W. H. Steier. Ring resonator-based electrooptic polymer traveling-wave modulator. *IEEE J. Lightwave Technol.*, 24, 3514–3519, 2006.
3. D. A. Cohen, M. Hossein-Zadeh, and A. F. J. Levi. High-Q microphotonicelectro-optic modulator. *Solid-State Electron*, 45(9), 1577–1589, 2001.
4. B. Bhola, H-C. Song, H. Tazawa, W. H. Steier. Polymer micro-resonator strain sensors. *Photon. Technol. Lett.*, 17, 867–869, 2005.
5. A. Savchenkov, A. B. Matsko, M. Mohageg, and L. Maleki. Ringdown spectroscopy of stimulated Raman scattering in a whispering gallery mode resonator. *Opt. Lett.*, 32, 497–499, 2007.
6. B. Schmidt, Q. Xu, J. Shakya, S. Manipatruni, and M. Lipson. Compact electro-optic modulator on silicon-on-insulator substrates using cavities with ultra-small modal volumes. *Opt. Express*, 15, 3140–3148, 2007.
7. M. Lipson. Compact electro-optic modulators on a silicon chip. *IEEE J. Sel. Top. Quant.*, 12(6), 1520–1526, 2006.
8. C. K. Madsen and G. Lenz. Optical all-pass filters for phase response design with applications for dispersion compensation. *IEEE Photon Tech. Lett.*, 10, 994–996, 1998.
9. J, K. Poon, L. Zhu, G. A. DeRose, and A. Yariv. Transmission and group delay of microring coupled-resonator optical waveguides. *Opt. Lett.*, 31, 456–458, 2006.
10. S. Xiao, M. H. Khan, H. Shen, and M. Qi. Multiple-channel silicon micro-resonator based filters for WDM applications. *Opt. Express*, 15, 7489–7498, 2007.
11. M. S. Nawrocka, T. Liu, X. Wang, and R. R. Panepucci. Tunable silicon microring resonator with wide free spectral range. *Appl. Phys. Lett.*, 89, 071110 2006.
12. J. E. Heebner and R. W. Boyd. 'Slow' and 'fast' light in resonator-coupled waveguides. *J. Mod. Opt.*, 49, 2629–2636, 2002.
13. J. E. Heebner, R. W. Boyd, and Q-H Park. SCISSOR solitons and other novel propagation effects in microresonator-modified waveguides. *J. Opt. Soc. Am. B*, 19, 722–731, 2002.
14. Yariv, Y. Xu, R. K. Lee, A. Scherer. Coupled-resonator optical waveguide: a proposal and analysis. *Opt. Lett.*, 24, 711–713, 1999.
15. Melloni, F. Morichetti and M. Martnelli. Linear and nonlinear pulse propagation in coupled resonator slow-wave optical structures. *Opt. Quant. Electron.*, 35, 365–378, 2003.
16. J. B. Khurgin. Expanding the bandwidth of slow light photonic devices based on coupled resonators. *Opt. Lett.*, 30, 513–515, 2005.
17. J. B. Khurgin. Optical buffers based on slow light in EIT media and coupled resonator structures—comparative analysis. *J. Opt. Soc. Am. B*, 22, 1062–1074, 2005.
18. M. Soljacic, S. G. Johnson, S. Fan, M. Inanescu, E. Ippen, and J. D. Joannopulos. Photonic-crystal slow-light enhancement of nonlinear phase sensitivity. *J. Opt. Soc. Am. B*, 19, 2052–2059, 2002.
19. J. B. Khurgin. Performance of nonlinear photonic crystal devices at high bit rates. *Opt. Lett.*, 30, 643–645, 2005.
20. Q. Xu, S. Manipatruni, B. Schmidt, J. Shakya, and M. Lipson. 12.5 Gbit/s carrier-injection-based silicon micro-ring silicon modulators. *Opt. Express*, 15(2), 430–436, 2007.
21. P. Millar, J. S. Aitchison, J. U. Kang, G. I. Stegeman, A. Villeneuve, G. T. Kennedy, and W. Sibbett. Nonlinear waveguide arrays in AlGaAs. *J. Opt. Soc. Am. B*, 14, 3224–3231, 1997.
22. D. A. Nikogosyan. Properties of Optical and Laser Related Materials. A Handbook. Wiley, NY, 1997.
23. J. B. Khurgin. Power dissipation in slow light devices: a comparative analysis. *Opt. Lett.*, 32, 163–165, 2007.

24. A. B. Matsko, A. A. Savchenkov, D. Strekalov, V. S. Ilchenko, and L. Maleki. Interference effects in lossy resonator chains. *J. Mod. Opt.*, 51, 2515–2522, 2004.
25. J. B. Khurgin. Dispersion and loss limitations on the performance of optical delay lines based on coupled resonant structures. *Opt. Lett.*, 32, 133–135, 2007.
26. F. Xia, L. Sekaric, M. O'Boyle, and Y. Vlasov. Coupled resonator optical waveguides based on silicon-on-insulator photonic wires. *Appl. Phys. Lett.*, 89, 041122, 2006.

13

Linear and Nonlinear Localization of Light in Optical Slow-Wave Structures

Shayan Mookherjea

University of California, San Diego

CONTENTS

13.1 Introduction

There are various aspects of localization which are important in condensed matter physics, including the metal-insulator (Anderson) transition, localization of charges in amorphous and disordered semiconductors, tunneling and breakdown in insulators, oscillations of the conductivity in inversion layers and field-effect transistors. In optics, light can be localized in k-space, as shown by coherent backscattering in aqueous solutions of polystyrene beads [1]. The width of the backscattering cone and the angle of its cusp are inversely proportional to l, the mean free path of photons. This called "weak localization" and is a precursor to strong, i.e., Anderson localization.

Localization of light has been demonstrated relatively recently, in three and two dimensions [2,3]. In such cases, recurrent multiple scattering of light in random media leads to strong localization if the Ioffe-Regel criterion is satisfied, $kl \leq 1$ where $k=2\pi/\lambda$ and l is the elastic mean free path of photons. Since the product kl diverges in both the short wavelength ($kl \sim 1/\lambda$) and long wavelength ($kl \sim \lambda^3$) limits, localization can take place only in a limited wavelength range [4].

Localization in a truly-1D or quasi-1D structure (in the latter case, the number of transverse modes $N = k^2 A/4\pi \gg 1$ where A is the cross-sectional area of the waveguide) is potentially simpler to achieve because of the restrictions on the propagation constant k_z imposed by the geometry, which makes it easier to realize sufficiently strong multiple scattering. In a long quasi-1D structure, energy injected at any point of the input is equally likely to emerge at any point of the output, and modes are completely mixed by the medium. A diffusion-theory based approach to the study of photon transport may be adopted. However, in a truly-1D structure, the diffusion time scale is absent, and there are only the ballistic propagation regime and the localized regime [5].

An optical slow-wave structure consists of a chain of resonant cavities (such as a defect cavity resonator in a photonic crystal slab, or a Fabry–Perot resonator) or other forms of resonators such as microrings, microdisks etc. [6,7]. In this context, localization is defined as an exponential maximum of the intensity in real (coordinate) space along the length of the waveguide. Confinement in the transverse plane is provided by the index contrast between the core and the cladding of the waveguide cross-section.

We will discuss here two distinct approaches to potentially localizing light in an optical slow-wave structure: (a) taking advantage of Anderson localization at the band edge in a disordered structure [8,9], or (b) utilizing the Kerr optical nonlinearity in conjunction with the unique dispersion relationship of the slow-wave structure to create slowly-moving or stopped soliton-like pulses [10,11]. (Very recently, researchers have also been investigating combinations of the two approaches—control over Anderson localization utilizing the optical Kerr effect in nonlinear coupled-waveguide structures [12].)

A typical realization of the optical slow-wave structure is the coupled-resonator optical waveguide (CROW), which is formed by placing optical resonators in a linear (or two/three dimensional) array, so as to guide light from one end of this chain to the other [13–15]. Energy transfer occurs because of nearest-neighbor interaction: the spatial overlap between the fields of adjacent unit cells. Although each unit cell could be custom-designed, which is done in the synthesis of optical filters with specified transfer functions [16], it is usually assumed that a slow-wave waveguide consists of nominally identical unit cells.

Since there is translational periodicity along the length of the waveguide, the modes of the structure can be described by Bloch functions [17]. In the presence of disorder, coherent back-scattering is restricted only to certain Bragg-selected channels, making the possibility of localization more likely [4]. We have fabricated such structures using a high index contrast material system (silicon for the waveguide core, and silicon dioxide or air for the waveguide cladding regions) which enhances the perturbation-induced coupling of modes—the coupling coefficient is (approximately) proportional to the difference in relative permittivity of the core and the cladding.

The dispersion relationship of an infinitely-long slow-wave waveguide can be found using tight-binding theory, as described below. For a finite-length structure, a matrix theory has been developed [18]. Many potential applications of slow-wave structures, such as slow light, enhanced optical nonlinearities etc., involve the use of pulses rather than continuous-wave (CW) light. Given an explicit relationship (albeit a nonlinear one) between ω and k, one can derive expressions describing pulse propagation [19]. We have derived such expressions in both the linear dispersion approximations [20] while taking into account the higher orders of dispersion [21]. The latter result is unusual in optical waveguide theory, as it is one of the rare cases when one can write down a nonperturbative propagation equation for any input envelope to *all* orders of dispersion, and therefore, pulse distortion can be exactly quantified.

13.2 Waveguiding Principles and the Dispersion Relationship

In this section, we describe mathematically how light propagates in a coupled-cavity slow-wave structure, and derive the waveguide dispersion relationship, which determines many of the optical transport properties. The dispersion relationship, and deviations from the ideal shape particularly in the band edge regions of the ω–k space, also determine the localization properties of light in both the linear and nonlinear regimes discussed in this Chapter.

13.2.1 Waveguide Mode

We assume the unit cells comprising the slow-wave waveguide are identical and lie along the z axis (unit vector $\mathbf{e}_z$) separated by a distance R. The dielectric coefficient for the periodic structure is written as $\varepsilon_{wg}(\mathbf{r})$ whereas the dielectric coefficient for a single resonator (isolated from other resonators) is written as $\varepsilon_{res}(\mathbf{r})$. The propagate ion constant k is defined according to the following: kR is the phase shift accumulated by the wave in propagating from the center of one unit cell of the period structure to the next.

Together with its time-evolution factor, the waveguide mode of the linear waveguide $\mathbf{E}_{wg}(\mathbf{r})$ at a particular propagation constant k is written as a linear combination of the individual eigenmodes $\mathbf{E}_{res}(\mathbf{r})$ of the resonators that comprise the structure [13]:

$$e^{-i\omega_k t}\mathbf{E}_{wg}(\mathbf{r}) = e^{-i\omega_k t}\sum_n e^{inR\,\mathbf{k}\cdot\mathbf{e}_z}\,\mathbf{E}_{res}(\mathbf{r}-nR\mathbf{e}_z), \tag{13.1}$$

where n is an index over the constituent resonators. Often, each constituent resonator has two (or more) mutually orthogonal eigenmodes, and then, each set of modes leads to a distinct waveguiding band [22]. Equation 13.1 has the Bloch form, and in drawing the dispersion relationship, we restrict the range of k to the first Brillouin zone, $|k|R<\pi$.

13.2.2 Dispersion Relationship

To find the dispersion relationship using the above Ansatz, we substitute Equation 13.1 into the wave equation,

$$\nabla^2\mathbf{E}_{wg}(\mathbf{r})+\varepsilon_{wg}\tfrac{\omega_k^2}{c^2}\mathbf{E}_{wg}= 0, \tag{13.2}$$

and use the fact that $\mathbf{E}_{res}$ obeys exactly the same equation but substituting ε_{res} in place of ε_{wg}, and Ω in place of ω_k. After some algebra [13], the dispersion relationship for a CROW waveguide mode is of the familiar tight-binding "cosine" form:

$$\omega_k=\Omega(1-\Delta\alpha/2)+\Omega k\,\cos[kR]\equiv\omega_0+\Delta\omega\,\cos[kR], \tag{13.3}$$

where the various parameters in the above equation all have physical meaning. Ω is the eigenfrequency of the individual resonators, and $\Delta\alpha$ and k are overlap integrals involving the individual resonator modes and the spatial variation of the dielectric constant, defined as

$$\Delta\alpha = \int d^3\mathbf{r}[\varepsilon_{\text{wg}}(\mathbf{r} - R\mathbf{e}_z) - \varepsilon_{\text{res}}(\mathbf{r} - R\mathbf{e}_z)]\,|\mathbf{E}_{\text{res}(\mathbf{r})}|^2,$$

$$\kappa = \int d^3\mathbf{r}[\varepsilon_{\text{res}}(\mathbf{r} - R\mathbf{e}_z) - \varepsilon_{\text{wg}}(\mathbf{r} - R\mathbf{e}_z)]\mathbf{E}_{\text{res}}(\mathbf{r}) \cdot \mathbf{E}_{\text{res}}(\mathbf{r} - R\mathbf{e}_z),$$

(13.4)

where, as before, ε_{res} is the dielectric coefficient of an individual resonator (in isolation), and ε_{wg} is the dielectric coefficient of the waveguide. $\Delta\alpha$ represents the shift of the eigenfrequency of a single resonator when it is coupled to its neighbors, and k is the coupling coefficient between two adjacent resonators. Consequently, the position and slope of the dispersion relationship are controlled by the geometric design of the waveguide through the overlap integrals defined in Equation 13.4.

The slowing factor for the waveguide describes how slow light is (at band-center) compared to a conventional waveguide,

$$S \equiv \frac{c/n_{\text{eff}}}{\text{max.}v_g} \approx \frac{1}{2\,|\,\kappa\,|\,m}$$

(13.5)

where $m = n_{\text{eff}}R/\lambda$ is the mode number expressing how many wavelengths λ/n_{eff} fit into a single period R. It is a constant in a particular design, and reflects the characteristic length scale of the structural elements, e.g., $m \approx 1$–10 for coupled Fabry–Perot resonators, whereas $m \approx 50$–100 for coupled microrings.

Based on Equation 13.3, the waveguiding bandwidth (defined as one-half of the end-to-end width) is

$$\Delta\omega = \frac{2\,|\,\kappa\,|\,mc}{n_{\text{eff}}R} = \frac{c}{Sn_{\text{eff}}R},$$

(13.6)

which shows that the slowing factor S and waveguide bandwidth $\Delta\omega$ are inversely proportional to each other, through their dependence on $|\kappa|$. Therefore, there is a fundamental tradeoff between how slow a pulse can propagate in a coupled-resonator optical slow-wave structure, and its transmission bandwidth. A representative figure-of-merit, defined as the product of the slowing factor and bandwidth, is inversely proportional to the length of a unit cell, R.

Values of $|\kappa|$ for typical coupled-resonator optical slow-wave structures are listed in Table 13.1. For coupled microresonators, a representative range for k is 0.05–0.25, with larger values resulting in a larger transmission bandwidth, and smaller values resulting in a larger slowing factor.

13.2.3 Tail of the Dispersion Relationship

Light in a slow-wave structure is slowest at the edges of the dispersion relationship, i.e., at $k=0$ or $\pm\pi/R$ in Equation 13.3. In principle, the group velocity should go to zero, but this does not occur because of disorder. A closer examination of the edge of the dispersion relationship of a weakly-disordered slow-wave structure (see Figure 13.1a) shows the existence of a band tail, and the slope of the dispersion curve (and hence v_g) is no longer zero [23].

TABLE 13.1

Typical Values of Microresonator Coupling Coefficients, Which are Used to Construct Coupled-Resonator Optical Slow-Wave Structures

Microrings:

6.5–9 µm radius Si on SiO2	$\lvert\kappa\rvert = 0.22 - 0.34$	Nature Photon. 1, 65–71, 2007
40 µm radius glass (Hydex)	$\lvert\kappa\rvert = 0.9 - 1.3\times10^{-2}$	Photon. Tech. Lett., 16, 2263–2265, 2004
60 µm radius polymer (PMMA)	$\lvert\kappa\rvert = 1.2\times10^{-1}$	Opt. Lett., 31, 456–458, 2006
Microspheres:		
1.4 µm polystyrene	$\lvert\kappa\rvert = 1.1\times10^{-3}$	Phys. Rev. B, 75, 245327, 2007
2–5 µm polystyrene	$\lvert\kappa\rvert = 2.8 - 3.5\times10^{-3}$	Phys. Rev. Lett., 82, 4623–4626, 1999
4.2 µm polystyrene	$\lvert\kappa\rvert = 1.3\times10^{-2}$	Phys. Rev. Lett., 94, 203905, 2005
2D Photonic crystal defects: (one missing hole, H1, even symmetry)		
One row between defects	$\lvert\kappa\rvert = 4.7\times10^{-2}$	Phys. Rev. Lett., 84, 2140–2143, 2000
Two rows between defects	$\lvert\kappa\rvert = 5.4\times10^{-3}$	J. Opt. Soc. Am., B, 17, 387–400, 2000
Three rows between defects	$\lvert\kappa\rvert = 3.7\times10^{-3}$	Phys. Rev. Lett., 92, 083901, 2004
	$\lvert\kappa\rvert = 1.3\times10^{-3}$	J. Opt. Soc. Am. B, 17, 387–400, 2000
Fabry–Perot and coupled waveguides:		
Silicon superlattice:	$\lvert\kappa\rvert = 1.23\times10^{-2}$	Appl. Phys. Lett., 88, 241103, 2006
Coupled GaAs lasers:	$\lvert\kappa\rvert = 6.03\times10^{-4}$	Appl. Phys. Lett., 46, 236–238, 1985
Coupled InGaAsP waveguides:	$\lvert\kappa\rvert = 5.2\times10^{-1}$	J. Opt. Soc. Am. B, 24, 2389–2393, 2007

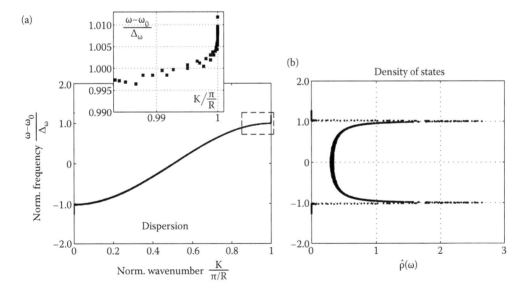

FIGURE 13.1

Dispersion and density of states for a weakly disordered slow-wave structure. (a) Dispersion relationship calculated *ab initio* for a slow-wave structure with 1% disorder in the nearest-neighbor unit cell coupling coefficients. For such weak disorder, the dispersion follows the tight-binding (cosine) function almost exactly, except at the band edge, shown by the inset. The band edge tail shows that $v_g \equiv dw/dk$ is non-zero. (b) In the presence of disorder, the density of states is calculated over the same range of frequencies, showing that the density of states, $\hat{\rho}(\omega)$, does not diverge to infinity at the band edges. (From Mookherjea, S., Park, S. J., Yang, S. S. H., and Bandaru, P. R. *Nature Photonics*, 2(2), 90–93, 2008. With permission.)

To understand why group velocity does not go to zero, it is necessary to consider the connection between group velocity and the density of states. In a weakly disordered slow-wave structure, the group velocity v_g is inversely proportional to the density of states, and is given by [23]:

$$v_g = \frac{1}{\hat{\rho}(\omega)} \frac{1}{2\pi/R} \left(1 + \frac{1}{NR} \frac{d\phi}{dk}\right), \tag{13.7}$$

where $\hat{\rho}$ is the optical density of states (normalized to unit integral over the waveguide band), and ϕ is the disorder-induced (scattering) phase shift. In a perfectly ordered structure, $\hat{\rho}(\omega_{edge}) \to \infty$ at the band edges [24] and thus $v_g \to 0$. Whereas, as shown in Figure 13.1b, in a disordered structure, $\hat{\rho}(\omega_{edge})$ only reaches a certain maximum value that depends on the statistics of variation of the elements of M—most importantly, the mean and standard deviation of the distribution of coupling coefficients, k and $\delta\kappa$, respectively [23].

In the presence of disorder, represented most conveniently by the standard deviation in the inter-unit-cell coupling coefficients $\delta\kappa$ normalized to its mean k, we have shown using *ab initio* simulations that [23]:

$$\frac{v_g \text{ at band center}}{v_g \text{ at band edge}} = 0.667 \log_{10}\left(\frac{\kappa}{\delta\kappa}\right) + 0.313. \tag{13.8}$$

Consequently, S_{be}, which is defined as the band edge slowing factor taking into account the effects of disorder, is

$$S_{be} \equiv \frac{c}{v_g \text{ at band edge}} = \frac{\lambda}{R} \frac{1}{(\delta\kappa^2 \cdot \kappa)^{1/3}}. \tag{13.9}$$

The band edge slowing factor is a measure of how much a slow-wave structure, or CROW, is free from disorder. Equation 13.9 shows that $S_{be} \to \infty$ (since v_g at band edge $\to 0$) if $\delta\kappa \to 0$. But, for a typical structure, if $\kappa = 10^{-2}$, $\delta\kappa = 5\%$ of k, and $R = 10\lambda$, then $S_{be} = 74$, a much more modest slowing factor. Experimental observations also indicate $S_{be} \approx 10$–100 [25,26].

A 1-cm long slow-wave structure which achieves maximum $S = 100$ may be replaced simply by a spool of fiber of length 1 cm $\times S \times n_{eff}/n_{fiber} \approx 2\,\text{m}$, which is considerably easier to implement in practice, and can be packaged into a fairly small volume. An alternate chip-scale method to achieve comparable slowing factors is described in Section 13.5. Therefore, using a coupled-resonator slow-wave structure only to slow down the propagation of light is unlikely to yield a commercially viable technology.

However, we may gain new insights from the study of fundamental physical phenomena related to electron transport within the context of optical slow-wave structures, such as Anderson localization. On-chip tunability of delay and localization at speeds in excess of achievable by thermal tuning of fiber Bragg gratings may yet offer a considerable advantage over the state-of-the-art.

13.3 Localization in the Presence of Disorder

The transmission spectrum of 1D slow-wave structures in the presence of disorder-induced Anderson localization shows many randomly distributed peaks in the transmission spectrum $T(\lambda)$ [9,27]. One way to check for localization is the following: since the localized modes decay exponentially with distance, the average of log T (λ) over many realizations of disorder decays linearly with the sample thickness, $<\log T> = -L/(\xi + \kappa_e^{-1})$ where ξ is a length scale called the localization length and κ_e^{-1} is an extinction coefficient (units of inverse length) that takes into account the total loss, including scattering losses and absorption. A slightly different expression, $<\log T> = -L(1/\xi + \kappa_e)$ has also been proposed [28]. Although the averaging is sometimes carried out over wavelength (assuming ergodicity) for the sake of experimental convenience—since the averaging over many realizations of disorder would require fabricating and measuring many samples—wavelength averaging is not suitable for dispersive structures such as the coupled-resonator waveguide.

Another way of examining localization is in the time domain: the Lorentzian line shape of a localized mode generates an asymmetric time response with an exponentially decaying tail [27]. Again, the highly dispersive nature of a coupled-resonator waveguide near the band edge makes this approach difficult for a slow-wave structure.

A third way of investigating localization, which has been used recently in several experiments [8,9,12] is to input light into the structure using a tunable CW laser and directly image the optical fields in a surface-normal configuration. Care should be taken not to perturb the fields strongly, e.g., by introduction of a metal-coated fiber tip in close vicinity of the waveguide. Here, localization is defined by the exponential maximum of the intensity in real (coordinate) space along the length of the waveguide. Confinement in the transverse plane is provided by the index contrast between the core and the cladding of the waveguide cross-section.

As we will discuss in this section, how disorder in the slow-wave structure can lead to localization of optical fields at frequencies near the band edges. This is shown in Figure 13.2. When there is no disorder, the spatial distribution of the field is

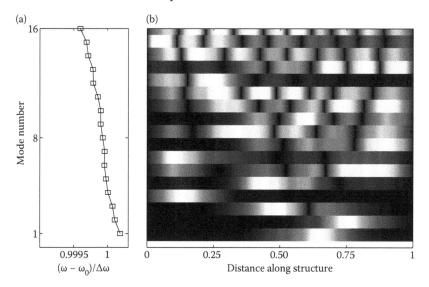

FIGURE 13.2
Spectral distribution of modes with disorder-induced localization. (a), Eigen-frequencies from a single Monte-Carlo calculation are labeled from the band edge inwards (1 = farthest detuning from band-center), showing that disorder creates states beyond the band edge of the ideal structure. The corresponding density of states is shown in Figure 13.1. (b) The corresponding field distributions along the (normalized) length of the slow-wave structure, showing that field distributions for frequencies near the band edge, at the bottom of the figure, are localized, and those inside the band, towards the top of the figure, are extended. (From Mookherjea, S., Park, S. J., Yang, S. S. H., and Bandaru, P. R. *Nature Photonics*, 2(2), 90–93, 2008. With permission.)

$$\langle z \,|\, \Psi_k \rangle \equiv \sum_{n=1}^{N} e^{-inkR} \, \mathbf{E}_{\text{single}}(\mathbf{r} - nR\hat{\mathbf{z}}) \tag{13.10}$$

i.e., a Bloch sum over the single-resonator fields $\mathbf{E}_{\text{single}}(\mathbf{r})$. ψ_k satisfies the generalized eigenvalue equation

$$\nabla \times \nabla \times |\psi_k\rangle = \varepsilon_{\text{wg}} E_k |\psi_k\rangle \tag{13.11}$$

where $E_k \equiv (\omega_k/c)^2$. Field distributions for modes near the band edge are shown in Figure 13.3. In the disorder-free case, the eigenfrequencies do not exceed $\omega_0 + \Delta\omega$, unlike in the disordered case shown in Figure 13.2.

The band edge modes do not couple to the input and output edges of the structure even in the ideal disorder-free case, as can be seen from the left and right edges of Figure 13.3b. But they can be excited by light already propagating in the waveguide at a nearby wavelength if the refractive index of the waveguide is changed, e.g., by the electro-optic or thermo-optic effect. Notice a change of less than 0.1% of the bandwidth of the dispersion relationship is sufficient; the dispersion bandwidth is itself on the order of a few nanometers in a typical CROW. In practice, transmission at the band edge may not fall by more than 10–20 dB in a typical structure, in part to the presence of substrate-guided or leaky modes, in addition to those considered in Equation 13.10.

We write the disorder as $\varepsilon_{\text{wg}}(\mathbf{r}) \rightarrow \varepsilon_{\text{wg}}(\mathbf{r}) + \delta\varepsilon_{\text{wg}}(\mathbf{r})$, and the new field distributions as $|\psi_k\rangle$, which satisfy

$$\nabla \times \nabla |\psi_k\rangle - \delta\varepsilon_{\text{wg}} E_k / \psi_k\rangle = \varepsilon_{\text{wg}} E_k |\psi_k\rangle. \tag{13.12}$$

Defining the operator $H \equiv \nabla \times (\nabla \times) - \delta\varepsilon_{\text{wg}} E_k$, the solution can be written as

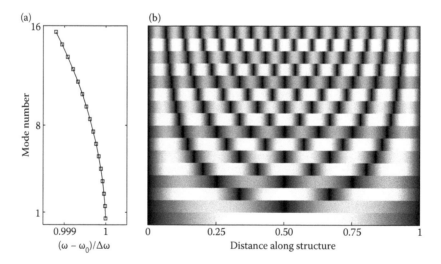

FIGURE 13.3
Spectral distribution of modes in the absence of disorder. (a) Eigen-frequencies from a single Monte-Carlo calculation are labeled from the band edge inwards (1=farthest detuning from band-center). In the absence of disorder, the eigenfrequencies approach but do not exceed the band edge $\omega_0 + \Delta\omega$ (compare with Figure 7.2a). (b) The corresponding field distributions along the (normalized) length of the slow-wave structure, which are similar to the amplitudes of the supermodes of waveguide arrays [43].

$$|\psi_k\rangle = |\Psi_k\rangle - (\varepsilon_{wg} - H)^{-1}\delta\varepsilon_{wg}E_k|\Psi_k\rangle, \qquad (13.13)$$

which may be verified by multiplying the left and right hand sides by $(\varepsilon_{wg}-H)$.

In terms of Green's function for the disordered structure, $G(z,z'; E_k)$, the solution may be written as

$$\psi_k(\mathbf{r}) = \Psi_k(\mathbf{r}) - \frac{\omega_k^2}{c^2}\sum_{\mathbf{r}'} G(\mathbf{r},\mathbf{r}'; E_k)\delta\varepsilon_{wg.}(\mathbf{r}')\Psi_k(\mathbf{r}'). \qquad (13.14)$$

$G(\mathbf{r},\mathbf{r}'; E_k)$ is not known exactly, of course, but may be found quasi-analytically from various well-known theories such as the average t-matrix approximation (ATA), or the coherent potential approximation (CPA) [24]. However, these methods do not work well in the region of our interest—the band edge, where the density of states is dominated by contributions from resonance or localized eigenstates, as the numerical results suggest.

To obtain a general idea of the behavior of Equation 13.14, one may approximate $G(z,z'; E_k)$ by $G_0(z,z'; E_k)$ for $z' \neq z$, where G_0 is Green's function for the unperturbed structure [24], and for $z'=z$, by $-\pi\rho(z,E)$, in terms of the local density of states, calculated numerically for the perturbed structure. Thus writing the disordered fields directly in terms of the perturbed density-of-states shows that those fields $\psi_k(\mathbf{r})$ which are most perturbed are those for which $\rho(E_k)$ is the most significantly affected—i.e., the band edge states.

Although it has been known theoretically that all the eigenmodes are localized for any degree of disorder (however small) if the structure is infinitely long [29], light may not be localized if the physical length of the slow-wave structure does not exceed the localization length. In a structure of finite length, localized modes are first observed near the band edges, as the modes at band-center are localized on a length-scale larger than the length of the system. To observe localization effects, fairly long slow-wave structures consisting of a large number of coupled resonators are needed, as is necessary anyway for practical applications of slow-wave structures as optical interconnects, which also require fairly long device lengths [30]. An estimate for how many resonators should be in the chain in order to observe localization can be obtained by the following argument: for a given value of N, the (dimensionless) energy separation δE between the two eigenmodes closest to the band edge is

$$\delta E = 2\kappa\left|\cos\frac{\pi N}{N+1} - \cos\frac{\pi(N-1)}{N+1}\right| \approx 3\kappa\pi^2/N^2 \qquad (13.15)$$

for large N, where N is the number of resonators in the chain, and k is the average coupling coefficient. Localization occurs when this energy separation is approximately equal to the energy perturbation due to disorder, which is $\approx \delta\kappa/\sqrt{N}$, after taking into account the effects of exchange narrowing [31]. The equality implies that $N=(3\pi^2\kappa/\delta\kappa)^{2/3}\approx 9(\kappa/\delta\kappa)^{2/3}$ which evaluates to 93 unit cells for 3% disorder, and 66 unit cells for 5% disorder. (Higher values of disorder may not obey the ballistic propagation model in this waveguide geometry.)

Three recent experiments on localization in 1D waveguide structures have fabricated waveguides consisting of (in two cases) 100 unit cells, in which each unit cell consisted of a lithographically defined resonator [8] or a sequence of side-coupled waveguides [12], and (in one case) about 250 unit cells of a single-line-defect photonic crystal waveguide [9].

One of these experiments [8] has demonstrated localization in a slow-wave structure (see Figure 13.4) consisting of a sequence of cuboidal resonators (the volume of a resonator

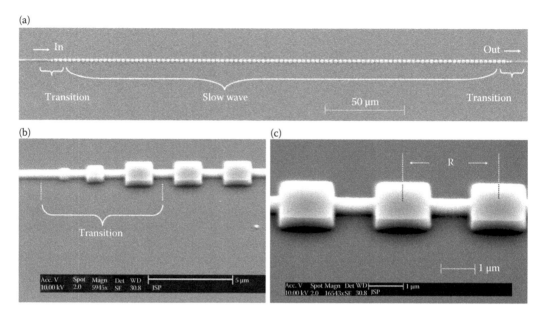

FIGURE 13.4

Slow-wave coupled-resonator waveguides. (a) The fabricated structure on an SOI chip consists of a single-mode waveguide loaded periodically with resonators, with two transition sections to better match impedances between the input/output sections and the slow-wave section. The top and side claddings are air, and the bottom cladding is silicon dioxide. (b) A magnified view of the transition section, and (c) a magnified view of the slow-wave section, with the periodicity R being 2.75 μm. (From Mookherjea, S., Park, S. J., Yang, S. S. H., and Bandaru, P. R. *Nature Photonics*, 2(2), 90–93, 2008. With permission.)

is ~1.1 μm³) with rounded edges to reduce scattering, periodically loading a single-mode optical waveguide. The operating principle is similar to the periodically-loaded transmission line in microwaves [32]. The photonic slow-wave structures were fabricated on a silicon-on-insulator (SOI) wafer using electron-beam lithography and dry-etching techniques. Upon excitation with a tunable CW laser, field profiles were recorded at wavelengths of interest near the band edge.

If we examine the field distributions corresponding to the band-tail in the dispersion relationship, such as the results of the simulation shown in Figure 13.2, we see that those field distributions are localized by disorder. From such profiles, a localization length may be defined as the root mean squared width of the intensity distribution for the localized modes. From Monte Carlo simulations of such profiles, it was observed that the localization length decreases exponentially with increasing disorder, as expected, but asymptotes to a finite non-zero lower limit (given by the offset of the exponential fit) which was calculated to be 5.6 unit cells for a finite-length structure consisting of a chain of 100 coupled resonators.

At the band edge wavelengths, the transmission spectrum decreases rapidly, as shown in Figure 13.5a. The presence of a small peak (~1.5 dB) in the transmission at the band edge has also been recently observed in a different slow-wave structure [33] and is conjectured to be related to enhanced coupling to the slow-light mode. The normalized localization length is calculated from the field profiles by numerically finding the root mean squared width of the intensity distribution, and dividing by the unit cell length. Figure 13.5c plots the localization length ($\hat{\xi}$) versus frequency ($\hat{\omega}$), both in normalized units, for the localized profiles shown in Figure 13.5b. The figure shows that $\hat{\xi} \propto \hat{\omega}^{-2}$, i.e. a straight line with slope –2 on a log-log plot [34], near the band edge, $\hat{\omega} \in (-0.2, 0)$, and is flat for positive

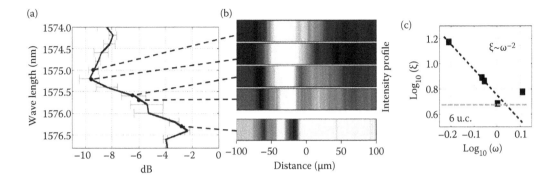

FIGURE 13.5

Experimental measurements of localization. (a) Measured transmission spectrum for the band edge wavelengths of a coupled-resonator structure as shown in Figure 13.4b, Measured spatial profiles of a representative extended field distribution at 1576.16 nm, and localized field distributions at 1575.64 nm, 1575.60 nm, 1575.10 nm, and 1575.00 nm (from bottom to top). (c) Log-log plot of the localization length $\hat{\xi}$ (normalized units) versus frequency $\hat{\omega}$ (normalized units), in agreement with the theoretically-calculated shape. (From Mookherjea, S., Park, S. J., Yang, S. S. H., and Bandaru, P. R. *Nature Photonics*, 2(2), 90–93, 2008. With permission.)

detuning from the frequency of maximum localization, in agreement with the theoretically expected behavior [28,29]. Based on simulations, we predicted that for a finite-length structure consisting of a chain of 100 coupled resonators, the minimum localization length should be approximately 5.6 unit cells. By way of comparison, from Figure 13.5c, the smallest localization length measured was 6 unit cells (17 μm). Note that unlike the theoretical calculations which assume an infinitely-long structure, "perfect" localization ($\log \hat{\xi} \rightarrow 0$) is not observed in a finite-length structure.

At the present time, given the fabrication difficulties in making long sequences of coupled resonators, all the three experiments carried out recently [8,9,12] have explored somewhat different aspects of localization. Future studies will undoubtedly converge to a more unified view of localization in slow-wave structures, and possible applications in creating high-Q states, or trapping and releasing light dynamically.

13.4 Nonlinear Localization

In Section 13.2, we discussed the dispersion relationship of an optical slow-wave structure consisting of a linear sequence of coupled unit cells. The dispersion relationship is nonlinear [21], and in particular, the higher-order dispersion terms are especially significant near the band edge.

13.4.1 Quadratic Dispersion at the Band Edge

Because the dispersion relationship is quadratic at the band edge, a coupled-resonator waveguide can cause significant pulse broadening. Conceptually unfolding the band at the edge of the band diagram, and performing the Taylor expansion of Equation 13.3 around $k=\pi/R$, we write $\omega = \bar{\omega} + (1/2)ak^2$ where k is now measured from π/R and $a = \Delta\omega R^2/2$. Given the field envelope $\varepsilon_0(z)$ at $t=0$, the envelope at a later time t is [21]

$$\varepsilon(z,t) = \frac{1}{2\pi} \int dK \,\tilde{\varepsilon}_0(K) \exp\left[\frac{i}{2}aK^2t - iKz\right], \tag{13.16}$$

where $\tilde{\varepsilon}_0(K) = \int dz \varepsilon_0(z) \exp(iKz)$. For a Gaussian envelope, $\varepsilon_0(z) = \exp[-z^2/2w_0^2]$, whose initial full-width at half-maximum is given by $w_{\mathrm{FWHM}} = 2[\sqrt{\ln 2}]w_0$, the initial width w_0 broadens with time,

$$w(t) = w_0\left[1 + \frac{t^2}{T_D^2}\right]^{1/2}, \tag{13.17}$$

where $T_D = w_0^2/|a| = 2(w_0/R)^2/|\Delta\omega|$, in terms of quantities defined in Equation 13.3.

Further aspects of pulse broadening and the role of dispersive terms are discussed elsewhere [35]. Here, we focus on the edges of the waveguiding band where the interplay of quadratic dispersion and self-phase modulation should give rise to soliton pulses.

In optical fibers and similar waveguides, the effects of (anomalous) group-velocity dispersion can be exactly balanced by the self-phase modulation induced by the Kerr effect, an intensity-dependent change in the refractive index of the material. This is the basis for the formation of the (fundamental) Schrödinger soliton in optical fibers, for instance. Here, we investigate the Kerr effect in coupled-resonator waveguides, with particular emphasis on determining whether self-phase modulation can compensate for the consequent distortion of the nonlinear dispersion relationship.

Such solutions would lead to the existence of envelopes that can exist or propagate without distortion in CROWs as eigensolutions of a nonlinear propagation equation (solitary waves and solitons). Whether such exact soliton pulse shapes exist or not, from a practical viewpoint (since disorder also plays a role), nonlinear propagation in such waveguides can be controlled, as in the linear cases analyzed earlier, by choosing (or changing) the structural properties of the waveguide, e.g., inter-resonator spacing, overlap integrals between adjacent resonator eigenmodes, and the Fourier spectrum of the initial excitation. Coupled-resonator waveguides therefore offer a wider range of design possibilities for the realization of all-optical information processing devices than available thus far.

13.4.2 The Nonlinear Evolution Equation

The field describing a pulse $\varepsilon(\mathbf{r},t)$ is written as a superposition of waveguide modes $\mathbf{E}_{\mathrm{wg}}(\mathbf{r})$ within the Brillouin zone, and using Equation 13.3,

$$\varepsilon(\mathbf{r},t) = e^{-i\omega_0 t} \int\limits_{-\pi/R}^{\pi/R} \frac{dK}{2\pi} \left[e^{-i\Delta\omega t \cos[(k_0+K)R]} c_{k_0+K}(t) \mathbf{E}_{\mathrm{wg}}(\mathbf{r})\right]. \tag{13.18}$$

Note that since nonlinear phenomena such as the Kerr effect change the relative weights of the eigenmodes as the waveform evolves with time, we have appropriately introduced a time-dependency in the superposition coefficients c_k appearing in Equation 13.18.

The exact nature in which these coefficients change with time is given by substituting the above expression into the nonlinear wave equation, written out with an explicit nonlinear polarization term describing the Kerr effect,

$$\nabla \times \nabla \times \varepsilon(\mathbf{r},t) - \mu \varepsilon_{\mathrm{wg}}(\mathbf{r})\frac{\partial^2}{\partial t^2}\varepsilon(\mathbf{r},t) = \mu \frac{\partial^2}{\partial t^2}\mathbf{P}_{\mathrm{NL}}(\mathbf{r},t) \tag{13.19}$$

where

$$\mathbf{P}_{NL}(\mathbf{r},t) = \frac{3}{4}\varepsilon_0\chi^{(3)}|\mathbf{E}(\mathbf{r},t)|^2\mathbf{E}(\mathbf{r},t). \tag{13.20}$$

in the instantaneous response approximation.

In simplifying the terms, we use the normalization

$$M\int d\mathbf{r}\,\varepsilon_{wg}(\mathbf{r})\sum_{m=1}^{M}|\mathbf{E}_{res}(\mathbf{r}-mR\mathbf{e}_z)|^2 = 1 \tag{13.21}$$

where the CROW waveguide comprises M resonators. A nonlinearity coefficient which we will need shortly is defined as

$$\gamma = 2n_0 n_2\,\varepsilon_0\,\omega_0\int d\mathbf{r}\sum_m|\mathbf{E}_{res}(\mathbf{r}-mR\mathbf{e}_z)|^4, \tag{13.22}$$

using the relationship $3\chi^{(3)}/8 = n_0 n_2$ [36], and we have ignored the dispersion (variation in ω) of γ. (Conversion to real world units follows conveniently from the equation $n^2 = n_0^2 + 2n_0 n_2|E|^2$ where n_0 is the linear refractive index.)

If we assume that $c_{k_0+K}(t)$ varies slowly over time intervals $\sim O(2\pi/\omega_0)$, as is usually the case, then we obtain

$$\frac{dc_{k_0+K}(t)}{dt} = i\gamma \int_{-\pi/R}^{\pi/R}\int \frac{dK_1}{2\pi}\frac{dK_2}{2\pi}\exp[-ik\Omega t\{-\cos[(k_0+K_1)R]+\cos[(k_0+K_2)R]$$

$$+[\{+\cos[(k_0+K_3)R]-\cos[(k_0+K)R]\}]\times c_{k_0+K_1}(t)^*c_{k_0+K_2}(t)c_{k_0+K_3}(t) \tag{13.23}$$

where $K_1+K=K_2+K_3$ and γ is the nonlinearity coefficient in the CROW geometry, described above.

Equation 13.23 can be written in a different basis set. The a_n's defined as

$$a_n(t) = \int_{-\pi/R}^{\pi/R}\frac{dK}{2\pi}c_{k_0+K}(t)\exp\left[in(k_0+K)R\right]\exp\left[i\left\{\frac{\Delta\alpha}{2}\Omega-\kappa\Omega\cos[(k_0+K)R]\right\}t\right] \tag{13.24}$$

are the coefficients that appear in the expansion of the field in terms of individual resonator modes, rather than the waveguide modes. The resultant differential equation is

$$i\frac{da_n}{dt}+\frac{\Delta\alpha}{2}\Omega a_n-\frac{\kappa}{2}\Omega(a_{n+1}+a_{n-1})+\gamma|a_n|^2 a_n=0. \tag{13.25}$$

Equation 13.25 is the nonlinear generalization of the linear evolution equation for $a_n(t)$ discussed later (Equation 13.45), along with a minor redefinition of the center frequency, represented by the second term, from one centered at an optical frequency to base-band (see Equation 13.24).

13.4.3 Time-Invariant Evolution

We separate the amplitude and phase of $c_{k_0+K}(t)$ as

$$c_{k_0+K}(t)=A_{k_0+K}(t)\exp[i\phi_{k_0+K}(t)] \tag{13.26}$$

We will look for solutions that retain their shape, i.e., $dA/dt=0$. Substituting Equation 13.26 into Equation 13.23 and separating the real and imaginary parts, we obtain a pair of equations,

$$\frac{dA_{k_0+K}}{dt} = -\gamma \iint \frac{dK_1}{2\pi} \frac{dK_2}{2\pi} A_{k_0+K_1} A_{k_0+K_2} A_{k_0+K_3} \sin \Phi, \tag{13.27}$$

$$\frac{d\phi_{k_0+K}}{dt} = \frac{\gamma}{A_{k_0+K}} \iint \frac{dK_1}{2\pi} \frac{dK_2}{2\pi} A_{k_0+K_1} A_{k_0+K_2} A_{k_0+K_3} \cos \Phi, \tag{13.28}$$

where Φ is defined as

$$\Phi \equiv -\{\phi_{k_0+K_1} - \kappa\Omega t\cos[(k_0+K_1)R]\} + \{\phi_{k_0+K_2} - \kappa\Omega t\cos[k_0+K_2)R] \tag{13.29}$$
$$+ \phi_{k_0+K_1} - \kappa\Omega t\cos[(k_0+K_3)R]\} + \{\phi_{k_0+K} - \kappa\Omega t\cos[(k_0+K)R]\}$$

Based on Equation 13.27, the A's will be independent of t if $\sin \Phi \equiv 0$ for all t. This implies that $\cos\Phi = 1$, and based on Equation 13.28, we take ϕ_{k_0+K} to be a linear function of t,

$$\phi_{k_0+K}(t) = a + bt + \kappa\Omega t\cos[(k_0+K)R], \tag{13.30}$$

where a and b are constants independent of t and k. We drop the constant a which represents a fixed phase that can be absorbed into the initial conditions. Substituting this form for $\phi_{k_0+K}(t)$ into Equation 13.28, we get

$$b + \kappa\Omega\cos[(k_0+K)R] = \frac{\gamma}{A_{k_0+K}} \int\limits_{-\pi/R}^{\pi/R} \int \frac{dK_1}{2\pi} \frac{dK_2}{2\pi} A_{k_0+K_1} A_{k_0+K_2} A_{k_0+K_3}. \tag{13.31}$$

13.4.4 The "Super-Resonant" Mode

In finding explicitly the solution to Equation 13.31, it is useful to remind ourselves that the basic physics lie in a balance between the phase modulation effects of the Kerr effect and (anomalous) group-velocity dispersion (GVD). The GVD term in the nonlinear Schrödinger equation appears as the coefficient of a second derivative term, which in the Fourier domain with the Fourier (frequency) variable K, translates to multiplication by $(iK)^2$.

In Equation 13.31, if we assume that k_0R is a multiple of 2π and $|KR| \ll 1$, then we may write $\cos[(k_0+K)R] \approx 1-(KR)^2/2$, which is the desired effective GVD term. Observe from the dispersion relationship, Equation 13.3, that ω_{k_0+K} is a quadratic function of K only at the edges of the Brillouin zone—where $d\omega_{k_0+K}/dK$ vanishes, i.e., the group velocity is zero. We expect, therefore, that the solutions of Equation 13.31 in this regime will be *stationary*, describing a localized state that is frozen in its initial ($t=0$) spatial distribution and does not propagate along the waveguide.

Using this approxmation, Equation 13.31 becomes

$$b + \kappa\Omega = \kappa\Omega\frac{(KR)^2}{2} + \frac{\gamma}{A_{k_0+K}} \int\limits_{-\pi/R}^{\pi/R} \int \frac{dK_1}{2\pi} \frac{dK_2}{2\pi} A_{k_0+K_1} A_{k_0+K_2} A_{k_0+K_3} \tag{13.32}$$

We assume that the A's are defined to be zero outside the regions of integration $-\pi/R$ and π/R so that the limits of integration can be taken as $-\infty$ to ∞. Equation 13.32 may then be solved [10],

$$A_{k_0+K} = A_{k_0+K}^{(0)} \operatorname{sech}(K/\bar{K}) \tag{13.33}$$

where $\bar{K}$ is a spectral width parameter whose relevance will become clear in the following discussion. Substituting Equation 13.33 into Equation 13.32, we get

$$b + \kappa\Omega = \kappa\Omega \frac{(KR)^2}{2} + 2\left[A_{k_0+K}^{(0)}\right]^2 \frac{\gamma}{(2\pi R)^2}\left[(KR)^2 + \left(\frac{\pi\bar{K}R}{2}\right)^2\right] \tag{13.34}$$

If b is to be independent of K, then we need

$$A_{k_0+K}^{(0)} = \sqrt{-(2\pi R)^2 \frac{\kappa\Omega}{4\gamma}}. \tag{13.35}$$

Since the left-hand side represents a real and positive number, we require that k as defined in Equation 13.4 be a negative number (as is physically expected from the meaning of $\varepsilon_{\mathrm{wg}}$ and $\varepsilon_{\mathrm{res}}$). This is equivalent to anomalous dispersion in optical fibers and similar waveguides.

Using Equations 13.30 and 13.35 in Equation 13.26, we write the final expression for $c_{k_0+K}(t)$,

$$c_{k_0+K}(t) = c_{k_0+K}(0)\exp\{-i\kappa\Omega t[1 + \pi^2(\bar{K}R)^2/8 - \cos(KR)]\}, \tag{13.36}$$

where

$$c_{k_0+K}(0) \equiv 2\pi R\sqrt{-\frac{\kappa\Omega}{4\gamma}}\operatorname{sech}(K/\bar{K}), \quad |KR| \leq \pi. \tag{13.37}$$

The field described by these coefficients is

$$\varepsilon(\mathbf{r},t) = e^{-i\omega_0 t} e^{-i\kappa\Omega t\left[1+\pi^2(\bar{K}R)^2/8\right]} \int\limits_{-\pi/R}^{\pi/R} \frac{dK}{2\pi} c_{k_0+K}(0)\mathbf{E}_{\mathrm{wg}}(\mathbf{r}) \tag{13.38}$$

In light of Equation 13.37, the integral on the second line of Equation 13.38 is not expressible in a simpler form. However, if $\bar{K}R \lesssim 1$, the hyperbolic secant function decays rapidly, and the limits of integration may be changed to $(-\infty,\infty)$. The integral then can be evaluated easily—the Fourier transform of a hyperbolic secant is itself a hyperbolic secant function. We derive the approximation,

$$\mathbf{E}(\mathbf{r},t) \approx e^{-i\omega_0 t} e^{-i\kappa\Omega t\left[1+\pi^2(\bar{K}R)^2/8\right]}\sqrt{-\frac{\kappa\Omega}{4\gamma}}\pi\bar{K}R\sum_n \operatorname{sech}\left(\frac{\pi\bar{K}}{2}nR\right)\mathbf{E}_{\mathrm{res}}(\mathbf{r}-nR\hat{z}) \tag{13.39}$$

The modulus of the amplitude $|\varepsilon(z,t=0)|$ normalized to its maximum value (in this approximation) is plotted in Figure 13.6. Values of the hyperbolic secant function in

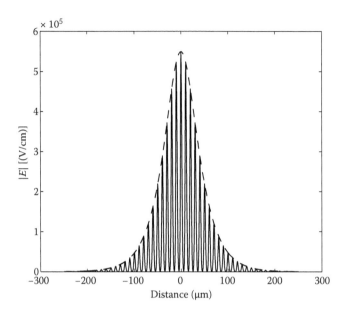

FIGURE 13.6
Calculation of a super-resonant field distribution, assuming that the individual resonator eigenmodes are Gaussian in shape with width 1 μm, with inter-resonator spacing $R=10$ μm, background refractive index $n_0=2.5$, nonlinear index $n_2=1.5\times10^{-13}$cm²/W and $\bar{K}=1/(5R)$. The input power is 1 Watt, and the cross-sectional dimensions are taken as 0.5 μm×0.5 μm. The inter-resonator coupling coefficient is assumed to be of magnitude $|\kappa|=0.01$. The dotted line is an envelope—a hyperbolic secant—connecting the excitation coefficients multiplying the individual resonator eigenmodes.

Equation 13.39 at nR (which has the dimensions of length) are the weights of the individual resonator eigenmodes. In this approximation, the envelope of these weights is a hyperbolic secant function whose width is inversely proportional to $\bar{K}$.

As we had expected from physical arguments, the envelope of $\varepsilon(\mathbf{r},t)$ is a stationary state that is independent of time: its spatial distribution at $t=0$ is maintained for all subsequent t. This is consistent with the observation that although the group velocity dispersion coefficient is nonzero, the group velocity itself is zero. We call the stationary state a super-resonant field since it is formed in a waveguide that itself comprises the coupling of individual (stationary) resonator modes. At the present time, the super-resonant mode of a coupled-resonator optical slow-wave structure has not been experimentally observed; however, it has been verified by simulations [37].

13.4.5 Nonlinear Anderson Localization

A recent experiment performed by Lahini et al. [12] has studied the interplay between disorder-induced Anderson localization and the optical Kerr nonlinearity in an array of 99 side-coupled optical waveguides patterned on an AlGaAs substrate. Light is injected into one or a few waveguides (typically about 10 waveguides for optimal localization) and can coherently tunnel between nearest neighbors as it propagates along z. The amplitude in the nth waveguide ($n=1, 2,...., N$) obeys the evolution equation

$$-i\frac{\partial U_n}{\partial z} = \beta_n U_n + C(U_{n+1}+U_{n-1})+\gamma\,|U_n|^2\,U_n \qquad (13.40)$$

where β_n is the propagation constant of the nth waveguide, C is the nearest-neighbor coupling coefficient (in units of inverse distance), and γ represents the effect of the Kerr nonlinear index.

Disorder is introduced by randomly changing the widths of each waveguide (a nominal value of the width is 4 μm), so that the parameters β_n become random in the range $\beta_0 \pm \Delta$. The measure of disorder is set by the ratio $\Delta/c = 1$. The input intensity was below the self-focusing limit so as to operate in the weakly nonlinear regime.

Recall from our earlier discussion that the two band edges of Equation 13.3 are characterized by $k=0$ and $k=\pm\pi/R$, which represent a relative phase shift $kR=0$ and $kR=\pm\pi$ between adjacent unit cells. It was observed by Lahini et al. that a positive Kerr nonlinearity tends to further localize those "flat phased" modes in which the phase-difference between nearest-neighbors is close to 0, and tends to delocalize those "staggered" modes in which the phase-difference between nearest-neighbors is close to π.

It was further observed, by injecting light of different intensities into a single waveguide and monitoring the evolution along z, that localization in a coupled 1D structure emerges out of the ballistic propagation regime without a signature of diffusive behavior [5], as would be observed in higher dimensions [2,3].

13.5 Cascaded Versus Nested Coupled-Resonator Structures

From the point of view of achievable delay (for modest values of S) in the presence of disorder, we show here that there is a "nested coupled-resonator" structure which may achieve the same levels of delay as a linear cascaded-resonator (i.e., CROW) delay line. The two structures are shown schematically in Figure 13.7. In particular, the nested structure we consider is formed by embedding a Fabry–Perot resonator inside a Gires–Tournois resonator as shown in Figure 13.7b. As the light bounces back and forth between the mirrors of the Gires–Tournois resonator, it traverses a Fabry–Perot resonator each time.

13.5.1 Slow Light in Fabry–Perot and Gires–Tournois Resonators

We first discuss the slowing of light in a single resonator, most conveniently taken to be of the Fabry–Perot type, i.e., two partially-reflective mirrors enclosing an air cavity, and then discuss the delay of a Fabry–Perot nested inside a Gires–Tournois resonator. The following three facts are noteworthy:

(a) The maximum group delay in transmission of a Fabry–Perot etalon of length d is obtained on resonance [38], and takes the value

$$T_{FP} = \left(\frac{1+r_2^2}{1-r_2^2}\right)T \tag{13.41}$$

where $\tau = 2n_{eff}d/c$ and r_2 is the (amplitude) reflection coefficient at the Fabry–Perot-to-air interface.

(b) On resonance, transmission through the Fabry–Perot etalon is 100%, i.e., all the light is transmitted, suffering only the phase shift $\phi_{FP} = \pi$ (or a multiple of π).

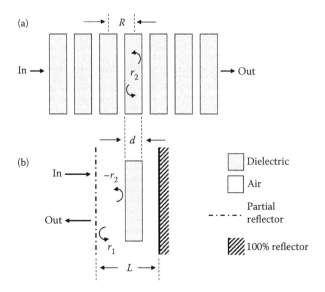

FIGURE 13.7
Cascaded and nested slow-wave architectures. (a) A cascaded resonator slow-wave architecture, consisting of identical Fabry–Perot etalons (thickness d) in a linear array with inter-resonator spacing R. (b) A nested resonator slow-wave architecture, consisting of a Fabry–Perot etalon inside a Gires–Tournois resonator of length L. The amplitude reflection coefficients at various air-dielectric interfaces are marked.

(c) The group delay of a Gires–Tournois resonator of length L is [38]

$$\tau_{GT} = \frac{\sigma}{1+(\sigma^2-1)\sin^2\phi_{GT}}\frac{d(2\phi_{GT})}{d\omega} \tag{13.42}$$

where $\phi_{GT}=\omega L/c$ is the one-pass phase accumulation, and $\sigma\equiv(1+|r_1|)/(1-|r_1|)$ in terms of the (amplitude) reflection coefficient at the front mirror, r_1 (see Figure 13.7b).

Combining these three facts (a–c), we can readily conclude that when a maximum-delay Fabry–Perot etalon is placed inside a Gires–Tournois resonator, as shown in Figure 1.7b, the only effect is to modify ϕ_{GT} which appears in Equation 13.42,

$$\phi_{GT} \rightarrow \phi_{FP+GT} = \frac{\omega}{c}(L-d)+\phi_{FP}. \tag{13.43}$$

Therefore, the group delay (upon reflection) of the nested structure shown in Figure 13.7b is

$$\tau_{FP+GT} = \frac{\sigma}{1+(\sigma^2-1)\sin^2\phi_{FP+GT}}\frac{d(2\phi_{FP+GT})}{d\omega}$$

$$= \underbrace{\frac{\sigma}{1+(\sigma^2-1)\sin^2\phi_{FP+GT}}}_{\equiv n}\left[\frac{2(L-d)}{c}+2\tau_{FP}\right]. \tag{13.44}$$

According to Equation 13.43, we can control ϕ_{FP+GT}, for example, by changing L by translating either of the mirrors comprising the Gires–Tournois etalon. In doing so, we can tune the pre-factor η in Equation 13.44 continuously between a minimum value $\approx 1/\sigma$ (antiresonance) to a maximum value $\approx \sigma$ (resonance). In other words, the group delay of the nested resonator structure (in reflection) can be σ times larger than twice the maximum group delay of the Fabry–Perot resonator alone (in transmission). The minimum group delay of the nested structure can also be made a factor of σ smaller, which may be useful in some applications.

13.5.2 Slow Light in the Coupled Fabry–Perot Structure

Having shown the versatility of the nested resonator architecture shown in Figure 13.7b for providing tunable optical delay, we ask the question, what length of coupled-resonator chain does it correspond to, for given values of the mirror reflectivities?

We assume, for simplicity, that the Fabry–Perot etalon shown in Figure 13.7b is located at the center of the Gires–Tournois resonator, and then model the corresponding unfolded chain of coupled Fabry–Perot etalons shown in Figure 13.7a using the tight-binding method. The resonators are indexed by $n=1, 2,\ldots, N$ and R is the distance between the centers of two adjacent etalons. An excitation (pulse) is described as a set of time-dependent coefficients $\{a_n(t)\}$, which represent the oscillation amplitudes of the individual etalons. Since each etalon is coupled symmetrically to its nearest neighbors, the evolution of the amplitudes is described by

$$i\frac{da_n}{at} + \Omega a_n + \Omega\frac{\kappa}{2\pi}(a_{n-1} + a_{n+1}) = 0 \tag{13.45}$$

where Ω is the resonant frequency of a single resonator and κ is the (dimensionless) nearest-neighbor coupling coefficient. Assuming periodic boundary conditions,[*] we guess a solution of the form

$$a_n(t) = \frac{1}{\sqrt{N}}e^{i(\omega t - nk_m R)} \tag{13.46}$$

where $k_m = m2\pi/(NR)$. Substituting Equation 13.46 into Equation 13.45, we obtain that

$$\frac{\omega_m}{\Omega} = 1 + \frac{\kappa}{\pi}\cos(k_m R), \quad m = 0, 1,\ldots, N-1 \tag{13.47}$$

are the N normalized eigenfrequencies of the chain. In the limit of large N, Equation 13.47 defines ω as a continuous function of k, which, in the first Brillouin zone, takes values from $-\pi/R$ to π/R. The group velocity is defined as $v_g(\omega) \equiv d\omega/dk$ from Equation 13.47.

---

[*] In assuming periodic boundary conditions, we rely on our physical intuition that the "bulk" optical properties within the waveguide are unaffected by the exact nature of the boundaries. In chains consisting of only a few unit cells, this need not be true [39]. Chains of resonators with gain may also be sensitive to boundary conditions [40].

A pulse propagates with minimum distortion if it is centered at mid-band, i.e., with carrier frequency $\omega = \Omega$. The time taken by such a pulse to propagate across M unit cells of the slow-wave structure is

$$\tau_{\text{CROW}}^{(M)} = \frac{MRn_{\text{eff}}}{|k|c}. \tag{13.48}$$

We ask: what value of M in Equation 13.48 equates $\tau_{\text{CROW}}^{(M)}$ with $\tau_{\text{FP–GT}}$ as derived in Equation 13.44? In other words, what length of a CROW structure achieves the same delay as the nested resonator?

We can obtain an expression for $|k|$ in terms of the mirror reflectivities and inter-etalon distances, using a transfer-matrix method [18,41]. It can be shown that $|k|$ lies within a range of values,

$$|\kappa| \in \left(\frac{1-|r_2|^2}{1+|r_2|^2}, 1 \right), \tag{13.49}$$

which correspond to the maximum and minimum group delays of the slow-wave structure, respectively. Using the minimum value of $|k|$ (for the largest possible delay), setting $R=L$, and thereafter equating Equation 13.48 and 13.44, we obtain

$$M = \sigma \frac{2d}{L} \left(2 + |\kappa| \frac{\tau_{\text{GT}}}{\tau} \right) \approx 2\sigma, \tag{13.50}$$

since typically, $|\kappa| \ll 1$, $\tau_{\text{GT}} \approx \tau$, and $2d \approx L$., As expected, M is proportional to σ, and therefore, the higher the quality factor of the Gires–Tournois resonator, the longer the length of the corresponding unfolded chain that achieves the same delay.

At the present time, the state-of-the-art in cascaded identical microrings is 12 sequentially coupled resonators [26]. (Longer chains of microrings have recently been reported [25] but their transmission spectra are complicated due to mis-matching of resonances in long chains of not-exactly-identical resonators.) On the other hand, achieving $\sigma = 6$ is fairly straightforward in a Gires–Tournois resonator, corresponding to an intensity reflection cofficient $|r_1|^2 = 51\% \approx 3$ dB at the front mirror. Therefore, for small delays, the obtainable delay from the nested structure would easily compare with the best cascaded micro-resonator structures demonstrated, which require considerably more effort to fabricate.

13.5.3 Advantages and Disadvantages of the Nested Architecture

We have argued that the same delay that can be achieved by cascading quite a large number of resonators can be achieved by nesting a single resonator inside a Gires–Tournois resonator. We summarize three advantages of the nested architecture.

1. The nested structure avoids some of the disorder-induced limitations of slow light in the unfolded coupled-resonator structure, whose performance greatly suffers if the inter-resonator spacings are not exactly identical [23]. Conceptually unfolding the nested resonator architecture into a chain of coupled resonators, it is clear that no matter what is the fine-structure of the unit-cell, the very same unit cell is repeated (approximately 2σ times, as we have shown in Equation 13.50). Thereby,

the disorder in coupling coefficients is overcome. Also, fabricating two resonators to meet target specifications is significantly easier than fabricating 12 or more identical resonators.

2. The nested structure is compact compared to the unfolded chain, occupying approximately $1/2\sigma$ times less physical length to achieve the same delay, and consuming correspondingly less dielectric material to fabricate.

3. The nested architecture is especially convenient for modulating the transmission properties of all the elements of a cascaded resonator chain at once, as is required for some applications of slow light [42]. Only a single resonator—the Fabry–Perot etalon in the center of the structure shown in Figure 13.7b—needs to be tuned, e.g., electro-optically or thermally. Furthermore, by varying the transmission of the front mirror of the Gires–Tournois etalon, one can dynamically tune the length of the equivalent unwrapped slow-wave structure in Figure 13.7a to realize a novel optical memory element whose storage capacity can be varied.

The main disadvantage of the nested architecture is that its enhanced group delay is obtained at the expense of reducing the bandwidth over which the delay peaks. Let us define $2\delta\omega^{(\text{nested})}$ as the separation between those frequencies at which $\tau_{\text{FP+GT}}$ falls to 50% of its peak value. Assuming $2\delta\omega^{(\text{nested})}$ is narrow enough so that the Fabry–Perot etalon remains near its resonance, we can use Equation 13.44 to find

$$\underbrace{\left(2\delta\omega^{(\text{nested})}\right)}_{\text{bandwidth}} \times \underbrace{\sigma\left(\frac{L-d}{c} + \tau_{\text{FP}}\right)}_{\text{half-delay}} = 2. \tag{13.51}$$

Using the minimum value of $|k|$ from Equation 13.49, we can re-write Equation 13.41 as $\tau_{\text{FP}} = n_{\text{eff}} d/(|\kappa|c)$ The bandwidth of the coupled-resonator chain is obtained from Equation 13.47 as

$$2\delta\omega^{(\text{chain})} = \frac{2|\kappa|c}{n_{\text{eff}}L} \tag{13.52}$$

(although usually only one-half of this bandwidth has approximately linear dispersion.) Therefore, we find

$$\frac{2\delta\omega^{(\text{nested})}}{2\delta\omega^{(\text{chain})}} = \frac{1}{\sigma\left(\frac{|\kappa|}{n_{\text{eff}}}\frac{L-d}{L} + \frac{d}{L}\right)} \approx \frac{L}{\sigma d} \text{ if } |\kappa| \ll 1. \tag{13.53}$$

Although this equation seems to suggest that the ratio of the bandwidths could approach unity if $d = L/\sigma$, such a configuration (thin Fabry–Perot etalon in a long Gires–Tournois resonator) results in the bandwidth of the nested structure being dominated by the Gires–Tournois resonator, rather than the Fabry–Perot etalon, and considerably reduces the value of M in Equation 13.50.

However, the bandwidth limitations can be circumvented by spectrally separating the input wavelengths using a diffraction grating, and using an array of nested-resonator devices to tune the delay in each narrow passband. Although we end up consuming the same amount of dielectric material and physical space as the slow-wave chain to regain the bandwidth, there is no longer any need to maintain phase-lock between the individual elements, as there would be in an ideal cascaded chain.

The significant attraction of the CROW (cascaded, rather than nested, resonator) architecture lies then in the new physical phenomena that can be observed in such structures, such as Anderson localization or Kerr-effect induced (solitonic) localization, which require a long unfolded optical path length to manifest themselves.

13.6 Summary

Recent research in slow-light has shown that it is difficult to achieve large slowing factors in practical devices. Each approach has its own limitations: the coupled-resonator optical waveguide (CROW) architecture is sensitive to fabrication errors which result in mismatching of resonances along the chain. This causes dephasing of the transmitted wave, and variations in the group delay, particularly at the band edges of the dispersion relationship, where light is expected to propagate slowly.

Disordered 1D structures are naturally prone to Anderson localization. Nano-lithographic tools, particularly the use of electron-beam lithography, allows the fabrication of slow-wave waveguides in the Mie scattering regime, where the dimension of a unit cell is comparable to the wavelength of light. Such structures can now be fabricated using materials of high index contrast, such as silicon and oxide or air, and with surface roughness on the order of a few nanometers. Most of the slow-wave structures demonstrated can also be fabricated using UV or deep-UV lithographic tools. Importantly, improvements in optical coupling techniques to nano-photonic waveguides of sub-micron dimensions, as well as the development of experimental methods to measure optical fields in situ have made experimental studies of optical localization in these waveguides possible.

Localization of light in a slow-wave structure can be investigated in both linear and nonlinear regimes. The former is closely connected with the usual theory of Anderson localization in disordered periodic structures, which has been an area of considerable interest in the last 20 years. Nonlinear localization can occur either in the perfectly-ordered case, or from the interplay of disorder and nonlinear self-focusing. The latter is of more practical interest, but has so far been demonstrated only in the quasi-1D case, where the structure is not single-moded in the transverse waveguide dimension. Single-mode waveguides are of more practical interest for applications.

Future studies of localization in slow-wave structures will investigate the role of diffusive processes (which are predicted to be absent in this geometry, but play a signature role in quasi-1D, 2D and 3D Anderson localization), and measure the time constants associated with localization. Control over localization, whether by optical, thermal, or electronic means, is also important. Since these waveguides are viewed as possibly playing the role of intra-chip interconnects in future micro-processor technologies, it is necessary to measure their dispersion and polarization properties, and at the same time, minimize both the absorption and coupling losses by proper design of the input and output transitions.

In summary, research on the coupled-resonator optical slow-wave structure is rapidly moving forward on both the theoretical and experimental fronts, and involves both basic physics and fundamental engineering topics. The waveguides display novel linear dispersion properties, interesting pulse propagation characteristics, and distinct nonlinear behavior. Using modern-day lithographic technology and commonly-used material systems which are compatible with the other developing areas of research in on-chip photonics, we can now fabricate and demonstrate slow-wave structures of considerable length, and gradually make the transition between laboratory demonstrations and practical applications.

Acknowledgments

The author is grateful to his students Jung S. Park, Michael L. Cooper, Marco A. Escobar, and Andrew Oh for assistance, and also to P. R. Bandaru, S. H. Yang, U. Levy, J. B. Khurgin, V. N. Astratov, J. E. Ford, G. C. Papen, and J. E. Sipe for useful discussions. This work was funded in part with support from the National Science Foundation (L. Goldberg, R. Hui and E. G. Johnson). The San Diego Supercomputing Center provided computational resources.

References

1. P.-E. Wolf and G. Maret. Weak localization and coherent backscattering of photons in disordered media. *Phys. Rev. Lett.*, 55(24): 2696–2699, 1985.
2. D. S. Wiersma, P. Bartolini, A. Lagendijk, and R. Righini. Localization of light in a disordered medium. *Nature*, 390: 671–673, 1997.
3. T. Schwartz, G. Bartal, S. Fishman, and M. Segev. Transport and Anderson localization in disordered two-dimensional photonic lattices. *Nature*, 446: 52–55, 2007.
4. S. John. Strong localization of photons in certain disordered dielectric superlattices. *Phys. Rev. Lett.*, 58: 2486–2489, 1987.
5. F. M. Izrailev, T. Kottos, A. Politi, and G. P. Tsironis. Evolution of wave packets in quasi-one-dimensional and one-dimensional random media: Diffusion versus localization. *Phys. Rev. E*, 55(5): 4951–4963, 1997.
6. A. Melloni, F. Morichetti, and M. Martinelli. Optical slow wave structures. *Optics and Photonics News*, 14: 44–48, 2003.
7. M. L. Povinelli, S. G. Johnson, and J. D. Joannopoulos. Slow-light, band-edge waveguides for tunable time delays. *Opt. Express*, 13(18): 7145–7159, 2005.
8. S. Mookherjea, J. S. Park, S.-H. Yang, and P. R. Bandaru. Localization in silicon nanophotonic slow-light waveguides. *Nature Photonics*, 2: 90–93, 2008.
9. J. Topolancik, B Ilic, and F. Vollmer. Experimental observation of strong photon localization in disordered photonic crystal waveguides. *Phys. Rev. Lett.*, 99: 253901, 2007.
10. S. Mookherjea and A. Yariv. Kerr-stabilized super-resonant modes in coupled-resonator optical waveguides. *Phys. Rev. E*, 66: 046610, 2002.
11. D. N. Christodoulides and N. K. Efremidis. Discrete temporal solitons along a chain of nonlinear coupled microcavities embedded in photonic crystals. *Opt. Lett.*, 27(8): 568–570, 2002.
12. Y. Lahini, A. Avidan, F. Pozzi, M. Sorel, R. Morandotti, D. N. Christodoulides, and Y. Silberberg. Anderson localization and nonlinearity in one-dimensional disordered photonic lattices. *Phys. Rev. Lett.*, 100: 013906, 2008.
13. A. Yariv, Y. Xu, R. K. Lee, and A. Scherer. Coupled-resonator optical waveguide: a proposal and analysis. *Opt. Lett.*, 24(11): 711–713, 1999.

14. M. Bayer, I. Gutbrod, J. P. Reithmaier, A. Forchel, T. L. Reinecke, P. A. Knipp, A. A. Dremin, and V. D. Kulakovskii. Optical modes in photonic molecules. *Phys. Rev. Lett.*, 81(12): 2582–2585, 1998.

15. S. Olivier, C. Smith, M. Rattier, H. Benisty, C. Weisbuch, T. Krauss, R. Houdré, and U. Oesterlé. Miniband transmission in a photonic crystal coupled-resonator optical waveguide. *Opt. Lett.*, 26(13): 1019–1021, 2001.

16. C. K. Madsen and J. H. Zhao. *Optical Filter Design and Analysis*. John Wiley & Sons, New York, 1999.

17. N. W. Ashcroft and N. D. Mermin. *Solid State Physics*. Harcourt, Fort Worth, TX, 1976.

18. J. K. S. Poon, J. Scheuer, S. Mookherjea, G. T. Paloczi, Y. Huang, and A. Yariv. Matrix analysis of microring coupled-resonator optical waveguides. *Opt. Express*, 12: 90–103, 2004.

19. S. Mookherjea and A. Yariv. Coupled resonator optical waveguides. *IEEE J. Select. Top. Quantum Electron.*, 8(3): 448–456, 2002.

20. S. Mookherjea and A. Yariv. Optical pulse propagation in the tight-binding approximation. *Opt. Express*, 9(2): 91–96, 2001.

21. S. Mookherjea, D. S. Cohen, and A. Yariv. Nonlinear dispersion in a coupled-resonator optical waveguide. *Opt. Lett.*, 27: 933–935, 2002.

22. Y. Xu, R. K. Lee, and A. Yariv. Propagation and second-harmonic generation of electromagnetic waves in a coupled-resonator optical waveguide. *J. Opt. Soc. Am. B*, 17(3): 387–400, 2000.

23. S. Mookherjea and A. Oh. Effect of disorder on slow light velocity in optical slow-wave structures. *Opt. Lett.*, 32: 289–291, 2007.

24. E. N. Economou. *Green's Functions in Quantum Physics*. 3rd Edition. Springer, Berlin, 2006.

25. F. Xia, L. Sekaric, and Y. Vlasov. Ultracompact optical buffers on a silicon chip. *Nature Photonics*, 1: 65–71, 2007.

26. J. K. S. Poon, L. Zhu, G. DeRose, and A. Yariv. Transmission and group delay of microring coupled-resonator optical waveguides. *Opt. Lett.*, 31: 456–458, 2006.

27. J. Bertolotti, S. Gottardo, D. Wiersma, M. Ghulinyan, and L. Pavesi. Optical necklace states in Anderson localized 1D systems. *Phys. Rev. Lett.*, 94: 113903, 2005.

28. A. R. McGurn, K. T. Christensen, F. M. Mueller, and A. A. Maradudin. Anderson localization in one-dimensional randomly disordered optical systems that are periodic on average. *Phys. Rev. B*, 47(20): 13120–13125, May 1993.

29. P. Sheng. *Introduction to Wave Scattering, Localization and Mesoscopic Phenomena*. Academic Press, San Diego, CA, 1995.

30. J. B. Khurgin. Optical buffers based on slow light in electromagnetically induced transparent media and coupled resonator structures: comparative analysis. *J. Opt. Soc. Am. B*, 22: 1062–1074, 2005.

31. F. Dominiguez-Adame and V. A. Malyshev. A simple approach to Anderson localization in one-dimensional disordered lattices. *Am. J. Phys.*, 72(2): 226–230, 2004.

32. J. C. Slater. *Microwave Electronics*. Dover, New York, 1969.

33. V. S. Volkov, S. I. Bozhevolnyi, L. H. Frandsen, and M. Kristensen. Direct observation of surface mode excitation and slow light coupling in photonic crystal waveguides. *Nano Lett.*, 7: 2341–2345, 2007.

34. G. Theodorou and Cohen. M. Extended states in a one-dimensional system with off-diagonal disorder. *Physical Review B*, 13: 4597–4601, 1976.

35. S. Mookherjea. Dispersion characteristics of coupled-resonator optical waveguides. *Opt. Lett.*, 30: 2406–2408, 2005.

36. G. P. Agrawal. *Nonlinear Fiber Optics*. Academic Press, San Diego, CA, 1989.

37. R. Iliew, U. Peschel, C. Etrich, and F. Lederer. Light propagation via coupled defects in photonic crystals. In *Conference on Lasers and Electro-optics, OSA Technical Digest, Postconference Edition*, volume 73 of OSA TOPS. OSA, Washington, DC, 191–192, 2002.

38. A. Yariv and P. Yeh. *Photonics*. 6th edition. Oxford University Press, New York, 2007.

39. Y.-H. Ye, J. Ding, D.-Y. Jeong, I. C. Khoo, and Q. M. Zhang. Finite-size effects on one-dimensional coupled-resonator optical waveguides. *Phys. Rev. E*, 69: 056604, 2004.

40. J. K. S. Poon and A. Yariv. Active coupled-resonator optical waveguides—part I: Gain enhancement and noise. *J. Opt. Soc. Am. B*, 24(9): 2378–2388, 2007.

41. G. T. Paloczi, Y. Huang, A. Yariv, and S. Mookherjea. Polymeric Mach-Zehnder interferometer using serially coupled microring resonators. *Optics Express*, 11: 2666–2671, 2003.

42. M. F. Yanik and S. Fan. Stopping light all optically. *Phys. Rev. Lett.*, 92: 083901, 2004.

43. A. Yariv. *Optical Electronics in Modern Communications*. 5th Edition. Oxford, New York, 1997.

Index